Grassland and Pasture Ecological Management and Utilization

Grassland and Pasture Ecological Management and Utilization

Editors

Kesi Liu
Xinqing Shao

MDPI

Basel • Beijing • Wuhan • Barcelona • Belgrade • Novi Sad • Cluj • Manchester

Editors
Kesi Liu
China Agricultural University
Beijing
China

Xinqing Shao
China Agricultural University
Beijing
China

Editorial Office
MDPI
St. Alban-Anlage 66
4052 Basel, Switzerland

This is a reprint of articles from the Special Issue published online in the open access journal *Agronomy* (ISSN 2073-4395) (available at: https://www.mdpi.com/journal/agronomy/special_issues/SBCMXE8235).

For citation purposes, cite each article independently as indicated on the article page online and as indicated below:

Lastname, A.A.; Lastname, B.B. Article Title. *Journal Name* **Year**, *Volume Number*, Page Range.

ISBN 978-3-7258-0167-1 (Hbk)
ISBN 978-3-7258-0168-8 (PDF)
doi.org/10.3390/books978-3-7258-0168-8

Cover image courtesy of Kesi Liu

Contents

About the Editors

Kesi Liu

Kesi Liu graduated with a Ph.D degree in Grassland Ecology and Management from the University of Florida in 2009. Liu is an Associated Professor of the Department of Grassland Resources and Ecology at China Agricultural University. He is working as an Executive Deputy Director of the Research Center of Grassland Engineering Technology and is a member of the Grassland Resources Committee of the China Natural Resources Society. He has more than 15 years of experience as a researcher dealing with the restoration of degraded grasslands and grassland multifunctional management. To date, Liu has published more than 60 scientific papers with referees.

Xinqing Shao

Xinqing Shao, Ph.D, is a Professor of the Department of Grassland Resources and Ecology at China Agricultural University. He is the Editor-in-Chief of the journal Grassland, Deputy Secretary General of the Chinese Grass Society, and Deputy Head of Guyuan grassland Ecosystem Experimental Station. His research focuses on grassland ecological management and restoration and biodiversity conservation. He has published 58 articles and authorized 5 invention patents.

agronomy

Editorial

Grassland Ecological Management and Utilization for Sustainability

Kesi Liu * and Xinqing Shao

Department of Grassland Resources and Ecology, China Agricultural University, Beijing 100193, China; shaoxinqing@163.com
* Correspondence: kliu@cau.edu.cn

1. Introduction

Grasslands, covering 40% of the land surface area [1], are important components of terrestrial ecosystems which provide multiple functions and services, including but not limited to livestock production, biodiversity conservation, the maintenance of soil and water, carbon sequestration, and habitats for wildlife [2,3]. To date, studies have made considerable contributions toward addressing major challenges and problems associated with identifying the internal mechanisms of various grasslands to provide ecological services and functions and develop knowledge-based strategies to effectively manage grasslands and restore degraded grasslands.

Many grasslands have been used as grazing lands for livestock production for centuries and have become key components of livestock production systems [4]. Animal grazing activities in livestock production systems can change the balance of the structure, composition, and functions of grassland or pasture ecosystems [5,6] and alter soil physiochemical properties and the enzyme activities of grasslands [7]. Moreover, grazing effects are likely to vary among response variables, intensity, and modes of grazing and abiotic conditions [8]. For example, light grazing is likely to induce mainly biotic shifts; moderate-to-heavy levels of grazing or prolonged intensive use, however, are likely to cause abiotic changes [9]. Currently, information on grazing grassland management is collected less widely than information on forests and cropland and tends to be of lower resolution and limited to a subset of regions and management practices [10].

Unfortunately, many grasslands are being degraded by the integration of internal drivers of grassland ecosystem fragility with external disturbances such as overgrazing, invasive species encroachment, and global climate change [11], resulting in a loss of biodiversity, water erosion intensification, carbon sequestration reduction, decreased grassland productivity, and reduced local human well-being. The degradation of grasslands can occur very rapidly, but the recovery of their multifunction is slow or does not occur at all [12]. Even so, grassland resilience has been extensively studied [13], and grassland restoration efforts are widely promoted. Consequently, many effective techniques to restore grasslands have been developed [14]. Despite this hope, significant confusion still exists regarding activities that constitute the ecological restoration of grasslands [15].

On-site studies have been conducted globally using mensurative and manipulative experiments; however, the pertinence management that optimizes livestock productivity and improves the ecosystem functions of grasslands still remains largely elusive. To understand the full potential of grasslands to deliver services and functions to society locally, regionally, and globally, further research is ongoing to address the uncertainty and context-dependency of grassland management and utilization and to explore the possible synergies and trade-offs of biodiversity conservation, climate mitigation, and food production in grassland ecosystems. This Special Issue can serve as a meaningful resource, providing valuable insights and evidence-based strategies that can help anyone interested in sustainable grassland management.

Citation: Liu, K.; Shao, X. Grassland Ecological Management and Utilization for Sustainability. *Agronomy* **2024**, *14*, 149. https://doi.org/10.3390/agronomy14010149

Received: 20 December 2023
Accepted: 8 January 2024
Published: 9 January 2024

2. Overview of the Special Issue

This Special Issue of *Agronomy* contains 22 research articles focused on grassland management and utilization, and these articles can be mainly grouped into three categories:

(1) The management and utilization of traditional grasslands [16–28].
(2) The restoration and management of degraded grasslands [29–34].
(3) Shrub encroachment in grasslands [35–37].

2.1. The Management and Utilization of Traditional Grasslands

The management and utilization of traditional grasslands involve many corresponding factors, including grassland species, grassland production, grassland grazing, grassland nutrient, and grassland health.

As for the management of forage species, Xing et al. [16] found that the precipitation is the key climatic factor restricting the distribution of Carex alatauensis on the Qinghai–Tibet Plateau. This finding will support the conservation and restoration planning process for this species on the Qinghai–Tibet Plateau. Jayasinghe et al. [17] investigated the factors influencing the expression of persistence in perennial ryegrass populations and found significant fixed effects of cultivar, endophyte, and environment and their interactions on the persistence traits of perennial ryegrass. Wu et al. [18] investigated the current and future distribution of orchardgrass-suitable areas globally and found that the areas suitable for habitats increased at higher latitudes while decreasing at lower latitudes as greenhouse gas emissions increased. Ferreira et al. [19] evaluated the forage mass and nutritional value of Guinea Massai grass in an open pasture or the silvopastoral system at different stages of development. These valuable insights will help balance economic development and ecological conservation goals to ensure the sustainable development of forage species and the stability of the ecosystem.

As for the management of grassland production, Li et al. [20] analyzed the structural and spatial characteristics of a grassland gross ecosystem product in karst desertification control. The results of this study can provide a reference for economic decision making regarding the management of grassland ecosystem services in karst areas with similar conditions and beyond. Meng et al. [21] comprehensively assessed the effects of multiple variables on the above-ground biomass (AGB) in managed grasslands in China and found the grassland AGB depends substantially on species, environments, and management practices.

As for grazing management, Temu et al. [22] assessed the sward structural responses of some native warm-season grasses to seasonal changes in harvest regimes and emphasized the importance of taking into consideration species' inherent morphological and physiological adaptations to grazing. Wang et al. [23] investigated the effects of various grazing intensities on the physicochemical properties and bacterial communities of soil in the desert steppe of the Inner Mongolia Autonomous Region.

As for the nutrient management of grasslands, Pan et al. [24] investigated the effects of grazing grassland, mowing grassland, and enclosed grassland on C, N, and P and their ecological stoichiometry in the plant–soil–microbe interaction in the artificial grassland of the karst desertification control area in Southern China and found that the chemical properties and stoichiometric characteristics of the plant–soil–microorganism interaction were significantly changed by different grassland use methods. Yang et al. [25] studied the short-term effects of N addition and mowing on the species diversity and biomass of a typical grassland in Inner Mongolia and found that mowing significantly increased species diversity. Species richness decreased significantly with an increasing N addition rate. Mowing can alleviate the negative effects of N addition on species richness.

As for the health evaluation and management of grasslands, Shi et al. [26] developed a sound warning system to diagnose the potential degradation risk of alpine grasslands, and this study is crucial for understanding the health level of alpine grassland and its further change trends and providing an important scientific basis for rational grazing. Abakumov et al. [27] studied the influence of land use type (pasture, vegetable garden, hayfield, or secondary afforestation) on key agrochemical parameters and the parameters of soil microbial biodiversity and found that the key factor regulating soil microbiome composition shifts was the duration and degree of the irreversibility of an agrogenic impact. Xiong et al. [28] analyzed 143 pertinent works on grassland ecological assets and ecological products and proposed insights into the enhancement of karst grassland ecosystem service functions based on three perspectives: a fragile environment, trade-off synergy, and service management. This study provides valuable insights for the development of regional ecological livestock and the scientific promotion of integrated desertification control.

2.2. The Restoration and Management of Degraded Grasslands

Ecosystem degradation has become a global issue which seriously affects the health of natural ecosystems and human well-being. Many measurements and methods have been investigated and evaluated by researchers for the restoration and management of degraded ecosystems. Hou et al. [29] used a conceptual framework of response–effect traits and the Community Assembly by Trait Selection model (CATS model) as a restoration strategy to achieve effective and efficient aims of restoration in degraded ecosystems.

Enclosure is a commonly used method of restoring degraded grasslands. Yang et al. [30] investigated the response of vegetation community characteristics to enclosure duration in a degraded alpine meadow in the Source Zone of the Yellow River and found that long-term enclosure (10 years) was observed to decrease the species diversity and nutrient utilization efficiency of alpine meadow vegetation. This finding is, to some extent, verified by the study of Liu et al. [31]. Liu et al. [31] evaluated the effects of enclosure on the vegetation characteristics of the main grassland types and found that different vegetation characteristics and grassland types showed different responses to enclosure duration; they suggested that management strategies for enclosed grasslands should be adjusted reasonably according to the type of grassland and the grassland management objectives in order to maintain or even improve the condition and services of grassland ecosystems.

The addition of objective materials is also a common way to retore degraded grasslands. Zhang et al. [32] explored how different strains of fungi affected plant growth and the community dynamics of different degraded levels of grasslands and found that using beneficial fungi (AMF and Trichoderma) for soil improvement and reducing harm from pathogenic Fusarium species (Fusarium boothii and Fusarium circinatum) to plant growth is of great significance for promoting the protection and management of grassland ecosystems, as well as for the restoration and recovery of grasslands. Li et al. [33] investigated the effects of effective microorganisms (EMs) and biochar addition on vegetation biomass, microorganisms, and soil properties in a degraded alpine grassland and found that the combination of the biochar and the EM addition had a synergistic effect on the restoration of degraded alpine grasslands. Chang et al. [34] used native dominant species combined with arbuscular mycorrhizal fungi (AMF) to recover grassland and restrain grassland degradation and found that various ratios of grass–legume mixtures plus AMF inoculation could be used to recover degraded grassland production and enhance grassland nutrient accumulation and stability.

2.3. Shrub Encroachment in Grasslands

Shrub encroachment in grasslands has received an increasing amount of attention in the context of climate change. Xie et al. [35] examined the responses of four stages of the Caragana shrub's life cycle to sandy habitats and found that sandy habitats promoted the population growth of Caragana shrubs during their whole life cycle and highlighted the significant role of sandy habitats in facilitating shrub encroachment in grasslands. Xie et al. [36] also evaluated whether sexual reproduction was the main mechanism for Caragana encroachment into grasslands and found that climatic aridity, grazing, and their combined effects had negative effects on the sexual reproduction of Caragana shrubs and that clonal reproduction might be of considerable importance for understanding the mechanism of shrub encroachment in grasslands. Manganyi et al. [37] evaluated the role of ruminants, particularly browsers, in the dispersal of woody plant seeds and found that after ingestion, shrub seeds were mostly still viable and might still be dispersed in the rangeland, leading to further bush encroachment.

3. Concluding Remarks

These findings in this Special Issue partly elucidate the inner mechanisms of change or propose tangible solutions to support the sustainable, rational use of grassland and pasture. In conclusion, effective ecological management and the sustainable utilization of grasslands and pastures are essential for maintaining their ecosystem services and ensuring long-term environmental sustainability. By adopting innovative approaches, integrated land use planning, and collaborative effective models, we can ensure the long-term sustainability of grasslands and pastures for the benefit of both nature and human society.

Author Contributions: K.L. and X.S. contributed equally during the development of this editorial. All authors have read and agreed to the published version of the manuscript.

Acknowledgments: We wish to thank all the authors' contributions to the success of this Special Issue. We also acknowledge the reviewers and editorial managers who assisted in the development of this Special Issue.

Conflicts of Interest: The authors declare no conflicts of interest.

References

1. Dixon, A.P.; Faber-Langendoen, D.; Josse, C.; Morrison, J.; Loucks, C.J. Distribution mapping of world grassland types. *J. Biogeogr.* **2014**, *41*, 2003–2019. [CrossRef]
2. Balvanera, P.; Pfisterer, A.B.; Buchmann, N.; He, J.S.; Nakashizuka, T.; Raffaelli, D.; Schmid, B. Quantifying the evidence for biodiversity effects on ecosystem functioning and services. *Ecol. Lett.* **2006**, *9*, 1146–1156. [CrossRef] [PubMed]
3. Bai, Y.; Francesca Cotrufo, M. Grassland soil carbon sequestration: Current understanding, challenges, and solutions. *Science* **2022**, *377*, 603–608. [CrossRef] [PubMed]
4. Erb, K.H.; Lauk, C.; Kastner, T.; Mayer, A.; Theurl, M.C.; Haberl, H. Exploring the biophysical option space for feeding the world without deforestation. *Nat. Commun.* **2016**, *7*, 11382. [CrossRef] [PubMed]
5. Liu, K.; Sollenberger, L.E.; Newman, Y.C.; Vendramini, J.M.B.; Interrante, S.M.; White-Leech, R. Grazing Management Effects on Productivity, Nutritive Value, and Persistence of 'Tifton 85' Bermudagrass. *Crop Sci.* **2011**, *51*, 353–360. [CrossRef]
6. White-Leech, R.; Liu, K.; Sollenberger, L.E.; Woodard, K.R.; Interrante, S.M. Excreta Deposition on Grassland Patches. I. Forage Harvested, Nutritive Value, and Nitrogen Recovery. *Crop Sci.* **2013**, *53*, 688–695. [CrossRef]
7. Zhang, X.R.; Zhang, W.Q.; Sai, X.; Chun, F.; Li, X.J.; Lu, X.X.; Wang, H.R. Grazing altered soil aggregates, nutrients and enzyme activities in a steppe of Inner Mongolia. *Soil Tillage Res.* **2022**, *219*, 105327. [CrossRef]
8. Eldridge, D.J.; Poore, A.G.B.; Ruiz-Colmenero, M.; Letnic, M.; Soliveres, S. Ecosystem structure, function, and composition in rangelands are negatively affected by livestock grazing. *Ecol. Appl.* **2016**, *26*, 1273–1283. [CrossRef]
9. Eldridge, D.J.; Soliveres, S.; Bowker, M.A.; Val, J. Grazing dampens the positive effects of shrub encroachment on ecosystem functions in a semi-arid woodland. *J. Appl. Ecol.* **2013**, *50*, 1028–1038. [CrossRef]
10. Conant, R.T.; Paustian, K. Grassland management activity data: Current sources and future needs. *Environ. Manag.* **2004**, *33*, 467–473. [CrossRef]
11. Stevens, N.; Erasmus, B.F.N.; Archibald, S.; Bond, W.J. Woody encroachment over 70 years in South African savannahs: Overgrazing, global change or extinction aftershock? *Philos. Trans. R. Soc. B Biol. Sci.* **2016**, *371*, 20150437. [CrossRef] [PubMed]
12. Veldman, J.W.; Buisson, E.; Durigan, G.; Fernandes, G.W.; Le Stradic, S.; Mahy, G.; Negreiros, D.; Overbeck, G.E.; Veldman, R.G.; Zaloumis, N.P.; et al. Toward an old-growth concept for grasslands, savannas, and woodlands. *Front. Ecol. Environ.* **2015**, *13*, 154–162. [CrossRef] [PubMed]
13. Jentsch, A.; Kreyling, J.; Elmer, M.; Gellesch, E.; Glaser, B.; Grant, K.; Hein, R.; Lara, M.; Mirzae, H.; Nadler, S.E.; et al. Climate extremes initiate ecosystem-regulating functions while maintaining productivity. *J. Ecol.* **2011**, *99*, 689–702. [CrossRef]
14. Li, J.; Shao, X.; Huang, D.; Liu, K.; Shang, J.; Zhang, Q.; Zhao, T.; Yang, X. Short-term biochar effect on soil physicochemical and microbiological properties of a degraded alpine grassland. *Pedosphere* **2022**, *32*, 426–437. [CrossRef]
15. Suding, K.; Higgs, E.; Palmer, M.; Callicott, J.B.; Anderson, C.B.; Baker, M.; Gutrich, J.J.; Hondula, K.L.; LaFevor, M.C.; Larson, B.M. Committing to ecological restoration. *Science* **2015**, *348*, 638–640. [CrossRef] [PubMed]
16. Xing, Y.; Shi, J.; De, K.; Wang, X.; Wang, W.; Ma, Y.; Zhang, H.; He, M.; Liu, Q. The Current Distribution of Carex alatauensis in the Qinghai–Tibet Plateau Estimated by MaxEnt. *Agronomy* **2023**, *13*, 564. [CrossRef]
17. Jayasinghe, C.; Jacobs, J.; Thomson, A.; Smith, K. Evaluation of the Relationship between Cultivar, Endophyte and Environment on the Expression of Persistence in Perennial Ryegrass Populations Using High-Throughput Phenotyping. *Agronomy* **2023**, *13*, 2292. [CrossRef]

18. Wu, J.; Yan, L.; Zhao, J.; Peng, J.; Xiong, Y.; Xiong, Y.; Ma, X. Modeling Climate Change Indicates Potential Shifts in the Global Distribution of Orchardgrass. *Agronomy* **2023**, *13*, 1985. [CrossRef]
19. Ferreira, M.R.; Cardoso, A.d.S.; Andrade, M.E.B.; Brito, T.R.; Ruggieri, A.C. How Are Warm-Season Pastures' Nutritive Value and Fermentation Characteristics Affected by Open Pasture, Silvopasture, and Sward Herbage Maturity? *Agronomy* **2023**, *13*, 1756. [CrossRef]
20. Li, Y.; Xiong, K.; Zhang, W.; Song, S.; Luo, L. Analyzing Characteristics of Grassland Gross Ecosystem Product to Inform Decision Making in the Karst Desertification Control. *Agronomy* **2023**, *13*, 1861. [CrossRef]
21. Meng, H.; Yang, J.; Sun, W.; Xiao, L.; Wang, G. Aboveground Biomass in China's Managed Grasslands and Their Responses to Environmental and Management Variations. *Agronomy* **2022**, *12*, 2913. [CrossRef]
22. Temu, V.W.; Kering, M.K. Compensatory Structural Growth Responses of Early-Succession Native Warm-Season Grass Stands to Defoliation Management. *Agronomy* **2023**, *13*, 1280. [CrossRef]
23. Wang, Y.; Guo, M.; Li, Y.; Yin, X.; Guo, J.; Wang, J. Responses of Soil Bacterial Communities and Chemical Properties to Grazing Regulation in Desert Steppe. *Agronomy* **2023**, *13*, 2817. [CrossRef]
24. Pan, G.; Song, S.; Wang, X.; Chi, Y. Using Ecological Stoichiometric Characteristics to Inform Grassland Management in the Karst Desertification Area. *Agronomy* **2023**, *13*, 1841. [CrossRef]
25. Yang, Z.; Minggagud, H.; Wang, Q.; Pan, H. Interacting Effects of Nitrogen Addition and Mowing on Plant Diversity and Biomass of a Typical Grassland in Inner Mongolia. *Agronomy* **2023**, *13*, 2125. [CrossRef]
26. Shi, H.; Liu, M.; Zhu, S.; Duan, Z.; Wu, R.; Quan, X.; Chen, M.; Zhang, J.; Qiao, Y. Construction of an EarlyWarning System Based on a Fuzzy Matter-Element Model for Diagnosing the Health of Alpine Grassland: A Case Study of Henan County, Qinghai, China. *Agronomy* **2023**, *13*, 2176. [CrossRef]
27. Abakumov, E.V.; Gladkov, G.V.; Kimeklis, A.K.; Andronov, E.E. The Microbiomes of Various Types of Abandoned Fallow Soils of South Taiga (Novgorod Region, Russian North-West). *Agronomy* **2023**, *13*, 2592. [CrossRef]
28. Xiong, K.; He, C.; Chi, Y. Research Progress on Grassland Eco-Assets and Eco-Products and Its Implications for the Enhancement of Ecosystem Service Function of Karst Desertification Control. *Agronomy* **2023**, *13*, 2394. [CrossRef]
29. Hou, J.; Wu, M.; Feng, H. Applying Trait-Based Modeling to Achieve Functional Targets during the Ecological Restoration of an Arid Mine Area. *Agronomy* **2022**, *12*, 2833. [CrossRef]
30. Yang, P.; Li, X.; Li, C.; Zhang, J. Effects of Long-Term Exclosure on Main Plant Functional Groups and Their Biochemical Properties in a Patchily Degraded Alpine Meadow in the Source Zone of the Yellow River, West China. *Agronomy* **2023**, *13*, 2781. [CrossRef]
31. Liu, C.; Li, H.; Liu, K.; Shao, X.; Huang, J.; Siri, M.; Feng, C.; Yang, X. Vegetation Characteristics of the Main Grassland Types in China Respond Differently to the Duration of Enclosure: A Meta-Analysis. *Agronomy* **2023**, *13*, 854. [CrossRef]
32. Zhang, Y.; Chang, J.; Xie, J.; Yang, L.; Sheteiwy, M.S.; Moustafa, A.-R.A.; Zaghloul, M.S.; Ren, H. The Impact of Root-Invasive Fungi on Dominant and Invasive Plant Species in Degraded Grassland at Nanshan Pasture. *Agronomy* **2023**, *13*, 1666. [CrossRef]
33. Li, J.; Li, H.; Shang, J.; Liu, K.; He, Y.; Shao, X. The Synergistic Effect of Biochar and Microorganisms Greatly Improves Vegetation and Microbial Structure of Degraded Alpine Grassland on Qinghai–Tibet Plateau. *Agronomy* **2023**, *13*, 2203. [CrossRef]
34. Chang, J.; Li, K.; Xie, J.; Zhang, Y.; Wang, S.; Ren, H.; Liu, M. Integrating Native Plant Mixtures and Arbuscular Mycorrhizal Fungi Inoculation Increases the Productivity of Degraded Grassland. *Agronomy* **2023**, *13*, 7. [CrossRef]
35. Xie, L.; Li, Y.; Guo, H.; Wang, C.; Chen, Q.; He, P.; Ma, C. Sandy Habitats Play an Important Role in Shrub Encroachment in Grasslands. *Agronomy* **2022**, *12*, 2858. [CrossRef]
36. Xie, L.; Li, Y.; Lin, M.; Guo, H.; Wang, Y.; Wang, L.; Ma, C. Sexual Reproduction Is Not Responsible for Caragana Shrub Encroachment in Grasslands. *Agronomy* **2023**, *13*, 1848. [CrossRef]
37. Manganyi, F.L.; Tjelele, J.; Mbatha, K.R.; Letsoalo, N.; Müller, F. The Potential for Endozoochorous Dispersal of Vachellia nilotica Seeds by Goats: Implications for Bush Encroachment. *Agronomy* **2023**, *13*, 1599. [CrossRef]

agronomy

Communication

The Current Distribution of *Carex alatauensis* in the Qinghai–Tibet Plateau Estimated by MaxEnt

Yunfei Xing [1,2,3], Jianjun Shi [1,2,3,4,*], Kejia De [1,2,*], Xiaoli Wang [1,2], Wei Wang [1,2], Yuan Ma [1,2], Hairong Zhang [1], Miaohua He [1] and Qingqing Liu [1]

1 Agriculture and Animal Husbandry College, Qinghai University, Xining 810016, China
2 Qinghai Academy of Animal Science and Veterinary Medicine, Xining 810018, China
3 Laboratory for Alpine Grassland Ecology in the Three-River-Source Region, Ministry of Education Key, Xining 810016, China
4 Qinghai Provincial Key Laboratory of Adaptive Management on Alpine Grassland, Xining 810016, China
* Correspondence: shjj0318@sina.com (J.S.); 18090900142@qhu.edu.cn (K.D.)

Abstract: Modeling the current distribution of and predicting suitable habitats for threatened species support the species conservation and restoration planning process. Therefore, the purpose of this study was to model the actual distribution and predict environmentally suitable habitats for *Carex alatauensis* S.R.Zhang 2015, a locally threatened native grass species on the Qinghai–Tibet Plateau. To realize this objective, based on the geographical samples within the natural distribution of *C. alatauensis*, the dominant climatic factors in its potential distribution range were analyzed using the maximum entropy (MaxEnt) model. The results showed that the average values of the area under the receiver operating characteristic curve (AUC) of the training data were 0.833 ± 0.044, which indicated that the accuracy of the MaxEnt model was pretty high for modeling potential distribution regions of *C. alatauensis*. The combined results from the Jackknife test and the presented contribution of environmental variables revealed that the annual precipitation, the growth season precipitation, and the precipitation of the driest month were the key climatic factors that restricted the distribution of *C. alatauensis* on the Qinghai–Tibet Plateau. It is predicted that the potential distribution area of *C. alatauensis* on the Qinghai–Tibet Plateau is 1.96×10^6 km^2, and the most suitable area is 3.7×10^5 km^2, mainly located in the Qilian Mountains, the Himalayas, and the Qingtanggula Mountains.

Keywords: *Carex alatauensis*; MaxEnt; species distribution model; Qinghai–Tibet Plateau; climate variable; habitat distribution

Citation: Xing, Y.; Shi, J.; De, K.; Wang, X.; Wang, W.; Ma, Y.; Zhang, H.; He, M.; Liu, Q. The Current Distribution of *Carex alatauensis* in the Qinghai–Tibet Plateau Estimated by MaxEnt. *Agronomy* **2023**, *13*, 564. https://doi.org/10.3390/agronomy13020564

Academic Editors: Kesi Liu and Xinqing Shao

Received: 16 December 2022
Revised: 14 February 2023
Accepted: 14 February 2023
Published: 16 February 2023

1. Introduction

The Qinghai–Tibet Plateau is called the third pole, with an average altitude of more than 4000 m and an area of 2.58 million km^2, accounting for 27% of China's land area [1–3]. Due to its unique geographical location and climatic conditions, it is considered a sensitive and vulnerable area for biodiversity [4]. The area of alpine meadows with *Carex alatauensis* S.R.Zhang 2015 as the dominant species is 480,000 km^2, accounting for 31% of the total grassland area of the Qinghai–Tibet Plateau and mainly distributed in the southern and eastern regions [5]. Alpine meadows not only provide production materials for the development of animal husbandry in the Qinghai–Tibet Plateau but also play important ecological service functions such as carbon pools, water conservation, climate regulation, and biodiversity protection [6–11].

C. alatauensis is a perennial herbaceous plant of the Carex genus with short rhizomes and subterranean shoots in cold mesophiles [12]. It is one of the dominant species in alpine meadows and is widely distributed in the Qinghai–Tibet Plateau. *C. alatauensis* has the characteristics of being able to adapt to cold and humid environments and having good palatability, high economic value, and resistance to water and soil erosion. It is a forage grass with excellent ecological and economic value in the Qinghai–Tibet Plateau [13]. Due to overuse and climate change, *C. alatauensis* meadows degenerate to form a "black beach" whereby irreversible changes take place in the structure and function of the ecosystem and the service and restoration functions of the ecosystem are weakened or even lost [14]. The degradation process of the alpine meadow ecosystem on the Qinghai–Tibet Plateau is quite different from that in the arid region due to its unique cold environment and wet characteristics. The phenotype of the small transpiration surface of *C. alatauensis* indicates the adaptability of this species to the cold and seasonally dry environment and may also be one of the factors for its wide distribution on the Qinghai–Tibet Plateau. The grass mat layer formed by its rhizome in the soil in alpine meadows is the main place for water conservation and carbon sinks [15,16]. Considering the dominant role of *C. alatauensis* in the alpine meadow ecosystem and the vulnerability of this species to climate change and human activities, this study is of great significance for the protection and management of the alpine meadow ecosystem. Since the alpine meadow area is also the source of major rivers in Asia [17], a better understanding of alpine vegetation distribution will help curb ecological problems.

The climate is closely related to the growth of plants and affects the geographical distribution of plants. Through the development of geographic information and the sharing of global data resources, the joint application of niche models and geographic information systems can accurately calculate the distribution of species and predict the habitat of species. The maximum entropy model (MaxEnt) is currently the most accurate and widely used model for studying species distribution and suitable habitat areas [18–20]. *C. alatauensis* is the dominant species in alpine meadows on the Qinghai–Tibet Plateau, and the analysis of the distribution of *C. alatauensis* is of great significance for species protection under climate change [21]. In this study, the MaxEnt model was used to predict the distribution area of *C. alatauensis* in the Qinghai–Tibet Plateau and to explore the relationship between the geographical distribution and the climate in order to clarify the dominant climatic factors affecting the distribution, and to combine the suitable grades for its potential, the geographical distribution range was visualized to provide theoretical and data support for the rational layout in degraded grassland ecological engineering.

2. Materials and Methods

2.1. Study Area

The Qinghai–Tibet Plateau (26°00′-39°47′ N, 73°19′-104°47′ E) is the highest plateau in the world and is known as the "roof of the world". It is located in the China–Tibet province and parts of Qinghai, Xinjiang, Gansu, Sichuan, and Yunnan and covers an area of approximately 2.58 million km^2. Topographically, it can be divided into six parts, including the Qiangtang Plateau, the South Tibet River Basin, the Tsaidam Basin, the Qilian Mountains, the Qinghai Plateau, and the Sichuan–Tibet Alpine Valley. Precipitation and temperature have clear regional distribution patterns. Annual precipitation increases gradually from 50 to 700 mm from the northwest to the southeast, and the annual temperature gradually increases from −15 to 20 °C from the northeast to the southeast [22]. The vegetation types on the Qinghai–Tibet Plateau are alpine meadows, alpine steppe, alpine shrub steppe, and desert grassland [23].

2.2. *Carex alatauensis* Distribution Data

According to the field investigation of *C. alatauensis* germplasm resources from 2019 to 2022, the Chinese Virtual Herbarium (https://www.cvh.ac.cn/ accessed on 12 August 2022) and the Global Biodiversity Information Facility database (https://www.gbif.org/ accessed on 12 August 2022), combined with local flora and literature reports, a total of 128 distribution data of *C. alatauensis* resources were obtained. Uncertain and repeated sampling points were removed, and finally, 90 distribution points that could execute the MaxEnt model program were retained; ArcGIS10.4.1 was used to draw the geographical distribution map of *C. alatauensis* (Figure 1). The map data come from the National Basic Geographic Information Center (http://www.ngcc.cn/ngcc/ accessed on 10 September 2022).

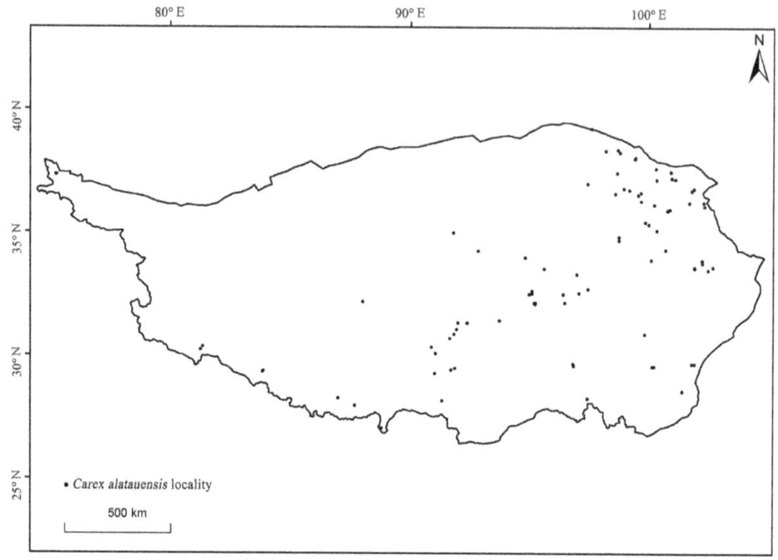

Figure 1. The occurrence sites of *Carex alatauensis* on the Qinghai–Tibet Plateau.

2.3. Climatic Data

The climate variable data come from the latest climate data layer provided by the WorldClim website (https://www.worldclim.org/ accessed on 20 August 2022). In January 2020, the website released climate data such as rainfall, temperature, solar radiation, and wind speed data from 1970 to 2000 and data on 19 bioclimatic variables (Table 1) [24]. Bioclimatic variables are derived from monthly temperature and rainfall values in order to generate more biologically meaningful variables. These are often used in species distribution modeling and related ecological modeling techniques. All of the climate variable data layers adopt a spatial resolution of 30″ (about 1 km).

2.4. Maximum Entropy Niche-Based Model

Import the existing distribution data of *C. alatauensis* collected above and the climate layer converted into ASCII format by the ArcGIS10.4.1 software into the MaxEnt3.4.1 software for calculation. During the operation, set the training set to 75%, the test set to 25%, the number of iterations to 10,000, and the repeated training to 10 to perform data simulation. Use the Jackknife module in MaxEnt3.4.1 software to calculate the contribution rate and important replacement value of each climate variable and to test the restriction strength of each climate variable on the geographical distribution. The accuracy of the MaxEnt model is tested by the receiver operating characteristic curve (ROC), represented by the area under the curve (AUC). The AUC value ranges from 0 to 1, and the larger the value, the more accurate the model's prediction. An AUC value less than 0.5 indicates that it is worse than the random simulation and the model simulation fails. If it is 0.5–0.7, this indicates that the model simulation is poor. If it is 0.7–0.9, this indicates that the model simulation effect is average. If it is greater than 0.9, this indicates that the model simulation accuracy is better, and it can more accurately reflect the potential distribution of the species [25].

Table 1. Bioclimatic variables' description.

Variable	Description
Bio1	Mean annual temperature
Bio2	Mean diurnal range
Bio3	Isothermality
Bio4	Temperature seasonality
Bio5	Max temperature of the warmest month
Bio6	Minimum temperature of the coldest month
Bio7	Temperature annual range
Bio8	Mean temperature of the wettest quarter
Bio9	Mean temperature of the driest quarter
Bio10	Mean temperature of the warmest quarter
Bio11	Mean temperature of the coldest quarter
Bio12	Annual precipitation
Bio13	Precipitation of the wettest month
Bio14	Precipitation of the driest month
Bio15	Precipitation seasonality
Bio16	Precipitation of the wettest quarter
Bio17	Precipitation of the driest quarter
Bio18	Precipitation of the warmest quarter
Bio19	Precipitation of the coldest quarter

Use the ArcGIS10.4.1 software to visualize the simulation results and divide the habitat suitability; according to the simulated moderate size of *C. alatauensis*, combined with the collected geographical distribution data, use the Jenks' natural breaks to divide it. There are four grades: the non-suitable region (<0.09), the suitable region (0.09–0.33), the sub-suitable region (0.33–0.62), and the most suitable region (>0.62). Then, use the Reclassify tool to calculate the distribution area corresponding to each grade.

3. Results

3.1. Accuracy of the Maxent Model

There are 90 geographical distribution coordinate data of *C. alatauensis*. In the construction of the maximum entropy model, 68 coordinate data were randomly selected as training samples, and the remaining 23 coordinate data were used as test samples. The results show that the AUC value of the model training set is 0.833 $\pm$ 0.044, which is greater than 0.8 and significantly greater than the AUC value of the random prediction distribution model (Figure 2). The simulation effect of the model is "good", with it meeting the model prediction accuracy requirements. This shows that the potential distribution area of *C. alatauensis* predicted by the maximum entropy model has good accuracy and the prediction results have high credibility.

3.2. The Variables' Contribution to Suitability

The MaxEnt model was run according to 19 environmental variables to evaluate the contribution rate of each climate factor (Table 2). The top five climatic factors with the highest contribution rate are the annual precipitation (Bio12, (42.31 $\pm$ 3.41)%), the mean temperature of the warmest quarter (Bio10, (14.63 $\pm$ 5.49)%), the mean diurnal range (Bio2, (12.62 $\pm$ 1.23)%), the precipitation of the driest month (Bio14, (7.38 $\pm$ 0.87)%), and the max temperature of the warmest month (Bio5, (5.17 $\pm$ 4.45)%), with a cumulative contribution rate of 82.11%. The Jackknife test scores for different climatic factors were analyzed by the Jackknife module (Figure 3). The results show that the top five climate factors are the annual precipitation (Bio12, 0.361), the precipitation of the wettest quarter (Bio16, 0.324), the precipitation of the warmest quarter (Bio18, 0.309), the precipitation of the wettest month (Bio13, 0.295), and the mean temperature of the warmest quarter (Bio10, 0.228).

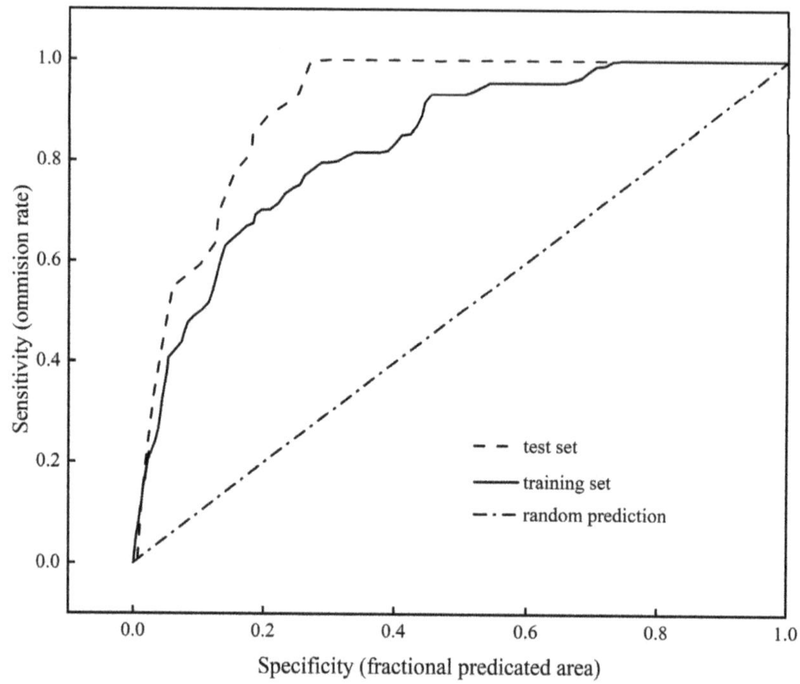

Figure 2. The receiver operating characteristic curve predicting the current potential distribution of *Carex alatauensis* based on the MaxEnt model.

Table 2. Contribution rate of each climatic factor for the distribution of *Carex alatauensis*.

Variables	Percent Contribution/%
Bio12	42.31 ± 3.41
Bio10	14.63 ± 5.49
Bio2	12.62 ± 1.23
Bio14	7.38 ± 0.87
Bio5	5.17 ± 4.45
Bio3	5.07 ± 1.61
Bio15	3.57 ± 1.08
Bio18	3.52 ± 2.28
Bio17	2.10 ± 0.95
Bio7	1.48 ± 1.05
Bio4	0.82 ± 0.49
Bio19	0.62 ± 0.16
Bio16	0.29 ± 0.18
Bio1	0.17 ± 0.21
Bio6	0.11 ± 0.14
Bio11	0.06 ± 0.15
Bio9	0.05 ± 0.05
Bio8	0
Bio13	0

3.3. The Variables' Response to Suitability

The response curves of the fitness degree of *C. alatauensis* were drawn for the above eight climatic factors. The existence probability of each climatic factor to *C. alatauensis* showed that with the increase in the climatic factor value, the existence probability showed a trend of first increasing rapidly and then decreasing slowly. In this study, the range of climate factors when the probability of existence is greater than 0.6 was used to represent the climate characteristics of the *C. alatauensis* distribution area. The climate characteristics of the distribution area of *C. alatauensis* are as follows: the mean diurnal range (Bio2) is 12.8–15.1 °C, the max temperature of the warmest month (Bio5) is 14.2–20.7 °C, the mean temperature of the warmest quarter (Bio10) is 7.7–14.3 °C, the annual precipitation (Bio12) is 300–651 mm, the precipitation of the wettest month (Bio13) is 69–143 mm, the precipitation of the driest month (Bio14) is 0–5 mm, the precipitation of the wettest quarter (Bio16) is 188–380 mm, and the precipitation of the warmest quarter (Bio18) is 187 mm~391 mm (Figure 4).

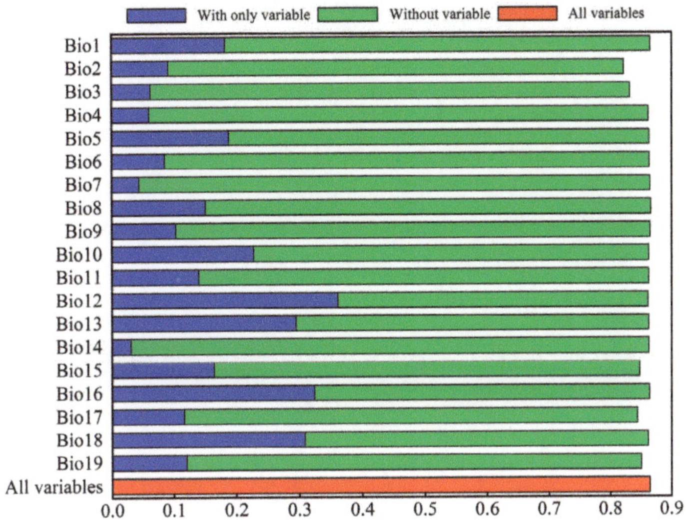

Figure 3. Jackknife test for climatic factors in the current potential distribution model of *Carex alatauensis*.

3.4. Potential Distribution of *Carex alatauensis*

The potential most suitable regions for *C. alatauensis* are mainly located in the Qilian Mountains, the Himalayas, and the Qingtanggula Mountains on the Qinghai–Tibet Plateau (Figure 5). Based on the MaxEnt model, the most suitable region for *C. alatauensis* is about 37.41×10^5 km^2 (Table 3). Most of the sub-suitable regions are located around the most suitable regions, such as near the Tanggula Mountains and the Nyainqentanglha Mountains. The area of the sub-suitable region is about 62.84×10^5 km^2. *C. alatauensis* has a wide range of suitable regions; except for the Qiangtang Plateau and the Tsaidam Basin, it is distributed in other areas, with an area of about 95.47×10^5 km^2.

Table 3. Potential distribution area of *Carex alatauensis*.

Classification	Simulation Area (km^2)
Most suitable region	37.41×10^5
Sub-suitable region	62.84×10^5

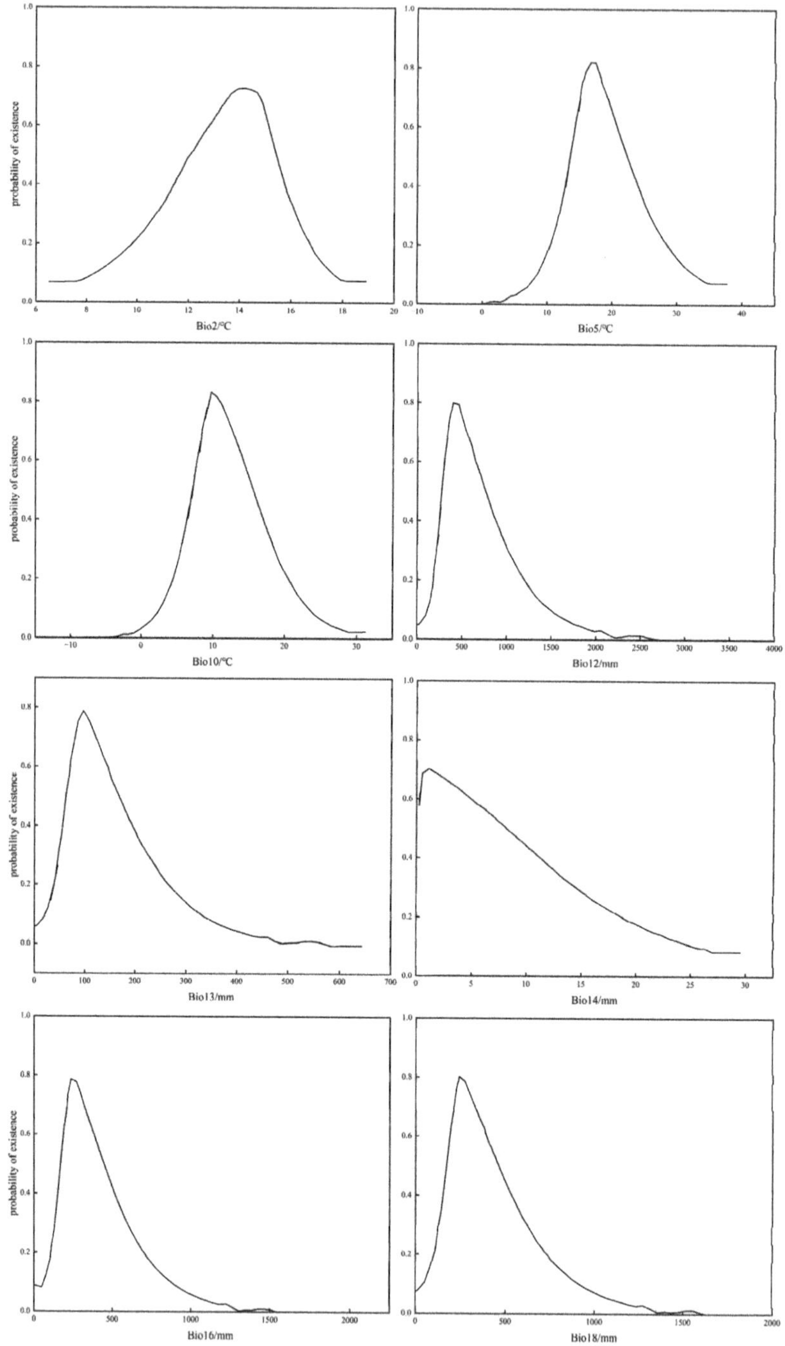

Figure 4. Probability response curves of the main climatic factors.

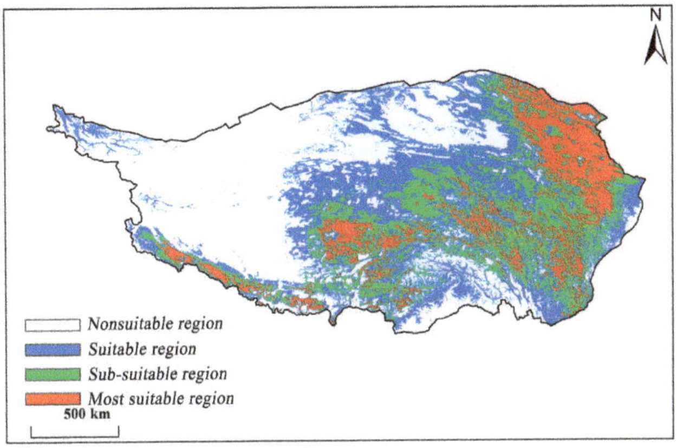

Figure 5. Potential distribution of *Carex alatauensis* based on MaxEnt.

4. Discussion

Climatic factors are one of the main factors restricting the geographical distribution of plants on a large scale, in which hydrothermal conditions play a leading role [26,27]. In this study, the MaxEnt model was used to simulate the potential distribution area of *C. alatauensis* on the Qinghai–Tibet Plateau. By integrating the contribution rate of each climate factor and the Jackknife test scores, it was shown that the main climatic factors affecting *C. alatauensis* are annual precipitation (Bio12), growth season precipitation (Bio16 for the precipitation of the wettest quarter, Bio18 for the precipitation of the warmest quarter, and Bio13 for the precipitation of the wettest month), heat (Bio10 for the mean temperature of the warmest quarter, Bio2 for the mean diurnal range, and Bio5 for the max temperature of the warmest month), and the precipitation of the driest month (Bio14). Annual precipitation, growth season precipitation, and the precipitation of the driest month have a greater impact on its distribution. Moisture conditions are the dominant climatic factors affecting the distribution of *C. alatauensis*, followed by heat conditions in the growing season [28,29]. Some studies have shown that temperature and moisture are the main limiting factors for the distribution of alpine species [30–32]. Other scholars believe that in the Qinghai–Tibet Plateau, temperature has a greater impact on *Cyperaceae* genus plants than precipitation [33]. However, *Cyperaceae* genus plants are mainly short and compact cushion plants. This special shape is beneficial for plants to absorb heat and reduce the impact of windy environments on plants, thereby improving their tolerance to alpine regions [34,35]. There are also some studies that show that water is the main limiting factor for the distribution of plateau plants. Song studied the changes in δ13C in the leaves of plants on the Qinghai–Tibet Plateau and found that precipitation is an important factor in regulating the ecosystem changes of the Qinghai–Tibet Plateau from southeast to northwest [36]. Growth season precipitation is also an important climatic factor restricting the growth of juniper (*Juniperus pingii var. wilsonii*) and birch (*Betula utilis*) [37,38]. According to the leaf area index satellite images of the National Oceanic and Atmospheric Administration in the United States, Xu found that rainfall is the main climatic factor affecting the interannual variation in the average vegetation cover in the entire Qinghai–Tibet Plateau. Higher temperatures lead to drought and insufficient rainfall in the northern Qinghai–Tibet Plateau; such arid climatic conditions are not conducive to increasing vegetation cover [39].

Among the 19 environmental variables adopted in the model, the mean temperature of the warmest quarter (Bio10) and the annual precipitation (Bio12) were the most important contributors to the habitat suitability distribution of *C. alatauensis*. For example, in the response curve of the mean temperature of the warmest quarter, *C. alatauensis* has lower temperature requirements (7.7–14.3 °C) in the warm season. Some studies have shown that C. *alatauensis* is a plant adapted to alpine meadows and likes a humid and cold environment [12]. Rainfall is one of the most important factors affecting the function and structure of plants and terrestrial ecosystems on the Qinghai–Tibet Plateau [21,40]. An area with an average annual precipitation of 300–651 mm is suitable for the growth of C. *alatauensis*, and the precipitation of the warmest quarter (Bio18) and the precipitation of the driest month (Bio14) will also affect the habitat suitability distribution.

The results showed that the total area of the potential distribution area was 1.96×10^6 km^2, of which the area of most suitable growth for potential distribution reached 3.7×10^5 km^2. The suitable distribution area of C. *alatauensis* simulated in this study is much higher than the total area of known alpine meadows [17,41]. The reasons may be as follows: (1) This study focuses on the distribution from the perspective of the relationship between climatic factors and species distribution, without considering the influence of distribution altitude, slope aspect, soil, land use type, and other factors. (2) The MaxEnt model simulates based on climate similarity and does not take into account other factors that are affected by the actual distribution of species, such as species growth characteristics, dispersal and migration capabilities, natural renewal capabilities, interactions between organisms, etc. [42,43]. (3) C. *alatauensis* is the dominant species in alpine meadows. Due to overgrazing and climate change, some alpine meadows have been degraded to *Ajania stenuifolia*, *Ligularia virgaurea*, *Pedicularis kansuensis*, etc., and have become black beach areas dominated by poisonous weeds [44]. These possible reasons will result in the predicted potential distribution area being larger than the actual known area. Nevertheless, the potential distribution areas of C. *alatauensis* obtained from the MaxEnt model all have similar climate characteristics, indicating that C. *alatauensis* has high adaptability, which can provide a certain basis for ecosystem restoration and the restoration of species selection in the abovementioned areas.

5. Conclusions

Through the analysis of the MaxEnt model, this study shows that water conditions are the dominant climatic factor affecting the distribution of C. *alatauensis*, followed by the heat conditions in the growing season. The range of dominant climatic factors in the suitable growth area has been clarified, among which the annual precipitation in the suitable growth area was 300–651 mm and the max temperature of the warmest month was 14.2–20.7 °C. It has been predicted that the potential distribution area of C. *alatauensis* on the Qinghai–Tibet Plateau is 1.96×10^6 km^2 and the most suitable area is 3.7×10^5 km^2, mainly located in the Qilian Mountains, the Himalayas, and the Qingtanggula Mountains.

Author Contributions: Conceptualization, Y.X., J.S., K.D., X.W. and Y.M.; data curation, Y.X., X.W., W.W., Y.M. and H.Z.; formal analysis, Y.X., J.S., K.D., W.W., Y.M. and H.Z.; investigation, Y.X., Y.M., H.Z., M.H. and Q.L.; methodology, Y.X., K.D. and X.W.; project administration, J.S.; software, Y.X. and X.W.; supervision, K.D. and X.W.; validation, J.S.; writing—original draft, Y.X.; writing—review and editing, Y.X., X.W. and Y.M. All authors have read and agreed to the published version of the manuscript.

Funding: This research was funded by the Second Comprehensive Scientific Investigation of Qinghai–Tibet Plateau, grant number 2019QZKK1002, and Qinghai Province High-End Innovative Talents Thousand Talents Plan, grant number 2019QHQRJH.

Data Availability Statement: The data that support the findings of this study are available from the corresponding author upon reasonable request.

Acknowledgments: The authors gratefully acknowledge the financial support of the Ministry of Science and Technology, China (2019QZKK1002), to conduct this research.

Conflicts of Interest: The authors declare no conflict of interest.

References

1. Cheng, G.; Wu, T. Responses of permafrost to climate change and their environmental significance, Qinghai-Tibet Plateau. *J. Geophys. Res. Atmos.* **2007**, *112*, F02S03. [CrossRef]
2. Harris, R.B. Rangeland degradation on the Qinghai-Tibetan plateau: A review of the evidence of its magnitude and causes. *J. Arid. Environ.* **2010**, *74*, 1–12. [CrossRef]
3. Zhang, Y.; Li, B.; Zhang, D. A discussion on the boundary and area of the Tibetan Plateau in China. *Geogr. Res.* **2002**, *21*, 1–8. [CrossRef]
4. Lu, Q.; Wu, S.; Zhao, D. Variations in Alpine Grassland Cover and Its Correlation with Climate Variables on the Qinghai-Tibet Plateau in 1982-2013. *Sci. Geogr. Sin.* **2017**, *37*, 292–300. [CrossRef]
5. Liao, G.; Jia, Y. *Rangeland Resources of China*; China Science and Technology Press: Beijing, China, 1996; pp. 489–492.
6. Raiesi, F.; Riahi, M. The influence of grazing exclosure on soil C stocks and dynamics, and ecological indicators in upland arid and semi-arid rangelands. *Ecol. Indic.* **2014**, *41*, 145–154. [CrossRef]
7. Luan, J.; Cui, L.; Xiang, C.; Wu, J.; Song, H.; Ma, Q.; Hu, Z. Different grazing removal exclosures effects on soil C stocks among alpine ecosystems in east Qinghai–Tibet Plateau. *Ecol. Eng.* **2014**, *64*, 262–268. [CrossRef]
8. Borer, E.T.; Seabloom, E.W.; Gruner, D.S.; Harpole, W.S.; Hillebrand, H.; Lind, E.M.; Adler, P.B.; Alberti, J.; Anderson, T.M.; Bakker, J.D.; et al. Herbivores and nutrients control grassland plant diversity via light limitation. *Nature* **2014**, *508*, 517–520. [CrossRef]
9. Yang, K.; Wu, H.; Qin, J.; Lin, C.; Tang, W.; Chen, Y. Recent climate changes over the Tibetan Plateau and their impacts on energy and water cycle: A review. *Glob. Planet. Chang.* **2014**, *112*, 79–91. [CrossRef]
10. Wen, L.; Dong, S.; Li, Y.; Wang, X.; Li, X.; Shi, J.; Dong, Q. The impact of land degradation on the C pools in alpine grasslands of the Qinghai-Tibet Plateau. *Plant and Soil* **2013**, *368*, 329–340. [CrossRef]
11. Cui, X.; Graf, H.-F. Recent land cover changes on the Tibetan Plateau: A review. *Clim. Chang.* **2009**, *94*, 47–61. [CrossRef]
12. Liang, S.; Dai, L.; Tang, Y.; Li, P. *Reipublicae Popularis Sinicae, Tomus 12*; Science Press: Beijing, China, 2000; pp. 35–41.
13. Liu, W.; Wang, F.; Yang, X.; Liu, Y.; Feng, B.; Yu, Y.; Zhang, C.; Cao, Q.; Dong, Q. Effects of the Traits of Reproductive and Vegetative Branches of Kobresia humilis under Different Herbivore Assemblage Grazing in Alpine Grassland. *Acta Agrestia Sin.* **2022**, *30*, 2231–2238. [CrossRef]
14. Wang, S. Vegetation degradation and protection strategy in the "Three rivers fountainhead"area in the Qinghai province. *Acta Prataculturae Sin.* **2003**, *19*, 1–9.
15. Schleuss, P.-M.; Heitkamp, F.; Sun, Y.; Miehe, G.; Xu, X.; Kuzyakov, Y. Nitrogen Uptake in an Alpine Kobresia Pasture on the Tibetan Plateau: Localization by 15N Labeling and Implications for a Vulnerable Ecosystem. *Ecosystems* **2015**, *18*, 946–957. [CrossRef]
16. Ingrisch, J.; Biermann, T.; Seeber, E.; Leipold, T.; Li, M.; Ma, Y.; Xu, X.; Miehe, G.; Guggenberger, G.; Foken, T.; et al. Carbon pools and fluxes in a Tibetan alpine Kobresia pygmaea pasture partitioned by coupled eddy-covariance measurements and $^{13}CO_2$ pulse labeling. *Sci. Total. Environ.* **2015**, *505*, 1213–1224. [CrossRef] [PubMed]
17. Miehe, G.; Schleuss, P.-M.; Seeber, E.; Babel, W.; Biermann, T.; Braendle, M.; Chen, F.; Coners, H.; Foken, T.; Gerken, T.; et al. The Kobresia pygmaea ecosystem of the Tibetan highlands—Origin, functioning and degradation of the world's largest pastoral alpine ecosystem: Kobresia pastures of Tibet. *Sci. Total. Environ.* **2019**, *648*, 754–771. [CrossRef] [PubMed]
18. Liu, D.; Lei, X.; Gao, W.; Guo, H.; Xie, Y.; Fu, L.; Lei, Y.; Li, Y.; Zhang, Z.; Tang, S. Mapping the potential distribution suitability of 16 tree species under climate change in northeastern China using Maxent modelling. *J. For. Res.* **2022**, *33*, 1739–1750. [CrossRef]
19. Kaky, E.; Nolan, V.; Alatawi, A.; Gilbert, F. A comparison between Ensemble and MaxEnt species distribution modelling approaches for conservation: A case study with Egyptian medicinal plants. *Ecol. Inform.* **2020**, *60*, 101150. [CrossRef]
20. Yan, X.; Wang, S.; Duan, Y.; Han, J.; Huang, D.; Zhou, J. Current and future distribution of the deciduous shrub Hydrangea macrophylla in China estimated by MaxEnt. *Ecol. Evol.* **2021**, *11*, 16099–16112. [CrossRef]
21. Hu, Z.; Guo, K.; Jin, S.; Pan, H. The influence of climatic changes on distribution pattern of six typical Kobresia species in Tibetan Plateau based on MaxEnt model and geographic information system. *Theor. Appl. Clim.* **2019**, *135*, 375–390. [CrossRef]
22. Jian, N. A Simulation of Biomes on the Tibetan Plateau and Their Responses to Global Climate Change. *Mt. Res. Dev.* **2000**, *20*, 80–89. [CrossRef]
23. Yue, P.P.; Lu, X.F.; Ye, R.R.; Zhang, C.X.; Yang, S.B.; Zhou, Y.B.; Peng, M. Distribution of Stipa purpurea steppe in the Northeastern Qinghai-Xizang Plateau (China). *Russ. J. Ecol.* **2011**, *42*, 50–56. [CrossRef]
24. Fick, S.E.; Hijmans, R.J. WorldClim 2: New 1-km spatial resolution climate surfaces for global land areas. *Int. J. Climatol.* **2017**, *37*, 4302–4315. [CrossRef]
25. Elith, J.H.; Graham, C.P.H.; Anderson, R.P.; Dudík, M.; Ferrier, S.; Guisan, A.J.; Hijmans, R.; Huettmann, F.; R. Leathwick, J.; Lehmann, A.; et al. Novel methods improve prediction of species' distributions from occurrence data. *Ecography* **2006**, *29*, 129–151. [CrossRef]
26. Rae, D.A.; Armbruster, W.S.; Edwards, M.E.; Svengård-Barre, M. Influence of microclimate and species interactions on the composition of plant and invertebrate communities in alpine northern Norway. *Acta Oecologica* **2006**, *29*, 266–282. [CrossRef]
27. Wang, Z.; Yang, G.; Yi, S.; Chen, S.; Wu, Z.; Guan, J.; Zhao, C.; Zhao, Q.; Ye, B. Effects of environmental factors on the distribution of plant communities in a semi-arid region of the Qinghai-Tibet Plateau. *Ecol. Res.* **2012**, *27*, 667–675. [CrossRef]
28. Li, J.; Zhang, F.; Lin, L.; Li, H.; Du, Y.; Li, Y.; Cao, G. Response of the plant community and soil water status to alpine Kobresia meadow degradation gradients on the Qinghai–Tibetan Plateau, China. *Ecol. Res.* **2015**, *30*, 589–596. [CrossRef]

29. Li, L. Study on the Suitable Distribution Mechanism and Response of the Kobresia Meadow in the Three-Rivers Source Region. Master's Thesis, Lanzhou University, Lanzhou, China, 2017.
30. Kennedy, A.D. Water as a Limiting Factor in the Antarctic Terrestrial Environment: A Biogeographical Synthesis. *Arct. Alp. Res.* **1993**, *25*, 308–315. [CrossRef]
31. Wu, L.; Wang, M.; Ouyang, H.; Cheng, S.; Song, M. Spatial Distribution Modelling of Kobresia pygmaea (Cyperaceae) on the Qinghai-Tibetan Plateau. *J. Resour. Ecol.* **2017**, *8*, 20–29. [CrossRef]
32. Boyce, R.L.; Clark, R.; Dawson, C. Factors Determining Alpine Species Distribution on Goliath Peak, Front Range, Colorado, U.S.A. *Arctic, Antarct. Alp. Res.* **2005**, *37*, 88–96. [CrossRef]
33. Zhou, Y.; Fan, J.; Zhang, W.; Harris, W.; Zhong, H.; Hu, Z.; Song, L. Factors influencing altitudinal patterns of C3 plant foliar carbon isotope composition of grasslands on the Qinghai-Tibet Plateau, China. *Alp. Bot.* **2011**, *121*, 79. [CrossRef]
34. Cavieres, L.A.; Quiroz, C.L.; Molina-Montenegro, M.A. Facilitation of the non-native Taraxacum officinale by native nurse cushion species in the high Andes of central Chile: Are there differences between nurses? *Funct. Ecol.* **2008**, *22*, 148–156. [CrossRef]
35. Liu, J.; Duan, Y.; Hao, G.; Ge, X.; Sun, H. Evolutionary history and underlying adaptation of alpine plants on the Qinghai–Tibet Plateau. *J. Syst. Evol.* **2014**, *52*, 241–249. [CrossRef]
36. Song, M.; Duan, D.; Chen, H.; Hu, Q.; Zhang, F.; Xu, X.; Tian, Y.; Ouyang, H.; Peng, C. Leaf δ13C reflects ecosystem patterns and responses of alpine plants to the environments on the Tibetan Plateau. *Ecography* **2008**, *31*, 499–508. [CrossRef]
37. Liang, E.; Lu, X.; Ren, P.; Li, X.; Zhu, L.; Eckstein, D. Annual increments of juniper dwarf shrubs above the tree line on the central Tibetan Plateau: A useful climatic proxy. *Ann. Bot.* **2012**, *109*, 721–728. [CrossRef] [PubMed]
38. Liang, E.; Dawadi, B.; Pederson, N.; Eckstein, D. Is the growth of birch at the upper timberline in the Himalayas limited by moisture or by temperature? *Ecology* **2014**, *95*, 2453–2465. [CrossRef]
39. Xu, X.; Chen, H.; Levy, J.K. Spatiotemporal vegetation cover variations in the Qinghai-Tibet Plateau under global climate change. *Sci. Bull.* **2008**, *53*, 915–922. [CrossRef]
40. Yang, Y.; Li, X.; Kong, X.; Ma, L.; Hu, X.; Yang, Y. Transcriptome analysis reveals diversified adaptation of Stipa purpurea along a drought gradient on the Tibetan Plateau. *Funct. Integr. Genom.* **2015**, *15*, 295–307. [CrossRef]
41. Genxu, W.; Ju, Q.; Guodong, C.; Yuanmin, L. Soil organic carbon pool of grassland soils on the Qinghai-Tibetan Plateau and its global implication. *Sci. Total. Environ.* **2002**, *291*, 207–217. [CrossRef]
42. Cao, X.; Wang, J.; Lu, S.; Zhang, X. Simulation of the potential distribution patterns of Picea crassifolia in climate change scenarios based on the maximum entropy (Maxent) model. *Acta Ecol. Sin.* **2019**, *39*, 5232–5240. [CrossRef]
43. Zhang, X.; Jiang, Y.; Bi, Y.; Liu, X.; Li, X.; Sun, T.; Chen, H.; Li, J. Identification of potential distribution area for Hippophae rhamnoides subsp. sinensis by the MaxEnt mode. *Acta Ecol. Sin.* **2022**, *42*, 1420–1428. [CrossRef]
44. Ma, Y.; Lang, B.; Li, Q.; Shi, J.; Dong, Q. Study on rehabilitating and rebuilding technologies for degenerated alpine meadow in the Changjiang and Yellow river source region. *Acta Prataculturae Sin.* **2002**, *19*, 1–5.

agronomy

MDPI

Article

Evaluation of the Relationship between Cultivar, Endophyte and Environment on the Expression of Persistence in Perennial Ryegrass Populations Using High-Throughput Phenotyping

Chinthaka Jayasinghe [1,*], Joe Jacobs [2,3], Anna Thomson [2] and Kevin Smith [1,3]

1 Agriculture Victoria, Hamilton Centre, Hamilton, VIC 3300, Australia
2 Agriculture Victoria, Ellinbank Centre, Ellinbank, VIC 3821, Australia
3 School of Agriculture, Food and Ecosystem Sciences, Faculty of Science, The University of Melbourne, Parkville, VIC 3010, Australia
* Correspondence: chinthaka.jayasinghe@agriculture.vic.gov.au

Abstract: Perennial ryegrass (*Lolium perenne* L.) is a commonly grown pasture species in temperate agriculture, mainly serving as a primary energy source for dairy cows. However, its limited persistence often leads to missed production potential and early resowing, especially in countries that experience summer drought, e.g., Australia and New Zealand. Therefore, understanding the factors influencing perennial ryegrass pasture persistence is crucial for sustainable land management and climate resilience in pasture-based animal production systems. Significant gaps in knowledge exist regarding the factors influencing pasture persistence, as the number of conducted studies in this area remains limited. This study aimed to investigate the factors influencing the expression of persistence in perennial ryegrass populations using airborne and ground-based sensors. A field experiment was conducted in the southwest region of Victoria, Australia, involving ten commercial perennial ryegrass cultivar–endophyte combinations in two different populations. Persistence was evaluated using sensor-based and conventional pasture measurements over two consecutive autumns. The results revealed significant fixed effects of cultivar, endophyte, and environment and their interactions on persistence traits of perennial ryegrass. Cultivars Alto, Samson, and One50 exhibited high levels of persistence when infected with novel endophyte strains. Furthermore, prolonged environmental stresses were found to drive directional selection within pasture populations. The findings emphasise the importance of selecting appropriate cultivar–endophyte combinations and early detection of signs of poor persistence to optimise sward longevity and financial returns from pasture-based animal production systems. This study fills a knowledge gap regarding the factors influencing pasture persistence and provides valuable insights for sustainable pasture management strategies.

Keywords: perennial ryegrass; persistence; endophyte; high-throughput phenotyping

Citation: Jayasinghe, C.; Jacobs, J.; Thomson, A.; Smith, K. Evaluation of the Relationship between Cultivar, Endophyte and Environment on the Expression of Persistence in Perennial Ryegrass Populations Using High-Throughput Phenotyping. *Agronomy* **2023**, *13*, 2292. https://doi.org/10.3390/agronomy13092292

Academic Editors: Kesi Liu and Xinqing Shao

Received: 10 August 2023
Revised: 27 August 2023
Accepted: 28 August 2023
Published: 30 August 2023

1. Introduction

Perennial ryegrass (*Lolium perenne* L.) is one of the major perennial pasture species used in global temperate agriculture due to its desirable traits, including producing high nutrient density for livestock and its ability to withstand frequent grazing pressure compared with other perennial pasture species [1,2]. Perennial ryegrass is native to southern Europe, temperate regions of Asia and northern Africa, and is used in the temperate regions of many other countries including Australia, New Zealand, and the United States [3]. Perennial ryegrass can survive for several years under defoliation pressure [3,4]. However, perennial ryegrass may not persist well in the majority of Australian and New Zealand livestock farms where they have rainfall <700 mm [5]. Temperature extremes, drought, pests, and diseases have all been identified as significant productivity-limiting factors of perennial pasture species [1]. Complex interactions between these environmental stressors influence plant growth and development [6], triggering protective mechanisms in perennial pasture species that enable plants to survive in a challenging environment. These protective mechanisms may include alterations in the expression of plant morphological, physiological, and biochemical traits. For example, as a response to drought stress, carbon dioxide assimilation rate of plants progressively decreases, resulting in reduced leaf size, stem extension, and root proliferation [7]. Persistence of perennial pastures species is the ability to maintain plant density and dry matter production throughout the life of the sward. Peak dry matter production per hectares during the first 4–7 years can be used as the threshold value of a sward to define perennial pasture persistence and identify the timing for pasture replacement or renovation [8].

Perennial ryegrass is a diploid species and exhibits outcrossing tendencies. The presence of a gametophytic, two-locus incompatibility system (SZ) at the stigmatic surface effectively restricts the occurrence of self-fertilization [9]. The out-breeding nature of perennial ryegrass shows a high degree of genetic diversity within and between populations [10]. As a result, there can be considerable overlap in the distribution of quantitative traits among wild populations [9]. Furthermore, cross-pollination in perennial ryegrass can lead to the production of heterozygous individuals and heterogeneous populations, resulting in high genetic variability within and between cultivars. For instance, the variation in heading date among individuals in a perennial ryegrass population can range from 35 days between the earliest and latest flowering individuals [11].

Perennial ryegrass has a symbiotic relationship with a fungal endophyte [12]. The endophyte colonises intercellular spaces in the above-ground tissues of healthy plants, typically near the base of the tillers, and is sustained by its host plant [13]. During the reproductive phase, the endophyte moves into seed heads and spreads to the next generation through seeds [12]. *Neotyphodium lolii* is a wild-type or standard endophyte strain that releases bioactive alkaloids into plant tissues when the host plant is under stress conditions such as water deficit and pest attack, which aid the plant's defense against these stress conditions [14]. Therefore, endophyte-infected plants may have greater persistence in a given environment. For example, endophyte-infected tillers in perennial ryegrass showed much lower Argentine stem weevil and African black beetle populations than in ryegrass tillers with low endophyte concentrations. Furthermore, a study found that wild-type endophyte-infected perennial ryegrass produced 38% greater dry matter, more leaf area, tillers, and roots than endophyte-free clones [15]. A recent study demonstrated that endophyte-infected ryegrass seeds have a higher seedling establishment, that may contribute to increased pasture DM yield and persistence of a sward through greater establishment [14].

Persistence of perennial ryegrass in temperate regions may similarly depend on various environmental factors, including rainfall distribution, temperature, and day length during the growing season [16]. However, other biotic and abiotic factors, such as defoliation, disease, and soil conditions, can also affect the persistence of perennial ryegrass in a dairy sward [1]. Extreme changes in these factors can create stress on individual plants, and those lacking the potential to resist such conditions may be eliminated from the sward. However, the impact of environmental factors on persistence can be managed at the initial stage.

Intensive rotational grazing, as used in dairy systems, may have both beneficial and destructive influences on the productivity and persistence of a sward [17]. The main detrimental effect of dairy grazing is defoliation, which forcefully removes available plant tissues, resulting in low water-soluble carbohydrate (WSC) reserves in the sward [18]. Defoliation can promote regrowth by enhancing light penetration, soil moisture, and nutrient availability for healthy tiller production [19]. However, the ability for shoot and root regrowth after defoliation may be limited by environmental stresses like prolonged dry and hot conditions. If the WSC reserves in the sward remain low during hot dry conditions, plants may die without entering conditional dormancy [20]. Moreover, increasing the frequency and intensity of grazing may reduce persistence and productivity in subsequent seasons, resulting in weed ingression to the sward.

Cultivar genomics of perennial ryegrass, endophyte, and environmental factors can interact and can result in adverse consequences for ryegrass persistence. For example, in the high rainfall zone of southern Australia, perennial grass ground cover is mainly influenced by growing season length, grazing method, and the interaction of stocking rate [21]. However, there needs to be more understanding of perennial ryegrass persistence and the main factors that affect it, as only a few studies have been conducted under real grazing conditions. These studies depended on traditional pasture measurements [22,23]. Investigating the potential factors that affect the expression of perennial ryegrass persistence may help in breeding cultivars with enhanced persistence and for farmers applying the proper mitigation practices to extend the life span of the dairy sward.

Traditional plant phenotyping methods have many limitations for large-scale plant screening, such as low repeatability, high labour cost, and user bias [24]. High-throughput phenotyping has emerged as a new technology in precision agriculture for non-destructive plant screening in controlled environments and under field conditions [25]. High-throughput phenotyping consists of utilising advanced sensors and computer vision-based pipelines for data acquisition and processing large data sets to extract specific plant traits from individual plants, plots or at the paddocks [26]. Field-based HTP platforms range from simple handheld devices to complex ground-based or space-borne systems, and the core element of remote sensing platforms is the sensors themselves. Sensors capture energy from specific segments of the electromagnetic spectrum (EMS), reflecting the morphological, physiological, and biochemical attributes of the target object [27]. Sensors can be classified into imaging and non-imaging types for precision agriculture in terms of imaging and non-imaging functionality, and image sensors have gained prominence due to advancements in image capture and processing technologies, facilitating efficient plant phenotyping [24]. A wide range of airborne and ground-based image sensors along with image-processing algorithms have shown high potential to estimate phenomic features related to pasture persistence, such as biomass [28], fractional ground cover [29], rate of pasture senescence [24,30], and pathogen infection [31]. The objective of this study was to evaluate the persistence of perennial ryegrass cultivars at plot scale and identify potential constraints on their expression of persistence using airborne multispectral and ground-based sound navigation and ranging (SONAR) and hyperspectral sensors.

2. Materials and Methods

2.1. Plant Materials for the Field Experiment

The plant material used in this study was obtained from an existing commercial study located in southwest Victoria, Australia (GPS coordinates: 38°14′28.8″ S 142°56′50.5″ E). The commercial study comprised 32 perennial ryegrass cultivars in four replicates, with all plots previously exposed to dairy grazing from 2013 for a further five years. Ten cultivars were subsequently selected for this experiment, representing combinations of four endophytes and four cultivar backgrounds. For each of the selected ten cultivars, one hundred soil cores with living perennial ryegrass tillers were randomly selected and excavated from each replicate using a 50 mm wide soil auger. Three healthy tillers were then randomly selected from each soil core and transferred to propagation seedling trays for field experiment planting. This group of tillers is referred to as "the selected population" in this experiment. Additionally, remnant seeds of the ten selected cultivars were sown in seedling trays, and approximately one hundred seedlings were grown from each cultivar. This group of seedlings is referred to as "the base population" of this experiment. Both the base and selected populations were then subjected to similar growing conditions to facilitate the growth of new tillers and roots until field experiment establishment.

2.2. Endophyte Status of Plant Materials

Genetic analysis was performed on the selected perennial ryegrass cultivars to test the presence of endophyte in plant materials in 2020. Three fully emerged young tillers from each plant was selected with 0.5–0.7 cm long stem at the base and sampled from the tiller using QIAGEN DNeasy 96 plant extraction kit. DNA was extracted from each cultivar using the MagAttract Plant DNA kit (Qiagen, Hilden, Germany) in an automated workflow with liquid handling platforms (Beckman Coulter, Brea, CA, USA). Each DNA sample was genotyped using a strain-specific Kompetitive Allele Specific PCR. (KASPTM) genotyping method was used to diagnose SNPs and KASP primers in this analysis. Each plate contained four positive controls as template controls (without DNA) for quality control. Each sample was tested for the presence of the expected endophyte. In addition, an immunoblot assay was used to quantify the presence of endophytes in the original populations in 2014 and the selected and base populations in 2021.

2.3. Experimental Design

The field experiment was established in May 2018 at the Hamilton Smart Farm, Agriculture Victoria Research, Hamilton, Victoria, Australia (GPS coordinates: 37°50′28.4″ S 142°04′16.9″ E). Hamilton receives its highest precipitation in July, August, and September, with an average annual temperature of 27 °C, and a mean annual precipitation of 660 mm. These weather conditions provide favourable environmental conditions for pasture growth and selection. As such, the region is considered an excellent location for conducting pasture research and breeding. Weather data were recorded throughout the experiment using an on-site meteorological station (Wireless Vantage Pro2™ Plus, Davis, CA, USA).

The experiment consisted of a completely randomised design, comprising four replicates. Each plot (3.5 m × 4.75 m) comprised four rows of the selected population and four rows of the base population, with both populations arranged adjacent to each other and spaced at 0.6 m intervals between rows. Twenty plants were established within each row, ensuring a spacing of 0.25 m between individual plants. This planting arrangement provided sufficient room for plant growth while fostering healthy competition among neighbouring plants.

2.4. Manual Pasture Measurements

In autumn 2019 and 2020, the number of surviving plants in each row was manually documented by an experienced pasture breeder. Destructive harvests were scheduled based on a predetermined criterion: when approximately 80% of the plants in each row reached the 3-leaf stage, indicating the development of three fully emerged leaves [32]. This stage was selected as an indicator of optimal growth and maturity for harvesting purposes. By employing this standardised approach, the timing of the harvests ensured consistency and allowed for accurate comparisons across the selected and base populations in each harvest. For each harvest, both the selected and base populations in each plot were mechanically harvested using a 21″ self-propelled lawn mower (Model: Honda HRX217K5HYUA) set at a cutting height of approximately 5 cm. The fresh weight of the harvested samples was subsequently measured using a precise compact scale manufactured by Mettler Toledo GmbH, Greifensee, Switzerland (Model: ICS6x5-1).

2.5. Sensor-Based Pasture Height

Prior to the destructive harvests, sensor-based pasture measurements were collected. This approach allowed for simultaneous data collections from sensors and the manual assessment at each harvest. The plant height of each row was recorded using a custom-modified side-by-side vehicle (Polaris Industries Inc., Medina, MN, USA). The vehicle was equipped with a set of integrated sensors, as illustrated in Figure 1. This innovative vehicle was developed by Agriculture Victoria [28], exclusively for the purpose of pasture height measurement. The vehicle was equipped with six ultrasonic sonar sensors (UNDK 30U6103/S14, Baumer group, Frauenfeld, Switzerland), mounted at the front of the vehicle, 0.6 m above the ground and in three rows 0.6 m apart, using a 1.45 m wide steel boom (Figure 1a). This configuration allowed for precise and accurate plant height measurement across the rows. Additionally, the vehicle was equipped with a Global Navigation Satellite System (GNSS) antenna (AG25, Trimble, Westminster, CO, USA) mounted on the rooftop. The GNSS antenna was connected to a real-time kinematics GNSS (RTK-GNSS) receiver

(FMX Integrated Display, Trimble, Westminster, CA, USA), enabling the generation of georeferenced and geolocated sensor data. This integration ensured that the recorded plant height measurements were associated with their corresponding geographic position. A datalogger (model: CR3000, Campbell Scientific, Inc., Logan, UT, USA) was attached to the back of the vehicle to record plant height data. The datalogger facilitated the collection and storage of plant height measurements, ensuring accurate and reliable data acquisition for subsequent analysis.

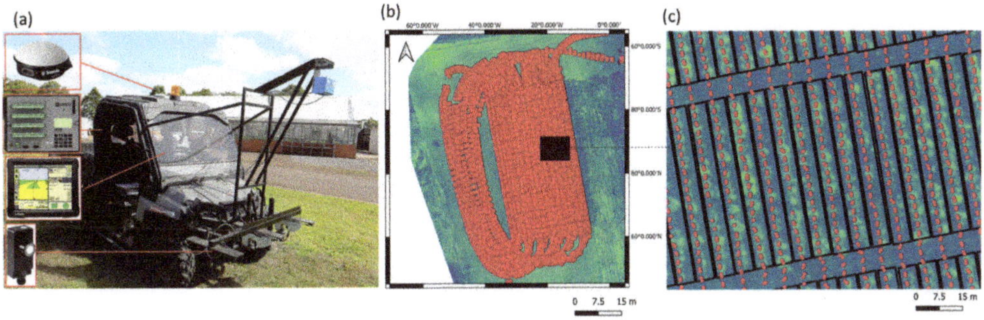

Figure 1. Simplified process of pasture plant height extraction (**a**) a modified side-by-side vehicle, (**b**) data extraction process in QGIS 3.12.3 (**c**) enlarged area of the dark square in image (**b**) where red colour dots represent georectified plant height and black polygons are row boundaries. Both point vector points and row plot polygons are overlayed on pseudocolour rendered NIR raster band.

The plant height data obtained from ultrasonic sonar sensors were processed using the methodology outlined by [28]. The georeferenced plant height measurements were autonomously recorded in the data logger and then transformed into a compatible format for third-party software using Loggernet 4.5 software (Campbell Scientific, Inc., Logan, UT, USA). To facilitate further analysis, the converted file containing the RTK-GNSS sensor data was georeferenced and transformed into the Universal Transverse Mercator (UTM) coordinate system using Microsoft Excel 2019 (https://www.microsoft.com/, accessed on 24 August 2020). The resulting file was saved in a comma-separated values (.csv) format for easy integration with other software. The UTM positions of each sensor measurement were calculated in QGIS 3.12.3 (QGIS Development Team, 2017, Raleigh, NC, USA; https://qgis.org/, accessed on 24 August 2020) utilising a pre-designed offset template within the HTP Geoprocessor 1.2 Python plugin. This allowed for precise spatial referencing of the plant height measurements. The data were extracted into a plot overlay in QGIS 3.12.3 through the processing steps provided by the HTP Geoprocessor 1.2 Python plugin (Figure 1b,c). This facilitated the creation of a vector file overlay representing the plant height measurements within the designated plan row.

2.6. Airborne Phenomic Data Acquisition

Prior to the destructive harvest, multispectral images were captured using a RedEdge-M camera (MicaSense, Inc., Seattle, WA, USA) mounted on a DJI Matrice 100 quadcopter (DJI Technology Co., Shenzhen, China). To facilitate the efficient collection of data, the Pix4D capture 4.9.0 application (Pix4D SA, Prilly, Switzerland; https://www.pix4d.com/, accessed on 2 February 2019) was utilised for the design, saving, and loading of pre-designed flight paths onto the quadcopter, automating the flight during data acquisition. Turning points of the flight path were positioned outside the area of interest to ensure comprehensive coverage of the experimental site. This ensured that the multispectral imaging process captured the entire experimental site. The flight mission was set with recommended overlap parameters for optimal image quality and accuracy. Specifically, 80% forward overlap and 75% sideways overlap were established for the RedEdge-M sensor.

During the flight, the quadcopter maintained a consistent altitude of 30 m above ground level while the ground speed was set at 6 ms^{-1} (21 km/h) to ensure the capture of stable and clear images. Calibration targets with known reflectance values (3%, 6%, 11%, 22%, and 33%) were placed on the ground within the sensor's field of view during image acquisition to facilitate accurate calibration and analysis. Furthermore, ground control points (GCPs) were strategically located within the experimental area. These GCPs played a crucial role in the georectification process of the airborne images during the image pre-processing steps.

Using Pix4D mapper 4.2.16 software (Pix4D SA, Prilly, Switzerland; https://www.pix4d.com/, accessed on 2 February 2019), digital terrain models (DTMs) and orthomosaic images were generated for each multispectral band, including green, blue, near-infrared (NIR), red, and red edge. During image pre-processing in Pix4D 4.2.16 software, the GCPs were used to ensure accurate georeferencing of the resulting DTM and orthomosaic images (Figure 2). The root mean square (RMS) error and ground sampling distance (GSD) of the orthomosaic images were approximately 2 cm in flight data (as detailed in Table 1). To establish a consistent coordinate reference system (CRS), the EPSG:32755–WGS 84/UTM zone 55S was adopted as the output coordinate system for both the DTM and orthomosaic images within the Pix4D workflow. The radiometric calibration of the orthomosaic images derived from the Pix4D workflow was conducted using QGIS 2.18.20 (QGIS Development Team, 2017, Raleigh, NC, USA; https://qgis.org/, accessed on 24 August 2020). The process involved employing the Zonal statistics plugin and raster calculator functions within QGIS. By applying radiometric calibration, the raw digital numbers (DNs) recorded by the RedEdge-M multispectral camera were converted into reflectance values, enabling further interpretation of the data. The resulting radiometrically corrected raster orthomosaics were subsequently employed in the phenomics pipeline to extract digital features of radiometrically calibrated images.

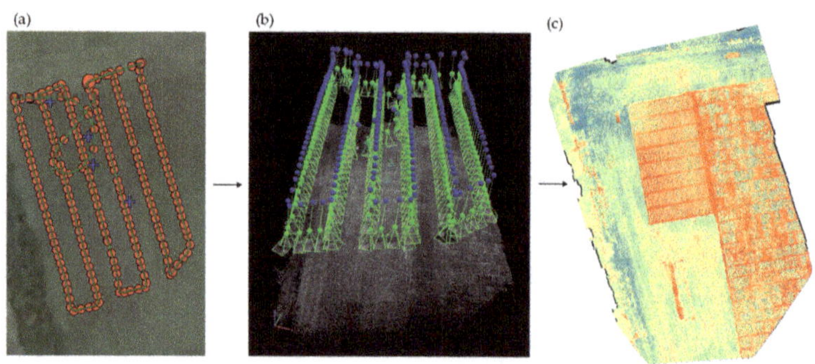

Figure 2. Simplified orthomosaic generating process using aerial multispectral images, (**a**) the flight path (**b**) image alignment and 3D point cloud along the flight path and (**c**) a georectified NIR orthomosaic raster where the band was rendered using pseudocolour (spectral) to visualise intensity variations in the NIR band across the experiment.

Table 1. Details of UAV image acquisition and orthomosaicing quality summary.

Image Acquisition Date	Image Overlap Forward/Side	Flight Speed (m/s)	Flight Height (m)	Mean RMS Error (m)	GSD (cm/pixels)
2019 Autumn	80%/75%	6	30	0.019	2.26
2020 Autumn	80%/75%	6	30	0.010	2.16

2.7. Vegetation Indices Extraction

The plot overlay, that followed the (CRS) EPSG:32755–WGS 84/UTM zone 55S, was digitised by importing calibrated orthomosaic bands into QGIS 2.1. (QGIS Development Team, 2017, Raleigh, NC, USA; https://qgis.org/, accessed on 24 August 2020). Orthomosaic images (raster format) of broadband vegetation indices (listed in Table 2.) were generated under the same CRS (EPSG:32755–WGS 84/UTM zone 55S) using the raster calculator. Reflectance values of generated vegetation indices were extracted through the Zonal statistics plugin. Reflectance values associated with the generated vegetation indices were extracted using the Zonal statistics plugin in QGIS 2.1.

Table 2. The list of vegetation indices derived from multispectral bands.

Vegetation Index	Abbreviation	Equation
Normalised Difference Vegetation Index	NDVI	(Rn − Rr)/(Rn + Rr) [33]
Green Normalised Difference Vegetation Index	GNDVI	(Rn − Rg)/(Rn + Rg) [34]
Red Edge Normalised Difference Vegetation Index	ReNDVI	(Rn − Rre)/(Rn + Rre)
Renormalised Difference Vegetation Index	RDVI	(Rn − Rr)/(Rn + Rr)$^{1/2}$ [35]
Soil Adjusted Vegetation Index	SAVI	(Rn − Rr)/(Rn + Rr + 0.5) × (1 + 0.5) [36]
Normalised green-red Difference Index	NGRDI	(Rg − Rr)/(Rg + Rr) [37]
Simple Ratio Index	SRI	Rn/Rr [38]
Red Edge Simple Ratio Index	ReSRI	Rn/Rre [39]
Green Simple Ratio Index	GSRI	Rn/Rg [40]
Green Leaf Index	GLI	(2 × Rg − Rr- Rb)/(2 × Rg + Rr + Rb) [41]
Chlorophyll Vegetation Index	CVI	Rn × Rr/Rg [42]
Normalised Green Intensity	NGI	Rg/(Rr + Rg + Rb) [43]
Infrared Percentage Vegetation Index	IPVI	Rn/(Rn + Rr) [44]
Visible Atmospherically Resistant Index	VARI	(Rn − Rr)/(Rr + Rg + Rb) [45]
Red Difference Index	RDI	Rn − Rr [46]
Green Difference Index	GDI	Rn − Rg [47]
Canopy Chlorophyll Concentration Index	CCCI	((Rn − Rre)/(Rn + Rre))/NDVI [48]
Core Red Edge Triangular Vegetation Index	CReTVI	100(Rn − Rre) − 10(Rn − Rg) [49]

Rn: reflectance of NIR band; Rr: reflectanceof red band; Rre: reflectance of red edge band; Rb: reflectance of blue band; and Rg: reflectance of green band.

2.8. Ground Cover Extraction from Multispectral Images

Ground cover extraction from the multispectral images was conducted using eCognition Developer 9.3.2 (Trimble Germany GmbH, Munich, Germany; http://www.ecognition.com/, accessed on 20 August 2020) with a predefined rule set, as illustrated in Figure 3. The plot overlay (vector file) and radiometrically corrected multispectral bands were imported into eCognition Developer 9.3.2 for further analysis. The rule set included image segmentation and ground cover classification, following previously described methodology [29]. The plot IDs from the plot overlay were used to reclassify the segmented objects using the "assign class by thematic layer" algorithm in eCognition Developer 9.3.2. The resulting green fraction data were exported as a comma-separated values (.csv) file using the "export object statistics" algorithm in eCognition Developer 9.3.2 for further analysis.

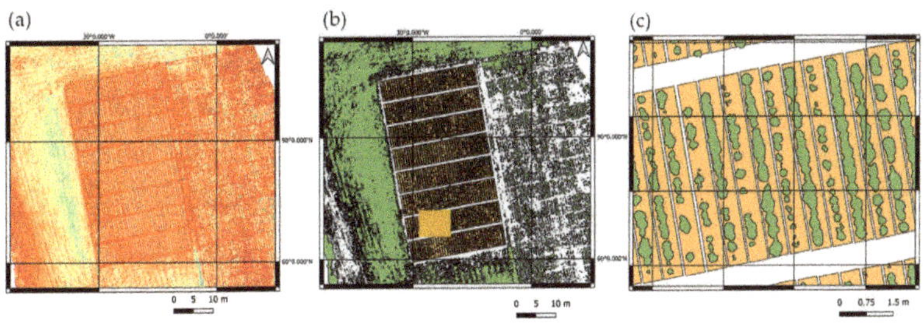

Figure 3. Perennial ryegrass ground cover extraction pipeline (**a**) radiometrically calibrated orthomosaic raster band where the band was rendered using pseudocolour (spectral) to visualise intensity variations across the study site, (**b**) green fraction classification in eCognition developer, (**c**) enlarged area of the light brown square in image (**b**) where green area represents living pasture ground cover and light brown polygons is boundaries of row plots.

2.9. Statistical Analysis

Interval plots of variables were generated using Minitab 19.1.1 (https://www.minitab.com, accessed on 20 August 2020). The most important VIs for representing perennial ryegrass persistence were selected by performing a correlation coefficient (Pearson's correlation coefficient) using pasture ground cover and plant surviving percentage. The effect of genotype, endophyte, and environment on the expression of selected variables was analysed using a linear mixed effect model approach using the following model

Variable = Fixed (Cultivar + Endophyte + Population type + Harvesting year) + all interaction + Random (Replicate + Row) + error

All variables were assumed to follow a normal distribution and denote interaction. All parameters were estimated using a residual maximum likelihood (REML) technique due to a missing cultivar–endophyte combination. Statistical analyses were performed in Genstat 18.2.0.18409 (https://www.vsni.co.uk, accessed on 20 August 2020).

3. Results

3.1. Meteorological Data

The monthly mean minimum and maximum temperature patterns exhibited a similar seasonal trend during the experimental period: May 2018–May 2020, as illustrated in Figure 4. The average maximum temperature from May 2018 to April 2019 was 27.9 °C. In the subsequent period from May 2019 to May 2020, the average maximum temperature slightly decreased to 27.1 °C. A comparable trend was also observed for the total seasonal rainfall during the different growing seasons. In winter, the total rainfall was 70.5 mm in 2018 and increased to 75.2 mm in 2019. In the early spring (September–October) period, the total rainfall was 43.9 mm in 2018 and raised to 62.8 mm in 2019 (Figure 4).

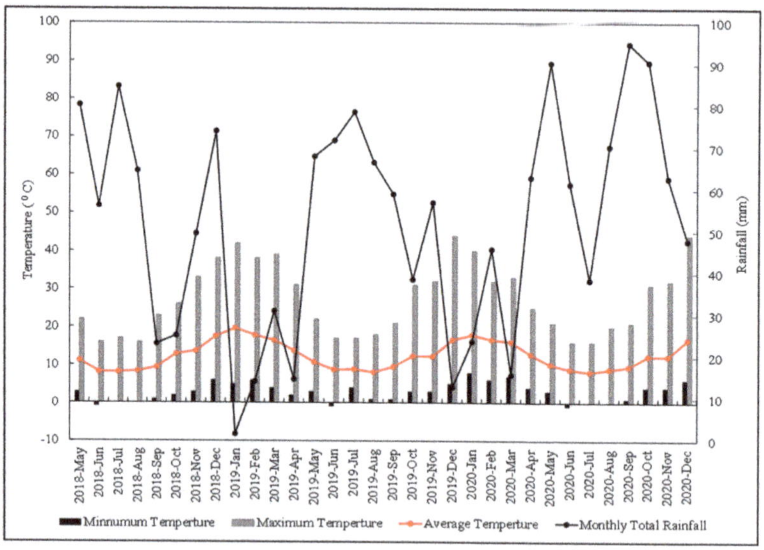

Figure 4. Changes in monthly mean, maximum and minimum temperate, and monthly total rainfall during the experimental period (May 2018—May 2020).

Similarly, in the late spring period (November), the total rainfall increased from 37.4 mm in 2018 to 47.8 mm in 2019. In autumn, the total rainfall was 37.9 mm in 2019 and raised to 60.9 mm in 2020. However, in summer, there was a decrease in the total rainfall from 29.9 mm in 2018 to 22.0 mm in 2019.

3.2. Endophyte Frequency

In 2014, immunoblot assay data showed that the endophyte frequency exceeded 60% for most cultivars. However, only 1% of One50 SE and Samson SE exhibited positive endophyte presence in their remnant seeds in 2020 (Figure 5a). In 2021, the selected populations showed a significantly higher endophyte frequency compared with their respective base populations, except for Samson Nil ($p < 0.001$; Figure 5b). This result was similar to the endophyte percentage observed in the original populations when tested in 2014, and a similar trend was observed in the base populations of all cultivars, except for base populations of Alto AR1, which had a higher frequency of 21%.

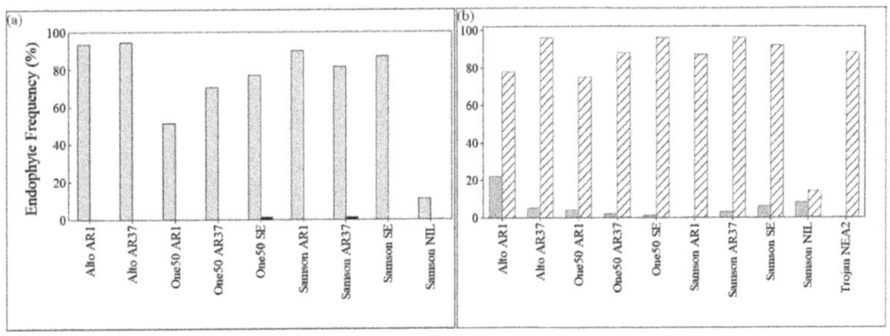

Figure 5. (**a**) Endophyte frequency of selected cultivars in original seed samples, tested in 2014 (■) and 2020 (▨), (**b**) endophyte frequency of perennial ryegrass populations in the base (■) and the selected population (▨) in each cultivar in 2021.

3.3. Manual Pasture Measurement

During the autumn harvest of 2019, a significant increase in dry matter yield was observed in the selected populations of Alto AR37, One50 AR37, One50 SE, Samson AR37, and Samson SE compared with their respective base populations ($p < 0.001$). This finding was consistent with the subsequent autumn harvest in 2020, where the selected populations of all cultivars displayed significantly higher dry matter yield compared with their base populations ($p < 0.001$). There were no significant differences in the base populations of all tested cultivars at the 2019 and 2020 autumn harvests ($p > 0.05$). However, the average of the base population for each cultivar experienced a significant reduction during the 2020 autumn harvest ($p > 0.05$; Figure 6).

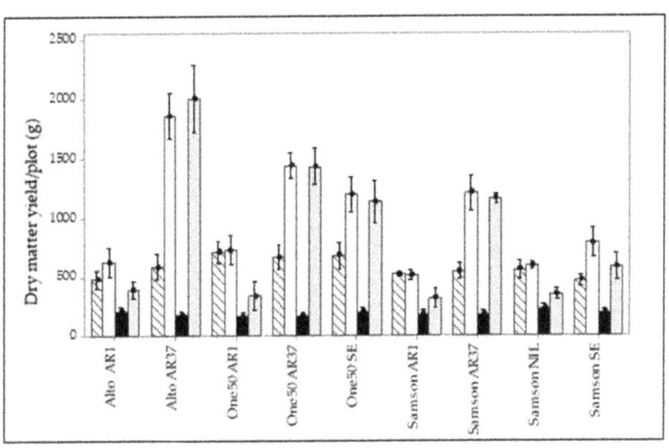

Figure 6. Plot level average DM yield in 2019 and 2020 autumn. Values are means (±SE) of n = 4, where ▨ base population of 2019 autumn, □ selected population of 2019 autumn, ■ base population of 2020 autumn, ■ selected population of 2020 autumn. Two side error bars represent the standard error of reported measurements at 95% confidence interval.

At the plot level, the dry matter yield of the selected populations exhibited significant variation among the cultivars in both 2019 and 2020 ($p < 0.001$). Alto AR37 consistently demonstrated the highest dry matter yield among all cultivars in both growing years, and this was not statistically significance in these two different harvest years ($p > 0.05$). However, in the 2020 autumn harvest, a significant decrease in average plot-level dry matter yield was observed for both selected and base populations of Alto AR1, One50 AR1, Samson AR1, and Samson Nil.

Except for Alto AR37, the surviving plant percentage of all selected cultivars did not significantly ($p > 0.05$) differ between the selected and base populations at the 2019 harvest. However, plant surviving percentage was significantly different among cultivars ($p < 0.001$). Moreover, Alto AR37 showed the highest surviving plant percentage, and Samson AR1 showed the lowest plant surviving percentage at the 2019 harvest. In 2020, the selected population of all cultivars showed significantly higher surviving plant percentages compared with the surviving plant percentage of the base population ($p < 0.001$), and the surviving plant percentage of the selected population was significantly different among all cultivars (Figure 7). Alto AR37 showed the highest plant surviving percentage in selected population at the 2020 autumn harvest. The base population of Alto AR37, One50 AR1 and One50 AR37 showed a similar trend, which was the lowest surviving percentage at the 2020 autumn harvest ($p > 0.05$).

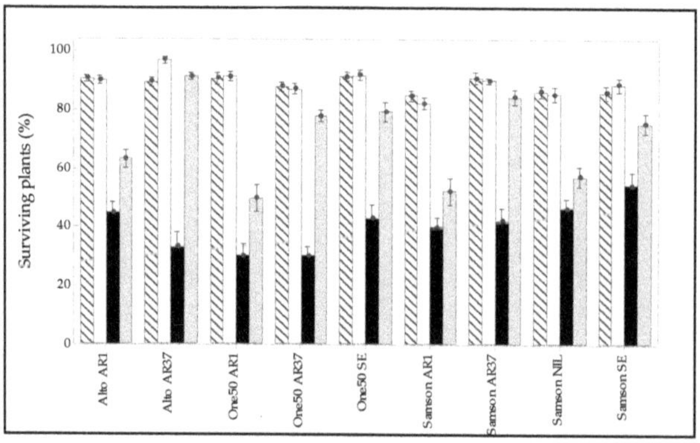

Figure 7. Row level plant surviving percentage, where ▨ base population of 2019 autumn, □ selected population of 2019 autumn, ▩ base population of 2020 autumn, ■ selected population of 2020 autumn. Two side error bars represent standard error of reported.

In the 2019 harvest, the surviving plant percentage of all cultivars, except for Alto AR37, was not significantly different ($p > 0.05$) between the selected and base populations. In contrast, during the 2020 harvest, the selected populations of all cultivars demonstrated significantly higher plant surviving percentages than those of their respective base populations, as depicted in Figure 7 ($p < 0.001$). Alto AR37 exhibited the highest surviving plant percentage within the selected population during the 2020 autumn harvest, and the base populations of Alto AR37, One50 AR1, and One50 AR37 displayed a similar trend, with the lowest plant surviving percentages observed at the 2020 autumn harvest ($p > 0.05$).

3.4. Sensor-Based Pasture Measurements

At the 2019 autumn harvest, the average row-level plant height of both the selected and base populations was found to be significantly lower compared with the row-level plant height observed at the 2020 autumn harvest ($p > 0.05$), as illustrated in Figure 8. However, the selected populations of Alto AR37, One50 AR37, and Samson AR37 exhibited significantly higher plant heights compared with their respective base populations when evaluated at the 2019 autumn harvest ($p < 0.001$). Similar trends in average plant height were observed for all other tested cultivars in both the selected and base populations during the same harvest. However, no significant differences in average row-level plant height were observed among Alto AR37, One50 AR37, and Samson AR37 within the selected population or the base population ($p > 0.05$) (Figure 8). The expression of fractional ground cover also showed the same trend in both selected and base populations (Figure 9).

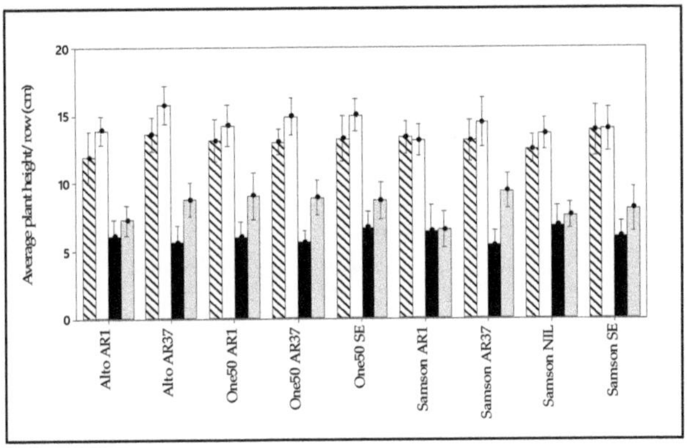

Figure 8. Row level average plant height, where ▨ base population of 2019 autumn, □ selected population of 2020 autumn, ■ base population of 2020 autumn, ▩ selected population of 2020 autumn. Two side error bars represent the standard error of reported measurements at 95% confidence interval.

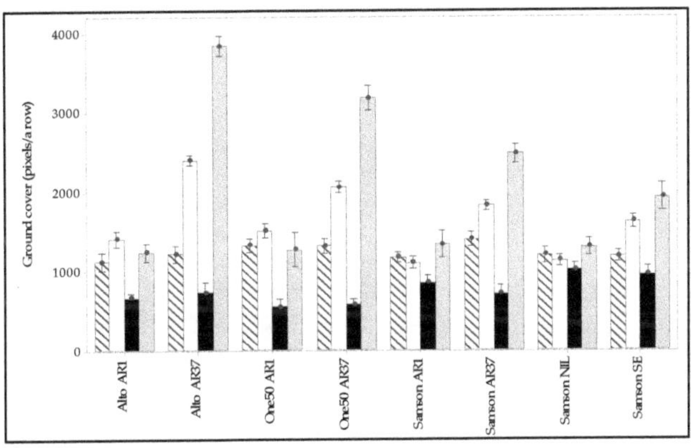

Figure 9. Row level plant ground cover, where ▨ base population of 2019 autumn, □ selected population of 2019 autumn, ■ base population of 2020 autumn, ▩ selected population of 2020 autumn. Two side error bars represent the standard error of reported measurements at a 95% confidence interval.

Analysis of preharvest multispectral images in 2019 revealed a positive correlation between all vegetation indices (VIs) and both row-level surviving plant percentage and pasture ground cover. However, the canopy chlorophyll concentration index (CCCI) did not show a significant correlation with either pasture ground cover or surviving plant percentage. The relationship between ground cover and VIs was stronger than the relationship between VIs and surviving plant percentage. Vegetation indices such as NDVI, GNDVI, RDVI, SAVI, SRI, IPVI, RDI, GDI, and CreTVI exhibited a robust positive relationship with ground cover at the 2019 harvest, with correlation coefficients (r) exceeding 0.95 ($p < 0.001$). This indicates a strong association between these VIs and the extent of ground cover, highlighting their potential as reliable indicators of vegetation density and coverage. A similar trend was observed in the analysis of 2020 autumn sensor-based data, including ground cover and plant surviving percentage. The relationship between VIs and ground cover at the 2020 harvest was even stronger than that observed in the 2019 harvest (Figure 10). These findings further emphasise the utility of VIs in assessing and quantifying ground cover, providing valuable insights into the overall vegetation dynamics and health of the experimental plots.

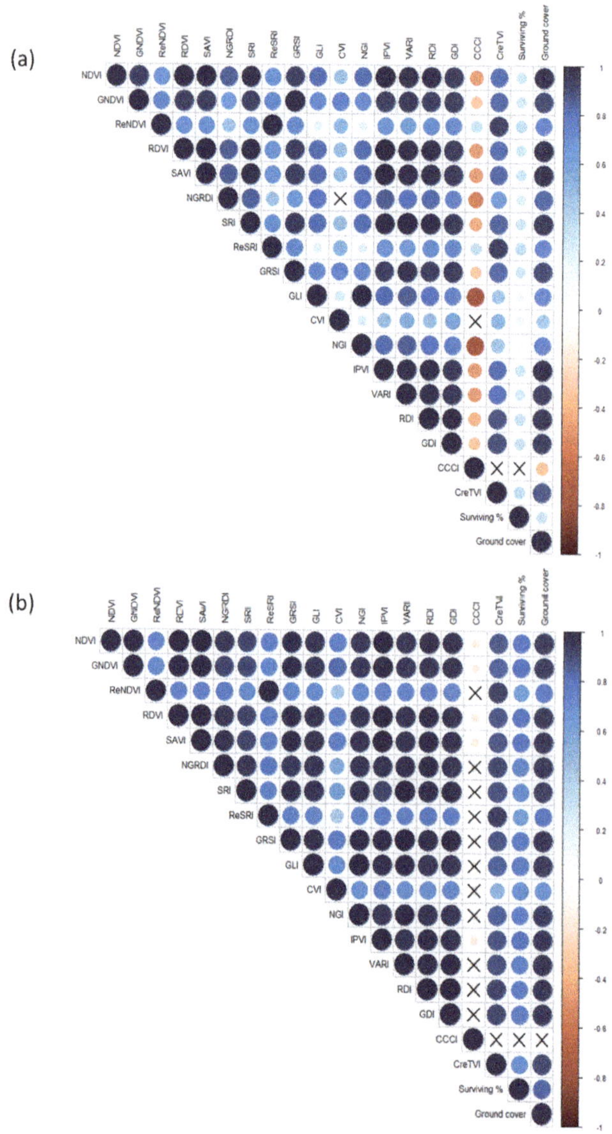

Figure 10. The correlogram for ground truth data (plant surviving % and ground cover) and multispectral vegetation indices; plot (**a**) 2019 autumn data (**b**) represents 2020 autumn data. Positive correlations are displayed in blue and negative correlations in red colour. The colour intensity and the size of the circle are proportional to the correlation coefficients and "×" indicates that no linear relationship exists between two variables.

3.5. Interaction of Endophyte, Cultivar, and Environment on Pasture Persistence

Table 3 provides an overview of the influence of four main factors, cultivar, endophyte, population type, and harvest year, on the expression of measured pasture traits. The results indicate that all four factors have significant main effects on dry matter yield, ground cover, surviving plant percentage, and several vegetation indices (SAVI, SRI, RDI, GLI, and CreTVI). However, the impact of the cultivar on plant height and the average row-level values of NDVI, GNDVI, RDVI, SAVI, and IPVI were not significant ($p > 0.05$). Furthermore, the expression of plant surviving percentage was observed to be independent of the influence of endophyte ($p = 0.07$), suggesting that endophyte presence or absence did not have a significant fixed effect on surviving plant number. Similarly, perennial ryegrass plant height was not influenced by the two-way interaction between endophyte and cultivar ($p = 0.336$), and a similar trend was observed for the two-way interaction between cultivar and population. However, the two-way interaction between endophyte and population significantly ($p < 0.05$) influenced all tested variables, both sensor-based and manual measurements (Figure 10).

Table 3. Summary of REML outcome for measured sensor-based and manual pasture measurements where cultivar, endophyte, population type, and harvest year were considered as fixed factors.

	Variable											
	HY	**PSP**	**GC**	**PH**	**NDVI**	**GNDVI**	**RDVI**	**SAVI**	**SRI**	**IPVI**	**RDI**	**GLI**
Host grass	<0.001	0.006	0.003	0.186	0.055	0.0943	0.057	0.049	<0.001	0.058	<0.001	0.027
Endophyte	<0.001	<0.001	<0.001	0.07	<0.001	<0.001	<0.001	<0.001	<0.001	<0.001	<0.001	<0.001
Population	<0.001	<0.001	<0.001	<0.001	<0.001	<0.001	<0.001	<0.001	<0.001	<0.001	<0.001	<0.001
Season	<0.001	<0.001	<0.001	<0.001	<0.001	<0.001	<0.001	<0.001	<0.001	<0.001	<0.001	<0.001
Host grass × Endophyte	<0.001	<0.0039	<0.001	0.336	<0.001	<0.001	<0.001	<0.001	<0.001	<0.001	<0.001	<0.001
Host grass × Population	<0.001	<0.001	0.012	0.069	<0.001	<0.001	<0.001	<0.001	<0.001	<0.001	<0.001	<0.001
Endophyte × Population	<0.001	<0.001	<0.001	0.047	<0.001	<0.001	<0.001	<0.001	<0.001	<0.001	<0.001	<0.001
Host grass × Season	0.28	0.001	0.001	0.749	0.055	0.154	0.055	0.059	0.063	0.055	0.262	0.474
Endophyte × Season	0.356	<0.001	<0.001	0.53	<0.001	<0.001	<0.001	<0.001	<0.001	<0.001	<0.001	0.003
Population × Season	<0.001	<0.001	<0.001	0.028	<0.001	<0.001	<0.001	<0.001	<0.001	<0.001	<0.001	<0.001
Host grass × Endophyte × Population	0.003	0.291	0.896	0.346	0.081	0.087	0.071	0.074	<0.001	0.071	0.262	0.027
Host grass × Endophyte × Season	0.831	0.212	<0.001	0.635	<0.001	<0.001	<0.001	<0.001	<0.001	<0.001	<0.001	0.273
Host grass × Population × Season	0.395	0.006	0.001	0.562	0.079	0.08	0.079	0.08	0.009	0.089	0.201	0.038
Endophyte × Population × Season	0.192	<0.001	<0.001	0.335	<0.001	<0.001	<0.001	<0.001	<0.001	<0.001	<0.001	<0.001

HY: herbage yield; PSP: plant surviving percentage; GC: ground cover; PH: plant height; abbreviations for vegetation indices are similar to Table 2. In cell show *p*-value for fixed effect or interaction effect at 95% confidence level. In REML analysis, F-tests are calculated using algebraic derivatives, ignoring fixed/boundary/singular variance parameters.

The impact of the two-way interaction between cultivar and harvest year was not significant for most variables, except for plant surviving percentage and plant height. Conversely, the two-way interaction between endophyte and harvest year significantly influenced the expression of all tested variables except for plant height. The interaction between population and harvest year also significantly influenced the expression of all tested variables ($p < 0.05$). Furthermore, the three-way interaction among cultivar, endophyte, population, and harvest year had an impact on all tested variables except for row-level surviving plant percentage. However, the strength and nature of this three-way interaction varied among the different variables (Table 3).

4. Discussion

4.1. Pasture Traits for Pasture Persistence Estimation

The aim of this experiment was to evaluate the impact of pasture cultivar, endophyte, and environment on the persistence of perennial ryegrass populations using sensor-based and manual pasture measurements. Persistence is a crucial trait in perennial pastures, typically assessed through field observations in the second or third year after sowing [50]. In the southeast region of Australia, where this study took place, perennial ryegrass experiences a period of no net growth during the summer months due to high temperatures and limited soil moisture. However, with the arrival of opening rainfall and the decline in temperature during autumn, active growth resumes in perennial ryegrass, providing an ideal time to evaluate persistence [30]. Therefore, pasture measurements collected after the autumn break were utilised to investigate the influence of cultivar, endophyte, and environment on perennial ryegrass persistence. The selected population in this experiment was derived from a pre-grazed sward that had experienced five previous summer periods before the start of this experiment. In contrast, the base population was obtained from remnant seeds of the selected population, with the plants experiencing two summer periods during the experiment. By using both the selected and base populations of the cultivars, this study aimed to examine the impact of short-term and long-term environmental factors on the expression of perennial ryegrass persistence within the same breeding line.

Persistence, along with productivity, is a critical characteristic of perennial pastures [51]. The assessment of pasture persistence can be determined through methods such as plant ground cover estimation or counting the number of plants in a defined area. Additionally, pasture height provides an approximate estimation of the available green feed in a paddock. However, pasture height may indicate plant surviving percentage or pasture density of a pasture population. Over time, a cultivar with poor persistence will experience a decline in annual pasture dry matter production due to the depletion of surviving plants [30]. As a result, there is a close relationship between pasture persistence and variables such as plant ground cover, dry matter yield, plant height, and the number of surviving plants in a sward [29]. Moreover, recent studies showed multispectral vegetation indices have great feasibility to assess persistence of perennial ryegrass at plot scale [24]. Therefore, this study used dry matter yield, fractional ground cover, plant surviving rate, and multispectral vegetation indices to evaluate the persistence of perennial ryegrass cultivars at plot scale and identify potential constraints on their expression of persistence.

The conventional methods employed for assessing pasture persistence within breeding trials and at the paddock scale present significant limitations, relying on destructive, labour-intensive, and non-digital approaches that hinder comprehensive analysis [24,29,30]. These constraints have made it challenging to efficiently phenotype plant traits, impeding the progress of pasture breeding programs. However, recent advances in sensor technology offer a promising avenue to address the challenge of plant phenotyping at scale. This experiment we were able to monitor persistence traits non-destructively throughout the experiment by utilising sensor-based platforms. These high-throughput phenotyping platforms enabled rapid and efficient screening of a large number of plots, maintaining data accuracy and consistency. As this research paper investigates the existing image-based high-throughput approaches for field phenotyping and evaluates their potential to replace conventional methods in pasture breeding programs, it contributes to the broader understanding of how such cutting-edge technologies can propel the advancement of agricultural practices. Therefore, in this experiment, along with high-throughput data from ground-based and airborne remote sensing platforms, the number of surviving plants, pasture height, pasture herbage yield, are used as indicators to assess the persistence of selected cultivars.

The presence of endophyte in perennial ryegrass may significantly improve the persistence of perennial ryegrass in temperate environments by enhancing protection from insect attack and water deficit conditions [52]. The fixed effect of endophyte infection of selected cultivar–endophyte combinations suggests that infection of endophyte may improve herbage dry matter yield, surviving plant percentage, and ground cover (Table 3). The selected population of cultivars with novel endophytes; Samson AR37 and One50 AR37, showed greater persistence than breeding lines with standard enodophytes: Samson SE and One50 SE. Our results suggest that novel endophytes such as AR37 may be more effective than standard endophytes (SEs) as these novel endophytes improved dry matter yield, resistance against biotic and abiotic stresses [53], and competition with other pasture species. Sutherland and Hoglund, 1989 [54], discovered that endophyte-infected ryegrass cultivars are more resistant to Argentine stem weevil and offered better persistence and higher dry matter yields than endophyte-free perennial ryegrasses. However, they are also more competitive against white clover and can reduce clover dry matter yield and persistence. Recent studies reported that there is a positive effect of novel endophyte on perennial ryegrass dry matter yield in the temperate region of New Zealand particularly during autumn and summer [14] and under subtropical conditions [53]. The findings of these recent studies agreed with the observed pattern of pasture measurements in our study. However, infection with endophyte in perennial ryegrass plants did not change the expression of some plant traits, such as plant height (Table 3).

4.2. Effect of the Environment on the Expression of Pasture Persistence

In the temperate region of Australia, pasture persistence is influenced by a combination of primary climatic conditions, defoliation, and diseases [15]. In this experiment, the base and selected population offered a long-term and short-term environmental effect on pasture persistence. Our results indicated that the population type had a significant fixed effect on the expression of all measured pasture measurements. Over time, the selected population may have undergone selection pressure, leading to the development of stronger genotypes that exhibit higher dry matter yields, greater ground cover, increased plant survival, and taller plant height. Adaption of the selected population explains its superior performance in terms of these phenotypic traits [55]. Additionally, the harvesting year was identified as a short-term environmental influence on the expression of plant phenotypic traits. The fixed effect of the harvest year was significant for all measured parameters (Table 3). Individual genotypes within the selected population may have developed adaptations to specific environmental conditions experienced during different harvesting years. The variations observed in terms of dry matter yield, ground cover, plant survival, and plant height in both the 2019 and 2020 autumn harvests may be attributed to differences in primary climatic conditions during the experiment. The distribution of rainfall in different seasons can significantly impact on pasture dry matter yield and plant survival [56]. Frequent rainfall events were recorded in the summer of 2019, potentially increasing soil moisture compared with the previous year. This change in environmental conditions may have triggered the activation of dormant ryegrass buds, leading to their premature death when exposed to subsequent hot and dry summer conditions. Failure to produce new buds during suitable growing conditions can result in a decrease in plant density [57]. The observed decrease in plant survival percentage of all cultivars at the 2020 autumn harvest may be influenced by fluctuations in primary climatic factors, including temperature and rainfall (Figure 4). Thus, changes in short-term environmental factors during the harvest year were significant contributors to the variation in perennial ryegrass persistence [58]. However, the impact of these environmental factors on the persistence of the selected populations was minor compared with their respective base populations. Nevertheless, the expression patterns of pasture traits in both populations may depend on the interactions among perennial ryegrass genotypes, endophytes, and environmental factors.

4.3. Interaction of Cultivar, Endophyte, and Environment

The REML analysis conducted in this experiment revealed an interaction between perennial ryegrass cultivars and endophytes in the expression of persistence within a given environment. The perennial ryegrass cultivars exhibited higher dry matter yields, ground cover, and plant survival percentages when interacting with endophytes or the environment. For instance, endophyte AR1 improved dry matter yield when paired with Alto but decreased dry matter yield when paired with Samson. Among the cultivars, the selected populations of Alto AR37, Samson AR37, and One50 AR37 showed the most stable cultivar–endophyte interactions in terms of persistence. The AR37 endophyte, that produces a complex of janthitrems and lacks the alkaloids found in wild-type endophytes [59] may contribute to enhance of persistence in a given environment. A recent study reported that the presence of the AR37 endophyte in the cultivar Alto increased the survival rate [60]. However, the expression of persistence in perennial ryegrass cultivars did not vary significantly due to rainfall and temperature variations during the growing periods. Therefore, of the endophytes tested in this study, AR37 endophyte appears to be the most suitable strain for enhancing persistence in the southwest region of Victoria.

The occurrence and performance of endophytes can be influenced by plant genotype and environmental factors, and the genomic information of endophytes can undergo changes through natural selection [55]. Thus, the observed persistence of perennial ryegrass can be influenced by the interaction between endophytes and the environment and between cultivars and the environment. Moreover, the persistence of perennial ryegrass is attributed to three-way interactions among cultivar, endophyte, and environmental factors, whether they are long-term or short-term in nature. These complex interactions highlight the multifaceted nature of persistence in perennial ryegrass and emphasise the need to consider both genetic and environmental factors in breeding and management practices.

5. Conclusions

The experiment aimed to investigate the factors influencing the persistence of perennial ryegrass in the temperate region of Australia, focusing on cultivar, endophyte, and environmental factors. The results revealed that the selected population exhibited higher dry matter yields, ground cover, and plant survival percentages compared with the base population, indicating the importance of long-term environmental effects. Additionally, the presence of the AR37 endophyte, with its unique janthitrem production, was associated with enhanced persistence in specific cultivars such as Alto, Samson, and One50. This study indicated that cultivars exhibit high levels of persistence when cultivars infect with novel endophyte strains. Therefore, this study revealed significant fixed effects of cultivar, endophyte, and environmental factors. Moreover, the interaction between cultivars, endophytes, and environmental factors played a significant role in the expression of persistence. Although variations in rainfall and temperature during growing periods did not significantly impact persistence, the AR37 endophyte emerged as the most suitable strain for enhancing perennial ryegrass persistence in the southwest region of Victoria. These findings highlight the complex interplay of genetic traits, endophyte interactions, and environmental conditions in determining the persistence of perennial ryegrass. This knowledge can guide breeding programs and management strategies to develop improved cultivars with enhanced persistence under diverse environmental conditions. Further research is warranted to unravel the underlying mechanisms of these interactions, aiding in the development of sustainable and resilient perennial ryegrass pastures.

Author Contributions: C.J.: data collection, data analysis, writing-original draft preparation, K.S. and J.J.: supervision, review and editing, A.T.: review and editing. All authors have read and agreed to the published version of the manuscript.

Funding: This research was funded by Dairy Australia, Gardiner Dairy Foundation, and Agriculture Victoria Research under Dairy Feedbase Project.

Data Availability Statement: Data sharing not applicable.

Acknowledgments: We thank DairyBio Hamilton team for their technical assistance during data collection and Kathryn Guthridge, Emma Ludlow, and Inoka Hettiarachchige for genetic analysis.

Conflicts of Interest: The authors declare no conflict of interest.

References

1. Waller, R.A.; Sale, P.W.G. Persistence and productivity of perennial ryegrass in sheep pastures in south-western Victoria: A review. *Aust. J. Exp. Agric.* **2001**, *41*, 117–144. [CrossRef]
2. Kemp, D.R.; Dowling, P.M. Towards sustainable temperate perennial pastures. *Aust. J. Exp. Agric.* **2000**, *40*, 125–132. [CrossRef]
3. Cunningham, P.J.; Blumenthal, M.J.; Anderson, M.W.; Prakash, K.S.; Leonforte, A. Perennial ryegrass improvement in Australia. *N. Z. J. Agric. Res.* **1994**, *37*, 295–310. [CrossRef]
4. Cunningham, P.; Graves, W.; Chakroun, M.; Mezni, M.; Saidi, S.; Ounejmate, M.; Porqueddu, C.; Reed, K. Novel perennial forage germplasm from North Africa and Sardinia. *Aust. Plant Introd. Rev.* **1997**, *27*, 13–46.

5. Fulkerson, W.; Sinclair, K.; Moore, K.; Rolfe, C. Management of *Lolium perenne/Trifolium repens* pastures in the subtropics. I. Effect of defoliation interval, seeding rate and application of N and lime. *Aust. J. Agric. Res.* **1993**, *44*, 1947–1958. [CrossRef]
6. Mantri, N.; Patade, V.; Penna, S.; Ford, R.; Pang, E. Abiotic Stress Responses in Plants: Present and Future. In *Abiotic Stress Responses in Plants: Metabolism, Productivity and Sustainability*; Ahmad, P., Prasad, M.N.V., Eds.; Springer: New York, NY, USA, 2012; pp. 1–19. [CrossRef]
7. Anjum, S.; Xie, X.-Y.; Wang, L.-C.; Saleem, M.; Man, C.; Lei, W. Morphological, physiological and biochemical responses of plants to drought stress. *Afr. J. Agric. Res.* **2011**, *6*, 2026–2032. [CrossRef]
8. Malcolm, B.; Smith, K.F.; Jacobs, J.L. Perennial pasture persistence: The economic perspective. *Crop Pasture Sci.* **2014**, *65*, 713–720. [CrossRef]
9. Bolaric, S.; Barth, S.; Melchinger, A.E.; Posselt, U.K. Molecular characterization of genetic diversity in European germplasm of perennial ryegrass. *Euphytica* **2005**, *146*, 39–44. [CrossRef]
10. Guthridge, K.M.; Dupal, M.P.; Kölliker, R.; Jones, E.S.; Smith, K.F.; Forster, J.W. AFLP analysis of genetic diversity within and between populations of perennial ryegrass (*Lolium perenne* L.). *Euphytica* **2001**, *122*, 191–201. [CrossRef]
11. Skøt, L.; Sanderson, R.; Thomas, A.; Skøt, K.; Thorogood, D.; Latypova, G.; Asp, T.; Armstead, I. Allelic Variation in the Perennial Ryegrass flowering locus T Gene Is Associated with Changes in Flowering Time across a Range of Populations. *Plant Physiol.* **2011**, *155*, 1013–1022. [CrossRef]
12. Thom, E.R.; Popay, A.J.; Hume, D.E.; Fletcher, L.R. Evaluating the performance of endophytes in farm systems to improve farmer outcomes—A review. *Crop Pasture Sci.* **2012**, *63*, 927–943. [CrossRef]
13. Hume, D.E.; Sewell, J.C. Agronomic advantages conferred by endophyte infection of perennial ryegrass (*Lolium perenne* L.) and tall fescue (*Festuca arundinacea* Schreb.) in Australia. *Crop Pasture Sci.* **2014**, *65*, 747–757. [CrossRef]
14. Popay, A.J.; Hume, D.E.; Davis, K.L.; Tapper, B.A. Interactions between endophyte (*Neotyphodium* spp.) and ploidy in hybrid and perennial ryegrass cultivars and their effects on Argentine stem weevil (*Listronotus bonariensis*). *N. Z. J. Agric. Res.* **2003**, *46*, 311–319. [CrossRef]
15. Latch, G.C.M. Physiological interactions of endophytic fungi and their hosts. Biotic stress tolerance imparted to grasses by endophytes. *Agric. Ecosyst. Environ.* **1993**, *44*, 143. [CrossRef]
16. Culvenor, R.A.; Simpson, R.J. Persistence traits in perennial pasture grasses: The case of phalaris. *Crop Pasture Sci.* **2014**, *65*, 1165–1176. [CrossRef]
17. Finch, H.J.S.; Samuel, A.M.; Lane, G.P.F. 21—Grazing management. In *Lockhart & Wiseman's Crop Husbandry Including Grassland*, 9th ed.; Finch, H.J.S., Samuel, A.M., Lane, G.P.F., Eds.; Woodhead Publishing: London, UK, 2014; pp. 499–512. [CrossRef]
18. Boschma, S.P.; Hill, M.J.; Scott, J.M.; Lutton, J.J. Carbohydrate reserves of perennial ryegrasses: Effect drought and defoliation intensity. In Proceedings of the Grasslands 2000—Proceedings of the XVIII International Grassland Congress, Winnipeg, MB, Saskatoon, SK, Canada, 8–19 June 1997; pp. 22–42.
19. Frank, D.A.; McNaughton, S.J.; Tracy, B.F. The Ecology of the Earth's Grazing Ecosystems: Profound functional similarities exist between the Serengeti and Yellowstone. *BioScience* **1998**, *48*, 513–521. [CrossRef]
20. Kemp, D.R.; Culvenor, R.A. Improving the grazing and drought tolerance of temperate perennial grasses. *N. Z. J. Agric. Res.* **1994**, *37*, 365–378. [CrossRef]
21. Sanford, P.; Cullen, B.R.; Dowling, P.M.; Chapman, D.F.; Garden, D.L.; Lodge, G.M.; Andrew, M.H.; Quigley, P.E.; Murphy, S.R.; King, W.M.; et al. SGS Pasture Theme: Effect of climate, soil factors and management on pasture production and stability across the high rainfall zone of southern Australia. *Aust. J. Exp. Agric.* **2003**, *43*, 945–959. [CrossRef]
22. Tozer, K.; BourdÃ´t, G.W.; Edwards, G.R. What factors lead to poor pasture persistence and weed ingress? *NZG Res. Pract. Ser.* **2011**, *15*, 129–137. [CrossRef]
23. Brougham, R.W. Some factors affecting the persistency of short-rotation ryegrass. *N. Z. J. Agric. Res.* **1961**, *4*, 516–522. [CrossRef]
24. Jayasinghe, C.; Badenhorst, P.; Jacobs, J.; Spangenberg, G.; Smith, K. Image-based high-throughput phenotyping for the estimation of persistence of perennial ryegrass (*Lolium perenne* L.)—A review. *Grass Forage Sci.* **2021**, *76*, 321–339. [CrossRef]
25. Pratap, A.; Tomar, R.; Kumar, J.; Pandey, V.R.; Mehandi, S.; Katiyar, P.K. High-Throughput Plant Phenotyping Platforms. In *Phenomics in Crop Plants: Trends, Options and Limitations*; Kumar, J., Pratap, A., Kumar, S., Eds.; Springer: New Delhi, India, 2015; pp. 285–296. [CrossRef]
26. Araus, J.L.; Cairns, J.E. Field high-throughput phenotyping: The new crop breeding frontier. *Trends Plant Sci.* **2014**, *19*, 52–61. [CrossRef]
27. Qiu, Q.; Sun, N.; Bai, H.; Wang, N.; Fan, Z.; Wang, Y.; Meng, Z.; Li, B.; Cong, Y. Field-Based High-Throughput Phenotyping for Maize Plant Using 3D LiDAR Point Cloud Generated With a "Phenomobile". *Front. Plant Sci.* **2019**, *10*, 554. [CrossRef]
28. Gebremedhin, A.; Badenhorst, P.; Wang, J.; Giri, K.; Spangenberg, G.; Smith, K. Development and Validation of a Model to Combine NDVI and Plant Height for High-Throughput Phenotyping of Herbage Yield in a Perennial Ryegrass Breeding Program. *Remote Sens.* **2019**, *11*, 2494. [CrossRef]
29. Jayasinghe, C.; Badenhorst, P.; Wang, J.; Jacobs, J.; Spangenberg, G.; Smith, K. An Object-Based Image Analysis Approach to Assess Persistence of Perennial Ryegrass (*Lolium perenne* L.) in Pasture Breeding. *Agronomy* **2019**, *9*, 501. [CrossRef]
30. Jayasinghe, C.; Badenhorst, P.; Jacobs, J.; Spangenberg, G.; Smith, K. High-Throughput Ground Cover Classification of Perennial Ryegrass (*Lolium perenne* L.) for the Estimation of Persistence in Pasture Breeding. *Agronomy* **2020**, *10*, 1206. [CrossRef]

31. Siebring, J.; Valente, J.; Domingues Franceschini, M.H.; Kamp, J.; Kooistra, L. Object-Based Image Analysis Applied to Low Altitude Aerial Imagery for Potato Plant Trait Retrieval and Pathogen Detection. *Sensors* **2019**, *19*, 5477. [CrossRef] [PubMed]
32. Donaghy, D.; Fulkerson, B. Principles for Developing an Effective Grazing Management System for Ryegrass-Based Pastures. 1999. Available online: https://www.dairyaustralia.com.au/resource-repository/2020/07/08/principles-for-developing-an-effective-grazing-management-system-for-ryegrass-based-pastures (accessed on 24 July 2019).
33. Alem, G.; Badenhorst, P.E.; Wang, J.; Spangenberg, G.C.; Smith, K.F. Prospects for Measurement of Dry Matter Yield in Forage Breeding Programs Using Sensor Technologies. *Agronomy* **2019**, *9*, 65.
34. Hunt, R.; Hively, W.; McCarty, G.; Daughtry, C.; Forrestal, P.; Kratochvil, R.; Carr, J.; Allen, N.; Fox-Rabinovitz, J.; Miller, C. NIR-Green-Blue High-Resolution Digital Images for Assessment of Winter Cover Crop Biomass. *GIScience Remote Sens.* **2011**, *48*, 86–98. [CrossRef]
35. Fu, Y.; Yang, G.; Wang, J.; Feng, H. A comparative analysis of spectral vegetation indices to estimate crop leaf area index. *Intell. Autom. Soft Comput.* **2013**, *19*, 315–326. [CrossRef]
36. Haboudane, D.; Miller, J.R.; Tremblay, N.; Zarco-Tejada, P.J.; Dextraze, L. Integrated narrow-band vegetation indices for prediction of crop chlorophyll content for application to precision agriculture. *Remote Sens. Environ.* **2002**, *81*, 416–426. [CrossRef]
37. Prabhakara, K.; Hively, W.D.; McCarty, G.W. Evaluating the relationship between biomass, percent groundcover and remote sensing indices across six winter cover crop fields in Maryland, United States. *Int. J. Appl. Earth Obs. Geoinf.* **2015**, *39*, 88–102. [CrossRef]
38. Jordan, C.F. Derivation of Leaf-Area Index from Quality of Light on the Forest Floor. *Ecology* **1969**, *50*, 663–666. [CrossRef]
39. Gitelson, A.A.; Viña, A.; Ciganda, V.; Rundquist, D.C.; Arkebauer, T.J. Remote estimation of canopy chlorophyll content in crops. *Geophys. Res. Lett.* **2005**, *32*, 688. [CrossRef]
40. Sripada, R.; Heiniger, R.; White, J.; Meijer, A. Aerial Color Infrared Photography for Determining Early In-Season Nitrogen Requirements in Corn. *Agron. J.* **2006**, *98*, 200. [CrossRef]
41. Louhaichi, M.; Borman, M.M.; Johnson, D.E. Spatially Located Platform and Aerial Photography for Documentation of Grazing Impacts on Wheat. *Geocarto Int.* **2001**, *16*, 65–70. [CrossRef]
42. Vincini, M.; Frazzi, E. Comparing narrow and broad-band vegetation indices to estimate leaf chlorophyll content in planophile crop canopies. *Precis. Agric.* **2011**, *12*, 334–344. [CrossRef]
43. Woebbecke, D.M.; Meyer, G.E.; Von Bargen, K.; Mortensen, D.A. Color Indices for Weed Identification Under Various Soil, Residue, and Lighting Conditions. *Trans. ASAE* **1995**, *38*, 259–269. [CrossRef]
44. Payero, J.O.; Christopher, N.; Wright, J.L. Comparison of eleven vegetation indices for estimating plant height of alfalfa and grass. *Appl. Eng. Agric.* **2004**, *20*, 57. [CrossRef]
45. Gitelson, A.A.; Kaufman, Y.J.; Stark, R.; Rundquist, D. Novel algorithms for remote estimation of vegetation fraction. *Remote Sens. Environ.* **2002**, *80*, 76–87. [CrossRef]
46. Tucker, C.J. Red and photographic infrared linear combinations for monitoring vegetation. *Remote Sens. Environ.* **1979**, *8*, 127–150. [CrossRef]
47. Sripada, R. Determining In-Season Nitrogen Requirements for Corn Using Aerial Color-Infrared Photography. 2005. Available online: https://repository.lib.ncsu.edu/handle/1840.16/4200 (accessed on 24 July 2019).
48. Jago, R.A.; Cutler, M.E.J.; Curran, P.J. Estimating Canopy Chlorophyll Concentration from Field and Airborne Spectra. *Remote Sens. Environ.* **1999**, *68*, 217–224. [CrossRef]
49. Chen, P.-F.; Nicolas, T.; Wang, J.-H.; Philippe, V.; Huang, W.-J.; Li, B.-G. New Index for Crop Canopy Fresh Biomass Estimation. *Spectrosc. Spectr. Anal.* **2010**, *30*, 512–517. [CrossRef]
50. Wilkins, P.W. Breeding perennial ryegrass for agriculture. *Euphytica* **1991**, *52*, 201–214. [CrossRef]
51. Cullen, B.R.; Rawnsley, R.P.; Eckard, R.J.; Christie, K.M.; Bell, M.J. Use of modelling to identify perennial ryegrass plant traits for future warmer and drier climates. *Crop Pasture Sci.* **2014**, *65*, 758–766. [CrossRef]
52. Hume, D.E.; Barker, D.J. Growth and Management of Endophytic Grasses in Pastoral Agriculture. In *Neotyphodium in Cool-Season Grasses*; John Wiley & Sons: Hoboken, NJ, USA, 2005; pp. 201–226. [CrossRef]
53. Lowe, K.F.; Bowdler, T.M.; Hume, D.E.; Casey, N.D.; Tapper, B.A. The effect of endophyte on the performance of irrigated perennial ryegrasses in subtropical Australia. *Aust. J. Agric. Res.* **2008**, *59*, 567–577. [CrossRef]
54. Sutherland, B.L.; Hoglund, J.H. Effcet of ryegrass containing the endophyte (*Acremonium lolii*), on the performance of associated white clover and subsequent crops. *Proc. N. Z. Grassl. Assoc.* **1989**, *50*, 265–269.
55. Cashman, P.; Gilliland, T.; McEvoy, M.; Watson, S.; O'Donovan, M. Changes in plant morphological expression in 12 perennial ryegrass cultivars following frequent and infrequent cutting management. *J. Agric. Sci.* **2015**, *154*, 456–471. [CrossRef]
56. Kaloki, P.; Trethowan, R.; Tan, D.K.Y. Effect of genotype × environment × management interactions on chickpea phenotypic stability. *Crop Pasture Sci.* **2019**, *70*, 453–462. [CrossRef]
57. Biddiscombe, E.; Rogers, A.; Maller, R. Summer dormancy, regeneration and persistence of perennial grasses in south-western Australia. *Aust. J. Exp. Agric.* **1977**, *17*, 795–801. [CrossRef]
58. Nie, Z.; Norton, M.R. Stress Tolerance and Persistence of Perennial Grasses: The Role of the Summer Dormancy Trait in Temperate Australia. *Crop Sci.* **2009**, *49*, 2405–2411. [CrossRef]

59. Tian, P.; Le, T.-N.; Ludlow, E.J.; Smith, K.F.; Forster, J.W.; Guthridge, K.M.; Spangenberg, G.C. Characterisation of novel perennial ryegrass host-Neotyphodium endophyte associations. *Crop Pasture Sci.* **2013**, *64*, 716–725. [CrossRef]
60. Faville, M.J.; Crush, J.R.; Hong, W.; Phillips, H.; Lee, J.M.; Chapman, D. Effects of pasture age on the genotype and phenotype of perennial ryegrass. *Grass Forage Sci.* **2020**, *75*, 135–144. [CrossRef]

agronomy MDPI

Article

Modeling Climate Change Indicates Potential Shifts in the Global Distribution of Orchardgrass

Jiqiang Wu [1,†], Lijun Yan [2,†], Junming Zhao [1], Jinghan Peng [1], Yi Xiong [1], Yanli Xiong [1] and Xiao Ma [1,*]

1 College of Grassland Science and Technology, Sichuan Agricultural University, Chengdu 611130, China; immortalwjq@outlook.com (J.W.); junmingzhao163@163.com (J.Z.); m17623220313@163.com (J.P.); xiongyi95@126.com (Y.X.); yanlimaster@126.com (Y.X.)
2 Sichuan Academy of Grassland Sciences, Chengdu 611743, China; yanlijun456@126.com
* Correspondence: maxiao@sicau.edu.cn
† These authors contributed equally to this work and should be considered co-first authors.

Abstract: Orchardgrass (*Dactylis glomerata* L.) is highly tolerant of shade, cold, and overwintering, making it an ideal species for grassland ecological restoration and livestock production. However, the genetic diversity of orchardgrass may be threatened by climate change. Using a Maximum Entropy (MaxEnt) model with the BCC-CSM2-MR global climate database and the Harmonized World Soil Database, we projected the current and future distribution of orchardgrass suitable areas globally. The predicted ecological thresholds for vital environmental factors were determined to be a temperature seasonality range of 411.50–1034.37 °C, a mean diurnal range of −0.88–10.69 °C, a maximum temperature of the warmest month of 22.21–35.45 °C, and precipitation of the coldest quarter of 116.56–825.40 mm. A range of AUC values from 0.914 to 0.922, indicating the accuracy of the prediction model. Our results indicate that the total area of current suitable habitats for orchardgrass was estimated to be 2133.01 × 10^4 km^2, it is dispersed unevenly over six continents. Additionally, the suitable areas of habitats increased in higher latitudes while decreasing in lower latitudes as greenhouse gas emissions increased. Therefore, efforts should be made to save places in the southern hemisphere that are in danger of becoming unsuitable, with the possibility of using northern America, China, and Europe in the future for conservation and extensive farming.

Keywords: bioclimatic; climate change; *Dactylis glomerata*; habitat shift; MaxEnt model; potential geographic distribution

Citation: Wu, J.; Yan, L.; Zhao, J.; Peng, J.; Xiong, Y.; Xiong, Y.; Ma, X. Modeling Climate Change Indicates Potential Shifts in the Global Distribution of Orchardgrass. *Agronomy* **2023**, *13*, 1985. https://doi.org/10.3390/agronomy13081985

Academic Editor: Saseendran S. Anapalli

Received: 20 June 2023
Revised: 22 July 2023
Accepted: 24 July 2023
Published: 27 July 2023

1. Introduction

Orchardgrass (*Dactylis glomerata* L.) is a perennial cold-season forage grass of the Poaceae family native to northern Africa, Europe, and some temperate regions of Asia [1,2]. It plays a crucial role in the generation of forage-based meat and dairy across temperate areas as one of the top four commercially significant forage types of grass grown globally [3,4]. Orchardgrass is known for its rapid growth, high biomass, high sugar content, strong shade tolerance, and adaptability to different environments [5–8]. Additionally, orchardgrass has been cultivated in North America for over two centuries and is currently one of the most widely cultivated grass, primarily for grass grazing and hay production [1]. Orchardgrass exhibits high genetic diversity, extensive geographical distribution, and varied habitat conditions, making it an excellent candidate grass for further genetic and ecological studies [2]. The temperature, moisture, and soil conditions are key factors influencing the growth and development of vegetation in orchardgrass [9–11]. Due to climate change, the regional climate patterns might change, and catastrophic climatic phenomena such as heat and droughts will become more frequent, leading to the extinction of species that cannot adapt to the environment or have a limited capacity to adjust [12]. Climate change will likely cause further harm to orchardgrass varieties and habitats in the future. Therefore, it is vital to improve the management and preservation of the main distribution regions of orchardgrass and safeguard critically threatened natural populations through in situ and ex-situ conservation measures.

Niches are habitats with the minimum thresholds necessary for survival [13]. The grassland niche is highly influenced by the surrounding environment, causing it to adapt or relocate in response to environmental changes. The Earth's temperature is projected to rise by 0.2 °C per decade due to greenhouse gas (GHG) emissions, with expected increases ranging from 0.3–1.7 °C at a minimum, up to 2.6–4.8 °C at a maximum for the twenty-first century [14]. As a widely distributed temperate perennial grass species, orchardgrass will inevitably struggle to move and colonize suitable habitats at a fast enough pace to cope with the predicted rapid climate change. This could result in elevated rates of species extinction, as well as diminished plant growth and yield due to the warmer temperatures [15,16]. As a result, there is a pressing need to determine the amount of climate change during the next few decades and evaluate its impact on specific indigenous grass habitats using a variety of techniques, which will develop future conservation and cultivation plans [17].

Species distribution modeling is a developing field of study that employs niche theory to derive the ecological requirements of specific species using mathematical models. These models combine environmental factors and occurrence data to provide a statistical or mechanical representation of the organism's probable distribution [18–20]. Currently, the most widely utilized niche models for species distribution are GARP (Genetic Algorithm for Rule-set Production) [21], MaxEnt (Maximum Entropy modeling) [22], Bioclim [23], Random Forest [24], and the Boosted Regression Tree [25]. Notably, MaxEnt, in accordance with the principle of maximum entropy [19,22,26], is frequently regarded as outperforming other species distribution models (SDMs) due to its strong toleration and precise forecasting in many model intercomparisons [27–29]. Researchers worldwide in the last decade have achieved significant success in applying species distribution models to issues such as protecting the diversity of rare animals and plants [30–33], estimating the dangers of invasive species [34–36], protecting marine ecosystem [37,38], predicting disaster distribution [39], and disease propagation [40,41], employing the MaxEnt model.

In this study, we used the MaxEnt model, combined with climatic factors, terrain factors, and soil factors, to predict the suitable area for orchardgrass. The four goals of this study were to: (1) assess the key external variables affecting the distribution of orchardgrass; (2) investigate the distribution of suitable areas for orchardgrass under present and future climate scenarios; (3) predict potential distribution shifts of orchardgrass; (4) pinpoint areas of habitat expansion and degradation for orchardgrass.

2. Materials and Methods

2.1. Data on Species Occurrence

To collect extensive data on the global natural distribution of orchardgrass, we searched multiple databases, including the Global Biodiversity Information Facility (GBIF; https://www.gbif.org/ (accessed on 13 May 2023)) and the Chinese Virtual Herbarium (CVH; https://www.cvh.ac.cn/ (accessed on 13 May 2023)) [42]. Furthermore, the scientific names of our target species (*Dactylis glomerata*) were used as search terms in a Web of Science (WOS; https://www.webofscience.com (accessed on 13 May 2023)) database search, and we recorded all the distribution sites that were mentioned in the literature. We conducted searches in the Chinese National Knowledge Infrastructure (CNKI; https://www.cnki.net/ (accessed on 13 May 2023)) database using the Chinese and scientific names of the target species [42]. This database is one of the most comprehensive databases in the Chinese scientific field and also records the distribution points of the target species. By utilizing these resources, we compiled a comprehensive list of orchardgrass occurrence records and a distribution map. (Figure 1; Table S1).

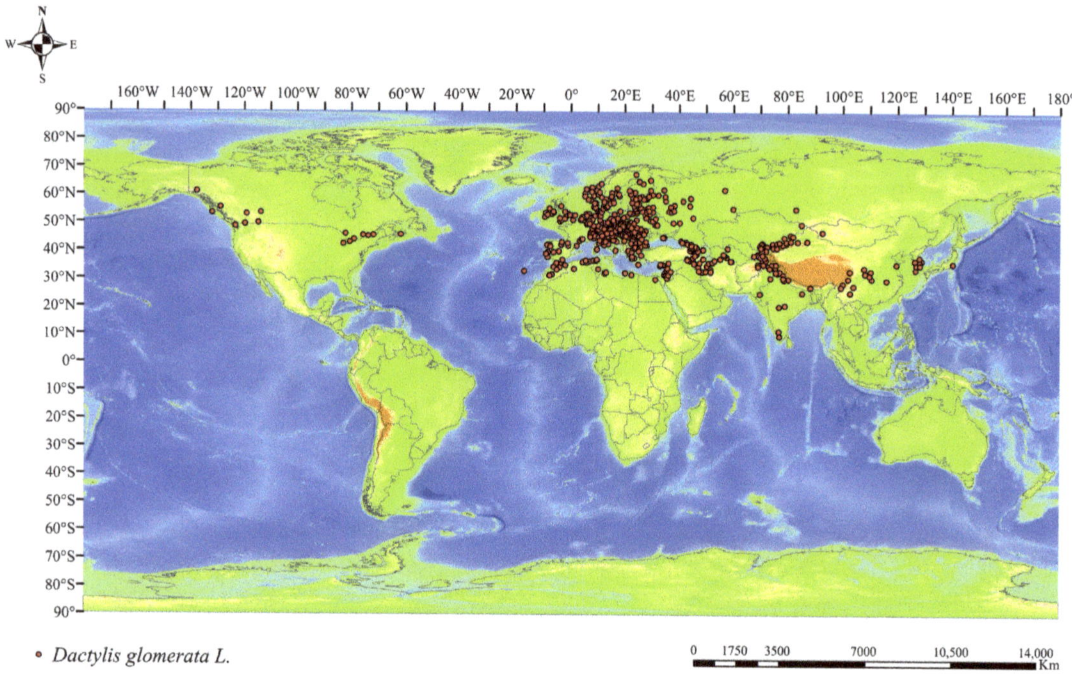

Figure 1. Records points of orchardgrass occurrence created in ESRI ArcMap 10.4.1.

The distribution data of our target species (orchardgrass) underwent several processing steps. Firstly, we convert the longitude and latitude values from sexagesimal to decimal. Secondly, we individually verified whether the reported latitudes and longitudes matched the corresponding collecting locations. In cases where the records did not match, we removed them from the dataset. We then implemented the ENMTools.pl version 1.0.4 software (https://github.com/danlwarren/ENMTools (accessed on 13 May 2023)) to reduce the impact of sampling bias by trimming the occurrence points to retain only one observation per 5 arc-min grid cell. The environmental data are the same for each grid cell [43]. In doing so, we obtained 522 orchardgrass occurrence points.

2.2. Environmental Variables

Climate data and terrain data for this study were obtained from the WorldClim database (https://www.worldclim.org/ (accessed on 13 May 2023)) and soil data were obtained from the Harmonised World Soil Database (https://www.fao.org/soils-portal/soil-survey/soil-maps-and-databases/harmified-world-soil-database-v12/ (accessed on 13 May 2023)) [44]. In total, 19 bioclimatic variables, 10 soil variables, and 1 terrain variable (Table 1) were used as candidates for constructing the MaxEnt model. The climatic data from the years 1970 to 2000 were used to depict the current climate, while the predicted future climate data came from the most recent WorldClim version 2.1 (https://www.worldclim.org/ (accessed on 13 May 2023)). Under the BCC-CSM2-MR climate model, we forecast alterations in suitable areas for the time periods 2041–2060 and 2081–2100 using the average values of three Shared Socio-economic Pathways (SSPs) [45]. The SSPs, which are anticipated to represent various ranges of GHG emission levels, included SSP126 (low GHG emissions), SSP370 (mid-GHG emissions), and SSP585 (high GHG emissions). In all, this study used 30 different environmental factors, such as soil type, altitude, and climate. We used the SDMtools resampling and transformation tool in ArcGIS (version 10.4.1) to resample the environmental data. The data was converted to a 5′ resolution and saved in

.asc format [46]. The resampling technique employed was nearest neighbor resampling, where the nearest pixel value from the original data was selected as the new pixel value to ensure data quality and accuracy.

Table 1. Meaning and units of environment variables used in this study. The variables in bold are the sixteen variables remaining after an iterative variable reduction process and are the variables used in the final climate envelope model.

Index	Variable	Source	Description	Unit
Bioclimatic variables	BIO1	Worldclim	Annual Mean Temperature	°C
	BIO2		**Mean diurnal range (mean of monthly (max temp–min temp))**	**°C**
	BIO3		Isothermality (BIO2/BIO7) (×100)	%
	BIO4		**Temperature seasonality (standard deviation × 100)**	**°C**
	BIO5		**Max temperature of the warmest month**	**°C**
	BIO6		Min temperature of the coldest month	°C
	BIO7		Temperature annual range (BIO5–BIO6)	°C
	BIO8		Mean temperature of wettest quarter	°C
	BIO9		Mean temperature of driest quarter	°C
	BIO10		Mean temperature of warmest quarter	°C
	BIO11		Mean temperature of coldest quarter	°C
	BIO12		Annual precipitation	mm
	BIO13		Precipitation of wettest month	mm
	BIO14		Precipitation of the driest month	mm
	BIO15		**Precipitation seasonality (coefficient of variation)**	**mm**
	BIO16		**Precipitation of the wettest quarter**	**mm**
	BIO17		**Precipitation of the driest quarter**	**mm**
	BIO18		Precipitation of the warmest quarter	mm
	BIO19		**Precipitation of the coldest quarter**	**mm**
Terrain variables	**Elevation**	Worldclim	**Elevation**	m
Soil variables	**ESP**	Harmonised World Soil Database	**Exchangeable sodium percentage**	—
	Gravel		**Volume percentage of gravel**	—
	OC		**Percentage of organic carbon**	—
	PH		**Soil reaction**	**mol·L^{-1}**
	AWC		**Available water capacity**	**g/kg**
	Bulk		**Cation exchange capacity**	**cmol (+)/kg**
	Clay		**Percentage of clay**	—
	Drainage		**Soil drainage class**	—
	CECS		Cation exchange capacity of the clay fraction	—
	Sand		Percentage of sand	—

ENMTools.pl was utilized to examine the correlation between the candidate variables (19 bioclimatic variables, 1 terrain variable, and 10 soil variables) in order to prevent multicollinearity among the variable inputs. The threshold of Pearson correlation coefficient was set at 0.75, and the correlated variables were removed accordingly [47]. Finally, we chose BIO2 (mean diurnal range), BIO4 (temperature seasonality), BIO5 (max temperature of the warmest month), BIO15 (precipitation seasonality), BIO16 (precipitation of wettest quarter), BIO17 (precipitation of driest quarter), BIO19 (precipitation of coldest quarter), ESP (Exchangeable sodium percentage), Gravel (Volume percentage of gravel), OC (percentage of organic carbon), AWC (available water capacity), Bulk (cation exchange capacity), Clay (percentage of clay), Drainage (soil drainage class), elevation, and PH, as the environmental factors to construct the model of species distribution (Figure 2, Table S2).

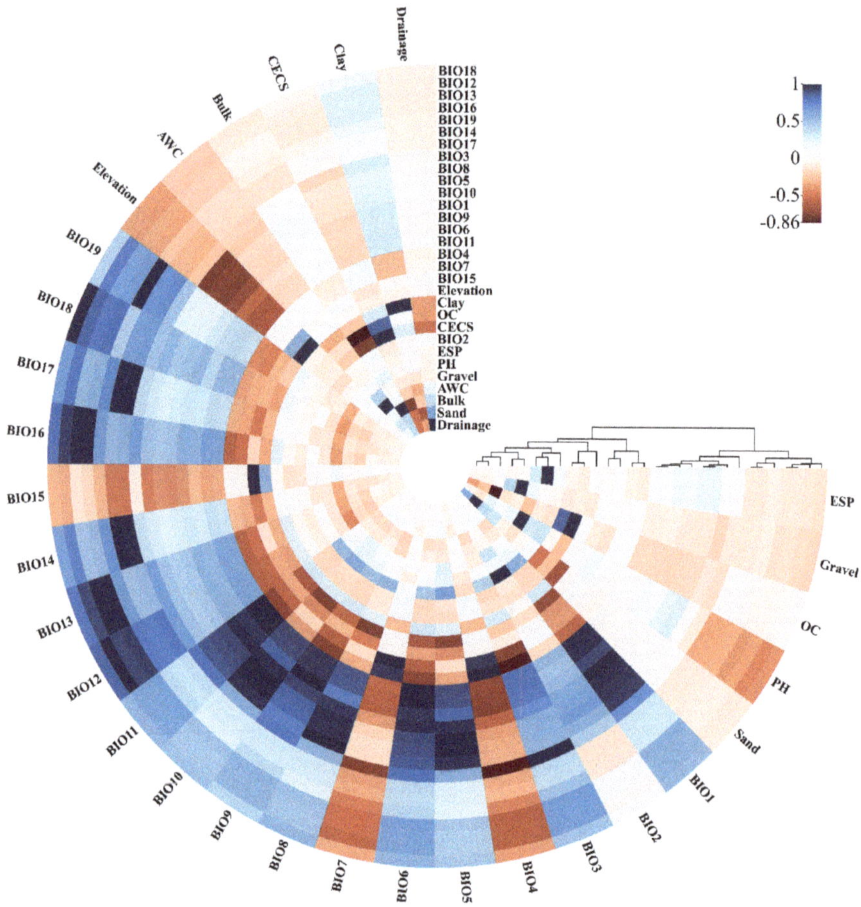

Figure 2. Pearson correlation coefficient of environmental variables.

2.3. Optimization of Model Parameters

We optimized the regularization multiplier and feature class parameters in R version 3.6.3 [48] using the Kuenm package (https://github.com/marlonecobos/kuenm (accessed on 13 May 2023)). This package is a prerequisite for building the species distribution model with MaxEnt version 3.4.4 (https://biodiversityinformatics.amnh.org/open_source/maxent/ (accessed on 13 May 2023)). The training set consisted of 75% of the data, which comprised of 392 occurrence points. It was determined that 1160 candidate models could be tested, representing all possible combinations of the 29 feature class combinations and the 40 regularization multiplier settings (0.1 to 4 with an interval of 0.1). The model's statistical significance, predictive performance, and complexity were evaluated in a sequential manner, with the partial ROC being assessed first, followed by the omission rates, and finally the AICc values. The statistical significance of the candidate models was the initial criterion for screening them. Second, to narrow the pool of models, omission rate requirements (i.e., 5% where feasible) were applied. In the final round of model selection, we chose models that had the lowest delta AICc values (<2) and met both the criteria of statistical significance and low omission rates.

2.4. Species Distribution Model

Construct species distribution models with MaxEnt 3.4.4 utilizing the above 16 environmental layers. Notably, the majority of earlier studies have concentrated on predicting suitable habitats in countries such as China and Pakistan using models that essentially use the standard parameters, leading to subpar model accuracy [49,50]. Therefore, to improve the accuracy of the model data, we employed the reg_mult function from the Kuenm package (https://github.com/marlonecobos/kuenm (accessed on 13 May 2023)). for model performance screening in our study. In addition to simulating regularization multipliers ranging from 0.1 to 4, the models with five types of feature categories (Hinge, Threshold, Linear, Quadratic, Product, and Threshold) also optimized their respective parameters [48,51]. Ten replicates were conducted for our analysis, with a maximum of 5000 iterations and 10,000 background points, encompassing the sixteen environmental layers mentioned earlier. The area under the receiver operating characteristic curve (AUC) was used to evaluate the model's performance [52]. An AUC value of 0.9–1.0 indicates a perfect prediction and a value of < 0.5 represents a random prediction [44]. Using the Jackknife test, we evaluated the relative contributions of each environmental variable [22]. Each grid cell has a probability of the model's output value being between 0 and 1, which can be read as a measure of relative suitability. The model outputs were categorized using the MTSPS (maximum test sensitivity plus specificity) threshold: grid cells with suitabilities above the MTSPS threshold were classified as potentially suitable habitat, while others were classified as unsuitable habitat [53]. We employed ArcGIS version 10.4.1 to create distribution maps, compute the percentage of potentially appropriate areas, and determine the average output of the BCC-CSM2-MR global model for the same SSP and identical period. The present study mainly summarizes the predicted mean results of the BCC-CSM2-MR global model under the different GHG emission scenarios in 2041–2060 and 2081–2100. The proportions of contracted and extended potentially suitable areas in orchardgrass, as well as changes in the potentially suitable areas of their centroids, were examined using SDMtoolbox version 2.4 software [54–56].

2.5. Estimation of Orchardgrass Distribution Area

In this study, we retrieved global land area data from the Food and Agriculture Organization of the United Nations (FAO) website (https://www.fao.org/ (accessed on 13 May 2023)) and used ArcGIS version 10.4.1 to divide it into suitable and unsuitable areas using the MTSPS method. By calculating the percentage of suitable and unsuitable grid cells out of the total grid cells and multiplying it by the corresponding area of each region, we can determine the area of suitable and unsuitable areas.

3. Results

3.1. Modeling of Species Distribution

The assessment revealed that the MaxEnt model with QPH (quadratic, product, and hinge) feature type parameters and a regularization multiplier of 2.6 was the most effective one that was used to proceed with further analysis. Under the current scenario, test AUC values from the final models' tenfold cross-validation are 0.919 (Figure 3), while the MaxEnt model's forecasts for potential orchardgrass habitats delivered positive outcomes, with a range of AUC values from 0.914 to 0.922 (Figure S1).

Internal jackknife testing of the MaxEnt model showed that temperature seasonality (Bio4, 34.9% of contribution), mean diurnal range (Bio2, 22.9% of contribution), the max temperature of the warmest month (Bio5, 7.8% of contribution), precipitation of coldest quarter (Bio19, 6.6% of contribution), contributed most to the Maxent model for orchardgrass, with an overall contribution of 72.2% (Table 2). Precipitation of wettest quarter (Bio16), PH, precipitation seasonality (Bio15), and other indicators made up 27.8% of the total contribution. Ecological thresholds for important environmental factors are known from environmental factor response curves (Figure S2): temperature seasonality (411.50–1034.37 °C), mean diurnal range (−0.88–10.69 °C), precipitation of the coldest quarter (116.56–825.40 mm), and max temperature of the warmest month (17.08–40.84 °C).

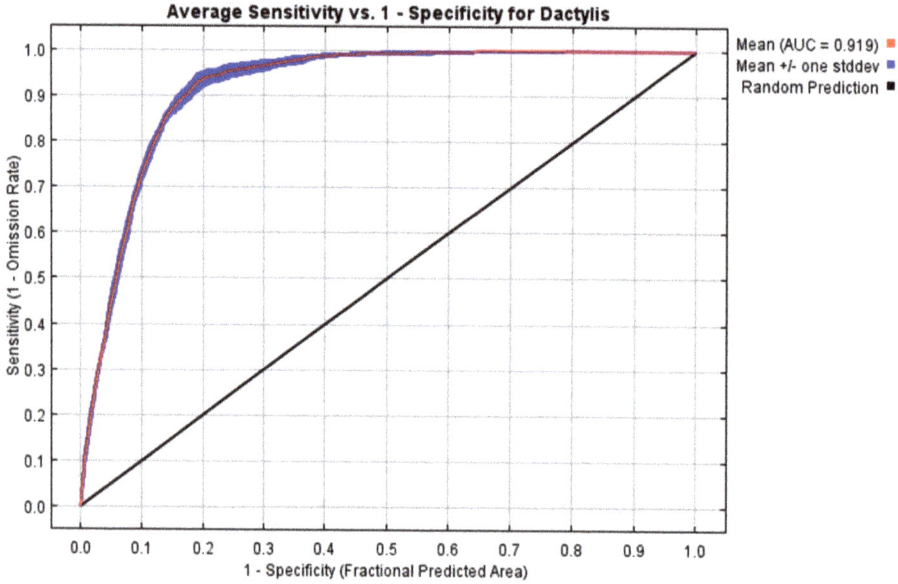

Figure 3. The final model of the receiver operating characteristic curve.

Table 2. Environmental factors contributing to the current suitable habitat for orchardgrass and their thresholds in MaxEnt models. Only the top four contributing environmental factors out of 16 are listed.

Code	Environmental Factor	Percent Contribution (%)	Suitable Threshold	Units
Bio4	temperature seasonality	34.9	411.50–1034.37	°C
Bio2	mean diurnal range	22.9	−0.88–10.69	°C
Bio5	max temperature of the warmest month	7.8	17.08–40.84	°C
Bio19	precipitation of the coldest quarter	6.6	116.56–825.40	mm

3.2. *Current Suitable Distribution for Orchardgrass*

Habitats for orchardgrass (Figure 4a) were widely distributed over three continents, primarily in western and southern North America, Europe, southwestern and southeastern Asia, and a few in southern South America, northern Africa, and southeastern Oceania, but not in Antarctica. The total appropriate habitat area was 2133.01 $\times$ 10^4 km^2, accounting for 14.19% of the world's land area.

Six continents had different suitable habitat distributions for orchardgrass (Figure 4b). Europe had the largest total area of appropriate habitat (754.79 $\times$ 10^4 km^2), covering most of the continent. Asia had the second biggest area, with a total appropriate habitat area of 597.64 $\times$ 10^4 km^2, largely in China, Turkey, and Iran. The appropriate habitat area for orchardgrass in North America was 247.80 $\times$ 10^4 km^2, largely found in coastal areas of the United States and Canada. Oceania has the smallest amount of suitable habitat for orchardgrass out of all the continents, with only 70.21 $\times$ 10^4 km^2 concentrated in the coastal areas of New Zealand and southeastern Australia. These findings support the current distribution pattern of orchardgrass and demonstrate the accuracy of using MaxEnt to predict species distribution patterns.

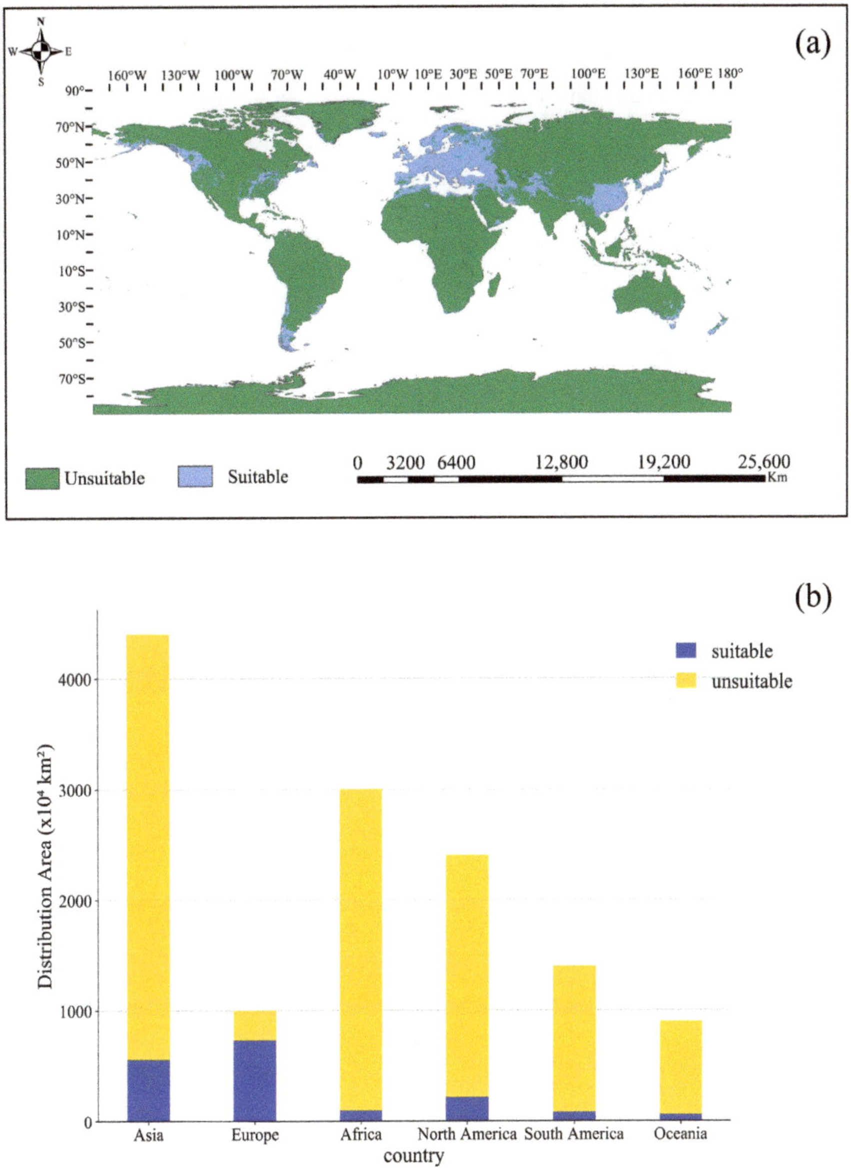

Figure 4. In accordance with current climate conditions, map (**a**) shows suitable habitat areas for orchardgrass. Created in ESRI ArcMap 10.4.1 map (**b**) shows a distribution area of orchardgrass habitats on six continents.

3.3. Potential Distribution of Orchardgrass under Future Climate Conditions

The study investigates the potential redistribution of orchardgrass habitats in the twenty-first century in response to climate change under three different scenarios. Based on a comparison of the present appropriate habitats (Figure 4a) with the predicted appropriate habitats for 2041–2060 and 2081–2100, we examined several tendencies that appeared under various climatic scenarios. (Figure 5). Total suitable habitat expanded from 75.91 $\times$ 10^4 km^2 (ssp245, 2041–2060) to 322.03 $\times$ 10^4 km^2 (ssp585, 2081–2100), while the contraction area varied from 182.04 × 10^4 km2 (ssp126, 2041–2060) to 378.50 × 10^4 km^2(ssp585, 2081–2100). Overall, the findings demonstrated that rising GHG emissions were significantly associated with a global shrinkage in the extent of appropriate habitats for orchardgrass. During the second half of the twenty-first century, the southern hemisphere, more notably South America, Central Africa, and Oceania had a particularly dramatic decline. The shrinking of Asia and South America is the most noteworthy at the end of the twenty-first century, owing to increasing GHG emissions.

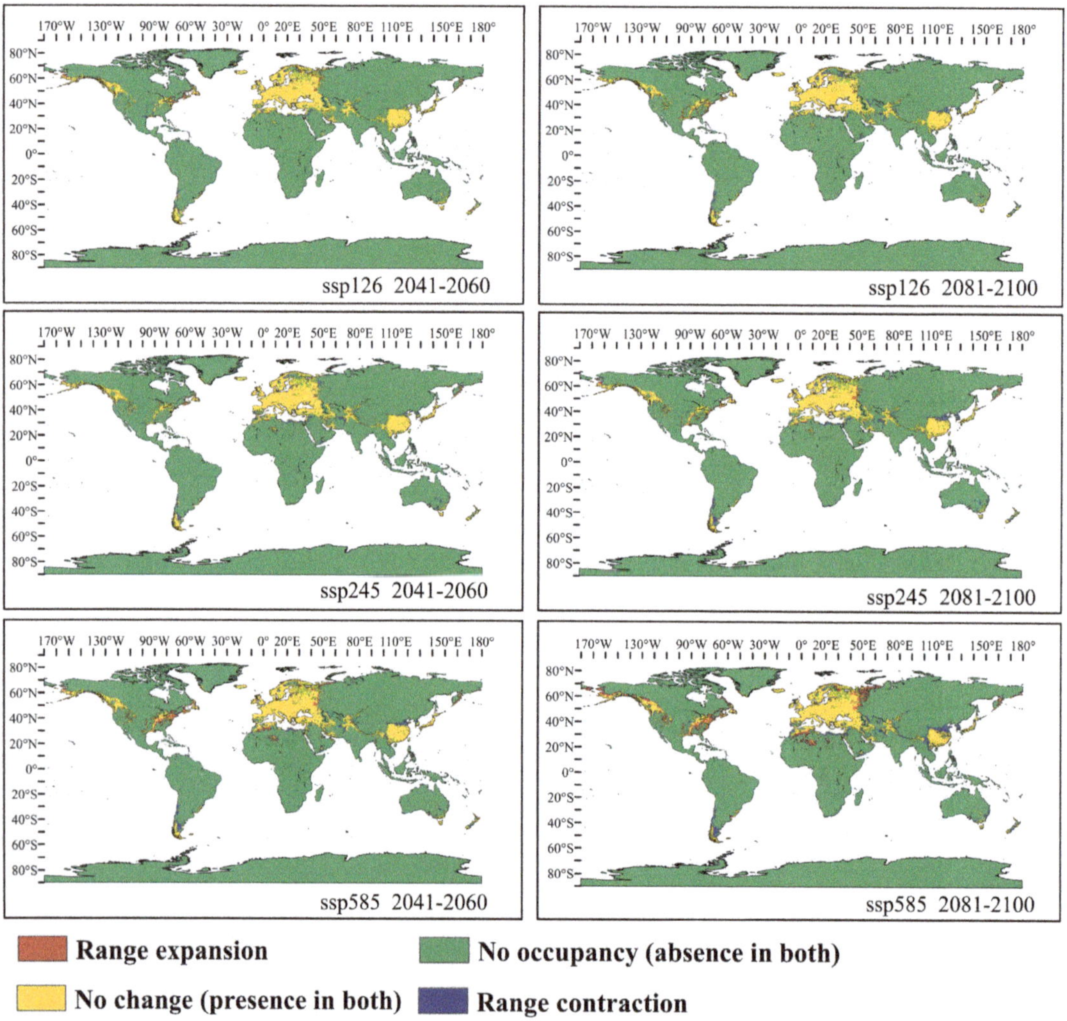

Figure 5. From years 2041 to 2100, changes in suitable habitats for orchardgrass under ssp126, ssp245, and ssp585 scenarios compared to the current distribution.

Significant expansion and contraction of suitable orchardgrass habitat towards the northwest were noted in all future climate scenarios, with significant differences between continents (Figure 6a, Table S3). Interestingly, appropriate growth regions in Asia, Europe, and North America stretch to high latitudes, with Asian habitat expansion ranging from 20.49 × 10^4 km^2 (ssp126, 2081–2100) to 38.13 × 10^4 km^2 (ss585, 2041–2060). European habitat expansion ranged from 6.24 × 10^4 km^2 (ssp245, 2041–2060) to 61.94 × 10^4 km^2

(ssp585, 2081–2100). In contrast, the size of the habitats in North America ranged from 33.58×10^4 km^2 (ssp245, 2041–2060) to 160.60×10^4 km^2 (ssp585, 2081–2100) (Table S4). North America has the largest area of suitable expansion. We observed that habitat expansion in Asia has occurred mainly in eastern Russia, eastern Kazakhstan, and eastern Afghanistan. The European habitat extension regions were mostly in western Russia, Finland, and Sweden, comparatively, the majority of the North American habitat extension regions were in the US and Canada. The study showed that the southern hemisphere's orchardgrass habitat is rapidly shrinking, with South American habitat areas declining by 16.74×10^4 km^2 (ssp126, 2041–2060) to 36.20×10^4 km^2 (ssp585, 2081–2100), which was 14.29 to 48.43 times the extent of expansion (4.04×10^3 km ^{2} [ssp126, 2041–2060] to 2.53×10^3 km^2 [ssp585, 2081–2100]). On the other hand, Oceania contracted by 14.70×10^4 km^2 (ssp124, 2041–2060) to 28.22×10^4 km^2 (ssp245, 2081–2100), which was 4.01 to 24.29 times the extent of expansion (1.16×10^4 km^2 [ssp245,2081–2100] to 3.94×10^4 km^2 [ssp126, 2081–2100]). African habitat areas contracted by 24.90×10^4 km^2 (ssp126, 2041–2060) to 33.57×10^4 km^2 (ssp126, 2081–2100), which was 0.80 to 5.70 times the extent of expansion (4.37×10^4 [ssp126, 2041–2060] to 72.78×10^4 km^2 [ssp585, 2081–2100]) (Table S4). The majority of the contracted regions were in Congo, Morocco, and Libya in Africa, Argentina and Chile in South America, and southeastern Australia and New Zealand in Oceania.

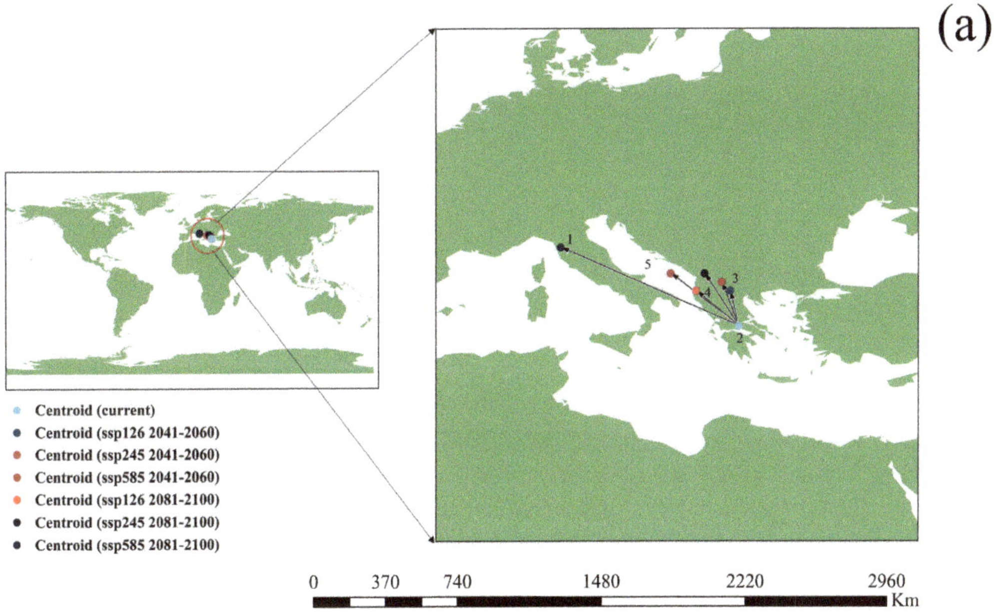

Figure 6. *Cont.*

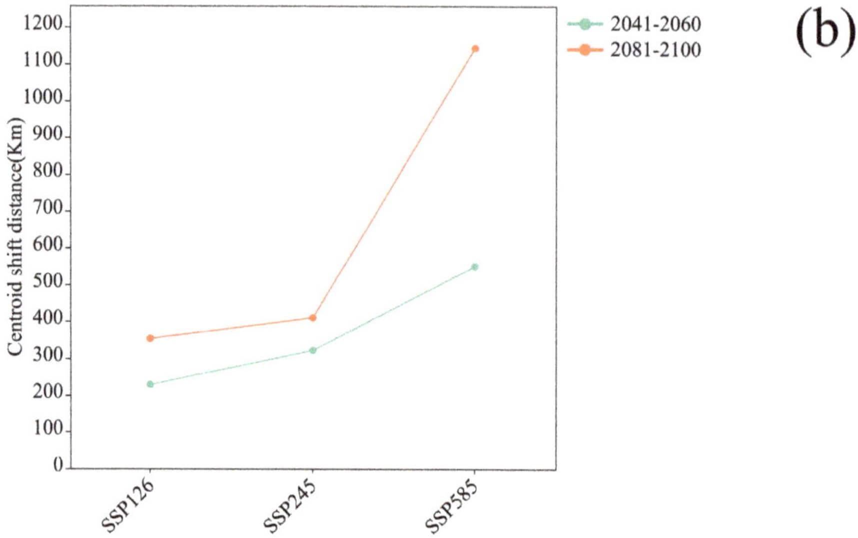

Figure 6. (**a**) Centroid changes for orchardgrass between current and future climate change scenarios. 1, Italy; 2, Greece; 3, Macedonia; 4, Republic of Albania; 5, Mediterranean Sea. (**b**) Centroid changes and Centroid shift distance under different GHG emission scenarios for orchardgrass.

4. Discussion

4.1. MaxEnt Modeling

MaxEnt, unlike other software programmes, can evaluate model performance using the under-the-curve (AUC) value of the receiver operating characteristic (ROC) curve [57], correct the sample deviation when obtaining data from a known distribution [58,59], and lessen spatial biases for the global distribution of species in the GBIF database [60]. AUC values for orchardgrass in this study, which were greater than the random AUC value of 0.50 and close to 1.00, demonstrated the reliability of the built-in model. Therefore, our model is able to accurately calculate the global distribution of orchardgrass habitat based on its performance.

Although the MaxEnt model possesses higher predictive performance and is widely used for predicting species distribution changes, it still has limitations such as high computational complexity, limited explanatory power, and susceptibility to the influence of low-resolution environmental data [61]. While climate and soil conditions are taken into account by the MaxEnt model, additional factors including adaptive ability, interspecific relationships, human development, and land utilization may also influence species distribution. In assessing model performance, AUC is sometimes not sufficient to assess model performance, and other metrics such as the Kappa coefficient and TSS are also important [62]. It is suggested that future studies may consider using more metrics to assess model performance. As a result, when all relevant factors are taken into account, we anticipate that orchardgrass suitable habitat will decline even more as a result of future global warming, and the reduction in habitat area may be more dramatic. However, the ensuing simulations might not always be more accurate if all variables are included in the model, because the impact of important variables might be reduced. Therefore, scientific selection of variables is needed to make the model more accurate and the results more meaningful.

4.2. Suitable Habitat Distribution Patterns of Orchardgrass under the Current Environment

Orchardgrass has a complicated distribution and diversity, most of which may be explained by its recent evolutionary, genomic histories, and migratory [63]. In general, response curves can be used to correlate species occurrence probability with major ecological factors. If a species is likely to have a probability of occurrence greater than 60%, the related ecological factor thresholds are appropriate for the existence of this species [26,64,65]. According to MaxEnt results and environmental factor response curves, the key environmental factors influencing distribution for orchardgrass in this study were temperature seasonality, mean diurnal range, maximum temperature of the warmest month, and precipitation of the coldest quarter. Findings from this study are consistent with those from previous studies [66,67].

According to our research, orchardgrass has a total suitable habitat area of 2133.01 × 10^4 km^2, with the majority of these areas being in Europe (including the Eastern European Plain, the Central European Plain, and the Western European Plain), Asia (including the Sichuan Basin, Yunnan-Guizhou Plateau, Northeast Plain, Middle and Lower Yangtze Plain, the North China Plain, and Japan), and North America (including the Cascade Mountain). This is consistent with the field distribution site of orchardgrass [68]. Since high-suitability locations are more beneficial to preserving orchardgrass variety than low-suitability areas, breeding, cultivation, and ex-situ conservation of orchardgrass species should be carried out in these locations.

4.3. Response of Suitable Habitats for Orchardgrass to Future Climate Change

Climate change, particularly worldwide warming, alters the pattern of precipitation distribution in addition to changing local temperatures. The distribution of these climatic factors will shift when the change is near to or over the adaptation threshold of the present plant development [69]. Our research demonstrates that different climate scenarios have different areas and distributions of orchardgrass suitable habitats, suggesting that the distribution of these habitats was impacted by climate change in spatially particular ways.

Among the 16 environmental variables adopted in the model, temperature seasonality and mean diurnal range made the greatest contributions to the distribution model for orchardgrass compared to other variables, indicating that these factors play important roles in its distribution. The probability of orchardgrass distribution increases and then decreases with the increase in temperature seasonality and mean diurnal range.

This result is supported by the fact that the climatic characteristics of an area act as key elements for population regeneration [70]. Orchardgrass is widely distributed in temperate and tropical regions of central and western Asia, temperate regions of southwestern Europe, and the Canary Islands of northern and western Africa [71]. Temperature seasonality refers to the magnitude of the temperature change between seasons [72]. In temperate and tropical regions of central and western Asia, where winters are cold and summers are hot, this temperature variation affects orchardgrass growth and dormancy cycles [73]. In the temperate regions of south-west Europe, where winters are warm and humid and summers are hot and dry, orchardgrass is well adapted to such environmental conditions [74,75]. This may explain the large values of temperature seasonality in orchardgrass. In contrast, in the Canary Islands in northern and western Africa, temperature seasonality is relatively small, and the warm and stable climate allows orchardgrass to grow and reproduce continuously. These connections help us better understand the ecological characteristics of orchardgrass and their interactions with the environment.

There is a certain relationship between plant growth and development and mean diurnal range [76]. The diurnal range is the difference between the daily maximum and minimum temperatures and is often used to describe the change in temperature between day and night [77]. Temperature regimes with a 5–15 °C amplitude enhanced seed germination percentages of orchardgrass, indicating that the conditional dormancy was released by these temperature regimes. Seeds germinated more rapidly under alternating temperatures than under constant temperatures. The dual effects of temperature on dormancy breaking and germination were accounted for by thermal time models based on alternating temperature regimes [78]. The impact of different diurnal temperature ranges on the growth of orchardgrass varies. Orchardgrass grown in the 21 °C day/13 °C night environment produced more aerial dry matter and had a larger leaf area compared to orchardgrass grown in the 29 °C day/31 °C night environment [79]. Under warm conditions (32 °C day/24 °C night), orchardgrass produces fewer flowering stems, while under cool or temperate conditions (18 °C day/10 °C night), it produces a higher number of flowering stems. Additionally, under these temperature conditions, orchardgrass exhibits the highest levels of growth rate, yield, tillering, leaf area, and dry matter allocation [80]. Therefore, the growth and adaptability of orchardgrass may be influenced by the diurnal temperature range in its surrounding environment. Understanding this relationship helps to gain deeper insights into the ecological adaptability of goosegrass under different environmental conditions and can provide valuable references and guidance for agricultural applications.

Our findings suggest suitable habitats for orchardgrass will experience considerable expansion and contraction across continents according to future climate scenarios. Orchardgrass is highly resistant to cold but shows cessation of growth at extreme temperatures [66]. With global warming, some high latitudes in the northern hemisphere may become suitable habitats for orchardgrass due to increased temperatures, whereas some low latitudes may face extended periods of extreme heat or dryness, posing a threat to the survival of orchardgrass. orchardgrass distribution continually changes north-westward, with European areas at its center, under future climatic circumstances. As the temperature goes up, the orchardgrass distribution area shifts farther away. The offset distances under different GHG emission scenarios in 2041–2060 are smaller than those in 2081–2100, and they reach a maximum of 1143.49 km under the high concentration emission scenario in 2081–2100 (Figure 6b). As a result, it is already evident that places at high latitudes are becoming more appropriate for orchardgrass, but as global warming progresses, orchardgrass habitats will generally decrease. Due to projected global warming, the southern hemisphere may no longer be suitable for orchardgrass, including South Africa and Angola in Africa, Chile and Argentina in South America, and southern Australia and New Zealand in Oceania. Surveys and collection of orchardgrass germplasms in these places are required to conserve the genetic variety of this plant, and some exceptional or unusual germplasm can be maintained in vitro for future use by asexual propagation. Areas such as Europe, the hilly and mountainous regions in southwest China and the Yangtze River Basin, as well as the northeast region of China and the Cascade Mountain Range, and Pacific Coast Ranges, which are mostly unaffected by climate change can serve as a base for future large-scale orchardgrass cultivation and usage, the preservation of local genetic resources, and other agricultural activities.

5. Conclusions

A MaxEnt model was successfully developed to represent the currently suitable areas for orchardgrass and predict the potential distribution of orchardgrass under future climate scenarios. The temperature seasonality and mean diurnal range were determined in this study as the ecological thresholds for the important environmental variables. Based on the results of the study, there are important implications for governmental departments to formulate relevant policies: firstly, protection and management, through the establishment of nature reserves, prohibiting destructive development and other measures to protect the ecological environment of orchardgrass; secondly, promotion of large-scale production, through financial support, technical training and other means to help farmers to carry out sustainable production of orchardgrass; thirdly, the formulation of policies and standards, including land-use planning, environmental protection regulations; Thirdly, to formulate policies and standards, including land use planning and environmental protection regulations, to protect and utilize orchardgrass; and lastly, to support research and monitoring, and to provide a scientific basis for policy adjustments and preventive measures by funding scientific research and monitoring work. These insights will help to balance economic development and ecological conservation goals to ensure the sustainable development of orchardgrass and the integrity of the ecosystem.

Supplementary Materials: The following supporting information can be downloaded at: https://www.mdpi.com/article/10.3390/agronomy13081985/s1, Table S1: The occurrence records of orchardgrass in the world; Table S2: Pearson correlation coefficient of environmental variables; Table S3: Projected potential suitable areas for orchardgrass under current and future climate scenarios; Table S4: The area change of suitable habitat of orchardgrass under different future climate scenarios on different continents; Figure S1: Receiver operating characteristic under future climate conditions; Figure S2: Response curves of the main predictors of orchardgrass citri occurrence probability. Curves show the mean response over ten replicate Maxent runs (red) and the mean ± 1 SD (blue, two shades for categorical variables).

Author Contributions: Methodology, J.W. and L.Y.; Software, J.W., L.Y., J.Z. and J.P.; Investigation, Y.X. (Yi Xiong); Resources, J.P., Y.X. (Yi Xiong) and Y.X. (Yanli Xiong); Data curation, J.W., L.Y., J.Z. and J.P.; Writing—original draft, J.W. and L.Y.; Writing—review & editing, X.M.; Visualization, J.W. and L.Y.; Supervision, X.M.; Funding acquisition, X.M. All authors have read and agreed to the published version of the manuscript.

Funding: This project was supported by the Sichuan Province "14th Five-Year Plan" Forage Breeding Research Project (2021YFYZ0013-2), Beef Innovation Team (SCCXTD-20-13), the National Modern Forages Industry Technology System (CARS-34) and Sichuan Forage Innovation Team Program (SCCXTD-2020-16).

Data Availability Statement: Please contact the first author with requests for data.

Conflicts of Interest: The authors declare no conflict of interest.

References

1. Hirata, M.; Yuyama, N.; Cai, H. Isolation and characterization of simple sequence repeat markers for the tetraploid forage grass *Dactylis glomerata*: Simple sequence repeat markers for Dactylis. *Plant Breed.* **2011**, *130*, 503–506. [CrossRef]
2. Xie, W.; Bushman, B.S.; Ma, Y.; West, M.S.; Robins, J.G.; Michaels, L.; Jensen, K.B.; Zhang, X.; Casler, M.D.; Stratton, S.D. Genetic diversity and variation in North American orchardgrass (*Dactylis glomerata* L.) cultivars and breeding lines. *Grassl. Sci.* **2014**, *60*, 185–193. [CrossRef]
3. Kole, C. (Ed.) *Wild Crop Relatives: Genomic and Breeding Resources: Millets and Grasses*; Springer: Berlin/Heidelberg, Germany, 2011. [CrossRef]
4. Wilkins, P.W.; Humphreys, M.O. Progress in breeding perennial forage grasses for temperate agriculture. *J. Agric. Sci.* **2003**, *140*, 129–150. [CrossRef]
5. Tronsmo, A.M. Resistance to Winter Stress Factors in Half-Sib Families of *Dactylis glomerata*, Tested in a Controlled Environment. *Acta Agric. Scand. Sect. B—Soil Plant Sci.* **1993**, *43*, 89–96. [CrossRef]
6. Turner, L.R.; Donaghy, D.J.; Lane, P.A.; Rawnsley, R.P. Distribution of Water-Soluble Carbohydrate Reserves in the Stubble of Prairie Grass and Orchardgrass Plants. *Agron. J.* **2007**, *99*, 591–594. [CrossRef]
7. Volaire, F. Seedling survival under drought differs between an annual (*Hordeum vulgare*) and a perennial grass (*Dactylis glomerata*). *New Phytol.* **2003**, *160*, 501–510. [CrossRef] [PubMed]
8. Volaire, F.; Conéjero, G.; Lelièvre, F. Drought survival and dehydration tolerance in *Dactylis glomerata* and *Poa bulbosa*. *Funct. Plant Biol.* **2001**, *28*, 743. [CrossRef]
9. Stout, W.L.; Jung, G.A. Influences of Soil Environment on Biomass and Nitrogen Accumulation Rates of Orchardgrass. *Agron. J.* **1992**, *84*, 1011–1019. [CrossRef]
10. Jensen, K.B.; Asay, K.H.; Waldron, B.L. Dry Matter Production of Orchardgrass and Perennial Ryegrass at Five Irrigation Levels. *Crop Sci.* **2001**, *41*, 479–487. [CrossRef]
11. Shaimi, N.; Kallida, R.; Volaire, F.; Saidi, N.; Faiz, C.A. Summer Dormancy and Drought Survival of Moroccan Ecotypes of Orchardgrass. *Crop Sci.* **2009**, *49*, 1416–1424. [CrossRef]
12. Pecl, G.T.; Araújo, M.B.; Bell, J.D.; Blanchard, J.; Bonebrake, T.C.; Chen, I.-C.; Clark, T.D.; Colwell, R.K.; Danielsen, F.; Evengård, B.; et al. Biodiversity redistribution under climate change: Impacts on ecosystems and human well-being. *Science* **2017**, *355*, eaai9214. [CrossRef]
13. Sanmartín, I. Biogeography: An Ecological and Evolutionary Approach, 7th edition. *Syst. Biol.* **2006**, *55*, 361–363. [CrossRef]
14. Intergovernmental Panel on Climate Change (Ed.) Evaluation of Climate Models. In *Climate Change 2013—The Physical Science Basis*; Cambridge University Press: Cambridge, UK, 2014; pp. 741–866. [CrossRef]
15. Hatfield, J.L.; Prueger, J.H. Temperature extremes: Effect on plant growth and development. *Weather Clim. Extrem.* **2015**, *10*, 4–10. [CrossRef]

16. Lawler, J.J. Climate Change Adaptation Strategies for Resource Management and Conservation Planning. *Ann. N. Y. Acad. Sci.* **2009**, *1162*, 79–98. [CrossRef]
17. Corlett, R.T.; Westcott, D.A. Will plant movements keep up with climate change? *Trends Ecol. Evol.* **2013**, *28*, 482–488. [CrossRef] [PubMed]
18. Araújo, M.B.; Peterson, A.T. Uses and misuses of bioclimatic envelope modeling. *Ecology* **2012**, *93*, 1527–1539. [CrossRef]
19. Elith, J.; Leathwick, J.R. Species Distribution Models: Ecological Explanation and Prediction Across Space and Time. *Annu. Rev. Ecol. Evol. Syst.* **2009**, *40*, 677–697. [CrossRef]
20. Gengping, Z.; Guoqing, L.; Wenjun, B.; Yubao, G. Ecological niche modeling and its applications in biodiversity conservation: Ecological niche modeling and its applications in biodiversity conservation. *Biodivers. Sci.* **2013**, *21*, 90–98. [CrossRef]
21. Stockwell, D. The GARP modelling system: Problems and solutions to automated spatial prediction. *Int. J. Geogr. Inf. Sci.* **1999**, *13*, 143–158. [CrossRef]
22. Phillips, S.J.; Anderson, R.P.; Schapire, R.E. Maximum entropy modeling of species geographic distributions. *Ecol. Model.* **2006**, *190*, 231–259. [CrossRef]
23. Beaumont, L.J.; Hughes, L.; Poulsen, M. Predicting species distributions: Use of climatic parameters in BIOCLIM and its impact on predictions of species' current and future distributions. *Ecol. Model.* **2005**, *186*, 251–270. [CrossRef]
24. Liaw, A.; Wiener, M. Classification and Regression by randomForest 2. *R News* **2002**, *2*, 18–22.
25. Elith, J.; Leathwick, J.R.; Hastie, T. A working guide to boosted regression trees. *J. Anim. Ecol.* **2008**, *77*, 802–813. [CrossRef] [PubMed]
26. Elith, J.; Phillips, S.J.; Hastie, T.; Dudík, M.; Chee, Y.E.; Yates, C.J. A statistical explanation of MaxEnt for ecologists: Statistical explanation of MaxEnt. *Divers. Distrib.* **2011**, *17*, 43–57. [CrossRef]
27. Li, G.; Du, S.; Wen, Z. Mapping the climatic suitable habitat of oriental arborvitae (*Platycladus orientalis*) for introduction and cultivation at a global scale. *Sci. Rep.* **2016**, *6*, 30009. [CrossRef] [PubMed]
28. Merow, C.; Smith, M.J.; Edwards, T.C.; Guisan, A.; McMahon, S.M.; Normand, S.; Thuiller, W.; Wüest, R.O.; Zimmermann, N.E.; Elith, J. What do we gain from simplicity versus complexity in species distribution models? *Ecography* **2014**, *37*, 1267–1281. [CrossRef]
29. Merow, C.; Smith, M.J.; Silander, J.A. A practical guide to MaxEnt for modeling species' distributions: What it does, and why inputs and settings matter. *Ecography* **2013**, *36*, 1058–1069. [CrossRef]
30. Guo, Y.; Zhao, Z.; Zhu, F.; Gao, B. The impact of global warming on the potential suitable planting area of *Pistacia chinensis* is limited. *Sci. Total Environ.* **2023**, *864*, 161007. [CrossRef]
31. Mohammadi, S.; Ebrahimi, E.; Shahriari Moghadam, M.; Bosso, L. Modelling current and future potential distributions of two desert jerboas under climate change in Iran. *Ecol. Inform.* **2019**, *52*, 7–13. [CrossRef]
32. Rong, Z.; Zhao, C., Liu, J.; Gao, Y.; Zang, F.; Guo, Z.; Mao, Y.; Wang, L. Modeling the Effect of Climate Change on the Potential Distribution of Qinghai Spruce (*Picea crassifolia* Kom.) in Qilian Mountains. *Forests* **2019**, *10*, 62. [CrossRef]
33. Shi, F.; Liu, S.; An, Y.; Sun, Y.; Zhao, S.; Liu, Y.; Li, M. Climatic factors and human disturbance influence ungulate species distribution on the Qinghai-Tibet Plateau. *Sci. Total Environ.* **2023**, *869*, 161681. [CrossRef]
34. Raffini, F.; Bertorelle, G.; Biello, R.; D'Urso, G.; Russo, D.; Bosso, L. From Nucleotides to Satellite Imagery: Approaches to Identify and Manage the Invasive Pathogen *Xylella fastidiosa* and Its Insect Vectors in Europe. *Sustainability* **2020**, *12*, 4508. [CrossRef]
35. Ramos, R.S.; Kumar, L.; Shabani, F.; Picanço, M.C. Risk of spread of tomato yellow leaf curl virus (TYLCV) in tomato crops under various climate change scenarios. *Agric. Syst.* **2019**, *173*, 524–535. [CrossRef]
36. Sultana, S.; Baumgartner, J.B.; Dominiak, B.C.; Royer, J.E.; Beaumont, L.J. Impacts of climate change on high priority fruit fly species in Australia. *PLoS ONE* **2020**, *15*, e0213820. [CrossRef] [PubMed]
37. Melo-Merino, S.M.; Reyes-Bonilla, H.; Lira-Noriega, A. Ecological niche models and species distribution models in marine environments: A literature review and spatial analysis of evidence. *Ecol. Model.* **2020**, *415*, 108837. [CrossRef]
38. Sterne, T.K.; Retchless, D.; Allee, R.; Highfield, W. Predictive modelling of mesophotic habitats in the north-western Gulf of Mexico. *Aquat. Conserv. Mar. Freshw. Ecosyst.* **2020**, *30*, 846–859. [CrossRef]
39. Convertino, M.; Annis, A.; Nardi, F. Information-theoretic portfolio decision model for optimal flood management. *Environ. Model. Softw.* **2019**, *119*, 258–274. [CrossRef]
40. Ardestani, E.G.; Mokhtari, A. Modeling the lumpy skin disease risk probability in central Zagros Mountains of Iran. *Prev. Vet. Med.* **2020**, *176*, 104887. [CrossRef] [PubMed]
41. Hanafi-Bojd, A.A.; Vatandoost, H.; Yaghoobi-Ershadi, M.R. Climate Change and the Risk of Malaria Transmission in Iran. *J. Med. Entomol.* **2020**, *57*, 50–64. [CrossRef]
42. Zhang, Q.-C.; Wang, J.-G.; Lei, Y.-H. Predicting Distribution of the Asian Longhorned Beetle, *Anoplophora glabripennis* (Coleoptera: Cerambycidae) and Its Natural Enemies in China. *Insects* **2022**, *13*, 687. [CrossRef]
43. Warren, D.L.; Glor, R.E.; Turelli, M. ENMTools: A toolbox for comparative studies of environmental niche models. *Ecography* **2010**, *33*, 607–611. [CrossRef]
44. Yu, X.; Tao, X.; Liao, J.; Liu, S.; Xu, L.; Yuan, S.; Zhang, Z.; Wang, F.; Deng, N.; Huang, J.; et al. Predicting potential cultivation region and paddy area for ratoon rice production in China using Maxent model. *Field Crops Res.* **2022**, *275*, 108372. [CrossRef]
45. Fick, S.E.; Hijmans, R.J. WorldClim 2: New 1-km spatial resolution climate surfaces for global land areas. *Int. J. Climatol.* **2017**, *37*, 4302–4315. [CrossRef]

46. Liu, J.; Wang, L.; Sun, C.; Xi, B.; Li, D.; Chen, Z.; He, Q.; Weng, X.; Jia, L. Global distribution of soapberries (*Sapindus* L.) habitats under current and future climate scenarios. *Sci. Rep.* **2021**, *11*, 19740. [CrossRef]
47. Santana, P.A.; Kumar, L.; Da Silva, R.S.; Pereira, J.L.; Picanço, M.C. Assessing the impact of climate change on the worldwide distribution of *Dalbulus maidis* (DeLong) using MaxEnt. *Pest Manag. Sci.* **2019**, *75*, 2706–2715. [CrossRef]
48. Cobos, M.E.; Peterson, A.T.; Barve, N.; Osorio-Olvera, L. Kuenm: An R package for detailed development of ecological niche models using Maxent. *PeerJ* **2019**, *7*, e6281. [CrossRef]
49. Peng, L.-P.; Cheng, F.-Y.; Hu, X.-G.; Mao, J.-F.; Xu, X.-X.; Zhong, Y.; Li, S.-Y.; Xian, H.-L. Modelling environmentally suitable areas for the potential introduction and cultivation of the emerging oil crop *Paeonia ostii* in China. *Sci. Rep.* **2019**, *9*, 3213. [CrossRef]
50. Zhang, K.; Zhang, Y.; Tao, J. Predicting the Potential Distribution of *Paeonia veitchii* (Paeoniaceae) in China by Incorporating Climate Change into a Maxent Model. *Forests* **2019**, *10*, 190. [CrossRef]
51. Li, D.; Li, Z.; Liu, Z.; Yang, Y.; Khoso, A.G.; Wang, L.; Liu, D. Climate change simulations revealed potentially drastic shifts in insect community structure and crop yields in China's farmland. *J. Pest Sci.* **2023**, *96*, 55–69. [CrossRef]
52. Fourcade, Y.; Engler, J.O.; Rödder, D.; Secondi, J. Mapping Species Distributions with MAXENT Using a Geographically Biased Sample of Presence Data: A Performance Assessment of Methods for Correcting Sampling Bias (Valentine, J.F., editor). *PLoS ONE* **2014**, *9*, e97122. [CrossRef] [PubMed]
53. Aidoo, O.F.; Souza, P.G.; da Silva, R.S.; Santana, P.A., Jr.; Picanço, M.C.; Kyerematen, R.; Sètamou, M.; Ekesi, S.; Borgemeister, C. Climate-induced range shifts of invasive species (*Diaphorina citri* Kuwayama). *Pest Manag. Sci.* **2022**, *78*, 2534–2549. [CrossRef] [PubMed]
54. Keenan, R.W.; Kruczek, M.E. The esterification of dolichol by rat liver microsomes. *Biochemistry* **1976**, *15*, 1586–1591. [CrossRef] [PubMed]
55. Esezobo, S.; Pilpel, N. Moisture and gelatin effects on the interparticle attractive forces and the compression behaviour of oxytetracycline formulations. *J. Pharm. Pharmacol.* **1977**, *29*, 75–81. [CrossRef]
56. Srivastava, V.; Roe, A.D.; Keena, M.A.; Hamelin, R.C.; Griess, V.C. Oh the places they'll go: Improving species distribution modelling for invasive forest pests in an uncertain world. *Biol. Invasions* **2021**, *23*, 297–349. [CrossRef]
57. Jiménez-Valverde, A. Insights into the area under the receiver operating characteristic curve (AUC) as a discrimination measure in species distribution modelling: Insights into the AUC. *Glob. Ecol. Biogeogr.* **2012**, *21*, 498–507. [CrossRef]
58. Kramer-Schadt, S.; Niedballa, J.; Pilgrim, J.D.; Schröder, B.; Lindenborn, J.; Reinfelder, V.; Stillfried, M.; Heckmann, I.; Scharf, A.K.; Augeri, D.M.; et al. The importance of correcting for sampling bias in MaxEnt species distribution models. *Divers. Distrib.* **2013**, *19*, 1366–1379. [CrossRef]
59. Syfert, M.M.; Smith, M.J.; Coomes, D.A. Correction: The Effects of Sampling Bias and Model Complexity on the Predictive Performance of MaxEnt Species Distribution Models. *PLoS ONE* **2013**, *8*, e55158. [CrossRef]
60. Beck, J.; Böller, M.; Erhardt, A.; Schwanghart, W. Spatial bias in the GBIF database and its effect on modeling species' geographic distributions. *Ecol. Inform.* **2014**, *19*, 10–15. [CrossRef]
61. Lissovsky, A.A.; Dudov, S.V. Species-Distribution Modeling: Advantages and Limitations of Its Application. 2. MaxEnt. *Biol. Bull. Rev.* **2021**, *11*, 265–275. [CrossRef]
62. West, A.M.; Kumar, S.; Brown, C.S.; Stohlgren, T.J.; Bromberg, J. Field validation of an invasive species Maxent model. *Ecol. Inform.* **2016**, *36*, 126–134. [CrossRef]
63. Stewart, A.V.; Ellison, N.W. Dactylis. In *Wild Crop Relatives: Genomic and Breeding Resources*; Kole, C., Ed.; Springer: Berlin/Heidelberg, Germany, 2011; pp. 73–87. [CrossRef]
64. Li, J.; Chang, H.; Liu, T.; Zhang, C. The potential geographical distribution of Haloxylon across Central Asia under climate change in the 21st century. *Agric. For. Meteorol.* **2019**, *275*, 243–254. [CrossRef]
65. Prevéy, J.S.; Parker, L.E.; Harrington, C.A.; Lamb, C.T.; Proctor, M.F. Climate change shifts in habitat suitability and phenology of huckleberry (*Vaccinium membranaceum*). *Agric. For. Meteorol.* **2020**, *280*, 107803. [CrossRef]
66. Davidson, J.L.; Milthorpe, F.L. The Effect of Temperature on the Growth of Cocksfoot (*Dactylis glomerata* L.). *Ann. Bot.* **1965**, *29*, 407–417. [CrossRef]
67. Ahmed, L.Q.; Escobar-Gutiérrez, A.J. Analysis of intra-specific variability of cocksfoot (*Dactylis glomerata* L.) in response to temperature during germination. *Acta Physiol. Plant* **2022**, *44*, 117. [CrossRef]
68. Sanada, Y.; Gras, M.-C.; van Santen, E. Cocksfoot. In *Fodder Crops and Amenity Grasses*; Boller, B., Posselt, U.K., Veronesi, F., Eds.; Springer: New York, NY, USA, 2010; pp. 317–328. [CrossRef]
69. Parmesan, C.; Yohe, G. A globally coherent fingerprint of climate change impacts across natural systems. *Nature* **2003**, *421*, 37–42. [CrossRef] [PubMed]
70. Hansen, W.D.; Braziunas, K.H.; Rammer, W.; Seidl, R.; Turner, M.G. It takes a few to tango: Changing climate and fire regimes can cause regeneration failure of two subalpine conifers. *Ecology* **2018**, *99*, 966–977. [CrossRef] [PubMed]
71. Jones, M.B.; Finnan, J.; Hodkinson, T.R. Morphological and physiological traits for higher biomass production in perennial rhizomatous grasses grown on marginal land. *GCB Bioenergy* **2015**, *7*, 375–385. [CrossRef]
72. Duan, J.; Ma, Z.; Wu, P.; Xoplaki, E.; Hegerl, G.; Li, L.; Schurer, A.; Guan, D.; Chen, L.; Duan, Y.; et al. Detection of human influences on temperature seasonality from the nineteenth century. *Nat. Sustain.* **2019**, *2*, 484–490. [CrossRef]
73. Archer, K.A.; Decker, A.M. Autumn-Accumulated Tall Fescue and Orchardgrass. I. Growth and Quality as Influenced by Nitrogen and Soil Temperature. *Agron. J.* **1977**, *69*, 601–605. [CrossRef]

74. Lumaret, R.; Borrill, M. Cytology, genetics, and evolution in the genus dactylis. *Crit. Rev. Plant Sci.* **1988**, *7*, 55–91. [CrossRef]
75. Fujimoto, F. Genetic Resources of Orchardgrass (*Dactylis glomerata* L.) and Related Subspecies from Warmer Regions. *Jpn. Agric. Res. Q.* **1993**, *27*, 106.
76. Geiger, D.R.; Servaites, J.C. Diurnal Regulation of Photosynthetic Carbon Metabolism in C3 Plants. *Annu. Rev. Plant. Physiol. Plant. Mol. Biol.* **1994**, *45*, 235–256. [CrossRef]
77. Parton, W.J.; Logan, J.A. A model for diurnal variation in soil and air temperature. *Agric. Meteorol.* **1981**, *23*, 205–216. [CrossRef]
78. Qiu, J.; Bai, Y.; Coulman, B.; Romo, J.T. Using thermal time models to predict seedling emergence of orchardgrass (*Dactylis glomerata* L.) under alternating temperature regimes. *Seed Sci. Res.* **2006**, *16*, 261–271. [CrossRef]
79. Taylor, T.H.; Cooper, J.P.; Treharne, K.J. Growth Response of Orchardgrass (*Dactylis glomerata* L.) to Different Light and Temperature Environments. I. Leaf Development and Senescence. *Crop Sci.* **1968**, *8*, 437–440. [CrossRef]
80. Knievel, D.P.; Smith, D. Influence of Cool and Warm Temperatures and Temperature Reversal at Inflorescence Emergence on Growth of Timothy, Orchardgrass, and Tall Fescue. *Agron. J.* **1973**, *65*, 378–383. [CrossRef]

agronomy

Article

How Are Warm-Season Pastures' Nutritive Value and Fermentation Characteristics Affected by Open Pasture, Silvopasture, and Sward Herbage Maturity?

Mariane Rodrigues Ferreira [1], Abmael da Silva Cardoso [1,2,*], Marina Elizabeth Barbosa Andrade [1], Thais Ribeiro Brito [1] and Ana Cláudia Ruggieri [1,*]

1 Department of Animal Sciences, Sao Paulo State University, Jaboticabal 14884-900, SP, Brazil; mariane.ferreira@unesp.br (M.R.F.); marina.elizabeth.15@hotmail.com (M.E.B.A.); tr.brito@unesp.br (T.R.B.)
2 Range Cattle Research and Education Center, University of Florida, Ona, FL 33865, USA
* Correspondence: adasilvacardoso@ufl.edu (A.d.S.C.); ana.ruggieri@unesp.br (A.C.R.)

Abstract: This study aimed to evaluate the forage mass and nutritional value (NV) of Guinea Massai grass (*Megathyrsus maximus* × *Megathyrsus infestum*) in an open pasture (OP) or the silvopastoral system (SPS) at different stages of development (SDs). The experimental design was completely randomized, with five replications. The treatments were distributed in a factorial scheme 2 × 4, corresponding to types of systems (OP and SPS) and four SD (vegetative stage—S1, pre-flowering—S2, full flowering—S3 and maturity after flowering—S4). There was no interaction of the system × stage of plant development in any of the variables evaluated. The production of forage mass and Guinea Massai grass morphological components did not differ ($p > 0.05$) between SPS and OP. However, they differed between the SDs ($p \leq 0.05$), with an increasing linear effect for forage mass, percentage of the stem, and dead material, and decreasing for leaf percentage and leaf: stem ratio. Only protein content differed ($p < 0.05$) among the chemical composition variables between culture systems. The shading caused by the SPS did not decrease the forage mass, NV, and in vitro dry matter digestibility of Guinea Massai grass compared to the system in OP. The advance of the SD of Guinea Massai grass increases the forage mass, with higher proportions of morphological components of low NV, such as stem and dead material. Silvopastoral system of Guinea Massai grass and eucalyptus is effective to prolongate the grazing season, with a greater amount of green leaves in the sward and better NV, without compromising herbage mass production in the forage growing season.

Keywords: Guinea Massai grass; forage management; integrated systems; nutritive value

Citation: Ferreira, M.R.; Cardoso, A.d.S.; Andrade, M.E.B.; Brito, T.R.; Ruggieri, A.C. How Are Warm-Season Pastures' Nutritive Value and Fermentation Characteristics Affected by Open Pasture, Silvopasture, and Sward Herbage Maturity?. *Agronomy* **2023**, *13*, 1756. https://doi.org/10.3390/agronomy13071756

Academic Editors: Kesi Liu and Xinqing Shao

Received: 6 May 2023
Revised: 22 June 2023
Accepted: 25 June 2023
Published: 28 June 2023

1. Introduction

Integrated systems are an option for efficient land use and intensification in ruminant production. The silvopastoral system (SPS), or forest livestock integration, is characterized by associating, in the same area, the forest component, the forage species, and the animal under grazing [1]. The combination of these components sustainably benefits the soil–plant–animal ecosystem.

The correct choice of the forestry and forage component plays a fundamental role in ensuring the success of the SPS implementation. Thus, the forest species should allow the luminous incidence to reach the forage canopy, must have low invasive potential, and provide marketable products [2]. In Brazil, the genus Eucalyptus has been highlighted for use in SPS because it presents desirable characteristics, besides being one of the most studied species for this purpose. On the other hand, the forage species must be adapted to the edaphoclimatic conditions of the region and have a good production capacity, be tolerant to the conditions generated by the system, especially shading [3], and meet the nutritional requirements of the animals. Thus, Guinea Massai grass (*Megathyrsus maximus*

× *Megathyrsus infestus*) becomes an option to make up the SPSs, especially because it tolerates 30–50% shading [4–6].

The shade created by the tree component alters the microclimate of the understory by reducing temperature and luminosity [7]. The reduction in temperature, in turn, slows the advance of the phenological stages of the plant, keeping it in more initial stages and of better nutritional value (NV) [8]. As the plant matures, especially in the reproductive phase, there is an increase in the thickness of cell walls, increasing fiber content, which reduces cell content and digestibility [9,10]. Thus, management conditions that keep the forage longer in the vegetative stage can benefit the production system.

Although it is known that there is a difference in forage NV in systems of open pasture (OP) and shaded [6,11–13], there is still little information about this variation in the different stages of development (SD) in integrated systems. Thus, the development of research evaluating tropical grasses in different SD plays a fundamental role in defining the ideal management in different systems.

We hypothesized that the shadow caused by the SPS would reduce the forage mass and improve the nutritive value of Guinea Massai grass compared to an OP. Additionally, that the more advanced the stage of development, the lower the nutritional value. Given the above, the objective was to evaluate the forage mass and NV of Guinea Massai grass during the different SDs maintained in the OP and SPS.

2. Materials and Methods

2.1. Experimental Area

The experiment was conducted in the Forage and Pasture sector at the São Paulo State University, Jaboticabal Campus, São Paulo, Brazil (21°14′ S and 48°17′ O, 598 masl), from February to June 2018, and from January to June 2019. The soil was classified as dystrophic yellow-red latosol [14]. In November 2017, soil analysis was performed, and there was no need for maintenance fertilization [15] (Table 1).

Table 1. Soil properties (depth 0–20) in the silvopastoral system (SPS) and open pasture.

Properties	SPS	OP
pH ($CaCl_2$)	5.3	5.4
Organic matter (g dm^{-3})	26	25
P (g dm^{-3})	38	36
S (g dm^{-3})	5	5
Ca (mmol$_c$ dm^{-3})	24	26
Mg (mmol$_c$ dm^{-3})	13	13
K (mmol$_c$ dm^{-3})	2.8	3.0
Al (mmol$_c$ dm^{-3})	0	0
H + Al (mmo$_{lc}$ dm^{-3})	31	28
Sum of bases (mmol$_c$ dm^{-3})	38.5	42.4
Cation exchange capability	69	70
Base saturation (V%)	56	60

According to the Köppen classification, the climate is of the type of Aw, described as tropical with a dry season from April to September and a concentration of rains in the summer months (October to March). The average annual precipitation is 1369 mm, and the average air temperature is 22.7 °C. Precipitation and temperature data were extracted from the data set of the area collection and Agrometeorology Department of Exact Sciences (Figure 1).

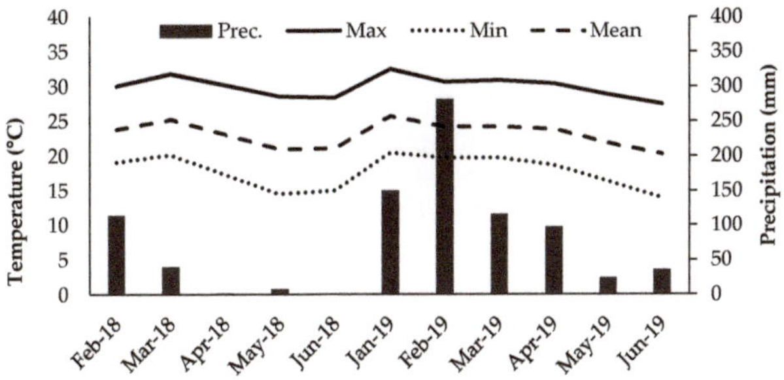

Figure 1. Accumulated precipitation (Prec.), mean, maximum (Max), and minimum (Min) temperatures recorded during the collection period, of the two experimental years (2018/2019) at Jaboticabal, São Paulo, Brazil.

2.2. Implementation of the Pasture Systems

The experimental area of 8000 m^2 comprised OP and SPS established in 2012 (800 m^2 per plot). Initially, in March 2012, the entire experimental area was established with Guinea Massai grass (*Megathyrsus maximus* × *Megathyrsus infestus*) integrated with maize (*Zea mays* L.). After grain harvesting, an irrigation system was used to aid in pasture formation (June to August, dry season in the region). In September 2012, the SPS plots were implanted, and seedlings of 1.2 m high of *Eucalyptus grandis* × *Eucalyptus urophylla* (*Eucalyptus urograndis*) were introduced to the system in single rows, in an East–West direction in the sets, one with spacing of 9 m × 1.5 m and the other 18 m × 1.5 m. Five years after implantation (2017), the tree component was thinning, and only the spacing of 18 m × 1.5 m was kept. This study was conducted in 2018/2019, six and seven years after the system implantation, respectively.

2.3. Treatments and Experimental Design

The experimental design was completely randomized, with five replications. The treatments were distributed in a factorial scheme 2 × 4, corresponding to two types of systems (SPS and OP), during four stages of development (SD) (vegetative stage—S1, pre-flowering—S2, full flowering—S3, and maturity after flowering—S4).

In January 2018, homogenization grazing was performed with dairy heifers (average weight of 350 kg), to achieve the sward residue height of 25 cm [6]. After grazing, five grazing exclusion cages of 1 m^2 were randomly distributed in each system to prevent the forage inside from being grazed. The cuts in each SD of Guinea Massai grass were performed according to the methodology described by Neel et al. [16]. In addition, Guinea Massai grass samples were collected for two consecutive years on the dates and cut-off intervals shown in Figure 2.

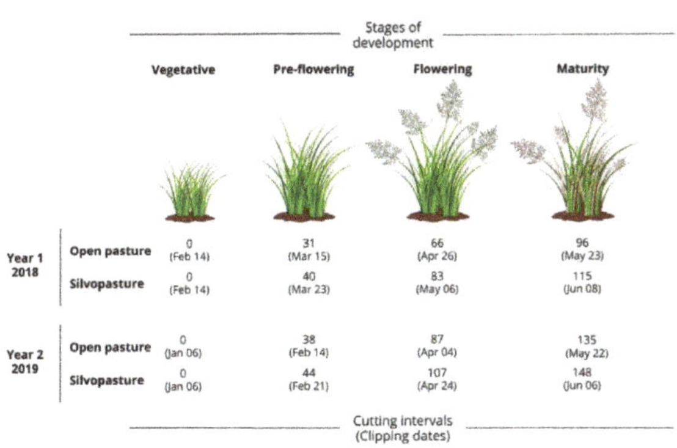

Figure 2. Dates and intervals between cuts in the different stages of development of Guinea Massai grass in open pasture and silvopastoral system at Jaboticabal, São Paulo, Brazil.

2.4. Evaluation of Forage Mass and Chemical Composition

A frame of 0.25 m^2 was used to delimit the sampling area within the cages. The forage samples in the four stages of grass development were collected at 5 cm of the soil. After collection, the material was weighed to obtain the green mass. The morphological components were separated into leaf blade (leaf), stem + sheath + reproductive structure (stem), and dead material obtained from a subsample. This subsample was dried in forced air circulation oven at 55 °C until constant weight, then the total dry mass of forage (kg DM ha^{-1}) and the proportion of its morphological components (% of DM) was calculated on a dry mass base at 55 °C.

Subsamples containing the whole plant were used to determine the chemical composition [17]. The chemical composition was expressed on a dry matter base of 105 °C. The dry matter content at 105 °C was quantified in an oven regulated at 105 °C, and the mineral matter (MM) by combustion at 600 °C [17]. Organic matter (OM) was calculated by the equation OM = 100 − MM. Neutral detergent fiber (NDF) and acid detergent fiber (ADF) concentrations were determined using procedures described by ANKOM Technology [18], and lignin (LIG) by acid hydrolysis. Hemicellulose was obtained by the difference between NDF content and ADF content, and cellulose by the difference between ADF and lignin content. Crude protein (CP) content was estimated using a LECO FP 528 nitrogen® io analyzer (Leco Corporation, St. Joseph, MI, USA).

2.5. Gas Production and In Vitro Degradability

In vitro grass production kinetics were evaluated in two consecutive trials, using 500 mg of sample and incubation vials of an internal volume of 125 mL [19,20], totaling eight treatments (four stages vegetative of forage × two systems) and five replications.

In each trial, rumen liquids used as inoculum (a pool of the rumen content of three animals) were collected from three fistulated lambs in the rumen-fed Guinea Massai grass hay, adapted for 15 days. The rumen liquids were collected before the bulky supply to the animals. Rumen fluids were collected proportionally for pool formation. The buffered inoculum was prepared by adding the ruminal liquid pool to the incubation medium (buffer), in a proportion of one part of the rumen fluid for every four parts of buffer solution [21]. Then, 75 mL of this solution was added to 125 mL vials filled with CO_2. The vials were sealed, measured for the holding of gases, and stored in a water bath at 39 °C for 48 h.

The gas pressure in the vials was measured at 3, 6, 9, 12, 24, and 48 h of fermentation, using a pressure transducer connected to a datalogger model GN200 (GN equipment, São Paulo, Brazil). The value recorded in the readings was converted to gas volume using the specific equation for laboratory conditions.

$$V = (5.4766 \times P) + 0.0934 \quad (1)$$

where V is the volume of gas in mL and P is the pressure measured in psi. The gas production was corrected by the average gas produced by the buffered inoculum and obtained from the Tifton 85 grass hay standard sample. In each assay, two weak ones containing standard samples and two vials with buffered inoculum were used. The production of gas accumulated during the 48 h was expressed in mL g^{-1} of OM.

The lag time (L) was estimated using the uncompartmentalized logistic model of Schofield et al. (1994) using the Equation (2):

$$Vt = Vf \times (1 + \exp(2 - 4 \times S \times (t - L))) - 1 \quad (2)$$

where Vf = final volume of accumulated gas (mL); S = degradability rate (h); t = time (h); and L = lag time (h).

The in vitro degradability of DM (DIVDM) was measured within 24 and 48 h of incubation. At the respective times, the containers were immediately immersed in ice, and, once open, the contents of the bottles were filtered on filter paper and the pH measured.

Then, the incubation residues were dried in a forced ventilation oven at 55 °C for 24 h and weighed on a precision scale. The difference between the forage weight calculated DIVDM and the residue after incubation was corrected for DM.

2.6. Statistical Analyses

Data from forage mass, morphological components, chemical composition, and DIVDM were submitted to variance analyses (ANOVA) according to the following model:

$$Yijk = \mu + Si + Ej + (SE)ij + \varepsilon ijk \quad (3)$$

where: μ = general mean; Si = system effect i; Ej = development stage effect j; $SEij$ = interaction effect between stage j and system i; εijk = random error associated with each observation.

The Shapiro–Wilk and Bartlett tests tested data for normality and homogeneity of variance, respectively. When significant, the systems were compared by the *t*-test, considering $p < 0.05$. When differences were observed between the SDs, orthogonal polynomial contrasts were used. All analyses were performed using the statistical program R (Version 4.0.2). The Tukey test compared the means, and differences were considered significant when $p < 0.05$.

3. Results

The adoption of the SPS compared to OP did not affect the forage mass and morphological components of Guinea Massai grass (Table 2). The production of forage mass and morphological components differed between SDs (Table 2). We observed a significant linear increase in the forage mass and the proportions of stem and dead material with the advance of plant maturity. On the contrary, the advance in the SD resulted in a linear reduction in the percentage of leaves and the leaf: stem ratio. Forage mass increased by 189% in S4 in relation to S1. The increase in forage mass in S4 comprised an increase of 261% in the proportion of stem and 233% in the proportion of dead material in plants compared to S1. However, the SD advance reduced the proportion of leaves in S4 by 48% in relation to S1.

Table 2. Forage mass and morphological components of Guinea Massai grass in open pasture (OP) and silvopastoral (SPS) systems (S) at different stages of development (SD) during 2018 and 2019 at Jaboticabal, São Paulo, Brazil. Values are means of five replicates of the S × DS interaction (n = 5).

Variable	Systems		SEM	SD				Effect	*p*-Value		
	OP	SPS		1	2	3	4		S	SD	S × SD
Forage mass [a]	5740	5704	2811	2931	4322	7006	8475	LIN	0.959	<0.001	0.736
Leaf (%)	58.74	58.14	11.33	83.41	65.96	42.80	42.99	LIN	0.826	<0.001	0.523
Stem (%)	20.56	21.32	10.23	8.25	15.65	29.31	29.71	LIN	0.762	<0.001	0.769
Dead (%) [b]	20.68	20.53	12.23	8.34	18.39	27.36	27.57	LIN	0.959	0.004	0.953
LSR [c]	5.10	5.19	3.09	11.98	5.23	1.76	1.99	LIN	0.909	<0.001	0.417

[a] in kg DM ha^{-1}; [b] dead material; [c] leaf ratio: stem; linear (LIN). Values of $p < 0.05$ differ, as per the *t*-test.

The interaction system × stage of plant development did not affect the variables of the chemical composition of Guinea Massai grass (Table 3). Except for the CP content, the pasture system did not significantly affect most of the chemically stable variables. The CP content in the pasture cultivated under the SPS was higher by 10% compared to the concentration of CP in the Guinea Massai grass in the OP. The plant development stage affected the forage chemical composition, except for the hemicellulose and cellulose content. These fractions comprised, on average, 34% of the DM of the forage, regardless of the SD evaluated. The forage NDF content showed a quadratic effect, with a higher content in S2. The contents of ADF and LIG showed a linear increase, with increments of 6% and 29% of S3 in relation to S1, respectively. The CP and ash contents showed a decreasing linear effect, with a reduction of 59% and 8% of S1 for S4, respectively.

Table 3. Chemical composition (g kg^{-1} DM) of Guinea Massai grass in open pasture (OP) and silvopastoral (SPS) systems (S) in different stages of development (SD) during 2018 and 2019 at Jaboticabal, São Paulo, Brazil. Values are means of five replicates of the S × DS interaction (n = 5).

Variable	System (S)		SEM	Stage of Development (SD)				Effect	*p*-Value		
	OP	SPS		1	2	3	4		S	DS	S × DS
CP [a]	76.5	84.4	14.5	118.7	81.3	65.8	48.0	LIN	0.023	<0.001	0.994
NDF [b]	742.4	736.6	22.3	736.1	753.8	736.1	732.0	QUA	0.243	0.020	0.250
ADF [c]	391.9	390.7	22.2	376.6	385.5	397.8	404.4	LIN	0.818	0.002	0.879
HEM [d]	353.9	345.7	31.9	353.8	357.5	342.0	345.5	NS	0.292	0.488	0.787
CEL [e]	347.4	345.2	20.0	347.9	344.6	350.4	342.5	NS	0.665	0.673	0.817
LIG [f]	29.4	34.0	10.8	28.9	25.3	34.8	37.4	LIN	0.074	0.004	0.406
Ash	101.2	102.6	11.9	10.58	110.1	94.6	97.3	LIN	0.649	<0.001	0.859

[a] Crude protein; [b] neutral detergent fiber; [c] acid detergent fiber; [d] hemicellulose; and [e] cellulose; [f] lignin, linear (LIN); quadratic (QUA); not significant (NS).

While there were a few significant effects for the main effect of S, and many for the main effect of DS, the S × SD interaction did not affect Vf and L (Table 4). The Vf of the forage cultivated in SPS was 6% higher than Vf provided by the OP. The lag time presented a similar response. The SPS presented an L of 84% higher than the forage cultivated in an OP (Table 4).

Table 4. Parameters of ruminal kinetics in vitro of Guinea Massai grass in open pasture and silvopastoral (SPS) systems during 2018 and 2019 at Jaboticabal, São Paulo, Brazil. Values are means of five replicates of the S × DS interaction (n = 5).

Variable	System (S)		SEM	Stage of Development (DS)				SEM	*p*-Value		
	Open Pasture	SPS		1	2	3	4		S	DS	S × DS
Vf (mL g^{-1} OM) [a]	158	168	3.89	196a	164b	140c	150bc	14	0.0389	<0.0001	
L (h) [b]	2.90	5.36	0.75	6.37a	6.79a	1.51b	1.85b	2.39	0.0213	0.0098	

[a] The final volume of gas produced in mL g^{-1} OM of organic matter, [b] lag time in hours. Means followed by the same letter on the line did not differ from each other by the Tukey test ($p \geq 0.05$).

As the SD of the forage plant increased, with the progress of the development stage, there was a reduction in Vf production. The maximum volume of gas from the fermentation of organic matter was observed during S1, which, when purchased with S4, was reduced by 23%. There was no difference in the colonization period between stages S1 and S2. In these stages, the lag time was higher by 74% compared to the other stages that did not differ from each other (Table 4).

The average pH was 7.0 and 6.9 in 24 h and 48 h, respectively. There was no effect of the interaction system × stage of development on DIVDM (Table 5). The pasture systems also did not affect DIVDM in incubation times of 24 h and 48 h. DIVDM in both evaluated times presented decreasing linear effects with the development of SD, with a reduction of approximately 34% from DIVDM in S1 to S4 in both incubation periods.

Table 5. In vitro degradability of dry matter (DIVDM; g kg^{-1} DM) of Guinea Massai grass in open pasture (OP) and silvopastoral (SPS) systems (S) at different stages of development (SD) after 24 h and 48 h of incubation during 2018 and 2019 at Jaboticabal, São Paulo, Brazil. Values are means of five replicates of the S × DS interaction (n = 5).

Variable	System (S)		SEM	Stage of Development (SD)				Effect	*p*-Value		
	Open Pasture	SPS		1	2	3	4		S	DS	S vs. DS
DIVDM 24 h	284.5	297.6	7.8	344.0	319.8	274.9	225.4	LIN	0.235	<0.001	0.127
DIVDM 48 h	447.9	453.8	7.1	552.4	502.0	394.0	354.8	LIN	0.562	<0.001	0.147

Linear (LIN). Values of $p \leq 0.05$ differ by the t-test.

4. Discussion

The similarity in the forage mass observed between OP and SPS (Table 2) suggests that the edaphoclimatic conditions imposed by the trees on the pasture in the understory were not limiting. This is possible due to the forage plant's ability to adapt to the reduction in the quantity and quality of sunlight [4–6]. Thus, it is inferable that, in a scenario similar to this research, eucalyptus can be used as a tree component in a spacing of 18 m between rows without the forage mass of Guinea Massai grass being affected. In a previous study in the same area of this research, Van Cleef et al. [22] observed that denser eucalyptus spacings (9 m between rows) reduce the forage mass of Guinea Massai grass. Lima et al. [13] found that tree age can also affect the development of grasses in SPS due to increased treetops and shading in the lower extract. The authors reported that between 6 and 7 years after planting, there was no reduction in the forage mass of *Urochloa decumbens*. The experimental period of this study occurred between the sixth and seventh year after planting eucalyptus seedlings, which may explain the similarity between the evaluated systems.

The absence of systems effect on forage mass (Table 2) ensures that the herbage allowance will not be a limiting factor for animal performance. However, while the forage mass was not different between systems, the level of shading may influence animal behavior regarding intake and rumination in the shade or full sunlight, especially if the paddocks were not fully shaded, which could cause animals to graze more briefly in the sun and lay down longer in the shade for thermal comfort. The difference between systems in CP and, potentially, lignin also would likely be a factor in animal performance. Moreover, the silvopastoral systems are known to provide thermal comfort to the animal [1], land use efficiency, and diversification of income sources within the property [23,24]. In this study, we did not measure the quality and quantity of sunlight. Further studies should measure the levels/extent of shading to better interpret the effect of the silvopastoral system on forage mass and nutritive value.

The chemical composition of Guinea Massai grass showed greater nutritive value in SPS due to the CP content in forage (Table 3). These results corroborate the responses of grasses grown in integrated production systems under conditions similar to this study's [6,11–13]. Kephart and Buxton [25] observed the highest protein content in plants submitted to shading, and the amount of nitrogen practically constant per cell results in a higher concentration of this nutrient. Moreover, according to these authors, the stress caused by shading can reduce the amount of photoassimilation for the development of the secondary cell wall, maintaining more cellular content, which contains higher protein content. Wilson [26] attributes the increase of protein in shaded environments to a higher soil moisture, which contributes to accelerating the degradation of organic matter and the cycling of nutrients in the soil and, consequently, greater utilization by plants. Neel et al. [16] relate this effect to the development of plants which, when submitted to shading, delay the time to reach physiological maturity. It is known that young plants have a higher protein content compared to those that are fully developed [13,22].

With the advance of the SD of Guinea Massai grass, there was a decrease in the percentage of leaves and, conversely, growth in the percentage of stem and dead material (Table 2). These results are in line with those observed by Caldeira [27] in Guinea Massai grass, Khral et al. [28] in different cultivars of giant missionary grass, and Peralta et al. [29] for cultivars of *Urochloa*. As the forage develops physiologically, moving from the vegetative to the reproductive stage, the stem elongates and prevents the generation of new leaves [30]. Advancing age resulted in a higher proportion of stems for two reasons. The first may be related to increased mass and self-shading level of the canopy. In these cases, the primary response of the plant is to lengthen the stem to expose the younger leaves to light. The second is related to flowering itself, which in Guinea Massai grass is induced by the reduction in daylength. Therefore, with the stimulation of flowering, the grass invested more energy in producing stems, reducing the leaf to stem ratio.

The advance of SD is responsible for the thickening and lignification of the cell wall, increasing the proportion of fiber constituents [31]. Young plant cells have only one outer layer, called the primary wall. However, as they mature, the secondary wall is formed, consisting mainly of structural carbohydrates, which attribute larger amounts of fiber to the forage in a more advanced SD [32]. A fact observed in this study is that the levels of ADF and LIG increased linearly with the advance of SD (Table 3), which corroborates the former studies [33,34]. The change in the wall and cell content proportion is also related to the decrease in CP content [26], due to the reduction in nitrogen availability due to the greater complexation of nitrogen compounds at the ADF fraction [33]. Another factor linked to the reduction in protein content is the decrease in hemicellulose, since it is in the leaves with higher concentrations of CP [35]. Low protein contents limit the animal's voluntary consumption, because nitrogen's low availability as a substrate for microbial synthesis reduces the activity of microorganisms in the rumen [35].

The in vitro gas production technique assumes the conversion of all major sources of carbohydrates into carbon dioxide, methane gas, and the reaction of volatile fatty acids [36]. It is known that the volume of gas produced during the in vitro fermentation process is related to the degradability of the substrate in the medium [13]. In this study, although a higher accumulated volume of gases (mL g^{-1} DM) was observed in Guinea Massai grass cultivated in SPS, no difference was observed in DIVDM compared to the OP (Table 4). This response suggests that the increase in gas production was not due solely to DIVDM, and that the lignin content in forage may have limited substrate degradation.

As for the stages of grass development, it was found that, with the advance of forage maturity, there was an increase in fiber content and a reduction in protein intake. This fact may have resulted in decreases in the maximum volume of gas produced during 48 h, and in the reduction in DIVDM (Table 5). It is known that the synchronism between energy and protein strongly influences the final volume of gas in vitro fermentation. In the same way, the lignification of structural tissue hinders the performance of digestive enzymes produced by ruminal microorganisms, reducing degradability [27,33]. Similar results were observed in Guinea Massai grass [31], Marandu palisade grass (*Urochloa brizantha*) [37], and different *Cynodon* cultivars [28], in which there was a reduction in degradability as the forage age progressed.

Different factors can affect microorganisms' colonization time and forage DIVDM. For example, the chemical composition, the physical characteristics of the cell wall, and the presence of readily-available carbohydrates are some determinant parameters in the lag time and degradability [35,38]. In this research, the forage cultivated in the SPS had a higher lag time (Table 4). This result may be due to the longer colonization time required by cellulolytic bacteria, since there was a tendency to increase the lignin content in the Guinea Massai grass submitted to SPS (Table 4), and cellulolytic bacteria populations have slower growth with a tendency to reduce; therefore, they require greater lag time.

Generally, plants in shaded environments tend to present a lower development speed and remain in a vegetative stage for longer [11,16]. In contrast, plants grown in the OP have a higher photosynthetic rate, providing accelerated growth [13]. Guinea Massai grass in SPS showed delayed physiological maturity compared to OP (Figure 2), prolonging the vegetative phase, and keeping the forage physiologically immature for a longer period [39]. This is a desirable feature in animal production systems, as the nutritional quality of forage generally declines with advancing age or during the reproductive phase [40].

5. Conclusions

The shading in the SPS did not decrease the forage mass, nutritive value, and value or DIVDM of Guinea Massai grass compared to the open pasture system, but it did increase the CP content. Forage mass averaged 5700 kg ha^{-1}, CP 8% of DM, and DIVDM 29% at 24h and 45% at 48h. The shading also induced Guinea Massai grass to remain vegetative for longer. The advance of the stage of development of forage results in a decrease in the proportion of leaves from 83 to 43%, crude protein from 11 to 4% content, and an increase in fiber fractions.

Author Contributions: Conceptualization, M.R.F., A.d.S.C. and A.C.R.; methodology, M.E.B.A.; formal analysis, M.R.F.; investigation, M.R.F., M.E.B.A. and T.R.B.; resources, A.C.R.; data curation, M.R.F.; writing—original draft preparation, M.R.F.; writing—review and editing, A.d.S.C. and A.C.R.; supervision, A.d.S.C. and A.C.R.; project administration, A.C.R.; funding acquisition, A.d.S.C. and A.C.R. All authors have read and agreed to the published version of the manuscript.

Funding: This research was funded by the Fundação de Amparo à Pesquisa do Estado de São Paulo (grant numbers: 2017/11274-5 and 2018/16273-0). The authors received scholarships from the Coordenação de Aperfeiçoamento do Pessoal de Nível Superior and the Conselho Nacional de desenvolvimento e Tecnologico.

Data Availability Statement: Data will be made available directly to the authors upon request.

Acknowledgments: This work was conducted as a part of the first author's PhD dissertation. We are grateful for the help of UnespFor (study group) for their assistance in conducting the experiments.

Conflicts of Interest: The authors declare no conflict of interest.

References

1. Oliveira, C.C.; Alves, F.V.; de Almeida, R.G.; Gamarra, É.L.; Villela, S.D.J.; Martins, P.G.M.D.A. Thermal comfort indices assessed in integrated production systems in the Brazilian savannah. *Agrofor. Syst.* **2018**, *92*, 1659–1672. [CrossRef]
2. Cardoso, A.S.; Alari, F.O.; de Andrade, N.; Nascimento, T.S.; van Cleef, F.D.O.S.; Alves, B.J.R.; Ruggieri, A.C.; Malheiros, E.B. Nitrous oxide emissions in silvopastoral systems: Key driving variables and seasonality. *Agric. For. Meteorol.* **2022**, *316*, 108851. [CrossRef]
3. de Oliveira, A.F.; Menezes, G.L.; Gonçalves, L.C.; de Araújo, V.E.; Ramirez, M.A.; Júnior, R.G.; Jayme, D.G.; Lana, Â.M.Q. Pasture traits and cattle performance in silvopastoral systems with Eucalyptus and Urochloa: Systematic review and meta-analysis. *Livest. Sci.* **2022**, *262*, 104973. [CrossRef]
4. Carvalho, C.A.B.D.; Silva, P.H.F.D.; Zanella, P.G.; Diavão, J.; Paiva, A.J. Morphogenetic and structural traits of tillers and herbage accumulation of Tanganyika grass under shading levels. *Rev. Bras. Saúde E Produção Anim.* **2022**, *23*, 1–6. [CrossRef]
5. Oliveira, E.P.D.; Silveira, L.P.D.O.; Teodoro, P.E.; Ascoli, F.G.; Torres, F.E. Effect of shading and seed encrusting on the development of *Panicum maximum* Jacq cultivars. *Biosci. J.* **2014**, *54*, 1682–1691.
6. Paciullo, D.S.C.; Gomide, C.D.M.; Castro, C.D.; Maurício, R.M.; Fernandes, P.B.; Morenz, M.J.F. Morphogenesis, biomass and nutritive value of *Panicum maximum* under different shade levels and fertilizer nitrogen rates. *Grass Forage Sci.* **2017**, *72*, 590–600. [CrossRef]
7. Almeida, E.M.; de Almeida, R.G.; Es, M.; de Freitas, P.V.D.X.; Ribeiro, F.M.; Fernandes, P.B.; Garcia, E.D.C. Silvopastoral systems: An approach on the integration of biotic and abiotic components. *Rev. Científica Rural* **2019**, *21*, 438–454.
8. Gobbi, K.F.; García, R.; Garcez Neto, A.F.; Pereira, O.G.; Rocha, G.C. Nutritional value of signalgrass and forage peanut submitted to shading. *Arch. Zootec.* **2010**, *59*, 379–390.
9. Rezaeifard, M.; Jafari, A.A.; Assareh, M.H. Effects of phenological stages on forage yield quality traits in cocksfoot (*Dactylis glomerata*). *J. Food Agric. Environ.* **2010**, *8*, 365–369.
10. Tesk, C.R.M.; Pedreira, B.C.; Pereira, D.H.; Pina, D.D.S.; Ramos, T.A.; Mombach, M.A. Impact of grazing management on forage qualitative characteristics: A review. *Sci. Electron. Arch.* **2018**, *11*, 188–197. [CrossRef]
11. Sousa, L.F.; Maurício, R.M.; Moreira, G.R.; Gonçalves, L.C.; Borges, I.; Pereira, L.G.R. Nutritional evaluation of "Braquiarão" grass in association with "Aroeira" trees in a silvopastoral system. *Agrofor. Syst.* **2010**, *79*, 189–199. [CrossRef]
12. Barros, J.S.; Meirelles, P.R.L.; Gomes, V.C.; Pariz, C.M.; Fachiolli, D.F.; Santana, E.A.R.; Costa, C.; Castilhos, C.A.; Souza, D.M. Nutritive value of xaraés palisade grass in three light intensities. *Arq. Bras. Med. Veterinária E Zootec.* **2019**, *71*, 1703–1711. [CrossRef]
13. Lima, M.A.; Paciullo, D.S.; Silva, F.F.; Morenz, M.J.; Gomide, C.A.; Rodrigues, R.A.; Bretas, I.L.; Chizzotti, F.H. Evaluation of a long-established silvopastoral *Brachiaria decumbens* system: Plant characteristics and feeding value for cattle. *Crop Pasture Sci.* **2019**, *70*, 814–825. [CrossRef]
14. Ruggieri, A.C.; Cardoso, A.S.; Ongaratto, F.; Casagrande, D.R.; Barbero, R.P.; Brito, L.D.F.; Oliveira, A.A.; Koscheck, J.F.W.; Reis, R.A. Grazing Intensity impacts on herbage mass, sward structure, greenhouse gas emissions, and animal performance: Analysis of brachiaria Pastureland. *Agronomy* **2020**, *10*, 1750. [CrossRef]
15. Raij, B.V.; Cantarella, H.; Quaggio, J.A.; Furlani, A.M.C. *Recomendações de Adubação e Calagem Para o Estado de São Paulo;* Instituto Agronômico/Fundação IAC: Campinas, Brazil, 1997; pp. 237–239.
16. Neel, J.P.S.; Felton, E.E.D.; Singh, S.; Sexstone, A.J.; Belesky, D.P. Open pasture, silvopasture and sward herbage maturity effects on nutritive value and fermentation characteristics of cool-season pasture. *Grass Forage Sci.* **2016**, *71*, 259–269. [CrossRef]
17. AOAC. *Official Methods of Analysis of the Association of Official Analytical Chemists*, 17th ed.; Association of Official Analytical Chemists: Gaithersburg, MD, USA, 2000.
18. ANKOM. *Acid Detergent Fiber in Feeds. Filter Bag Technique (for A200, A200I)*; Ankom Technology Method 8; Ankom Technology Corp.: Macedon, NY, USA, 2006.

19. Theodorou, M.K.; Williams, B.A.; Dhanoa, M.S.; McAllan, A.B.; France, J.A. Simple gas production method using a pressure transducer to determine the fermentation kinetics of ruminant feeds. *Anim. Feed Sci. Technol.* **1994**, *48*, 185–197. [CrossRef]
20. Mauricio, R.M.; Mould, F.L.; Dhanoa, M.S.; Owen, E.; Channa, K.S.; Theodorou, M.K. A semi-automated in vitro gas production technique for ruminant feedstuff evaluation. *Anim. Feed Sci. Technol.* **1999**, *79*, 321–330. [CrossRef]
21. Goering, H.K.; Van Soest, P.J. *Forage Fiber Analysis: Apparatus Reagents, Procedures, and Some Applications*; Agricultural Handbook 379; U.S. Agricultural Research Service: Washington, DC, USA, 1970; 20p.
22. van Cleef, F.O.S.; van Cleef, E.H.C.B.; Longhini, V.Z.; Nascimento, T.S.; Ezequiel, J.M.B.; Ruggieri, A.C. Feedlot performance, carcass characteristics, and meat characteristics of lambs grown under silvopastoral systems. *Can. J. Anim. Sci.* **2019**, *100*, 385–388. [CrossRef]
23. Cardoso, A.S.; Longhini, V.Z.; Berça, A.S.; Ongaratto, F.; Siniscalchi, D.; Reis, R.A.; Ruggieri, A.C. Pasture management and greenhouse gases emissions. *Biosci. J.* **2022**, *38*, e38099. [CrossRef]
24. Pezzopane, J.R.M.; Bosi, C.; Nicodemo, M.L.F.; Santos, P.M.; Cruz, P.G.D.; Parmejiani, R.S. Microclimate and soil moisture in a silvopastoral system in southeastern Brazil. *Bragantia* **2015**, *74*, 110–119. [CrossRef]
25. Kephart, K.D.; Buxton, D.R. Forage quality responses of C3 and C4 perennial grasses to shade. *Crop Sci.* **1993**, *33*, 831–837. [CrossRef]
26. Wilson, J.R. Cell wall characteristics in relation to forage digestion by ruminants. *J. Agric. Sci.* **1994**, *122*, 173–182. [CrossRef]
27. Caldeira, R.R. Avaliação da Produção e da Qualidade do Capim *Panicum maximum* cv. Massai. Master Thesis, Universidade de Brasília, Brasília, Brazil, 2016.
28. Krahl, G.; Baldissera, T.C.; Pinto, C.E.; Garagorry, F.C.; Werner, S.S.; Lopes, C.F.; Ribeiro Filho, H.M.N. Can pasture defoliation management targets be similar under full sun and shaded conditions? *Crop Pasture Sci.* **2022**, *74*, 259–269. [CrossRef]
29. Peralta, M.D.L.A.; Garcia, A.R.R.; Olmedo, F.M.; Mariela, P.M.; Aniano Aguirre, H.; Gasga Perez, L. Physical-chemical quality of Urochloa grasses in different phenological stages. *Int. J. Agric. Environ. Biores.* **2020**, *5*, 162–171. [CrossRef]
30. Hodgson, J. *Grazing Management—Science into Practice*; Longman Scientific & Technical: Essex, New Zealand, 1990; 203p.
31. Garcez, B.S.; Alves, A.A.; Macedo, E.D.O.; Santos, C.M.D.; Araújo, D.L.D.C.; Lacerda, M.D.S.B. Ruminal degradation of Panicum grasses in three post-regrowth ages. *Ciênc. Anim. Bras.* **2020**, *21*, 1–8. [CrossRef]
32. Asaadi, A.M.; Yazdi, A.K. Phenological stage effects on forage quality of four forbs species. *J. Food Agric. Environ.* **2011**, *9*, 380–384.
33. Garcez, B.S.; Alves, A.A.; Araújo, D.L.C.; Lacerda, M.D.S.B.; Souza, L.G.C.; Carvalho, L.F. Ruminal degradability of colonião grass (*Panicum maximum* Jacq. cv. Colonião) at three post-regrowth ages. *Acta Vet. Bras.* **2016**, *10*, 130–134. [CrossRef]
34. Monção, F.P.; Costa, M.A.M.S.; Sampaio, J.P.; Gomes, M.; Leal, D.B.; Maranhão, C.M.A.; Chamone, J.M.A. Yield and nutritional value of BRS Capiaçu grass at different regrowth ages. *Semin. Ciências Agrárias* **2019**, *40*, 2045–2056. [CrossRef]
35. Van Soest, P.J. *Nutritional Ecology of the Ruminant*; Cornell University Press: New York, NY, USA, 1994; 476p.
36. Santo, A.X.; Silva, L.D.F.; Lançanova, J.A.C.; Ribeiro, E.L.A.; Mizubuti, I.Y.; Fortaleza, A.P.S.; Henz, E.L.; Júnior, F.L.M. Fractionation of carbohydrates, protein and "in vitro" ruminal degradation kinetics by gas production technique of supplementary rations containing sunflower cake. *Arq. Bras. Med. Veterinária E Zootec.* **2017**, *69*, 234–242. [CrossRef]
37. Velásquez, P.A.T.; Berchielli, T.T.; Reis, R.A.; Rivera, A.R.; Dian, P.H.M.; Teixeira, I.A.M.D.A. Chemical composition, fractionation of carbohydrates and crude protein and *in vitro* digestibility on tropical forages in the different cutting ages. *Rev. Bras. Zootec.* **2010**, *39*, 1206–1213. [CrossRef]
38. Xue, Z.; Liu, N.; Wang, Y.; Yang, H.; Wei, Y.; Moriel, P.; Palmer, E.; Zhang, Y. Combining orchardgrass and alfalfa: Effects of forage ratios on in vitro rumen degradation and fermentation characteristics of silage compared with hay. *Animals* **2019**, *10*, 59. [CrossRef] [PubMed]
39. Lopes, C.M.; Paciullo, D.S.C.; Araújo, S.A.C.; Gomide, C.A.M.; Morenz, M.J.F.; Villela, S.D.J. Herbage mass, morphological composition, and nutritive value of signalgrass, submitted to shading and fertilization levels. *Arq. Bras. Med. Veterinária E Zootec.* **2017**, *69*, 225–233. [CrossRef]
40. Collins, M.; Newman, Y.C. Forage quality. In *Forages*, 7th ed.; Collins, M., Nelson, C.J., Moore, K.J., Barnes, R.F., Eds.; John Wiley & Sons: Hoboken, NJ, USA, 2018; Volume 1, pp. 269–285.

agronomy

MDPI

Article

Analyzing Characteristics of Grassland Gross Ecosystem Product to Inform Decision Making in the Karst Desertification Control

Yongyao Li [1,2], Kangning Xiong [1,*], Wenfang Zhang [1], Shuzhen Song [1] and Lu Luo [1]

1 School of Karst Science, Engineering Laboratory for Karst Desertification Control and Eco-Industry of Guizhou Province, Guizhou Normal University, Guiyang 550001, China; 21030170044@gznu.edu.cn (Y.L.); 222100170580@gznu.edu.cn (W.Z.); 15010170764@gznu.edu.cn (S.S.); 20030170039@gznu.edu.cn (L.L.)
2 Bijie High-Tech Institute, Science and Technology Bureau of Bijie, Bijie 551700, China
* Correspondence: xiongkn@gznu.edu.cn

Abstract: Synergistically enhancing and realizing the value of grassland ecosystem services (ES) for economic activity is an important but challenging task for achieving sustainability in the karst desertification control (KDC). However, how to use grassland ES value characteristics in the KDC to make decisions on ES improvement, human well-being enhancement, and sustainable development remains unclear. In this paper, we took the contiguous region of karst desertification in Yunnan-Guangxi-Guizhou, China, a global hotspot, as the study area. Based on the valuation of the gross ecosystem product (GEP) and county economic intensity, we analyzed the structural and spatial characteristics of grassland GEP in the KDC using spatial analysis methods. We found that: (1) the grassland GEP in the KDC is mainly distributed in counties with low economic intensity (86.05% of the total number of counties) and vulnerable to losses caused by the livelihood of farmers; (2) the grassland GEP in the KDC is spatially small and scattered (the geographic concentration lies between 0.015 and 0.237), which makes it difficult to form industrial scale advantages; (3) the public product index (66.22–96.77%) and industry scale concentration (97.87–99.86%) of grassland GEP in the KDC are high, and most of the GEP is difficult to transform on the private market. Based on our findings, we proposed three corresponding recommendations for economic decision-making. The results of this study can provide a reference for economic decision-making regarding the management of grassland ES in karst areas with similar conditions and beyond.

Keywords: gross ecosystem product; valuation; grassland; karst desertification control; economic decision

Citation: Li, Y.; Xiong, K.; Zhang, W.; Song, S.; Luo, L. Analyzing Characteristics of Grassland Gross Ecosystem Product to Inform Decision Making in the Karst Desertification Control. *Agronomy* **2023**, *13*, 1861. https://doi.org/10.3390/agronomy13071861

Academic Editors: Kesi Liu and Xinqing Shao

Received: 8 June 2023
Revised: 10 July 2023
Accepted: 11 July 2023
Published: 14 July 2023

1. Introduction

Karst regions cover approximately 20% of the world's dry ice-free land [1]. Various types of karst have been widely developed in China. It covers about 36% of the national territory [2]. The South China karst, centered on the Guizhou Plateau, is the largest area of exposed carbonate rocks amoung the three main concentrated karst areas in the world, with a karst area of 555,000 km^2 [3]. Karst ecosystems have a binary three-dimensional spatial structure coupled with unique hydrological and geological conditions [4], forming a fragile, sensitive, and vulnerable ecological environment [5]. In the context of high population pressure and limited livelihoods, the ecosystem is unduly disturbed by traditional industries [6], resulting in the globally representative and typical ecosystem problem of karst desertification [7,8]. Karst desertification is an extreme manifestation of land degradation [9]. It affects the supply of ES and forms a poverty trap [10], which in turn threatens human well-being.

As a pioneer community ecosystem for alleviating severe karst desertification and restoring degraded land [11], grasslands not only provide an ecological security shield for regional sustainable development but also contribute to household livelihoods through forage supply and grazing [12,13]. Since the 1990s, grassland establishment and herbivorous animal husbandry development have been regarded as the main projects for controlling karst desertification [14]. For example, in the early stages of the KDC, the whole society generally attached importance to the livestock support services provided by grasslands [15] due to the fact that these services can be traded on private markets or contribute to livelihoods for local residents. In the context of an understanding that the pursuit of ecological material goods is dominant, grass cultivation and livestock breeding became the main modes of KDC. Accordingly, it not only promoted the greening of vegetation [16–18] but also improved the diversity of household livelihoods [19]. As research has progressed, researchers have generally recognized that karst is a complex ecosystem, and its internal elements are not a single cause-and-effect relationship but a complex feedback [20]. Therefore, in the later period of the KDC, the structure, function, and process of karst ecosystems are of great concern by relevant scholars [21]. Correspondingly, the objectives of engineering measures for the management of karst desertification are gradually oriented towards the overall improvement of grassland ES. For example, increasing species diversity in grasslands by selecting pioneer species (e.g., legumes) and optimizing community configuration [22,23]; stimulating plant growth and increasing herbaceous biomass by adding nutrients (e.g., N, P, and K) during grassland establishment and management [24]; and improving the health index of grassland ecosystems through engineering measures such as returning farmland to grassland or establishing artificial grasslands [25]. After decades of engineering management, a special grassland ecosystem (hereinafter referred to as grassland in the KDC) has been formed correspondingly. Now, in response to the major needs of the United Nations Decade on Ecosystem Restoration (2021–2030) [26], the objectives of ecosystem conservation and restoration of grassland in the KDC have shifted from the initial pursuit of vegetation greening to the synergistically comprehensive enhancement of ES and human well-being [27]. Quantifying the contribution of the ES value of grasslands in the KDC to people and incorporating it into economic decision making is an important step towards achieving this goal.

In previous research, researchers have generally accounted for the physical quantities of individual ES in karst areas based on biophysical and monetary measures. For example, the accounting cases of water yield, soil conservation, and carbon sequestration in typical karst areas in southwest China [28–30]. Meanwhile, some scholars have also used the value transfer method to evaluate the value of each karst ES. For instance, Hu et al. [31], Wang et al. [32], and Zhang et al. [33] took the karst-Beibu Gulf of southwest Guangxi, the Yunnan-Guizhou Plateau, and China, respectively, as the study areas to estimate the ES value of karst ecosystems. From the perspective of karst area to global grassland ecosystem, researchers have widely discussed the monetary value of grassland ES from the aspects of index system, value accounting method, influencing factors, and land-use management. Richter et al. summarized 29 indicators and 85 methods of grassland ES value accounting [34]. Chen et al. discussed the influencing factors by analyzing the temporal and spatial changes based on grassland ES value accounting [35]. Rao et al. calculated the ES value of grasslands in semi-arid areas to guide land-use management [36]. From these typical cases, researchers universally apply comparative analysis of ES values, either vertically (e.g., time, space) or horizontally (e.g., compared with GDP), to inform management decisions on ES and land use by analyzing ES trade-offs and synergistic relationships, driving mechanisms, etc. However, a thematic valuation of grassland ES values in the KDC is rare, and no researchers have analyzed the structural and spatial characteristics of the ES values of grasslands in the KDC and proposed valuable economic decision suggestions. More notably, previous valuations have focused on ecological attributes and paid insufficient attention to economic attributes and their utility. In addition, decisions have not taken into account local economic strengths, resulting in the suspension of economic decisions to improve ES and human well-being.

In this study, we took the concentrated contiguous area of karst desertification in Yunnan-Guangxi-Guizhou, China as the study area, where the regional characteristics and engineering measures are typical and representative globally. According to the background of the study area, we propose the following theoretical hypotheses of the characteristics of grassland ES value in the KDC (Figure 1): (1) The ES value of grassland in the KDC was mainly concentrated in the mountainous area far from the city, and the farther from the city, the higher the ES value. (2) For the structure and classification of grassland ES value in the KDC, the value of regulation services accounts for the majority proportion (the value is difficult to translate through the market due to its public nature and externalities), and the value of ecological material goods and services that can be directly traded based on the private market accounts for a small proportion. (3) Owing to the small scattered landscape pattern of grassland in the KDC, the geographical concentration of ES value is low, and it is difficult to form an industrial scale advantage. According to the theoretical assumptions, we applied the GEP accounting framework, a method to assess the value of nature's contribution to economic activity [37,38], to quantify the value of grassland ES in the KDC. On the basis of accounting for the counties' economic intensity, we analyzed the structural and special characteristics of the grassland GEP using the geospatial analysis method and proposed some economic decision recommendations for the management of the grassland ES in the KDC.

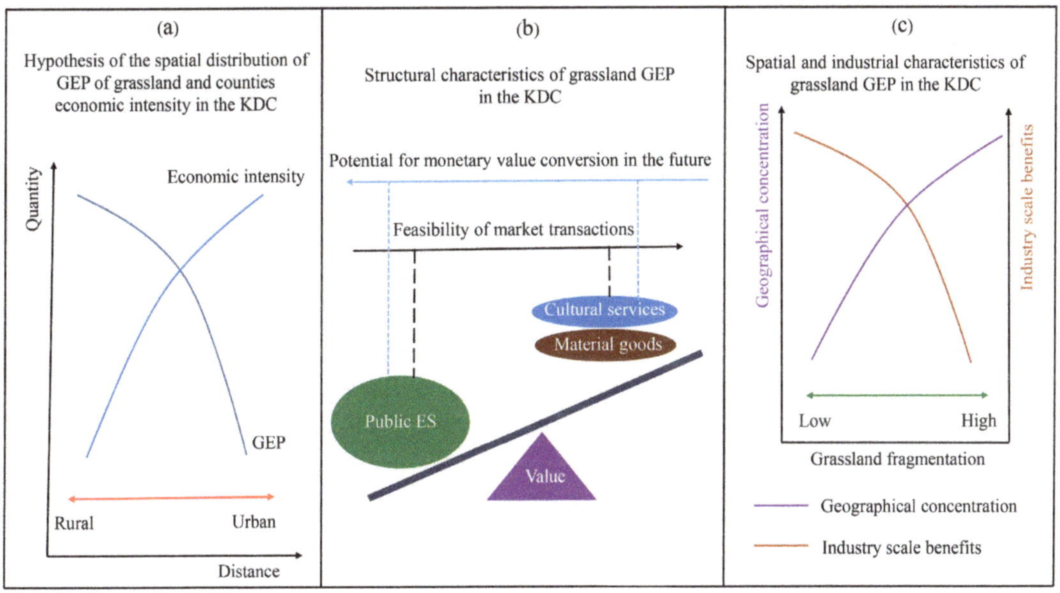

Figure 1. Theoretical hypothesis of GEP characteristics of grassland in the KDC.

Through the GEP valuation of grasslands in the KDC, researchers and stakeholders can better understand the economic value of grassland ES in the KDC. More importantly, we proposed differentiated recommendations for managing grassland ES in the KDC in relation to county economic strength, which can improve the feasibility of economic decision-making processes. At the same time, this study shifts from the previous bio-physical assessment that focused on ecological attributes and their intrinsic values to an assessment that focuses on economic attributes and their utility values, while bridging the knowledge gap between previous studies on ES management decisions that lacked actionability because they did not consider local input potential. Going forward, this study can provide theoretical references for the implementation of major projects to protect and restore grasslands in ecologically fragile areas (especially karst areas with the same ecological and environmental conditions), the realization of the grassland eco-products value, and rural revitalization.

2. Materials and Methods

In this paper, we first selected the globally representative and typical hot study area of the KDC. On this basis, we followed the logical threads of basic geospatial data and socioeconomic collection—GEP and economic intensity index accounting—GEP characteristic analysis—decision making (Figure 2). In particular, we applied national and local standards (or specifications) to calculate the GEP of grasslands in the KDC. Finally, we used SPSS 2019 (version 19.0; SPSS, Chicago, IL, USA) to analyze the indicators and used ArcGIS 10.3 (Redwood City, CA, USA) and Originlab Origin 2021 (Northampton, MA, USA) to visualize the data.

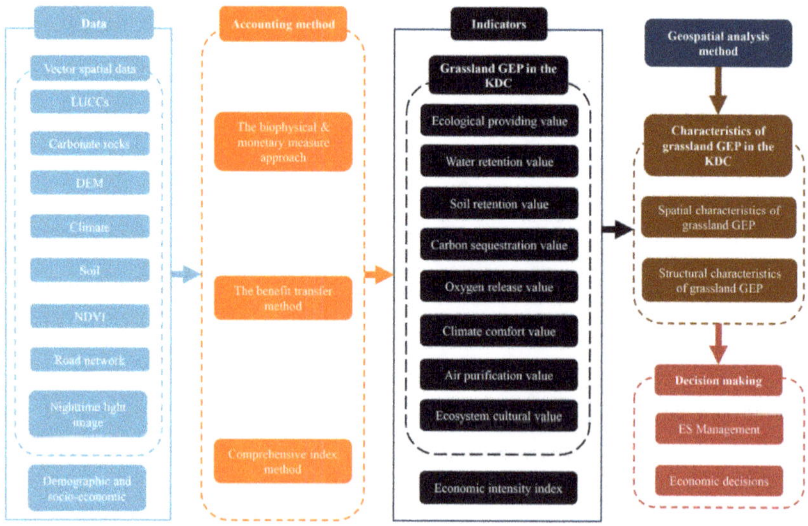

Figure 2. The flow chart of research methodology and logical threads.

2.1. Research Areas

The concentrated contiguous area of karst desertification in Yunnan-Guangxi-Guizhou is located at the intersection of Yunnan Province, Guangxi Zhuang Autonomous Region, and Guizhou Province, China (97°21′–117°19′ E and 20°13′–34°19′ N) (Figure 3). It includes 91 counties (among which 5 counties have no karst grasslands, and the following data of grassland ES in KDC are calculated based on 86 counties). In the region, most of the sites are located in the southeastern part of the Yunnan-Guizhou Plateau. It is a transition zone to the Guangxi Basin and borders Vietnam to the south, which is a typical tectonic plateau mountainous area. It includes six topographic zones, namely, karst plateau, karst fault basin, karst peak cluster depression, karst peak forest plain, karst gorge, and karst trough valley. At the same time, this area is also an important ecological functional area in the Pearl River and Yangtze River basins. The study area covers 228,000 km^2, of which 135,300 km^2 is karst area, accounting for 59.34% of the total area; karst grassland is 22,000 km^2, accounting for 16.26% of the karst area and 9.7% of the total area. The region is a tropical and subtropical monsoon zone. Due to the altitude difference in the region, the climate has the characteristics of both low-latitude monsoon and mountain plain climates, with annual precipitation ranging from 880 to 1991 mm, mainly concentrated in the months of April–September. The population of the region is 24.6 million, and the GDP reaches CNY 398.89 billion in 2020.

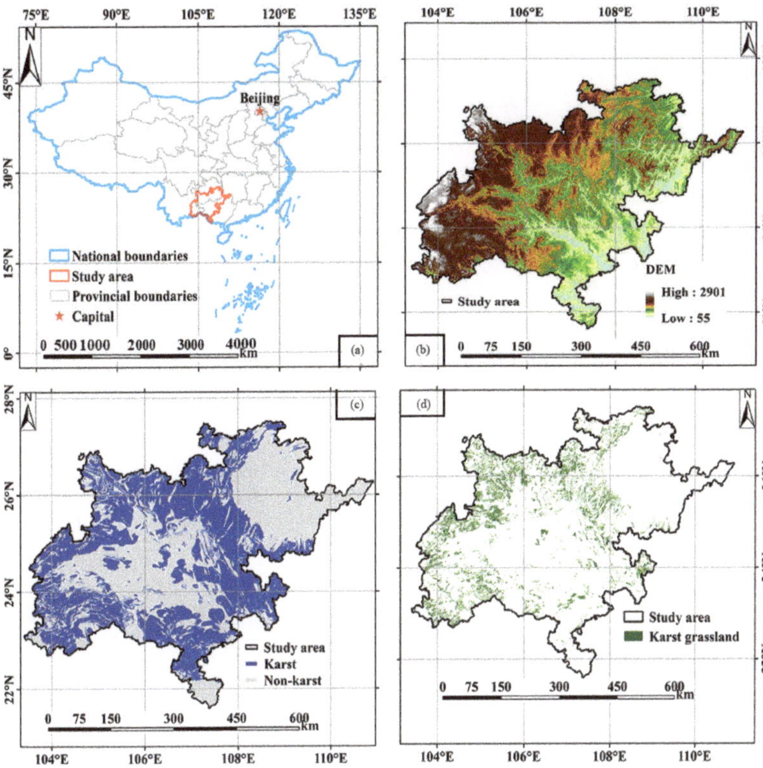

Figure 3. Study area: (**a**) Location of study area; (**b**) the DEM of the study area; (**c**) Karst distribution; and (**d**) Grassland distribution in the KDC area.

In recent years, the study area has carried out a number of national projects to combat karst desertification and alleviate poverty. For example, in February 2008, China's State Council approved the "Comprehensive Management Plan for the Control Karst Desertification in the Karst Region (2006–2015)", which plans to carry out comprehensive management in 100 pilot counties in eight provinces in southern China during the Eleventh Five-Year Plan. Pilot projects were carried out in 38 counties in the study area [39]. Based on this, the Chinese government implemented the "Thirteenth Five-Year Plan for the Construction of a Comprehensive Management Project for Karst Desertification" in 2016 [40]. After two phases of karst desertification management, all karst counties (karst area of more than 30% of the national territory) in the study area were fully covered. At the same time, the Chinese government has implemented the "Plan for Regional Development and Poverty Alleviation in the Contiguous Karst Desertification Region of Yunnan-Guangxi-Guizhou (2011–2020)" [41]. By the end of 2020, all counties in the study area were lifted out of poverty.

The grassland ecosystem of the concentrated contiguous area of karst desertification in Yunnan-Guangxi-Guizhou, China, not only has the basic attributes of fragility, sensitivity, and vulnerability, but is also characterized by high population pressure, numerous ethnic minorities, and lagging industrial development [42]. What's more, the spatial heterogeneity of the region is extremely significant. Compared to other non-karst regions, especially in relation to globally important rangelands, the grassland ecosystems in this region are more complex and the ES value is more difficult to realize. Many issues about the value realization of grassland ES globally (especially in karst areas) can be answered in this region. It can be seen that the study area is typical and representative.

2.2. Data Source

The data and sources are shown in Table 1.

Table 1. Data, resolution, and sources in study area.

Data	Resolution	Source
Land Use (grassland)	30 m	Resource and Environment Science and Data Center, https://www.resdc.cn (accessed on 13 December 2022)
Carbonate rocks	1:500,000	Karst Science Data Center, Institute of Geochemistry Chinese Academy of Sciences
DEM	30 m	Resource and Environment Science and Data Center, https://www.resdc.cn (accessed on 5 December 2022)
Meteorological data	1 km	(1) National Meteorological Science Data Center, https://data.cma.cn (accessed on 3 December 2022). (2) National Qinghai-Tibet Plateau Scientific Data Center, http://data.tpdc.ac.cn/zh-hans/ (accessed on 15 December 2022). (3) National Earth System Science Data Center, https://www.geodata.cn (accessed on 5 December 2022).
Soil	1 km	Harmonized World Soil Database (HWSD) http://www.ncdc.ac.cn/portal/metadata/fcd60332-575c-4209-ae9c-d1efc1f424d4 (accessed on 5 December 2022)
NPP	1 km	Resource and Environment Science and Data Center, https://www.resdc.cn (accessed on 5 December 2022)
Administrative Boundaries	--	Resource and Environment Science and Data Center, https://www.resdc.cn (accessed on 5 December 2022)
Road network data	1 km	Geographic remote sensing ecological network platform, www.gisrs.cn (accessed on 7 May 2023)
Demographic and socioeconomic data	--	(1) Statistical Yearbook (Province, City, County, District); (2) Government Statistical Bulletin of National Economic and Social Development (County, District); (3) Annual Government Work Report (County, District); (4) Seventh National Census Bulletin (China); (5) Missing Data: Obtained by the linear interpolation method. Prioritization hierarchy of the same indicator acquisition: (1) > (2) > (3) > (5)
Nighttime light	0.44 km	National earth system science data center, https://www.resdc.cn (accessed on 8 December 2022)

2.3. Evaluation and Analysis Methods

2.3.1. Measuring GEP of Grassland in the KDC

- Ecosystem providing value

Considering that the grazing output value of grassland in the KDC is not based on natural boundary statistics, the theoretical value of grazing yield of karst grassland is used as the basis for accounting. The formula is:

$$V_M = E_P \times A_G \tag{1}$$

where V_M is the value of material goods of grassland in the KDC (CNY); E_P refers to the value of ecological material goods per unit of grassland area (CNY·hm^{-2}). According to the research results of Shen et al. [43], we take 1.07 cattle/hm^2. The price of cattle is based on the "National Agricultural Product Cost-Benefit Compilation-2021" [44], which is calculated at 16,713.15 CNY per cattle. A_G is grassland area (hm^2).

- Ecosystem regulation services value

1. Water retention value

Estimation of supply. Following the principle of water balance, water retention was calculated using the InVEST model (with two modules for water yield and water retention). The formula for measuring water yield can be expressed as follows:

$$Y_x = (1 - \frac{AET_x}{P_x}) \times P_x \tag{2}$$

where Y_x (mm·yr^{-1}) is the water yield for pixel x, AET_x (mm·yr^{-1}) is the actual annual evapotranspiration for pixel x, and P_x (mm·yr^{-1}) is the annual precipitation for pixel x.

Since actual annual evapotranspiration is difficult to measure accurately, the evapotranspiration portion of the water balance, $\frac{AET_i}{P_i}$, is based on an expression of the Budyko curve.

$$\frac{AET_x}{P_x} = 1 + \frac{PET_x}{P_x} - \left[1 + \left(\frac{PET_x}{P_x}\right)^{W_x}\right]^{\frac{1}{W_x}} \tag{3}$$

$$W_x = Z \times \frac{AWC_x}{P_x} + 1.25 \tag{4}$$

$$PET_x = K_{c_x} \times ET_0 \tag{5}$$

Here, PET_x is potential evapotranspiration (mm·yr^{-1}); W_x is a non-physical parameter that represents natural climate-soil properties; Z is a dimensionless constant ranging from 1 to 30 and follows local precipitation and hydrogeological characteristics, AWC_x is plant-available water content (mm); ET_0 is the reference evapotranspiration of pixel x and represents the coefficient of vegetation evapotranspiration associated with pixel x' and K_{c_x} is the evapotranspiration coefficient for pixel x.

Then, water retention is calculated using the following formula:

$$WR = \min(1, \frac{249}{Velocity}) \times \min(1, \frac{0.9 \times TI}{3}) \times \min(1, \frac{Ksat}{300}) \times Y_x \tag{6}$$

$$TI = \log(\frac{DrainageArea}{SoilDepth \times Percent\_slope}) \tag{7}$$

where WR represents water retention (ton·yr^{-1}); $Velocity$ is velocity coefficient; TI refers to topographic index; $Ksat$ is soil saturated hydraulic conductivity; Y_x is water yield (mm·yr^{-1}); $DrainageArea$ is catchment area of pixel x; $SoilDepth$ is soil depth; and $Percent\_slope$ is percent slope.

Value accounting. The calculation formula is:

$$V_{wr} = P_{wr} \times WR \tag{8}$$

where V_{wr} represents the total value of water retention (CNY); P_{wr} is the price of water resources, using the average water price (2.28 CNY·ton^{-1}) of 31 provincial capitals in Mainland China in 2020 with data from the China Water Price Network, www.h20-china.com/price/ (accessed on 13 January 2023); and WR represents water retention (ton·yr^{-1}).

2. Soil retention value

Estimation of supply. Soil retention refers to the ability of an ecosystem to regulate soil erosion and sediment interception, which plays a key role in maintaining land productivity. It can be calculated by using the revised universal soil loss equation (RUSLE), and the formula can be expressed as:

$$Q_{sr} = R \times K \times L \times S \times (1 - C \times P) \tag{9}$$

Q_{sr} is soil retention (t/a), R, K, L, S, C and P is rainfall erosivity, soil erodibility factor, terrain factor, vegetation cover and management factor, and soil and factors for water conservation measures, respectively. The calculation methods of R, K, L, S, C, and P refers to Xu et al. [45].

Value accounting. Using an alternative cost method. The calculation formula is:

$$V_{SC} = V_1 + V_2 \tag{10}$$

$$V_1 = (Q_{SR}/\rho) \times r \times \beta_1 \tag{11}$$

$$V_2 = \sum (Q_{OI}/r_i) \times \beta_{4i} \tag{12}$$

$$Q_{OI} = Q_{SR} \times C_i \tag{13}$$

where V_{SC} is the total value of soil conservation of grassland in the KDC (CNY); V_1 and V_2 are the values of sedimentation reduction and fertility retention, respectively (CNY); Q_{SR} is the amount of soil retention (ton); and ρ is the soil capacity. Referring to "the Technical Guide to Accounting for Gross Ecosystem Product" (hereinafter referred to as "GEP specification") [46], the average soil capacity of Yunnan, Guizhou, and Guangxi, China was taken to be 1.23 $t \cdot m^{-3}$. r refers to the percentage of sediment lost to soil erosion that is silted up in reservoirs, rivers, and lakes (%). Referring to "the specification of grassland ecosystem service function assessment" (hereinafter referred to as "grassland specification") [47], we taken 0.24; β_1 is the cost of the unit storage capacity project, taken to be 42 $CNY \cdot ton^{-1}$; Q_{OI} is annual fertilizer retention ($ton \cdot yr^{-1}$); C_i is the N, P, K, and organic matter content of the soil (%); r_i is N, P, K, and organic matter content in fertilizer and organic matter (according to the "grassland specification" 14%, 15%, 50%, and 100% were taken, respectively); and β_2 is the value of chemical fertilizer and organic matter ($CNY \cdot ton^{-1}$). According to the "grassland specification": diammonium phosphate contains 14% of nitrogen and 15.01% of phosphorus and potassium chloride contains 50% potassium content. The price of diammonium phosphate is 3060 $CNY \cdot t^{-1}$, the price of potassium chloride fertilizer is 2350 $CNY \cdot t^{-1}$, and the price of organic matter is 855 $CNY \cdot t^{-1}$.

3. Carbon sequestration and storage value

Estimation of supply. The net ecosystem productivity method was used. The formula is:

$$Q_{tCO_2} = M_{CO_2} \times NEP \tag{14}$$

$$NEP = \alpha \times NPP \tag{15}$$

where Q_{TCO_2} is the carbon sequestration in grasslands in the KDC (ton); $M_{CO_2}/M_C = 44/12$, which is the coefficient of conversion of C to CO_2; NPP is net primary productivity ($t \cdot C \cdot yr^{-1}$); and α is the conversion coefficient between NEP and NPP, which is assumed to be 0.1532 according to the "GEP Specification".

Value accounting. The calculation formula is:

$$V_{CO_2} = Q_{tCO_2} \times P_{CO_2} \tag{16}$$

V_{CO_2} is the value of carbon sequestration of GKD (CNY); Q_{tCO_2} is the amount of carbon sequestered by grassland in the KDC (ton); and P_{CO_2} is the market price of carbon trading. The average price (33.6 $CNY \cdot t^{-1}$) of eight Chinese trading markets (Beijing, Fujian, Guangdong, Hubei, Shanghai, Shenzheng, Tianjin, and Chongqing) was taken from China Carbon Trading Network (www.tanjiaoyi.com (accessed on 6 January 2023)) in 2020.

4. Oxygen release value

Estimation of supply. According to the chemical equation of photosynthesis, for every 1 mole of CO_2 produced and absorbed, the plant releases 1 mole of O_2. The calculation formula is:

$$Q_{op} = M_{O_2}/M_{CO_2} \times Q_{CO_2} \tag{17}$$

where Q_{op} is the amount of oxygen released from grasslands in the GKD ($ton \cdot O_2 \cdot yr^{-1}$); $M_{O_2}/M_{CO_2} = 32/44$, which is the coefficient of conversion of CO_2 to O_2; and Q_{CO_2} is the amount of carbon sequestered by grasslands in the KDC ($ton \cdot C \cdot yr^{-1}$).

Value accounting. The market value method was used. The calculation formula is:

$$V_{op} = Q_{op} \times C_O \tag{18}$$

V_{op} is the value of oxygen release ($CNY \cdot yr^{-1}$); Q_{op} is the amount of oxygen released ($t \cdot O_2 \cdot yr^{-1}$); and C_O is the price of industrial oxygen production, taken to be 1462.35 $CNY \cdot ton^{-1}$ (data from "Grassland Specification").

5. Climate comfort value

Estimation of supply. The total energy consumed by ecosystem evapotranspiration was chosen as the valuation index for the physical extent of climate regulation services provided by grasslands in the KDC. The formula is:

$$E_{pt} = EEP \times S \times D \times 10^6 / (3600 \times r) \tag{19}$$

E_{pt} is the total energy consumed by transpiration and evapotranspiration on the grassland in the KDC ($kW \cdot h \cdot yr^{-1}$); EEP is the heat consumption per unit area of transpiration on the grassland in the KDC ($kJ \cdot m^{-2} \cdot d^{-1}$); S, D is the area of the grassland ecosystem, the number of days with a daily maximum temperature of 26 °C or more, respectively; and r is the air-conditioning energy efficiency ratio, taken to be 3.0, dimensionless.

Value accounting. The value of climate comfort of grasslands in the KDC is calculated using the alternative cost method, i.e., the cost of electricity needed to artificially regulate temperature and humidity. The formula is:

$$V_{pt} = E_{pt} \times P_e \tag{20}$$

V_{pt} is the climate comfort value of grassland in the KDC ($CNY \cdot yr^{-1}$); E_{pt} is the total energy consumed by grassland in the KDC to regulate climate ($kW \cdot h \cdot yr^{-1}$); P_e is the local consumer electricity price, using the average electricity price of Yuannan, Guangxi, and Guizhou, China, in 2020, data is from China's electricity price check system, http://cep.cec.org.cn/ (accessed on 11 January 2023), taken to be 0.51 $CNY \cdot kW^{-1} \cdot h^{-1}$.

6. Air purification value

Estimation of supply. Considering the characteristics of small and scattered grasslands in the KDC, the air purification capacity of the grassland ecosystem was chosen to estimate the physical volume. The calculation method is:

$$Q_{ap} = \sum_{i=1}^{n} Q_i \times A \tag{21}$$

where Q_{ap} is the amount of air purification in grassland in the KDC ($ton \cdot yr^{-1}$); n is the number of air pollutant categories; i is an air pollutant category, $I = 1, 2, 3, \ldots n$; and Q_i is the cleaning purification volume of air pollutants of category i per unit area ($t \cdot km^{-2} \cdot yr^{-1}$). According to the "Gross Eco-products accounting specification (hereinafter referred to as "eco-products specification") [34], SO_2, NO_x, and dust purification were 2.94 $t \cdot km^{-2} \cdot yr^{-1}$, 1.57 $t \cdot km^{-2} \cdot yr^{-1}$, and 8.41 $t \cdot km^{-2} \cdot yr^{-1}$, respectively. A is the area of grassland in the KDC (km^2).

Value accounting. The value of air purification of grasslands in the KDC was accounted for using the substitution cost method (i.e., the cost of industrial treatment of air pollutants). The calculation formula is:

$$V_{ap} = Q_{ap} \times c_i \tag{22}$$

V_{ap} is the value of air purification of grassland in the KDC ($CNY \cdot yr^{-1}$); Q_{ap} is the amount of air purification of grassland in the KDC ($ton \cdot yr^{-1}$); and c_i is the treatment cost unit of air pollutant category i ($CNY \cdot ton^{-1}$), according to "grassland specification". The treatment cost of SO_2, NO_x, and dust are 1.28 $CNY \cdot kg^{-1}$, 1.41 $CNY \cdot kg^{-1}$, and 0.33 $CNY \cdot kg^{-1}$, respectively.

- Ecosystem cultural services value

Since the cultural value of non-karst grasslands is not included in the calculation, it is difficult to accurately distinguish the cultural value of grassland in the KDC independently. Here, the equivalent factor method is used for the calculation. The formula is:

$$V_{cp} = A_i \times D \times E_{vpa} \quad (23)$$

$$D = S_r \times F_r + S_w \times F_w + S_c \times F_c \quad (24)$$

where V_{cp} is the cultural service value of grassland in the KDC (CNY); A_i is the grassland area of the i-th county in the KDC (hm^2); D is the cultural service value equivalent per unit area of grassland in the KDC ($CNY \cdot hm^{-2}$); S_r, S_w, and S_c are rice, wheat, and corn as the percentage of the sown area of the three crops in 2020, respectively (%), taken from China's annual average data from 2016 to 2020 and data from "China Statistical Yearbook"; F_r, F_w, and F_c represent the average profit per unit area in 2020 for rice, wheat, and corn, respectively ($CNY \cdot hm^{-2}$), taken from China's annual average data from 2016 to 2020 and data from "National Agricultural Product Cost-Benefit Compilation-2021"; and E_{vpa} is the landscape's aesthetic value equivalent per unit area of grassland, based on the research results of Xie et al. [48], taken to be 0.96.

2.3.2. Evaluation of Counties Economic Intensity

- Structure of indicator system

In accordance with the purpose of the indicator system, which is used to reflect the degree of regional economic development, and in conjunction with the possibility of data acquisition, the following 15 indicators (Table 2) were used as indicators of economic intensity at different levels of the county.

Table 2. The indicators system and weights of counties economic intensity.

Number	Indicators	Unit	Weight
X1	Economic density	10,000 CNY/km^2	0.1057
X2	Per capita GDP	CNY/capita	0.0317
X3	The proportion of the output value of secondary and tertiary industries in GDP	%	0.0200
X4	Local general public budget revenue per capita	CNY/capita	0.0743
X5	General public budget expenditure per capita	CNY/capita	0.0228
X6	Per capita household savings deposit balance	CNY/capita	0.0465
X7	Per capita year-end loan balances of financial institutions	CNY/capita	0.0698
X8	Number of industrial enterprises above designated size per 10,000 people	Num/10,000 capita	0.0952
X9	Number of fixed-line telephone users per 100 people	Households/100 capita	0.0601
X10	Urbanization rate	%	0.0407
X11	Per capita total retail sales of consumer goods	CNY/capita	0.0619
X12	Per capita disposable income of urban households	CNY/capita	0.0253
X13	Per capita disposable income of rural households	CNY/capita	0.0658
X14	Nighttime light index	--	0.0394
X15	Highway density	Km/km^2	0.2406

- Standardization of data.

Since the units of the indicators are not uniform and there is a lack of comparability among the indicators, the entropy value method was used to standardize the 15 indicators. Since the selected indicators are all positive indicators, the following formula is used for calculation and processing.

$$X'_{ij} = \frac{X_{ij} - \min(X_{ij})}{\max(X_{ij}) - \min(X_{ij})} \quad (25)$$

where X'_{ij} is the standardized value of the j-th indicator for the i-th county; X_{ij} is the actual value of the j-th indicator for the i-th county; $\max(X_{ij})$ is the maximum value of the j-th indicator in the county; and $\min(X_{ij})$ is the minimum value of the j-th indicator in the county.

- Determination of weights

The entropy value method was used for calculation, and the calculation process was as follows:

First, calculate the weight of the j-th indicator for the i-th county.

$$Y_{ij} = \frac{X_{ij}}{\sum\limits_{i=1}^{n} X_{ij}} \tag{26}$$

Second, account for the entropy value of the *j*-th indicator.

$$e_i = -k\sum_{i=1}^{m} Y_{ij} \ln Y_{ij} \tag{27}$$

Third, determine the weights of each indicator.

$$W_j = \frac{1-e_i}{\sum\limits_{i=1}^{n} 1-e_i} \tag{28}$$

Finally, perform a composite index calculation as follows:

$$I_i = \sum_{j=1}^{n} W_j \times X_{ij} \tag{29}$$

In Equations (26)–(29) above, Y_{ij} denotes the weight of the *j*-th indicator of the *i*-th county, *n* indicates the number of counties, e_i is the entropy value, $k = \frac{1}{\ln m}$, m is the number of indicators, W_j is the indicator weight, and I_i denotes the economic intensity index of the *i*-th county.

The specific index weights are shown in Table 2. Among the 15 indicators, the economic density and highway density have a higher weight because the higher the density of the county economy, the more frequent the economic activity, and the more vigorous the demand and consumption of ES. At the same time, there are also funds, technology, and labor inputs to protect the ecological environment; in particular, it is beneficial to promote the transformation of ES into ecological funds and realize the value of the grassland ES in the KDC. Highway density is an important index that reflects transportation accessibility. The higher the highway density, the faster and more convenient the material flow, information flow and capital flow in the county, and the more beneficial it is to improve the value realization rate of the grassland ES in the KDC.

2.3.3. Characteristic Quantification of Grassland GEP in the KDC Area

- Structural characteristic of grassland GEP in the KDC area

Public eco-products index. It is used to reflect the size of the proportion of public eco-products. The larger the share, the more prominent the regional ecological function, the lower the degree of marketability, and the more necessary it is for both the government and market instruments to increase the value realization rate of ES. The formula is:

$$R_{EVR} = \frac{ERV}{GEP} \tag{30}$$

where R_{EVR} and ERV refer to the public eco-products index and ecosystem regulation services, respectively.

Industrial scale concentration index. It is used to reflect the degree of industry aggregation. The formula is:

$$CR_5 = \frac{\sum_{i=1}^{5} GEP_i}{GEP} \tag{31}$$

where CR_5 is the concentration index of the industrial scale (%), which is the ratio between the value of the five largest categories of ES and the GEP; $\sum_{i=1}^{5} GEP_i$ is the sum of the values of the five largest ES (10,000 CNY); and GEP is total value of grassland ES in the GKD (10,000 CNY).

- Geospatial characteristics of grassland GEP in the KDC

We use geographic concentration as a metric. The Herfindahl index is used to reflect the spatial concentration of the GEP. The more concentrated the spatial distribution of the GEP, the more limited the relative beneficiaries. However, for two regions with comparable total GEP, the smaller the Herfindahl index, the more beneficial it is to improve the equalization and effectiveness of ecological well-being.

$$G_h = \sum_{k=1}^{n} Sk^2 - \frac{1}{K} \quad (32)$$

where G_h is the Herfindahl index; S_K is the share of GEP in the total GEP for the k-th region; and K is number of regions.

2.3.4. Spatial Autocorrelation Analysis

The Moran index is used to analyze the autocorrelation of the counties in the study area, and the formula is as follows:

$$Moran'sI = \frac{\sum_j^n \sum_{j=i}^n (x_i - \overline{x})(x_j - \overline{x})}{S^2 \sum_j^n \sum_{j=i}^n w_{ij}} \quad (33)$$

where w_{ij} is the proximity relationship between county i and county j; n is the number of counties; x_i is the value of the ith county; x_j is the value of the jth county; and $Moran'sI$ takes values between −1 and 1, and negative, positive, and 0 represent negative correlation, positive correlation, and no correlation, respectively.

2.3.5. Indicator Analysis Method

The comparative mean method was used in IBM SPSS Statistics 19.0 (version 19.0; SPSS, Chicago, IL, USA) to analyze the distribution dynamics of the industrial characteristics of the grassland ES in the KDC, with indicators including mean, standard deviation, median, skewness, and kurtosis.

3. Results

3.1. Spatial Distribution of Grassland GEP and County Economic Intensity in the KDC

In 2020, the GEP of the grassland in the KDC of the study area reached CNY 499.81 billion. Among the 86 counties, the GEP lies between CNY 0.07 and 25.52 billion (Figure 4a), with a highly positive skewed, low kurtosis distribution, and most counties (80.23%) were below CNY 10 billion (Table 3). In terms of spatial distribution, the GEP was mainly distributed in the central and western regions of the study area, with the trend of West > Central > East. This distribution trend is closely related to the distribution of grasslands in the KDC. Its western fringe areas have a higher GEP value, such as Pingbian Miao Autonomous County (CNY 25.52 billion), Shuicheng District (CNY 25.42 billion), and Maguan County (CNY 23.65 billion). The eastern fringe areas have a lower GEP value, such as Jinping County (CNY 0.022 billion), Ziyuan County (CNY 0.013 billion), and Jianhe County (CNY 0.007 billion), etc. Moreover, the GEP increases with outward distance, centering on the location of the city government.

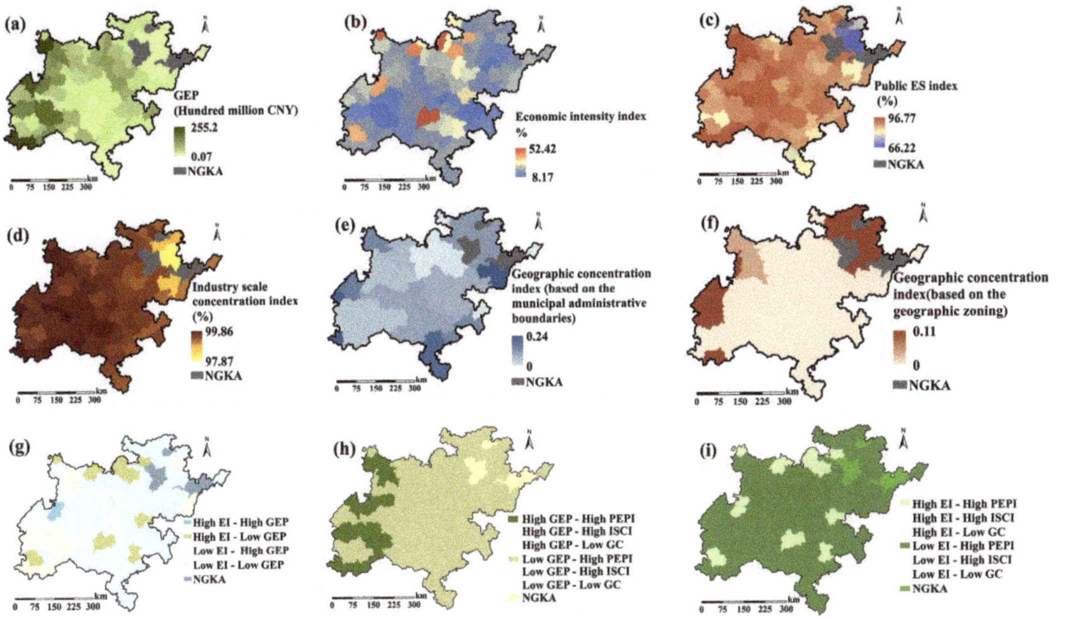

Figure 4. Spatial distribution of grassland GEP in the KDC and its characteristics (NGKA: no grassland in the karst area, EI: economic intensity, PEPI: public eco-product index, ISCI: industrial scale concentration index, and GC: geographical concentration; (**a**), Distribution of GEP; (**b**), Distribution of EI; (**c**), Distribution of public ES index; (**d**), Distribution of ISCI; (**e**), Distribution of GC (based on the municipal administrative boundaries); (**f**), Distribution of GC index (based on the geographic zoning); (**g**), Spatial coupling of GEP and EI; (**h**), Spatial coupling between GEP and PEPI, ISCI, GC; (**i**), Spatial coupling between EI and PEPI, ISCI, GC).

Table 3. Indicators analysis of grassland GEP in the KDC.

Indicators	Mean	Variation Range	Median	Skewness	Kurtosis
GEP (0.1 billion CNY)	58.12	255.13	39.84	1.58	2.37
County economic intensity (%)	20.74	44.24	18.49	1.58	2.64
Public eco-product index (%)	89.83	30.55	90.86	−2.36	7.32
Industrial scale concentration index (%)	99.48	1.99	99.56	−2.73	10.03
Geographical concentration (based on the municipal administrative boundaries)	0.069	0.24	0.06	1.41	1.56
Geographical concentration (based on the geographic zoning)	0.037	0.11	0.03	2.68	0.81

In 2020, the counties' economic intensity in the study area ranged from 8% to 53% (Figure 4b). The distribution is highly positively skewed and has low kurtosis. Among them, the majority of counties have an economic intensity below 30% (Table 3). Counties with strong economies are distributed in a dotted pattern, mostly counties where municipal administrative districts are located, such as Zhongshan District (52.41%), Youjiang District (51.45%), Xixiu District (43.65%), etc., or in areas closer to provincial capitals, such as Longli County (52.42%). Economically underdeveloped regions are mainly located in the southwest, southeast, and northeast of the study area, such as Guangnan County (8.18%), Du'an County (8.2%), Congjiang County (9.33%), etc. In general, it basically forms a distribution status quo with the municipal administrative center as the radiance, and the farther from the administrative center, the less economically developed.

The median line was used as the basis for classification, and the economic intensity of the counties and the GEP of grassland in the KDC were divided into four combination types. Among the 86 counties, there are 64 counties with low economic intensity—low GEP and 10 counties with low economic intensity—high GEP, accounting for 74.42% and 11.63% of the total number of counties, respectively. These two types of counties account for 86.05% of the total number of counties. There are 11 counties with high economic intensity—low GEP and 1 county with high economic intensity—high GEP, representing 12.79% and 1.16% of the total number of counties, respectively (Figure 5). Spatially, counties with low economic and low intensity GEP are widely distributed, and other types of counties are distributed at the edge of the study area (Figure 4g).

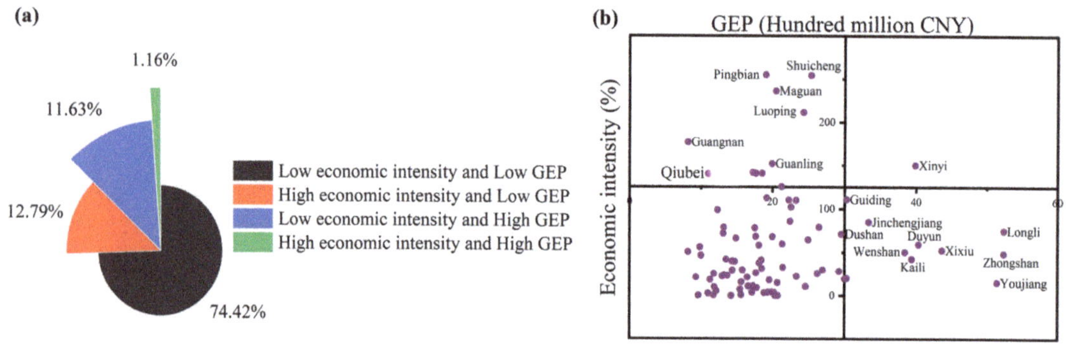

Figure 5. Spatial coupling relationship between the GEP of grassland and county economic intensity in the KDC (**a**), Structural distribution of different types of GEP and county economic intensity combinations; (**b**), Spatial distribution of different combinations of GEP and county economic intensity).

Based on the perspective of spatial autocorrelation, the Moran indices of GEP and economic intensity of stone desertification management grasslands were 0.46 and 0.129, respectively, indicating that there was a spatial aggregation effect. The hotspot area of GEP was the western part of the study area, and the hotspot area of economic intensity was distributed in a point pattern. The GEP and economic intensity were spatially misaligned (Figure 6). It can be seen that the grassland ES in the KDC mainly comes from areas with lower economic intensity, and this distribution pattern is basically consistent with the theoretical hypothesis of Figure 1a.

3.2. *Structural Characteristics of Grassland GEP in the KDC*

The public ES index of grassland in the KDC of the study area ranges from 66.22% to 96.77% (Figure 4c), with a negatively skewed and low kurtosis distribution (Table 3), and most counties (88.37%) above 85% (Figure 7a). It supports the scientific conjecture of Figure 1b. Overall, there was little variation between counties. Spatially, counties with a high public eco-product index are relatively concentrated, while counties with a relatively low index are mainly found in the outlying areas of the study area. Counties with a relatively low index include Jinping County (66.22%), Liping County (69.51%) and Jianhe County (71.99%), etc. Counties with a high index include Pingbian Miao Autonomous County (96.77%), Qinglong County (95.93%), and Shuicheng District (95.88%), etc.

In terms of the index of industry scale concentration of grassland GEP in the KDC, all counties in the study area were consistently above 97% (Figure 4d), with a negatively skewed, low kurtosis distribution (Table 3). In addition, most counties (94.19%) were above 99% (Figure 7b). The reason is that the value among GEP components is mainly concentrated in five ES: soil retention, material goods, climate regulation, carbon sequestration, and oxygen release. Among the five ES with the largest value, the largest proportion is accounted for by ecological regulation services (63.73–96.79%) and a smaller proportion is accounted for by provisioning services (3.21–34.27%).

In terms of the distribution of the industrial structure of grassland ES with GEP and county economic intensity in the KDC, the study area basically formed a distribution trend of high public eco-product index, low GEP, and low economic intensity (Figure 4h,i). Counties with a high public ES index and high GEP were mainly distributed in the western part of the study area, such as Pingbian Miao autonomous county, Qubei county, and Shuicheng district. The counties with a high publicness ecological product index and high economic intensity are distributed in a dotted pattern, which is basically consistent with the spatial distribution of high economic intensity.

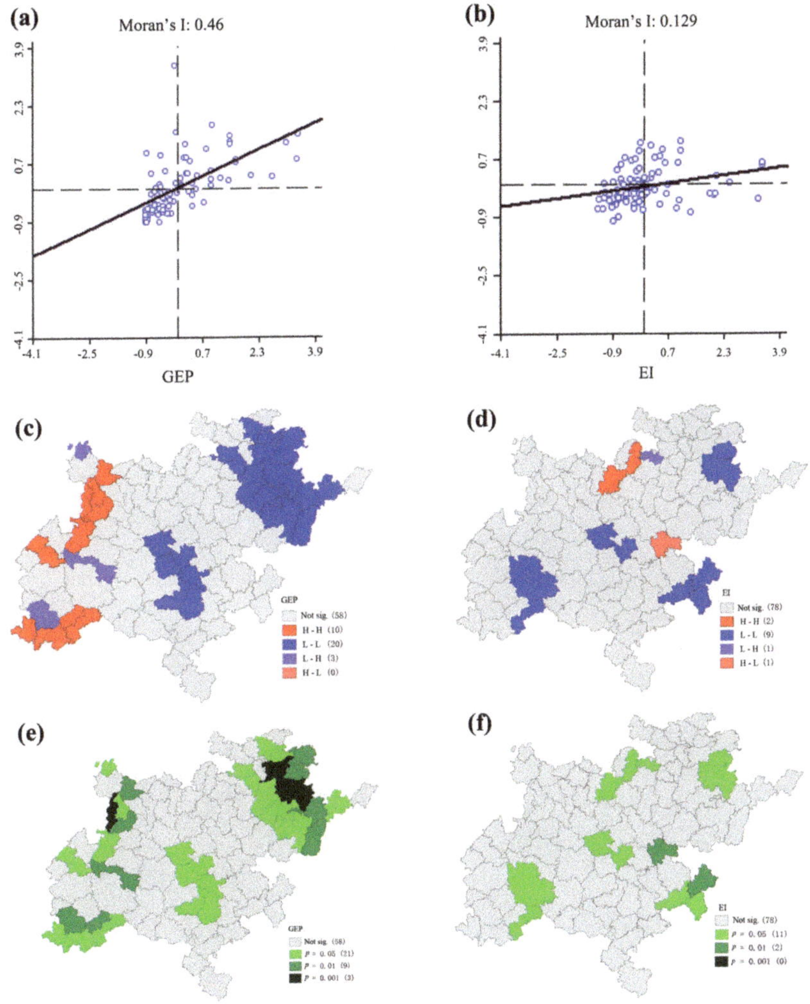

Figure 6. Moran index of GEP and economic intensity in the study area. (**a**) Moran index of GEP in the KDC; (**b**) Moran index of economic intensity; (**c**) Spatial autocorrelation distribution of GEP; (**d**) Spatial autocorrelation distribution of economic intensity; (**e**) Spatial autocorrelation significance of GEP; and (**f**) Spatial autocorrelation significance of economic intensity. EI: economic intensity; Not sig.: Not significant; H-H: High-High; L-L: Low-Low; L-H: Low-High; and H-L: High-Low.

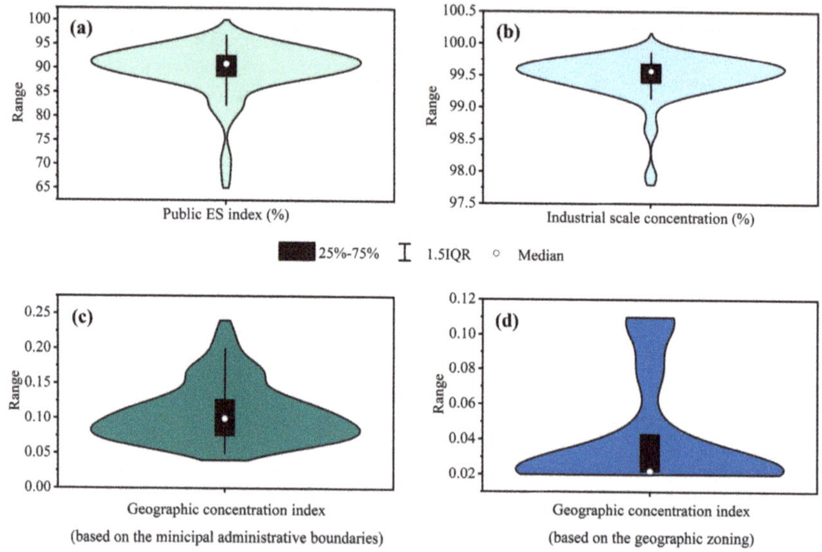

Figure 7. Structural and spatial characteristics of grassland ES in the KDC.

3.3. *Spatial Characteristics of Grassland GEP in the KDC*

The geographic concentration of grassland GEP in the KDC of the study area is generally low, with a positively skewed, low kurtosis distribution (Table 3). Based on the municipal administrative boundaries, the geographical concentration ranges from 0.015 to 0.237. Among them, Liuzhou city (0.237), Honghe Hani and Yi Autonomous Prefecture (0.178), and Chongzuo City (0.11) are relatively higher, and Wenshan Zhuang and Miao Autonomous Prefecture (0.042), Qianxinan Buyi and Miao Autonomous Prefecture (0.037), and Qiannan Buyi and Miao Autonomous Prefecture (0.015) are lower. The overall trend is decreasing from the periphery to the center (Figure 4e). Based on geomorphological zoning, the geographic concentration of grassland ES in the KDC of the study area ranges from 0.022 to 0.108, with karst trough valley (0.108) > karst faulted basin (0.079) > karst gorge (0.043) > karst plateau (0.024) > karst peak cluster depression (0.022) (Figure 4f). The small geographical concentration of grassland ES in the KDC verifies its significant spatial variability and unevenness (Figure 7c,d). This leads to a low scale of grassland eco-industry, which conforms to the theoretical hypothesis of Figure 1c.

In terms of the relationship between the industrial geographic of grassland ES in the KDC and the GEP and the counties' economic intensity, the distribution pattern is consistent with the results of Section 3.2 (Figure 4h,i). Therefore, in a comprehensive view, the grassland ES in the KDC have typical industrial characteristics of low economic intensity, low GEP, high public ES index, and low geographic concentration, which increased the complexity, difficult, and long-term nature of transforming the economic benefits.

4. Discussion

Scientifically measuring the value of nature to people [49] and incorporating ES and natural capital into mainstream economic policy is an important step toward improving ES and human well-being [50–52]. However, this presupposes an understanding of the three elements of an ES. The first is physical quantity and value quantity from ES to nature's contribution (including positive and negative) to people [53]. Second, according to the classification of ES [54], the structural proportions of ES value should be considered, and the feasibility of private market-based transactions should be clarified. Finally, the spatial patterns of an ES should be fully considered. In particular, the flow and enhancement pathways from an ES to ecological capital and human well-being should be explored in the context of local socioeconomic systems [55,56]. Accordingly, focusing on karst desertification areas, we should use the typical characteristics of the grassland GEP in the KDC to inform economic decisions for achieving inclusive and sustainable development by managing ES (Figure 8).

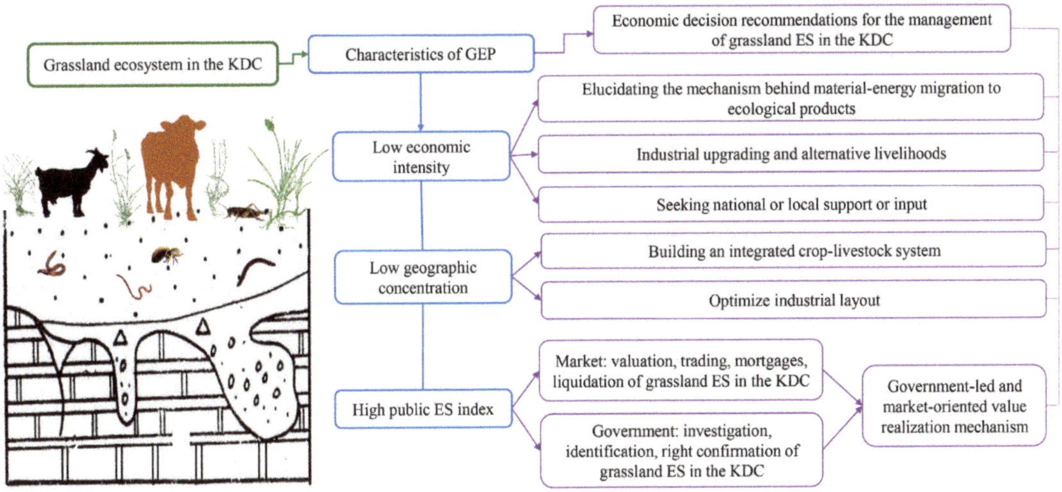

Figure 8. Recommendations for the management of the grassland ES in the KDC.

4.1. Spatial Coupling Relationship between Grassland GEP in the KDC and County Economic Intensity

Grassland GEP in the KDC mainly come from rural and remote areas with relatively less developed economies and traditional industries. At the same time, the region has a background of high population pressure [57]. Farmers and herders are prone to overgrazing based on their livelihood needs. Moreover, grasslands tend to develop into arable lands [58]. This leads to the destruction and degradation of grasslands, which in turn affects the provision of ESs, resulting in the loss of natural capital. This feature is consistent with the arid and semi-arid temperate steppes and alpine grasslands of Xinjiang, Inner Mongolia, China, Central Asia region [59–62]. However, it is different from the economically developed ranches of Western Europe, the Central Great Plains of the United States, Australia, and the Pampas of Latin America. This corroborates the findings of Ganguli and O'Rourke [63].

Due to this, it is essential to improve the quality and efficiency of the grassland ES in the KDC and to seek alternative livelihoods for farmers. Since most of the grassland patches in the KDC are located in economically underdeveloped remote rural areas, coupled with large populations and high livelihood pressures [64], the survival and development of farmers is an inevitable reality. After decades of KDC and socioeconomic development, the current main contradiction in the region has changed to the gap between the supply of quality eco-products and human needs for a beautiful ecological environment. Therefore, we should elucidate the migration mechanism of rock, soil, water, and gas elements to animals and plants in karst grasslands with different terrains, lithologies, climates, grades of karst desertification, and social ecosystems in order to provide theoretical support for improving the supply capacity of high-quality ES. Furthermore, we should develop derivative industries, extend the ecological industrial chain through industrial integration and development, and increase the value added of ecological products. Accordingly, we should also promote labor migration and flexible employment to increase the income of farmers and herders. Equally important, it is critical to seek national or local policy and financial support in agriculture, science and technology, the environment, poverty alleviation, etc., to enhance the endogenous driving force for development.

4.2. Industrial Spatial Characteristics of Grassland GEP in the KDC

Grassland in the KDC is an ecosystem controlled by special geological conditions (e.g., carbonate rocks) [65]. The complex and unique hydrogeological structure results in an extremely fragile ecological environment [66]. Due to the control of karst degradation ecosystem structure, function, and process [67], grasslands in the KDC show fragmented, small, and scattered spatial patterns with significant non-equilibrium and heterogeneous characteristics in space [68,69]. This is a distinctive feature that distinguishes the grasslands in the KDC from other grassland ecosystems, such as tropical and subtropical savannas, temperate steppes, and arctic-alpine grasslands with concentrated and cohesive features in Eurasia (e.g., Himalayas of South Asia and China, Afghanistan—Xinjiang—Inner Mongolia, China, five Central Asian countries, and Russia), South America, and East Africa.

Influenced by the low spatial concentration of grassland GEP in the KDC, the costs of industrial development are high, and the benefits of economies of scale are low. However, the region has a dominant agroecosystem and is rich in biological resources [70,71]. Still, from November to February, much of the arable land is available for grazing or forage production. Moreover, refined agricultural management (e.g., weed mowing) can provide sufficient forage resources for livestock development within the context of population benefits, and the potential of converting grain to forage can compensate for the lack of grassland resources in karst areas. Thus, by optimizing the structure of grass resources, cereals, warp, and forage crops, we can increase feed sources and ensure feed supply. Moreover, a reasonable industrial layout can be designed by developing and using local resources (e.g., flora, fauna, climate, and land), with the market playing a crucial role in resource allocation.

4.3. Structural Characteristics of the Grassland GEP in the KDC

Publicity and externality are the distinctive features of grassland GEP in the KDC. The value of regulating services of grassland ES in the KDC accounts for more than 60% of the GEP, which is largely consistent with the structural characteristics of ES values such as forests and wetlands, etc., [72,73]. Due to the typical public and external characteristics of GEP, most grassland GEPs in the KDC have the dilemma of being difficult to define, quantify, trade, and transfer [74,75]. On the one hand, the non-competitive and non-exclusive characteristics of grasslands in the KDC have led to market failures. Due to the "tragedy of the commons", the stakeholders of grassland in the KDC area cannot achieve the appropriate returns, and their enthusiasm is inhibited. On the other hand, since any active subject can possess more resources without paying or at a lower cost, economic behavior evolves in the direction of over-cultivation of grasslands, overgrazing, and random emissions, which leads to the destruction of resources and the environment, thus threatening the livelihood of farmers and herders. It can be seen that the publicness and externalities of grassland GEPs in the KDC are important shortcomings that restrict the improvement of ES and human well-being, as well as the public issues facing non-karst grasslands globally. This issue is not only unique to grasslands in the KDC, but also to non-karst grasslands and other terrestrial ecosystems.

Despite the dilemma that the ES value of grasslands in the KDC is difficult to transform through the market, it is also important to recognize the potential of transforming public ES into economic benefits. Public ES are a difficult and painful point of value realization and have the potential to be explored to improve the value realization rate of ES. This needs to be supported by the application of two extremely important instruments: government and market [76,77]. On the one hand, it may be most effective to integrate the management of ES into the existing government policy framework [78]. At the same time, a public ES requires the government to be the largest provider and consumer [79]. It is equally important for the government to fully play its management role and strengthen its services in identifying and managing resources of a public nature to create the conditions for ecosystem goods and services to enter the market. On the other hand, the market is an important lever in regulating the supply and demand of ES [80]. The crucial role of the market in resource allocation should be fully realized to achieve the goal that ES providers receive benefits and consumers pay. It is highlighted that governments and markets are not two instruments used independently but are mutually supportive and complementary [81]. The absence of either instrument can lead to a significant reduction in the realization rate of the ES value. For this reason, it is necessary to create a government-led and market-oriented mechanism to improve the value realization rate of grassland ES in the KDC.

5. Conclusions

In this paper, the contiguous area of karst desertification in Yunnan-Guangxi-Guizhou, China, were taken as the study area, and the GEP of grassland in the KDC and the economic intensity of counties were calculated. At the same time, we analyzed the structural and spatial characteristics of grassland GEP in the KDC using geospatial analysis methods, and based on this, the corresponding economic decision proposals are put forward. We found that the GEP of grasslands in the KDC is unevenly distributed, mainly in the areas with low economic intensity of counties. The socio-ecological system of grasslands in KDC is similar to that of temperate grasslands in northern China and Central Asia and alpine meadows in the Qinghai-Tibet Plateau, but different from that of economically developed pastures in European countries, the United States, Australia, and Argentina. Meanwhile, due to the small and scattered landscape pattern, the geographical concentration of grassland in the KDC is low (0.015–0.237). It is difficult to expand the industrial scale and form industrial advantages. This characteristic differs from the globally concentrated and continuous grasslands in non-karst areas. In addition, the high index of public ES (66.22–96.77%) of grasslands in the KDC is basically consistent with the distribution trend of service values of major ecosystems globally and faces the dilemma of difficult transformation of market economic benefits. According to the structural and spatial characteristics of grassland GEP in the KDC, we should elucidate the mechanism behind material–energy migration to ecological products, develop an integrated crop–livestock system, seek alternative livelihoods, and use both government and market instruments to promote the monetary value transformation of public ES.

Focusing on the contiguous region of karst desertification in Yunnan-Guangxi-Guizhou, China, we point out the current shortcomings of improving grassland ES and enhancing human well-being in the KDC by analyzing the characteristics of GEP. Accordingly, we propose a two-pronged strategy recommendation. In contrast to previous physical and spatial analyses that focused on ecological attributes, we considered not only the economic value of nature to people, but also economic intensity and spatial patterns. This makes our recommendations more targeted and operational. The research results can provide theoretical support to inform economic decisions for ecosystem protection and restoration, value realization of ecosystem products, urban–rural integration development, and rural revitalization in the KDC areas with similar man–earth areal systems (e.g., Southeast Qinghai-Tibet Plateau—Southwest China—South Central Peninsula region) or even other ecologically fragile areas around the world.

Author Contributions: Conceptualization, Data curation, Formal analysis, Investigation, Methodology, and Drawing, Y.L.; Writing—review and editing, Funding acquisition, and Supervision, K.X.; Methodology and Drawing, W.Z.; Writing—review and editing, Formal analysis, S.S.; Methodology, L.L. All authors have read and agreed to the published version of the manuscript.

Funding: This research was funded by the Key Science and Technology Program of Guizhou Province: Poverty Alleviation Model and Technology Demonstration for Eco-industries Derived from the karst desertification control (Grant No. 5411 2017 QKHPTRC); the China Overseas Expertise Introduction Program for Discipline Innovation: Overseas Expertise Introduction Center for South China Karst Eco-environment Discipline Innovation (Grant No. D17016); and the Philosophy and Social Science Planning Key Project of Guizhou Province: Research on Ecological Civilization and Guizhou New Road in Ecological Civilization Construction (Grant No. 21GZZB43).

Data Availability Statement: Not applicable.

Acknowledgments: The authors thank the reviewers and editor for their insightful comments and constructive suggestions.

Conflicts of Interest: The authors declare no conflict of interest.

Abbreviations

ES	Ecosystem services
KDC	Karst desertification control
GEP	gross ecosystem product
NGKA	no grassland in the karst area
EI	economic intensity
PEPI	public eco-product index
ISCI	industrial scale concentration index
GC	geographical concentration

References

1. White, W.B.; Culver, D.C.; Herman, J.S.; Kane, T.C.; Mylroie, J.E. Karst Lands. The dissolution of carbonate rock produces unique landscapes and poses significant hydrological and environmental concerns. *Am. Sci.* **1995**, *83*, 450–459.
2. Jiang, Z.C.; Lian, Y.Q.; Qin, X.Q. Rocky desertification in Southwest China: Impacts, causes, and restoration. *Earth Sci. Rev.* **2014**, *132*, 1–12. [CrossRef]
3. Sweeting, M.M. Refections on the development of Karst geomorphology in Europe and a comparison with its development in China. *Z. Geomorphol.* **1993**, *37*, 127–136.
4. LeGrand, H.E. Hydrological and ecological problems of karst regions: Hydrological actions on limestone regions cause distinctive ecological problems. *Science* **1973**, *179*, 859–864. [CrossRef] [PubMed]
5. Ford, D.; Williams, P.W. *Karst Hydrogeology and Geomorphology*; John Willy and Sons: Chichester, UK, 2007; pp. 471–472.
6. Wang, S.J.; Liu, Q.M.; Zhang, D.F. Karst rocky desertification in southwestern China: Geomorphology, land use, impact and rehabilitation. *Land Degrad. Dev.* **2004**, *15*, 115–121. [CrossRef]
7. Xiong, K.N.; Li, P.; Zhou, Z.F.; An, Y.; Lyu, T.; Lan, A.J. *Remote Sensing of Karst Rocky Desertification—A Typical Research of GIS: Taking Guizhou Province as an Example*; Geology Publishing House: Beijing, China, 2002; pp. 17–23.
8. Yuan, D.X. Rock desertification in the subtropical karst of South China. *Z. Geomorphol.* **1997**, *108*, 81–90.
9. Li, Y.B.; Yu, M.; Zhang, H.; Xie, Y.X. From expansion to shrinkage-exploring the evolution and transition of karst rocky desertification in karst mountainous areas of Southwest China. *Land Degrad. Dev.* **2022**. [CrossRef]
10. Zuo, T.A.; Zhang, F.T.; Zhang, J.Y.; Gao, L.; Yu, S.J. Rocky desertification poverty in Southwest China: Progress, challenges and enlightenment to rural revitalization. *J. Geogr. Sci.* **2022**, *32*, 1357–1382. [CrossRef]
11. Qi, X.K.; Wang, K.L.; Zhang, C.H. Effectiveness of ecological restoration projects in a karst region of southwest China assessed using vegetation succession mapping. *Ecol. Eng.* **2013**, *54*, 245–253. [CrossRef]
12. Zhu, X.Y.; Zheng, J.H.; An, Y.; Xin, X.P.; Xu, D.W.; Yan, R.R.; Xu, L.J.; Shen, B.B.; Hou, L.L. Grassland ecosystem progress: A review and bibliometric analysis based on research publication over the last three decades. *Agronomy* **2023**, *13*, 614. [CrossRef]
13. Kang, B.Y.; Shao, Q.Y.; Xu, H.K.; Jiang, F.Y.; Wei, X.T.; Shao, X.Q. Research on grassland ecosystem service value in China under climate change based on meta-analysis: A case study of Qinghai province. *Int. J. Clim. Chang. Str.* **2020**, *12*, 617–637. [CrossRef]
14. Qiu, S.J.; Peng, J.; Zheng, H.N.; Xu, Z.H.; Meersmans, J. How can massive ecological restoration programs interplay with social-ecological systems? A review of research in the South China karst region. *Sci. Total Environ.* **2021**, *807*, 150723. [CrossRef]
15. Song, S.Z.; Xiong, K.N.; Chi, Y.K.; Shen, X.Y.; Guo, T.; Lu, N.N. Research progress and prospect of grassland and establishment and ecological animal husbandry in karst rocky desertification area. *Fresen. Environ. Bull.* **2018**, *27*, 7017–7030.
16. Chen, C.; Park, T.; Wang, X.H.; Piao, S.L.; Xu, B.D.; Chaturvedi, R.K.; Fuchs, R.; Brovkin, V.; Ciais, P.; Fensholt, R.; et al. China and India lead in greening of the world through land-use management. *Nat. Sustain.* **2019**, *2*, 122–129. [CrossRef]
17. Qiao, Y.N.; Jiang, Y.J.; Zhang, C.Y. Contribution of karst ecological restoration engineering to vegetation greening in southwest China during recent decade. *Ecol. Indic.* **2021**, *121*, 107081. [CrossRef]
18. Tong, X.W.; Brandt, M.; Yue, Y.M.; Horion, S.; Wang, K.L.; De Keersmaecker, W.; Tian, F.; Schurgers, G.; Xiao, X.M.; Luo, Y.Q.; et al. Increased vegetation growth and carbon stock in China karst via ecological engineering. *Nat. Sustain.* **2018**, *1*, 44–50. [CrossRef]
19. Zhang, J.Y.; Dai, M.H.; Wang, L.C.; Su, W.C. Household livelihood change under the rocky desertification control project in karst areas, Southwest China. *Land Use Policy* **2016**, *56*, 8–15. [CrossRef]
20. Tian, Y.C.; Wang, S.J.; Bai, X.Y.; Luo, G.J.; Xu, Y. Trade-offs among ecosystem services in a typical Karst watershed, SW China. *Sci. Total Environ.* **2016**, *566*, 1297–1308. [CrossRef]
21. Wang, K.L.; Zhang, C.H.; Chen, H.S.; Yue, Y.M.; Zhang, W.; Zhang, M.Y.; Qi, X.K.; Fu, Z.Y. Karst landscapes of China: Patterns, ecosystem processes and services. *Landsc. Ecol.* **2019**, *34*, 2743–2763. [CrossRef]
22. Song, S.Z.; Xiong, K.N.; Chi, Y.K.; He, C.; Fang, J.Z.; He, S.Y. Effect of cultivated pastures on soil bacterial communities in the karst rocky desertification area. *Front. Microbiol.* **2022**, *13*, 922989. [CrossRef] [PubMed]

23. Xiao, D.; Tan, Y.J.; Liu, X.; Yang, R.; Zhang, W.; He, X.Y.; Xu, Z.H.; Wang, K.L. Responses of soil diazotrophs to legume species and density in a karst grassland, southwest China. *Agr. Ecosyst. Environ.* **2020**, *288*, 106707. [CrossRef]
24. Liu, C.C.; Liu, Y.G.; Guo, K.; Qiao, X.G.; Zhao, H.W.; Wang, S.J.; Zhang, L.; Cai, X.L. Effects of nitrogen, phosphorus and potassium addition on the productivity of a karst grassland: Plant functional group and community perspectives. *Ecol. Eng.* **2018**, *117*, 84–95. [CrossRef]
25. Liao, C.J.; Yue, Y.M.; Wang, K.L.; Fensholt, R.; Tong, X.W.; Brandt, M. Ecological restoration enhances ecosystem health in the karst regions of southwest China. *Ecol. Indic.* **2018**, *90*, 416–425. [CrossRef]
26. United Nations Decade on Ecosystem Restoration (2021–2030). Available online: https://undocs.org/A/RES/73/284 (accessed on 28 April 2023).
27. Wang, K.L.; Yue, Y.M.; Chen, H.S.; Zeng, F.P. Mechanisms and realization pathways for integration of scientific poverty alleviation and ecosystem services enhancement. *Bull. Chin. Acad. Sci.* **2020**, *35*, 1264–1272.
28. Lang, Y.Q.; Song, W. Quantifying and mapping the responses of selected ecosystem services to projected land use changes. *Ecol. Indic.* **2019**, *102*, 186–198. [CrossRef]
29. Peng, J.; Tian, L.; Zhang, Z.M.; Zhao, Y.; Green, S.M.; Quine, T.A.; Liu, H.Y.; Meersmans, J. Distinguishing the impacts of land use and climate change on ecosystem services in a karst landscape in China. *Ecosyst. Serv.* **2020**, *46*, 101199. [CrossRef]
30. Zhang, M.Y.; Wang, K.L.; Liu, H.Y.; Zhang, C.H.; Yue, Y.M.; Qi, X.K. Effect of ecological engineering projects on ecosystem services in a karst region: A case study of northwest Guangxi, China. *J. Clean. Prod.* **2018**, *183*, 831–842. [CrossRef]
31. Hu, Z.Y.; Wang, S.J.; Bai, X.Y.; Luo, G.J.; Li, Q.; Wu, L.H.; Yang, Y.J.; Tian, S.Q.; Li, C.Y.; Deng, Y.H. Changes in ecosystem service values in karst areas of China. *Agr. Ecosyst. Environ.* **2020**, *301*, 107026. [CrossRef]
32. Wang, Z.J.; Liu, S.J.; Li, J.H.; Pan, C.; Wu, J.L.; Ran, J.; Su, Y. Remarkable improvement of ecosystem service values promoted by land use/land cover changes on the Yungui Plateau of China during 2001–2020. *Ecol. Indic.* **2022**, *142*, 109303. [CrossRef]
33. Zhang, L.L.; Hu, B.Q.; Zhang, Z.; Liang, G.D. Research on the spatiotemporal evolution and mechanism of ecosystem service value in the mountain-river-sea transition zone based on "production-living-ecological space"—Taking the Karst-Beibu Gulf in Southwest Guangxi, China as an example. *Ecol. Indic.* **2023**, *148*, 109889. [CrossRef]
34. Richter, F.; Jan, P.; El Benni, N.; Lüscher, A.; Buchmann, N.; Klaus, V.H. A guide to assess and value ecosystem services of grasslands. *Ecosyst. Serv.* **2021**, *52*, 101376.
35. Cheng, W.; Shen, B.B.; Xin, X.P.; Gu, Q.; Guo, T. Spatiotemporal variations of grassland ecosystem service value and its influencing factors in inner Mongolia, China. *Agronomy* **2022**, *12*, 2090. [CrossRef]
36. Rao, Y.X.; Zhou, M.; Ou, G.L.; Dai, D.Y.; Zhang, L.; Zhang, Z.; Nie, X.; Yang, C. Integrating ecosystem services value for sustainable land-use management in semi-arid region. *J. Clean. Prod.* **2018**, *186*, 662–672. [CrossRef]
37. NDRM (National Development and Reform Commission); NBS (National Bureau of Statistics). *Gross Eco-Products Accounting Specification*; People's Publishing House: Beijing, China, 2022; pp. 1–201.
38. Ouyang, Z.Y.; Song, C.S.; Zheng, H.; Polasky, S.; Xiao, Y.; Bateman, I.J.; Liu, J.G.; Ruckelshaus, M.; Shi, F.Q.; Xiao, Y.; et al. Using gross ecosystem product (GEP) to value nature in decision making. *Proc. Natl. Acad. Sci. USA* **2020**, *117*, 14593–14601. [CrossRef] [PubMed]
39. Du, Y. *Harmony with Nature—Exploration and Practice of Comprehensive Management of Karst Desertification*; Chinese Forestry Publishing House: Beijing, China, 2011; pp. 333–345.
40. Thirteenth Five-Year Plan for the Construction of the Comprehensive Management Project of Karst Desertification. Available online: https://www.ndrc.gov.cn/xxgk/zcfb/ghwb/201604/t20160422_962180.html?state=123&code=&state=123 (accessed on 27 April 2023).
41. Launch Meeting on Regional Development and Poverty Alleviation in Yunnan-Guangxi-Guizhou Karst Desertification Concentrated Contiguous Area. Available online: http://www.gov.cn/ldhd/2012-06/28/content_2172407.htm (accessed on 28 April 2023).
42. Chen, Q.W.; Lu, S.X.; Xiong, K.N.; Zhao, R. Coupling analysis on ecological environment fragility and poverty in South China Karst. *Environ. Res.* **2021**, *201*, 111650. [CrossRef]
43. Shen, X.Y.; Xiong, K.N.; Zhang, J.B.; Zhao, K.; Chi, Y.K. *Research on Mineral Nutrition of Forage Plants in Karst Areas of Southern China*; China Agricultural Press: Beijing, China, 2021; pp. 8–12.
44. NDRCPD (National Development and Reform Commission Price Division); NDRCPCSC (National Development and Reform Commission Price Cost Survey Center). *National Agricultural Cost-Benefit Information Compendium—2021*; China Statistics Press: Beijing, China, 2021; pp. 265–266.
45. Xu, G.Y.; Xiong, K.N.; Shu, T.; Shi, Y.J.; Chen, L.S.; Zheng, L.L.; Fan, H.X.; Zhao, Z.M.; Yang, Z.H. Bundling evaluating changes in ecosystem service under karst rocky desertification restoration: Projects a case study of Huajiang-Guanling, Guizhou province, Southwest China. *Environ. Earth Sci.* **2022**, *81*, 302. [CrossRef]
46. The Technical Guideline on Gross Ecosystem Product (GEP). Available online: http://www.caep.org.cn/zclm/sthjyjjhszx/zxdt_21932/202010/t20201029_805419.shtml (accessed on 27 April 2023).
47. Grassland Ecosystem Service Function Assessment Specification. Available online: https://dbba.sacinfo.org.cn/stdDetail/f3cdabdedc60927fcb42bdf3ab66ec7b996d274a3c88502230d36c273570d6f4 (accessed on 2 April 2023).
48. Xie, G.D.; Zhang, C.X.; Zhang, L.M.; Chen, W.H.; Li, S.M. Improvement of the evaluation method for ecosystem service value based on per unit area. *J. Nat. Resour.* **2015**, *30*, 1243–1254.

49. Liu, Y.X.; Fu, B.J.; Wang, S.; Rhodes, J.R.; Li, Y.; Zhao, W.W.; Li, C.J.; Zhou, S.; Wang, C.X. Global assessment of nature's contribution to people. *Sci. Bull.* **2023**, *68*, 424–435. [CrossRef]
50. Costanza, R.; de Groot, R.; Braat, L.; Kubiszewski, I.; Fioramonti, L.; Sutton, P.; Farber, S.; Grasso, M. Twenty years of ecosystem services: How far have we come and how far do we still need to go? *Ecosyst. Serv.* **2017**, *28*, 1–16. [CrossRef]
51. Daily, G.C.; Ruckelshaus, M. 25 years of valuing ecosystems in decision-making. *Nature* **2022**, *606*, 465–466. [CrossRef]
52. Guerry, A.D.; Polasky, S.; Lubchenco, J.; Chaplin-Kramer, R.; Daily, G.C.; Griffin, R.; Ruckelshaus, M.; Bateman, I.J.; Duraiappah, A.; Elmqvist, T.; et al. Natural capital and ecosystem services informing decisions: From promise to practice. *Proc. Natl. Acad. Sci. USA* **2015**, *112*, 7348–7355. [CrossRef]
53. Díaz, S.; Pascual, U.; Stenseke, M.; Martín-López, B.; Watson, R.T.; Molnár, Z.; Hill, R.; Chan, K.M.A.; Baste, I.A.; Brauman, K.A.; et al. Assessing nature's contributions to people. *Science* **2018**, *359*, 270–272. [CrossRef]
54. La Notte, A.; D'Amato, D.; Mäkinen, H.; Paracchini, M.L.; Liquete, C.; Egoh, B.; Geneletti, D.; Crossman, N.D. Ecosystem services classification: A systems ecology perspective of the cascade framework. *Ecol. Indic.* **2017**, *74*, 392–402. [CrossRef]
55. Bennett, E.M.; Cramer, W.; Begossi, A.; Cundill, G.; Díaz, S.; Egoh, B.N.; Geijzendorffer, I.R.; Krug, C.B.; Lavorel, S.; Lazos, E.; et al. Linking biodiversity, ecosystem services, and human well-being: Three challenges for designing research for sustainability. *Curr. Opin. Env. Sust.* **2015**, *14*, 76–85. [CrossRef]
56. Daily, G.C.; Matson, P.A. Ecosystem services: From theory to implementation. *Proc. Natl. Acad. Sci. USA* **2008**, *105*, 9455–9456. [CrossRef] [PubMed]
57. Xu, E.Q.; Zhang, H.Q.; Li, M.X. Mining spatial information to investigate the evolution of karst rocky desertification and its human driving forces in Changshun, China. *Sci. Total Environ.* **2013**, *458*, 419–426. [CrossRef] [PubMed]
58. Zhang, S.H.; Xiong, K.N.; Qin, Y.; Min, X.Y.; Xiao, J. Evolution and determinants of ecosystem services: Insights from South China karst. *Ecol. Indic.* **2021**, *133*, 108437. [CrossRef]
59. Bi, X.; Li, B.; Fu, Q.; Fan, Y.; Ma, L.X.; Yang, Z.H.; Nan, B.; Dai, X.H.; Zhang, X.S. Effects of grazing exclusion on the grassland ecosystems of mountain meadows and temperate typical steppe in a mountain-basin system in Central Asia's arid regions, China. *Sci. Total Environ.* **2018**, *630*, 254–263. [CrossRef]
60. Liu, C.; Siri, M.J.; Li, H.; Ren, C.; Huang, J.; Feng, C.L.; Liu, K.S. Drought is threatening plant growth and soil nutrients of grassland ecosystems: A meta-analysis. *Ecol. Evol.* **2023**, *13*, e10092. [CrossRef]
61. Dong, S.K.; Kassam, K.S.; Tourrand, J.F.; Boone, R.B. *Building Resilience of Human-Natural Systems of Pastoralism in the Developing World: Interdisciplinary Perspectives*; Springer International Publishing: Cham, Switzerland, 2016; p. 1.
62. Ren, H.Y.; Eviner, V.T.; Gui, W.Y.; Wilson, G.W.; Cobb, A.B.; Yang, G.W.; Zhang, Y.J.; Hu, S.J.; Bai, Y.F. Livestock grazing regulates ecosystem multifunctionality in semi-arid grassland. *Funct. Ecol.* **2018**, *32*, 2790–2800. [CrossRef]
63. Ganguli, A.C.; O'Rourke, M.E. How vulnerable are rangelands to grazing? *Science* **2022**, *378*, 834. [CrossRef] [PubMed]
64. Yan, X.; Cai, Y.L. Multi-scale anthropogenic driving forces of karst rocky desertification in Southwest China. *Land Degrad. Dev.* **2015**, *26*, 193–200. [CrossRef]
65. Cao, J.H.; Yuan, D.X.; Tong, L.Q.; Azim, M.; Yang, H.; Huang, F. An overview of karst ecosystem in southwest China: Current state and future management. *J. Resour. Ecol.* **2015**, *6*, 247–256.
66. Grime, J.P.; Brown, V.K.; Thompson, K.; Masters, G.J.; Hillier, S.H.; Clarke, I.P.; Askew, A.P.; Corker, D.; Kielty, J.P. The response of two contrasting limestone grasslands to simulated climate change. *Science* **2000**, *289*, 762–765. [CrossRef] [PubMed]
67. Xiao, J.; Xiong, K.N. A review of agroforestry ecosystem services and its enlightenment on the ecosystem improvement of rocky desertification control. *Sci. Total Environ.* **2022**, *852*, 158538.
68. Bátori, Z.; Valkó, O.; Vojtkó, A.; Tölgyesi, C.; Farkas, T.; Frei, K.; Hábenczyus, A.A.; Tóth, Á.; Li, G.; Rádai, Z.; et al. Environmental heterogeneity increases the conservation value of small natural features in karst landscapes. *Sci. Total Environ.* **2023**, *872*, 162120. [CrossRef] [PubMed]
69. Gao, J.B.; Zuo, L.Y.; Liu, W.L. Environmental determinants impacting the spatial heterogeneity of karst ecosystem services in Southwest China. *Land Degrad. Dev.* **2021**, *32*, 1718–1731. [CrossRef]
70. Ren, J.Z. Agro-grassland system is the proper way for the karst area in Southwest China. *Pratac. Sci.* **2008**, *9*, 26–30.
71. Cheng, X.M.; Liu, W.D.; Zhou, J.H.; Wang, Z.Z.; Zhang, S.Q.; Liao, S.X. Extraction of mountain grasslands in Yunnan, China, from sentinel-2 data during the optimal phenological period using feature optimization. *Agronomy* **2022**, *8*, 1948. [CrossRef]
72. Costanza, R.; de Groot, R.; Sutton, P.; van der Ploeg, S.; Anderson, S.J.; Kubiszewski, I.; Farber, S.; de Groot, R.; Brander, L.; van der Ploeg, S.; et al. Global estimates of the value of ecosystems and their services in monetary units. *Ecosyst. Serv.* **2012**, *1*, 50–61.
73. Turner, R.K. Changes in the global value of ecosystem services. *Global Environ. Chang.* **2014**, *26*, 152–158.
74. Gao, X.L.; Xu, W.H.; Hou, Y.; Ouyang, Z.Y. Market-based instruments for ecosystem services: Framework and case study in Lishui City, China. *Ecosyst. Health Sustain.* **2020**, *6*, 1835445. [CrossRef]
75. Liu, H.F.; Hou, L.L.; Kang, N.N.; Nan, Z.B.; Huang, J.K. A meta-regression analysis of the economic value of grassland ecosystem services in China. *Ecol. Indic.* **2022**, *138*, 108793. [CrossRef]
76. DeFries, R.; Nagendra, H. Ecosystem management as a wicked problem. *Science* **2017**, *356*, 265–270. [CrossRef]
77. Gomez-Baggethun, E.; Muradian, R. In markets we trust? Setting the boundaries of Market-Based Instruments in ecosystem services governance. *Ecol. Econ.* **2015**, *117*, 217–224. [CrossRef]
78. Joppa, L.N.; Boyd, J.W.; Duke, C.S.; Hampton, S.; Jackson, S.T.; Jacobs, K.L.; Kassam, K.A.S.; Mooney, H.A.; Ogden, L.A.; Ruckelshaus, M.; et al. Government: Plan for ecosystem services. *Science* **2016**, *351*, 1037. [CrossRef]

79. Adamowicz, W.; Calderon-Etter, L.; Entem, A.; Fenichel, E.P.; Hall, J.S.; Lloyd-Smith, P.; Ogden, F.L.; Regina, J.A.; Rad, M.R.; Stallard, R.F. Assessing ecological infrastructure investments. *Proc. Natl. Acad. Sci. USA* **2019**, *116*, 5254–5261. [CrossRef]
80. Sheng, J.C.; Qiu, W.G.; Han, X. China's PES-like horizontal eco-compensation program: Combining market-oriented mechanisms and government interventions. *Ecosyst. Serv.* **2020**, *45*, 101164. [CrossRef]
81. Hahn, T.; McDermott, C.; Ituarte-Lima, C.; Schultz, M.; Green, T.; Tuvendal, M. Purposes and degrees of commodification: Economic instruments for biodiversity and ecosystem services need not rely on markets or monetary valuation. *Ecosyst. Serv.* **2015**, *16*, 74–82. [CrossRef]

agronomy

MDPI

Article

Aboveground Biomass in China's Managed Grasslands and Their Responses to Environmental and Management Variations

Huimei Meng [1,†], Jingrui Yang [2,3,†], Wenjuan Sun [4], Liujun Xiao [5] and Guocheng Wang [6,7,*]

[1] College of Science, Tibet University, Lhasa 850000, China
[2] State Key Laboratory of Soil and Sustainable Agriculture, Institute of Soil Science, Chinese Academy of Sciences, Nanjing 210008, China
[3] University of Chinese Academy of Sciences, Beijing 100049, China
[4] State Key Laboratory of Vegetation and Environmental Change, Institute of Botany, Chinese Academy of Sciences, Beijing 100093, China
[5] Institute of Agriculture Remote Sensing and Information Technology, College of Environmental Resource Sciences, Zhejiang University, Hangzhou 310058, China
[6] LAPC, Institute of Atmospheric Physics, Chinese Academy of Sciences, Beijing 100029, China
[7] Key Laboratory of Agricultural Remote Sensing and Information System, Hangzhou 310058, China
* Correspondence: wanggc@mail.iap.ac.cn
† These two authors contributed equally to this work.

Abstract: Aboveground biomass (AGB) in managed grasslands can vary across a suite of environmental and management conditions; however, there lacks a quantitative assessment at the national scale of China. Although the potential effects of individual drivers (e.g., species, nutrient fertilization, and water management) have been examined in China's managed grasslands, no attempts have been made to comprehensively assess the effects of multiple variables on AGB. Using a meta-data analysis approach, we created a database composed of AGB and associated attributes of managed grasslands in China. The database was used to assess the responses of AGB to anthropogenic factors, in addition to a suite of natural variables including climate, soil, and topography. The average AGB in managed grasslands of China is approximately 630 g m^{-2} of dry matter, ranging from 55 to 2172 g m^{-2} (95% confidence interval). *Medicago sativa* is the most widely planted species in China's managed grasslands, followed by *Elymus dahuricus* and *Bromus japonicus*. The national average AGB of these three species was around 692, 530, and 856 g m^{-2}, respectively. For each species, AGB shows a large discrepancy across different places. In general, grassland AGB depends substantially on species, environments, and management practices. The dependence can be well described by a linear mixed-effects regression in which a series of biotic and abiotic factors are used as predictors. We highlight that establishing managed grassland can potentially contribute to not only AGB enhancement, but also grassland restoration on degraded natural grasslands.

Keywords: aboveground biomass; managed grasslands; plant species; management; climate; soil; topography

Citation: Meng, H.; Yang, J.; Sun, W.; Xiao, L.; Wang, G. Aboveground Biomass in China's Managed Grasslands and Their Responses to Environmental and Management Variations. *Agronomy* **2022**, *12*, 2913. https://doi.org/10.3390/agronomy12122913

Academic Editors: Kesi Liu and Xinqing Shao

Received: 25 October 2022
Accepted: 21 November 2022
Published: 22 November 2022

1. Introduction

Covering nearly 40% of the Earth's surface, grassland is the world's largest terrestrial ecosystem [1], providing food and ecosystem services and contributing substantively to regulating the global carbon cycle and climate change [2]. Due to either anthropogenic activities (e.g., cultivation and overgrazing) or climate change (e.g., drought), grasslands have been degrading across the world, leading to substantial decreases in the production of forages and livestock [2,3]. To combat terrestrial ecosystem degradation, the United Nations (UN) General Assembly declared 2021–2030 the "UN Decade on Ecosystem Restoration", which outlined the urgent need for global restoration of degraded lands, including grasslands.

China has the world's second largest area of grasslands, a majority of which has been degrading to some extent due to both climate change and/or anthropogenic activities since more than half a century ago [4–6]. Grassland degradation is characterized mainly by decreases in plant species diversity, plant cover, and productivity. Synchronously occurring with degradation, grassland biodiversity and productivity have also been widely declining [7,8]. China's grassland aboveground biomass (AGB) was generally 30–50% lower than that of approximately 60 to 70 years ago [9]. Decreases in grassland AGB pose massive threats to food security and ecosystem health. To address this issue, the establishment of managed grasslands by planting grass species with high quality and quantity forage production and/or supplying additional nutrients and water has widely been recommended [10,11]. However, to date, managed grasslands make up only approximately 3% of China's total grassland area, which is substantially lower than those in developed regions such as Europe and Australia. Overall, to meet the growing needs of forage and livestock production, the establishment of managed grasslands is becoming an increasing trend [11].

Over the past two decades, a number of managed grassland experiments aimed at assessing the responses of plant productivity to environmental and anthropogenic factors have been conducted across China (Appendix B). Specifically, in this study, managed grasslands are defined as the grasslands managed by introducing species with high-quality and high-quantity forage production combined with a series of recommended management practices, such as nutrient fertilization and irrigation [11]. Most existing field-scale managed grassland experiments, however, focused on the effect of a single factor, such as plant species [12,13] and nutrient fertilization [14,15], on AGB. It is well known that grassland productivity is coregulated by both natural and anthropogenic attributes, such as irrigation and climatic, edaphic, and topographic conditions [16–19]. However, the regulating effects of these drivers have seldom been assessed [9], leaving a gap in knowledge on the dynamics of grassland AGB under changing environments and/or management. Moreover, potential forage production through establishing managed grasslands remains unclear at the national scale, which prevents policy makers and local shepherds from projecting the expected benefits from establishing managed grasslands.

In this study, we conducted a data synthesis of 101 publications to collate AGB observations in China's managed grasslands. By employing a suite of environmental and management covariates, we further clarified the regulating effects of these predictors on managed grassland AGB in China.

2. Materials and Methods

2.1. Data Compilation

Using the keywords of *aboveground biomass*, *productivity*, *production*, *China*, and *managed grasslands* (or *managed pasture*), peer-reviewed publications reporting AGB measurements in managed grasslands in China were collected by searching Web of Science (WoS; since the publication year of 1990) to construct a synthesis dataset during October–December 2021. The online datasets used for the literature search in WoS included the WoS Core Collection and Chinese Science Citation Database (articles written in Chinese). We ultimately obtained 101 studies (19 in English and 82 in Chinese) by screening the return publications using the following criteria: (1) managed grasslands were involved; (2) grass species were specified; (3) amount of AGB was reported; (4) experimental locations with precise latitude and longitude coordinates were specified; (5) observation year was directly reported; (6) if fertilization was adopted, type and amount of fertilizers [e.g., nitrogen (N), phosphorus (P) and potassium (K)] were reported; and (7) water management (i.e., rainfed or irrigation) was reported or could be obtained through personal communications. We finally obtained 864 individual AGB measurements widely distributed in China's main grassland areas (Figure 1). In all the managed grassland experiments included in our study (Appendix B), zero-grazing was adopted. In general, the standing tissue of grasses was cut and harvested manually for measuring AGB during its peak amount for a year (e.g., July or August). All other agronomy regarding fertilization and water management for each experiment was summarized in a dataset, which is publicly obtainable via this link: https:

//figshare.com/articles/dataset/AGB_of_managed_grasslands_in_China/19641654 (accessed on 1 July 2022).

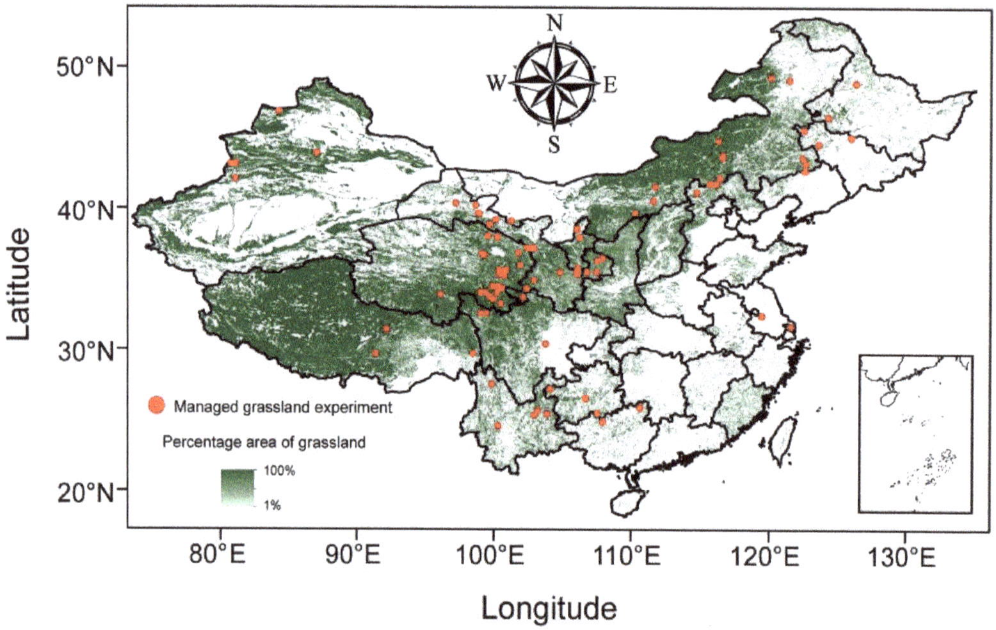

Figure 1. Locations of managed grassland experimental sites collected from literature reviews.

We obtained a set of global and/or national layers of environmental covariates, including edaphic, climatic, and topographic variables (Table A1), as potential predictors of managed grassland AGB. These variables have widely been used to assess the possible regulation effects of environment on changes in grassland aboveground biomass [20] and terrestrial carbon cycles [21]. Here, 10 soil physical and chemical properties (Table A1) were obtained from the ISRIC-WISE soil profile database [22] with a spatial resolution of 1 km^2. We also obtained 16 topographic attributes with the same resolution as the WISE database from Amatulli, et al. [23]. In addition, we determined 19 bioclimatic attributes (T1–T11 and P1–P8; Table A1) quantifying biologically meaningful variables using monthly maximum and minimum temperature and precipitation [24] at each location in the observation year of the experiment. Specifically, for the observation year, we first extracted the monthly maximum and minimum temperature and precipitation from Peng, et al. [25] using the location information of each measurement. Then, the 19 bioclimatic variables (Table A1) were calculated using the *biovars* function in the R package *dismo*. Global gridded rasters of these 19 bioclimatic attributes (representing the period of 1980–2000) with a spatial resolution of 1 km^2 were derived from WorldClim [24]. More details of these global spatial covariate layers are described in Table A1. The collated AGB measurements and their associated predictors are documented and publicly available from: https://figshare.com/articles/dataset/AGB_of_managed_grasslands_in_China/19641654 (accessed on 1 July 2022). The spatial distribution of grasslands was obtained from the National Land Cover DataSets (NLCD) of China developed from Landsat TM digital images [26].

2.2. Drivers of Managed Grassland AGB

We first produced boxplots characterizing the mean, median, and interquartile range of AGB among different groups of environmental and management attributes. Then, we used linear mixed-effects regression (LMER) to examine the relationship of a suite of predicting variables with AGB. We assumed that AGB is potentially associated with soil properties, climatic factors, topographic conditions (Table A1), plant species, and management practices. In fitting the LMER, plant species was treated as a random effect, i.e., the coefficients of other predictor variables were modified by species, and all numerical variables were standardized to unit variance; therefore, the absolute magnitude of the coefficients for the predictor variables reflected their relative importance [27]. A principal component analysis (PCA) was applied to eliminate potential correlations in the 10 edaphic variables, 19 climatic variables, and 16 topographic variables (Table A1). The most important principal components (PCs) with variances greater than 1 were retained for the regression [28]. PCA and LMER were performed using *procomp* in the R package *stats* and *lmer* in the R package *arm*, respectively, in R 4.0.3 [29].

Furthermore, we performed a machine learning-based regression (i.e., a random forest model) to assess the drivers of AGB. Before fitting the regression, the variance inflation factor (VIF) was calculated and used to minimize the multicollinearity of environmental covariates. Specifically, the variables with a VIF value larger than 10 were excluded from further regressions. Treating the remaining covariates together with species, fertilization, and irrigation regimes as independent variables, and AGB as a dependent variable, we then fitted the machine learning-based model, which inherently quantifies the variable importance of predictors using importance scores for each predictor in the regression.

3. Results

3.1. AGB in China's Managed Grasslands

In total, 16 dominant plant species were identified in these studies (Figure A1). In general, the top four species with the highest frequencies identified among the 107 studies, including *Medicago sativa, Elymus dahuricus*, and *Bromus japonicus*, were generally more widespread than the remaining species (i.e., *Poa pratensis, Leymus chinensis, Elymus sibiricus, Agropyron cristatum, Lolium perenne, Onobrychis viciifolia, Dactylis glomerata, Festuca ovina, Trifolium repens, Astragalus adsurgens*, and *Phleum pratense*; Figures A1 and A2).

The data synthesis suggested that, by averaging across the 101 sites in China (Figure 1), AGB in managed grasslands was estimated to be 630 g m^{-2} of dry matter (ranging from 55 to 2172 g m^{-2}, lower and upper limit of 95% confidence interval). Among the four most widely distributed species (Figure A1), *Bromus japonicus* had the highest average AGB (856 g m^{-2}), followed by *Medicago sativa* (692 g m^{-2}) and *Elymus dahuricus* (530 g m^{-2}, Figure 2). A large variability existed in the observed AGB (Figure 2). For example, the uncertainty (expressed as CV, i.e., standard deviation divided by mean) of observed AGB for *Medicago sativa* was 96% (data not shown).

Regardless of other factors, such as species and water management, AGB under fertilization (e.g., nitrogen, phosphorus, and potassium) was on average 50% higher than that under zero fertilization (Figure 3a). Similarly, the adoption of irrigation enhanced the average AGB by approximately 100% compared with that under rainfed conditions (Figure 3b). When pooling all data together, we found that AGB was generally higher in regions with higher mean temperatures during the plant growing seasons (i.e., April–October; Figure A3a), while the correlation between AGB and P_G (accumulated precipitation during the growing season) was much weaker (Figure A3b). By excluding the impacts of fertilization and irrigation, we found that the AGB of *Medicago sativa*, for example, was generally higher in regions with a warmer and wetter growing season (Figure 3c,d).

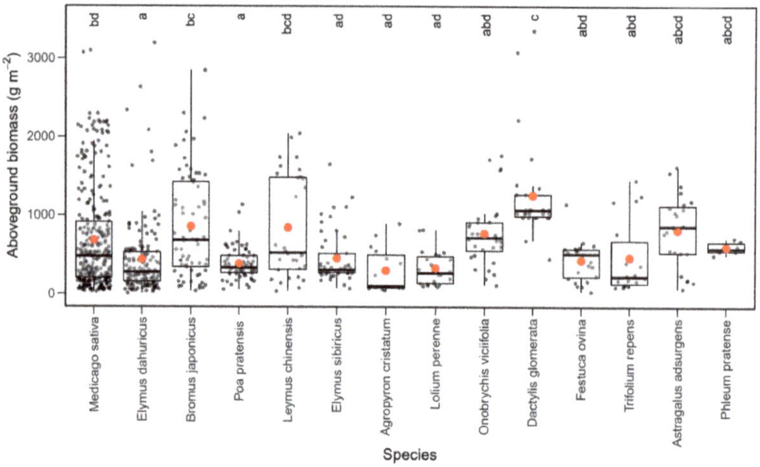

Figure 2. Observed AGB of different grass species planted in managed grasslands. Red dots show the average, and boxplots show the median and interquartile range with whiskers extending to 1.5 times the interquartile range. Different letters above the boxes indicate significant differences ($p < 0.05$) between the AGB of different species.

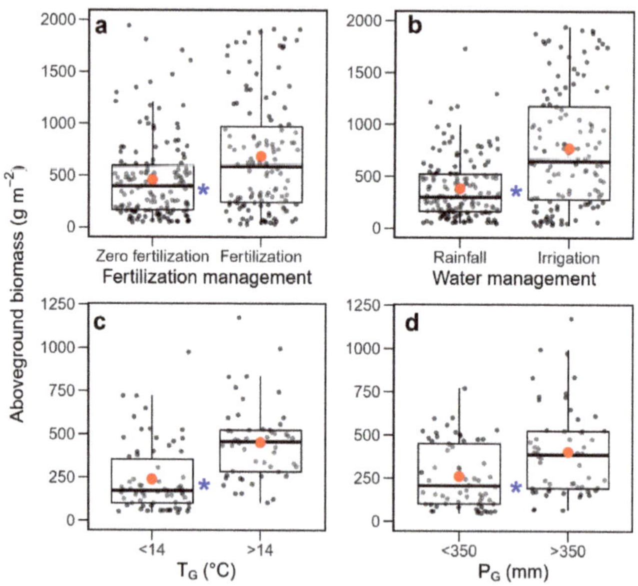

Figure 3. AGB as impacted by management and environmental factors. (**a**): fertilization; (**b**): water management; (**c**): mean temperature during plant growing seasons (April–October; TG); (**d**): accumulated precipitation during plant growing seasons (April–October; PG). Red dots show the average, and the boxplots show the median and interquartile range with whiskers extending to 1.5 times the interquartile range. Blue stars between two boxes of AGB indicate significant differences ($p < 0.05$) between the two groups of data as determined by t test. Observed AGBs of all species were analyzed in (a) and (b), while only AGB of *Medicago sativa* under zero fertilization and rainfall conditions were used in (c) and (d).

3.2. Comprehensive Assessment of Drivers on Managed Grassland AGB

We further introduced three sets of environmental factors, i.e., soil, climate, and topography. Principal component analysis (PCA) suggested that the first two soil principal components (PCs), four climate PCs, and four topography PCs could explain 71% (Figure 4a), 94% (Figure 4c), and 77% (Figure 4e) of the variances in the 10 soil attributes, 19 climate variables, and 16 topography properties (Table A1), respectively. For the first two PCs of soil, the most important contributing variables were ORGC (organic carbon) and CLPC (clay content) (Figures 4b and A4a). For the first four PCs of climate, the most important contributing attributes were P1 (annual precipitation), T10 (mean temperature of warmest quarter), T3 (isothermality), and P4 (precipitation seasonality) (Figures 4d and A4b). For the first four PCs of topography, the most important contributing variables were TRI (topographic position index), Northness, PCURV (profile curvature), and Aspectsine (aspect sine) (Figures 4f and A4c).

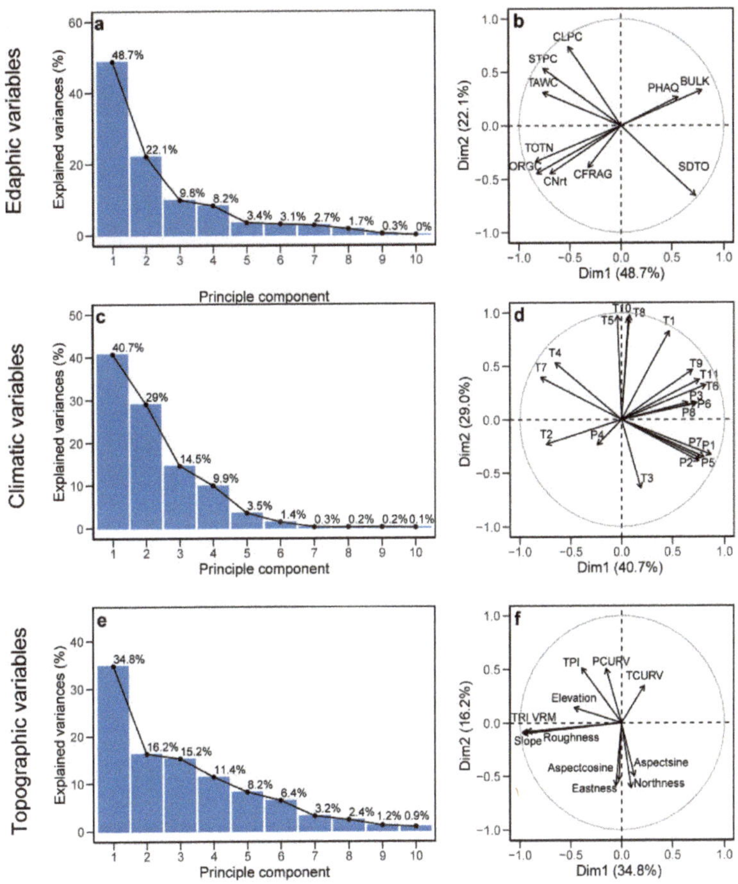

Figure 4. Principal component analysis of edaphic ((**a**) and (**b**); first row), climatic ((**c**) and (**d**); second row) and topographic ((**e**) and (**f**); third row) variables at the managed grassland sites. See Table A1 for detailed descriptions of the variables. The first column shows the loadings of each environmental variable to the top two most important principal components (PCs); the second column shows the percentage of explained variances of the first ten PCs.

Using these PCs as predictors, together with species, which were treated as random effects, and nutrient fertilization attributes as co-predictors, a linear mixed-effects regression (LMER) was fitted to the observed AGB (Figure 5). On average, the fitted LMER explained 68% of the variances in AGB (Figure 5). In general, AGB was significantly and positively correlated with nitrogen fertilization (Figure 5). AGB was also positively and significantly correlated with the first two PCs of climate variables (Figure 5). Thus, annual precipitation (P1, the most contributing variable of PC1 of climate, Figure A4b) and mean temperature of the warmest quarter (T10, the most contributing variable of PC2 of climate, Figure A4b) were generally positively correlated with AGB. The first PC of soil was generally negatively correlated with AGB (Figure 5). Variations in AGB were also regulated by topographic factors, and AGB was significantly correlated with the first three PCs of topography (Figure 5).

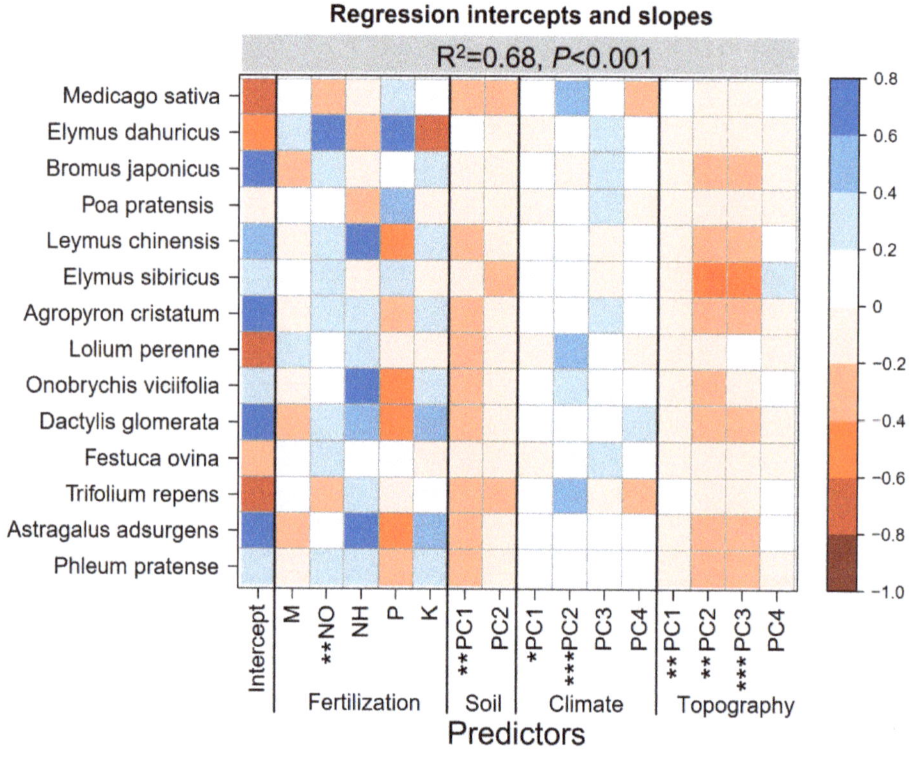

Figure 5. Coefficients of the fitted linear mixed-effects regression (LMER) in simulating AGB. Intercept is the intercept of the LMER. M, NO, NH, P, and K are the amounts of different fertilizers (manure, nitrate nitrogen, ammonium nitrogen, phosphorus, and potassium, respectively) applied during a plant growing season. PC1, PC2, PC3, and PC4 are the most important principal components (PCs) of different groups of driving factors (i.e., soil, climate, and topography). The R^2 for the LMER model is 0.68. *, ** and *** under the predicting variables indicate that the coefficients are statistically significant at the levels of $p < 0.05$, $p < 0.01$ and $p < 0.001$, respectively. Detailed principal component analyses on the predictor variables are presented in Figure A4.

The fitted machine learning-based model (i.e., random forest) indicated that 57% of the variances in AGB can be explained by the species, management of fertilization and irrigation, and the environmental attributes selected in Section 2.2 (Figure 6a). As indicated by the random forest model, three climatic variables (i.e., T8, P2, and T4) are the top three most important regulators of AGB (Figure 6b).

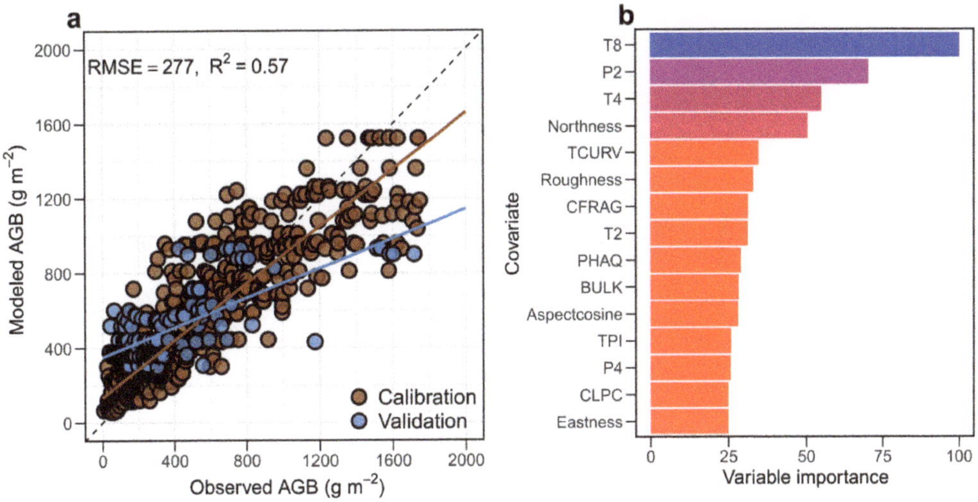

Figure 6. Performance of the fitted random forest model to predict aboveground biomass (AGB) (**a**) and the relative importance of the top 15 most important variables for predicting AGB (**b**). See Table A1 for detailed descriptions of the variables.

4. Discussion

Managed grasslands in China are mainly distributed in the northern and western regions, from Inner Mongolia to the Qinghai–Tibetan Plateau (Figure 1). These regions are generally characterized by a relatively cold and/or dry climate due to high altitudes and/or more northern latitudes [30,31]. As such, both temperature and precipitation were found to be positively correlated with grassland AGB (Figure 3c,d). As expected, fertilization significantly increased AGB compared with zero fertilization (Figure 3a), which is consistent with existing findings that addition of nutrients (e.g., nitrogen, phosphorus, and potassium) can enhance the nutritional quality of plant tissues, thereby promoting grassland AGB [32–34]. This is because in the world's most terrestrial ecosystem [35,36], including China's grasslands [37,38], plant productivity is widely acknowledged to be nutrient-limited. Apart from nutrients, water availability is another strong constraint on plant productivity in global terrestrial ecosystems, particularly in grasslands [39,40]. This can help to explain the general promoting effect of water irrigation on AGB (Figure 3b). Moreover, in a system with sufficient water supply, the impact of precipitation on AGB can be eliminated [41], which underpins our results that, when irrigation was involved, precipitation seemed to have limited influence on AGB (Figure A3b). In addition, our results demonstrated the coregulating effects of edaphic and topographic attributes on AGB (Figures 5 and 6b), which are generally comparable with the findings in the literature [42–44].

Regardless of plant species, AGB in managed grasslands averaged approximately 630 g m^{-2} of dry matter, ranging from 6 to 7 times China's national natural grassland AGB [45–48]. Although grasslands account for approximately 40% of China's land [4,49], only around 3% of the grassland area is under managed conditions (i.e., managed grasslands), which is substantially lower than those in developed regions such as Europe, Australia and New Zealand [11]. In addition, improving management practices and species varieties can help to enhance AGB. It has been reported that *Medicago sativa* is one of the world's most popular species due to its high forage productivity, high nutritional quality, and wide adaptability to different climatic and edaphic conditions [50]. Our results indicate that, on average, the AGB of *Medicago sativa* can reach approximately 700 g m^{-2} (Figure 2), i.e., ~7 Mg ha^{-1}, which is only half the production of that in developed countries, such as the USA [51]. This could be attributed to the fact that most managed grassland experiments in China (Appendix B) were conducted on existing degraded lands with relatively poorer soil nutrient condition.

Despite the positive effect on AGB, management of grasslands at large scales should be undertaken with caution due to the possible negative consequences on other ecosystem functionalities. For example, growing evidence has suggested that nutrient enrichment due to fertilization can lead to widespread decreases in grassland biodiversity [52], which is deemed another key characteristic of grassland degradation [53]. Moreover, conversion from natural lands to managed grasslands can possibly result in significant losses of ecosystem carbon stock. In natural grasslands, root biomass, rather than AGB, constitutes the majority of the total plant biomass carbon stocks [45,47]. Furthermore, the largest carbon reservoir of grassland is soil, containing more than 90% of the carbon in the whole grassland system [54]. A number of studies have indicated that conversion from natural grasslands to managed grasslands would not only reduce root biomass [55–57] but also substantially decrease the soil carbon pool [58,59], thereby causing a net warming effect on climate. In contrast, in regions that have already suffered from degradation, the establishment of managed grasslands can significantly enhance both root biomass productivity and soil carbon content [60,61]. On this basis, we suggest that priority of managed grassland establishment should be given to grasslands that have been severely degraded to benefit not only forage production but also ecosystem restoration. In addition, since managed grassland is in general established in areas with soil degradation possibly caused by water limitation, we highlight a need for introducing the plant species with high drought tolerance and/or deep rooting capacity in the future.

5. Conclusions

We comprehensively and quantitatively assessed the aboveground biomass (AGB) of dominant plant species in China's managed grasslands. We found that the establishment of managed grasslands via practices such as introducing advanced species, fertilization, and irrigation can potentially increase aboveground biomass. The magnitude of enhanced AGB through establishing managed grasslands is associated with a series of biotic and abiotic factors, such as species and climatic, edaphic, and topographic attributes. We highlight the need for a more extensive establishment of managed grasslands, particularly in areas suffering degradation of natural grasslands, thereby favoring not only livestock production, but also grassland restoration. Apart from improving management practices, introducing plant species with drought tolerance and/or deep rooting capacity may also contribute to productivity enhancement and grassland restoration.

Author Contributions: Conceptualization, G.W. and W.S.; methodology, H.M. and J.Y.; formal analysis, H.M., L.X. and G.W.; investigation, G.W. and L.X., J.Y.; data curation, J.Y.; writing—original draft preparation, G.W.; writing—review and editing, W.S.; funding acquisition, W.S. All authors have read and agreed to the published version of the manuscript.

Funding: This study was financially supported by the Strategic Priority Research Program of the Chinese Academy of Sciences (Grant No. XDA26010103) and the Major Program for Basic Research Project of Yunnan Province (Grant No. 202101BC070002).

Data Availability Statement: All data used in this paper is publicly obtainable via this link: https://figshare.com/articles/dataset/AGB_of_managed_grasslands_in_China/19641654 (accessed on 1 July 2022).

Acknowledgments: We thank Zhongkui Luo from Zhejiang University and Zhangcai Qin from Sun Yat-sen University for their help on editing the MS.

Conflicts of Interest: The authors declare no conflict of interest.

Appendix A

Table A1. Environmental covariates.

Covariates	Code	Description	Unit
Edaphic variables	CFRAG	Coarse fragments (>2 mm)	%
	BULK	Bulk density	g cm^{-3}
	ORGC	Organic carbon	g kg^{-1}
	SDTO	Sand content	%
	CLPC	Clay content	%
	STPC	Silt content	%
	TAWC	Available water capacity	cm m^{-1}
	TOTN	Total nitrogen	g kg^{-1}
	CNrt	C:N ratio	-
	PHAQ	pH measured in H_2O	-
Bioclimatic variables	T1	Annual mean temperature	°C
	T2	Mean diurnal range	°C
	T3	Isothermality (T2/T7 × 100)	%
	T4	Temperature seasonality (standard deviation × 100)	°C
	T5	Max temperature of warmest month	°C
	T6	Min temperature of coldest month	°C
	T7	Temperature annual range (T5–T6)	°C
	T8	Mean temperature of wettest quarter	°C
	T9	Mean temperature of direst quarter	°C
	T10	Mean temperature of warmest quarter	°C
	T11	Mean temperature of coldest quarter	°C
	P1	Annual precipitation	mm
	P2	Precipitation of wettest month	mm
	P3	Precipitation of driest month	mm
	P4	Precipitation seasonality (coefficient of variation)	%
	P5	Precipitation of wettest quarter	mm
	P6	Precipitation of driest quarter	mm
	P7	Precipitation of warmest quarter	mm
	P8	Precipitation of coldest quarter	mm
Topographic variables	Elevation	Elevation	m
	Roughness	Roughness	-
	TRI	Terrain Ruggedness Index	-
	TPI	Topographic Position Index	-
	VRM	Vector Ruggedness Measure	-
	Aspectcosine	Aspect Cosine	-
	Aspectsine	Aspect Sine	-
	Slope	Slope	-
	Eastness	Index from −1 to 1 of how east or west a site faces	-
	Northness	Index from −1 to 1 of how north a site faces	-
	PCURV	Profile curvature	°
	TCURV	Tangential curvature	°
	dx	First order partial derivative (E-W slope)	-
	dy	First order partial derivative (N-S slope)	-
	dxx	Second order partial derivative (E-W slope)	-
	dyy	Second order partial derivative (N-S slope)	-

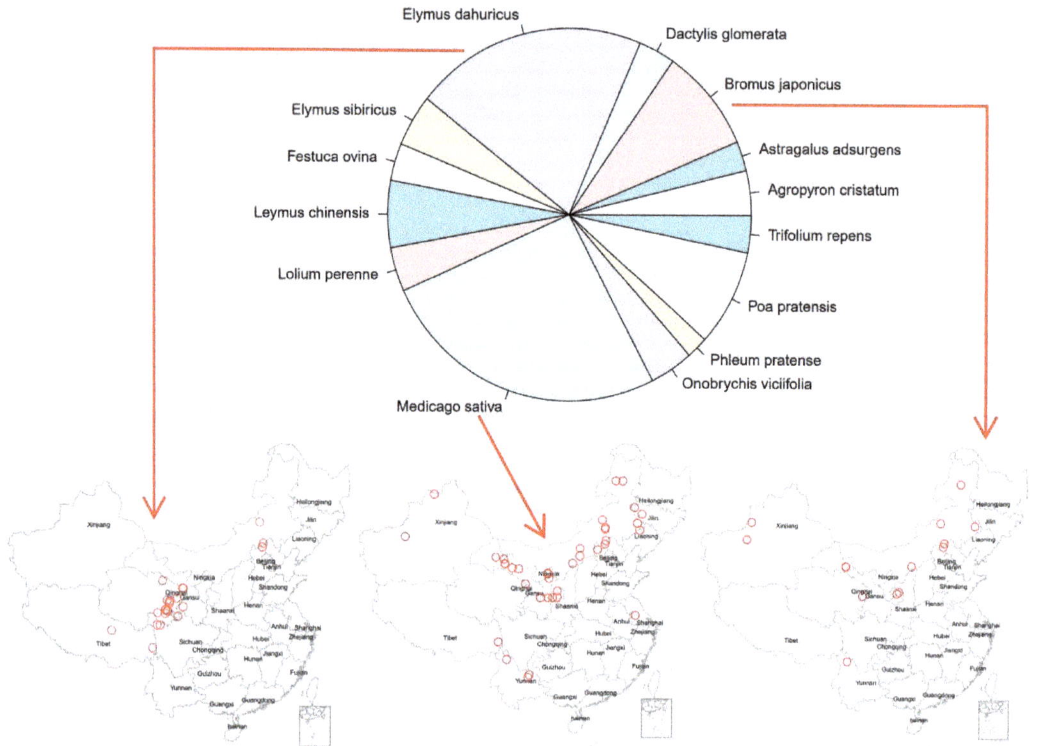

Figure A1. Proportions of plant species and spatial distribution of the top three species that were most commonly found in the 101 literature articles. Spatial distribution of the remaining 11 species is presented in Figure A2.

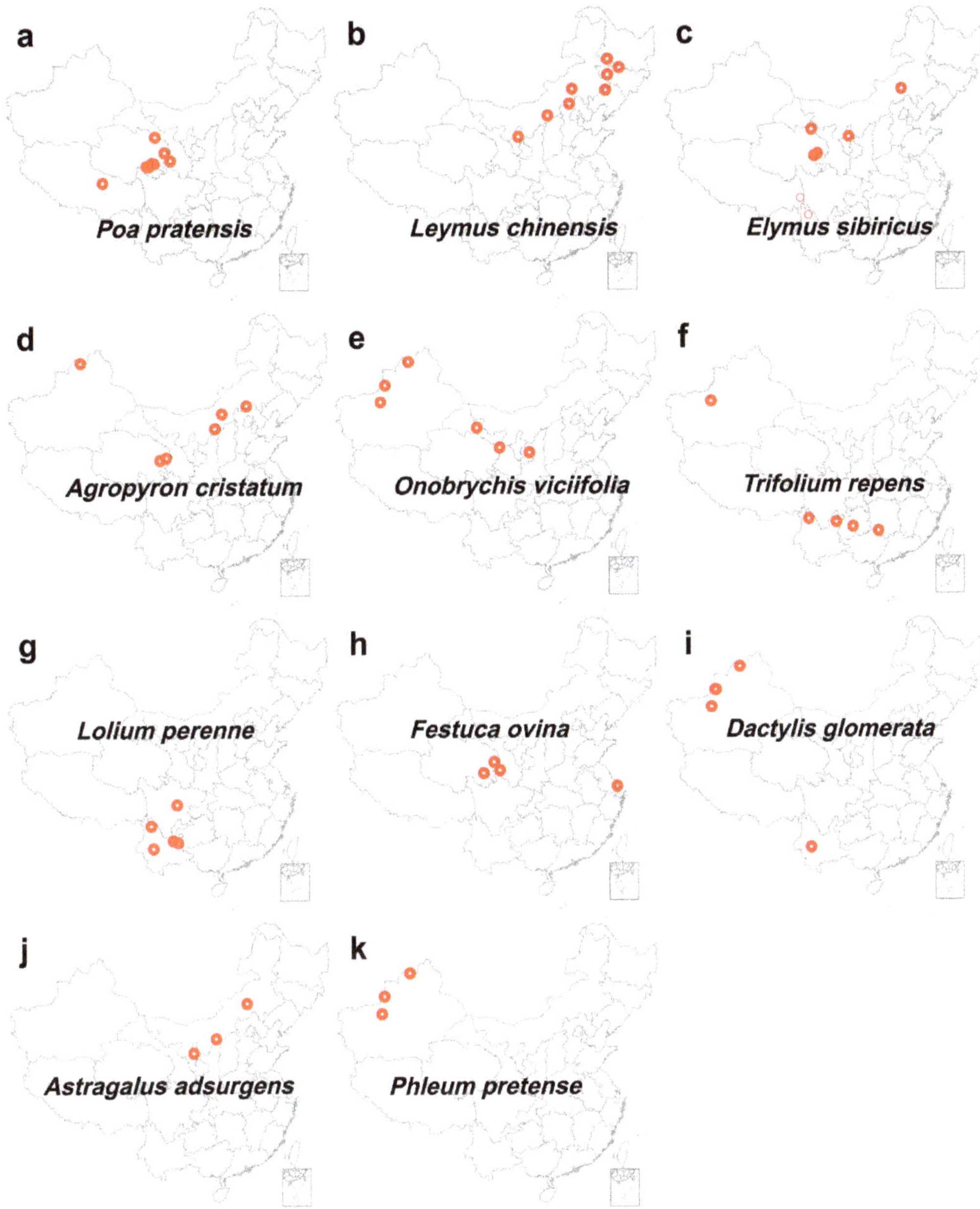

Figure A2. Spatial distribution of the 11 plant species other than those top three species presented in Figure A1.

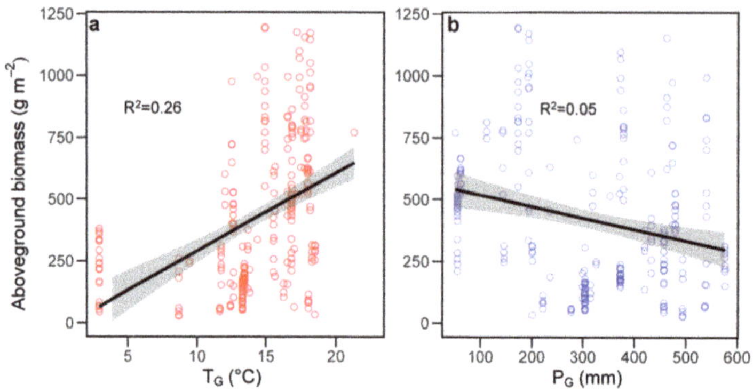

Figure A3. Relationship between aboveground biomass and two climatic attributes: (**a**): mean temperature during plant growing seasons (April–October; T_G), (**b**): accumulated precipitation during plant growing seasons (April–October; P_G). Data for all management such as fertilization and water were included.

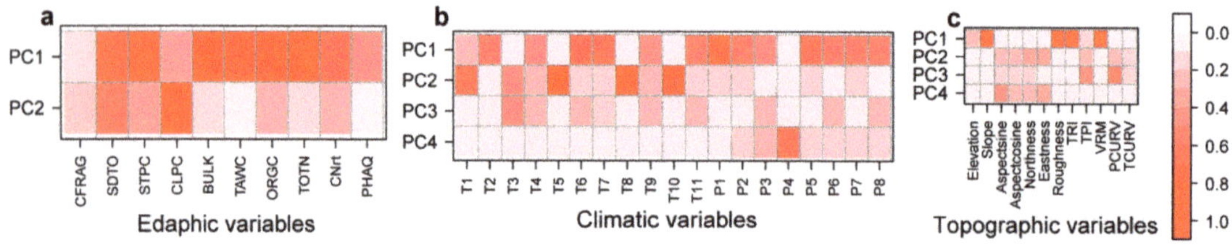

Figure A4. Contributions of each variable to the most *n* important principal components (PCs) of edaphic variables (**a**), climatic variables (**b**), and topographic variables (**c**). *n* is the number of PCs retained based on Kaiser's criterion.

Appendix B. References for Extracting AGB in China's Managed Grasslands

(1) Chen, F., Xia, H. and Qin, X.-j., 2019. Effect of mixture sowing on biomass allocation in the artificially-planted pastures, Southeastern Tibetan. Journal of Mountain Science, 16(1): 54–63.

(2) Chi, Y., Xiong, K., Dong, Y. and Zhang, J., 2015. Community characteristics of white clover and Ophiopogon japonicus mixed planting system under forest in karst area. Jiangsu Agricultural Sciences, 43(1): 114–116 (in Chinese).

(3) Chu, X., Shan, G., Bi, Y., Xue, S. and Kuang, C., 2012. Production performance and persistence of ten introduced alfalfa varieties. Pratacultural Science, 29(4): 610-614 (in Chinese).

(4) Dong, W. et al., 2010. The study on dynamics of aboveground biomass and nutrients of perennial sowing grassland in a 'Grain for Green' area. Pratacultural Science, 27(2): 54–58 (in Chinese).

(5) Fang, Y., Huang, Z., Cui, Z., He, H.H. and Liu, Y., 2021. Trade-offs between forage crop productivity and soil nutrients for different ages of alfalfa grassland. Land Degradation & Development, 32(1): 374–386.

(6) Feng, B., Liu, Z., Baoyin, T. and Tang, H., 2021. Effects of mowing frequency on community characteristics of dry mixed sowing artificial pasture. Chinese Journal of Grassland, 43(03): 10–18 (in Chinese).

(7) Feng, M. et al., 2016. Effects of water and nitrogen fertilizer on biomass distribution and water use efficiency of alfalfa (Medicago sativa) in Hexi Corridor. Chinese Journal of Eco-Agriculture, 24(12): 1623–1632 (in Chinese).

(8) Gao, C., Han, G., Wang, Z., Li, Z. and Ren, H., 2017. Carbon sequestration effect of different artificial grasslands in Inner Mongolia desert grassland area. Chinese Journal of Grassland, 39(4): 81–85 (in Chinese).
(9) Geng, W. et al., 2000. Study on establishing or improving pasture with phytocide tillage. Grassland of China, (5): 41–45 (in Chinese).
(10) Guan, H., Fan, J. and Li, Y., 2019. The impact of different introduced artificial grassland species combinations on community biomass and species diversity in temperate steppe of the Qinghai-Tibetan Plateau. Acta Prataculturae Sinica, 28(9): 192 (in Chinese).
(11) Han, D., Xu, Z., Ai, L. and Han, L., 2009. Effect of fertilization on the forage biomass and quality of aged leafy Elymus sibiricus. Plant Nutrition and Fertilizer Science, (6): 1486–1490 (in Chinese).
(12) He, F. et al., 2019. Variation characteristics of CO_2 fluxes of Elymus nutans artificial grassland for a planting cycle in Agro-pastoral transition area of Sanjiangyuan. Ecology and Environmental Sciences, 28(5): 918–929 (in Chinese).
(13) He, H. et al., 2018. The asymptotic response of soil water holding capacity along restoration duration of artificial grasslands from degraded alpine meadows in the Three River Sources, Qinghai–Tibetan Plateau, China. Ecological research, 33(5): 1001–1010.
(14) He, X., Yang, H., Ma, Y. and He, S., 2016. Study on biomass dynamic changes of Poa pratensis L.cv. Qinghai. Chinese Qinghai Journal of Animal and Veterinary Sciences, 46(1): 13-16 (in Chinese).
(15) He, Y., Wang, H. and Qi, B., 2014. Productivity dynamic of Festuca sinensis pasture built on 'bare soil land' degraded meadow during 5years on Tibetan plateau. Pratacultural Science, 31(1): 200-204 (in Chinese).
(16) Hou, X. et al., 2015. Study on the community structure characteristics and soil physicochemical property of Poa pratensis L.cv.Qinghai single artificial grasslands in different ages. Chinese Journal of Grassland, (1): 65-69 (in Chinese).
(17) Hu, D. et al., 2017. Effects of fertilization on yield and quality of Leymus chinensis cultivated grassland. Chinese Journal of Grassland, 39(1): 35-41 (in Chinese).
(18) Hu, L., Zi, H., Luo, X., Lerdau, M. and Wang, C., 2021. Root dynamics along a restoration chronosequence of revegetated grasslands in degraded alpine meadows of the Qinghai-Tibetan Plateau, China. Land Degradation & Development, 32(13): 3561-3572.
(19) Huang, Z. et al., 2018. Soil water storage deficit of alfalfa (Medicago sativa) grasslands along ages in arid area (China). Field Crops Research, 221: 1-6.
(20) Ji, B. et al., 2020. Soil carbon storage characteristics of alfalfa (Medicago sativa) artificial grasslands in the semi-arid hilly gully region of the loess plateau, China. Russian Journal of Ecology, 51(5): 466-476.
(21) Ji, W. et al., 2012. Suitability of several alfalfa varieties to alpine regions in the Qinghai-Tibetan Plateau. Pratacultural Science, 29(7): 1137-1141 (in Chinese).
(22) Jing, M., Wang, Y., Ma, S. and Li, S., 2019. Optimum conditions of Poa pratensis L.cv.Qinghai cultivation in Qilian Mountain. Acta Agriculturae Boreali-occidentalis Sinica, 28(1): 31-40 (in Chinese).
(23) Li, H., Zhu, C., An, S., Ma, J. and Deng, H., 2013. Changes of community characteristics in the legume/grass pasture in different year after establishment. Pratacultural Science, 30(3): 430-435 (in Chinese).
(24) Li, J.-H., Xu, D.-H. and Wang, G., 2008. Weed inhibition by sowing legume species in early succession of abandoned fields on Loess Plateau, China. Acta Oecologica, 33(1): 10-14.
(25) Li, Q., Zhang, H., Huang, Y. and Zhou, D., 2019a. Forage nitrogen yield and soil nitrogen in artificial grasslands with varied Medicago seedling proportion. Archives of Agronomy and Soil Science, 66(1): 110-125.

(26) Li, S. et al., 2019b. Effect of different grazing intensity on plant community and soil property in cultivated grassland in Weining, Guizhou Province. Grassland and Turf, 39(4): 19-24 (in Chinese).
(27) Li, X. et al., 2020. Effect of stubble height on yield and quality of alfalfa artificial grassland. China Herbivore Science, 40(06): 43-45+54 (in Chinese).
(28) Li, Y.-Y., Dong, S.-K., Wen, L., Wang, X.-X. and Wu, Y., 2014. Soil carbon and nitrogen pools and their relationship to plant and soil dynamics of degraded and artificially restored grasslands of the Qinghai–Tibetan Plateau. Geoderma, 213: 178-184.
(29) Li, Z., Hou, F. and An, Y., 2011. Effects of grazing and light on productivity of artificial pasture of understory. Pratacultural Science, 28(3): 414-419 (in Chinese).
(30) Lian, P.Y., Zeng, D.H., Liu, J.Y., Ding, F. and Wu, Z.W., 2012. Impact of land-use change on carbon stocks in meadow steppe of Northeast China, Applied Mechanics and Materials. Trans Tech Publ, pp. 262-268.
(31) Liu, D. et al., 2009. Impact of enclosure on community characters of sowed Elymus nutans grassland in "black soil Land". Pratacultural Science, 26(10): 59-66 (in Chinese).
(32) Liu, M. and Baoyin, T., 2004. An experiment on mix-sowing of Elymus Sibiricus and Medicago Varia. Grassland of China, 26(1): 22-27 (in Chinese).
(33) Liu, M. et al., 2016. Effects of Legume-grass mixed sowing on forage grass yield and quality in artificial grassland. Arid Zone Research, 33(1): 179-185 (in Chinese).
(34) Liu, Y. et al., 2019. Vegetation biomass allocation relationships in the different utilization patterns in alpine meadow. Grassland and Turf, 39(6): 58-65 (in Chinese).
(35) Luo, S. et al., 2018. Vegetation community of Poa pratensis L. cv. Qinghai cultivated grassland with different growth years. Journal of Grassland and Forage Science, (5): 24-29 (in Chinese).
(36) Luo, Z. et al., 2015. Soil moisture and alfalfa productivity response from different years of growth on the Loess Plateau of central Gansu. Acta Prataculturae Sinica, (1): 31-38 (in Chinese).
(37) Ma, L., Yang, H., Pan, Z. and Rong, Y., 2020. In situ measurements and meta-analysis reveal that land-use changes combined with low nitrogen application promote methane uptake by temperate grasslands in China. Science of the Total Environment, 706: 136048.
(38) Ma, Y. et al., 2003. Study on Cultivation and Tame of Poa Paratensis. Chinese Qinghai Journal of Animal and Veterinary Sciences, 33(3): 6-7 (in Chinese).
(39) Ma, Z., Zhang, C., Zhou, H.K., Yao, B.Q. and Zhao, X.Q., 2017. Role of seed bank in establishment of single and mixed-sowing artificial grasslands of Tibetan Plateau. Polish Journal of Ecology, 65(4): 334-344.
(40) Min, X., Ma, Y., Li, S. and Wang, Y., 2014. Effects of sheep manure on productivity and nutrition of soil for Poa pratensis cv. Qinghai pasture. Pratacultural Science, (6): 1039-1044 (in Chinese).
(41) Mu, Y., Liu, Y., Tian, F.-P., Chang, X.-F. and Wu, G.-L., 2016. Influence of artificial grassland restoration on soil carbon pool in an arid mining land. Journal of soil science and plant nutrition, 16(4): 890-900.
(42) Pan, L. et al., 2012. Effects of fertilizers and sowing rates on growth characteristics and forage yields of alfalfa in Yangzhou region. Acta Prataculturae Sinica, 20(6): 1099-1104 (in Chinese).
(43) Qi, Y., Zhu, L. and Xu, X., 2015. Effects of mixed combination and proportion between leguminous and graminaceous forages grown in Ningxia desert steppe. Pratacultural Science, 32(9): 1463-1472 (in Chinese).
(44) Qin, J. et al., 2019. Effects of grazing intensity in spring on forage growth in Poa pratensis L. cv. Qinghai artificial grassland in Qilian mountains. Journal of Qinghai University, 37(4): 1-6 (in Chinese).
(45) Ren, W., Zhang, Z., Lin, C., Yang, Q. and Shen, Y., 2020. Effects of mixed seeding ratio on biomass allocation and competition of Onobrychis viciifolia and Elymus nutans

under cold conditions in the Tianzhu alpine region. Pratacultural Science, 37(10): 2035-2048 (in Chinese).
(46) Ruan, K. et al., 2013. Experiment on adaptive selection and cultivation of artificial grassland in Liangshan Tianchi. Hubei Journal of Animal and Veterinary Sciences, 34(12): 88-90 (in Chinese).
(47) Shan, G. et al., 2016. Adaptability evaluation of 12 perennial forage cultivars in alpine region of Diqing. Pratacultural Science, 33(9): 1793-1800 (in Chinese).
(48) Sun, H., 2000. Studies on the degradation succession of mixed artificial grassland. Grassland of China, (2): 8-14 (in Chinese).
(49) Tai, J., Zhang, L. and Yang, H., 2009. Effect of different planting years on the yield of alfalfa and content of N, P, K in soil. Pratacultural Science, 26(12): 82-86 (in Chinese).
(50) Tang, L., Dang, X., Liu, G., Shao, C. and Xue, S., 2014. Response of artificial grassland carbon stock to management in mountain region of Southern Ningxia, China. Chinese Geographical Science, 24(4): 436-443.
(51) Tao, Z.W. et al., 2021. Effects of different artificial planting schemes on invasive weeds. Global Ecology and Conservation, 28: e01651.
(52) Wang, C. et al., 2009. Community succession of differently aged artificial grasslands and their soil nutrient changes in three rivers' source regions in Qinghai, China. Chin J Appl Environ Biol, 15(6): 737-744 (in Chinese).
(53) Wang, C. et al., 2020. Variation characteristics of plant communities in black soil shoals with different establishment years in the source area of the Yellow River. Pratacultural Science, 37(12): 2422-2430 (in Chinese).
(54) Wang, D. et al., 2018. Response of microbial C, N and respiration characteristics to sowing rates in alfalfa cultivation grasslands in Hulunber, Inner Mongolia. Acta Prataculturae Sinica, 27(3): 135-143 (in Chinese).
(55) Wang, L., 2012. The study of the alfalfa cultivation experiment in the artificial grass, Western Jilin. Chinese Agricultural Mechanization, (4): 165-167 (in Chinese).
(56) Wang, L., Bi, Y., Ma, Y. and Shi, J., 2007. Study of the optimal combination of mixed seeding artificial pasture in "Black Soil Type" degraded meadow. Journal of Qinghai University, 25(3): 1-5 (in Chinese).
(57) Wang, P. et al., 2021. Effects of conservation seeding on the yield and quality of perennial mowing mixed grassland in alpine area. Acta Prataculturae Sinica, 29(08): 1818-1827 (in Chinese).
(58) Wang, Q. et al., 2017. Effects of sowing depth on botanical characters and production performance of Alfalfa (Medicago sativa) under different soil salt concentration. Chinese Journal of Grassland, 39(1): 19-26 (in Chinese).
(59) Wang, Q., Wang, S., Chen, W. and Chen, B., 2016. Effects of fertilization on the yield, quality and economic benefits of alfalfa in low-production croplands of Minle. Pratacultural Science, 33(2): 230-239 (in Chinese).
(60) Wang, S., Hao, M., Pu, Q. and Wu, Z., 2014. Ecological and productive succession process of a cultivated alfalfa grassland community on Loess Plateau. Acta Prataculturae Sinica, 23(6): 1-10 (in Chinese).
(61) Wang, S. et al., 2015. Effects of sowing quantity and stubble height on the yield and quality of Elymus sibiricus. Pratacultural Science, (1): 107-113 (in Chinese).
(62) Wang, X. et al., 2021d. Effects of perennial Legume-Gramineae mixtures on forage yield and quality in the Hexi Corridor Region. Pratacultural Science, 38(7): 1339-1350 (in Chinese).
(63) Wen, L., Jinlan, W., Xiaojiao, Z., Shangli, S. and Wenxia, C., 2018. Effect of degradation and rebuilding of artificial grasslands on soil respiration and carbon and nitrogen pools on an alpine meadow of the Qinghai-Tibetan Plateau. Ecological Engineering, 111: 134-142.
(64) Wen, Z. and Hui, Y., 1996. Establishment of rape-forage mixture pasture in Bashang Plateau Region. Grassland of China, (1): 27-30 (in Chinese).

(65) Wu, G.L., Liu, Z.H., Zhang, L., Hu, T.M. and Chen, J.-M., 2010. Effects of artificial grassland establishment on soil nutrients and carbon properties in a black-soil-type degraded grassland. Plant and soil, 333(1): 469-479.
(66) Wu, H. et al., 2019. Sediment addition and legume cultivation result in sustainable, long-term increases in ecosystem functions of sandy grasslands. Land Degradation & Development, 30(14): 1667-1676.
(67) Xiao, X. et al., 2021. Effects of utilization modes, planting patterns, and nitrogen applications on the yield and quality of perennial forage. Pratacultural Science, 38(4): 703-715 (in Chinese).
(68) Xing, Y. et al., 2020. Characteristics of plant community and soil organic carbon and nitrogen in artificial grassland with different establishment years. Acta Prataculturae Sinica, 28(2): 521-528 (in Chinese).
(69) Xu, K., Wu, X., Xie, Y. and Yang, J., 2013. Main factors controlling soil CO_2 efflux from alfalfa fields of different standing ages in an arid region. Ecology and Environmental Sciences, 22(10): 1671-1677 (in Chinese).
(70) Xu, R., Chang, S. and Jia, Q., 2020. Effects of nitrogen application and utilization methods on yield, quality and water use of grass-legume mixed grassland in Loess Plateau. Acta Prataculturae Sinica, 28(6): 1744-1755 (in Chinese).
(71) Xu, S. et al., 2000a. Study of the pasture improvement by CaMgP application. Pratacultural Science, 17(12): 13-17 (in Chinese).
(72) Xu, S. et al., 2000b. Renewing deteriorated pasture without cultivation. Pratacultural Science(z1): 10-12 (in Chinese).
(73) Xu, Z. et al., 2018. Effects of grazing methods on vegetation productivity and carbon density in artificial grassland. Journal of Northern Agriculture, 4(46): 110-117 (in Chinese).
(74) Yang, H. et al., 2011. Dynamics of vegetation characteristics and soil physical properties of Poa pratensis cv. Qinghai. Pratacultural Science, 28(6): 910-914 (in Chinese).
(75) Yang, K. et al., 2015a. Effects of nitrogen application on Phleum pretense pasture's forage yield and quality. Pratacultural Science, 32(12): 2071-2077 (in Chinese).
(76) Yang, X., Li, P., Dong, C., Ma, X. and Gou, W., 2020. Studies on the dynamics of above-ground biomass and nutritive value of annual ryegrass and common oat mixtures. Acta Prataculturae Sinica, 28(1): 149-158 (in Chinese).
(77) Yang, X., Li, X., Jin, L. and Sun, H., 2019. Effectiveness of different artificial restoration measures for soil and vegetation recovery on coal mine tailings in an alpine area. Acta Prataculturae Sinica, 28(3): 1-11 (in Chinese).
(78) Yang, X., Wang, C., Zi, H. and Liu, M., 2015b. Soil microbial community structure characteristics in artificial grassland with different cultivation years in the headwater region of Three Rivers, China. Chin J Appl Environ Biol, 21(2): 341-349 (in Chinese).
(79) Yang, Y., Zhou, W., Lian, X. and Kou, S., 2018. R. Comprehensive evaluation of production characteristics of annual forage crops based on principal component analysis: A case study of Zhuanglang in Gansu. Pratacultural Science, 35(6): 1503-1509 (in Chinese).
(80) Yang, Z., Wang, D., Liu, Y., Zhu, Y. and Wu, G., 2015c. Soil moisture and infiltration characteristics for artificial pasture planted on opencast coal mining tailings. Acta Prataculturae Sinica, 24(12): 29-37 (in Chinese).
(81) Yao, Z., Li, J. and Song, L., 2020a. Study on the effect of cross sowing of elymus nutans and medicago sativain different periods. Acta Prataculturae Sinica, 28(5): 1454-1459 (in Chinese).
(82) Yao, Z. et al., 2020b. Study on the leguminous-gramineous grass mixed sowing in Bashang area of Zhangjiakou. Acta Prataculturae Sinica, 28(4): 1076-1082 (in Chinese).
(83) Yu, X. et al., 2014. Plant community characteristics of plateau pika rathole hazard zone in head waters region of three rivers. Journal of Gansu Agricultural University, 49(3): 107-112 (in Chinese).

(84) Zhai, X. et al., 2019. Stoichiometric characteristics of different agroecosystems under the same climatic conditions in the agropastoral ecotone of northern China. Soil Research, 57(8): 875-882.
(85) Zhang, H. et al., 2021. Effects of water regulation on yield, quality and water use of mixed artificial grassland. Water Resources Planning and Design, (4): 63-69 (in Chinese).
(86) Zhang, J. et al., 2010. Effects of clipping on forage yield and quality of mixed-sown artificial grassland in Yangtze and Yellow river headwater region. Pratacultural Science, 27(1): 92-96 (in Chinese).
(87) Zhang, J., Zhang, J., Wang, Y. and Han, T., 2014a. Adaptability of introduced species for improvement of degraded alpine grassland in Gannan areas, China. Pratacultural Science, 31(4): 744-753 (in Chinese).
(88) Zhang, L. et al., 2012. Relationships of dominant species root activity, plant community characteristics and soil micro-environment in artificial grassland over different cultivation periods. Acta Prataculturae Sinica, 21(5): 185-194 (in Chinese).
(89) Zhang, Q., Jing, Y., Yang, X., Jiang, L. and Li, F., 2014b. Preliminary study on mix-sowed pasture in eastern section of Tal Barker Taishan mountain. Pratacultural Science, 31(5): 943-948 (in Chinese).
(90) Zhang, R. et al., 2018. Characteristics of biomass carbon density of degraded natural grassland and artificial grassland in the "Three-River Headwaters" Region. Journal of Natural Resources, 33(2): 185-194 (in Chinese).
(91) Zhang, X. et al., 2019. Effects of fertilization rate on forage yield and water use efficiency of artificial grassland in an alpine arid area. Scientia Agricultura Sinica, 52(8): 1368-1379 (in Chinese).
(92) Zhang, X. et al., 2020. Study on plant community structure characteristics of alpine mixed-seeding grassland in Three Rivers Source Regions. Acta Prataculturae Sinica, 28(4): 1090-1099 (in Chinese).
(93) Zhang, X., Zhu, J. and Ding, H., 2014c. Effects of different mixed sowing patterns on productivity of legume/grass mixture. Grassland and Turf, 34(1): 44-48 (in Chinese).
(94) Zhang, Y. and Zhang, L., 2006. A Study on forage yield dynamics of Nedicago varia/Bromus inermis mixture and single grassland. Chinese Journal of Grassland, 28(5): 23-28 (in Chinese).
(95) Zhang, Z., Xiong, Y. and Pan, Q., 2009. Study on the mix-sowing of five gramineous grasses and Medicago varia in pastoral agronomy area. Grassland and Turf, 2009(6): 15-19 (in Chinese).
(96) Zhao, Q. et al., 2013. Effect of different grazing systems on nutrition of artificial pasture and sheep weight. Chinese Journal of Grassland, 35(1): 67-72 (in Chinese).
(97) Zhao, W. et al., 2020. Effects of vegetation reconstruction on vegetation and microbial community characteristics of black soil beach grassland. Ecology and Environmental Sciences, 29(1): 71-80 (in Chinese).
(98) Zhao, Y., Chai, Q., Chen, Y. and Zhu, W., 2008. Improvement and utilization of saline-alkali grassland in Hexi Corridor. Pratacultural Science, 25(2): 21-25 (in Chinese).
(99) Zheng, W., Zhu, J., Jianaerguli, Li, H. and Zhang, J., 2011. Effects of different mixed sowing patterns on production performance of legume-grass mixture. Chinese Journal of Grassland, 33(5): 45-52 (in Chinese).
(100) Zhou, J., 2020. Artificial grassland productivity of native grass in north of Tibet. Tibet Journal of Agricultural Sciences, 42(S1): 69-72 (in Chinese).
(101) Zhou, Z. et al., 2003. Experimental study on mixed sowing of brome and alfalfa. Inner Mongolia Prataculture, 15(3): 43-44 (in Chinese).

References

1. Suttie, J.M.; Reynolds, S.G.; Batello, C. *Grasslands of the World*; Food & Agriculture Org: Rome, Italy, 2005; Volume 34.
2. Bardgett, R.D.; Bullock, J.M.; Lavorel, S.; Manning, P.; Schaffner, U.; Ostle, N.; Chomel, M.; Durigan, G.; Fry, E.L.; Johnson, D. Combatting global grassland degradation. *Nat. Rev. Earth Environ.* **2021**, *2*, 720–735. [CrossRef]
3. Gibbs, H.; Salmon, J.M. Mapping the world's degraded lands. *Appl. Geogr.* **2015**, *57*, 12–21. [CrossRef]
4. Kang, L.; Han, X.; Zhang, Z.; Sun, O.J. Grassland ecosystems in China: Review of current knowledge and research advancement. *Philos. Trans. R. Soc. B Biol. Sci.* **2007**, *362*, 997–1008. [CrossRef] [PubMed]
5. Cao, J.; Adamowski, J.F.; Deo, R.C.; Xu, X.; Gong, Y.; Feng, Q. Grassland degradation on the Qinghai-Tibetan Plateau: Reevaluation of causative factors. *Rangel. Ecol. Manag.* **2019**, *72*, 988–995. [CrossRef]
6. Qi, J.; Chen, J.; Wan, S.; Ai, L. Understanding the coupled natural and human systems in Dryland East Asia. *Environ. Res. Lett.* **2012**, *7*, 015202. [CrossRef]
7. Wu, G.L.; Ren, G.H.; Dong, Q.M.; Shi, J.J.; Wang, Y.L. Above-and Belowground Response along Degradation Gradient in an Alpine Grassland of the Qinghai-T ibetan Plateau. *CLEAN–Soil Air Water* **2014**, *42*, 319–323. [CrossRef]
8. Xu, H.P.; Zhang, J.; Pang, X.P.; Wang, Q.; Zhang, W.N.; Wang, J.; Guo, Z.G. Responses of plant productivity and soil nutrient concentrations to different alpine grassland degradation levels. *Environ. Monit. Assess.* **2019**, *191*, 678. [CrossRef]
9. Han, J.; Zhang, Y.; Wang, C.; Bai, W.; Wang, Y.; Han, G.; Li, L. Rangeland degradation and restoration management in China. *Rangel. J.* **2008**, *30*, 233–239. [CrossRef]
10. Zhou, H.; Zhao, X.; Tang, Y.; Gu, S.; Zhou, L. Alpine grassland degradation and its control in the source region of the Yangtze and Yellow Rivers, China. *Grassl. Sci.* **2005**, *51*, 191–203. [CrossRef]
11. Fang, J.; Bai, Y.; Li, L.; Jiang, G.; Huang, J.; Huang, Z.; Zhang, W.; Gao, S. Scientific basis and practical ways for sustainable development of China's pasture regions. *Chin. Sci. Bull.* **2016**, *61*, 155–164. (In Chinese)
12. Shan, G.; Zhang, M.; Liao, X.; Zhong, S.; Zhou, P.; Xue, S. Adaptability evaluation of 12 perennial forage cultivars in alpine region of Diqing. *Pratacultural Sci.* **2016**, *33*, 1793–1800. (In Chinese with English abstract)
13. Chen, F.; Xia, H.; Qin, X.-J. Effect of mixture sowing on biomass allocation in the artificially-planted pastures, Southeastern Tibetan. *J. Mt. Sci.* **2019**, *16*, 54–63. [CrossRef]
14. Han, D.; Xu, Z.; Ai, L.; Han, L. Effect of fertilization on the forage biomass and quality of aged leafy Elymus sibiricus. *Plant Nutr. Fertil. Sci.* **2009**, *15*, 1486–1490. (In Chinese with English abstract)
15. Pan, L.; Wei, Z.; Wu, Z.; Zhang, D.; Zheng, X.; Chen, F.; Liu, Q.; Li, W. Effects of Fertilizers and Sowing Rates on Growth Characteristics and Forage Yields of Alfalfa in Yangzhou Region. *Acta Prataculturae Sin.* **2012**, *20*, 1099–1104. (In Chinese with English abstract)
16. Zavalloni, C.; Vicca, S.; Büscher, M.; de la Providencia, I.E.; Dupré de Boulois, H.; Declerck, S.; Nijs, I.; Ceulemans, R. Exposure to warming and CO_2 enrichment promotes greater above-ground biomass, nitrogen, phosphorus and arbuscular mycorrhizal colonization in newly established grasslands. *Plant Soil* **2012**, *359*, 121–136. [CrossRef]
17. Yuan, Z.-Q.; Fang, C.; Zhang, R.; Li, F.-M.; Javaid, M.M.; Janssens, I.A. Topographic influences on soil properties and aboveground biomass in lucerne-rich vegetation in a semi-arid environment. *Geoderma* **2019**, *344*, 137–143. [CrossRef]
18. Xiao, Y.; Zhang, J.; Jia, T.T.; Pang, X.P.; Guo, Z.G. Effects of alternate furrow irrigation on the biomass and quality of alfalfa (*Medicago sativa*). *Agric. Water Manag.* **2015**, *161*, 147–154. [CrossRef]
19. Cui, H.; Sun, W.; Delgado-Baquerizo, M.; Song, W.; Ma, J.-Y.; Wang, K.; Ling, X. Phosphorus addition regulates the responses of soil multifunctionality to nitrogen over-fertilization in a temperate grassland. *Plant Soil* **2020**, *473*, 73–87. [CrossRef]
20. Wang, G.; Luo, Z.; Huang, Y.; Sun, W.; Wei, Y.; Xiao, L.; Deng, X.; Zhu, J.; Li, T.; Zhang, W. Simulating the spatiotemporal variations in aboveground biomass in Inner Mongolian grasslands under environmental changes. *Atmos. Chem. Phys.* **2021**, *21*, 3059–3071. [CrossRef]
21. Luo, Z.; Rossel, R.A.V.; Shi, Z. Distinct controls over the temporal dynamics of soil carbon fractions after land use change. *Glob. Chang. Biol.* **2020**, *26*, 4614–4625. [CrossRef]
22. Batjes, N.H. Harmonized soil property values for broad-scale modelling (WISE30sec) with estimates of global soil carbon stocks. *Geoderma* **2016**, *269*, 61–68. [CrossRef]
23. Amatulli, G.; Domisch, S.; Tuanmu, M.-N.; Parmentier, B.; Ranipeta, A.; Malczyk, J.; Jetz, W. A suite of global, cross-scale topographic variables for environmental and biodiversity modeling. *Sci. Data* **2018**, *5*, 180040. [CrossRef] [PubMed]
24. Fick, S.E.; Hijmans, R.J. WorldClim 2: New 1-km spatial resolution climate surfaces for global land areas. *Int. J. Climatol.* **2017**, *37*, 4302–4315. [CrossRef]
25. Peng, S.; Ding, Y.; Liu, W.; Li, Z. 1 km monthly temperature and precipitation dataset for China from 1901 to 2017. *Earth Syst. Sci. Data* **2019**, *11*, 1931–1946. [CrossRef]
26. Liu, J.Y.; Liu, M.L.; Zhuang, D.F.; Zhang, Z.X.; Deng, X.Z. Study on spatial pattern of land-use change in China during 1995–2000. *Sci. China Ser. D-Earth Sci.* **2003**, *46*, 373–384.
27. Gelman, A.; Hill, J. *Data Analysis Using Regression and Multilevel/Hierarchical Models*; Cambridge University Press: Cambridge, UK, 2006.
28. Kaiser, H.F. The Application of Electronic Computers to Factor Analysis. *Educ. Psychol. Meas.* **1960**, *20*, 141–151. [CrossRef]
29. R Development Core Team. *R: A Language and Environment for Statistical Computing*; R Foundation for Statistical Computing: Vienna, Austria, 2022.

30. Hu, Q.; Pan, F.; Pan, X.; Zhang, D.; Li, Q.; Pan, Z.; Wei, Y. Spatial analysis of climate change in Inner Mongolia during 1961–2012, China. *Appl. Geogr.* **2015**, *60*, 254–260. [CrossRef]
31. Yang, K.; Wu, H.; Qin, J.; Lin, C.; Tang, W.; Chen, Y. Recent climate changes over the Tibetan Plateau and their impacts on energy and water cycle: A review. *Glob. Planet. Chang.* **2014**, *112*, 79–91. [CrossRef]
32. Fay, P.A.; Prober, S.M.; Harpole, W.S.; Knops, J.M.H.; Bakker, J.D.; Borer, E.T.; Lind, E.M.; MacDougall, A.S.; Seabloom, E.W.; Wragg, P.D.; et al. Grassland productivity limited by multiple nutrients. *Nat. Plants* **2015**, *1*, 15080. [CrossRef]
33. Firn, J.; McGree, J.M.; Harvey, E.; Flores-Moreno, H.; Schütz, M.; Buckley, Y.M.; Borer, E.T.; Seabloom, E.W.; La Pierre, K.J.; MacDougall, A.M.; et al. Leaf nutrients, not specific leaf area, are consistent indicators of elevated nutrient inputs. *Nat. Ecol. Evol.* **2019**, *3*, 400–406. [CrossRef]
34. Xu, X.; Liu, H.; Song, Z.; Wang, W.; Hu, G.; Qi, Z. Response of aboveground biomass and diversity to nitrogen addition along a degradation gradient in the Inner Mongolian steppe, China. *Sci. Rep.* **2015**, *5*, 10284. [CrossRef]
35. Elser, J.J.; Bracken, M.E.; Cleland, E.E.; Gruner, D.S.; Harpole, W.S.; Hillebrand, H.; Ngai, J.T.; Seabloom, E.W.; Shurin, J.B.; Smith, J.E. Global analysis of nitrogen and phosphorus limitation of primary producers in freshwater, marine and terrestrial ecosystems. *Ecol. Lett.* **2007**, *10*, 1135–1142. [CrossRef]
36. LeBauer, D.S.; Treseder, K.K. Nitrogen limitation of net primary productivity in terrestrial ecosystems is globally distributed. *Ecology* **2008**, *89*, 371–379. [CrossRef]
37. Li, J.; Lin, S.; Taube, F.; Pan, Q.; Dittert, K. Above and belowground net primary productivity of grassland influenced by supplemental water and nitrogen in Inner Mongolia. *Plant Soil* **2011**, *340*, 253–264. [CrossRef]
38. Dai, L.; Ke, X.; Du, Y.; Zhang, F.; Li, Y.; Li, Q.; Lin, L.; Peng, C.; Shu, K.; Cao, G. Nitrogen controls the net primary production of an alpine Kobresia meadow in the northern Qinghai-Tibet Plateau. *Ecol. Evol.* **2019**, *9*, 8865–8875. [CrossRef]
39. Sala, O.E.; Parton, W.J.; Joyce, L.; Lauenroth, W. Primary production of the central grassland region of the United States. *Ecology* **1988**, *69*, 40–45. [CrossRef]
40. Knapp, A.K.; Smith, M.D. Variation among biomes in temporal dynamics of aboveground primary production. *Science* **2001**, *291*, 481–484. [CrossRef]
41. Bradford, J.B.; Lauenroth, W.K.; Burke, I.C.; Paruelo, J.M. The influence of climate, soils, weather, and land use on primary production and biomass seasonality in the US Great Plains. *Ecosystems* **2006**, *9*, 934–950. [CrossRef]
42. Ji, C.J.; Yang, Y.H.; Han, W.X.; He, Y.F.; Smith, J.; Smith, P. Climatic and edaphic controls on soil pH in alpine grasslands on the Tibetan Plateau, China: A quantitative analysis. *Pedosphere* **2014**, *24*, 39–44. [CrossRef]
43. Xie, Y.; Sha, Z.; Yu, M.; Bai, Y.; Zhang, L. A comparison of two models with Landsat data for estimating above ground grassland biomass in Inner Mongolia, China. *Ecol. Model.* **2009**, *220*, 1810–1818. [CrossRef]
44. Sun, J.; Cheng, G.; Li, W. Meta-analysis of relationships between environmental factors and aboveground biomass in the alpine grassland on the Tibetan Plateau. *Biogeosciences* **2013**, *10*, 1707–1715. [CrossRef]
45. Fang, J.Y.; Yang, Y.H.; Ma, W.H.; Mohammat, A.; Shen, H.H. Ecosystem carbon stocks and their changes in China's grasslands. *Sci. China-Life Sci.* **2010**, *53*, 757–765. [CrossRef] [PubMed]
46. Fang, J.; Guo, Z.; Piao, S.; Chen, A. Terrestrial vegetation carbon sinks in China, 1981–2000. *Sci. China Earth Sci.* **2007**, *50*, 1341–1350. [CrossRef]
47. Piao, S.; Fang, J.; Zhou, L.; Tan, K.; Tao, S. Changes in biomass carbon stocks in China's grasslands between 1982 and 1999. *Glob. Biogeochem. Cycles* **2007**, *21*, GB2002. [CrossRef]
48. Ni, J. Forage yield-based carbon storage in grasslands of China. *Clim. Chang.* **2004**, *67*, 237–246. [CrossRef]
49. Shen, H.; Zhu, Y.; Zhao, X.; Geng, X.; Gao, S.; Fang, J. Grassland area, biomass and productivity in China: A literature survey and model evaluation. *Chin. Sci. Bull.* **2016**, *61*, 139–154. (In Chinese with English abstract)
50. Brink, G.; Sanderson, M.; Casler, M. Grass and legume effects on nutritive value of complex forage mixtures. *Crop Sci.* **2015**, *55*, 1329–1337. [CrossRef]
51. Lindenmayer, R.B.; Hansen, N.C.; Brummer, J.; Pritchett, J.G. Deficit Irrigation of Alfalfa for Water-Savings in the Great Plains and Intermountain West: A Review and Analysis of the Literature. *Agron. J.* **2011**, *103*, 45–50. [CrossRef]
52. Stevens, C.J.; Dise, N.B.; Mountford, J.O.; Gowing, D.J. Impact of nitrogen deposition on the species richness of grasslands. *Science* **2004**, *303*, 1876–1879. [CrossRef]
53. Allan, E.; Manning, P.; Alt, F.; Binkenstein, J.; Blaser, S.; Blüthgen, N.; Böhm, S.; Grassein, F.; Hölzel, N.; Klaus, V.H. Land use intensification alters ecosystem multifunctionality via loss of biodiversity and changes to functional composition. *Ecol. Lett.* **2015**, *18*, 834–843. [CrossRef]
54. Yang, Y.; Fang, J.; Ma, W.; Smith, P.; Mohammat, A.; Wang, S.; Wang, W. Soil carbon stock and its changes in northern China's grasslands from 1980s to 2000s. *Glob. Chang. Biol.* **2010**, *16*, 3036–3047. [CrossRef]
55. Glover, J.D.; Culman, S.W.; DuPont, S.T.; Broussard, W.; Young, L.; Mangan, M.E.; Mai, J.G.; Crews, T.E.; DeHaan, L.R.; Buckley, D.H. Harvested perennial grasslands provide ecological benchmarks for agricultural sustainability. *Agric. Ecosyst. Environ.* **2010**, *137*, 3–12. [CrossRef]
56. DuPont, S.T.; Culman, S.W.; Ferris, H.; Buckley, D.H.; Glover, J.D. No-tillage conversion of harvested perennial grassland to annual cropland reduces root biomass, decreases active carbon stocks, and impacts soil biota. *Agric. Ecosyst. Environ.* **2010**, *137*, 25–32. [CrossRef]

57. Guan, Z.-H.; Li, X.G.; Wang, L.; Mou, X.M.; Kuzyakov, Y. Conversion of Tibetan grasslands to croplands decreases accumulation of microbially synthesized compounds in soil. *Soil Biol. Biochem.* **2018**, *123*, 10–20. [CrossRef]
58. Post, W.M.; Kwon, K.C. Soil carbon sequestration and land-use change: Processes and potential. *Glob. Change Biol.* **2000**, *6*, 317–327. [CrossRef]
59. Chang, J.; Ciais, P.; Gasser, T.; Smith, P.; Herrero, M.; Havlík, P.; Obersteiner, M.; Guenet, B.; Goll, D.S.; Li, W. Climate warming from managed grasslands cancels the cooling effect of carbon sinks in sparsely grazed and natural grasslands. *Nat. Commun.* **2021**, *12*, 118. [CrossRef]
60. Li, Y.-Y.; Dong, S.-K.; Wen, L.; Wang, X.-X.; Wu, Y. Soil carbon and nitrogen pools and their relationship to plant and soil dynamics of degraded and artificially restored grasslands of the Qinghai–Tibetan Plateau. *Geoderma* **2014**, *213*, 178–184. [CrossRef]
61. Wen, L.; Jinlan, W.; Xiaojiao, Z.; Shangli, S.; Wenxia, C. Effect of degradation and rebuilding of artificial grasslands on soil respiration and carbon and nitrogen pools on an alpine meadow of the Qinghai-Tibetan Plateau. *Ecol. Eng.* **2018**, *111*, 134–142. [CrossRef]

agronomy

MDPI

Article

Compensatory Structural Growth Responses of Early-Succession Native Warm-Season Grass Stands to Defoliation Management

Vitalis W. Temu * and Maru K. Kering

Agricultural Research Station, Virginia State University, Petersburg, VA 23806, USA; mkering@vsu.edu
* Correspondence: vtemu@vsu.edu; Tel.: +1-804-524-6717; Fax: +1-804-524-5186

Abstract: There is a growing recognition of the significance of unique morphological and physiological adaptation of native warm-season grasses (NWSG) of North America as summer forage resources and major grassland ecosystem components. Defoliation management plays a major role in ensuring eco-friendly utilization of grassland natural resources. To assess sward structural responses of big bluestem (BB, *Andropogon gerardii* Vitman), eastern gamagrass (GG, *Tripsacum dactyloides* L.), indiangrass (IG, *Sorghastrum nutans* L. Nash), and switchgrass (SG, *Panicum virgatum* L.) stands to seasonal changes in harvest regimes, a five-year forage harvesting trial was conducted, in a randomized complete block design, at Virginia State University's research farm. Vegetation structural response attributes (sward-height, canopy closure, stand density and basal cover) of newly established the NWSG stands to second year changes in harvest regimes were monitored. In 2013, 64 plots of year-old stands of transplanted BB, GG, IG, and SG separated by ≥120-cm alleys were cut once in early-August and mid-November to suppress weeds and promote tillering. Starting June 2014, each plot had three 1.5-m wide side-by-side harvest-strips cut once-, twice-, or thrice year^{-1} (frequencies) ending mid-Oct for four consecutive years followed by a single mid-summer harvest in 26 June 2018, using a forage plot-harvester. In 2015, harvest frequencies for the three- and single-cut strips, in plots 32–64, were switched/flipped once and never reverted. Data was recorded on four early-summer and late-fall sward heights, from each strip at 60-cm intervals before the first and the last harvest, each year. early-spring basal- and canopy-diameter, for mid-April 2015 and 2016, concurrent early-spring canopy light interception, using the LI-191 Line Quantum Sensor, and season-end visual obstruction heights, for stand density in 2016 and 2017. All regrowth sward-heights showed effects of harvest frequency and exhibited compensatory structural responses to the change in harvest regimes. Basal and canopy diameters tended to be greater for the single-cut strips that were previously cut thrice.

Keywords: defoliation frequency; forage; sward-height; structure; canopy; light interception; habitat; warm-season; big bluestem; gamagrass; indiangrass; switchgrass; compensatory

Citation: Temu, V.W.; Kering, M.K. Compensatory Structural Growth Responses of Early-Succession Native Warm-Season Grass Stands to Defoliation Management. *Agronomy* **2023**, *13*, 1280. https://doi.org/10.3390/agronomy13051280

Academic Editors: Kesi Liu and Xinqing Shao

Received: 28 March 2023
Revised: 28 April 2023
Accepted: 28 April 2023
Published: 29 April 2023

1. Introduction

A growing recognition of the unique suitability of native warm-season grasses (NWSG) of North America for various economic and ecological uses has generated interests in management strategies for sustainable utilization. As forage plants, most NWSGs often grow better and persist longer under hot and drought growing conditions than their exotic counterparts such as bermudagrass (*Cynodon dactylon* L. Pers), bahiagrass (*Paspalum notatum* Flueggé), and dallisgrass (*Paspalum dilatatum* Poir.). Being morphologically and physiologically tolerant to harsh growing conditions also makes them suitable candidates for various ecosystem services such as wildlife habitat, soil conservation, bioenergy production, carbon sequestration, stabilizing stream banks, riparian buffers and filtering off sediments from runoff waters in agricultural landscapes. In many ways, NWSGs have shown the potential to play a role towards alleviating some major global challenges of food and/energy insecurity, climate change and environmental sustainability. Thus far, five mostly researched NWSG species—big bluestem (BB, *Andropogon gerardii* Vitman), eastern gamagrass (EG, *Tripsacum dactyloides* L.), indiangrass (IG, *Sorghastrum nutans* L. Nash), little bluestem (LB, *Schizachyrium scoparium* Michx. Nash), switchgrass (SG, *Panicum virgatum* L.) are promising and have shown desirable attributes for this multiple use [1]. Regardless of the long established economic and environmental benefits of NWSGs in managed grasslands, their adoption remains very low. Hindrances to their incorporation into forage systems, for example, are mostly associated with high costs of establishment and inability to withstand grazing when compared with conventional/exotic forage grasses [2]. The inability for farmers to realize the well-promoted unique ecological advantages that NWSGs have over the conventional forage grasses remains a challenge.

In many aspects, these deterrents to faster incorporation of NWSGs into forage systems seem mostly rooted in misperceptions around their stand persistency, forage quality and proneness to weeds pressure and bush encroachment. These are, usually, negative impacts of severe defoliation and deferential growth responses of the plant components and also reduction in leaf area coupled with changes in respiratory and growth rates, as well as carbon allocation patterns [3]. Because of their unique morphological features and physiological characteristics, NWSG stands require appropriate defoliation management to sustain high forage productivity. To be appropriate, decisions on the timing, intensity, and frequency of defoliation events must be mindful of their potential effects to critical plant growth responses. For tall-growing bunch grasses, defoliation usually removes the hormonal dominancy of reproductive tillers and that stimulates bud break and production of the relatively leafier vegetative tillers [4,5]. The rate at which plants may repair the damaged tissues and restore their lost photosynthetic capacities is dependent on the proportion of functional leaf area retained and timing and the time between successive defoliation events [3,6–8].

Depending on the management objective(s), mainly, biomass production, soil conservation and/or ecosystem services, relevant responses may include changes in tiller numbers, stand density, vegetation structure, species composition and spatial distribution, and growth performance amongst others. Proper defoliation management of NWSG stands must allow enough time for recovery growth and also minimize excessive losses of or damages to growing points, which happen to be high above the ground and mostly beyond the common cutting heights [3,9,10]. As an adaptation to grazing, regrowth of defoliated NWSGs usually involves apportioning more resources towards faster regrowth and repair or even compensate for sustained physical damages or loss of meristematic tissues [11]. However, species morphological adaptations to tissue damage differ and are reflected in a plants' tolerance to defoliation and respective growth responses [12–14]. The degree to which plants may compensate for tissue damages or loss is also influenced by local factors such as; precipitation, defoliation intensity and history [15–18]. Therefore, appropriate defoliation management of mixed stands must also consider how the most dominant species may respond to stresses induced under the prevailing weather conditions.

Where grassland management is interested more in ecosystem services, monitoring the stand responses to defoliation need to focus on changes in vegetation structure, species composition and the proportion of ground covered by plant biomass. Vegetation structural changes include the Canopy architecture, which affects the physiological functions of forage plants, the forage quality on offer to grazing animals [19], such processes as photosynthesis, transpiration, cell enlargement, and species competition in a plant community, root growth, soil moisture retention, soil health and residue decomposition [20]. For a plant community, canopy structure is the spatial arrangement of the above-ground organs in it, which includes leaves and other photosynthetic organs as well as the stems and branches for their support and strategic positioning. Commonly used parameters for monitoring changes in vegetation structure include canopy and basal diameters [21], sward-height that is mostly preferred in the management of grazing system due to its strong influence on bite dimensions [22] and canopy light interception, which varies positively with leafiness and forage quality [23]. Canopy structure has indirect influence on soil moisture and temperature, affects root growth, soil moisture retention, biomass residue composition, and other soil microbial processes [20,24]. These canopy structural attributes have implications on species composition, ground-level air temperatures and suitability of the stand for wildlife habitat. In a grazing system, canopy architecture affects both the physiological functions of forage plants and their forage quality [19].

Another parameter is visual obstruction, which is a non-destructive tool for estimating herbaceous standing crop in tallgrass prairie [25]. As a technique for monitoring standing crop and vegetation height and density on grasslands, visual obstruction, which integrates plant-height and density, is simple, cost effective, and provides pertinent information for both livestock and wildlife management purposes [26]. In plant communities, these structural changes have impact on species composition whose functional traits influence the efficiency of resource use by the biological system [27]. The resulting spatial and temporal changes in sward structure also have implications on their wildlife habitat qualities for different types of grassland birds and small mammals. For ground-nesting birds, the most habitat quality features include visibility while foraging, vigilance to predators, concealment, and easy mobility through the stands [28,29]. While a second-year switching of harvest frequencies between the single- and three-cut NWSG stands have exhibited compensatory forage yield responses [30], information on its associated impact on wildlife habitat quality is not clear. This study, therefore, was established to generate data that may help in developing sustainable defoliation management strategies for young NWSG stands for dual use summer forage production and wildlife habitat. The study focused on NWSGs' sward height, canopy closure, stand density, and ground cover attribute responses to a second year change in harvest frequencies. It was hypothesized that reducing the number of harvests per year would result in faster regrowth rates enough to compensate for prior-year losses in stand vigor that will also reflect in vegetation structure.

2. Materials and Methods

2.1. Location and Field Preparations

The study was conducted at Randolph Farm–Virginia State University's research and demonstration farm located in Chesterfield county, Virginia at 37°13′43″ N; 77°26′22″ W, about 45 m above sea level. The soil at the farm is Bourne series fine sandy loam (mixed, semiactive, thermic Typic Fragiudults) with low organic matter content. No fertilizers were applied to the NWSGs for the duration of this study. By the summer of 2013, the study area had a 20-year June, July, and August average precipitation of 92, 113, and 121 mm with day temperatures of 30.2, 32.1, and 31.2 °C, respectively [31]. The mean monthly temperatures and precipitation amounts around the study area from April through October for the 2012 through 2018 production years are shown in Figures 1a and 1b, respectively. In 2013, 64 plots (roughly 6-m W × 7-m L) of year-old transplanted BB, GG, IG, and SG stands separated by ≥120-cm alleys received a first cut in early-August and second one in mid-November to suppress annual weeds. The seedlings were raised in the greenhouse from seeds, U.S. ecotype for GG and NC ecotypes for the others, supplied by Ernst Conservation Seeds Inc., Meadville, PA, USA. At planting, the seedlings were spaced 30 × 45 cm within and between rows, respectively. The plots were in eight 8-plot rows, in which each NWSG species was assigned two, in a randomized complete block design. Starting early-June of 2014, three 1.5-m wide side-by-side strips in each plot were cut once-, twice-, or thrice per year (harvest frequencies). After four consecutive forage harvesting years, a fifth-year single mid-summer harvest followed on 26 June 2018. Harvest dates for the entire study are summarized in Table 1, by year and treatment. A CIBUS F Plot Forage Harvester (Wintersteiger Ag, Dimmelstrasse, Austria) with a 120-cm cutting width cutting height set at 18-cm was used. During the first through fourth year, harvesting for the three-cut systems happened in early-June, late-July to early-Aug, and late-September to mid-October. In most cases, the same first and last harvest dates were used for the two-cut systems except when weather and logistical problems impacted operations. The last harvest date for the three- and two-cut systems was also basically the same for the one-cut per year system. In the second harvest-yearn, however, the harvest frequencies were flipped once for the single-

and three-cut strips in plots 33–64, but not in the other 32 plots (Figure 2). The flipping of harvest frequencies was never reverted throughout the study. Having the flipped and not flipped plots in separate blocks was necessary to avoid potential shading effect of single-cut strips in one row on the three-cut strips of a neighboring row. For the same reason, the middle alley separating the two blocks was also made 60-cm wider than the 120-cm alleys separating plots, within blocks. To facilitate machine operations on different scheduled harvest dates and to limit associated disturbance to the harvest strips, end-to-end aligned strips within each 8-plot row were assigned the same harvest frequency.

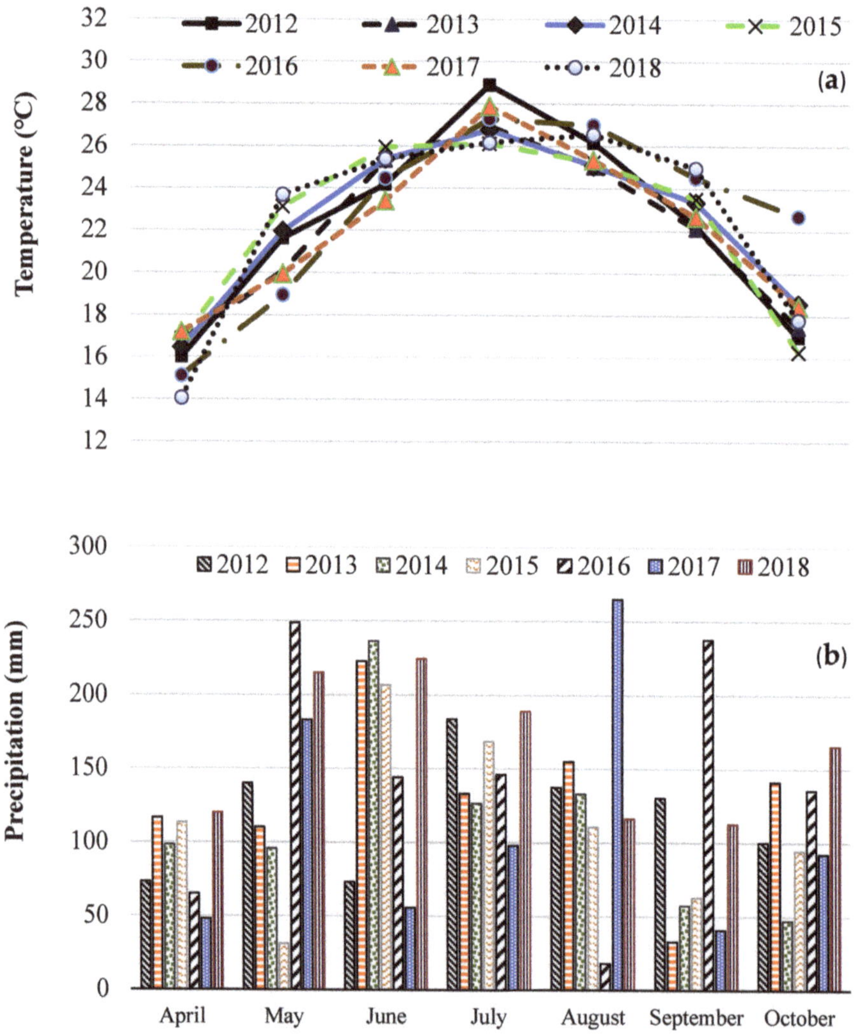

Figure 1. Mean monthly temperatures (**a**), top, and precipitation (**b**), below, recorded at nearby National Weather Service Stations (Hopewell and Petersburg, VA, USA) from April through October during the 2012–2018 production years. Adopted from Temu, et. al. (2022) [29].

Table 1. Actual harvest dates for three different cutting frequencies (cuts year^{-1}) on newly established NWSG stands recorded from 2013 to 2018.

		Harvest Dates by Harvest Regime		
Year	**Cuts**	**Three Cuts**	**Two Cuts**	**Single Cut**
2013 *	1st 2nd	5 June 2013 18 November 2013	5 June 2013 18 November 2013	5 June 2013 18 November 2013
1st (2014)	1st 2nd 3rd	24 June 2014 29 July 2014 12 September 2014	23 July 2014 12 September 2014	14 September 2014
2nd (2015)	1st 2nd 3rd	14 June 2015 31 July 2015 14 October 2015	18 June 2015 14 October 2015	14 October 2015
3rd (2016)	1st 2nd 3rd	18 June 2016 05 August 2016 24 October 2016	18 June 2016 24October 2016	24 October 2016
4th (2017)	1st 2nd 3rd	19 June 2017 17 August 2017 17 October 2017	29 June 2017 18 October 2017	18 October 2017
5th (2018) *	N/A	26 June 2018	26 June 2018	26 June 2018

* All plots experienced the same pre- and post-treatment harvests in 2013 and 2018, respectively; N/A = Not applicable.

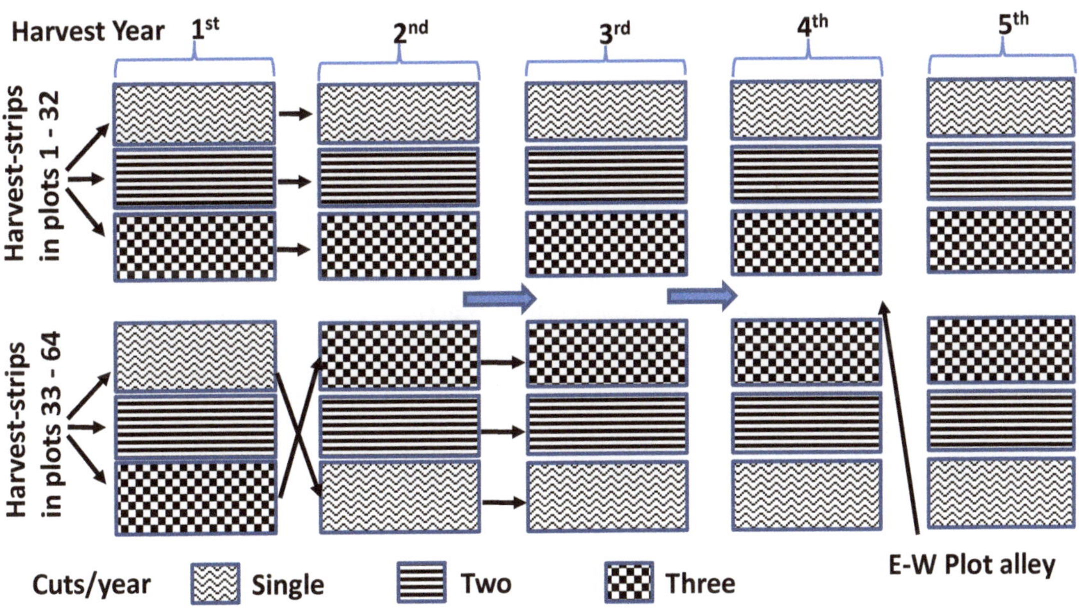

Figure 2. Relative arrangement of harvest-strips within a plot assigned to one, two, or three cuts/year, with harvest regimes being flipped between the single- and three-cut strips after the 1st year–2014, for plots 32–64 and all plots receiving a single late-June harvest in the 5th year–2018, adopted from Temu, et. al. (2022) [29].

2.2. *Vegetation Measurements*

A day or two before the first and the last harvest of the year, four early-summer and late-fall sward height (cm) measurements were recorded from every harvest strip at 60-cm intervals. A sward height reading was recorded as the highest point above ground at which a meter stick, held horizontally above the sward and perpendicular to a vertical Robel pole, touched at least two native grass leaves on separate rows. For the data analysis, the sward-height readings were entered as four-point averages. Around mid-April of the 2015 and 2016 harvest years, NWSG early-spring basal diameter (BD) measurements at about 2.5 cm above soil surface (crown's widest cross-sectional distance, cm) of three inner-row stubbles $\geq$ 1-m apart were recorded from each harvest strip. About a month later, three respective canopy diameter (CD) measurements (widest horizontal distance between canopy edges) of inner-row clumps perpendicular to the plot length orientation, spaced $\geq$ 1-m apart, were also recorded. From the measured BD and CD values, a CD:BD ratio was calculated. Due to gradual merging of crowns, within rows, individual BD and CD readings for the fourth and fifth harvest years were considered misleading in assessing response to the second-year flipping of harvest frequencies.

The treatment effects assessment was also done on canopy closures in the regrowth stands, during both 2015 and 2016. For that, early-spring instantaneous photosynthetically active solar radiation (PAR) intercepted by the vegetation layer in each harvest-strip was recorded between the 12:00 and 14:00 h. The day-time for the PAR readings was intended to minimize likely distortions that the recorder's shadow or tall plants in neighboring strips might have on the actual proportions of the intercepted solar radiation. For each PAR reading above the canopy (PARa), five matching readings, $\geq$1-m apart, of that reaching the ground surface beneath (PARb) were also recorded. From the PAR readings, average light interception was calculated as = $\sum$ [(PARa $-$ PARb)/PARa] $\times$ 100/5, [32]. The PAR measurements, µmol m^{-2} s^{-1}, were taken using the LI-191 Line Quantum Sensor (LI-COR 2000, LICOR, Lincoln, NE, USA).

During the 2016 and 2017 growing seasons, Visual Obstruction height (VOH) measurements as indicators of how the second harvest-year change in harvest frequencies may have influenced the subsequent stand density, that also has implications on visibility-related wildlife habitat qualities, was also done. As a technique, VOH is simple, cost effective, and provides pertinent information to both livestock and wildlife management purposes [26]. The VOH method is also considered an effective, non-destructive tool for estimating herbaceous standing crop in tallgrass prairies [25]. In the current study, VOH was recorded as the height at which a naked eye on the opposite end of the 6-m long harvest-strip could sight the lowest unobstructed mark on a graduated pole through the stand, when viewed at about a meter above the ground. Concurrently, canopy height measurements were also recorded using a modified Robel pole and a meter rule.

2.3. Data Analysis

The data were organized and subjected to analysis of variance (ANOVA) as a RCBD with, the year of assessment, defoliation management (system), species, and harvest frequency (cuts) as fixed effects. For the analysis, the windows-based SAS software 9.4 (SAS Institute Inc., Cary, NC, USA) was used. Because of significant year, species and treatment interactions, multiple ANOVA procedures were done for each year separately to compare treatments within species and harvest regimes. The ANOVA first compared the single-, two-, and three-cut systems on their sward structure attributes within year, species and harvest regimes (flipped or same). Then, a separate ANOVA compared the harvest regimes within year, species and harvest frequency. Respectively, the probabilities of difference from each run were used for the means comparison within year, species and harvest frequency or harvest regime. Means were compared by the Fisher's Least Significant Difference test at α = 0.05.

3. Results and Discussion

The ANOVA results showed highly significant ($p < 0.001$) main effects of year, species and harvest frequency as well as their two- and three-way interactions on the measured sward heights (Table 2).

Table 2. The ANOVA F and p values for the main and interaction effects of Year, harvest regime (System), Species, and harvest frequency (Cuts) on early- and late-season sward heights of young native warm-season grass stands harvested once, twice, and thrice year^{-1} recorded from 2014 to 2018.

Source	Early-Season Heights			Late-Season Heights		
	DF	Fα	p > Fα	DF	Fα	p > Fα
Model	119	49.63	<0.001	95	300.75	<0.001
Year	4	133.37	<0.001	3	452.26	<0.001
System	1	9.60	0.002	1	0.55	0.458
Year × System	4	13.37	<0.001	3	35.95	<0.001
Species	3	727.32	<0.001	3	1562.85	<0.001
Year × Species	12	98.54	<0.001	9	24.93	<0.001
System × Species	3	6.93	0.001	3	17.14	<0.001
Year × System × Species	12	8.41	<0.001	9	3.83	<0.001
Cuts	2	551.60	<0.001	2	8889.27	<0.001
Year × Cuts	8	38.16	<0.001	6	215.31	<0.001
System × Cuts	2	13.41	<0.001	2	26.11	<0.001
Year × System × Cuts	8	2.21	0.025	6	1.38	0.218
Species × Cuts	6	32.83	<0.001	6	338.78	<0.001
Year × Species × Cuts	24	6.56	<0.001	18	46.20	<0.001
System × Species × Cuts	6	0.71	0.641	6	5.83	<0.001
Year × System × Species × Cuts	24	0.49	0.982	18	4.27	<0.001
Error	840			672		
Corrected Total	959			767		

DF = degrees of freedom; p > Fα = probability of difference between means within species.

As well, ANOVA for the two- and three-cut systems showed highly significant main effects of year, species, harvest frequency, and harvest timing on light interception, visual obstruction, as well as basal and canopy diameters (Table 3). The interaction effects, however, were equally strong for some attributes but not observed on others.

Table 3. The ANOVA p values for the main and interaction effects of Year, harvest regime (System), Species, and harvest frequency (Cuts) on the percent light interception (PARi), basal diameter (CD), canopy diameter (BD) readings and CD:BD ratio recorded in May 2015 and 2016, and visual obstruction heights (VOH) recorded in October 2016 and 2017, from young native warm-season grass stands harvested once, twice, and thrice year^{-1}.

Source	DF	p > Fα				
		PARi	VOH	BD	CD	CBDR
Model	47	<0.001	<0.001	<0.001	<0.001	<0.001
Year	1	<0.001	0.003	<0.001	<0.001	<0.001
System	3	<0.001	<0.001	<0.001	0.001	0.001
Year × System	3	<0.001	0.153	0.049	<0.001	<0.001
Species	2	<0.001	<0.001	<0.001	<0.001	<0.001
Year × Species	2	0.160	<0.001	<0.001	<0.001	<0.001
System × Species	6	<0.001	<0.001	0.937	0.575	0.151
Year × System × Species	6	0.229	0.081	<0.059	0.001	0.002
Cuts	1	<0.001	0.607	0.096	<0.001	<0.001
Year × Cuts	1	<0.001	0.107	0.154	<0.001	0.004
Manage × Cuts	3	<0.001	0.058	0.370	0.575	0.454
Year × System × Cuts	3	0.020	0.894	0.699	0.357	0.832
Species × Cuts	2	0.014	0.277	0.041	<0.001	0.041
Year × Species × Cuts	2	0.204	0.031	<0.001	0.022	0.374
System × Species × Cuts	6	0.643	0.923	0.109	0.229	0.038
Year × System × Species × Cuts	6	0.927	0.249	0.840	0.715	0.637
Error	336					
Corrected Total	383					

DF = degrees of freedom; p > Fα = probability of difference between means within species.

Because of significant factor interactions, treatment means within-species and for each harvest regime are presented, separately, for each year.

3.1. Early-Summer Sward-Heights

During the 2014 growing season, the newly established NWSG stands, which had already received two common cuts in July and November 2013, experienced a second harvest year. As expected, the early-summer sward heights, which ranged from about 42 cm (BB) to 65 cm (IG), Table 4, reflected differences in the species growth responses to the common defoliation management during stand establishment. The observed sward-height differences were also consistent with two reported mechanisms used by plants to cope with herbivory; the ability to reduce their probability of being grazed or the increase in ability to recover following grazing [33]. It is also reported that a plant's response to defoliation management is related mainly to the species morphological and physiological characteristics [34]. However, the 2015 early-season sward-height records, that followed the first season under the single- two- and three-cut year^{-1} harvest regime showed significant defoliation treatment effects ($p < 0.001$) (Figure 3). Within species, all strips harvested thrice the previous year (2014) had the shortest sward-heights, in 2015 that ranged from about 36.4 cm in SG to an average of 55.9 cm in GG plots.

Figure 3. Top; Aerial view of 64 native warm-season grass plots showing relative arrangement of single-, two-, and three-cut strips (X1, X2, X3) in the block harvested the same consecutively and the one where the single- and three-cut strips were flipped during the second year. Bottom; early-summer regrowth sward-height measurement in the side-by-side harvest strips.

Table 4. Effects of a seasonal change (Same vs. Flipped) in harvest frequency (Cuts) on subsequent early-summer and late-fall † sward heights of young NWSG stands cut once, twice or thrice year^{-1} recorded from September 2014 to October 2018.

Year	Cuts	Species and Harvest Regime							
		Big Bluestem		Gamagrass		Indiangrass		Switchgrass	
		Same	Flipped	Same	Flipped	Same	Flipped	Same	Flipped
		Early-summer Sward Heights							
		----- cm -----							
2014	Twice	43.3 A§	42.4 A§	53.1 A	56.2 A	65.4 A	63.8 A	53.7 A	50.8 A
	Thrice	-	-	-	-	-	-	-	-
2015	Once	62.4 aA	62.4 aA	77.9 aA	75.1 aA	44.8 aA	45.4 aA	66.1 aA	67.2 aA
	Twice	52.3 bA	52.2 bA	69.8 bA	63.0 bB	44.6 aA	43.1 aA	42.4 bB	45.3 bA
	Thrice	45.7 cA	45.1 cA	55.9 cA	57.3 cA	39.1 bA	38.3 bA	36.4 cB	39.9 cA
	$p > \alpha$ #	<0.001	<0.001	<0.001	<0.001	<0.001	<0.001	<0.001	<0.001
2016	Once	67.6 aA	65.1 aA	82.3 aA	80.8 aA	47.9 aA	49.4 aA	73.2 aA	67.5 aB
	Twice	53.5 bA	57.1 bA	69.0 bA	72.0 bA	39.6 bB	44.7 bA	58.3 bB	60.5 bA
	Thrice	44.7 cB	52.2 bA	59.2 cA	62.9 cA	33.1 cA	35.8 cA	44.9 cA	51.4 cA
	$p > \alpha$	<0.001	<0.001	<0.001	<0.001	<0.001	<0.001	<0.001	<0.001
2017	Once	60.1 aA	56.6 aA	75.4 aA	71.2 aA	39.2 aA	38.0 aA	71.2 aA	65.4 aB
	Twice	46.5 bA	45.4 bA	68.0 bA	66.6 aA	34.4 bA	35.6 aA	60.5 bA	61.6 bA
	Thrice	41.9 bA	40.4 cA	59.4 cA	59.1 bA	36.1a bA	37.7 aA	45.7cB	52.2 cA
	$p > \alpha$	<0.001	<0.001	<0.001	<0.001	0.027	0.289	<0.001	<0.001
2018	Once	86.0 aA	80.9 aA	87.5 aA	84.2 aA	46.6 aA	39.9 bB	76.4 aA	73.2 aA
	Twice	61.8 bA	58.8 bA	74.6 bA	75.3a bA	44.7a bA	41.9a bA	61.8 bA	61.2 bA
	Thrice	52.7 cA	56.0 bA	67.7 bA	70.4 bA	42.3 bA	44.5 aA	46.0 cA	50.1 cA
	$p > \alpha$	<0.001	<0.001	<0.001	<0.001	0.022	0.018	<0.001	<0.001
		Late-fall Sward Heights							
2014	Once	171.5 aA	159.6 aA	70.9 aB	96.3 aA	197.9 aA	173.1aB	194.4 aA	174.1 aB
	Twice	69.8 bA	61.1 bA	64.5 aA	65.5 bA	69.2 bA	63.5 bA	91.0 bA	95.4 bA
	Thrice	47.4 cA	49.9 bA	49.9 bB	57.3 bA	66.2 bA	58.5 bA	85.6 bA	86.2 cA
	$p > \alpha$	<0.001	<0.001	<0.001	<0.001	<0.001	<0.001	<0.001	<0.001
2015	Once	188.9 aA	165.8 aB	111.6 aA	103.9 aB	187.2 aA	194.2 aA	206.1 aA	185.0 aB
	Twice	165.0 bA	145.5 bB	82.7 bA	87.2 bA	157.9 bA	160.0 bA	152.2 bB	158.9 bA
	Thrice	30.1 cB	40.1 cA	45.3 cB	50.9 cA	96.5 cA	90.2 cA	57.5 cB	66.7 cA
	$p > \alpha$	<0.001	<0.001	<0.001	<0.001	<0.001	<0.001	<0.001	<0.001
2016	Once	183.7 aA	182.6 aA	104.0 aB	118.1 aA	183.4aB	194.6 aA	208.5 aA	205.7 aA
	Twice	120.7 bB	141.3 bA	86.2 bB	108.0 bA	156.5 bA	164.2 bA	150.8 bA	155.6 bA
	Thrice	29.4 cB	41.3 cA	50.5cB	69.9 cA	92.6 cA	110.1 cA	54.7cB	76.3 cA
	$p > \alpha$	<0.001	<0.001	<0.001	<0.001	<0.001	<0.001	<0.001	<0.001
2017	Once	186.7 aA	167.9 aB	90.8 aA	91.0 aA	179.6 aA	168.7 aB	207.9 aA	207.6 aA
	Twice	89.8 bA	70.4 bB	69.4 bA	71.6 bA	148.1 bA	134.0 bB	122.7 bA	121.5 bA
	Thrice	32.9 cB	37.7 cA	41.5 cA	44.6 cA	42.3 cA	41.4 cA	45.6 cA	47.7 cA
	$p > \alpha$	<0.001	<0.001	<0.001	<0.001	<0.001	<0.001	<0.001	<0.001

NWSG = big bluestem—*Andropogon gerardii*, Gamagrass—*Tripsacum dactyloides*, Indiangrass—*Soghastrum nutans*, and Switchgrass—*Panicum virgatum*; † the height at which a horizontal meter stick held across a harvest strip touched at list three topmost leaves of the NWSG bunches; § Means of the same grass followed by the same lowercase letter within-, or uppercase letter between-columns are not statistically different at $\alpha = 0.05$; # Probability of mean difference between harvest frequencies, within a year. Note: There was no treatment effect on sward-height records for May 2014, the same year the harvest frequencies were first imposed.

The sward-heights for these three-cut strips were shorter than their respective single-cut values by about 14% for IG, 26.5% for BB and GG and up to 43% for SG. The observed shorter sward-heights for the three-cut strips was actually in agreement with reported aftermath reflection of severity and frequency of defoliation history in *Dactylis glomerata* stands [35]. Similar responses to defoliation are also reported from a grazing trial on elephant grass (*Pennisetum purpureum*) with regrowth sward-heights being shorter for the severely grazed stands compared to those grazed lightly [36]. As well, regrowth in the more severely grazed stands (70% defoliation) showed reduced leaf appearance, lower stem and leaf elongation rates, and higher proportions of sheaths than leaf lamina, unlike those defoliated at 50%. According to [37], leaf sheaths contribute less than 5% of canopy photosynthesis, which may explain the associated lower rates of leaf elongation in the more severely defoliated stands. In the respective two-cut strips, the swards were significantly taller ($p < 0.001$) than in the three- but shorter than the single-cuts. However, in the IG plots, differences between the two- and single-cut sward-heights were not significant ($p > 0.05$). This indicates that, for the two-cut forage harvest systems, the early-season growing conditions were good enough for the NWSG stands to compensate for differences in severity of the prior-year defoliation tissue-damages. The timing and frequency of prior-years defoliation often reflect in aftermath growth performance including DM yield, bud and tiller numbers for the NWSGs of North America [16,38]. Up to 60% greater herbage yields have been reported from unclipped wester wheatgrass (*Pascopyrum smithii*, Rydb.) compared to their counterparts subjected to multiple defoliations [39]. For the IG, however, the single- and two-cut strips had comparable sward-heights as the three-cuts in BB plots implying that its early-season resources were probably diverted to other growth components rather soon. That also suggests that its relatively more spread crowns were less prone to self-shading and, effectively, their tillers attained dependable photosynthetic capacities rather faster.

After the 2015 flipping of harvest frequencies between the single- and three-cut strips, the early-season sward heights recorded in summer of 2016, (Table 4), tended to be consistently greater in the three-cut strips that were previously cut once year^{-1} than those cut thrice all along, although the differences were only significant for BB. In the same assessment, sward-heights in the corresponding single-cut strips that flipped from three cuts year^{-1} were statistically similar to those cut only once each year, except in SG plots. Growth performance of the flipped single-cut strips, through to the 2018 assessment, also exhibited the ability to compensate for prior losses in growth vigor, except in IG plots (Figure 3). For IG, the corresponding early-summer sward height values only came close to, but never reached 50 cm. Such species differences in response to shared defoliation management are often attributable to respective morphological and physiological characteristics that can translate into different regrowth rates [34]. Similar differences in species compensatory dynamics, in a plant community, have been credited to increases in the abundance of one species with a decrease in another [27]. By the 2018 harvest-year and within species, the 2014 single-cut strips that in 2015 flipped into three cuts year^{-1} were statistically the same height as those harvested thrice all along ($p < 0.001$). However, numerically, the flipped strips maintained an edge over the ones that were not. The observed growth responses to the change in harvest regimes indicate that flipping the harvest frequencies can strategically induce conducive growth responses for both forage biomass and sward structures of managed NWSG stands. In fact, in both the two- and three-cut strips, the mean sward-heights exceeded the 30–40 cm minimum grass height requirement for good ground-nesting bird habitat [28]. The single- and two-cut strips will also provide good winter cover for a variety of wildlife, including small mammals. A practical flipping-of-harvest-regimes scenario could involve dividing a NWSG field into halves where one section is mowed thrice year^{-1}, for hay, and the other only once year^{-1} as feedstock in alternate years. During the summer, small mammals and grassland birds can forage and find cover in the three-cut stands while the single-cut year^{-1} section may also provide forage for herbivores such as the white-tailed deer (*Odocoileus virginianus*).

3.2. Late-Fall Sward-Heights

Also, in Table 4, are the results of a season-end assessment of the effect of changing harvest frequencies on the NWSG aftermath sward structures. Except in IG plots, the sward-heights were significantly taller ($p < 0.05$) in all three-cut strips that were previously cut once year^{-1} than those receiving three cuts year^{-1} since 2014. In 2015, the late-fall sward-heights were about 6-, 9-, and 10-cm taller for the flipped three-cut strips in GG, SG, and BB plots, respectively, than their non-flipped counterparts. The observed superiority in late-fall sward-heights of the flipped three-cut strips was actually clearer in the 2016 when values were from 27% greater for BB to 29% for both GG and SG, respectively. For the same species, the 2017 late-fall swards still tended to be taller in the flipped three-cut strips than in those cut thrice for three consecutive years, but only significantly so in BB plots ($p < 0.001$). Usually, multiple cuts per year remove apical dominance and may encourage tillering [40–42] and also reduce root growth and root reserves [42]. Repeated cutting is reported to reduce tiller structural carbohydrate contents, which has an effect on tiller growth vigor [43]. The greater height in the plants flipped from single-cut may be attributed to better root reserves that allowed more vibrant growth compare to those in a continuous three-cut year^{-1} regime. Also, there is a potential that the plants transitioning to a single-cut from a three-cut year^{-1} had fewer tillers hence reduced competition for resources that allowed them better overall growth.

Among the strips cut only once year^{-1}, the 2015 late-fall sward-heights were 8–23 cm taller in the strips that flipped from the three-cut year^{-1} regime. In fact, the relative increases in late-fall sward-heights between 2015 and 2016 also showed abilities for the NWSG stands to compensate for tissue-damages sustained during the three-cut year^{-1} harvest regime. For the strips that were maintained on the same harvest regime, the increases in mean late-fall sward-heights were only about 12 and 17 cm for SG and BB, respectively, and not more than 41 cm for GG. In comparison, the sward-heights for strips that flipped from three- to a single-cut year^{-1} regime equaled 2–3-fold increases from ~50 cm to ~86 cm in 2014 to the range 104–206 cm in 2015. However, a within-year comparison between the growth performances of the flipped and not flipped single-cut strips showed that, except for IG, a single year's rest is not enough for the flipping effect to close the respective sward-height differences. As evidenced here, following two rest-years, the 2016 late-fall sward-heights of all strips whose harvest frequencies were flipped from three- to single-cut year^{-1} in 2015 matched or exceeded their counterparts, significantly ($p < 0.05$). During the first year upon transitioning to one- from a three-cut per year, the plants had enough time to build carbohydrate reserves in their tiller buds and roots. Such reserves serve to promote prolific regrowth and sustained growth later as previously reported [43].

3.3. Basal Diameter

The assessment of growth responses of the NWSG stands to defoliation management was also based on crown/basal diameters (BD) recorded in early-spring whose results are summarized in Table 5. There was significant year × species × harvest regime interactions affecting the measured BDs and so the results are discussed separately, for each factor. As in Figure 4, the mean BD for each species were generally greater in 2016 than in 2015. Those harvested thrice year^{-1} had BDs with lower magnitudes and which were significant ($p = 0.05$), in most cases. This agrees with previous findings where more intensely grazed pastures showed reduced basal areas [44]. An earlier study on grass growth under grazing found that periodic heavy grazing during the growing season restricted basal-area growth [45]. In a bunch-grass clipping frequency trial with *Arrhenatherum elatius*, [21], data showed that the crown diameters of the less frequently defoliated plants were greater than those defoliated more frequently. Similarly, it is reported that grazing up to 80% forage biomass removal resulted in significant reduction in live basal cover [46]. Within species, the BD values tended to be greater in the strips whose harvest frequencies were flipped than those harvested the same, consecutively. However, the observed flipping effects were only significant among the three-cuts in SG plots (Table 5). Within a harvest regime, the

BD differences due to harvest frequencies were not in any consistent pattern and mostly numerical, except for the flipped strips in BB plots. The observed differences in species response to the harvest regimes agree with earlier reports [34] that attributed responses to species morphological and physiological characteristics.

Table 5. Effects of a seasonal change (Same vs. Flipped) in harvest regimes † on early-spring and -summer basal and canopy diameters and their ratios §, respectively, in native warm-season grass stands harvested once, twice or thrice in the first year with or without a frequency switch between the one- and three-cut strips during the second-year.

Year	Cuts	Species and Harvest Regime							
		Big Bluestem		Gamagrass		Indiangrass		Switchgrass	
		Same	Flipped	Same	Flipped	Same	Flipped	Same	Flipped
		Basal							
		-cm-							
2015	Once	21.7 a	26.2 a	19.1 aA	19.8 bA	19.4 aA	21.3 aA	18.6 aA	21.0 aA
	Twice	17.4 aA	18.5 bA	19.4 aB	29.8 aA	18.4 aA	22.7 aA	17.3 aA	18.8 aA
	Thrice	17.2 aA	19.6 bA	24.4 aA	28.7 aA	19.4 aA	21.8 aA	17.2 aB	21.1 aA
	$p > \alpha$	0.134	<0.001	0.257	<0.001	0.853	0.828	0.541	0.455
2016	Once	26.8 aA	27.6 bA	39.7 aA	36.9 aA	28.1 aA	30.2 aA	25.1 aA	28.1 aA
	Twice	28.2 aA	29.5 abA	37.4 aA	38.6 aA	30.2 aA	32.2 aA	27.5 aA	29.4 aA
	Thrice	28.3 aA	31.3 aA	37.9 aA	37.5 aA	32.2 aA	34.0 aA	26.4 aB	31.1 aA
	$p > \alpha$	0.718	0.052	0.783	0.727	0.231	0.157	0.244	0.169
		Canopy							
		-cm-							
2015	Once	46.3 aA	39.8 aA	27.7 bA	30.3 aA	20.1 bA	21.8 bA	29.1 aA	24.6 aB
	Twice	37.0 abA	32.5 bA	33.3 aA	34.2 aA	26.5 aA	25.1 bA	18.7 bA	19.5 bA
	Thrice	35.3 bA	30.6 bA	37.0 aA	31.7 aA	28.8 aA	30.1 aA	21.1 bA	24.2 aA
	$p > \alpha$	0.101	0.035	0.257	0.332	0.022	0.007	<0.001	0.007
2016	Once	57.1 aB	63.3 aA	63.7 abB	81.6 aA	52.8 bA	55.6 aA	55.2 aA	58.5 aA
	Twice	58.9 aA	60.2 aA	70.1 aA	78.1 aA	60.5 aA	58.0 aA	52.0 aB	59.3 aA
	Thrice	49.0 bA	53.1 bA	56.2 bB	70.0 bA	52.7 b	52.5 a	44.9 bB	50.1 bA
	$p > \alpha$	<0.001	0.002	0.009	<0.001	0.069	0.310	0.001	<0.001
		Canopy: Basal Diameter							
		Ration-							
2015	Once	2.40 aA	1.56 aB	1.49 aA	1.59 aA	1.09 bA	1.04 aA	1.61 aA	1.23 aB
	Twice	2.23 aA	1.78 aA	2.08 aA	1.17 bA	1.48 aA	1.19 aA	1.10 bA	1.09 aA
	Thrice	2.14 aA	1.57 aB	1.59 aA	1.11 bB	1.50 aA	1.45 aA	1.24 bA	1.16 aA
	$p > \alpha$	0.858	0.408	0.316	0.005	0.038	0.141	0.008	0.655
2016	Once	2.14 aA	2.31 aA	1.63 abB	2.23 aA	1.93 aA	1.85 aA	2.22 aA	2.10 aA
	Twice	2.10 aA	2.05 aA	1.90 aA	2.05 abA	2.05 aA	1.84 aA	1.91 bA	2.04 aA
	Thrice	1.80 bA	1.72 bA	1.52 bB	1.88 bA	1.67 aA	1.55 aA	1.71 bA	1.62 bA
	$p > \alpha$	0.054	0.001	0.045	0.021	0.163	0.112	0.007	<0.001

NWSG = big bluestem—*Andropogon gerardii*, Gamagrass—*Tripsacum dactyloides*, Indiangrass—*Soghastrum nutans*, and Switchgrass—*Panicum virgatum*; † harvesting once, twice or thrice a year with or without a second-year switch between the single and three cuts assignments. § horizontal distance between opposite outer edges of the clump base or its canopy of the same NWSG bunch, average of three each. The same lowercase letter within-, or uppercase letter between-columns are not statistically different at $\alpha = 0.05$

3.4. Canopy Diameter

To assess how defoliation management might be reflected in vegetation structural responses, changes in species CD values is also helpful. In the current study, the early-summer CD values showed significant year × species × harvest regime interactions and so the results are presented separately by year, species and harvest regimes (Table 5). In most cases, the NWSG clumps cut once year^{-1} had greater CD values than their counterparts cut twice or thrice year^{-1}. A similar observation with 20% greater CD in ungrazed bunchgrasses compared with their grazed counterparts is reported [47]. In the current study, each species also registered significantly greater CD values in 2016 than in 2015 (Figure 4). This observation might be linked to progressive growth performance among the grasses as the stand matures. From the 2015 data, differences due to prior-year harvest frequencies were only significant in SG plots, among the single-cut strips, with those previously cut once year^{-1} performing better than their three-cut counterparts. In 2016, however, plants in the strips that flipped from three- to a single-cut year^{-1}, exhibited significantly greater CDs in BB and GG, but not IG or SG plots. Among the strips cut thrice year^{-1}, the same was true in GG and SG, but not BB or IG plots. Similar differences in species characteristics and physiological and morphological responses to frequency and intensity of defoliation are reported [34].

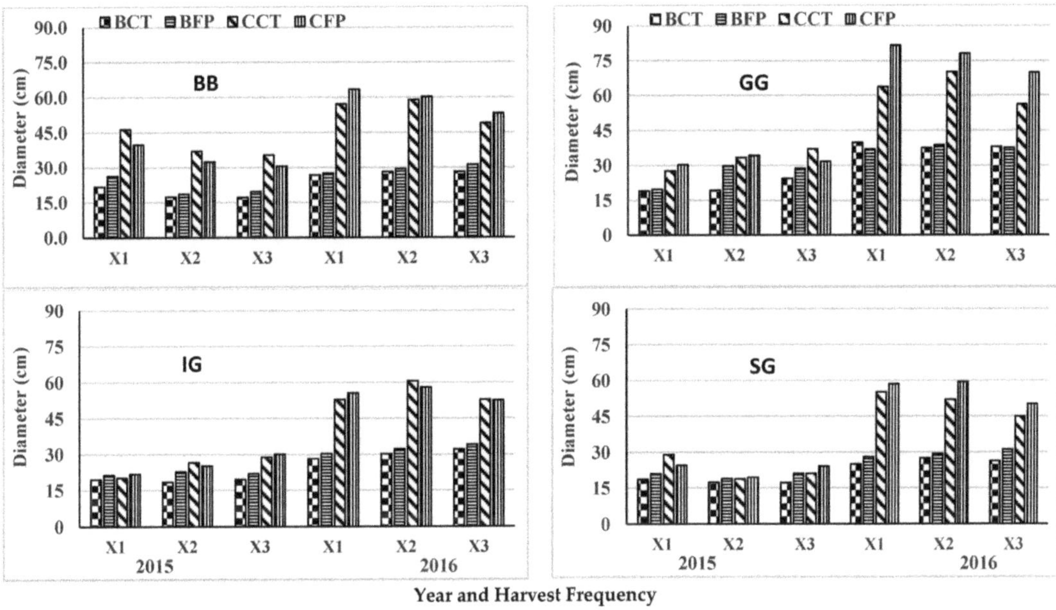

Figure 4. Early-spring basal- (B) and canopy (C) diameters of native warm-season grass [(BB) big bluestem—*Andropogon gerardii*, (GG) Gamagrass—*Tripsacum dactyloides*, (IG) Indiangrass—*Soghastrum nutans*, and (SG) Switchgrass—*Panicum virgatum*] clumps as affected by being harvested once—X1, twice—X2, or thrice—X3 year^{-1}, continuously (CT) or with the harvest regimes for the X1 and X3 harvest strips flipped during the second year as recorded in 2015 and 2016. Note: The 32 plots that did not have harvest regimes flipped were in a separate block from the other 32 that did.

Although the recorded 2016 mean CD values were consistently greater than in 2015, the data also showed species differences due to the flipping of harvest frequencies (Figure 4). Unlike the BD values, the 2015 CDs differed significantly between the single- and three-cut strips, except in the GG plots where cut frequencies flipped. Also, it is worth noting the species difference in the recorded CD, and that, in 2015, IG bunches in the single-cut strips had smaller CDs than those cut thrice year^{-1}. Significant CD differences between the single- and three-cut strips were also observed in 2016 except in the IG plots in which harvest frequencies were flipped. Stems under the one-cut year^{-1} regime were taller and more outward leaning plus, though not determined, had leaves that appeared longer and broader. These may have contributed to the observed differences in CD between harvest regimes. That is so because swards submitted to more severe grazing often have higher proportions of sheaths than leaves lamina while, in pastures with 50% defoliation, higher leaf-area and leaf elongation rates are more likely [37]. Others studies assessing defoliation effects on range species have reported 60% higher herbage yields by unclipped western wheat grass (*Pascopyrum smithii*, Rydb.) than their counterparts receiving multiple defoliations [39].

3.5. Canopy-to-Basal Diameter Ratios

While changes in BD, in young NWSG stands, indicate dynamics in tiller numbers, the regrowth CDs may provide information on how the former relates to species proportional ground cover between harvest events. The statistical analysis of the within-treatment CD:BD ratios showed species and year differences ($p < 0.004$) in proportional CD and BD responses to the changes in harvest regimes (Table 5). In 2015, only the BB clumps in the strips harvested the same, consistently, had canopies that were twice as wide as their crowns. However, in 2016, that was true for all single-cut BB and GG strips and their corresponding SG strips with prior three-cut year^{-1} experience. Also, in 2015, and within species, canopies in the single-cut BB and SG clumps outsized their crowns in the strips harvested the same all along than those flipped from three-cuts. Still, in 2016, differences among the single-cut strips were only significant in GG plots where clumps in the strips that were previously cut thrice year^{-1} had greater CD:BD ratios than those cut once only. Besides, year differences in growing conditions and timing of operations, the observed similarity in CD:BD ratios may be attributable to compensatory NWSG growth responses in the strips with prior multiple defoliation experience. Similar increases in CDs for less frequently defoliated bunch grasses are reported [21].

On the current data (Table 5), the higher CD:BD ratios also indicate preferential resource allocation for the expansion and elongation of residual leaf and stem tissues relative to the growth of new tillers. The dissimilarities in CD:BD ratios are attributable to species differences in compensatory growth responses to the change in defoliation regimes. That often involves changes in respiratory rates, growth rates, and carbon allocation patterns [3]. Specific responses to defoliation regimes are rooted in morphological adaptations to physical tissue-damages, which increase plants' tolerance to defoliation [12,13,15]. By large, these species growth responses to defoliation, strategically seek to repair the tissue-damages, and restore lost physiological functions. That often involves increased tiller density in the more frequently harvested stands and reduced vegetative growth on the less frequently harvested ones [17,38,48]. Depending on the severity of sustained injuries, recovering NWSGs usually exhibit preferential allocation of energy reserves to replace lost leaf area and/or increase photosynthesis rates on the residual or regrowth tissues [17,48].

Year and species differences in CB:CD ratios were also observed among the three-cut strips. In 2015, the CB:CD ratios tended to be greater in the strips harvested thrice every year than those previously harvested only once year^{-1}. However, significant differences were only present in the BB and GG plots. In 2016, however, the BD:CD difference was only significant in GG plots where the flipped previous single-cut strips outnumbered those harvested thrice year^{-1}, consistently. Species morphological differences may explain the demonstrated ability for GG, whose stems are more prostrate oriented and with larger proportion of growing points more likely to escape defoliation physical tissue-damages, to be less impacted by frequent harvesting. This agrees with previous reports on insensitivity of GG to frequent defoliation [49]. The portrayed species differences in compensatory growth potentials underscore the importance of species-specific defoliation management.

3.6. Sward Structure

The effects of flipping the harvest frequencies on subsequent NWSG performance were also assessed based on changes in sward structural features associated with canopy light interception and stand density. The proportions of instantaneous PAR above the canopy intercepted by the vegetation layer, in 2015 and 2016, are summarized by harvest regimes, within species (Table 6). In the same table, mean VOH, within-species and for the same harvest regime are presented.

Table 6. Effects of a seasonal change (Same vs. Flipped) in harvest frequency (Cuts) on subsequent early-season canopy closure recorded in May 2015 and 2016, and late-season stand density of young NWSG stands based on respective light interceptions † and visual obstruction heights § recorded in October 2016 and 2017.

Year	Cuts	Species and Harvest Regime							
		Big Bluestem		Gamagrass		Indiangrass		Switchgrass	
		Same	Flipped	Same	Flipped	Same	Flipped	Same	Flipped
		Light Interception							
		-%-							
2015	Once	93.5 aA *	93.0 aA	79.6 aB	89.0 aA	76.7 aA	82.0 aA	79.2 aA	84.4 aA
	Twice	86.2 bA	83.6 bA	66.9 bB	80.8 bA	67.0 bA	73.0 bA	48.1 bB	62.1 bA
	Thrice	81.2 bA	79.4 bA	58.2 cB	73.9 cA	55.5 bA	62.0 cA	38.9 cB	55.4 bA
	$p > \alpha$ #	<0.001	0.011	0.007	<0.001	0.003	<0.001	<0.001	<0.001
2016	Once	94.8 aA	93.1 aA	89.8 aA	88.7 aA	85.9 aA	89.9 aA	93.3 aA	92.5 aA
	Twice	85.5 bA	81.6 bA	75.3 bA	77.0 bA	80.3 bA	78.4 bA	74.3 bA	71.2 bA
	Thrice	77.2 cA	81.6 bA	65.4 cB	75.9 bA	74.3 cA	78.3 bA	62.2 cA	68.0 bA
	$p > \alpha$	<0.001	0.009	<0.001	<0.001	<0.001	0.002	<0.001	<0.001
		Visual Obstruction Height							
		-cm-							
2016	Once	186.3 aA	169.7 aA	81.4 aA	78.7 aA	181.9 aA	172.9 aA	208.7 aA	210.0 aA
	Twice	73.5 bA	85.0 bA	69.4 bA	71.2 aA	111.4 bA	132.8 bA	136.5 bA	148.9 bA
	Thrice	21.3 cA	18.0 cB	35.3 cA	35.0 bA	29.6 cA	28.9 cA	32.7 cB	41.4 cA
	$p > \alpha$	<0.001	<0.001	<0.001	<0.001	<0.001	<0.001	<0.001	<0.001
2017	Once	176.0 aA	172.3 aA	93.1 aA	72.0 aA	166.1 aA	173.7 aA	205.4 aA	210.0 aA
	Twice	73.7 bA	56.6 bB	67.7 aA	58.5 bA	120.1 bA	110.4 bA	102.1 bA	110.1 bA
	Thrice	29.9 cA	26.9 cA	34.2 bA	35.3 cA	35.1 cA	28.1 cB	34.5 cA	39.4 cA
	$p > \alpha$	<0.001	<0.001	0.002	<0.001	<0.001	<0.001	<0.001	<0.001

NWSG = big bluestem—*Andropogon gerardii*, Gamagrass—*Tripsacum dactyloides*, Indiangrass—*Soghastrum nutans*, and Switchgrass—*Panicum virgatum*; † The proportions of photosynthetically active radiation measured above the canopy that did not reach the ground vertically below the canopy; § The height at which a white card held against a Robel pole was invisible, through the vegetation, to the naked eye positioned about a meter above ground; * Means of the same grass followed by the same lowercase letter within-, or uppercase letter between-columns are not statistically different at α = 0.05. # Probability of mean difference between harvest frequencies within a year.

3.6.1. Canopy Light Interception

As expected, light interception values recorded in 2015 were greater for the single cuts than their respective two- and three-cuts year^{-1} (Table 6). The fact that the percentage PAR interception in all flipped single-cut strips either matched or slightly exceeded that of their non-flipped counterparts is an indication of the positive impact of allowing plants long recovery rests to repair their damaged tissues. Species differences in the ability to restore the lost photosynthetic surface may explain the significantly greater values, nearly 10-points margin, observed among the single-cuts in the GG plots. Likewise, for GG and SG, the percentage PAR interception, among the three-cut strips, were statistically greater for the strips that had previously received a single-cut year^{-1} than those on a second three-cut year $^{-1}$ cycle. Logically, the previous three cuts year^{-1} may have stimulated more tiller buds and, that lead to greater stem densities. That seems the most plausible scenario because, under such an open stand, the red/far-red ratio also reported to increase tiller/stem formation remains high [50]. And as the system transitions from a three- to a single-cut, the increase in tiller numbers due to previous defoliation management allowed for a more robust plant growth that resulted in an improved canopy structure and increased PAR interception. The observed changes can be attributed to the defoliation effects on canopy density that can profoundly influence the quality of radiation and the irradiant flux density received by the plants [51]. As roots recovered from effects of repeated defoliation in previous years, there was potentially greater root density and better uptake of soil water and nutrients, which allowed for fast growth. In similar patterns, the 2016 early-season values for PAR interception mostly outnumbered their respective 2015 records, but differences due to previous defoliation experiences were only significant among the three-cut GG strips. The observed similarities in canopy closure attributes among the single- or three-cut strips that experienced the same or flipped harvest regimes, as indicated by larger PAR values, may also have resulted from a gap-filling effect of annual weeds. The early-season stands often have high proportions of annual weeds, which also include tall-growing broadleaf ones occupying voids between the bunch grasses, as reported in [30]. With respect to wildlife habitat quality features, bunched grasses interspersed with legumes and forbs with insect-attracting wild flowers are among desirable attributes of a good habitat for bobwhites [29].

3.6.2. Stand Density

Mean sward structural response assessment of the NWSG stands to the seasonal changes in harvest frequencies was also based on season-end VOH measurements in 2016 and 2017. There was significant year × species × frequency interaction on the measured VOH (Table 3). The results are presented and discussed separately (Table 6) by year, species and harvest regimes. Among the single-cut strips and, both in 2016 and 2017, VOH within species showed no effect due to the flipping of harvest frequencies ($p > 0.05$). The 2016 values averaged 178, 80, 177, and 209 cm for BB, GG, IG and SG, respectively, with their corresponding 2017 averages as ~174, 82, 170, and 208 cm. As an indicator of stand density, the recorded VOH values also seemed more reflective of the NWSG stem densities and leafiness than the presence of annual weeds. That may also explain their notable closeness, in trend, to the respective late-fall sward-heights (Table 4). The data also suggests that, in both years, the single-cut strips that were previously cut thrice, somehow compensated for their losses in stand vigor and hence the lack of difference due to prior harvest regimes. This better performance of the flipped three-cut strips against their non-flipped counterparts are consistent with the asserted prior-year defoliation-induced increases in tiller buds coupled with greater buildup of energy reserves [30], as the frequency changes from three- to a single-cut year^{-1}. In plant communities, compensatory dynamics are often manifested in different ways and, may involve increases in the abundance of one species at the expense of another [27]. In more frequently harvested stands, compensatory growth responses often result in increases in tiller density and delayed preferential allocation of resources to root growth [17,48]. That will logically boost the regrowth stand density, as reflected in the recorded VOHs, in the current study.

In assessing the NWSG structural response to the changes in harvest regimes, the three cut VOH values were also compared (Table 6). During the 2016, clear three-cut VOH differences due to prior harvest frequencies were detected in the BB (21.3 vs. 18.0 cm) and SG (32.7 vs. 41.4 cm), but not the GG (35.1 cm) or IG (29.2 cm) data. While the BB strips flipped from a single-cut year^{-1} trailed their not-flipped counterparts in VOH, the opposite was true for SG. During the 2017, a flipping effect, among the three-cut strips was only significant in the IG (35.1 vs. 28.1 cm) plots. For the 2017 data, the three-cut VOH values were 3- and 5-cm greater in the not-flipped BB and the flipped SG strips, respectively. However, while these differences observed for BB and SG were insignificant, the 7-cm difference for IG was significant. Largely, the dissimilarities in the recovery stand density originate from differences among species in the susceptibility of shoot growing points to physical damages and how they, preferentially, allocate resources to new tillers vs regrowth on residual tissues. Likewise, yearly differences in growing conditions can impact the proportions of annual weeds in the stands. These differential structural responses to defoliation management have implications on such ecosystem services as ground cover, wildlife habitat quality—food and shelter, and airflow through the stands. Depending on the species composition and the desired season-end ecosystem services, appropriate strategies on the frequency and timing of harvesting operations can be implemented.

In both years, the season-end VOH means were about 2–3-folds greater ($p < 0.001$) in the single- than the three-cut GG strips, 9-folds so for BB in 2016 and five to six times in all other plots (Table 6). The VOH values for the single-cuts also outnumbered the two-cuts, significantly ($p < 0.001$), except in 2016 flipped and the 2017 not-flipped GG strips. All strips cut twice year^{-1} still had significantly greater VOH values compared to their respective three-cut counterparts. These results are consistent with typical NWSG growth responses to defoliation associated with greater tiller densities in the absence of apical dominance [40,41]. Frequently harvested grasses often produce tillers with reduced structural carbohydrate contents, which negatively impacts their growth vigor [43]. Although the two-cut strips experienced no flipping of harvest frequencies, the 2016 single- and three-cut BB, IG, and SG stands in the flipped-strips block had, numerically, greater VOH values, about 10-units taller, than those in plots harvested the same, continuously. This better performance of the two-cut strips in the flipped-plots block is mostly attributable to the resulting better sunlight environment as they all bordered the relatively shorter three-cut strips to the North. In the not-flipped block, however, the matching two-cut strips would have experienced relatively more shading from the taller single-cut strips to their North. Considering also the temporal microclimate implications that this may have on the associated wildlife habitat attributes, the importance of field layout, orientation, and the timing of operations to strategic defoliation management of NWSG stands could not be clearer.

4. Conclusions

The data has shown that strategic defoliation management of NWSG stands for forage biomass and or feedstock production may have species-specific practical implications on subsequent wildlife habitat quality and other ecosystem services. The demonstrated differences in stand structural responses to the changes in harvest regimes also shows the importance of taking into consideration species inherent morphological and physiological adaptations to grazing, when planning sustainable defoliation management of mixed NWSG stands. The data also shows that each of the NWSGs has the potential for developing a multipurpose summer forage system that may also provide wildlife shelter from extreme weather conditions.

Based on the observed differences in the sward structural responses to defoliation (sward-heights, canopy closure, percent light interception and visual obstruction) due to species and changes in harvest regimes, strategic defoliation management of NWSG fields for both forage biomass production and wildlife habitat must aim at creating spatial-structural heterogeneity, which may involve partitioning the field into adjacent sections so one can be repeatedly mowed for forage while the other provides critical wildlife habitat needs.

The observed lack of effect due to flipping of the harvest frequencies on the regrowth sward-heights demonstrated the inherent ability for the NWSGs to exhibit compensatory structural responses to changes in defoliation management. The flipping effects on stand density and canopy closure as indicated by the mean PARi and VOH readings further underscored the importance of using a combination of response variables for evaluating potential impact of defoliation management practices on wildlife habitat quality and other ecosystem services.

Author Contributions: Conceptualization, visualization, supervision, project administration, funding acquisition, formal analysis, investigation, and writing—original draft preparation, V.W.T.; methodology, validation, writing—review and editing, V.W.T. and M.K.K. All authors have read and agreed to the published version of the manuscript.

Funding: This research was funded by USDA-NIFA through the EVANS ALLEN program.

Data Availability Statement: Not Applicable.

Acknowledgments: Throughout the study and during the manuscript preparation, the project team received tremendous administrative and logistical support from the management of the Agricultural Research Station (ARS) at Virginia State University for which the authors are very grateful. As well, the technical support of the plant science and small ruminant research technicians (Amanda Miller, Joshquinn Andrews, and Ryan Mason) for research plots maintenance and data collection and that of ARS research students (Bianca Jacques, Courtney Epps, Ariel Coleman, David Johnson, Christos Galanopoulos and Christopher Copeland), who took measurements and typed data, are acknowledged. This article is a publication No. 390 of the ARS at Virginia State University.

Conflicts of Interest: The authors declare no conflict of interest. The funders had no role in the design of the study; in the collection, analyses, or interpretation of data; in the writing of the manuscript; or in the decision to publish the result.

References

1. Temu, V.W.; Rude, B.J.; Baldwin, B.S. Yield response of native warm-season forage grasses to harvest intervals and durations in mixed stands. *Agronomy* **2014**, *4*, 90–107. [CrossRef]
2. Caswell, H.; Reed, F.; Stephenson, S.N.; Werner, P.A. Photosynthetic pathways and selective herbivory: A hypothesis. *Am. Nat.* **1973**, *107*, 465–480. [CrossRef]
3. Ferraro, D.O.; Oesterheld, M. Effect of defoliation on grass growth. a quantitative review. *Oikos* **2002**, *98*, 125–133. [CrossRef]
4. Manske, L.L.; Ske, L.L. *General Description of Grass Growth and Development and Defoliation Resistance Mechanisms*; Range Management Report DREC 98-1022; NDSU Dickinson Research Extension Center: Dickinson, ND, USA, 1998.
5. Garay, A.H. Tiller size/density compensation in perennial ryegrass miniature swards subject to differing defoliation heights and a proposed productivity index. *Grass Forage Sci.* **1999**, *54*, 347–356. [CrossRef]
6. Slepetys, J.; Šterne, D. The productivity and persistency of pure and mixed forage legume swards. *Latv. J. Agron.* **2008**, *11*, 276–281.
7. Harris, W. Pasture as an Ecosystem. In *Pastures, Their Ecology and Management*; Langer, R.H.M., Ed.; Oxford University Press: Oakdale, CA, USA, 1990; pp. 75–131.
8. Lee, W.G.; Fenner, M.; Loughnan, A.; Lloyd, K.M. Long-term effects of defoliation: Incomplete recovery of a New Zealand alpine tussock grass, *Chionochloa pallens*, after 20 years. *J. Appl. Ecol.* **2000**, *37*, 348–355. [CrossRef]
9. Briske, D.D.; Richards, J.H. *Plant Responses to Defoliation: A Physiological, Morphological and Demographic Evaluation. Wildland Plants: Physiological Ecology and Developmental Morphology*; Society for Range Manage: Denver, CO, USA, 1995; pp. 635–710.
10. Dahl, B.E. Developmental Morphology of Plants. In *Wildland Plants: Physiological Ecology and Developmental Morphology*, 1st ed.; Bedunah, D.J., Sosebee, R.E., Eds.; Society for Range Manage: Denver, CO, USA, 1995; pp. 22–58.
11. Noy-Meir, I. Compensating growth of grazed plants and its relevance to the use of rangelands. *Ecol. Appl.* **1993**, *3*, 32–34. [CrossRef]
12. Anderson, V.J.; Briske, D.D. Herbivore-induced species replacement in grasslands: Is it driven by herbivory tolerance or avoidance? *Ecol. Appl.* **1995**, *5*, 1014–1024. [CrossRef]
13. Cullen, B.R.; Chapman, D.F.; Quigley, P.E. Comparative defoliation tolerance of temperate perennial grasses. *Grass Forage Sci.* **2006**, *61*, 405–412. [CrossRef]
14. Loeser, M.R.; Crews, T.E.; Sisk, T.D. Defoliation increased above-ground productivity in a semi-arid grassland. *J. Range Manag.* **2004**, *57*, 442–447. [CrossRef]
15. Smith, S.E. Variation in response to defoliation between populations of *Bouteloua curtipendula* var. caespitosa (Poaceae) with different livestock grazing histories. *Am.J. Bot.* **1998**, *85*, 1266–1272. [CrossRef]
16. Mullahey, J.J.; Waller, S.S.; Moser, L.E. Defoliation effects on production and morphological development of little bluestem. *J. Range Manag.* **1990**, *43*, 497–500. [CrossRef]
17. Turner, C.L.; Seastedt, T.R.; Dyer, M.I. Maximization of aboveground grassland production: The role of defoliation frequency, intensity, and history. *Ecol. Applic.* **1993**, *3*, 175–186. [CrossRef]
18. Woodis, J.E.; Jackson, R.D. The effects of clipping height and frequency on net primary production of Andropogon gerardii (C4 grass) and Bromus inermis (C3 grass) in greenhouse experiments. *Grass Forage Sci.* **2008**, *63*, 458–466. [CrossRef]
19. Nelson, C.J.; Moser, L.E. Plant Factors Affecting Forage Quality. In *Forage Quality, Evaluation, and Utilization*; ASA-CSSA-SSSA: Madison, WI, USA, 1994; pp. 115–154.
20. Campbell, G.S.; Norman, J.M. The Description and Measurement of Plant Canopy Structure. In *Plant Canopies: Their Growth, Form and Function*; Russell, G., Marshall, B., Jarvis, P.G., Eds.; Cambridge University Press: Cambridge, UK, 1989; pp. 1–20.
21. D'angelo, G.H.; Postulka, E.B.; Ferrari, L. Infrequent and intense defoliation benefits dry-matter accumulation and persistence of clipped Arrhenatherum elatius. *Grass Forage Sci.* **2005**, *60*, 17–24. [CrossRef]
22. Hodgson, J. *Grazing Management. Science Into Practice*; Longman Group UK Ltd.: Harlow, UK, 1990; p. 203.
23. Difante, G.D.S.; Nascimento Júnior, D.D.; Euclides, V.P.B.; Silva, S.C.D.; Barbosa, R.A.; Gonçalves, W.V. Sward structure and nutritive value of tanzania guineagrass subjected to rotational stocking managements. *Rev. Bras. De Zootec.* **2009**, *38*, 9–19. [CrossRef]

24. Norman, J.M.; Campbell, G.S. Canopy Structure. In *Plant Physiological Ecology*; Springer: Dordrecht, The Netherlands, 1989; pp. 301–325.
25. Vermeire, L.T.; Gillen, R.L. Estimating herbage standing crop with visual obstruction in tallgrass prairie. *J. Range Manag.* **2001**, *54*, 57–60. [CrossRef]
26. Benkobi, L.; Uresk, D.W.; Schenbeck, G.; King, R.M. Protocol for monitoring standing crop in grasslands using visual obstruction. *Rangel. Ecol. Manag. J. Range Manag. Arch.* **2000**, *53*, 627–633.
27. Gonzalez, A.; Loreau, M. The causes and consequences of compensatory dynamics in ecological communities. *Annu. Rev. Ecol. Evol. Syst.* **2009**, *40*, 393–414. [CrossRef]
28. Lusk, J.J.; Smith, S.G.; Fuhlendorf, S.D.; Guthery, F.S. Factors influencing northern bobwhite nest-site selection and fate. *J. Wildl. Manag.* **2006**, *70*, 564–571. [CrossRef]
29. Jones, J.; Coggin, D.S.; Cummins, J.L.; Hill, J. Restoring and Managing Native Prairies. In *A Handbook for Mississippi Landowners*; Wildlife Mississippi: Starkville, MS, USA, 2007.
30. Temu, V.W.; Rutto, L.K.; Kering, M.K. Compensatory Yield Responses of Young Native Warm-Season Grass Stands to Seasonal Changes in Harvest Frequencies. *Agronomy* **2022**, *12*, 2761. [CrossRef]
31. NOAA Satellite and Information Service. *Drought Termination and Amelioration*; National Climatic Data Center: Asheville, NC, USA, 2017.
32. Pedreira, C.G.; Sollenberger, L.E.; Mislevy, P. Botanical composition, light interception, and carbohydrate reserve status of grazed 'Florakirk' Bermudagrass. *Agron. J.* **2000**, *92*, 194–199.
33. Briske, D.D. Developmental Morphology and Physiology f Grasses. In *Grazing Management: An Ecological Perspective*; Heitschmidt, R.K., Stuth, J.W., Eds.; Timber Press: Portland, OR, USA, 1991; pp. 85–108.
34. Belesky, D.P.; Fedders, J.M. Defoliation effects on seasonal production and growth rate of cool-season grasses. *Agron. J.* **1994**, *86*, 38–45. [CrossRef]
35. Garcia, F.; Carrère, P.; Soussana, J.F.; Baumont, R. How do severity and frequency of grazing affect sward characteristics and the choices of sheep during the grazing season? *Grass Forage Sci.* **2003**, *58*, 138–150. [CrossRef]
36. Rodolfo, G.R.; Schmitt, D.; Dias, K.M.; Sbrissia, A.F. Levels of defoliation and regrowth dynamics in elephant grass swards. *Ciência Rural.* **2015**, *45*, 1299–1304. [CrossRef]
37. Parsons, A.J.; Leafe, E.L.; Collett, B.; Stiles, W. The physiology of grass production under grazing. I. Characteristics of leaf and canopy photosynthesis of continuously-grazed swards. *J. Appl. Ecol.* **1983**, *20*, 117–126. Available online: https://www.jstor.org/stable/2403380 (accessed on 27 March 2023). [CrossRef]
38. Mullahey, J.J.; Waller, S.S.; Moser, L.E. Defoliation effects on yield and bud and tiller numbers of two Sandhills grasses. *J. Range Manag.* **1991**, *44*, 241–245. [CrossRef]
39. Buwai, M.; Trlica, M.J. Multiple defoliation effects on herbage yield, vigor, and total nonstructural carbohydrates of five range species. *J. Range Manag.* **1977**, *30*, 164–171. [CrossRef]
40. Richards, J.H.; Mueller, J.H.; Mott, J.J. Tillering in tussock grasses in relation to defoliation and apical bud removaL. *Ann. Bot.* **1988**, *62*, 173–179. [CrossRef]
41. Murphy, J.S.; Briske, D.D. Regulation of tillering by apical dominance: Chronology, interpretive value, and current perspectives. *J. Range Manage.* **1992**, *45*, 419–429. [CrossRef]
42. Holland, E.A.; Detling, J.K. Plant response to herbivory and belowground nitrogen cycling. *Ecology* **1990**, *71*, 1040–1049. [CrossRef]
43. Benot, M.-L.; Morvan-Bertrand, A.; Mony, C.; Huet, J.; Sulmon, C.; Decau, M.-L.; Prud'homme, M.-P.; Bonis, A. Grazing intensity modulates carbohydrate storage pattern in five grass species from temperate grasslands. *Acta Oecologica* **2019**, *95*, 108–115. [CrossRef]
44. Ash, A.J.; Corfield, J.P.; Mclvor, J.G.; Ksiksi, T.S. Grazing management in tropical savannas: Utilization and rest strategies to manipulate rangeland condition. *Rangel. Ecol. Manag.* **2011**, *64*, 223–239. [CrossRef]
45. Eckert, R.E.; Spencer, J.S. Growth and reproduction of grasses heavily grazed under rest-rotation management. *Rangel. Ecol. Manag. J. Range Manag. Arch.* **1987**, *40*, 156–159. [CrossRef]
46. Guevara, J.C.; Stasi, C.R.; Estevez, O.R. Effect of cattle grazing on range perennial grasses in the Mendoza plain, Argentina. *J. Arid Environ.* **1996**, *34*, 205–213. [CrossRef]
47. Davies, K.W.; Boyd, C.S.; Bates, J.D. Eighty years of grazing by cattle modifies sagebrush and bunchgrass structure. *Rangel. Ecol. Manag.* **2018**, *71*, 275–280. [CrossRef]
48. Gutman, M.; Noy-Meir, I.; Pluda, D.; Seligman, N.A.; Rothman, S.; Sternberg, M. Biomass partitioning following defoliation of annual and perennial Mediterranean grasses. *Conserv. Ecol.* **2002**, *5*. Available online: http://www.consecol.org/vol5/iss2/art1 (accessed on 27 March 2023).
49. Peterson, P.R.; Hutton, S.J.; Abaye, A.O.; Wolf, D.D.; Benson, G.B. Eastern gamagrass responses to defoliation management. In Proceedings of the 2nd Eastern Native Grass Symposium, Baltimore, MD, USA, 17–19 November 1999; pp. 253–258.

50. Ballaré, C.L.; Scopel, A.L.; Sánchez, R.A. Photocontrol of stem elongation in plant neighbourhoods: Effects of photon fluence rate under natural conditions of radiation. *Plant Cell Environ.* **1991**, *14*, 57–65. [CrossRef]
51. Frank, A.B.; Hofman, L. Light quality and stem numbers in cool-season forage grasses. *Crop Sci.* **1994**, *34*, 468–473. [CrossRef]

agronomy

Article

Responses of Soil Bacterial Communities and Chemical Properties to Grazing Regulation in Desert Steppe

Yue Wang [1,2], Mishan Guo [3,*], Yongfu Li [3], Xiaolin Yin [3], Jianying Guo [1,4] and Jing Wang [2]

[1] Yinshanbeilu Grassland Eco-Hydrology National Observation and Research Station, China Institute of Water Resources and Hydropower Research, Beijing 100038, China
[2] College of Resources and Environmental Economics, Inner Mongolia University of Finance and Economics, Hohhot 010070, China
[3] Department of Sediment Research, China Institute of Water Resources and Hydropower Research, Beijing 100048, China
[4] Institute of Water Resources for Pastoral Area, Ministry of Water Resources, Hohhot 010020, China
* Correspondence: guomsh@iwhr.com

Abstract: Due to the region's social economy and historical culture, rough grazing has led to unresolved grassland-based ecological problems in Northern China. Soil microorganisms are essential structural and functional components of underground ecosystems, and the effects of various grazing intensities on the physicochemical properties and bacterial communities of soil are unclear. A stocking density regulation experiment was carried out in the desert steppe of the Inner Mongolia Autonomous Region. In the study area, four grazing intensities were set, namely, the enclosure control group (CK), light grazing, moderate grazing, and heavy grazing. Field investigations and 16S rRNA sequencing were used to compare and analyze the characteristics of soil bacterial community structures and their correlations with soil nutrient factors under different grazing intensities. The experiment showed the following results: (1) The Shannon, Simpson, and Pielou indices of the light grazing group were significantly higher than those of the CK ($p < 0.05$), and the indices of the moderate and heavy grazing groups were lower than those of the CK, but the difference was not significant, and there were no significant differences in the Chao1 index between each group. (2) Acidobacteria, Actinobacteria, Proteobacteria, and Chloroflexi were the main bacterial phyla. (3) With the exception of soil organic matter and available potassium, which had significant negative correlations with the Shannon index ($p < 0.05$), other soil factors had no significant correlation with the soil bacterial diversity. (4) The contents of soil organic matter, total phosphorus, alkali-hydrolyzed nitrogen, available phosphorus, and available potassium influenced the differences between soil bacterial communities under different grazing intensities.

Keywords: graze; soil bacteria; desert steppe; soil properties

Citation: Wang, Y.; Guo, M.; Li, Y.; Yin, X.; Guo, J.; Wang, J. Responses of Soil Bacterial Communities and Chemical Properties to Grazing Regulation in Desert Steppe. *Agronomy* **2023**, *13*, 2817. https://doi.org/10.3390/agronomy13112817

Academic Editor: Tony Vancov

Received: 17 October 2023
Revised: 6 November 2023
Accepted: 13 November 2023
Published: 15 November 2023

1. Introduction

Soil microorganisms commonly participate in natural energy flows and material cycles, while their microbial diversity boosts the versatility of terrestrial ecosystems because of these microorganisms' large populations and wide distributions [1]. Soil microbes have wide-ranging distributions and functions, meaning that their diversity maintenance mechanisms and responses to environmental changes have been widely studied in the field of ecological restoration [2,3]. Soil microbes are sensitive to both natural climate change and anthropogenic land use change, with both processes having a great impact on their community structure and diversity [4].

Grasslands play an irreplaceable ecological role in soil conservation, climate regulation, and biodiversity maintenance in the construction and protection of ecological environments [5,6], and grassland environments account for more than 40% of the land area of China [7]. Desert steppe, as one of the most important land types in Northwestern China, also supports local animal husbandry practices. However, due to the long-term high intensity grazing and human exploitation of this environment, many of the desert steppes have been significantly degraded [8]. The main manifestations of degradation are desertification, soil fertility reduction, a decrease in biodiversity and a decline in soil health [9]. The safety of the ecological environment and the sustainability of livestock husbandry are seriously threatened in China. In order to reverse these trends, measures for ecologically managing grassland, such as enclosure and the prohibition of grazing, thus returning farmland to grassland, and transplanting and replanting trees, have been employed, and these measures have achieved obvious improvements in the health of ecosystems [10]. However, most of these measures require years to deliver positive results, and the process of natural recovery is relatively slow, which somewhat limits the productive capacity of grasslands. Therefore, on the premise of achieving ecological governance objectives, many scholars have considered how to sustainably develop animal husbandry and create scientific grazing methods that maximize productive potential while ensuring ecological security.

The study of the influence of grazing on grassland ecosystems has mostly focused on the physical and chemical properties of soil, and there are many indices of these properties which can reflect soil conditions. Soil microorganisms are an important group to study as they are highly sensitive to environmental changes. They can be used as characteristic factors to evaluate and regulate the health of soil ecosystems [11]. It is generally believed that grazing reduces the biomass and diversity of plants and has a negative impact on the microbial community, but the input of manure can also introduce and stimulate some microbial populations. Therefore, we hypothesized that (1) grazing reduced soil microbial diversity compared to the level of diversity present in enclosed grassland, (2) the stocking density had a significant effect on the soil microbial community, and (3) changes in the soil microbial community were closely related to changes in the physical and chemical properties of soil caused by grazing. In addition, the study of the multi-layer correlation between the microbial community, the physical and chemical properties of soil, and the stocking density can help researchers to understand the effects of grazing on soil ecosystems with regard to material transfer, which could provide a scientific basis for guiding the regulation of grazing and support research into the microbial measures of ecological restoration and protection of grassland environments.

2. Material and Methods

2.1. Study Site

The study site was located in Siziwang County (41°46′43″ N, 111°54′52″ E) in Ulanqab, Inner Mongolia Autonomous Region. The area is characterized by a mid-temperate continental climate, the site's altitude is 1459 m, the average annual temperature is 3–4 °C, the average annual rainfall is 240–320 mm, and the average annual evaporation rate is 2900–3000 mm. The frost-free period lasts for about 108 days. The study area was located in a desert steppe environment, where *Stipa breviflora* Griseb. was a group species and *Artemisia frigida* Willd. and *Cleistogenes songorica* Roshev. were dominant species, and the main soil type was sandy chestnut soil.

2.2. Sampling

Based on the grazing control experimental plots set up in 2002, the sampling areas for each group were established, having stocking densities of 0, 0.80, 1.92, and 2.71 [sheep/(hm^2·a)], respectively. These groups were referred to as the fencing-in group (CK), light grazing group (LG), medium grazing group (MG), and heavy grazing group (HG), respectively. The grazing period took place from June to December per year. There were three repeated plots of 4.40 hm^2 for each treatment group, resulting in a total of 12 plots.

Sample collections were conducted in December 2022, which represented the end of the grazing period. The soil samples were collected in 0–10 cm soil layers from each plot following the removal of loose litter, herbs, and a humus layer. A random sampling method was adopted to collect soil samples in each plot. Every three soil samples were mixed into one composite sample, and for each plot, three repeated composite samples were tested. These samples were put into sealed bags and numbered. The 36 soil samples were kept in a portable incubator at 4 °C for less than 8 h, before being stored at −80 °C. Each sample was divided into two parts to allow us to analyze the soil's chemical properties and bacterial community structure.

2.3. Soil Chemical Properties Analysis

Roots and stone debris were removed from the air-dried soil samples. The presence of soil organic matter (SOM) was determined via potassium dichromate oxidation [12]; the contents of total nitrogen (TN) and alkali-hydrolyzed nitrogen (AN) were determined via the Kjeldahl nitrogen determination and alkali-hydrolytic diffusion methods, respectively [13]; and the contents of soil total phosphorus (TP) and available phosphorus (AP) were determined via molybdenum-antimony and scandium colorimetry, respectively [14]. The contents of oil total potassium (TK) and available potassium (AK) were determined via the alkaline melting and ammonium acetate extraction methods [15], respectively.

2.4. Molecular Characterization of the Soil Bacteria

The composite soil samples were thoroughly mixed before DNA extraction took place. Next, 2 g soil was used to extract DNA via a Powersoil DNA Isolation Kit (MoBio, Carlsbad, CA, USA) [16]. The V3–V4 regions of the 16S rRNA gene were amplified using the primers 338F: ACTCCTACGGGAGGCAGCAG and 806R: GGACTACHVGGGTWTCTAAT. The PCR products were mixed and tested via 2% agarose gel electrophoresis. The PCR products were gelled and recovered using an AxyPrep DNA Gel Extraction Kit (AXYGEN, Corning, NY, USA). The qualified PCR products were used to construct and sequence the Illumina Miseq library, with this step being carried out by Allwegene Technology Inc., Beijing, China.

The original sequences were double-ended sequence data. We filtered the sequences with read tail mass values of less than 20. We set a window of 50 bp: if the average mass value in the window was lower than 20, the back-end bases were cut off from the window, and the reads below 50 bp were filtered following a quality control procedure. Then, we used the overlap of PE sequencing to merge pairs of sequences into a single sequence. After this step, the raw tag data were obtained upon removing the barcode and primer and performing splicing, and the high-quality sequence clean tags were obtained after further removing the chimeric and short sequences. To minimize the sample variation-related error and maximize sample coverage, the data sizes of all samples were homogenized to 16,104 sequences. The remaining sequences were classified as an operational classification unit (OTU) using Uclust (Version 1.2.22, http://www.drive5.com/uclust/downloads1_2_22q.html, accessed on 15th December 2022), with all sequences having more than 97% similarity [17].

2.5. Bioinformation Analysis and Statistical Analysis

The Chao1, Simpson, Shannon, and Pielou indices were used to characterize the diversity and evenness of the soil bacterial community, and they were computed using the vegan package in R (version 3.4.3) [18–21]. We used the nonparametric Kruskal–Wallis test to test three or more sets of data, with differential OTUs ($p < 0.05$), that originated from multiple larger sets of data generated in multiple biological replicates. The similarities and differences between bacterial communities were obtained via nonmetric multidimensional scaling analysis (NMDS) and analysis of similarities (ANOSIMs). The correlation between soil bacterial community and soil chemical properties was expressed on the NMDS map via envfit. The Mantel test was calculated using the Euclidian distances between the soil bacterial communities and the Bayesian distances between the soil properties of different stocking density groups. Redundancy analysis (RDA) of the correlation between soil bacteria and soil properties was performed using Canoco 5.0 for Windows. The diversity indices, the Kruskal–Wallis test and the above similarity and heterogeneity test were performed and the OTU clustering heatmap was generated via R. Other histograms were created via Origin 2022 (OriginLab Corporation, Northampton, MA, USA).

Differences in the soil properties (SOM, TN, AN, TP, AP, TK, AK) and bacterial diversity indices of the various stocking density groups were compared by performing one-way ANOVAs and post hoc Tukey tests ($p < 0.05$). The correlation between the soil bacterial diversity and the soil properties was tested via Pearson's correlation coefficient. All the above-mentioned mathematical statistical analyses were conducted via SPSS 26.0 (IBM, Chicago, IL, USA).

3. Results

3.1. Soil Properties under Different Stocking Density Groups

Soil properties were affected by stocking density, but there was no significant difference between these effects (Table 1). The TN, TP, and AP contents of the grazing groups were lower than those of the fencing-in group. The contents of other soil properties were highest in the heavy grazing group and lowest in the medium grazing group.

Table 1. Effects of different grazing intensities on chemical properties of soil.

	SOM (g/kg)	TN (g/kg)	TP (g/kg)	TK (g/kg)	AN (mg/kg)	AP (mg/kg)	AK (mg/kg)
CK	24.62 ± 1.43a	1.61 ± 0.12a	0.46 ± 0.02a	28.93 ± 2.11a	70.98 ± 6.28a	4.82 ± 0.59a	169.22 ± 16.66a
LG	24.70 ± 0.91a	1.60 ± 0.06a	0.44 ± 0.01a	29.21 ± 3.86a	77.31 ± 19.54a	3.88 ± 0.93a	157.33 ± 12.88a
MG	24.49 ± 4.35a	1.52 ± 0.18a	0.44 ± 0.02a	27.46 ± 0.76a	70.84 ± 14.37a	3.63 ± 0.58a	151.97 ± 45.02a
HG	25.70 ± 0.43a	1.61 ± 0.06a	0.43 ± 0.00a	30.09 ± 1.04a	95.42 ± 13.79a	4.44 ± 1.28a	196.72 ± 8.44a

Note: The same letter being present in the same column means that no significant difference was noted. CK—fencing-in group; LG—light grazing group; MG—medium grazing group; HG—heavy grazing group; SOM—soil organic matter; TN—soil total nitrogen; A—soil alkali-hydrolyzed nitrogen; TP—soil total phosphorus; AP—soil-available phosphorus; TK—soil total potassium; AK—soil-available potassium.

3.2. The Composition and Structure of the Soil Bacterial Community under Different Grazing Intensities

The soil bacterial diversity indices determined under different grazing intensities in the desert steppe environment are shown in Figure 1. The Shannon and Simpson indices of the light grazing group were significantly higher than those of the medium and heavy grazing groups ($p < 0.05$). The Pielou index of the fencing-in group was significantly lower than that of the light grazing group ($p < 0.05$), but there was no significant difference between the first group and the other two groups ($p > 0.05$). The Pielou index of the medium grazing group was significantly lower than those of the light and heavy grazing groups ($p < 0.05$). The values of the Chao1 index differed between the four groups, albeit not to a significant extent ($p > 0.05$).

A total of 4746 OTUs were collected, and they were further identified as 36 phyla, 95 classes, 116 orders, 206 families, and 277 genera. The main bacterial groups (relative abundance $\geq$ 1%) are shown in Figure 2. Chloroflexi, Acidobacteria, Proteobacteria, and Chloroflexi were the dominant groups. These groups had average relative abundances of 36.64% ± 2.65%, 24.89% ± 4.44%, 12.25% ± 1.48%, and 11.46% ± 1.30%, respectively. The relative abundance of Acidobacteria and Proteobacteria decreased in correlation with the increase in the stocking density, while Actinobacteria showed the opposite pattern; the relative abundance of Chloromycetes initially increased, before decreasing in correlation with the stocking density. Moreover, at the class level, the populations with the three greatest relative abundances were Blastocatellia, Subgroup_6 (belongs to Acidobacteria) and Alphaproteobacteria, while at the order level, they were Blastocatellales, Rubrobacterales, and Rhizobiales. At the family level, the populations with the three greatest relative abundances were Blastocatellaceae-Subgroup-4, Rubrobacteriaceae, and Gemmatimonadaceae, while at the genus level, they were RB41, Rubrobacter, and Krasilnikovia.

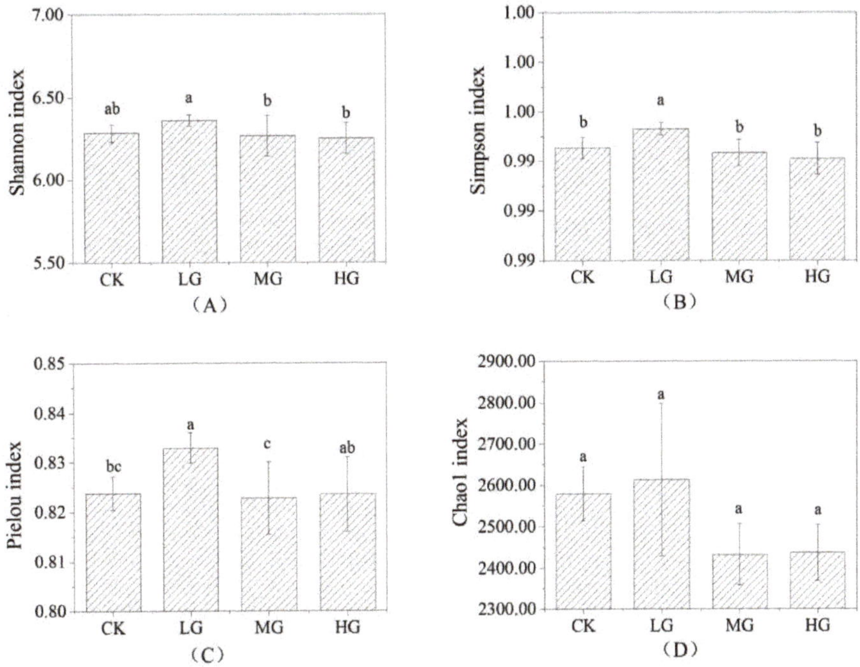

Figure 1. Soil bacterial community diversity indices under different stocking densities. Letters indicate significant differences for each parameter ($p < 0.05$). Diversity index includes Shannon (**A**), Simpson (**B**), Pielou (**C**), and Chao1 (**D**) indices. CK—fencing-in group; LG—light grazing group; MG—medium grazing group; HG—heavy grazing group.

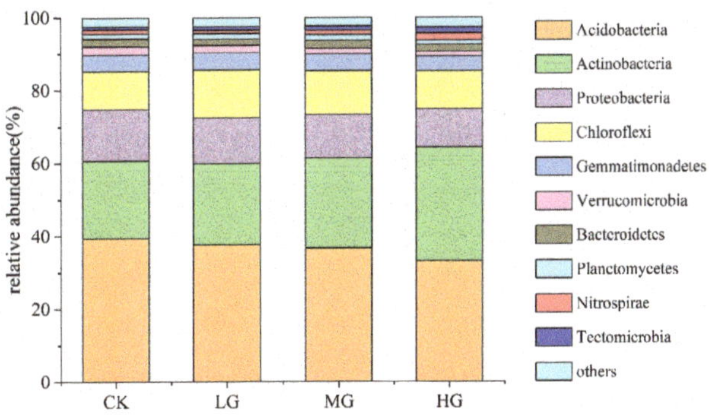

Figure 2. Soil bacterial community composition under different stocking densities (phylum level). Note: only the bacterial phyla for which the relative abundance was greater than 1% are listed here, and those with lower values were counted as other elements. CK—fencing-in group; LG—light grazing group; MG—medium grazing group; HG—heavy grazing group.

The NMDS analyses of soil bacterial communities under different grazing intensities are shown in Figure 3. As shown in Figure 3A, all the sample plots with the same color showed a state of aggregation except for the medium grazing group, although there was no obvious boundary between them. The dispersion of the medium grazing group indicated that the soil bacterial communities in this group greatly varied. The close distance between the heavy grazing group and the fencing-in group indicated that their soil bacterial compositions were more similar to one another than to those of other groups.

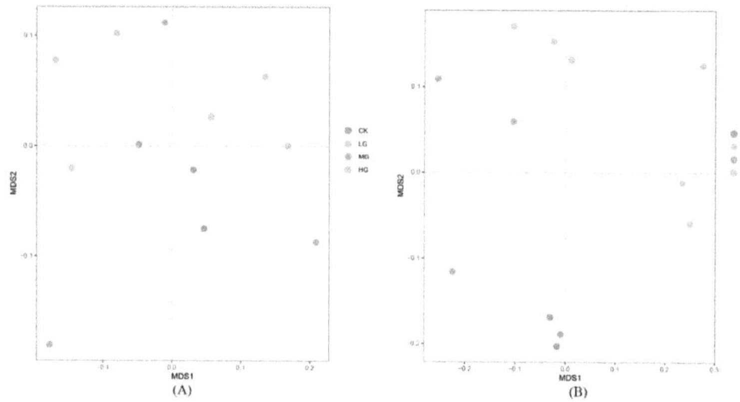

Figure 3. Analysis of nonmetric multidimensional scaling in soil bacterial communities under different grazing intensities, considering the overall OTUs and differential OTUs. Note: Figure (**A**) was generated based on overall OTUs, while Figure (**B**) was generated based on differential OTUs.

We used the Kruskal–Wallis test to obtain the differences in OTU levels between multiple sets of data with several biological replicates: Figure 3B shows the results of the NMDS analyses of the differential OTUs in each group determined via the Kruskal–Wallis test. Each group was separated from the others, indicating that their differential OTUs had great heterogeneity. The distance between the plots in the medium grazing group was still large, indicating that the variation in the OTUs between the plots in the same group was great; this result was consistent with the rule outlined in Figure 3A, given the overall situation. In the fencing-in group, the plots had stronger cohesion, indicating that the differential OTUs in this group did not undergo significant change.

3.3. Correlations between Soil Bacteria and Soil Properties under Different Grazing Intensities

In Table 2, the Shannon index has the strongest and most significant negative correlation with soil organic matter and available potassium content ($p < 0.05$). The Simpson, Pielou, and Chao1 indices were not correlated with the soil properties.

Table 2. Pearson's correlation coefficients for soil bacterial diversity and soil properties under different grazing intensities.

	SOM	TN	TP	TK	AN	AP	AK
Shannon	−0.652 *	−0.263	−0.053	0.231	−0.151	−0.308	−0.604 *
Simpson	−0.433	−0.012	0.117	0.162	−0.222	−0.328	−0.495
Pielou	−0.453	−0.120	−0.031	0.314	−0.031	−0.469	−0.467
Chao1	−0.141	0.010	0.297	−0.245	−0.186	0.454	−0.068

Note: the asterisk indicates a significant difference ($p < 0.05$). SOM—soil organic matter; TN—soil total nitrogen; AN—soil alkali-hydrolyzed nitrogen; TP—soil total phosphorus; AP—soil-available phosphorus; TK—soil total potassium; AK—soil-available potassium.

Among soil properties, only the soil organic matter content was significantly correlated with the soil bacterial community composition ($r = 0.4353$, $p < 0.05$), as determined via the Mantel test (Table 3). The other correlations were relatively weak and insignificant.

Table 3. Correlations between the soil bacterial communities under different grazing intensities and soil properties, as determined via the Mantel test.

	SOM	TN	TP	TK	AN	AP	AK
r	0.4353	0.2983	−0.1221	0.2819	−0.0122	−0.0840	0.2790
p	**0.019**	0.085	0.675	0.052	0.514	0.671	0.076

Note: the bold font represents the significant difference ($p < 0.05$). SOM—soil organic matter; TN—soil total nitrogen; AN—soil alkali-hydrolyzed nitrogen; TP—soil total phosphorus; AP—soil-available phosphorus; TK—soil total potassium; AK—soil-available potassium.

At the OTU level, the soil properties could explain 70.96% of the rate of variation in the soil bacterial community determined via the redundancy analysis (Figure 4). The total contents of phosphorus, available phosphorus, available potassium, and soil organic matter had greater effects on the soil bacterial community structure. The fencing-in group and the grazing groups were horizontally distributed on both sides of the axis, showing the differences in soil bacterial community structure caused by grazing. The fencing-in group was strong correlated with the contents of total phosphorus and available phosphorus. The heavy grazing group was strongly correlated with the contents of soil organic matter and alkali-hydrolyzed nitrogen. At the bacterial phyla level, the contents of soil organic matter, total nitrogen, available potassium, alkali-hydrolyzed nitrogen, and total phosphorus had strong plasticizing effects on the distribution of the soil bacterial phyla. The contents of soil total phosphorus and nitrogen showed a strong positive correlation with most bacteria, such as Proteobacteria and Acidobacteria. In contrast, the contents of soil total potassium, alkali-hydrolyzable nitrogen, and available phosphorus showed strong positive correlations with Nitrospirae, Tectomicrobia, and Actinobacteria.

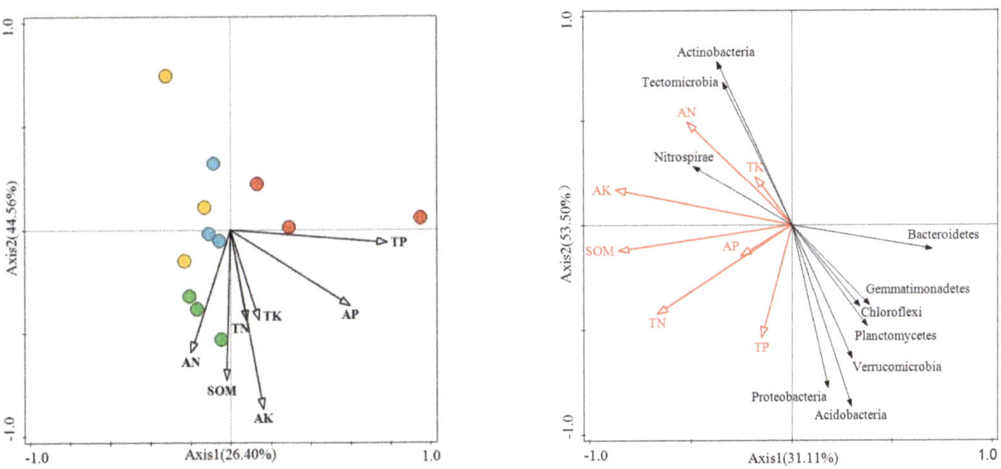

Figure 4. Redundancy analysis (RDA) of the soil bacterial structure and the soil properties under different grazing intensities, based on the full OTUs and the phylum level. Note 1: the left figure was the result based on all OTUs, while the right figure was based on the bacterial phylum level. Note 2: in the left figure, the red points represent CK, the blue points represent LG, the yellow points represent MG, and the green points represent HG. SOM—soil organic matter; TN—soil total nitrogen; AN—soil alkali-hydrolyzed nitrogen; TP—soil total phosphorus; AP—soil-available phosphorus; TK—soil total potassium; AK—soil-available potassium.

4. Discussion

4.1. Effects of Stocking Density on Soil Nutrient Content

Although the contents of total and available nutrients were affected by the change in the stocking density, there was no significant difference. There is no uniform rule in the research into the effects of stocking density on soil nutrient content. According to a four-year grazing regulation experiment conducted in the farming–grazing transitional zone of Shanxi Province, there was no significant difference in the contents of soil organic carbon and nitrogen between different grazing intensities in a single month [22]. The influence of grazing on grassland soil nutrients was generally considered to be caused by animals' nibbling on above-ground plants and litter, as well as the change in the physical properties of soil, such as porosity reduction or the mechanical destruction of the soil's structure caused by trampling by animals, and the addition of animal excrement; all these factors impacted soil nutrient storage and cycling [23,24]. The contents of organic matter, total potassium, and alkali-hydrolyzed nitrogen in the light grazing group were increased compared to the fencing-in group. Many previous studies have verified this observation, and moderate trampling disturbance could improve grassland productivity and carbon sequestration [25]. In our study, almost all soil properties had the lowest results in the medium grazing groups. The overcompensated growth of plants in the medium grazing group may be the reason for the decrease in the soil organic matter and nutrients [26]. However, the increase in nutrients in the heavy grazing group may be due to a large cumulative amount of animal excreta, as the nutrient elements in excreta exceeded the dynamic balance of the nutrient cycle in the ecosystem but were not transported outwards, meaning that they were stored in the surface soil. It is also believed that the comprehensive influence of animals in the heavy grazing group would lead to an increase in the plant root–shoot ratio, which stimulated the accumulation of root biomass and, thus, increased the nutrient elements released by root litter into the soil [27].

4.2. Responses of Soil Bacterial Diversity and Community Structure to Stocking Density

The α-diversity indices of the light grazing group were higher than those of the fencing-in group, but the difference between them was not significant. Compared to the fencing-in group, the excrement produced via moderate grazing and plant residue, as well as litters chewed and trampled by animals, were more easily decomposed, providing rich carbon sources for the growth of soil bacteria, thus improving the bacterial diversity [28]. The Shannon and Simpson indices of the medium and heavy grazing groups were significantly lower than those of the light grazing group. Intensive grazing would lead to the degradation of above-ground vegetation and a reduction in soil fertility, thus inhibiting the growth of some bacteria. A significant decrease in soil bacterial diversity was found in a 64-year intensive grazing experiment [29]. In the grazing gradient (control, light, medium, heavy and overgrazing) tests in Inner Mongolia, the soil bacterial α-diversity index increased and then decreased, which was similar to the changing trend of the results of our study [30]. Scholars also found that heavy grazing and control groups had no significant effects on the α-diversity indices [31]. There were several reasons for this differentiation between results, including the differences in environmental conditions, vegetation status, grazing animals, and grazing experiment designs in different study areas [32,33].

Acidobacteria, Actinomycetes, and Proteobacteria were the most common soil bacterial phyla, and their relative abundances in soil varied depending on the climatic conditions, soil types, vegetation types, and land use patterns [34]. Previous studies showed that grazing decreased the relative abundances of dominant bacteria compared to the control group, but the difference was not significant [35]. Acidobacteria and Proteobacteria showed such changes in our study. However, this rule was not consistent, as the dominant bacteria in different grasslands showed various trends in correlation with the increase in grazing intensity [36]. In the NMDS analysis of the overall bacterial community structure, the sample sites of different groups had a certain degree of cohesion, but the boundary with other groups was not obvious, indicating that the soil bacterial community of each group had a certain degree of overlap. However, the NMDS analysis results of different OTUs indicated that the bacteria that were significantly affected by stocking density belonged to nondominant species, which accounted for a small proportion of the soil bacterial community. Blastocatellia (belongs to Acidobacteria), Thermomicrobia (belongs to Chloroflexi), and Subgroup_6 (belongs to Acidobacteria) class were the three OTUs with the largest abundances of different OTUs. The response of the bacterial community structure to grazing was affected by the species of animal and the grazing time, as well as other factors. In the grassland environment, with *Leymus chinensis* being the dominant species in Jilin Province, the soil bacterial community structures had significant differences, which were caused by the grazing intensity [37].

4.3. The Correlation between Stocking Density, Soil Nutrient Content, and Soil Bacteria

Influencing the soil organic carbon transfer was one of the most important ways that grazing could change the soil's bacterial community structure [38]. This measure showed that the soil bacterial α-diversity was closely related to soil organic matter content (SOC) and soil water content under different grazing intensities [30]. In our study, the Shannon index was only negatively correlated with soil organic matter and available potassium content among the soil properties ($p < 0.05$). The Mantel test showed a significant positive correlation between the soil organic matter content and the bacterial community ($p < 0.05$). The soil bacterial community structure was significantly positively correlated with the soil contents of organic carbon and total nitrogen in alpine grassland ecosystems in northern Tibet [39]. The variations in the above results may be caused by climatic and vegetation conditions. In terms of the eutrophic hypothesis, eutrophic bacteria were positively correlated with soil organic carbon, and the accumulation of soil organic carbon reduced the relative abundance of oligotrophic bacteria [40]. In our results, there was no significant change in soil organic carbon content between the grazing groups ($p > 0.05$), and it varied in the following manner: HG > LG > CK > MG. However, from the analysis of the overall range of numerical change, the results related to Acidobacteria, which is one of the representatives of oligotrophic bacteria, were consistent with this viewpoint. Meanwhile, the negative correlation between Acidobacteria and soil organic carbon was also explained by some scholars based on the principle of soil carbon mineralization [41]. Proteobacteria are defined as eutrophic bacteria with a high carbon utilization rate, and the high soil contents of carbon and nitrogen are conducive to their growth [42]. However, as one of the dominant phyla in this study, Proteobacteria did not conform to this hypothesis. This may be caused by the fact that our study was not a simple soil nutrient control experiment, instead overlaying the influence of grazing factors. Some bacteria in Actinomycetes have a photosynthetic autotrophic ability to fix CO_2 [43], which made their relative abundances exhibit an increasing trend, even when the grazing intensity increased and the above-ground vegetation decreased. As determined via RDA analysis, the close positive correlation between the heavy grazing group and soil organic matter may be due to the dominant position of Actinomyces.

Compared to the accumulation of soil nutrients brought about via natural restoration in the fencing-in group, the increase in soil nitrogen and phosphorus in the heavy grazing group mainly occurred due to the input of animal feces [44]. Soil bacteria participate in nitrogen fixation, nitrification, denitrification, and other processes in the ecosystem. The content of soil-available nitrogen had a significant positive correlation with Nitrospirae, while it had a negative correlation with Acidobacteria. These findings were contrary to the results of previous studies that stated that there was a significant positive correlation between Acidobacteria and nitrogen because of the significant use of ammonium nitrogen and nitrate nitrogen as nitrogen sources [45]. Soil-available phosphorus had little correlation with the various bacterial phyla. It has been shown that bacteria are far less sensitive to soil phosphorus utilization than fungi, and there is no obvious correlation between the available phosphorus and bacterial abundance [46]. The strong correlation between soil phosphorus and the soil bacterial community structure, as determined via RDA analysis, may be due to the indirect effect that phosphorus has on the above-ground plant and soil fungal community, before acting on the bacterial community [47].

The correlation between soil nutrient content and soil bacteria under the influence of grazing was complex and changeable. More experiments are needed, such as controlling fecal input and more combined regulation of the stocking density and time, to explore the correlation between these factors and provide a more effective and accurate theoretical basis for studying actual grazing activities.

5. Conclusions

The soil bacterial communities present in the desert steppe environment under different stocking densities were identified as 36 phyla, 95 classes, 116 orders, 206 families, and 277 genera. Acidobacteria, Actinobacteria, Proteobacteria, and Chloroflexi were the dominant phyla. Light grazing increased the bacterial α-diversity indices of the soil, while medium and heavy grazing decreased these indices. There were some differences in soil bacterial communities between the different grazing intensities, and the variability in the soil bacterial communities was greater in the medium grazing group. In our study, soil chemical properties did not significantly change in response to the stocking density. Only soil organic matter and available potassium had significant negative correlations with the soil bacterial Shannon index, while the contents of soil organic matter, total phosphorus, alkali-hydrolyzed nitrogen, available phosphorus, and available potassium played strong roles in shaping the bacterial community differences and taxa composition. It would be unwise for governments to pursue desert steppe land restoration by completely banning grazing for a long time, given its complex relationship with soil microbial characterization. Nonetheless, we will continue to conduct multi-year observations and analyses to further study the relationship between soil microbes and grazing.

Author Contributions: Y.W.: Conceptualization, software, formal analysis, investigation, resources, data curation, writing—original draft preparation, project administration, project administration M.G.: Conceptualization, methodology, validation, writing—original draft preparation, writing—review and editing Y.L.: software, visualization X.Y.: validation, supervision J.G.: methodology, resources, supervision J.W.: formal analysis, investigation, visualization. All authors have read and agreed to the published version of the manuscript.

Funding: This study was supported by Yinshanbeilu Grassland Eco-hydrology National Observation and Research Station, China Institute of Water Resources and Hydropower Research (Grant NO. YSS202117); National Natural Science Foundation of China (32260279); and the Natural Science Foundation of Inner Mongolia (2020BS04007, 2020BS03001).

Institutional Review Board Statement: Not applicable.

Informed Consent Statement: Not applicable.

Data Availability Statement: The original contributions presented in the study are included in the article; further inquiries can be directed to the corresponding author.

Conflicts of Interest: The authors declare that they have no conflict of interest to report regarding the present study.

References

1. Delgado-Baquerizo, M.; Maestre, F.T.; Reich, P.B.; Jeffries, T.C.; Gaitan, J.J.; Encinar, D.; Berdugo, M.; Campbell, C.D.; Singh, B.K. Microbial diversity drives multifunctionality in terrestrial ecosystems. *Nat. Commun.* **2016**, *7*, 10541. [CrossRef]
2. Geisen, S.; Wall, D.H.; van der Putten, W.H. Challenges and opportunities for soil biodiversity in the anthropocene. *Curr. Biol.* **2019**, *29*, 1036–1044. [CrossRef] [PubMed]
3. Zhang, G.; Zhao, Z.; Yin, X.; Zhu, Y. Impacts of biochars on bacterial community shifts and biodegradation of antibiotics in an agricultural soil during short-term incubation. *Sci. Total Environ.* **2021**, *771*, 144751. [CrossRef] [PubMed]
4. Zhang, Z.C.; Shi, Z.Q.; Yang, J.Y.; Hao, B.H.; Hao, L.J.; Diao, F.W.; Wang, L.X.; Bao, Z.H.; Guo, W. A new strategy for evaluating the improvement effectiveness of degraded soil based on the synergy and diversity of microbial ecological function. *Ecol. Indic.* **2021**, *120*, 106917. [CrossRef]
5. Xue, Y.; Bai, X.; Zhao, C.; Tan, Q.; Li, Y.; Luo, G.; Long, M. Spring photosynthetic phenology of Chinese vegetation in response to climate change and its impact on net primary productivity. *Agric. For. Meteorol.* **2023**, *342*, 109734. [CrossRef]
6. Qiu, D.; Zhu, G.; Lin, X.; Jiao, Y.; Lu, S.; Liu, J.; Chen, L. Dissipation and movement of soil water in artificial forest in arid oasis areas: Cognition based on stable isotopes. *CATENA* **2023**, *228*, 107178. [CrossRef]

7. Pan, D.R.; Yan, H.W.; Li, Q.; Liu, D.Y.; Liu, X.N.; Zhang, D.G.; Han, T.H.; Sun, B.; Jiang, J.C. Loss of grassland ecosystem service values based on potential vegetation in China. *Rangel. J.* **2022**, *43*, 363–375. [CrossRef]
8. Li, M.Y.; Li, X.B.; Liu, S.Y.; Li, X.; Lyu, X.; Dang, D.L.; Dou, H. Ecosystem services under different grazing intensities in typical grasslands in Inner Mongolia and their relationships. *Glob. Ecol. Conserv.* **2021**, *26*, e01526. [CrossRef]
9. Akiyama, T.; Kawamura, K. Grassland degradation in China: Methods of monitoring, management and restoration. *Grassl. Sci.* **2007**, *53*, 1–17. [CrossRef]
10. Zhang, H.Y.; Fan, J.W.; Shao, Q.Q.; Zhang, Y.X. Ecosystem dynamics in the 'Returning Rangeland to Grassland' programs, China. *Acta Prataculturae Sin.* **2016**, *25*, 1–15.
11. Coban, O.; Deyn, G.B.; van der Ploeg, M. Soil microbiota as game-changers in restoration of degraded lands. *Science* **2022**, *375*, abe0725. [CrossRef] [PubMed]
12. Walkley, A.; Black, I.A. An examination of the degtjareff method for determining soil organic matter, and a proposed modification of the chromic acid Titration method. *Soil Sci.* **1934**, *37*, 29–38. [CrossRef]
13. Mason, C.J.; Edwards, M.; Riby, P.G.; Coe, G. The use of microwaves in the acceleration of digestion and colour development in the determination of total Kjeldahl nitrogen in soil. *Analyst* **1999**, *124*, 1719–1726. [CrossRef]
14. Bao, S. *Soil and Agricultural Chemistry Analysis*, 3rd ed.; Agriculture Press of China: Beijing, China, 2000; p. 11.
15. Liu, J.S.; Ma, Q.; Hui, X.L.; Ran, J.Y.; Ma, Q.X.; Wang, X.; Wang, Z.H. Long-term high-P fertilizer input decreased the total bacterial diversity but not *phoD*-harboring bacteria in wheat rhizosphere soil with available-P deficiency. *Soil Biol. Biochem.* **2020**, *149*, 107918. [CrossRef]
16. Edgar, R. Search and Clustering Orders of Magnitude Faster than BLAST. *Bioinformatics* **2010**, *26*, 2460–2461. [CrossRef]
17. Caporaso, J.G.; Lauber, C.L.; Walters, W.A.; Berg-Lyons, D.; Huntley, J.; Fierer, N.; Gormley, J.A.; Smith, G.; Knight, R. Ultra-high-throughput microbial community analysis on the Illumina HiSeq and MiSeq platforms. *ISME J.* **2012**, *6*, 1621–1624. [CrossRef]
18. Shanon, C.E. A mathematical theory of communications. *Bell Syst. Tech. J.* **1948**, *27*, 379–423. [CrossRef]
19. Simpson, E.H. Measurement of diversity. *Nature* **1949**, *168*, 668. [CrossRef]
20. Chao, A. Non-parametric estimation of the number of classes in a population. *Scand. J. Stat.* **1984**, *11*, 265–270.
21. Pielou, E.C. Species-diversity and pattern-diversity in the study of ecological succession. *J. Theor. Biol.* **1966**, *10*, 370–383. [CrossRef]
22. Haynes, R.; Williams, P.H. Nutrient Cycling and Soil Fertility in the Grazed Pasture Ecosystem. *Adv. Agron.* **1993**, *49*, 119–199.
23. Ma, W.M.; Ding, K.Y.; Zhong, W.L. Comparison of soil carbon and nitrogen stocks at grazing-excluded and yak grazed alpine meadow sites in Qinghai-Tibetan Plateau, China. *Ecol. Eng.* **2016**, *87*, 203–211. [CrossRef]
24. Rakkar, M.K.; Blanco-Canqui, H. Grazing of crop residues: Impacts on soils and crop production. *Agric. Ecosyst. Environ.* **2018**, *258*, 71–90. [CrossRef]
25. Zhu, G.Y.; Deng, L.; Zhang, X.B.; Shangguan, Z.P. Effects of grazing exclusion on plant community and soil physicochemical properties in a desert steppe on the Loess Plateau, China. *Ecol. Eng.* **2016**, *90*, 372–381. [CrossRef]
26. Stewart Ibarra, A.; Frank, D. Short sampling intervals reveal very rapid root turnover in temperate grassland. *Oecologia* **2008**, *157*, 453–458. [CrossRef]
27. Zhou, X.; Wang, J.; Hao, Y.; Wang, Y. Intermediate grazing intensities by sheep increase soil bacterial diversities in an Inner Mongolian steppe. *Biol. Fert. Soils* **2010**, *46*, 817–824. [CrossRef]
28. Zhang, Y.T.; Gao, X.L.; Hao, X.Y.; Alexander, T.; Shi, X.J.; Jin, L.; Thomas, B.W. Heavy grazing over 64 years reduced soil bacterial diversity in the foothills of the Rocky Mountains, Canada. *Appl. Soil Ecol.* **2020**, *147*, 103361. [CrossRef]
29. Wang, Z.; Jiang, S.Y.; Struik, P.C.; Wang, H.; Jin, K.; Wu, R.; Na, R.; Mu, H.B.; Ta, N. Plant and soil responses to grazing intensity drive changes in the soil microbiome in a desert steppe. *Plant Soil* **2022**, *491*, 219–237. [CrossRef]
30. Li, Y.M.; Wang, S.P.; Jiang, L.L.; Zhang, L.R.; Cui, S.J.; Meng, F.D.; Wang, Q.; Li, X.; Zhou, Y. Changes of soil microbial community under different degraded gradients of alpine meadow. *Agr. Ecosyst. Environ.* **2016**, *222*, 213–222. [CrossRef]
31. Xu, S.; Silveira, M.; Inglett, K.; Sollenberger, L.; Gerber, S. Soil microbial community responses to long-term land use intensification in subtropical grazing lands. *Geoderma* **2017**, *293*, 73–81. [CrossRef]
32. Wang, M.M.; Wang, S.P.; Wu, L.W.; Xu, D.P.; Lin, Q.Y.; Yi, G.H.; Li, X.Z.; Zhou, J.Z.; Yang, Y.F. Evaluating the lingering effect of livestock grazing on functional potentials of microbial communities in Tibetan grassland soils. *Plant Soil* **2016**, *407*, 385–399. [CrossRef]
33. Baker, K.L.; Langenheder, S.; Nicol, G.W.; Ricketts, D.; Killham, K.; Campbell, C.D.; Prosser, J.I. Environmental and spatial characterisation of bacterial community composition in soil to inform sampling strategies. *Soil Biol. Biochem.* **2009**, *41*, 2292–2298. [CrossRef]
34. Yin, Y.L.; Wang, Y.Q.; Li, S.X.; Liu, Y.; Zhao, W.; Ma, Y.S.; Bao, G.S. Effects of enclosing on soil microbial community diversity and soil stoichiometric characteristics in a degraded alpine meadow. *J. Appl. Ecol.* **2019**, *30*, 127–136.
35. Le Roux, X.; Poly, F.; Currey, P.; Commeaux, C.; Hai, B.; Nicol, G.W.; Prosser, J.I.; Schloter, M.; Attard, E.; Klumpp, K. Effects of aboveground grazing on coupling among nitrifier activity, abundance and community structure. *ISME J.* **2008**, *2*, 221–232. [CrossRef] [PubMed]
36. Qu, T.B.; Du, W.C.; Yuan, X.; Yang, Z.M.; Liu, D.B.; Wang, D.L.; Yu, L.J. Impacts of grazing intensity and plant community composition on soil bacterial community diversity in a steppe grassland. *PLoS ONE* **2016**, *11*, e0159680. [CrossRef]

37. Xun, W.B.; Yan, R.R.; Ren, Y.; Jin, D.Y.; Xiong, W.; Zhang, G.S.; Cui, Z.L.; Xin, X.P.; Zhang, R.F. Grazing-induced microbiome alterations drive soil organic carbon turnover and productivity in meadow steppe. *Microbiome* **2018**, *6*, 170. [CrossRef]
38. Gao, F.; Wang, B.; Shi, Y.X.; Zhang, G.X.; Wang, J.; Si, G.C.; Han, C.H.; Yuan, Y.L.; Hu, Z. The response of alpine grasslands ecosystem in the north Tibet to short-term enclosure. *Acta Ecol. Sin.* **2017**, *37*, 4366–4374.
39. Wang, Z.; Li, X.L.; Ji, B.M.; Struik, P.C.; Jin, K.; Tang, S.M. Coupling between the responses of plants, soil, and microorganisms following grazing exclusion in an cvergrazed grassland. *Front. Plant Sci.* **2021**, *12*, 640789. [CrossRef]
40. Xun, W.B.; Zhao, J.; Xue, C.; Zhang, G.S.; Ran, W.; Wang, B.R.; Shen, Q.R.; Zhang, R.F. Significant alteration of soil bacterial communities and organic carbon decomposition by different long-term fertilization management conditions of extremely low-productivity arable soil in South China. *Environ. Microbiol.* **2016**, *18*, 1907–1917. [CrossRef]
41. Thomson, B.C.; Ostle, N.; McNamara, N.; Bailey, M.J.; Whiteley, A.S.; Griffiths, R.I. Vegetation affects the relative abundances of dominant soil bacterial taxa and soil respiration rates in an upland grassland soil. *Microb. Ecol.* **2010**, *59*, 335–343. [CrossRef]
42. He, Z.L.; Piceno, Y.; Deng, Y.; Xu, M.Y.; Lu, Z.M.; DeSantis, T.; Andersen, G.; Hobbie, S.E.; Reich, P.B.; Zhou, J.Z. The phylogenetic composition and structure of soil microbial communities shifts in response to elevated carbon dioxide. *ISME J.* **2011**, *6*, 259–272. [CrossRef] [PubMed]
43. Liu, J.; Li, L.; Ji, L.; Li, Y.; Liu, J.; Li, F. Divergent effects of grazing versus mowing on plant nutrients in typical steppe grasslands of Inner Mongolia. *J. Plant Ecol.* **2022**, *16*, rtac032. [CrossRef]
44. Melo, V.; Barros, L.; Silva, M.; Veloso, T.; Senwo, Z.; Matos, K.; Nunes, T.K.O. Soil bacterial diversities and response to deforestation, land use and burning in North Amazon, Brazil. *Appl. Soil Ecol.* **2021**, *158*, 103775. [CrossRef]
45. Wang, Z.; Zhang, J.X.; Yang, X.L.; Huang, X.; Chen, S.L.; Qiao, Y.M. Characteristics of soil microbial diversity in different patches of alpine meadow. *Acta Agrestia Sin.* **2021**, *29*, 1916–1926.
46. Adair, K.L.; Wratten, S.; Lear, G. Soil phosphorus depletion and shifts in plant communities change bacterial community structure in a long-term grassland management trial. *Env. Microbiol. Rep.* **2013**, *5*, 404–413. [CrossRef] [PubMed]
47. John, M.K. Colorimetric determination in soil and plant material with ascorbic acid. *Soil Sci.* **1970**, *109*, 214–220. [CrossRef]

agronomy

Article

Using Ecological Stoichimetric Characteristies to Inform Grassland Management in the Karst Desertification Area

Guochang Pan, Shuzhen Song *, Xueling Wang and Yongkuan Chi *

School of Teacher Education, State Engineering Technology Institute for Karst Desertification Control, Guizhou Normal University, Guiyang 550001, China; 100050489@gznu.edu.cn (G.P.); 222200171830@gznu.edu.cn (X.W.)
* Correspondence: 15010170764@gznu.edu.cn (S.S.); 201907002@gznu.edu.cn (Y.C.)

Abstract: C, N and P play an important indicator role in explaining the material cycles and elemental balances of living and non-living systems. In order to control karst rocky desertification, China has established a large number of artificial grasslands for the development of herbivorous animal husbandry, which has played an important role in ecological restoration and economic development. However, the effects of different use patterns on the ecological stoichiometry of the carbon (C), nitrogen (N) and phosphorus (P) of the grassland plant–soil–microorganism are not clear. In this study, the effects of grazing grassland (GG), mowing grassland (MG) and enclosed grassland (EG) on C, N and P and their ecological stoichiometry in the artificial grassland plant–soil–microbe were investigated in the karst desertification control area in Southern China. The results showed that (1) the C content was EG > GG > MG. The N content was GG > EG > MG, while the P content was MG > GG > EG. C:N, C:P and N:P were shown as EG > GG > MG. The plant N:P was more than 20, indicating a P deficiency and limitation, especially in EG. (2) The content of C and P in soil was EG > GG > MG. The N content was GG > EG > MG. The soil C:N showed EG > MG > GG, while C:P and N:P were shown as MG > GG > EG. The soil N:P ratios were all less than 14, indicating that all of them had an obvious N limitation. (3) Soil microbial biomass carbon (MBC) was GG > MG > EG. Soil microbial biomass nitrogen (MBN) was GG > EG > MG. Soil microbial biomass phosphorus (MBP) showed EG > GG > MG. MBC:MBN was MG > EG > GG. MBC:MBP was MG > EG > GG. MBN:MBP was GG > MG > EG. The MBN:MBP in GG and MG was greater than 9.6, which is P-limited, while the MBN:MBP in EG is less than 8.9, which is N-limited. (4) Plant C and N were significantly correlated with soil C and N, but plant P was significantly negatively correlated with soil P, while MBP was significantly positively correlated with soil TP. Soil microorganisms had the tendency to assimilate available P in GG and MG treatments, but the potential of releasing P from mineralized soil organic matter was higher in EG treatment. The results showed that the chemical properties and stoichiometric characteristics of the plant–soil–microorganism were significantly changed by different grassland-use methods, which provided scientific guidance for the management of C, N and P elements and the further optimization of soil microbial environment for artificial grassland in the karst rocky desertification area.

Keywords: artificial grasslands; karst desertification; ecological restoration; carbon; nitrogen; phosphorus; soil microbial biomass; ecological stoichiometry

Citation: Pan, G.; Song, S.; Wang, X.; Chi, Y. Using Ecological Stoichimetric Characteristies to Inform Grassland Management in the Karst Desertification Area. *Agronomy* **2023**, *13*, 1841. https://doi.org/10.3390/agronomy13071841

Academic Editors: Kesi Liu and Xinqing Shao

Received: 18 June 2023
Revised: 11 July 2023
Accepted: 11 July 2023
Published: 12 July 2023

1. Introduction

Nitrogen (N) and phosphorus (P) are the most essential nutrients affecting plant growth and development, and they are the main limiting elements in terrestrial ecosystems [1]. At the same time, N and P elements will affect the fixation of plant carbon (C), and C, N and P interact with each other to regulate plant physiological activities [2]. Soil is the basis for plant growth, and the nutrients needed for plant growth mainly come from soil, while microorganisms are important drivers of the soil nutrient cycle, which can reflect the growth conditions of organisms [3]. Differences in soil parent material, microbial activities, plant types, litter return and human utilization methods result in the obvious different characteristics of plants, soil and microorganisms, thus increasing the complexity of their relationship with the environment [4,5]. C, N and P play important indicative roles in explaining the material cycle and elemental balance of biological and abiotic systems [6,7]. Therefore, it is of great significance for understanding the coupling relationship between the material cycling process of the biological system and environment to study the dynamic balance of C, N and P nutrients in the "plant–soil–microbe" system.

Ecological stoichiometry emphasizes the balance of C, N and P ratios and focuses on revealing the quantitative relationship and law of element interaction and process balance in an ecosystem [8]. Ecological stoichiometry is a useful tool to reveal the ability of plants, soils and microorganisms to maintain a relatively constant nutrient composition in response to changes in available soil resources [9] and to understand the variation of nutrients and their interactions in the plant–soil–microorganisms [10]. Therefore, understanding the nutrient balance in the plant–soil–microorganism is very important for the restoration of degraded ecosystems. Studies have shown that C, N and P in the plant–soil–microorganisms are distinctly affected by different ecological processes, such as grassland degradation and mining recovery [11,12], so their ratio changes are highly valued globally [13]. At present, the study of ecological stoichiometry on vegetation mainly focuses on regional and ecosystem scales, and the research results are mostly on forest, grassland and wetland ecosystems [14]. These studies also showed that the ecological stoichiometry of C, N and P in plants and soil had spatial variability in different regions and habitats. For perennial plants (e.g., grassland plants), their different utilization methods will result in differences in C assimilation and accumulation capacity, nutrient absorption efficiency and litter return ability, which will definitely lead to changes in the C, N and P contents of plants, soils and soil microorganisms. Therefore, it is necessary to study the ecological stoichiometry of plant–soil–microbe relationships from the perspective of different utilization methods [5].

Karst landform accounts for about 15% of the total global land area and is home to more than 1 billion people [15]. However, due to its special geological structure, under the interference of natural and human activities, soil erosion is serious, and it is easy to cause a landscape similar to desertification, namely karst desertification [16]. Previous studies in the karst region of Southern China have shown that the "Grain for Green" project (refers to the control mode of stopping cultivation of sloping farmland, changing it into artificial grassland establishment, restoring vegetation and controlling soil erosion in order to prevent and control soil erosion) is an important measure to rapidly repair the damages to the ecological environment in the karst desertification area [17] and that it is of great significance to promote ecological reconstruction and economic development. The implementation of a large number of rocky desertification control engineering projects will inevitably lead to differences in soil nutrient consumption and return and make the changes of C, N, P nutrient cycle in the "plant–soil–microorganism" system, while ecological stoichiometry can reveal its internal relationship, thus guiding practice. The most important utilization methods of artificial grasslands are grazing and mowing, and different utilization methods may change the long-tern evolution of soil–plant–microbial nutrient relationships in grassland [18], thus exerting far-reaching impacts on ecosystem functions [19]. Currently, the studies of ecological stoichiometry in karst areas mainly focus on forests or different land uses [20–24], while the research on ecological stoichiometry characteristics, change law and internal correlation of grassland plants, soil and microorganisms under different measures of rocky desertification control project is still insufficient. Therefore, it is urgent to clarify the ecological stoichiometry characteristics of the soil–plant–microorganism and their inner correlations, revealing the nutrient utilization methods in grassland under different utilization methods in the karst desertification areas to provide a scientific basis for the restoration and reconstruction of degraded karst ecosystems. For this purpose, we proposed a scientific hypothesis: is there an effect of different utilization patterns on the ecological stoichiometry characteristics of the soil–plant–microorganisms in the grasslands of karst rocky desertification areas? To address the hypothesis, this study took the artificial grassland under different utilization methods (grazing, mowing and enclosure) in the karst desertification area as the research object and studied the contents of C, N P and their ecological stoichiometry characteristics and correlation of the plant–soil–microorganism under different grassland utilization methods in order to provide theoretical support for the sustainable utilization and ecological restoration of artificial grassland in the karst desertification control area.

2. Materials and Methods

2.1. Research Area

The study area is located in Salaxi Town and Yejiao Township, Qixingguan District, Bijie City, Guizhou Province, China (105°02′01′–105°08′09′ E, 27°11′36′–27°16′51′ N), belonging to the Upper Liuchong River Basin of Wujiang River, which is a karst desertification control with typical karst plateau–mountain and potential/light karst desertification. Karst is widely distributed in the study area, and the karst desertification area in the study area is 55.931 km^2, accounting for 64.93% of the total area of the demonstration area. The soil is mainly yellow loam, which was mainly planted by traditional agriculture such as corn and potato for a long time, and soil erosion is serious, so the soil thickness is generally about 20–50 cm. The study area has a subtropical monsoon climate zone, with an average annual rainfall of about 1000 mm, average annual temperature of about 12 °C, frost-free period of 245 days and average annual sunshine hours of 1360 h. Due to the influence of the karst geological structure, the storage of groundwater is extremely complex, so it is difficult to utilize and exploit groundwater resources. The main production and living waters are spring and surface water, with sufficient water in summer but serious water shortage in the dry season. The vegetation of trees and shrubs is dominated by *Pinus yunnanensis*, *Rhododendron simsii*, *Pyracantha fortuneana*, *Cyclobalanopsis glauca*, *Rosa roxburghii* and *Juglans regia*. Herbaceous plants include *Artemisia lavandulaefolia*, *Chenopodium glaucum*, *Stellaria media*, *Digitaria sanguinalis*, *Trifolium repens*, *Trifolium pratense*, *Lolium perenne*, *Dactylis glomerata* and *Bromus catharticus*.

2.2. Sample Plot Setting

In April 2012, the research team established an artificial grassland by mixed seeding in the study area, namely *Lolium perenne* + *Dactylis glomerata* + *Trifolium repens*, with the seed quantity of 2:2:1. The grassland has been mainly used for free grazing after the establishment, and the carrying capacity was 600 m^2/sheep unit. In order to reveal the difference of ecological stoichiometry characteristics of grassland plant–soil–microorganism under different use patterns, we established three different types of grassland usage in August 2019, namely grazing grassland (GG), mowing grassland (MG) and enclosed grassland (EG), among which EG was the control group. Three replicates were set for each plot, and the distance between the plot boundaries was more than 50 cm. Each plot of enclosed grassland was about 100 m^2, and none shall be used for any purpose.

Each plot of grazing grassland was about 3000 m^2. The average grazing quantity of each plot in grazing grassland was 5 (basically consistent with the local grazing situation), and the grazing livestock were Guizhou semi-fine wool sheep, about 1 year old. The grazing time was 300 days per year, except in extreme weather.

Each plot of mowing grassland was about 3000 m^2. When the grass height reached about 40 cm, the mowing was carried out, and it was cut 4 times a year; the stubble height was about 5 cm.

2.3. Sample Collection

In mid-August 2021, three 1 m × 1 m quadrates were set up in each sample plot, and the aboveground plants in the small quadrate were harvested with a stubble height of 5 cm. The plant samples harvested in the quadrat of each plot were fully mixed and evenly bagged back to the laboratory, and a total of 9 plant samples were obtained. In order to reduce the spatial heterogeneity of the soil, the S-shaped sampling method was used in each plot afterward, and 15 sampling points were uniformly set. After removing the litter layer on the soil surface, soil samples (0–10 cm) were collected with a soil drill. Soil samples from 15 sampling points in each plot were mixed into one sample and brought back to the laboratory. A total of 9 soil samples were obtained. The plant samples brought back were inactivated at 105 °C for 30 min and then dried at 65 °C to constant weight. After drying, the contents of C, N and P of the plants were determined after passing through a 2 mm sieve. The impurities in the soil samples were removed and divided into two parts. One part of the soil samples was naturally air-dried indoors and then passed through a 2 mm sieve for the determination of soil C, N and P contents. The other part was placed in dry ice at −78.5 °C and returned to the laboratory for the determination of soil microbial C, N and P contents.

2.4. Determination of Samples

According to the method of Hu et al. [14], the contents of organic carbon, total nitrogen and total phosphorus in the soil and plants were determined. The organic carbon content and total nitrogen content of plants and soil were determined by an automatic elemental analyzer (FlashSmart, Thermo Fisher, Waltham, WA, USA). The total phosphorus content of plants was digested with H_2SO_4-H_2O_2, and the total phosphorus (TP) of soil was fused with NaOH and colored with molybdenum–antimony resistance and then determined with an ultraviolet spectrophotometer (Specord 200 PLUS, Analytik, Jena, Germany). Soil microbial biomass carbon (MBC) was measured via the $CHCl_3$ fumigation-K_2SO_4 method, soil microbial biomass nitrogen (MBN) was measured via the $CHCl_3$ fumigation–K_2SO_4 extraction–nitrogen automatic analyzer method and soil microbial biomass phosphorus (MBP) was measured via the $CHCl_3$ fumigation–$NaHCO_3$ extraction–Pi determination–Pi correction method [14,25]. The stoichiometric characteristics of carbon, nitrogen and phosphorus and of plants, soils and microorganisms were expressed by mass ratio.

2.5. Data Processing

Excel 2013 was used to make the statistics and preliminary analysis of the experimental data. A one-way ANOVA, LSD multiple comparison and Person correlation analysis were conducted in SPSS 22 to study the effects of different land-use patterns on the ecological stoichiometry characteristics of grassland soil–plant–microorganism, and the figures were plotted with Origin 2018.

3. Results

3.1. C, N and P Contents and Stoichiometric Ratios of Grassland Plants under Different Utilization Methods

According to Table 1, the C content of grassland plants under the three different utilization methods was EG > GG > MG, and the C content of plants under the three treatments was significantly different ($p < 0.05$). The N content of plants was GG > EG > MG, and the N content of plants in GG was significantly higher than that in MG and EG ($p < 0.05$), but there was no significant difference between MG and EG. The results showed that EG increased the C content, GG increased the N content, and MG increased the P content.

Table 1. C, N and P contents and stoichiometric ratios of grassland plants under different utilization methods.

Utilization Methods	C/(g/kg)	N/(g/kg)	P/(g/kg)	C:N	C:P	N:P
Grazing grassland	429.56 ± 9.92 b	21.47 ± 0.16 a	0.81 ± 0.08 a	20.01 ± 0.61 c	531.74 ± 42.68 b	26.61 ± 2.62 b
Mowing grassland	406.70 ± 5.25 c	18.54 ± 0.13 bc	0.87 ± 0.10 a	21.93 ± 0.22 b	474.63 ± 61.02 bc	21.63 ± 2.67 bc
Enclosed grassland	472.45 ± 12.90 a	19.79 ± 0.03 b	0.42 ± 0.02 b	23.88 ± 0.68 a	1133.09 ± 40.40 a	47.49 ± 2.38 a

Mean value (mean ± standard error, n = 3). Different lowercase letters indicate the difference between treatments reaching a significant level ($p < 0.05$).

The C:N of grassland plants under three different utilization methods showed EG > MG > GG, and there were significant differences among the three treatments ($p < 0.05$). The C:P and N:P showed EG > GG > MG, and the EG treatment was significantly higher than GG and MG ($p < 0.05$). The average values of the C:N, C:P and N:P of grassland plants under three different utilization methods were 21.94, 713.15 and 31.93, respectively.

3.2. C, N and P Contents and Stoichiometric Ratio of Grassland Soil under Different Utilization Methods

According to Table 2, the C content of grassland soil under three different utilization methods was shown as EG > GG > MG, and the C content of soil in EG was significantly higher than that in MG ($p < 0.05$). The N content of soil was GG > EG > MG, and the N content of soil in GG was significantly higher than that in MG and EG ($p < 0.05$). The P content of soil was EG > GG > MG, and the difference among the three treatments was significant ($p < 0.05$). The average values of C, N and P contents in grassland soil under three different utilization methods were 19.77 g/kg, 1.92 g/kg and 0.88 g/kg, respectively. The results showed that EG increased the C and P content, while GG increased the N content.

Table 2. C, N and P contents and stoichiometric ratios of grassland soil under different utilization methods.

Utilization Methods	C/(g/kg)	N/(g/kg)	P/(g/kg)	C:N	C:P	N:P
Grazing grassland	19.48 ± 1.12 ab	2.19 ± 0.09 a	0.85 ± 0.05 b	8.91 ± 0.97 c	22.78 ± 0.83 b	2.58 ± 0.36 b
Mowing grassland	17.37 ± 0.42 b	1.74 ± 0.10 bc	0.53 ± 0.01 c	10.04 ± 0.67 b	32.88 ± 0.79 a	3.29 ± 0.27 a
Enclosed grassland	22.45 ± 1.39 a	1.83 ± 0.09 b	1.26 ± 0.020 a	12.31 ± 1.91 a	17.91 ± 2.07 c	1.46 ± 0.08 c

Mean value (mean ± standard error, n = 3). Different lowercase letters indicate the difference between treatments reaching a significant level ($p < 0.05$).

The C:N of grassland soil under the three different utilization methods was EG > MG > GG, and the difference among the three treatments was significant ($p < 0.05$). The C:P and N:P showed MG > GG > EG, and there were significant differences among the three treatments ($p < 0.05$). The average values of the C:N, C:P and N:P of grassland soil under three different utilization methods were 10.42, 24.52 and 2.44, respectively.

3.3. MBC, MBN and MBP Contents and Stoichiometric Ratio of Grassland Soil under Different Utilization Methods

According to Table 3, the MBC, MBN and MBP contents of grassland soil under three different utilization methods are significantly different. The average values of MBC, MBN and MBP of grassland soil under three different utilization methods were 312.25 mg/kg, 106.97 mg/kg and 12.42 mg/kg, respectively. The MBC content of grassland soil was GG > MG > EG, and GG was significantly higher than EG ($p < 0.05$), but there was no significant difference between GG and MG or between MG and EG. The MBN content was GG > EG > MG, and there were significant differences among the three treatments ($p < 0.05$). The MBP content showed EG > GG > MG, and there were significant differences among the three treatments ($p < 0.05$). The results showed that GG increased the content of MBC and MBN, while EG increased the MBP content.

Table 3. C, N and P contents and stoichiometric ratios of soil microorganisms in grassland under different utilization methods.

Utilization Methods	MBC (mg/kg)	MBN (mg/kg)	MBP (mg/kg)	MBC:MBN	MBC:MBP	MBN:MBP
Grazing grassland	350.24 ± 37.56 a	142.46 ± 10.45 a	10.95 ± 37.56 b	2.48 ± 0.41 bc	32.34 ± 5.93 b	13.11 ± 1.81 a
Mowing grassland	315.61 ± 31.44 ab	81.53 ± 7.23 c	7.50 ± 0.20 c	3.88 ± 0.32 a	42.09 ± 3.98 a	10.86 ± 0.71 a
Enclosed grassland	270.89 ± 11.44 b	96.93 ± 10.14 b	18.82 ± 0.65 a	2.81 ± 0.19 b	14.39 ± 0.20 c	5.14 ± 0.37 b

Mean value (mean ± standard error, n = 3). Different lowercase letters indicate the difference between treatments reaching a significant level ($p < 0.05$).

The stoichiometric ratios of soil microorganisms in grassland under three different utilization methods were also significantly different. The MBC:MBN was MG > EG > GG, and there were significant differences among the three treatments ($p < 0.05$). The MBC:MBP showed MG > GG > EG, and there were significant differences among the three treatments ($p < 0.05$). The MBN:MBP showed GG > MG > EG, but there was no significant difference between GG and MG ($p > 0.05$).

3.4. Correlation Analysis of C, N, P Contents and Stoichiometric Ratios in Plant–Soil–Microorganism

The correlation analysis results showed that SOC was significantly positively correlated with TP, plant C, MBP, soil C:N, plant C:P and plant N:P. It was negatively correlated with plant P, soil C:P, soil N:P and MBC:MBP (Figure 1). TN was positively correlated with plant N and MBN and negatively correlated with MBC:MBN. TP was significantly positively correlated with plant C, MBP, plant C:P and plant N:P. It was significantly negatively correlated with plant P, soil C:P, soil N:P, MBC:MBP and MBN:MBP. Plant C was significantly positively correlated with MBP, plant C:N, plant C:P and plant N:P. It was significantly negatively correlated with plant P, soil C:P, soil N:P, MBC:MBP and MBN:MBP. Plant N was positively correlated with MBN and negatively correlated with MBC:MBN. Plant P was significantly positively correlated with soil C:P, soil N:P, MBC:MBP and MBN:MBP. It was significantly negatively correlated with MBP, soil C:N, plant C:N, plant C:P and plant N:P. MBC was positively correlated with MBC:MBP and MBN:MBP and negatively correlated with soil C:N and plant C:N. MBN was negatively correlated with MBC:MBN. MBP was significantly positively correlated with plant C:N, plant C:P and plant N:P. It was significantly negatively correlated with soil C:P, soil N:P, MBC:MBP and MBN:MBP.

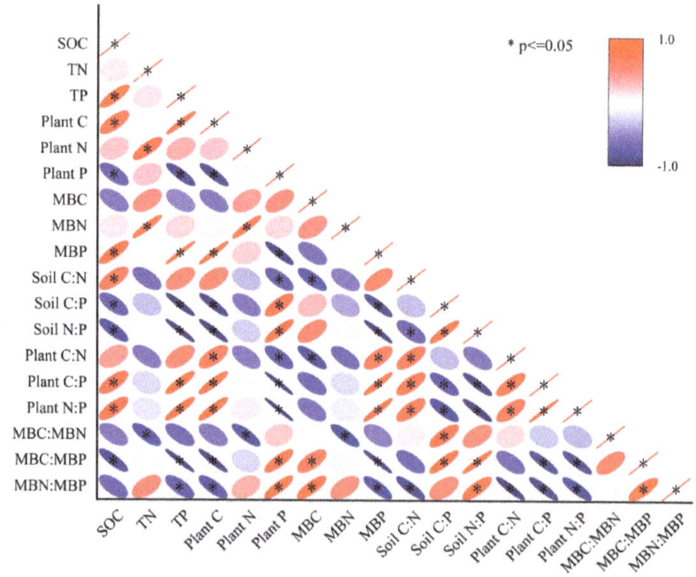

Figure 1. Correlation analysis of plant–soil–microorganism in grassland under different utilization methods.

Soil C:N was significantly positively correlated with plant C:N, plant C:P and plant N:P and significantly negatively correlated with soil N:P and MBN:MBP. Soil C:P was significantly positively correlated with soil N:P, MBC:MBN and MBC:MBP. It was significantly positively correlated with plant C:P and plant N:P. Soil N:P was significantly positively correlated with MBC:MBP and MBN:MBP. It was significantly negatively correlated with plant C:P and plant N:P. Plant C:N was positively correlated with plant C:P and plant N:P, and it was negatively correlated with MBN:MBP ($p < 0.05$). Plant C:N was positively correlated with plant C:P and plant N:P, and it was negatively correlated with MBN:MBP. Plant C:P was positively correlated with plant N:P, and it was negatively correlated with MBC:MBP and MBN:MBP. Plant C:P was positively correlated with plant N:P, and it was negatively correlated with MBC:MBP and MBN:MBP. Plant N:P was negatively correlated with MBC:MBP and MBN:MBP. There was a significant positive correlation between MBC:MBP and MBN:MBP.

4. Discussion

4.1. Plant C, N and P Content and Stoichiometric Ratio

Previous studies have shown that the contents of C, N and P in plants are often different under different utilization patterns or at different growth stages [26–30]. In this study, the plant C content was highest in EG and lowest in MG, showing EG > GG > MG. N and P are important components that affect the synthesis of proteins, nucleic acids, enzymes and other important compounds in plants, and they also affect all physiological activities of a plant's life cycle [27]. Forage requires large amounts of protein and nucleic acids to support the rapid growth of plant tissues due to being harvested and mowed by livestock, so the P content of forage in GG and MG was significantly higher than that in EG ($p < 0.05$). The plant N content in GG was significantly higher than the other two treatments due to factors such as the excretion of feces and urine from grazing livestock, while the MG, which was mowed and utilized for a long time, had the lowest N content ($p < 0.05$). The forage in EG was degraded because of not being used for a long time, and the nutrients such as P were diluted; thus, as the P content decreases, the pasture biomass decreases, the tissue structure is stabilized and the C content increases obviously. C mainly exists in the form of organic matter in the plant, and it accumulates in the process of forage growth, which was also the reason why the C content in EG was the highest and the MG content decreased rapidly, which might be related to the long-term removal of a large amount of C from forage; meanwhile, the MG treatment was the lowest, and this result might be related to the increase of long-term unused C content, which is consistent with the previous research results [31,32].

The C:N, C:P and N:P ratios of plants can indicate the ability of plants to assimilate C for nutrient absorption and the nutrient limitation of plants. Some studies showed that grazing reduces some senescent tissues of plants through livestock feeding, and the growth rate of young tissues is faster, resulting in a lower leaf C:N [33], which is consistent with the findings of this paper; that is, GG significantly reduced plant C:N compared with EG ($p < 0.05$). At the same time, GG increased the C:P and P:N, a result which is consistent with previous research results [34]. In terrestrial ecosystems, N and P are the main nutrient elements that limit plant growth, and the N:P can effectively predict the nutrient limitations of individual plants or ecosystems [35]. Some studies have suggested that when plant N:P is <10 or 14, plant growth is limited by N, and when plant N:P is > 16 or 20, plant growth is limited by P [36,37]. He et al. studied the main grassland vegetation types in China and concluded that the grassland vegetation in China is generally P-limited [37], and Han et al. also reached the same conclusion by studying 753 species of higher terrestrial plants in China [4]. In this study, the N:P values in three treatments were all greater than 20, meaning that they exceeded the average value of N:P in China herbaceous (13.5) [4]. The N:P in EG was the highest, and that in MG was the lowest; EG was 1.79 times and 2.20 times of GG and MG, respectively, indicating that P restriction exists in the three treatments, and the P restriction in EG is the most serious, while the P restriction in MG is the least. We believe that this may be related to the fact that the grass on the ground of MG is cleaned after each cutting, the relatively smooth ground is more conducive to the decomposition of soil P, and the decomposition and turnover rate of soil P is higher than that of N, results which are basically consistent with the previous research results of our team [38]. However, the factors that affect the nutrient limitation of plants are complex and diverse, and the threshold values of the N and P elements are different in different ecosystem types, different regions and different evolution stages [33]. Using the plant N:P ratio to characterize the limiting effect of N and P nutrients is more important to reflect the relative size and mutual transformation trend of N and P elements, and its significance is mainly used as an indication. To refine the threshold value of N and P elements in specific grassland ecosystems in karst areas, further in-depth verification is needed.

4.2. Soil C, N and P Nutrient Content and Stoichiometric Ratio

Soil nutrients are an important source of plant nutrients, and different utilization methods often lead to changes in soil nutrient status, which affects plants' absorption of soil C, N and P, resulting in changes in plant C, N and P stoichiometric ratios and microbial communities and ultimately affecting the stability of the grassland ecosystem [33,39]. In this study, we found some differences in soil nutrients under the three different treatments. EG may be more conducive to SOC accumulation, and its content was 1.15 times that of GG and 1.29 times that of MG, a result which may be related to the long-term enclosure to restore the vegetation characteristics and maintain or improve the soil carbon sequestration capacity, as is consistent with the previous research results [40]. Soil nitrogen is the largest mineral element absorbed by plants from soil, and it is also a major limiting factor for vegetation growth in grassland [41], as it can affect community diversity and ecological function [42]. In this study, the TN content of grassland soil in GG was 2.19 g/kg, which was 1.26 times higher than that of MG and 1.20 times higher than that of EG. This may be related to the excretion of livestock, indicating that the manure of grazing livestock can promote the improvement of TN in grassland soil to a certain extent, which is consistent with the previous results of our research group [38,43]. In this study, the TP content in EG was the highest, which was 1.48 times of GG and 2.38 times of MG. To a certain extent, it also indicated that enclosure had a promoting effect on the increase of soil TP, a result which is consistent with the research in the desert grassland of Northwest China [44]. It may be mainly related to the fact that EG can increase the P content of soil through the cycle of litter compared with GG and MG by human disturbance. However, some studies suggest that after 6–8 years of fencing, there is no difference in soil P content between unfenced grassland and fenced grassland, a result which may be related to the time of fencing [45]. Some studies also consider that with the increase of fence years, the adverse effects of fences become increasingly prominent, which may gradually reduce the content of P [46]. Therefore, Ahmad et al. suggested that biological and abiotic environmental and human factors together determine the changes of soil C, N and P contents [47]. The results of this study also showed that different artificial utilization methods increased the heterogeneity of the grassland soil environment and affected the development trend of soil C, N, P and other elements. Therefore, corresponding restoration strategies should be formulated to adapt to grassland utilization in ecological restoration areas to improve ecosystem productivity.

The ecological stoichiometric ratio of soil C, N and P is a key index to estimate soil quality [48]. Soil C:N and C:P are commonly used to evaluate the availability of soil N and P. Generally, the lower the C:N and C:P, the higher the content of soil available N and P [49]. In this study, the average values of C:N in GG, MG and EG were 8.91, 10.04 and 12.31, respectively. Compared with the average value of C:N of Chinese land surface soil (14.4) [50], EG was closer to the average level of China. The higher the soil C:N ratio, the slower the decomposition rate of SOC, which could maintain soil fertility for a long time. Therefore, the decomposition rate of SOC in EG was lower than that in GG and MG, and the available N and P contents were relatively lower. Some studies suggested that when the C:P is > 300, soil nutrients are net fixed; when the C:P is <200, the P element experiences net mineralization; and when the C:P is 200 < C:P < 300, the soluble phosphorus concentration has little change [51]. In this study, the C:P in GG, MG and EG was 22.78, 32.88 and 17.91, respectively, far lower than the average value in China (136) [52], also indicating that the net mineralization of soil P element in MG is stronger than that in GG and EG, which may be related to the long-term removal of forage material by MG. Plant growth requires the soil to accelerate the provision of P, which intensifies the net mineralization of P and the abundance of P in the soil, a result which is consistent with the previous research results [53]. Soil N:P is an important predictor of nutrient limitation. It is generally believed that plant growth is limited by N when the N:P is <14, plant growth is limited by P when the N:P is >16, and plant growth is limited by both N and P when 14 < N:P < 16 [54]. In this study, the N:P of grassland soil was less than 14 under three different utilization modes, indicating that there was an obvious N limitation, a result which is consistent with previous similar research results [54–57]. However, some studies suggest that soil nutrients in karst areas are mainly restricted by the P element [58]. The reason for the low N:P in EG may be that the higher soil P element has net mineralization and a moderately abundant N element, which reduces its ratio, a result which is consistent with that of the study of Wan et al. [59]. The main reason for the higher N:P ratio in MG than that in GG may be that the higher net mineralization of P in MG soil counteracts the advantage of N emission from livestock manure in GG. Therefore, we believe that the reasons for the limitation of soil N:P may be related to environmental and human factors such as soil type, vegetation characteristics and utilization methods. Meanwhile, the rational application of N fertilizer should be considered under the three different grassland uses in order to alleviate the limitation of N on soil nutrients and promote the balance of soil nutrients.

4.3. Microorganisms C, N and P Content and Stoichiometric Ratio

Microorganisms are the most active components in the soil ecosystem [60]. Soil microorganisms affect the decomposition rate of soil organic matter by adjusting the productivity of mineralization enzymes for soil nutrient availability, which is an ideal indicator to measure soil quality [61]. In this study, there were some differences in soil microbial biomass under different grassland-utilization methods. The MBC in GG and MG under human intervention was higher than that in EG, while the MBN in EG was between that of GG and MG, and the MBP in EG was the highest, which may be closely related to the nutrient demand of grassland and the nutrient supply of soil under different utilization methods.

Soil microorganisms can adapt to changes of the surrounding environment by adjusting their ecological stoichiometric ratio and resource-utilization efficiency [62]. The ecological stoichiometry of soil microorganisms is significantly affected by the ecological stoichiometry of soil [63], which can reflect the microbial community's structure and microbial activity, and it is an important indicator to determine the direction of microbial activity, the release of organic matter nutrients [64] and the change of the microbial community's structure [65]. In this study, the MBC:MBN ranged from 2.48 to 3.88 under three different grassland utilization modes and was significantly lower than the global average range (4.6–10.3) [66]; it was also lower than the average value of desert grassland soil in China (8.55 ± 3.79) [67]. This may be due to the fact that the study area is located in the karst rocky desertification area, and the surface engineering water shortage is serious, resulting in low soil MBC and MBN content, coupled with phosphate mineralization and other factors, thus reducing the ratio of MBC:MBN. There were also significant differences in the MBC:MBP under the three different grassland-utilization modes ($p < 0.05$), with the highest being in MG and the lowest in EG, which may be related to soil water content and soil texture to a certain extent. MG is a relatively strong human disturbance due to the removal of all aboveground plants, as this will reduce the mineralization rate and content of P, only to meet the needs of microorganisms, thus leading to the decrease of MBC:MBP, a result which is consistent with the research results of degraded desert grassland [40]. The MBN:MBP can be used to evaluate the nutrient restriction of soil [68]. Gonzalez-Chavez et al. consider that soil is limited by P when MBN:MBP > 9.6 [69], and soil is limited by N when MBN:MBP < 8.9 [70]. In this study, the MBN:MBP in GG and MG were 13.11 and 10.86, respectively, indicating that P is limited. Meanwhile, the MBN:MBP in EG was only 5.14, indicating that its N is limited; moreover, the value is much higher than the average value of MBN:MBP in the desert grassland of China (0.20 ± 0.20) [67]. The research results also showed that there were some differences in the MBC:MBN, MBC:MBP and MBN:MBP of grassland under different utilization methods, and there were also differences in nutrient requirements for forage growth under different utilization patterns, which ultimately reflected the different microbial activity and material metabolism.

4.4. Plant–Soil–Microbial Stoichiometric Correlation Analysis

The chemical elements of plant–soil–microorganism circulate and flow with each other [25]. In this study, plant C and N were significantly correlated with soil C and N ($p < 0.05$), but plant P was significantly negatively correlated with soil P ($p < 0.05$), indicating that soil P supply has little influence or even a negative influence on plant P absorption. Meanwhile, MBP was significantly positively correlated with soil P ($p < 0.05$), indicating that the P of microbial biomass is more closely related to soil P, as is consistent with the results of Hu et al. [14]. The average content of soil P in the three treatments in this study was 0.88 g/kg, which is not significantly different from that in the previous studies [5,14,71], but the P content of plants (0.70 g/kg) was higher than that in previous studies [14], indicating that the absorption and utilization rate of soil P by forage in the study area is not too low, which may be related to forage species and planting years. Current studies have shown that the ratio of MBC:MBP can be used as an index to measure the potential of releasing P from soil organic matter by microbial mineralization or absorbing P from soil. The ratio of MBC:MBP is generally between 7 and 30. The smaller the ratio is, the greater the potential of microorganisms to release P in mineralized soil organic matter is, and MBP can supplement the soil available phosphorus pool [72]. The higher the ratio of MBC:MBP, the more it indicates that soil microorganisms have a tendency to assimilate soil available phosphorus, and it is easy for microorganisms to compete with crops to absorb soil available phosphorus, which has a strong phosphorus fixation potential [72]. Our results showed that the MBC:MBP in GG, MG and EG was 32.34, 42.09 and 14.39, respectively. In contrast to the those in EG, soil microorganisms have a tendency to assimilate soil available phosphorus in GG and MG, making it easy to compete with crops to absorb soil available phosphorus; CC and MG also have a strong phosphorus fixation potential. Meanwhile, the potential of releasing P from mineralized soil organic matter by soil microorganisms was higher in EG, and MBP could supplement the available P pool of the soil. The reason for this phenomenon may be related to utilization measures such as mowing and grazing, which also fully reflect the complexity of the "plant–soil–microbial" system in the nutrient cycle under different utilization methods. Therefore, we should further deepen the research on the transformation relationship among plants, soils and microbials in the future; enhance the ecosystem service function; and provide theoretical support for the artificial grassland in karst fragile ecological areas to cope with the pressure brought on by global climate change.

5. Conclusions

(1) There were significant differences in C, N, P and the ecological stoichiometric characteristics of plants under different grassland utilization methods. Among the three treatments, the EG treatment had the highest plant C content, while the MG treatment had the lowest. The plant P of GG and MG was significantly higher than that of EG. The plant N content in GG was significantly higher than that in MG and EG. The ratio of plant N:P in the three treatments was higher than 20, and P deficiency and P limitation existed in the three treatments. The P limitation of EG was the most serious, and that of MG was the least.

(2) There were some differences in soil nutrients and ecological stoichiometry among the three treatments. EG may be more beneficial to accumulate SOC and increase TP content. The C:N in EG was higher than that in GG and MG, and the decomposition rate of SOC in EG was slower, which was beneficial to the long-term maintenance

of soil fertility. The C:P ratio of MG was the highest, and the net mineralization of P in MG was stronger than that of GG and EG. Soil N:P ratios under three different grassland-use patterns were all lower than 14, indicating that there was obvious N limitation.

(3) There were some differences in soil microbial biomass under different grassland-utilization methods. The MBC of GG and MG was higher than that of EG, but the MBN of EG was between that of GG and MG, and the MBP of EG was the highest. The MBC:MBN under the three different grassland-utilization methods ranged from 2.48 to 3.88, which was lower than the global average range. The MBC:MBP also showed significant differences, with MG being the highest and EG the lowest. The MBN:MBP ratios of GG and MG were both higher than 9.6, indicating that they were P-limited, while the MBN:MBP of EG was lower than 8.9, indicating that it was N-limited.

(4) Plant C and N were significantly correlated with soil C and N ($p < 0.05$). Plant P was significantly negatively correlated with soil P ($p < 0.05$), but MBP was significantly positively correlated with soil TP ($p < 0.05$). The results of MBC:MBP showed that soil microorganisms in GG and MG had a tendency to assimilate soil available phosphorus, and there was a tendency for microorganisms to compete with crops to absorb soil available phosphorus, which had a strong phosphorus fixation potential. Meanwhile, soil microorganisms in EG had a greater potential to release P from mineralized soil organic matter, and MBP could supplement soil available phosphorus pool.

In summary, this study described the element cycle and nutrient limitation from the perspective of the plant–soil–microorganism system, which can provide some basis for the fertilization management of artificial grassland in the karst rocky desertification area, such as necessary fertilization management or other measures to supplement elements according to the limitation of N and P, so as to solve the element limitation problem. However, due to the relationship between the experimental time, this study is still not systematic and perfect; the later period should be long-term observation to form systematic research results to provide the necessary theoretical support for the scientific management of grasslands in the ecological restoration area.

Author Contributions: Conceptualization, G.P. and S.S.; methodology, X.W. and Y.C.; software, X.W. and Y.C.; validation, Y.C., S.S. and G.P.; formal analysis, X.W. and Y.C.; investigation, G.P.; resources, Y.C.; data curation, X.W.; writing—original draft preparation, G.P.; writing—review and editing, S.S.; visualization, S.S.; supervision, Y.C. and G.P.; project administration, Y.C.; funding acquisition, Y.C. and G.P. All authors have read and agreed to the published version of the manuscript.

Funding: This research was funded by the Natural Science Research Project of Education Department of Guizhou Province (Qianjiaohe KY Zi (2022) 157), the Academic New Seedling Fund Project of Guizhou Normal University (Qianshi Xinmiao B15) and Special Research Project on Cadre Training for Rural Revitalization in Cooperation between East and West of Tongren City in 2022 (12004/0622051).

Data Availability Statement: Not applicable.

Acknowledgments: The authors thank the reviewers and editor for their insightful comments and constructive suggestions.

Conflicts of Interest: The authors declare no conflict of interest.

References

1. Reich, P.B.; Walters, M.B.; Ellsworth, D.S. From tropics to tundra: Global convergence in plant functioning. *Proc. Natl. Acad. Sci. USA* **1997**, *94*, 13730–13734. [CrossRef] [PubMed]
2. Reich, P.B.; Tjoelker, M.G.; Machado, J.L.; Oleksyn, J. Universal scaling of respiratory metabolism, size and nitrogen in plants. *Nature* **2006**, *439*, 457–461. [CrossRef] [PubMed]
3. Wang, C.; Hou, Y.; Zheng, R.L.; Hu, Y.X.; Li, X.N. Plant diversity and nitrogen addition affect the architecture of plant–soil–microbe stoichiometric networks. *Plant Soil* **2023**. [CrossRef]
4. Han, W.X.; Fang, J.Y.; Reich, P.B.; Ian Woodward, F.; Wang, Z.H. Biogeography and variability of eleven mineral elements in plant leaves across gradients of climate, soil and plant functional type in China. *Ecol. Lett.* **2011**, *14*, 788–796. [CrossRef]

5. Song, S.Z.; Wang, X.L.; He, C.; Chi, Y.K. Effects of utilization methods on C, N, P rate and enzyme activity of artificial grassland in karst desertification area. *Agronomy* **2023**, *13*, 1368. [CrossRef]
6. Sun, B.; Barnes, A.D. Editorial: Soil-root-microbe interactions promote soil and plant health. *Front. Microbiol.* **2023**, *14*, 1155234. [CrossRef]
7. Chen, Z.J.; Li, Y.; Hu, M.; Xiong, Y.W.; Huang, Q.Z.; Jin, S.; Huang, G.H. Lignite bioorganic fertilizer enhanced microbial co-occurrence network stability and plant–microbe interactions in saline-sodic soil. *Sci. Total Environ.* **2023**, *879*, 163113. [CrossRef]
8. Elser, J.J.; Fagan, W.F.; Kerkhof, A.J.; Swenson, N.G.; Enquist, B.J. Biological stoichiometry of plant production: Metabolism, scaling and ecological response to global change. *New Phytol.* **2010**, *186*, 593–608. [CrossRef]
9. Zhang, C.P.; Li, Q.; Feng, R.Q.; Zhang, Z.H.; Yang, Y.F.; Liu, J. C:N:P stoichiometry of plant-soil-microbe in the secondary succession of zokor-made mounds on Qinghai-Tibet Plateau. *Environ. Res.* **2023**, *222*, 115333. [CrossRef]
10. Isles, P.D.F. The misuse of ratios in ecological stoichiometry. *Ecology* **2020**, *101*, e03153. [CrossRef]
11. Oldroyd, G.E.D.; Leyser, O.A. Plant's diet, surviving in a variable nutrient environment. *Science* **2020**, *368*, eaba0196. [CrossRef] [PubMed]
12. Wang, Y.; Ren, Z.; Ma, P.; Wang, Z.; Elser, J.J. Effects of grassland degradation on ecological stoichiometry of soil ecosystems on the Qinghai-Tibet Plateau. *Sci. Total Environ.* **2020**, *722*, 137910. [CrossRef] [PubMed]
13. Cleveland, C.C.; Liptzin, D. C:N:P stoichiometry in soil: Is there a "Redfield ratio" for the microbial biomass? *Biogeochemistry* **2007**, *85*, 235–252. [CrossRef]
14. Hu, P.L.; Wang, K.L.; Zeng, Z.X.; Zhang, H.; Li, S.S.; Song, X.J. Ecological stoichiometric characteristics of plants, soil, and microbes of *Pennisetum purpureum* cv. Guimu-1 pastures at different rehabilitation ages in a karst rocky desertification region. *Acta Ecol. Sin.* **2017**, *37*, 896–905.
15. Garousi, F.; Shan, Z.J.; Ni, K.; Yang, H.; Shan, J.; Cao, J.H.; Jiang, Z.C.; Yang, J.L.; Zhu, T.B.; Müller, C. Decreased inorganic N supply capacity and turnover in calcareous soil under degraded rubber plantation in the tropical karst region. *Geoderma* **2021**, *381*, 114754. [CrossRef]
16. Xiong, K.N.; Chi, Y.K.; Shen, X.Y. Research on photosynthetic leguminous forage in the karst rocky desertification regions of Southwestern China. *Pol. J. Environ. Stud.* **2017**, *26*, 2319–2329. [CrossRef]
17. Wang, K.L.; Yue, Y.M.; Chen, H.S.; Zeng, F.P. Mechanisms and realization pathways for integration of scientifiic poverty alleviation and ecosystem services enhancement. *Bull. Chin. Acad. Sci.* **2020**, *35*, 1264–1272.
18. Pei, G.T.; Sun, J.F.; He, T.X.; Hu, B.Q. Effects of long-term human disturbances on soil microbial diversity and community structure in a karst grassland ecosystem of northwestern Guangxi, China. *Chin. J. Plant Ecol.* **2021**, *45*, 74–84. [CrossRef]
19. Liu, M.; Bai, X.Y.; Tan, Q.; Luo, G.J.; Zhao, C.W.; Wu, L.H.; Hu, Z.Y.; Ran, C.; Deng, Y.H. Monitoring impacts of ecological engineering on ecosystem services with geospatial techniques in karst areas of SW China. *Geocarto. Int.* **2022**, *7*, 5091–5115. [CrossRef]
20. Lu, M.; Liu, K.; Zhang, L.; Zeng, F.; Song, T.; Peng, W.; Du, H. Stoichiometric variation in soil carbon, nitrogen, and phosphorus following cropland conversion to forest in Southwest China. *Forests* **2022**, *13*, 1155. [CrossRef]
21. Wen, L.; Li, D.; Xiao, X.; Tang, H.M. Alterations in soil microbial phospholipid fatty acid profile with soil depth following cropland conversion in karst region, southwest China. *Environ. Sci. Pollut. Res.* **2023**, *30*, 1502–1519. [CrossRef] [PubMed]
22. Song, M.; Peng, W.X.; Du, H.; Xu, Q.G. Responses of soil and microbial C:N:P stoichiometry to vegetation succession in a karst region of Southwest China. *Forests* **2019**, *10*, 755. [CrossRef]
23. Chen, H.J.; Peng, W.X.; Du, H.; Song, T.Q.; Zeng, F.P.; Wang, F. Effect of different grain for green approaches on soil bacterial community in a karst region. *Front. Microbiol.* **2020**, *11*, 577242. [PubMed]
24. Xiao, S.Z.; He, J.H.; Zeng, C.; Wang, J.L. Soil chemical properties under various land-use types in the karst area with a case study in Shibing County of China. *J. Chem.* **2021**, *2021*, 5523060. [CrossRef]
25. Lin, X.G. *Principles and Methods of Soil Microbiology Research*; Higher Education Press: Beijing, China, 2010.
26. Bradshaw, C.; Kautsky, U.; Kumblad, L. Ecological stoichiometry and multi-element transfer in a coastal ecosystem. *Ecosystems* **2012**, *15*, 591–603. [CrossRef]
27. Chen, X.; Chen, H.Y.H.; Chang, S.X. Meta-analysis shows that plant mixtures increase soil phosphorus availability and plant productivity in diverse ecosystems. *Nat. Ecol. Evol.* **2022**, *6*, 1112–1121. [CrossRef]
28. Kohmann, M.M.; Silveira, M.L.; da Silva Cardoso, A.; Bracho, R. Short-term impacts of prescribed fire on C, N, and P dynamics in a subtropical rangeland. *Plant Soil* **2023**. [CrossRef]
29. Dai, M.; Wang, T.; Wang, Y.; Xu, J. Effects of warming and phosphorus enrichment on the C:N:P stoichiometry of *Potamogeton crispus* organs. *Front. Plant Sci.* **2022**, *13*, 814255. [CrossRef]
30. Taggart, C.B.; Muir, J.P.; Brady, J.A.; Kan, E.; Mitchell, A.B.; Obayomi, O. Impacts of biochar on *Trifolium incarnatum* and *Lolium multiflorum*: Soil nutrient retention and loss in sandy loam amended with dairy manure. *Agronomy* **2023**, *13*, 26. [CrossRef]
31. Xu, H.; Qu, Q.; Wang, Z.; Xue, S.; Xu, Z. Plant-soil-enzyme C-N-P stoichiometry and microbial nutrient limitation responses to plant-soil feedbacks during community succession: A 3-year pot experiment in China. *Front. Plant Sci.* **2022**, *13*, 1009886. [CrossRef]
32. Wang, J.; He, K.; Bao, G.; He, L.; Xu, H.; Song, M.; Jin, L.; Wei, X. Effects of increased phosphorus fertilizer on C, N, and P stoichiometry in different organs of bluegrass (*Poa* L.) at different growth stages. *Front. Ecol. Evol.* **2023**, *11*, 1092840. [CrossRef]

33. Xie, L.L.; Wang, X.L.; Ma, Y.; MA, Y.S.; Wang, Y.L.; Zhou, X.B. Effect of the no-grazing practice in regreening period on the quality and stoichiometric ratio of C, N, P of plant community in alpine meadow. *Acta Agrestia Sin.* **2023**, *31*, 1454–1460.
34. Wang, M.; Zhang, Y.; Li, R.Q.; Xin, X.P.; Zhu, X.Y.; Cao, J.; Zhou, Z.Y.; Yan, R.R. Effects of Grazing Disturbance on the Stoichiometry of Nitrogen and Phosphorus in Plant Organs of *Leymus chinensis* Meadow Steppe. *Sci. Agric. Sin.* **2022**, *55*, 1371–1384.
35. Tessier, J.T.; Raynal, D.J. Use of nitrogen to phosphorus ratios in plant tissue as an indicator of nutrient limitation and nitrogen saturation. *J. Appl. Ecol.* **2003**, *40*, 523–534. [CrossRef]
36. Güsewell, S. N:P ratios in terrestrial plants: Variation and functional significance. *New Phytol* **2004**, *164*, 243–266. [CrossRef]
37. He, J.S.; Wang, L.; Flynn, D.F.B.; Wang, X.P.; Ma, W.H.; Fang, J.Y. Leaf nitrogen:phosphorus stoichiometry across Chinese grassland biomes. *Oecologia* **2008**, *155*, 301–310. [CrossRef] [PubMed]
38. Song, S.Z. Study on Coupling of Degraded Grassland Improvement and Semi-House Feeding Cattle and Sheep in the Karst Rocky Desertification Area. Master's Thesis, Guizhou Normal University, Guiyang, China, 2019.
39. Peco, B.; Navarro, E.; Carmona, C.P.; Medina, N.G.; Marques, M.J. Effects of grazing abandonment on soil multifunctionality: The role of plant functional traits. *Agr. Ecosyst. Environ.* **2017**, *249*, 215–225. [CrossRef]
40. Li, S.Y.; Cui, Y.X.; Sun, Z.J.; Liu, H.X.; Ye, H.W. Effect of grazing exclusion on soil organic carbon and stoichiometry characteristics of soil microbial biomass in sagebrush desert. *Acta Pratacul. Sin.* **2023**, *32*, 58–70.
41. Hali, A.; Sun, Z.J.; He, P.X.; Liu, H.X. Effects of grazing exclusion on soil nitrogen content and its component characteristics in sagebrush desert grassland. *J. Soil Wat. Conserv.* **2022**, *36*, 222–230.
42. Li, W.; Su, T.; Shen, Y.; Ma, H.; Zhou, Y.; Lu, Q.; Wang, G.; Liu, Z.; Li, J. Effects of warming seasonal rotational grazing on plant communities' structure and diversity in desert steppe. *Ecol. Evol.* **2023**, *13*, e9748. [CrossRef]
43. Chi, Y.K.; Song, S.Z.; Zhao, D.G.; Wu, J.H. Effects of fertilization on plant-soil ecological stoichiometric characteristics of degraded artificial grassland in karst area. *Fresen. Environ. Bull.* **2022**, *31*, 10190–10198.
44. Fan, D.D.; Liu, Y.J.; Cao, H.L.; Chen, H.; Kong, W.D.; Li, X.Z. On the effect of fencing on physicochemical property and microbial community of grassland soils. *Sci. Technol. Rev.* **2022**, *40*, 41–51.
45. Lu, X.; Yan, Y.; Sun, J.; Zhang, X.; Chen, Y.; Wang, X.; Cheng, G. Short-term grazing exclusion has no impact on soil properties and nutrients of degraded alpine grassland in Tibet, China. *Solid Earth* **2015**, *6*, 1195–1205. [CrossRef]
46. Wu, G.L.; Du, G.Z.; Liu, Z.H.; Thirgood, S. Effect of fencing and grazing on a Kobresia-dominated meadow in the QinghaiTibetan Plateau. *Plant Soil* **2009**, *319*, 115–126. [CrossRef]
47. Ahmad, E.H.; Demisie, W.; Zhang, M. Effects of land use on concentrations and chemical forms of phosphorus in different-size aggregates. *Eurasian Soil Sci.* **2017**, *50*, 1435–1443. [CrossRef]
48. Wang, S.Q.; Yu, G.R. Ecological stoichiometry characteristics of ecosystem carbon, nitrogen and phosphorus elements. *Acta Ecol. Sin.* **2008**, *28*, 3937–3947.
49. Shu, M.; Jiang, T.; Wang, D.L.; Lian, Z.; Tang, J.X.; Kong, T.; Xu, Y.Y.; Han, X.M. Soil ecological stoichiometry under the planted of *Pinus sylvestris* var. mongolica forests with different stand ages in the Horqin Sandy Land. *Arid Zone Res.* **2018**, *35*, 789–795.
50. Tian, H.Q.; Chen, G.S.; Zhang, C.; Melillo, J.M.; Hall, C.A.S. Pattern and variation of C:N:P rations in China's soils: Synthesis of observational data. *Biogeochemistry* **2010**, *98*, 139–151. [CrossRef]
51. Jia, Y.; Xu, B.C.; Li, F.M.; Wang, X.L. Availability and contributions of soil phosphorus to forage production of seeded alfalfa in semiarid Loess Plateau. *Acta Ecol. Sin.* **2007**, *27*, 42–47.
52. Zhu, Q.L.; Xing, X.Y.; Zhang, H.; An, S.S. Soil ecological stoichiometry under different vegetation area on loess hilly gully region. *Acta Ecol. Sin.* **2013**, *33*, 4674–4682.
53. He, J.; Wei, Q.Q.; Zhong, Y.X.; Luo, L.L. Eco-stoichiometry characteristics of grape fields with different planting years in the east foothills of Helan Mountain. *Agric. Res. Arid Areas* **2020**, *38*, 23–30.
54. Zhao, D.; Li, F.; Yang, Q.; Wang, R.; Song, Y.; Tao, Y. The influence of different types of urban land use on soil microbial biomass and functional diversity in Beijing, China. *Soil Use Manag.* **2013**, *29*, 230–239. [CrossRef]
55. Wu, L.F.; Wang, Z.Q.; Wang, Y.; Liu, Y.G.; Yang, B.; Zhang, Y.F. Relationship between soil C, N, P stoichiometric characteristics and enzyme activity in karst plateau soils with different degree of rocky desertification. *Ecol. Environ. Sci.* **2019**, *28*, 2332–2340.
56. Yang, X.W.; Liu, J.; Hou, M.Y.; Cheng, X.M.; Huang, X.X. Enzyme activities and stoichiometric characteristics of rhizometric and non-rhizosphere soil in different ancient tea gardens in Fengqing County. *Chin. J. Appl. Eco.* **2023**, *29*, 1368.
57. Tian, J.; Sheng, M.Y.; Wang, P.; Wen, P.C. Influence of land use change on litter and soil C, N, P stoichiometric characteristics and soil enzyme activity in karst ecosystem, Southwest China. *Environ. Sci.* **2019**, *40*, 4278–4286.
58. Liu, L.B.; Zhong, Q.L.; Ni, J. Ecosystem C:N:P stoichiometry and storages of a secondary plateau-surface karst forest in Guizhou Province, southwestern China. *Acta Ecol. Sin.* **2019**, *39*, 8606–8614.
59. Wan, F.; Meng, Z.J.; Dang, X.H.; Wang, R.D.; Zhang, H.M. C, N and P ecological stoichiometry characteristics of a Stipa species plant–soil system subject to grazing exclusion in a desert steppe. *Acta Pratacul. Sin.* **2020**, *29*, 49–55.
60. Zhang, J.H.; Li, M.X.; Xu, L.; Zhu, J.X.; Dai, G.H.; He, N.P. C:N:P stoichiometry in terrestrial ecosystems in China. *Sci. Total Environ.* **2021**, *795*, 148849. [CrossRef]
61. Arunachalam, A.; Pandey, H.N. Ecosystem restoration of jhum fallows in northeast india: Microbial C and N along altitudinal and successional gradients. *Restor. Ecol.* **2003**, *11*, 168–173. [CrossRef]
62. Manzoni, S.; Trofymow, J.A.; Jackson, R.B.; Porporato, A. Stoichiometric controls on carbon, nitrogen, and phosphorus dynamics in decomposing litter. *Ecol. Monogr.* **2010**, *80*, 89–106. [CrossRef]

63. Mooshammer, M.; Wanek, W.; Zechmeister-Boltenstern, S.; Richter, A. Stoichiometric imbalances between terrestrial decomposer communities and their resources: Mechanisms and implications of microbial adaptations to their resources. *Front. Microbiol.* **2014**, *5*, 22. [CrossRef]
64. Heuch, C.; Weig, A.; Spohn, M. Soil microbial biomass C:N:P stoichiometry and microbial use of organic phosphorus. *Soil Biol. Biochem.* **2015**, *85*, 119–129. [CrossRef]
65. Li, J.W.; Liu, Y.L.; Hai, X.Y.; Shangguan, Z.P.; Deng, L. Dynamics of soil microbial C:N:P stoichiometry and its driving mechanisms following natural vegetation restoration after farmland abandonment. *Sci. Total. Environ.* **2019**, *693*, 133613. [CrossRef] [PubMed]
66. Tischer, A.; Potthast, K.; Hamer, U. Land-use and soil depth affect resource and microbial stoichiometry in a tropical mountain rainforest region of southern Ecuador. *Oecologia* **2014**, *175*, 375–393. [CrossRef] [PubMed]
67. Pan, Y.; Fang, F.; Tang, H. Patterns and internal stability of carbon, nitrogen, and phosphorus in soils and soil microbial biomass in terrestrial ecosystems in China: A Data Synthesis. *Forests* **2021**, *12*, 1544. [CrossRef]
68. Hartman, W.H.; Richardson, C.J. Differential nutrient limitation of soil microbial biomass and metabolic quotients(qCO_2): Is there a biological stoichiometry of soil microbes. *PLoS ONE* **2013**, *8*, 57127. [CrossRef]
69. González-Chávez, M.C.A.; Aitkenhead-Peterson, J.A.; Gentry, T.J.; Zuberer, D.; Hons, F.; Loeppert, R. Soil microbial community, C, N, and P responses to long-term tillage and crop rotation. *Soil Till. Res.* **2016**, *106*, 285–293. [CrossRef]
70. Wang, B.R.; Yang, J.J.; An, S.S.; Zhang, H.X.; Bai, X.J. Effects of vegetation and topography features on ecological stoichiometry of soil and soil microbial biomass in the hilly-gully region of the Loess Plateau, China. *Chin. J. Appl. Ecol.* **2018**, *29*, 247–259.
71. Yu, Y.F.; Peng, W.X.; Song, T.Q.; Zeng, F.P.; Wang, K.L.; Wen, L.; Fan, F.J. Stoichiometric characteristics of plant and soil C, N and P in different forest types in depressions between karst hills, southwest China. *Chin. J. Appl. Ecol.* **2014**, *25*, 947–954.
72. Peng, P.Q.; Zhang, W.J.; Tong, C.L.; Qiu, S.J.; Zhang, W.C. Soil C, N and P contents and their relationships with soil physical properties in wetlands of Dongting Lake floodplain. *Chin. J. Appl. Ecol.* **2005**, *16*, 1872–1878.

agronomy

MDPI

Article

Interacting Effects of Nitrogen Addition and Mowing on Plant Diversity and Biomass of a Typical Grassland in Inner Mongolia

Zhaoping Yang [1,*,†], Hugjiltu Minggagud [2,†], Qian Wang [1] and Hongyuan Pan [2]

1 Jiangsu Collaborative Innovation Center of Atmospheric Environment and Equipment Technology, School of Environmental Science and Engineering, Nanjing University of Information Science & Technology, Nanjing 210044, China
2 School of Ecology and Environment, Inner Mongolia University, Hohhot 010021, China
* Correspondence: yangzp04@163.com
† These authors contributed equally to this work.

Abstract: It is unclear how N addition and mowing interactively affect grassland community structure and function. We studied the short-term effects of N addition and mowing on the species diversity and biomass of a typical grassland in Inner Mongolia using a split-plot design, with the whole plot representing the mowing treatment and the subplot representing the N addition rate. Three stubble heights were set at 10 cm (M2), 6 cm (M3) and 3 cm (M4), and a no-mowing treatment (M1) was also included. N addition consisted of five rates, i.e., 0 (N1), 2 (N2), 5 (N3), 10 (N4) and 20 (N5) g N $m^2.yr^{-1}$. Mowing significantly increased species diversity. Mowing with a stubble height of 10 cm (M2) significantly increased both species biomass and community biomass. Heavy mowing (3 cm stubble height, M4) weakened the difference in *C. squarrosa* biomass caused by N addition. Species richness decreased, while community biomass and *Leymus chinensis* biomass increased significantly with increasing N addition rate. Mowing alleviated the negative effects of N addition on species richness, and this effect was influenced by stubble height. Community biomass and *L. chinensis* biomass tended to be stable when N addition was greater than 10 g N $m^2.yr^{-1}$. The N saturation threshold of *C. squarrosa* biomass was much lower than that of community biomass and *L. chinensis* biomass. Species richness was negatively correlated with the Gini coefficient and litter production, which indicated that light competition and litter accumulation were important factors affecting the decrease in species richness in our study.

Keywords: nitrogen addition; mowing; species diversity; biomass; Inner Mongolia

Citation: Yang, Z.; Minggagud, H.; Wang, Q.; Pan, H. Interacting Effects of Nitrogen Addition and Mowing on Plant Diversity and Biomass of a Typical Grassland in Inner Mongolia. *Agronomy* **2023**, *13*, 2125. https://doi.org/10.3390/agronomy13082125

Academic Editor: Lili Jiang

Received: 1 July 2023
Revised: 8 August 2023
Accepted: 11 August 2023
Published: 14 August 2023

1. Introduction

Since the industrial revolution, human activities such as fossil fuel combustion and agricultural production have led to a substantial increase in reactive nitrogen (N) entering the terrestrial ecosystem [1,2]. It is estimated that the amount of N deposition caused by human activities has increased from 31.6 Tg per year in 1860 to 103 Tg per year in the 1990s and is expected to increase to 195 Tg per year by 2050 [3]. Many studies have indicated that the enhancement of N deposition could profoundly impact the structure, function and processes of ecosystems [4,5]. The response of ecosystems to N deposition has always been a popular research topic in the field of global change. Under the background of the continuous increase in global N deposition, exploring the effects of N addition on plant community structure and function will be beneficial for better understanding ecosystem responses to global change.

As one of the most widely distributed vegetation types worldwide, grassland accounts for approximately 20% of the global land surface [6] and plays an important role in supporting livestock production and providing ecosystem services such as sand fixation, soil and water conservation and biodiversity conservation [7,8]. N is a key factor limiting the productivity of grassland ecosystems [4]. N enrichment can directly increase the availability of soil N and subsequently alleviate N limitation to plant growth [9,10], thereby changing plant community structure and function [11,12]. The response of soil N availability and the community light environment to nitrogen enrichment is a gradually changing process, resulting in phase-dependent community changes with the continuous input of reactive N [13]. Therefore, exploring the short-term effects of N addition on grassland community biomass and diversity is critical for comprehensively evaluating the effects of N enrichment on grassland ecosystems. Most studies have indicated that N enrichment results in a decline in species diversity [14–16]; however, some studies have also shown that N enrichment leads to positive or no change in diversity [17–19]. The changes in community biomass and species diversity and their underlying mechanisms vary depending on the grassland type, experiment duration and N deposition rate [14,16,20]. Given these uncertainties, it is necessary to verify the response of grassland community biomass and diversity to N addition.

Mowing for hay and silage is both a widely used human disturbance and a pasture management activity in grassland ecosystems across the world [21,22]. Mowing can change community structure and function by removing aboveground biomass, altering structural heterogeneity and decreasing litter accumulation [23,24]. Moderate mowing can improve species diversity and biomass, while heavy mowing can decrease productivity and thus threaten ecosystem stability [25,26]. Stubble height is a key factor affecting mowing effects on community structure and function [22]. A study carried out in a semiarid steppe showed that plant biomass production increased under light mowing (stubble height > 12 cm) but decreased under heavy mowing (stubble height < 6 cm) [22]. Suitable stubble height may increase light availability [25], thereby increasing species diversity by promoting the growth of subdominant species and forbs [26]. For example, one previous study indicated that an 8 cm stubble height significantly increased the Shannon diversity index of a typical grassland in Inner Mongolia. Another study, however, reported that there was no significant difference in Shannon index and evenness index among stubble heights [22]. In general, the effects of stubble height on community structure and function are still unclear.

The N addition can promote light interception by upper-layer tall grasses and thereby decrease the light availability for lower-layer plants; however, mowing can initially enhance the light availability of subdominant species [26,27]. N addition causes an increase in litter accumulation, while mowing can remove litter [26,28], which produces the opposite effects on seed germination and seedling regeneration. The decrease in N accumulation caused by the mowing-induced removal of aboveground biomass would weaken the soil N availability enhanced by N enrichment [29–31]. However, it is unclear how N addition and mowing interactively affect grassland community structure and function. The uncertainty related to the mechanisms of how the plant community responds to N addition and mowing requires further investigation to fill our knowledge gap.

The typical grasslands in Inner Mongolia account for approximately 10.5% of the total grassland area in China and are an important livestock production base and green ecological barrier in northern China [32,33]. Continuous input of reactive N will inevitably change the plant community structure and function of this typical grassland [34]. Apart from N enrichment, mowing is a widely used practice in this temperate grassland [22]. Therefore, the objective of this study was to determine the independent and joint effects of N addition and mowing on the plant biomass and species diversity of a typical grassland in Inner Mongolia.

2. Materials and Methods

2.1. Study Site

Our study was conducted in a typical steppe region located in the Grassland Ecology Research Station of Inner Mongolia University (44°09′~44°18′ N, 116°12′~116°30′ E, 1160 m a.s.l). The climate is temperate and semiarid with a dry spring and wet summer. The long-term (1971–2010) mean annual temperature was 2.8 °C, with the lowest monthly mean temperature in January (−19.03 °C) and highest in July (21.38 °C). The annual mean precipitation is 272 mm, 87% of which falls in the plant growing season from May to September. The soil is a sandy loam light chestnut soil according to the Chinese soil taxonomy classification, which is equivalent to Calcic-orthic Aridisol in the United States Soil Taxonomy classification system [22]. Mean soil pH of the top 10 cm is 8.4. Soil organic carbon, total N and total phosphorus concentrations for the top 10 cm are 12.8, 1.7 and 0.3 g kg^{-1}, respectively. The vegetation is dominated by native rhizomatous perennial C3 grass, i.e., *Leymus chinensis* and native perennial C4 bunchgrass, i.e., *Cleistogenes squarrosa*, which together account for more than 80% of the total aboveground biomass. The experimental grassland with fairly uniform vegetation appeared topographically and floristically similar. According to background investigation, the species richness, Shannon-Wiener index and evenness index ranged from 6 to 10 averaging 8, 0.32 to 0.63 averaging 0.47 and 0.15 to 0.34 averaging 0.23, respectively.

2.2. Experimental Design and Sampling

The experiment was established in 2018 using a split-plot design, with the whole plot being the mowing treatment and the subplot being N addition. The mowing treatment included a no-mowing control (M1) and mowing at stubble heights of 10 cm (M2), 6 cm (M3) and 3 cm (M4). Mowing treatments were implemented once a year at the time of peak plant biomass (mid-August). N addition consisted of five rates, i.e., 0 (N1), 2 (N2), 5 (N3), 10 (N4) and 20 (N5) g N $m^2.yr^{-1}$. There were 20 experimental treatments in total, with 5 replicate blocks for each treatment, resulting in 100 treatment plots in total. Each plot was 5 m × 5 m in area, separated by 1 m rows. Starting in 2018, $CO(NH_2)_2$ was applied to the soil surface in each N3-N5 plot twice a year with 50% of N applied at each time. The first nitrogen application time was the early growing season (8–10 May), and the second application time was the middle of the growing season (3–5 July). For the N2 plot, the level of 2 g N $m^2.yr^{-1}$ was too low to split and applied once a year at the first application time.

In 2020, just before annual mowing at the time of peak biomass (middle–late August), one 1 m × 1 m quadrat was randomly placed within 5 m × 5 m each plot for vegetation measurement. The quadrat was placed at least 50 cm inside the border of each plot to avoid edge effects. Individuals of all vascular plant species were identified in quadrats, and the height and density of each species were measured. All vascular plants in the quadrat were clipped and sorted by species. Litter was collected from each quadrat after clipping. The plant samples were oven-dried at 65 °C for 48 h to a constant weight and then weighed.

2.3. Data Analysis and Statistics

For each plot, we determined the vascular plant species richness (S), Shannon-Wiener index (H') and evenness index (J') as $H' = -\sum pi \times \ln pi$ and $J' = H'/\ln S$, where pi is the proportion of the total abundance of the ith species. Community density was abundance of vascular plants per unit area. Community biomass was calculated by adding the dry weights of all vascular plants. The asymmetry of plant height can reflect species competition ability for light resources. The Gini coefficient (G) was calculated based on the plant height of each species to quantify the intensity of interspecific competition in the community [35]. The Gini coefficient was calculated as follows:

$$G = \frac{\sum_{i=1}^{n} \sum_{j=1}^{n} |Xi - Xj|}{2n\overline{X}}$$

where Xi and Xj represent the plant height of the ith species and jth species, respectively. $\overline{X}$ is the mean height of all species and n is the number of species for each plot.

Statistical analyses were conducted using the GLM procedure. Treatment effects on community density, species diversity, biomass, litter production and the Gini coefficient were analyzed by univariate ANOVA with mowing and nitrogen addition as the fixed factors and with block as random factor. If the interaction term was nonsignificant, then it was removed from the statistical model, and the main factor effects were reanalyzed using one-way ANOVAs. Data were tested for normality using the Kolmogorov–Smirnov test and for equality of error variance using Levene's test. Duncan's test was used for multiple comparisons in all cases of significant treatment effects. Regression models with N as a continuous variable were used to determine the general relationship between N addition and the response variable, i.e., response variable = Intercept + Slope × Ln (N). The relationships between species richness and litter production and the Gini coefficient were analyzed using linear regression. In all analyses, a difference was considered significant at $p < 0.05$. All analyses were conducted using version 20.0 of SPSS software (SPSS Inc., Chicago, IL, USA).

3. Results

3.1. Plant Diversity

Mowing treatment had significant impacts on the plant diversity of a typical steppe in Inner Mongolia ($p < 0.05$) (Table 1, Figure 1A–C). There was no significant difference in the species richness between mowing at 10 cm (M2) and 3 cm (M4) treatments ($p > 0.05$), but they were significantly lower compared with mowing at 3 cm (M3) treatment ($p < 0.05$). In no-mowing control (M1) and mowing at 10 cm (M2) treatments, the species richness with the rate of 20 g N $m^2.yr^{-1}$ (N5) was significantly lower than that with the rates of 0 (N1) and 2 g N $m^2.yr^{-1}$ (N2). Mowing at a stubble height of 10 cm (M2) increased plant diversity as measured by the Shannon-Wiener index more than mowing at a stubble height of 3 cm (M4) ($p < 0.05$), and there was no significant difference in Shannon-Wiener diversity between these treatments and mowing at 6 cm stubble height (M3) ($p > 0.05$). In no-mowing treatment (M1), the Shannon-Wiener diversity with the rate of 20 g N $m^2.yr^{-1}$ was significantly lower than that with no N addition (N1). No significant difference in Pielou evenness existed between mowing with 3–10 cm stubble height (M2-M4) ($p > 0.05$), but they were significantly higher than that in no-mowing control treatment (M1) ($p < 0.05$). In all mowing treatments, N addition did not significantly alter the species evenness.

Table 1. ANOVA of the mowing effects, N addition and their interactions on community response variables.

	Source of Variance							
	Whole Plot				Sub Plot			
	B df (4, 12)		M df (3, 12)		N df (4, 64)		M * N df (12, 64)	
	F-Test	*p*	*F*-Test	*p*	*F*-Test	*p*	*F*-Test	*p*
Species richness	1.03	0.43	4.41	0.03	10.63	0.00	0.52	0.89
Shannon–Wiener index	0.07	0.99	11.68	<0.001	1.45	0.23	0.80	0.65
Pielou evenness index	0.17	0.95	4.85	0.02	0.50	0.73	0.69	0.75
Community biomass	0.21	0.93	6.21	0.01	26.24	0.00	2.29	0.02
Litter production	0.51	0.74	145.12	<0.001	66.00	<0.001	3.92	<0.001
L. chinensis biomass	0.97	0.46	6.20	0.01	16.14	0.00	1.04	0.43
C. squarrosa biomass	0.40	0.81	7.96	0.00	8.78	0.00	2.98	0.00
Gini coefficient	1.60	0.32	5.52	0.01	4.56	0.01	0.40	0.96
Community density	10.39	0.02	8.69	0.00	14.27	0.00	0.61	0.82

Note: B, block; M, mowing; N, nitrogen addition; * interaction; df, degree of freedom.

The species richness with the level of 20 g N $m^2.yr^{-1}$ (N5) was significantly lower than that with no N addition (N1) ($p < 0.05$), and there was no significant difference in species richness among the rates of 2-10 N g N $m^2.yr^{-1}$ (N2-N4) ($p > 0.05$) (Figure 1D). Species richness showed a significant decreasing trend with increasing N addition rate ($p < 0.001$). N addition did not significantly affect Shannon-Wiener diversity (Table 1, Figure 1E) or Pielou evenness (Table 1, Figure 1F) ($p > 0.05$).

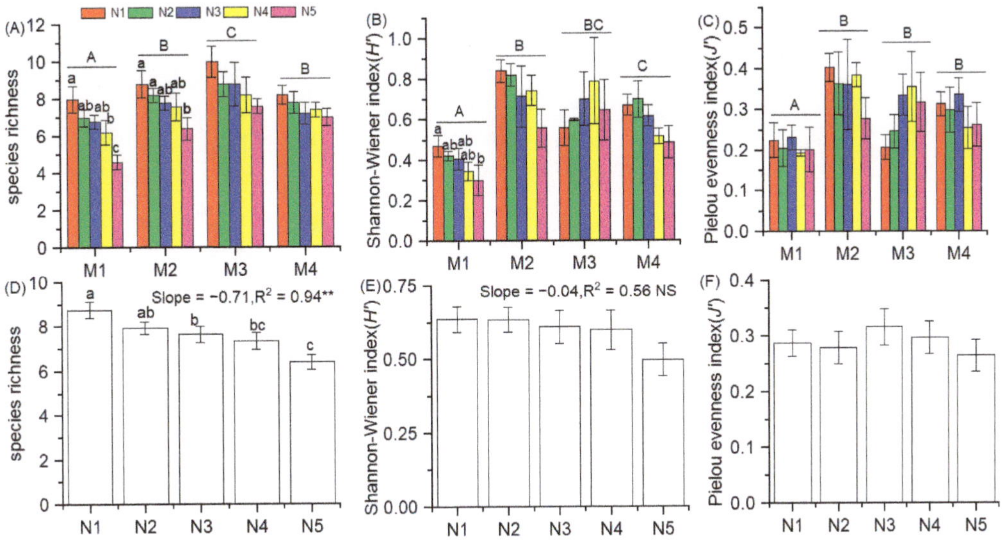

Figure 1. Effects of mowing and N addition on plant diversity. The values are mean ± SE. For (**A–C**), different lowercase letters under the line denote significant differences among the N addition treatments with $p < 0.05$; different capital letters above the line denote significant differences among mowing treatments with $p < 0.05$. For (**D–F**), different lowercase letters denote significant differences among the treatments with $p < 0.05$; regression parameters were estimated for species richness and Shannon-Wiener index using log-linear model with N treatment as a continuous predictor, i.e., species richness or Shannon-Wiener index = Intercept + Slope × Ln (N); significant differences are reported as NS, $p > 0.05$; ** $p < 0.01$. M1, no-mowing control; M2, M3 and M4 are mowing at stubble heights of 10 cm, 6 cm and 3 cm, respectively; N1, N2, N3, N4 and N5 are N addition rate with 0, 2, 5, 10 and 20 g N $m^2.yr^{-1}$, respectively.

3.2. Biomass

The biomass of community, *L. chinensis* and *C. squarrosa* was significantly affected by the N addition rate and mowing ($p < 0.05$) (Table 1, Figure 2A–C). The mowing and N addition rates interactively affected the community biomass and biomass of *C. squarrosa*. The biomass of community, *L. chinensis* and *C. squarrosa* in mowing at 10 cm (M2) treatment were the highest and were significantly higher than those in no-mowing (M1) and mowing at 3 cm (M4) treatments ($p < 0.05$) (Figure 2A–C). In all mowing treatments, no significant differences in community biomass and *L. chinensis* biomass between the rates of 10 g N $m^2.yr^{-1}$ (N4) and 20 g N $m^2.yr^{-1}$ (N5) treatments were observed ($p > 0.05$), and they were both significantly higher than those with the no N addition (N1) treatment ($p < 0.05$) (Figure 2A,B). In the no-mowing (M1) and mowing at 6–10 cm (M2 and M3) treatments, there was no significant difference in *C. squarrosa* biomass between the rates of 10 g N $m^2.yr^{-1}$ (N4) and 20 g N $m^2.yr^{-1}$ (N5) ($p > 0.05$), and they were both lower than that with the rate of 2 g N $m^2.yr^{-1}$ (N2) ($p < 0.05$) (Figure 2C). No significant difference in *C. squarrosa* biomass was found in any of the nitrogen addition rates in mowing at 3 cm (M4) treatment ($p > 0.05$) (Figure 2C).

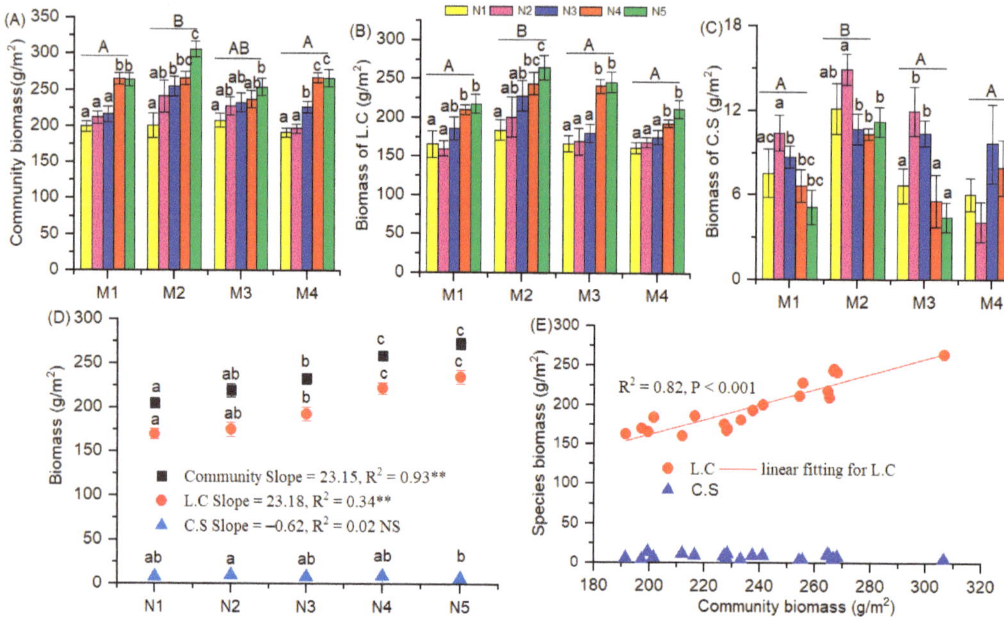

Figure 2. Effects of mowing and nitrogen addition on biomass and the relationship between community biomass and species biomass. The values are mean ± SE. For (**A–C**), different lowercase letters under the line denote significant differences among the N addition treatments with $p < 0.05$; different capital letters above the line denote significant differences among the mowing treatments with $p < 0.05$. For (**D**), different lowercase letters under the line denote significant differences among the treatments with $p < 0.05$; regression parameters were estimated for biomass using a log-linear model with N treatment as a continuous predictor, i.e., biomass = Intercept + Slope × Ln (N); significant differences are reported as NS, $p > 0.05$; ** $p < 0.01$. L.C: *Leymus chinensis*, C.S: *Cleistogenes squarrosa*. M1, no-mowing control; M2, M3 and M4 are mowing at stubble heights of 10 cm, 6 cm and 3 cm, respectively; N1, N2, N3, N4 and N5 are N addition rate with 0, 2, 5, 10 and 20 g N $m^2.yr^{-1}$, respectively. (**E**) indicates the relationship between community biomass and species biomass.

The community biomass with rate 5 g N $m^2.yr^{-1}$ (N3) was significantly lower than that with the rates of 10 g N $m^2.yr^{-1}$ (N4) and 20 g N $m^2.yr^{-1}$ (N5), but significantly higher than that with no N addition (N1) ($p < 0.05$) (Figure 2D). The response of *L. chinensis* biomass to the N addition rate was consistent with that of community biomass (Figure 2D). The plot with the rate of 2 g N $m^2.yr^{-1}$ (N2) had the highest *C. squarrosa* biomass, which was significantly higher than that in the plot with the rate of 20 g N $m^2.yr^{-1}$ (N5) ($p < 0.05$) (Figure 2D). Both community biomass and *L. chinensis* biomass showed a significant increasing trend with increasing N addition rate ($p < 0.05$) (Figure 2D). Although *C. squarrosa* biomass presented a decreasing trend with increasing N addition rate, this trend was not statistically significant ($p > 0.05$). The results of the linear regressions showed that the community biomass was significantly positively correlated with *L. chinensis* biomass ($p < 0.05$) (Figure 2E).

3.3. Gini Coefficient

The effect of mowing on the Gini coefficient was highly significant ($p < 0.05$) (Table 1). In no-mowing (M1) and mowing at 3 cm (M4) treatments, the Gini coefficient with a rate of 20 g N $m^2.yr^{-1}$ (N5) was significantly higher than that with no N addition (N1) ($p < 0.05$), and no significant differences existed among the rates of 2-10 N g N $m^2.yr^{-1}$ (N2-N4) ($p > 0.05$) (Figure 3A). N addition did not significantly change the Gini coefficient when mowing at 10 cm (M2) and 6 cm (M3) ($p > 0.05$) (Figure 3A). The Gini coefficient in mowing at 6 cm (M3) treatment was significantly lower than that in the no-mowing (M1) treatment ($p < 0.05$) and was not significantly different from those in the mowing at 10 cm (M2) and 3 cm (M4) treatments ($p > 0.05$) (Figure 3A). The main effect of the N addition rate had a significant impact on the Gini coefficient ($p < 0.05$) (Table 1, Figure 3B). No significant differences in the Gini coefficient were detected among the rate of 0–5 g N $m^2.yr^{-1}$ (N1-N3) ($p > 0.05$) (Figure 3B), and they were all significantly lower than that with the rate of 20 g N $m^2.yr^{-1}$ (N5) ($p < 0.05$). The Gini coefficient showed a significant increasing trend with the increasing rate of N addition ($R^2 = 0.14$, $p < 0.01$) (Figure 3B).

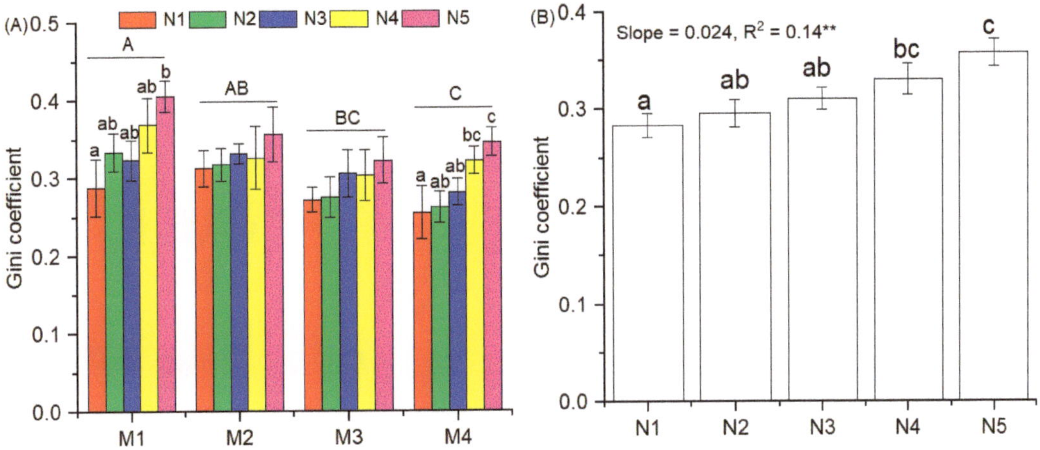

Figure 3. Effects of mowing and N addition on the Gini coefficient. The values are mean ± SE. For (**A**), different lowercase letters under the line denote significant differences among nitrogen addition treatments with $p < 0.05$; different capital letters above the line denote significant differences among mowing treatments with $p < 0.05$. For (**B**), different lowercase letters under the line denote significant differences among treatments with $p < 0.05$; regression parameters were estimated for Gini coefficient using log-linear model with N treatment as a continuous predictor, i.e., Gini coefficient = Intercept + Slope × Ln (N); significant differences are reported as ** $p < 0.01$. M1, no-mowing control; M2, M3 and M4 are mowing at stubble heights of 10 cm, 6 cm and 3 cm, respectively; N1, N2, N3, N4 and N5 are N addition rate with 0, 2, 5, 10 and 20 g N $m^2.yr^{-1}$, respectively.

3.4. Community Density and Litter

The main effects of mowing and N addition rate had significant effects on community density and litter, and their interactions had a significant effect on litter (Table 1, Figure 4). No significant difference in community density was found between the no-mowing (M1) and mowing at 10 cm (M2) treatments ($p > 0.05$), which were significantly different from that in the mowing at 3 cm (M4) treatment ($p < 0.05$) (Figure 4A). There was no significant difference in the community density among the different N addition rates in no-mowing (M1) treatment ($p > 0.05$) (Figure 4A), while the community density with a rate of 20 g N $m^2.yr^{-1}$ (N5) was significantly higher than that with the rates of 0 (N1) and 2 g N $m^2.yr^{-1}$ (N2) in the other mowing treatments ($p < 0.05$) (Figure 4A). The community densities with the rates of 10 g N $m^2.yr^{-1}$ (N4) and 20 g N $m^2.yr^{-1}$ (N5) were significantly higher than that in the other N addition rates ($p < 0.05$) (Figure 4C). No significant differences in community density were observed between the rates of 2–5 g N $m^2.yr^{-1}$ (N2-N3) or between the rates of 5–10 g N $m^2.yr^{-1}$ (N3-N4) ($p > 0.05$) (Figure 4C).

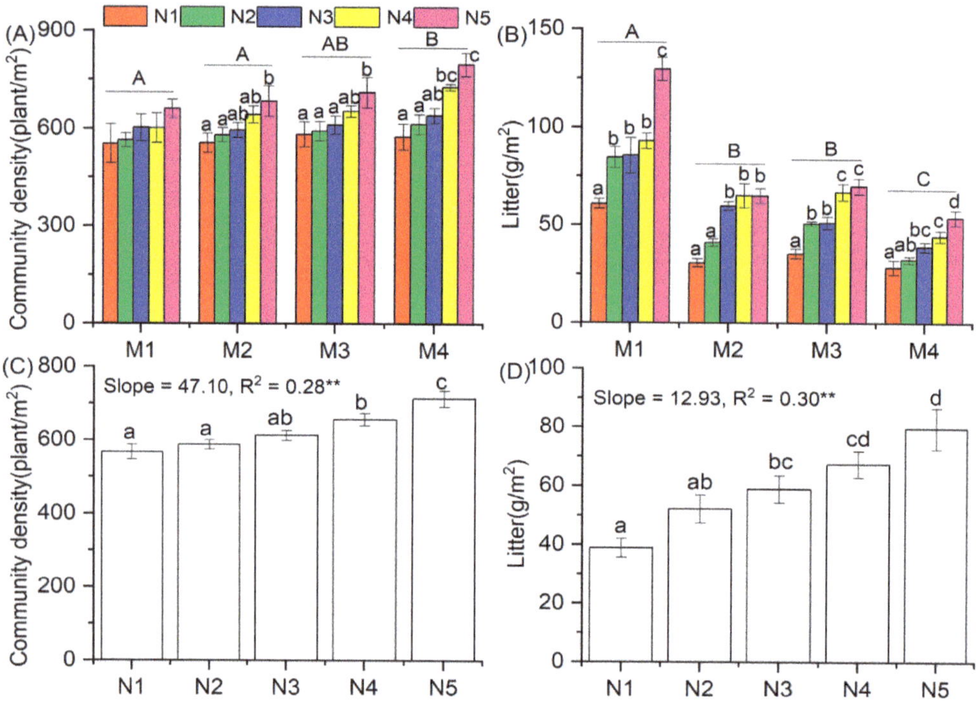

Figure 4. Effects of mowing and N addition on community density and litter. The values are mean ± SE. For (**A**,**B**), different lowercase letters under the line denote significant differences among N addition treatments with $p < 0.05$; different capital letters above the line denote significant differences among mowing treatments with $p < 0.05$. For (**C**,**D**), different lowercase letters under the line denote significant differences among treatments with $p < 0.05$; regression parameters were estimated for community density and litter using log-linear model with N treatment as a continuous predictor, i.e., community density or litter = Intercept + Slope × Ln (N); significant differences are reported as ** $p < 0.01$. M1, no-mowing control; M2, M3 and M4 are mowing at stubble heights of 10 cm, 6 cm and 3 cm, respectively; N1, N2, N3, N4 and N5 are N addition rate with 0, 2, 5, 10 and 20 g N $m^2.yr^{-1}$, respectively.

No significant difference in litter production existed between the mowing at 10 cm (M2) and 6 cm (M3) treatments ($p > 0.05$), and they were significantly lower than that in the no-mowing (M1) treatment and significantly higher than the litter production in mowing at 3 cm (M4) treatment ($p < 0.05$) (Figure 4B). In the mowing at 10 cm (M2) and 6 cm (M3) treatments, no significant differences in litter production existed between the rates of 10 (N4) and 20 g N $m^2.yr^{-1}$ (N5) ($p < 0.05$). The litter production with the rate of 10 g N $m^2.yr^{-1}$ (N4) and 20 g N $m^2.yr^{-1}$ (N5) was not significantly different ($p > 0.05$), but was significantly higher than that with the rates of 0 (N1) and 2 g N $m^2.yr^{-1}$ (N2) ($p < 0.05$) (Figure 4D). In the regression analyses, the N addition rate had a significant positive linear relationship with community density ($R^2 = 0.28$, $p < 0.01$) and litter ($R^2 = 0.30$, $p < 0.01$).

3.5. Relationship between Species Richness and the Gini Coefficient and Litter

The Gini coefficient was significantly negatively correlated with species richness ($R^2 = 0.76$, $p < 0.001$, Figure 5A). Our results also showed that species richness was significantly negatively correlated with litter ($R^2 = 0.63$, $p < 0.001$, Figure 5B).

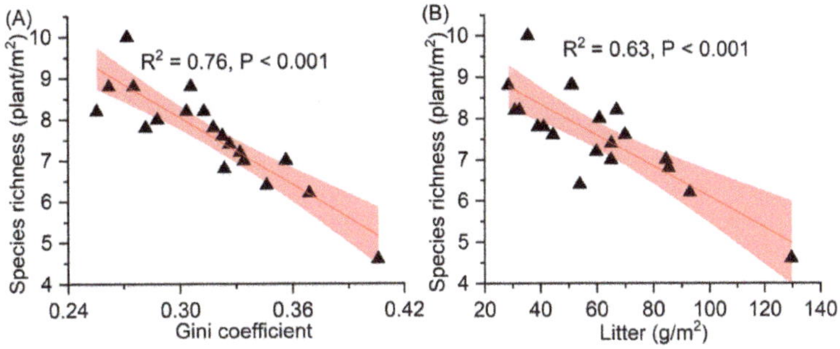

Figure 5. Relationship between species richness and the Gini coefficient (**A**) and litter (**B**).

4. Discussion

The effects of exogenous nitrogen input on plant species diversity have always been a popular topic in the field of global change. Most studies have shown that nitrogen enrichment will lead to a decline in species diversity [12,36–38]; however, some studies have shown that N enrichment has positive or neutral effects on species diversity [17–19]. As N addition increases, species richness decreases significantly, which is consistent with existing research results [36,38]. Various mechanisms have been proposed to explain the decrease in species richness caused by nitrogen enrichment, such as light competition [27,39], regeneration constraints and soil acidification [38,40–42]. Some studies have confirmed that nitrogen addition can shift plant competition from underground nutrient resources to aboveground light resources [43]. There was a significant positive correlation between the N addition rate and Gini coefficient ($p < 0.05$) (Figure 3B), indicating that the competition for light resources among species intensified with increasing N addition rate. The intensification of aboveground light resource competition in this study occurred for the following reasons. First, as the dominant species in our experimental grassland, *L. chinensis* is an upper-layer tall grass and has an efficient N utilization strategy [44]. Therefore, the increase in soil N availability caused by N addition promoted its rapid growth. With an increasing N addition rate, the interception of light resources by *L. chinensis* was enhanced, which gradually reduced the light availability of middle- and lower-layer plants [26,27]. Second, the significant increase in litter caused by N addition reduced the light availability of lower-layer vegetation in the community through the shading effect (Figure 4D). Third, N addition significantly increased the community density (Figure 4C), which reduced the light transmittance of the community. The light efficiency for the lower-layer dwarf or shade-intolerant species gradually decreased with increasing N rate, and finally, they were excluded from the community. Therefore, our results highlight that light competition was an important reason for the significant decrease in species richness due to the increase in the N addition rate in this study [25–27]. Species richness had a significant negative correlation with litter production (Figure 5B), which indicates that the increase in litter production was an important factor that caused a significant decrease in species richness [26,28]. The increase in litter production can inhibit seed germination and seedling colonization, which may also have been an important factor in the negative effect of N addition on species richness in this study. Numerous studies have shown that N enrichment will significantly reduce the Shannon-Wiener diversity index [12,36] and species evenness [45,46], which is different from our results. The short experimental duration may have been the reason why the Shannon-Wiener diversity index and species evenness among the N addition rates were not significant.

The enhanced species richness that occurred with mowing is in line with the results of studies conducted in threatened temperate grasslands [47]. The increase of species richness with mowing may be ascribed to mowing's positive effects on subdominant species and germination rates via increased light availability [25,26]. In this study, mowing reduced the Gini coefficient (Figure 3A), indicating that mowing decreased the intensity of light competition among species in the community, which was beneficial for promoting the survival of light-demanding seedlings and low-growth species and led to an increase in species richness. In this study, mowing increased the species richness, Shannon-Wiener diversity index and evenness index (Figure 1A–C), demonstrating that short-term mowing was beneficial for maintaining the species diversity of the community. Research has confirmed that litter accumulation will have a negative impact on species diversity because it will inhibit seedling establishment of some target species and reduce the survival rate of seedlings [26,28]. Mowing significantly reduced litter production (Figure 4B), which contributed to improving forb richness, thus leading to a significant increase in community species richness. Mowing weakens the inhibition of upper-layer tall plants on middle- and lower-layer plants [25], which causes mowing treatment to significantly improve species evenness.

Species richness was significantly different among the N addition rates in no-mowing (M1) and mowing at 10 cm (M2) treatments; however, there was no significant difference among the N addition rates in mowing at 6 cm (M3) and 3 cm (M4) treatments (Figure 1A). Significant differences in the Shannon-Wiener diversity indices among N addition rates only existed in no-mowing (M1) treatment (Figure 1B). Our results demonstrated that mowing alleviated the negative effects of N addition on species richness and Shannon-Wiener diversity, which is consistent with the research results carried out in European grasslands [48,49]. Therefore, mowing under the background of nitrogen deposition can maintain the relative stability of species richness in typical grasslands. In addition, our results further confirm that decreased mowing to reduce the negative effects of nitrogen addition on species diversity is influenced by stubble height.

N is one of the key factors limiting the productivity of grassland ecosystems [50]. Many studies have shown that N addition can increase soil N availability, alleviate nutrient constraints on plant growth, and promote the growth of plant aboveground parts. In our study, the community biomass increased significantly with increasing N addition (Figure 2D). As an N-exploitative species, *L. chinensis* has a flexible N absorption strategy and can adjust its N uptake preference as nutrient availability changes [44]. With the increase in N addition, the rhizome of *L. chinensis* will extend more ramets to use nutrients more effectively and maintain its dominance in the community, resulting in a significant increase in *L. chinensis* biomass. The significant positive correlation between *L. chinensis* biomass and community biomass indicated that the increase in community biomass was mainly due to the contribution of *L. chinensis* (Figure 2E). Our results indicated that the community biomass and dominant species biomass increased first and then tended to be stable with increasing N addition rate (Figure 2D). Previous studies have shown that there may be a threshold for aboveground productivity with increasing N addition rate, beyond which N fertilizer effects on aboveground productivity will be weakened [51]. Community biomass and *L. chinensis* biomass tended to be stable when N deposition was greater than 10 g N $m^2.yr^{-1}$ (Figure 2D), indicating that the saturation threshold of grassland ecosystem productivity in response to the N addition rate in this study was 10 g N $m^2.yr^{-1}$, which is consistent with results from a previous nitrogen addition experiment on a typical grassland in Inner Mongolia [14]. *C. squarrosa* is a perennial bunchgrass and belongs to the middle- and lower-layer dominant species in our experimental grassland. The increase in the N addition rate reduced the light availability of the middle and lower layers of plants [26,27], thereby limiting the growth of *C. squarrosa* and resulting in its biomass not significantly changing or even showing a downward trend with increasing N addition rate (Figure 2D). With a conservative nutrient acquisition strategy, *C. squarrosa* exhibited a low plasticity of N uptake under a changing soil nutrient environment [44], which may be another important reason why its biomass did not change with increasing N addition rate.

Previous studies have shown that mowing may stimulate plant growth through compensatory growth [52]. Mowed plants may exhibit equal, under- or overcompensatory growth depending on stubble height [53]. In this study, compared with no mowing, with mowing, both species biomass and community biomass at 10 cm stubble height showed overcompensatory growth, resulting in the species biomass and community biomass in mowing at 10 cm (M2) treatment being higher than that in no-mowing (M1) treatment (Figure 2A–C). Mowing at a 10 cm stubble height can remove the apical meristem and senescent plant tissues of upper-layer tall plants that stimulate the growth of collateral branches and formation of new tissues. At the same time, light mowing can improve the light conditions of middle- and lower-layer plants [54], thus increasing their biomass. Therefore, mowing at 10 cm (M2) treatment significantly increased the biomass of *L. chinensis* and *C. squarrosa* compared with no-mowing (M1) treatment. Due to the increase in the biomass of the upper-layer tall plants and the middle- and lower-layer plants, the community biomass increased significantly (Figure 2A). When the stubble height was 3–6 cm, *L. chinensis* and *C. squarrosa* had equal compensatory growth, which may explain why there was no significant difference between community biomass in the M1 treatment and community biomass in the M3 and M4 treatments.

In mowing at 3 cm (M4) treatment, no significant difference existed in *C. squarrosa* biomass among the N addition rates (Figure 2C), which indicated that heavy mowing (i.e., 3 cm stubble height) could weaken the difference in *C. squarrosa* biomass caused by the N addition treatment. *C. squarrosa* biomass in no-mowing (M1), mowing at 10 cm (M2) and 6 cm (M3) treatments reached the maximum value with level of 2 g N $m^2.yr^{-1}$ (N2) and then decreased (Figure 2C), which reflected that the N saturation threshold of *C. squarrosa* biomass was much lower than that of community biomass and *L. chinensis* biomass. When the N addition rate was greater than 2 g N $m^2.yr^{-1}$, the changes in the *L. chinensis* biomass and *C. squarrosa* biomass showed an opposite trend (Figure 2B,C). We speculate that there are two reasons for the continuous decline in *C. squarrosa* from the level of 2 g N $m^2.yr^{-1}$ (N2) to 20 g N $m^2.yr^{-1}$ (N5). *C. squarrosa* has a conservative N acquisition strategy [44], and the increase in the N addition rate did not promote the increase in *C. squarrosa* biomass. The continuous increase in *L. chinensis* biomass gradually enhanced the shading effect on *C. squarrosa* [26], which is another reason for this result. With the increase in the N addition rate, the space released by the decrease in *C. squarrosa* biomass was occupied by *L. chinensis* to enhance its community dominance under the interference of N addition.

This study concludes that light availability is a key factor affecting the response of species richness to N addition and mowing based on the measurement results of the Gini coefficient. In future work, community light intensity should be measured to further elucidate the role of light limitation in reducing species richness under N addition and mowing interference. Similarly, it is necessary to further clarify the relationship between seed germination, seedling colonization and litter production to determine how litter accumulation regulates the response of species richness to N addition and mowing. Our results show that mowing and N addition both increase community biomass, and they have a synergistic effect on changes in ecosystem productivity. However, studies on different types of grassland ecosystems show that long-term mowing will lead to a decrease in community biomass [29]. N addition over a long period will lead to community biomass first increasing and then reaching saturation [14,55]. Theoretically, community biomass will be in a relatively stable state under N addition and mowing interference. Therefore, long-term monitoring and research are needed to determine the time threshold of community biomass stability. Long-term N enrichment will lead to soil acidification [38,42], which can be alleviated by mowing to remove N from the ecosystem [30,31]. Stubble height is the key factor affecting the amount of nutrients removed [22]. A suitable stubble height that can eliminate soil acidification and maintain ecosystem stability under future conditions of N deposition is important to study. N addition and mowing both affect soil nitrogen availability [56,57], and the response of plants to soil N availability is species-specific [44,58]. The existing studies show that N addition can promote the growth of nitrogen-loving plants, thus excluding forbs that are insensitive to soil N availability. Therefore, clarifying the N acquisition strategy and plasticity of species is the key approach for explaining the influence of N enrichment and mowing on species richness and community biomass.

5. Conclusions

Our results show that mowing significantly increased species diversity. Species richness decreased significantly with increasing N addition rate. Short-term N addition did not significantly affect Shannon-Wiener diversity and Pielou evenness. Mowing at a stubble height of 10 cm significantly increased the biomass of community, *L. chinensis* and *C. squarrosa*, indicating that light mowing can improve ecosystem productivity. Community biomass and *L. chinensis* biomass showed a significant increasing trend with increasing N addition rate. Species richness was significantly negatively correlated with litter production and the Gini coefficient, which demonstrated that light competition and litter accumulation were important reasons for the significant decrease in species richness in this study. Our results also demonstrated that mowing can alleviate the negative effects of N addition on species richness, and mowing to alleviate the negative effects of N addition on species diversity is influenced by stubble height. The response mechanism of the grassland community to the interaction between N addition and mowing and the suitable stubble height maintaining ecosystem stability under future conditions of N deposition need further study.

Author Contributions: Z.Y. and H.M. designed the research and conducted the field investigation. Z.Y. wrote the original draft. Q.W. and H.P. analyzed the data. All authors have read and agreed to the published version of the manuscript.

Funding: This study was funded by the National Natural Science Foundation of China (31870413).

Data Availability Statement: Please contact the first author with requests for data.

Acknowledgments: Special thanks are given to Baoyin Taogetao for his valuable help in experiment design. This experimental site has been managed by Grassland Ecology Research Station of Inner Mongolia University. Many thanks are expressed to the anonymous reviewers for their helpful suggestions.

Conflicts of Interest: The authors declare no conflict of interest.

References

1. Vitousek, P.M.; Aber, J.D.; Howarth, R.W.; Likens, G.E.; Matson, P.A.; Schindler, D.W.; Schlesinger, W.H.; Tilman, D.G. Human alteration of the global nitrogen cycle: Sources and consequences. *Ecol. Appl.* **1997**, *7*, 737–750. [CrossRef]
2. Galloway, J.N.; Bleeker, A.; Erisman, J.W. The human creation and use of reactive nitrogen: A global and regional perspective. *Annu. Rev. Environ. Resour.* **2021**, *46*, 255–288. [CrossRef]
3. Galloway, J.N.; Dentener, F.J.; Capone, D.G.; Boyer, E.W.; Howarth, R.W.; Seitizinger, S.P.; Asner, G.P.; Cleveland, C.C.; Green, P.A.; Holland, E.A.; et al. Nitrogen cycles: Past, present and future. *Biogeochemistry* **2004**, *70*, 153–226. [CrossRef]
4. Elser, J.J.; Bracken, M.E.S.; Cleland, E.E.; Gruner, D.S.; Harpole, W.S.; Hillebrand, H.; Ngai, J.T.; Seabloom, E.W.; Shurin, J.B.; Smith, J.E. Global analysis of nitrogen and phosphorus limitation of primary producers in freshwater, marine and terrestrial ecosystems. *Ecol. Lett.* **2007**, *10*, 1135–1142. [CrossRef]
5. Pierik, M.; van Ruijven, J.; Bezemer, T.M.; Geerts, R.H.E.M.; Berendse, F. Recovery of plant species richness during long-term fertilization of a species-rich grassland. *Ecology* **2011**, *92*, 1393–1398. [CrossRef] [PubMed]
6. Scurlock, J.; Hall, D.O. The global carbon sink: A grassland perspective. *Glob. Chang. Biol.* **1998**, *4*, 229–233. [CrossRef]
7. Havstad, K.M.; Peters, D.P.C.; Skaggs, R.; Brown, J.; Bestelmeyer, B.; Fredrickson, E.; Herrick, J.; Wright, J. Ecological services to and from rangelands of the United States. *Ecol. Econ.* **2007**, *64*, 261–268. [CrossRef]
8. Schirpke, U.; Kohler, M.; Leitinger, G.; Fontana, V.; Tasser, E.; Tappeiner, U. Future impacts of changing land-use and climate on ecosystem services of mountain grassland and their resilience. *Ecosyst. Serv.* **2017**, *26*, 79–94. [CrossRef]
9. Chapman, S.K.; Devine, K.A.; Curran, C.; Jones, R.O.; Gilliam, F.S. Impacts of Soil Nitrogen and Carbon Additions on Forest Understory Communities with a Long Nitrogen Deposition History. *Ecosystems* **2016**, *19*, 142–154. [CrossRef]
10. Stevens, C.J.; Lind, E.M.; Hautier, Y.; Harpole, W.S.; Borer, E.T.; Hobbie, S.; Seabloom, E.W.; Ladwig, L.; Bakker, J.D.; Chu, C.J.; et al. Anthropogenic nitrogen deposition predicts local grassland primary production worldwide. *Ecology* **2015**, *96*, 1459–1465. [CrossRef]
11. Tognetti, P.M.; Prober, S.M.; Báez, S.; Chaneton, E.J.; Firn, J.; Risch, A.C.; Schuetz, M.; Simonsen, A.K.; Yahdjian, L.; Borer, E.T.; et al. Negative effects of nitrogen override positive effects of phosphorus on grassland legumes worldwide. *Proc. Natl. Acad. Sci. USA* **2021**, *118*, e2023718118. [CrossRef]
12. Isbell, F.; Reich, P.B.; Tilman, D.; Hobbie, S.E.; Polasky, S.; Binder, S. Nutrient enrichment, biodiversity loss, and consequent declines in ecosystem productivity. *Proc. Natl. Acad. Sci. USA* **2013**, *110*, 11911–11916. [CrossRef] [PubMed]

13. Tian, Q.; Yang, L.; Ma, P.; Zhou, H.; Liu, N.; Bai, W.; Wang, H.; Ren, L.; Lu, P.; Han, W.; et al. Below-ground-mediated and phase-dependent processes drive nitrogen-evoked community changes in grasslands. *J. Ecol.* **2020**, *108*, 1874–1887. [CrossRef]
14. Bai, Y.; Wu, J.; Clark, C.M.; Naeem, S.; Pan, Q.; Huang, J.; Zhang, L.; Han, X. Tradeoffs and thresholds in the effects of nitrogen addition on biodiversity and ecosystem functioning: Evidence from inner Mongolia Grasslands. *Glob. Chang. Biol.* **2010**, *16*, 358–372. [CrossRef]
15. Stevens, C.J.; Dise, N.B.; Mountford, J.O.; Gowing, D.J. Impact of nitrogen deposition on the species richness of grasslands. *Science* **2004**, *303*, 1876–1879. [CrossRef] [PubMed]
16. Humbert, J.Y.; Dwyer, J.M.; Andrey, A.; Arlettaz, R. Impacts of nitrogen addition on plant biodiversity in mountain grasslands depend on dose, application duration and climate: A systematic review. *Glob. Chang. Biol.* **2016**, *22*, 110–120. [CrossRef]
17. Goldberg, D.E.; Miller, T.E. Effects of different resource additions on species diversity in an annual plant community. *Ecology* **1990**, *71*, 213–225. [CrossRef]
18. Huberty, L.E.; Gross, K.L.; Miller, C.J. Effects of nitrogen addition on successional dynamics and species diversity in Michigan old-fields. *J. Ecol.* **1998**, *86*, 794–803. [CrossRef]
19. Zheng, H.P.; Chen, Z.X.; Wang, S.R.; Niu, J.Y. Effects of fertilizer on plant diversity and productivity of desertified alpine grassland at Maqu, Gansu. *Acta Pratacult. Sin.* **2007**, *16*, 34–39, (In Chinese with English Abstract).
20. Yang, Q.; Wang, W.; Zeng, H. Effects of nitrogen addition on the plant diversity and biomass of degraded grasslands of NeiMongol, China. *Chin. J. Plant Ecol.* **2018**, *42*, 430–441. [CrossRef]
21. Hou, S.L.; Yin, J.X.; Sistla, S.; Yang, J.J.; Sun, Y.; Li, Y.Y.; Lü, X.T.; Han, X.G. Long-term mowing did not alter the impacts of nitrogen deposition on litter quality in a temperate steppe. *Ecol. Eng.* **2017**, *102*, 404–410. [CrossRef]
22. Yang, Z.P.; Hugjiltu, M.; Baoyin, T.; Li, F.Y. Plant production decreases whereas nutrients concentration increases in response to the decrease of mowing stubble height. *J. Environ. Manag.* **2020**, *253*, 109745. [CrossRef]
23. Kotas, P.; Choma, M.; Šantrůčková, H.; Lepš, J.; Tříska, J.; Kaštovská, E. Linking above- and belowground responses to 16 years of fertilization, mowing, and removal of the dominant species in a temperate grassland. *Ecosystems* **2017**, *20*, 354–367. [CrossRef]
24. Blüthgen, N.; Dormann, C.F.; Prati, D.; Klaus, V.H.; Kleinebecker, T.; Hölzel, N.; Alt, F.; Boch, S.; Gockel, S.; Hemp, A.; et al. A quantitative index of land-use intensity in grasslands: Integrating mowing, grazing and fertilization. *Basic Appl. Ecol.* **2012**, *13*, 207–220. [CrossRef]
25. Borer, E.T.; Seabloom, E.W.; Gruner, D.S.; Harpole, W.S.; Hillebrand, H.; Lind, E.M.; Adler, P.B.; Alberti, J.; Anderson, T.M.; Bakker, J.D.; et al. Herbivores and nutrients control grassland plant diversity via light limitation. *Nature* **2014**, *508*, 517–520. [CrossRef]
26. Socher, S.A.; Prati, D.; Boch, S.; Müller, J.; Klaus, V.H.; Hölzel, N.; Fischer, M. Direct and productivity-mediated indirect effects of fertilization, mowing and grazing on grassland species richness. *J. Ecol.* **2012**, *100*, 1391–1399. [CrossRef]
27. Hautier, Y.; Niklaus, A.P.; Hector, A. Competition for light causes plant biodiversity loss after eutrophication. *Science* **2009**, *324*, 636–638. [CrossRef]
28. Foster, B.L.; Gross, K.L. Species richness in a successional grassland: Effects of nitrogen enrichment and plant litter. *Ecology* **1998**, *79*, 2593–2602. [CrossRef]
29. Maron, J.L.; Jefferies, R.L. Restoring enriched grasslands: Effects of mowing on species richness, productivity, and nitrogen retention. *Ecol. Appl.* **2001**, *11*, 1088–1100. [CrossRef]
30. Hiernaux, P.; Turner, M.D. The effect of clipping on growth and nutrient uptake of Sahelian annual rangelands. *J. Appl. Ecol.* **1996**, *33*, 387–399. [CrossRef]
31. Barker, C.G.; Power, S.A.; Bell, J.N.B.; Orme, C.D.L. Effects of habitat management on heathland response to atmospheric nitrogen deposition. *Biol. Conserv.* **2004**, *120*, 41–52. [CrossRef]
32. Li, Z.Q.; Liu, Z.G.; Chen, Z.Z.; Yang, Z.G. The effects of climate changes on the productivity in the Inner Mongolia steppe of China. *Acta Pratacult. Sin.* **2003**, *12*, 4–10, (In Chinese with English Abstract).
33. Pan, Q.M.; Xue, J.G.; Tao, J.; Xu, M.Y.; Zhang, W.H. Current status of grassland degradation and measures for grassland restoration in northern China. *Chin. Sci. Bull.* **2018**, *63*, 1642–1650. [CrossRef]
34. Lü, C.Q.; Tian, H.Q. Spatial and temporal patterns of nitrogen deposition in China: Synthesis of observational data. *J. Geophys. Res.* **2007**, *112*, 10–15. [CrossRef]
35. Ren, Z.W.; Li, Q.; Chu, C.J.; Zhao, L.Q.; Zhang, J.Q.; Ai, D.X.C.; Yang, Y.B.; Wang, G. Effects of resource additions on species richness and ANPP in an alpine meadow community. *J. Plant Ecol.* **2010**, *3*, 25–31. [CrossRef]
36. Niu, D.C.; Yuan, X.B.; Cease, A.J.; Wen, H.Y.; Zhang, C.P.; Fu, H.; Elser, J.J. The impact of nitrogen enrichment on grassland ecosystem stability depends on nitrogen addition level. *Sci. Total Environ.* **2017**, *618*, 1529–1538. [CrossRef]
37. Clark, C.M.; Tilman, D. Loss of plant species after chronic low-level nitrogen deposition to prairie grasslands. *Nature* **2008**, *451*, 712–715. [CrossRef]
38. Zhang, Y.H.; Lü, X.T.; Isbell, F.; Stevens, C.; Han, X.; He, N.P.; Zhang, G.M.; Yu, Q.; Huang, J.H.; Han, X.G. Rapid plant species loss at high rates and at low frequency of N addition in temperate steppe. *Glob. Chang. Biol.* **2014**, *20*, 3520–3529. [CrossRef]
39. Newman, E.I. Competition and diversity in herbaceous vegetation. *Nature* **1973**, *244*, 310. [CrossRef]
40. Henry, M.; Stevens, H.; Bunker, D.E.; Schnitzer, S.A.; Carson, W.P. Establishment limitation reduces species recruitment and species richness as soil resources rise. *J. Ecol.* **2004**, *92*, 339–347. [CrossRef]
41. Tilman, D. Species richness of experimental productivity gradients: How important is colonization limitation? *Ecology* **1993**, *74*, 2179–2191. [CrossRef]

42. Tian, D.S.; Niu, S.L. A global analysis of soil acidification caused by nitrogen addition. *Environ. Res. Lett.* **2015**, *10*, 024019. [CrossRef]
43. Dickson, T.L.; Foster, B.L. Fertilization decreases plant biodiversity even when light is not limiting. *Ecol. Lett.* **2011**, *14*, 380–388. [CrossRef] [PubMed]
44. Yang, Z.P.; Li, Y.L.; Wang, Y.D.; Cheng, J.W.; Li, F.Y.H. Preferences for different nitrogen forms in three dominant plants in a semi-arid grassland under different grazing intensities. *Agric. Ecosyst. Environ.* **2022**, *333*, 107959. [CrossRef]
45. Avolio, M.L.; Koerner, S.E.; La Pierre, K.J.; Wilcox, K.R.; Wilson, G.W.T.; Smith, M.D.; Collins, S.L. Changes in plant community composition, not diversity, during a decade of nitrogen and phosphorus additions drive above-ground productivity in a tallgrass prairie. *J. Ecol.* **2014**, *102*, 1649–1660. [CrossRef]
46. van den Berg, L.J.L.; Vergeer, P.; Tim, C.G.R.; Smart, S.M.; Guest, D.; Ashmore, M.R. Direct and indirect effects of nitrogen deposition on species composition change in calcareous grasslands. *Glob. Chang. Biol.* **2011**, *17*, 1871–1883. [CrossRef]
47. Smith, A.L.; Barrett, R.L.; Milner, R.N.C. Annual mowing maintains plant diversity in threatened temperate grasslands. *Appl. Veg. Sci.* **2017**, *12*, 3218–3221. [CrossRef]
48. Socher, S.A.; Prati, D.; Boch, S.; Müller, J.; Baumbach, H.; Gockel, S.; Hemp, A.; Schöning, I.; Wells, K.; Kalko, E.K.V.; et al. Interacting effects of fertilization, mowing and grazing on plant species diversity of 1500 grasslands in Germany differ between regions. *Basic Appl. Ecol.* **2013**, *14*, 126–136. [CrossRef]
49. Jones, L.; Stevens, C.; Rowe, E.C.; Payne, R.; Caporn, S.J.M.; Evans, C.D.; Field, C.; Dale, S. Can on-site management mitigate nitrogen deposition impacts in non-wooded habitats? *Biol. Conserv.* **2016**, *212*, 464–475. [CrossRef]
50. Lebauer, D.S.; Treseder, K.K. Nitrogen limitation of net primary productivity in terrestrial ecosystems is globally distributed. *Ecology* **2008**, *89*, 371–379. [CrossRef]
51. Niu, S.; Classen, A.T.; Dukes, J.S.; Kardol, P.; Liu, L.; Luo, Y.; Rustad, L.; Sun, J.; Tang, J.; Templer, P.H.; et al. Global patterns and substrate-based mechanisms of the terrestrial nitrogen cycle. *Ecol. Lett.* **2016**, *19*, 697–709. [CrossRef] [PubMed]
52. Zhao, W.; Chen, S.P.; Lin, G.H. Compensatory growth responses to clipping defoliation in Leymus chinensis (Poaceae) under nutrient addition and water deficiency conditions. *Plant Ecol.* **2008**, *196*, 85–99. [CrossRef]
53. Liu, Y.; Pan, Q.; Liu, H.; Bai, Y.; Simmons, M.; Dittert, K.; Han, X. Plant responses following grazing removal at different stocking rates in an Inner Mongolia grassland ecosystem. *Plant Soil.* **2011**, *340*, 199–213. [CrossRef]
54. Keddy, P.A.; Twolan-Sstrutt, L.; Shipley, B. Experimental evidence that interspecific competitive asymmetry increases with soil production. *Oikos* **1997**, *80*, 253–256. [CrossRef]
55. Yang, G.J.; Hautier, Y.; Zhang, Z.J.; Lü, X.T.; Han, X.G. Decoupled responses of above- and below-ground stability of productivity to nitrogen addition at the local and larger spatial scale. *Glob. Chang. Biol.* **2022**, *28*, 2711–2720. [CrossRef]
56. Cheng, Y.; Wang, J.; Chang, S.X.; Cai, Z.; Mueller, C.; Zhang, J. Nitrogen deposition affects both net and gross soil nitrogen transformations in forest ecosystems: A review. *Environ. Pollut.* **2019**, *244*, 608–616. [CrossRef]
57. Mikola, J.; Setälä, H.; Virkajärvi, P.; Saarijärvi, K.; Ilmarinen, K.; Voigt, W.; Vestberg, M. Defoliation and patchy nutrient return drive grazing effects on plant and soil properties in a dairy cow pasture. *Ecol. Monogr.* **2009**, *79*, 221–244. [CrossRef]
58. Houle, D.; Moore, J.D.; Ouimet, R.; Marty, C. Tree species partition N uptake by soil depth in boreal forests. *Ecology* **2014**, *95*, 1127–1133. [CrossRef]

agronomy

MDPI

Article

Construction of an Early Warning System Based on a Fuzzy Matter-Element Model for Diagnosing the Health of Alpine Grassland: A Case Study of Henan County, Qinghai, China

Huilan Shi [1], Mengping Liu [1,2], Shihai Zhu [1], Zhonghua Duan [3], Rongrong Wu [1], Xiaolong Quan [3], Mengci Chen [3], Jiexue Zhang [1] and Youming Qiao [3,*]

[1] College of Eco-Environmental Engineering, Qinghai University, Xining 810016, China; hlshi7701@126.com (H.S.); mengping0919@163.com (M.L.); zsh696000@163.com (S.Z.); wuluobo55rong@163.com (R.W.); jxzhang@163.com (J.Z.)
[2] Bureau of Ecology and Environment of Puyang, Puyang 457005, China
[3] State Key Laboratory of Plateau Ecology and Agriculture, Qinghai University, Xining 810016, China; zhonghuaduan@sina.com (Z.D.); quanxl@126.com (X.Q.); chenmengci@126.com (M.C.)
* Correspondence: ymqiao@aliyun.com

Abstract: To maintain alpine grassland in a healthy and sustainable state, a sound warning system was developed to diagnose its potential degradation risk. Data related to grassland quality (six indicators), habitat (six indicators), and eco-carrying capacity (three indicators) at eight sampling plots were collected from Henan Mongol Autonomous County of West China in 2014 and 2017, representing five types of grassland and three grazing styles. Compared to the warning level in 2014, alpine grassland had a higher warning level in 2017, demonstrating the degradation of grassland ecosystems. *Kobresia tibetica* exhibited the lowest level of warning, while *Kobresia humilis* had the highest, indicating its corresponding safety and unsafety under the environmental change. Grassland quality is the most important index for grassland health, and soil total carbon and available phosphorus are the most important indices of habitat quality, which finally greatly influence the warning level of alpine grassland. Further analysis results suggested that winter grazing is beneficial for the health of grassland, and moderate grazing can accelerate the self-recovery of the alpine grassland due to the increase in organic matter. This study is crucial for understanding the health level of alpine grassland and its further change trends, and providing an important scientific basis for rational grazing.

Keywords: alpine grassland; early warning system; grassland type; grazing style; fuzzy matter-element model; analytic hierarchy process

Citation: Shi, H.; Liu, M.; Zhu, S.; Duan, Z.; Wu, R.; Quan, X.; Chen, M.; Zhang, J.; Qiao, Y. Construction of an Early Warning System Based on a Fuzzy Matter-Element Model for Diagnosing the Health of Alpine Grassland: A Case Study of Henan County, Qinghai, China. *Agronomy* **2023**, *13*, 2176. https://doi.org/10.3390/agronomy13082176

Academic Editor: Qingzhu Gao

Received: 25 June 2023
Revised: 11 August 2023
Accepted: 16 August 2023
Published: 19 August 2023

1. Introduction

Grassland is a renewable natural resource that provides various important eco-services [1]. The alpine grassland ecosystem on the vast Qinghai-Tibet Plateau not only protects the land surface, but also supports livestock grazing. Owing to the harsh natural environment, this ecosystem is particularly sensitive to disturbances and prone to degradation [2]. With population growth, the grassland ecosystem is facing a mounting pressure of overgrazing, which makes it highly vulnerable to degradation [3]. In recent years, due to overgrazing and climate warming [4], the alpine grassland ecosystem has shown signs of degraded functionality [5,6], declined grass coverage, and reduced biomass [7,8]. In order to prevent the grassland from further degradation, it is rather important to comprehensively and realistically assess its health so that appropriate measures can be put in place to protect it. An early warning system is a multi-functional network being able to assess capacities and identify gaps and opportunities in comprehensively evaluating all kinds of natural hazards. It can also be used for the evaluation, prediction, and warning about the potential destruction to the grassland eco-environment triggered by excessive resource exploitation [9]. Pioneering research on early warning systems mainly focused on the principles and methods of ecological early warning on environmental disasters and environmental pollution [10]. It has also been used to study the early warning signal to mitigate abrupt climate change [11]. Early warning systems have been developed to protect endangered species [12], predict rainfall-induced land sliding [13], and monitor and manage land resources [14]. In addition, early warning systems have also been developed to detect and predict long-term marine pollution [15], monitor wetland water eutrophication and eco-environment [16,17], and assess the ecological risk of wetlands for water scarcity [18].

In spite of its proved utility in managing diverse natural hazards and conserving natural resources [19], however, no early-warning systems have been constructed for effectively managing alpine grassland on the Qinghai-Tibet Plateau that has been degraded even severely with reduced ecological functions [20]. While the mechanism of grassland degradation has been studied in depth [3,20], inadequate attention has been paid to diagnose grassland health and the effect that different grazing styles have upon it. In order to protect this vulnerable resource, it is important to assess its current potential risk of degradation, which can be achieved by constructing an early warning system. With the assistance of this system, it is possible to determine the warning level of the grassland. This system can also provide an early warning of potential risks of degradation, based on which measures regarding how the grassland should be managed properly can be prescribed. However, a rare study has developed an early warning system for predicting the health of grassland, especially its potential for degradation.

A fuzzy matter-element model based on matter-element analysis is used to describe the fuzziness of objects. This is a method that falls between mathematics and experimentation. Fuzzy-matter elements have been widely used in fields such as environmental safety assessment, comprehensive evaluation of water quality, ecological environment warning, and sustainable utilization comprehensive evaluation due to their ability to effectively address the impact of fuzzy uncertainty in evaluation criteria and overall objectivity in evaluation [21–23].

The objective of this study is to develop an ecological early warning system for diagnosing the health of alpine grassland and assessing the impact of different grazing styles on its resilience to degradation. Specifically, it develops a structured model for an early warning system applying a fuzzy matter-element model. This early warning system helps to (1) reveal the risk level of degradation for different types of grassland under varying grazing styles, (2) assess the effectiveness of various indicators in the early warning system in different eco-environments, and (3) evaluate the current vulnerability of five types of grassland in the study area on the Qinghai-Tibet Plateau, a geographic area with widespread grassland resources that have shown signs of degradation.

Study area: The study site is located in Henan Mongol Autonomous County (100°34′–102°28′ E, 34°04′–36°10′ N) in the south-eastern Qinghai Province of western China (Figure 1). The County has a total area of 6997.45 km^2 and a mean elevation of 3600 m. It is the largest organic animal husbandry production base in China, with horses, sheep, and yak as the main stock. It has an alpine continental climate with a large diurnal temperature range. The annual temperature averages between −0.4 and −2.6 °C. The warm season is very short, resulting in a short growing period of about 150 days per annum. Annual rainfall varies between 591.1 and 615.5 mm, concentrating mostly between May and October. Annual evaporation mounts to 1349.70 mm, leading to a water deficit.

Land use in the study area is primarily grazing, with alpine grassland totaling 63.12 × 10^4 ha (93.76% of the total land area). The alpine grassland comprises four subcategories of alpine meadow, shrub meadow, marsh meadow, and mountain meadow. The common plant species are *Kobresia humilis*, *Potentilla fruticose*, *Kobresia tibetica*, and *Elymus nutans*, of which *Kobresia humilis*—the dominant species—has a short rhizome and a height of 3–10 cm.

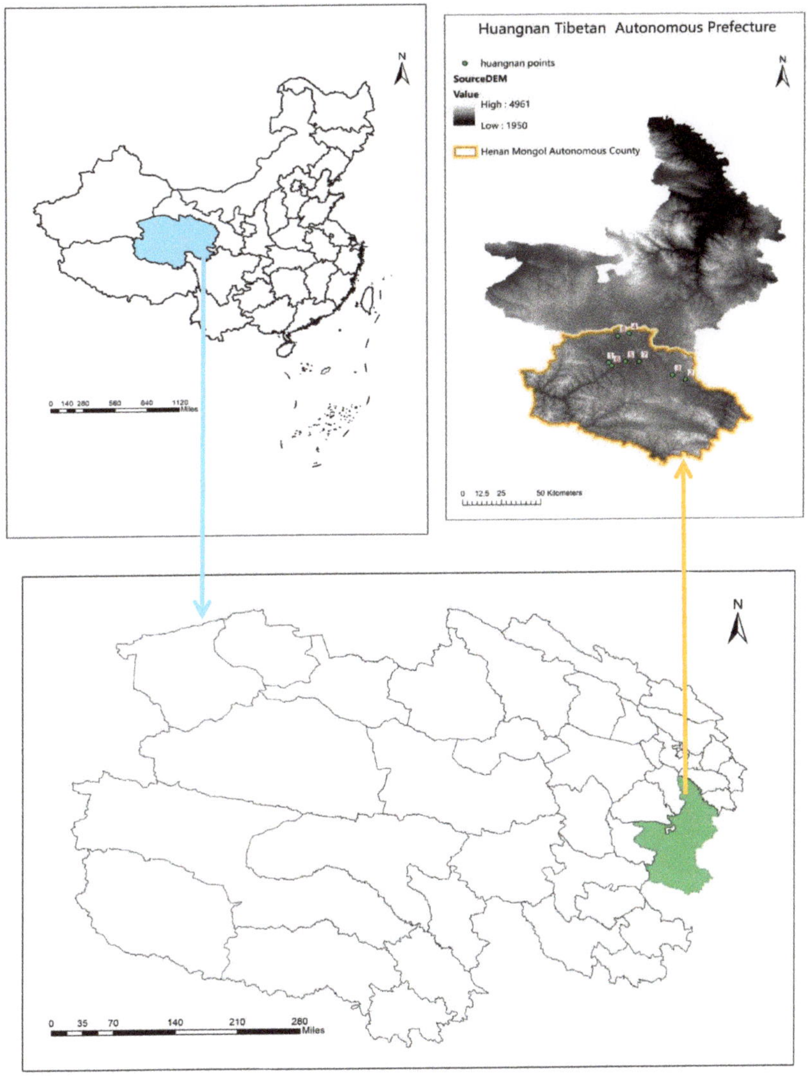

Figure 1. Location of Henan Mongol Autonomous County (orange) in Huangnan Tibetan Autonomous Prefecture (green) and Qinghai Province (sky blue).

As one of the main organic grazing bases in China, it faces an aggravated conflict between protection of the grazing resource and its utilization for grazing. Consequently, the grassland has shown signs of degradation. Different types of grassland have different levels of degradation risk owing to variable grazing pressure. In order to remedy the negative effect of grazing, various grassland management strategies have been implemented. However, their impact on grassland degradation thresholds remains unknown. There is a dire need for constructing an ecological early warning system to comprehensively reflect the multi-functions of various types of grassland so that safe management practices can be applied to the grassland system to achieve sustainable grazing.

2. Research Methodology

2.1. Field Data Collection

Field samples were collected from eight plots of 1 km by 1 km in size in 2014 and 2017 (Figure 1). They have a slope gradient mostly within 5–10% but can be as steep as 15–25%. These plots cover five types of grassland, including *shrub Potentilla fruticosa* (PF) and forbs species as *Elymus nutas* (EN), *Kobresia humilis* (KH), *Ligularia birgaurea* (LB), and *Kobresia tibetica* (KT), all of which are grazed in one of three styles (winter, summer, and no grazing) (Table 1). These plots also capture different proportions of vegetative cover. As shown in the table, each site has a unique community of dominant species and grazing style. The community structure and characteristics in each sampling plot were surveyed along three parallel transects in August when the biomass was maximum [24]. Each transect contains five sampling plots of 0.5 m × 0.5 m in size (changed to 1 m by 1 m for shrubbery vegetation). Within these randomly selected plots, aboveground biomass of all plants was measured by clipping the vegetation to the ground surface as closely as possible. Vegetation cover was estimated within 10 spatially adjoining plots at each site. Forage quantity was measured in three randomly selected plots [24]. The fresh grass was dried at 70 °C for 48 h in an oven to measure the biomass [25].

Table 1. Properties of the eight sampling plots.

Type of Grassland		Longitude (E)	Latitude (N)	Altitude (m)	Species Richness (S)	Veg Coverage (%)	Veg Height (cm)	Dominant Species	Grazing Style
PH	M1	101°24′41″	34°41′53″	3541	48	99.0	60	*Potentilla fruticosa* (0.051)	Winter Pasture (WP)
	M6	101°25′48″	34°40′42″	3518	21	82.0	35	*Potentilla fruticosa* (0.225)	Summer Pasture (SP)
EN	M2	101°51′53″	34°35′36″	3590	28	94.3	75	*Elymus nutans* (0.108)	Winter Pasture (WP)
	M3	101°47′17″	34°37′16″	3636	16	75.5	55	*Elymus nutans* (0.150)	Summer Pasture (SP)
	M7	101°35′27″	34°42′11″	3542	35	79.5	70	*Elymus nutans* (0.182)	No Grazing (NGP)
KH	M4	101°31′51″	34°52′21″	3610	28	92.1	3	*Kobresia humilis* (0.079)	Summer Pasture (SP)
LB	M5	101°30′37″	34°42′15″	3509	30	88.0	28	*Ligularia birgaurea* (0.114)	No Grazing (NGP)
KT	M8	101°27′57″	34°51′22″	3588	33	98.3	55	*Kobresia tibetica* (0.180)	Winter Pasture (WP)

Soil samples were collected with an earth drill of 40 mm in inner diameter (φ) at three depths of 0~10 cm (top soil), 10~20 cm (top-sub soil), and 20~30 cm (sub soil) within each plot after plants and surface litters had been removed. The soil samples collected at three drills were mixed in a plastic bag and taken to the lab where they were dried at 105 °C for 48 h. Afterwards, soil properties such as C_{SOC} (potassium dichromate oxidation with external Heating) [26], C_{TN} (Kjeldahl procedure followed by colorimetric analysis) [27], C_{AN} (Alkaline hydrolysis diffusion method) [28], C_{TP} (the sodium bicarbonate alkali digestion method and molybdenum antimony colorimetry) [29], and C_{AP} (0.5 mol·L^{-1} $NaHCO_3$ solution) were analyzed.

2.2. Indicators

The ecological early warning system needs to reflect the eco-carrying capacity and plant and soil characteristics, which can be achieved by including the most appropriate and important indicators of grassland ecosystem health. Commonly considered indicators include vegetation properties, soil properties, and grazing intensity [30]. In addition to these factors, this study also takes into consideration topography, grazing style, grassland type, and grassland eco-carrying capacity. In total, 15 indicators were included in the early warning system: bare ground proportions, vegetation coverage, biomass, grass layer thickness, plant diversity, forage production and quality, height of dominant species, soil total nitrogen, available phosphorus, potassium, moisture, soil slope, and grazing styles, all of which have been extensively used in classification and grading system of grassland [31,32].

Due to the lack of literature on this topic, the ecological early warning system was constructed jointly based on theoretical analysis [5,33,34] and experts' opinions. Specifically, the analytical hierarchy process (AHP) [35] was used to construct the system comprising three layers (target, criterion, and indicator). The target layer (A) aims to derive the level of warning on the potential risk of grassland degradation (Table 2). The criterion layer (B) contains three categories of grassland quality (B_1), habitat index (B_2), and eco-carrying capacity (B_3). The indicator layer (C) contains 15 indicators termed as C_1–C_{15} (Table 2).

Table 2. The indicators selected in the early warning system.

Target Layer (A)	Criterion Layer (B)		Indicator Layer (C)		
	Weight			Factor	Unit
Derive the level of warning on the potential risk of grassland degradation	0.630	B1—grassland quality	C_1	Bare ground proportion	%
			C_2	Turf layer thickness	cm
			C_3	Above ground biomass	g/m^2
			C_4	Vegetation coverage	%
			C_5	Height of dominant species	cm
			C_6	Plant richness	Number of species
	0.218	B2—habitat quality	C_7	Total carbon	g/Kg
			C_8	Available phosphorus	g/Kg
			C_9	Total nitrogen	mg/Kg
			C_{10}	Available potassium	mg/Kg
			C_{11}	Moisture	%
			C_{12}	Slope	°
	0.151	B3—eco-carrying capacity	C_{13}	Grazing intensity	sheep-unit/hm^2.year
			C_{14}	Forage production	RMB/hm^2
			C_{15}	Fine forage ratio	%

2.3. Weighting Indicators

The 15 selected indicators must be assigned the appropriate weights in accordance with their importance using the fuzzy matter-element model. It can deal with fuzzy objects by precise digital means, and make a more scientific, reasonable, and close-to-reality representation of the information. This is commonly achieved through AHP because of its multi-functionality. This decision-making approach can aid in the solution to complex multiple criteria problems. This effective and practical approach can reach complex and unstructured decisions by weighting the indicators selected for the early warning system.

AHP was used to process the collected data and to construct the early warning system in three steps. The first was to determine the relative weight W (varying between 1, 3, 5, 7, and 9) by comparing a pair of indicators (Supplementary Table S1). A larger weight suggests that one of the indicators is increasingly more important than the other. The comparison between any two potential pairs resulted in an evaluation matrix [36]. The second was to determine the relative importance of the elements. The third was to test for consistency using the consistency index (*C.I.*); it was calculated using the following formula:

$$C.I. = \frac{\lambda_{max}}{n-1} \quad (1)$$

where n refers to the dimension of the evaluation matrix (n = 15) and $\lambda_{max} = n$. The larger the *C.I.* value is, the more consistent the evaluation matrix is. If $C.R. \leq 0.1$, the evaluation matrix is considered consistent [34] (Supplementary Table S5). The revised value *R.I.* is the random consistency index; it is introduced according to the dimension of the evaluation matrix. The mean random consistency index *R.I.* values depend on 1–15 order average consistency index used in this study (Supplementary Table S2). The *C.R.* is taken as an index as follows,

$$C.R. = \frac{C.I.}{R.I.} \quad (2)$$

2.4. Data Processing—Fuzzy Matter Element Standardization

AHP was implemented using a fuzzy matter element that is used to generate the standard fuzzy matter element R_{mn} (m = 15 indicators, n = 8 sites). This fuzzy value matrix can be expressed as follows:

$$R_{mn} = \begin{bmatrix} & M_1 & \cdots & M_m \\ c_1 & u_{11} & \cdots & u_{m1} \\ \vdots & \vdots & \cdots & \vdots \\ c_n & u_{1n} & \cdots & u_{mn} \end{bmatrix} \quad (3)$$

This matrix was used to derive a composite prosperity index I_m by multiplying the W matrix (15 × 8) derived using the AHP method, namely,

$$I_{\mathrm{m}} = W_{\mathrm{m}} \times R_{\mathrm{mn}} = (I_1, \quad I_2, \quad \cdots, \quad I_m) \quad (4)$$

where m = 1, 2, 3, . . ., 8 (8 sampling plots).

2.5. Grading the Severity of Warning

The derived composite prosperity index value $\boldsymbol{I}$ with a range of 0–10 was converted to five warning levels of no warning, light warning, medium warning, serious warning, and extreme warning, at an interval of 2 (Table 3). The actual threshold value S_i (for the calculation of this value, refer to Supplementary Table S3) was compared with these theoretical threshold values in the table to determine its warning level and to produce a comprehensive and objective ecological warning for the alpine grassland.

Table 3. The corresponding relations of alert degree, booming exponents, and index scope.

Warning Degree	No Warning	Light Warning	Medium Warning	Serious Warning	Extreme Warning
Prosperity index	[0, 2]	[2, 4]	[4, 6]	[6, 8]	[8, 10]
Range of indicators	$<S_4$	$[S_4, S_3]$	$[S_3, S_2]$	$[S_2, S_1]$	$>S_1$

Note: $S_1 > S_2 > S_3 > S_4$ indicates the critical value of different warning ranges.

On the basis of fuzzy matter-element matrix, combined with the range of the prosperity index value, each standard early warning index and standardized fuzzy matter-element matrix was calculated. If the calculated standard early warning index is <0, it is set to 0; if the value is >10, it is set to 10 [36].

3. Results

3.1. Standardized Fuzzy Matter-Element Matrix of the Alpine Grassland

The standard fuzzy matter-element matrix of 2014 is calculated as below:

$$R_{mn} = \begin{bmatrix} & M_1 & M_2 & M_3 & M_4 & M_5 & M_6 & M_7 & M_8 \\ C_1 & 0.20 & 1.14 & 7.90 & 1.58 & 3.40 & 5.60 & 7.10 & 0.34 \\ C_2 & 8.00 & 2.00 & 8.67 & 6.00 & 8.00 & 1.67 & 2.00 & 1.00 \\ C_3 & 0 & 0 & 0 & 0 & 0 & 1.98 & 0 & 0 \\ C_4 & 1.80 & 1.90 & 4.90 & 1.95 & 2.40 & 3.60 & 4.10 & 1.82 \\ C_5 & 0 & 0 & 0 & 7.27 & 0 & 0 & 0 & 0 \\ C_6 & 0 & 0.27 & 1.87 & 0.27 & 0 & 1.20 & 0 & 0 \\ C_7 & 1.48 & 1.25 & 1.64 & 3.24 & 2.77 & 1.53 & 3.44 & 0.40 \\ C_8 & 5.97 & 7.00 & 7.66 & 7.65 & 7.80 & 5.09 & 8.38 & 5.25 \\ C_9 & 8.96 & 8.13 & 8.69 & 7.97 & 8.52 & 8.15 & 8.51 & 8.52 \\ C_{10} & 1.64 & 1.73 & 0.95 & 1.59 & 1.99 & 1.93 & 5.48 & 0.16 \\ C_{11} & 0.69 & 1.61 & 4.63 & 2.91 & 2.64 & 1.71 & 3.71 & 0.90 \\ C_{12} & 6.60 & 2.00 & 3.40 & 2.60 & 2.80 & 5.40 & 3.40 & 2.20 \\ C_{13} & 3.33 & 2.00 & 2.00 & 1.11 & 6.00 & 4.00 & 8.22 & 1.11 \\ C_{14} & 4.40 & 1.60 & 1.96 & 6.40 & 4.40 & 8.20 & 6.20 & 2.32 \\ C_{15} & 9.80 & 2.47 & 9.86 & 8.64 & 9.79 & 5.7 & 7.64 & 1.99 \end{bmatrix} \quad (5)$$

The standard fuzzy matter-element matrix in 2017 is calculated as below:

$$R_{mn} = \begin{bmatrix} & M_1 & M_2 & M_3 & M_4 & M_5 & M_6 & M_7 & M_8 \\ C_1 & 1.68 & 1.12 & 0.8 & 2.12 & 0.74 & 0.66 & 7.2 & 0 \\ C_2 & 8.33 & 1.67 & 8.00 & 8.00 & 7.00 & 1.50 & 5.00 & 4.00 \\ C_3 & 0 & 0 & 0 & 0 & 0 & 0 & 0 & 0 \\ C_4 & 1.96 & 1.90 & 1.90 & 2.07 & 1.86 & 1.85 & 4.67 & 1.78 \\ C_5 & 0 & 0 & 0 & 6.70 & 0 & 0 & 0 & 0 \\ C_6 & 1.73 & 2.40 & 2.40 & 5.00 & 4.00 & 2.80 & 3.60 & 3.60 \\ C_7 & 1.07 & 0.90 & 1.24 & 2.01 & 3.46 & 1.38 & 1.86 & 0.87 \\ C_8 & 7.17 & 5.87 & 5.5 & 6.97 & 7.1 & 5.83 & 6 & 6.87 \\ C_9 & 8.68 & 8.23 & 8.20 & 9.02 & 9.04 & 8.89 & 8.52 & 7.73 \\ C_{10} & 5.47 & 2.11 & 2.00 & 2.10 & 1.71 & 3.22 & 1.18 & 2.15 \\ C_{11} & 1.69 & 2.97 & 1.63 & 3.48 & 2.71 & 4.77 & 1.98 & 1.71 \\ C_{12} & 6.60 & 2.00 & 3.40 & 2.60 & 2.80 & 5.40 & 3.40 & 2.20 \\ C_{13} & 4.67 & 2.67 & 2.70 & 0 & 0 & 4.67 & 0 & 0 \\ C_{14} & 5.20 & 1.64 & 2.60 & 6.4 & 4.8 & 8.3 & 7.4 & 4.8 \\ C_{15} & 1.74 & 4.64 & 1.61 & 8.1 & 7.82 & 1.80 & 1.41 & 3.18 \end{bmatrix} \quad (6)$$

The standardized fuzzy matter-element and the composite prosperity index (Table 3) indicate that the eight typical alpine grasslands (M1–M8) in the study area in 2014 generally showed no warning in terms of their aboveground biomass. Sample plot 6 had light warning, indicating that the vegetation growth in the sample plots was generally good, and the plant diversity was the light warning and below. The content of soil total nitrogen is medium warning and above. The contents of available phosphorus in soil are all heavy warning or huge warning, which indicates that the vegetation habitat is poor and the soil nutrient is low. The proportion of fine herbage is in the light warning only in the sample plot 2 and 8; the rest are in the middle warning and above, which indicates that except for plots 2 and plots 8, the forage quality is lower and there were more poisonous weeds.

3.2. Importance of All Indicators

The weight of all the 15 evaluation indicators considered for possible inclusion in the early warning index system passed the random consistency test (Supplementary Table S4) as the single ranking *C.I.* value and the total sorting *C.R.* value are both smaller than 0.1, suggesting that the ranking results are acceptable. The experts' judgments are consistent with the importance of each index. Hence, all the 15 indicators are retained in the constructed early warning system.

Of the three criterion layers, the most important layer is the quality of grassland (B_1) that receives a weight of 0.630, followed by habitat quality (B_2) (weight = 0.218), and eco-carrying capacity (B_3) (weight = 0.151) (Table 4). Therefore, the quality of grassland is the absolute determinant of whether the grassland faces the risk of degradation, given that its weight is almost twice higher than the combined weight of the other two layers. This outcome is attributable likely to the fact that the habitat quality can account for the eco-carrying capacity of the grassland partially.

Table 4. Weights of evaluation matrix.

	Parameter							
		B_1	C_1	C_2	C_3	C_4	C_5	C_6
B_1-C	$\overline{W_l^b}$	2.289	1.246	0.888	0.342	0.525	1.414	3.448
	$\overline{W_i^b}$	0.630	0.11	0.066	0.028	0.042	0.114	0.278
		B_2	C_7	C_8	C_9	C_{10}	C_{11}	C_{12}
B_2-C	$\overline{W_l^b}$	0.794	2.928	0.293	0.567	1.763	1.399	0.833
	$\overline{W_i^b}$	0.218	0.082	0.008	0.016	0.049	0.039	0.023
		B_3	C_{13}	C_{14}	C_{15}			
B_3-C	$\overline{W_l^b}$	0.550	2.466	0.405	3.271			
	$\overline{W_i^b}$	0.151	0.022	0.012	0.110			

The most critical indicator in the grassland quality layer is the number of species or species richness C_6. It has a weight of 0.278, the highest among all the six indicators in this layer. Moreover, this weight is more than twice higher than the next most important indicator, height of predominant species C_5 (0.114). This indicator is nearly as important as the proportion of bare ground C_1 (0.101). Jointly, C_5 and C_1 can account for the quantity of grassland biomass to a certain degree. This is because bare ground can lead to a loss of nutrients from the grassland through soil erosion and result ultimately in the grassland being degraded, even directly to a serious level. The side effect of bare ground is a reduced yield of fine forage, which makes the grassland sensitive to grazing intensity. A lower plant community exposes the ground to more evaporation and hence a lower soil moisture that is detrimental to grassland health. In contrast, the other three indicators of turf thickness C_2, aboveground biomass C_3, and vegetation coverage C_4 are much less important, probably because their effect has been taken into consideration indirectly via the remaining three indicators.

Of the six habitat quality indicators, total carbon C_7 (0.082) is the most significant, followed by available potassium C_{10} (0.049) and moisture C_{11} (0.039). In comparison, available phosphorus C_8, soil total nitrogen C_9, and plot slope (C_{12}) are much less important. This is because a higher biomass on the ground will lead to more decomposition of plant litters. Moisture is rather critical to the health of grasses as it affects whether and how quickly the grass can absorb nutrients from the soil.

3.3. Level of Early Warning by Grassland Type

As shown in Figure 2, in 2014 all the five types of grassland except KT had a composite prosperity index larger than 2, the threshold for light warning. In contrast, KH had the highest warning index value of 3.4. There is no noticeable difference among the other three types of grassland in their composite prosperity index. The lowest warning index for KT is explained by its change from a former *Kobresia* marsh to alpine grassland. The relative high warning level of KH is accounted for mainly by the dwarf stature (about 2–3 cm in height) of the grass. In addition, it is also the backbone and dominant grassland that is grazed regularly. Overall, the general level of warning indicates that caution must be exercised in grazing the grassland rationally to prevent it from evolving to the moderate warning level.

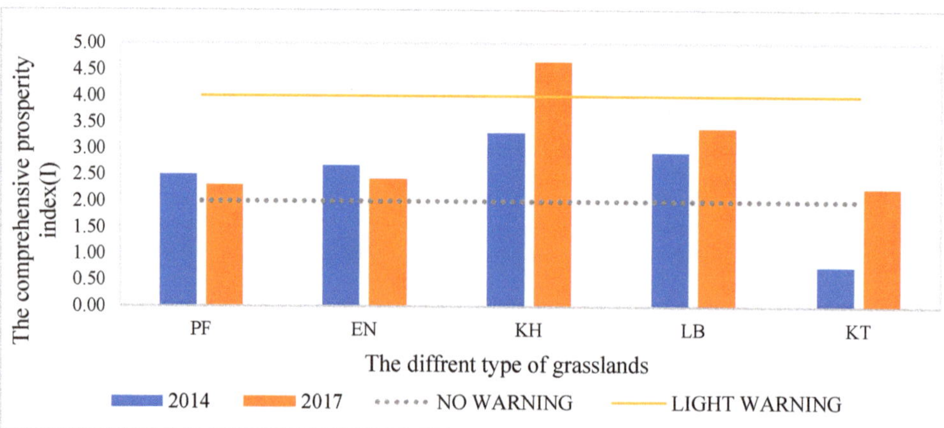

Figure 2. The composite prosperity index value of the five types of grassland in 2014 and 2017 (PF—Potentilla fruticosa shrub, EN—Elymus nutans, KH—Kobresia humilis, LB—forbs, and KT—Kobresia tibetica).

However, the index value changed differently among the five types of grassland from 2014 to 2017 (Figure 2). Only two of them (PF and EN) had a lower warning level in 2017 than in 2014. The decrease in PF's index value is attributed to the marked increase in the height of the fine dominant species of grasses and fine forage, in spite of the fact that its moisture level and aboveground biomass both decreased (Table 5). The decrease in the proportion of bare ground also contributed to the lessened level of warning. Similarly, EN also had a lower proportion of bare ground in 2017 than in 2014. Its soil nitrogen and carbon both increased, despite the highest drop in its aboveground biomass among all the five types of grassland, and a decrease in the portion of fine forage.

Table 5. Change in the 15 indicators among the five types of grassland during 2014–2017.

Indicator	Factor	Change Degree (%)				
		PF	EN	KH	LB	KT
C1	Bareground proportion	−38.42	−35.70	30.38	−69.17	147.06
C2	Turf layer thickness	0.00	10.71	−25.00	16.67	−7.69
C3	Aboveground biomass	−20.86	−72.75	−68.66	−28.49	−38.18
C4	Vegetation coverage	4.03	7.26	−2.61	9.43	−2.54
C5	Height of dominant species	21.48	2.76	4.88	0.00	7.09
C6	Plant diversity	−57.14	−50.63	−67.86	−66.67	−72.73
C7	Total carbon	11.26	18.61	18.19	−9.57	13.99
C8	Available phosphorus	−21.12	49.43	14.49	18.18	41.92
C9	Total nitrogen	−15.93	7.76	−51.46	−35.22	−1.05
C10	Available potassium	−30.74	2.91	−18.04	14.05	39.99
C11	Moisture	−35.41	15.94	−7.99	−0.95	14.37
C12	Slope	0.00	0.00	0.00	0.00	0.00
C13	Grazing intensity	13.04	−47.37	20.00	−100.00	−400.00
C14	Forage production	−8.93	−7.86	0.00	−5.26	−14.29
C15	Fine forage ratio	102.74	92.99	43.57	925.00	42.68

Note: Change degree—The degree of the indicators change, PF—Mean values of indicators for *Potentilla fruticosa* under winter pasture and summer pasture, EN—Mean values of indicators for *Elymus nutans* under no grazing, winter pasture, and summer pasture.

Of the five types of grassland, KT suffered the biggest increase in warning level that more than doubled within three years. Such a drastic increase is accounted for by the change in its grassland type from the former marshy grassland to alpine grassland through artificial planting of the *Elymus* plant species, leading to an increase in grass cover and height as well as biomass. However, this intervention decreased the portion of *Kobresia* plant species that is resistant to degradation owing to its well-developed root system. Although this intervention increased the average height of grasses and vegetation cover, quite a number of indicators actually decreased in their prosperity index value. In spite of this, however, its vulnerability to degradation actually increased due to its shallower roots. Accompanying these changes is a reduction in aboveground biomass, biodiversity, soil nutrients (total C and P) and moisture, and fine forage.

The worsened index value of KH is related closely to its reduced above-ground biomass, biodiversity, fine species of grass, and vegetation cover (Table 5). Also associated with these changes are decreases in some soil properties. The most important indicators to the observed change are turf layer thickness and total nitrogen. LB grassland suffered a decrease in biomass and grass diversity and soil total P, even though its plant coverage increased while its bare ground decreased.

Of the five types of grassland, *Kobresia* grassland has the highest composite prosperity index value while *Elymus* grassland has the lowest, mainly because plant height makes the most contribution to the composite prosperity index (Figure 2). Of all types of *Kobresia* grassland, its composite prosperity index all increased from 2014 to 2017, irrespective of the exact sub-type of grassland. For instance, *Kobresia humilis* with a thin turf layer is more sensitive to external disturbance. However, the composite prosperity index of *Kobresia tibetica* grassland was lower in 2014, but this value increased the most after artificial planting of *Elymus*. Therefore, appropriate external intervention is beneficial to the restoration of grassland ecosystem. The composite prosperity index of *Elymus* grassland is variable in 2014 and 2017. Overall, the average prosperity index increased by 0.27 from 2014 to 2017, indicating that the grasslands urgently need strengthened management.

3.4. Level of Early Warning by Grazing Style

In total, there are three grazing styles: winter pasture (WP), summer pasture (SP), and no grazing pasture (NGP). In 2014, all three grazing styles had a composite prosperity index value less than 4 (Figure 3) while WP had a value below 2. Therefore, compared with grasslands under grazing and utilization in winter, the impact of grazing methods on grassland stability is relatively small. Both SP and NGP had a similar level of light warning with a composite index value around 2.8. Thus, grazing style affects the composite prosperity index, and the level of degradation risk. From 2014 to 2017, all three grazing styles had a larger composite prosperity index value (Figure 3). In particular, WP had the largest increase (35.1%), followed by SP (10.33%). In contrast, NGP had the lowest increase of only 9.3%. These changes are related closely to the change in the properties of the relevant indicators. For instance, WP suffered a decrease in aboveground biomass, biodiversity, and soil available K, in spite of the increase in the height of fine species and soil available P (Table 6). For SP, its aboveground biomass, biodiversity, and soil available K all decreased. However, such decreases did not translate into a lower prosperity index value because its turf thickness, vegetation cover, and the portion of fine grasses all increased while its bare ground portion decreased. For NGP, above ground biomass, biodiversity, forage production and soil total P all decreased. Therefore, for NGP grassland, the most important factors to the change in its composite prosperity index are related to the vegetation conditions.

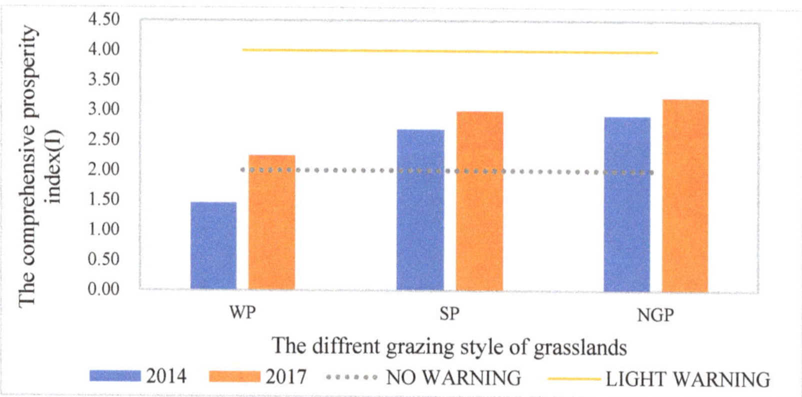

Figure 3. The composite prosperity index of grassland by grazing style in 2014 and 2017 (WP stands for winter pasture, SP stands for summer pasture, NGP stands for no grazing pasture).

Table 6. Change in the 15 indicators among the three types of grazing.

		Value Difference			Change Degree (%)		
		WP	**SP**	**NGP**	**WP**	**SP**	**NGP**
C1	Bareground proportion	−1.87	10.93	2.90	−17.08	99.89	26.53
C12	Slope	0.00	0.00	0.00	0.00	0.00	0.00
C13	Grazing intensity	−0.13	−0.10	1.75	−7.62	−5.71	0.00
C2	Turf layer thickness	0.00	−0.33	0.00	0.00	−100.00	0.00
C3	Above ground biomass	183.40	113.99	85.26	100.00	62.15	46.49
C4	Vegetation coverage	1.87	−10.93	−2.90	17.08	−100.00	−26.53
C5	Height of dominant species	−4.13	0.27	−1.00	−100.00	6.46	−24.21
C6	Plant diversity	22.67	9.67	22.00	99.99	42.64	97.04

Table 6. *Cont.*

		Value Difference			Change Degree (%)		
		WP	SP	NGP	WP	SP	NGP
C7	Total carbon	−1.93	−5.76	−3.31	−33.56	−100.00	−57.38
C8	Available phosphorus	0.86	−0.42	−1.59	54.09	−26.62	100.00
C9	Total nitrogen	−0.81	1.09	0.67	−74.19	100.00	61.56
C10	Available potassium	88.80	49.73	−75.44	100.00	56.00	−84.95
C11	Moisture	16.33	0.62	−4.29	100.00	3.80	−26.24
C14	Forage production	3.67	1.67	4.00	91.67	41.67	100.00
C15	Fine forage ratio	−5.33	−44.54	−32.72	−11.97	−100.00	−73.46

Note: Value difference—The value of the indicators change from 2014 to 2017; Change degree—the value/the $value_{max}$.

4. Discussion

4.1. Grassland Type and Its Best Way of Use

The stable states of plant communities have been described as high and low steady states [37], and community stable states have been characterized by species turnover rate in community succession. Based on the theory of "ecological monoclimax" or "succession", the system must overcome the threshold to transfer from one state to another. During early warning, the community state that can be identified and the factors threshold value is difficult to cross is called the stable state when early warning is given [38], so the detectability of early warning is very important [39]. The index characterizing species turnover rate is the number of species, the variation of which is very important, and the index chosen in this study is C6 (Plant diversity) with a weighting index of 0.278.

In the study, 3 types of grazing methods, 5 types of alpine scrub and grassland, and 15 indicator factors were selected to construct a grassland ecological early warning system, which comprehensively considered the effects of grassland types, utilization methods, and early warning indicator factors on the ecological early warning of alpine grassland.

From the study, it can be seen that from 2014 to 2017, the composite prosperity index value all showed an increasing trend, with the greatest degree of change in winter pastures.

There is a complex relationship between the type of grassland and its best way of use or grazing style due to its varying composite prosperity index. In 2014, the composite prosperity index of the five types of grassland and the three types of grazing follows the sequence of *Kobresia tibetica* (winter pasture) < *Elymus* (winter pasture) < forbs (no grazing), *Kobresia humilis* (summer pasture) < *Elymus* (summer pasture). In 2017, this order changed to *Elymus* (winter pasture) < *Elymus* (summer pasture) < *Kobresia tibetica* (winter pasture) < forbs (no grazing) < *Kobresia humilis* (summer pasture). This change in sequence demonstrates that the style of grazing exerts a critical impact on the sustainable grazing of a given type of grassland. According to Figure 3, summer pasture had a significantly lower composite prosperity index in 2017 than in 2014 because the ecological recovery benefits showed up only within three years following the sowing of the *Elymus* grass (summer pasture). Thus, in order to reap the benefits, both the existing intervention intensity and grazing style should be maintained. In the winter pasture of *Kobresia tibetica* grassland, artificial construction of *Elymus* has seriously altered the natural state of the ecological system. Different grazing styles should be applied to different types of grassland to maximize the economic return from the grassland without the negative impact on its quality and long-term sustainability. *Potentilla fruticosa* shrub is best to be used as winter pasture to make the grazing more sustainable, whereas the vegetation of Elymus nutans grassland is low, but the forage is high in nutritional value, which is most suitable for as summer pasture.

4.2. Influence of Grazing Intensity

Under natural conditions, the steady-state transition period is relatively long. When disturbed, the system will undergo gentle and continuous changes in the current steady state. When the disturbance intensity is large enough, the system can be pushed to cross the unstable equilibrium point and fall to another steady state [40]; when the disturbance reaches a certain regime, alternative stable states occur; when the disturbance is greater than the steady-state threshold, a jump between different steady states occurs [41].

The threshold is different depending on the configuration of the community stable state, and the grazing pressure threshold is different in herbaceous grassland and shrub communities [42]. The mono-climax communities which are dominated by single species that are strongly resistant to changes caused by grazing are inconsistently affected by grazing closures on community stable state [43]. Intense disturbances such as heavy grazing can remove dominant species from the dwarf grassland [44]. The large-scale Artemisia annua shrub vegetation, which has been in a stable state for 10 years, lacks response to grazing [45]. The steady state of desert grassland shrub communities is difficult to break [46].

Since it is difficult to control grazing intensity during the study period, this indicator was not included in the early warning system. However, it has been taken into consideration partially through the indicator of grassland eco-carrying capacity. Judging from its highest weight (0.022) in the criterion layer (Table 2), it can be seen that grazing is fundamental to the grassland warning level. This is because more intense grazing will lead directly to a reduction in aboveground biomass, vegetation cover, and change in soil physical properties (e.g., soil compactness), indirectly increasing soil available P [47].

The study shows (Table 6) that the effects of winter grazing, summer grazing, and no grazing on the health status of alpine grasslands were considered in this study. However, the impact of grazing intensity on the health of alpine grasslands was not set, and there was no experiment conducted to determine whether summer grazing avoided the grass rejuvenation period (from April to May each year). Therefore, further research is needed to construct and apply an ecological warning system for alpine grasslands under grazing interference.

4.3. Impact of Human Intervention

Different ecosystem structures and functions correspond to different stable states, and the internal stability of ecosystems can be represented by the "attraction domain" and elasticity. If the "attraction domain" is wide, the system elasticity is large, and vice versa, the system elasticity is small [48]. When the attraction domain of a stable system narrows, small disturbances can also drive the stable state into another attraction domain, leading to significant changes in the ecosystem [49]. Ecosystem mutations have a hysteresis effect, and a greater disturbance is required to return to the original state [50]. Alpine grassland ecosystems are multistable systems. Irreversibility makes the detection of major ecosystem mutations of practical significance, and regime determination and ecological early warning become the focus of homeostatic applications. Accurate estimation of biophysical and chemical covariates of grassland vegetation to realize the dynamic detection of spatial and temporal changes in grassland is an important basis for ecological early warning systems for grassland.

Whether a grassland is grazed or not affected the grassland condition as the composite prosperity index of non-grazing pasture also increased, just like grazed pasture. This demonstrates that a moderate level of grazing is actually conducive to maintaining the health of the grassland. We should consider the modified warning model to adjust the classification standard of total carbon and available phosphorus and adapt to the alpine grassland ecosystem [51].

In this study, the early warning model constructed by the fuzzy object element method and the comprehensive prosperity index was used to evaluate the health status of the grassland and the warning status in 2014 and 2017, and to analyze the changes in the health status of different alpine grasslands and the degree of changes in the early warning and alert changes from the specific changes in the 15 index factors, which not only achieves the purpose of the health evaluation and ecological early warning, but also provides the healthy management of grassland ecosystems with a method of screening the early warning factors, the specific numerical indicators of the changes in the warning factors, and the methods of health management for each warning factor to be performed. It also provides a method for the health management of grassland ecosystems to identify the early warning factors, the specific numerical indexes of the changes of the warning factors, and the health management for the early warning factors. It will also provide references for the selection of sensitivity indicators for grassland ecosystem resilience in future studies, as well as quantitative research on the theory of steady-state critical slowdown and the critical transition [52] of early warning ecosystems.

5. Conclusions

The ecological early warning system constructed in this study for diagnosing grassland health is comprehensive and sound as it comprises a total of 15 indicators related to grassland quality, habitat, and its ecological carrying capacity in three layers (target, criterion, and indicator). It can paint a holistic picture about the true risk of grassland degradation in the study area that cannot be generated from each of the indicators individually. This system is able to yield warning levels of grassland degradation and diagnose its change over a period of time when grassland grazing style may have changed. Grassland quality is the most important to grassland health, followed by habitat quality among the three layers. In turn, grassland habit quality is subject the most to total carbon and available potassium of the soil. Of the five types of grassland, K. Tibetic had the lowest composite prosperity index, and hence the lowest level of warning (safe level), while all other four types of grassland had a higher but very similar level of warning (light) in 2014. However, in 2017, *K. tabetic*'s index increased the most (198%). *K. humilis*'s index had the next highest increase to reach the moderate warning level. All three grazing styles had a higher index in 2017 than in 2014. In particular, the winter pasture had the largest increase (35.1%), followed by the spring pasture (10.33%). In contrast, the no-grazing pasture had the lowest increase of only 9.3% to reach the light warning level. The fact that no indicators are equally important to all five types of grassland and the three grazing styles shows that grazing practices should vary with the conditions of the grassland so as to reap the most benefits from the resource without compromising its sustainability.

The fuzzy matter-element model for ecological warning of alpine grassland not only inherits the advantages of classical matter-element analysis such as clear concepts, strong logic, simple calculation process, and high resolution of evaluation results, but also fully considers the uncertainty of weights, comprehensively evaluates grassland conditions, reflects the trend of alpine grassland or reverse succession, and provides protection for targeted health management of grasslands [53]. However, there may be subjectivity when setting weights, so the model application requires a complete rule base [21].

Supplementary Materials: The following supporting information can be downloaded at: https://www.mdpi.com/article/10.3390/agronomy13082176/s1, Table S1: The weights and steps of each evaluation factor; Table S2: Mean random consistency index values; Table S3: Thresholds values (Si) of individual indicators Ci (Uij) used in calculating the level of warning; Table S4: The formula used to standardize Uij to Xij based on the level of warning judged from; Table S5: Test of the consistency of the indicators' weights.

Author Contributions: Conceptualization, Investigation, Writing—review & editing, H.S.; Software, Writing-original draft, M.L.; Methodology, Investigation, S.Z.; Formal analysis, Z.D.; Investigation, R.W.; Data curation, X.Q.; Formal analysis, M.C.; Writing—original draft, J.Z.; Supervision, Funding acquisition, Y.Q. All authors have read and agreed to the published version of the manuscript.

Funding: This work was funded by the Basic Research Project of Science and Technology Department of Qinghai Province (2021-ZJ-715); Project to promote scientific research cooperation and high-level talent training with CANZ and Latin America (No. [2021]109); Famous Teachers cultivation project of Qinghai University (2019-2022); First-class discipline "Ecosystem Management" in Sanjiangyuan of Qinghai University (ZYHX-202208).

Data Availability Statement: The data presented in this study are available on request from the corresponding author.

Conflicts of Interest: The authors declare no conflict of interest.

References

1. Zhang, J.N.; Zhang, J.G. Grazing strategies and suggestions for sustainable utilization of alpine meadow resources on the Qinghai Tibet Plateau. *Chin. Herbiv. Sci.* **2017**, *37*, 63–67.
2. Wu, G.L.; Du, G.Z. Discussion on the restoration and sustainable development of degraded alpine grassland ecosystem in Qinghai Tibet Plateau. *Nat. J.* **2007**, *29*, 159–164.
3. Li, X.L.; Gao, J.; Brierley, G.; Qiao, Y.M.; Zhang, J.; Yang, Y.W. Rangeland degradation on the Qinghai—Tibet Plateau: Implications for rehabilitation. *Land Degrad. Dev.* **2013**, *80*, 72–80. [CrossRef]
4. Song, Y.; Long, J.; Wang, H.B. Vegetation Changes along the Qinghai-Tibet Plateau Engineering Corridor since 2000 Induced by Climate Change and Human Activities. *Remote Sens.* **2018**, *10*, 95. [CrossRef]
5. Zhao, X.Q.; Zhou, H.K. Eco-environmental degradation, vegetation regeneration and sustainable development in the headwaters of three rivers on Tibetan Plateau. *China Acad. J. Electron. Publ. House* **2005**, *20*, 471–476.
6. Hu, L.; Wang, C.T.; Wang, G.X.; Ma, L.; Liu, W.; Xiang, Z.Y. Changes in the activities of soil enzymes and microbial community structure at different degradation successional stages of alpine meadows in the headwater region of three rivers, China. *Acta Prataculturae Sin.* **2014**, *23*, 8–19.
7. Wang, G.X.; Li, Y.S.; Wang, Y.B.; Shen, Y.P. Impacts of alpine ecosystem and climate changes on surface runoff in the headwaters of the Yangtze River. *J. Glaciol. Geocryol.* **2007**, *29*, 159–168.
8. Li, N.; Wang, G.X.; Liu, G.S.; Liu, Y.; Sun, X.Y. The ecological implications of land use change in the source regions of the Yangtze and Yellow Rivers, China. *Reg. Environ. Chang.* **2013**, *13*, 1099–1108.
9. Soyza, A.G.; Whitford, W.G.; Herrick, J.E.; Van Zee, J.W.; Havstad, K.M. Early warning indicators of desertification: Examples of tests in the Chihuahuan Desert. *J. Arid. Environ.* **1998**, *39*, 101–112. [CrossRef]
10. Fu, B.j. Principle and method of regional ecological environment early warning. *Resour. Dev. Prot.* **1991**, *7*, 138–141.
11. Dakos, V.; Scheffer, M.; Van Nes, E.H.; Held, H. Slowing down as an early warning signal for abrupt climate change. *Proc. Natl. Acad. Sci. USA* **2008**, *105*, 14308–14312. [CrossRef]
12. Drake, J.M.; Griffen, B.D. Early warning signals of extinction in deteriorating environments. *Nature* **2010**, *467*, 456–459. [CrossRef]
13. Capparelli, G.; Tiranti, D. Application of the MoniFLaIR early warning system for rainfall-induced landslides in Piedmont region (Italy). *Landslides* **2010**, *7*, 401–410. [CrossRef]
14. Wen, D.L.; Pan, H. Construction of monitoring and early warning system for cold soil black soil environment. *Environ. Sci. Manag.* **2010**, *35*, 124–126.
15. Hackett, B.; Comerma, E.; Daniel, P. Marine pollution monitoring and prediction. *Oceanography* **2009**, *22*, 168–175. [CrossRef]
16. Wright, A.L. Wetland Eutrophication: Early Warning Biogeochemical Indicators. *Soil Water Sci.* **2016**, *304*, 1–3. [CrossRef]
17. Van Dam, R.A.; Camilleri, C.; Finlayson, C.M. The potential of rapid assessment techniques as early warning indicators of wetland degradation: A review. *Environ. Toxicol.* **2015**, *13*, 297–312. [CrossRef]
18. Qin, Y.; Yang, Z.; Yang, W. Ecological risk assessment for water scarcity in China's Yellow River Delta Wetland. *Stoch. Environ. Res. Risk Assess.* **2011**, *25*, 697–771. [CrossRef]
19. Romañach, S.S.; Benscoter, A.M.; Brandt, L.A. Value-focused framework for defining landscape-scale conservation targets. *J. Nat. Conserv.* **2016**, *32*, 53–61. [CrossRef]
20. Shang, Z.H.; Dong, Q.M.; Shi, J.J.; Zhou, H.K.; Dong, S.K.; Shao, X.Q.; Li, S.X.; Wang, Y.L.; Ma, Y.S.; Ding, L.M.; et al. Research progress in recent ten years of ecological restoration for 'Black Soil Land' degraded grassland on Tibetan plateau-concurrently discuss of ecological restoration in Sangjiangyuan region. *Acta Agrestia Sin.* **2018**, *26*, 1–21.
21. Liu, D.J.; Zou, Z.H. Water quality evaluation based on improved fuzzy matter-element method. *J. Environ. Sci.* **2012**, *24*, 1210–1216. [CrossRef] [PubMed]
22. Zhu, J.; Tao, X. Evaluation of Land Ecological Safety Based on Fuzzy Matter-Element Theory. *J. Comput.* **2011**, *6*, 2639–2646. [CrossRef]
23. Li, L.; Xu, Y.L.; Jiang, S.H.; Wu, Q.M. Evaluation of Urban Vitality Based on Fuzzy Matter-Element Model. *Geogr. Geo-Inf. Sci.* **2010**, *26*, 73–77.
24. Ren, J.Z. *Methods of Grassland Science Research*; China Agriculture Press: Beijing, China, 1998.
25. You, Q.G.; Xian, X.; Peng, F.; Dong, S.Y. Alpine meadow degradation effect on soil thermal and hydraulic properties and its environmental impacts. *J. Desert Res.* **2015**, *35*, 1183–1192.
26. Cai, X.B.; Zhou, J. Spatial-temporal variation of soil organic carbon and its relations to soil physical properties in degraded alpine grasslands. *Chin. J. Appl. Ecol.* **2009**, *20*, 2639–2645.
27. Wei, J.; Jiang, P.; Yu, D.Y.; Wu, G.; Fu, H.W. Distribution patterns of vegetation biomass and nutrients bio-cycle in alpine tundra ecosystem on Changbai Mountains. *Northeast. China J. For. Res.* **2007**, *18*, 271–278. [CrossRef]
28. Saccone, P.; Cirel, J.; Pages, J.P.; Brun, J.J.; Michalet, R. Ecological resistance to Acer negundo invasion in a European riparian forest: Relative importance of environmental and biotic drivers. *Appl. Veg. Sci.* **2013**, *16*, 184–192. [CrossRef]

29. Cao, Y.Z.; Wang, X.D.; Lu, X.Y.; Yan, Y.; Fan, J.H. Soil organic carbon and nutrients along an alpine grassland transect across Northern Tibet. *J. Mt. Sci.* **2013**, *10*, 564–573. [CrossRef]
30. Hou, F.J.; Li, G.; Chang, S.H. Physiological indicators of grazing grassland health management. *J. Appl. Ecol.* **2002**, *13*, 1049–1053.
31. Liu, J.Y.; Xu, X.L.; Shao, Q.Q. Grassland degradation in the "Three-River Headwaters" region, Qinghai Province. *J. Geogr. Sci.* **2008**, *18*, 259–273. [CrossRef]
32. Dong, Q.M.; Ma, Y.S.; Xu, C.J.; Shi, J.J.; Wang, Y.; Wang, Y.L.; Sheng, L.; Li, S.X. Study of Classification and Gradation, Restoration of Black-Soil Beach Degraded Grassland in the Headwaters of Three Rivers. *Acta Agrestia Sin.* **2015**, *23*, 441–447.
33. Yang, W.C. *Health Evaluation of Qinghai Tibet Alpine Meadow Ecosystem: A Case Study of Chengduo County*; Gansu Agricultural University: Lanzhou, China, 2010.
34. Liu, Z. *Evaluation of Ecological Carrying Capacity of Qilian Mountains Nature Reserve*; Nanjing Normal University: Nanjing, China, 2004.
35. Partovi, F.Y. Determining what to benchmark: An analytic hierarchy process approach. *Int. J. Oper. Prod. Manag.* **1994**, *14*, 25–39. [CrossRef]
36. Jing, P. The Fuzzy Matter-Element Model for Forewarning Analysis of Regional Water Resources Sustainable Development. *China Rural Water Hydropower* **2005**, *8*, 22–24.
37. Murray, R.B.; Mayland, H.F.; Van Soest, P.J. *Growth and Nutritional Value to Cattle of Grasses on Cheatgrass Range in Southern Idaho*; Department of Agriculture, Forest Service, Intermountain Forest and Range Experiment Station: Ogden, UT, USA, 1978.
38. Xu, C.; Wang, H.J.; Liu, Q.X.; Wang, B. Alternative stable states and tipping points of ecosystems. *Biodivers. Sci.* **2020**, *28*, 1417–1430. [CrossRef]
39. Dakos, V.; Van Nes, E.H.; Donangelo, R.; Fort, H.; Scheffer, M. Spatial correlation as leading indicator of catastrophic shifts. *Theor. Ecol.* **2010**, *3*, 163–174. [CrossRef]
40. Sutherland, W.J.; Freckleton, R.P.; Godfray, H.C.J.; Beissinger, S.R.; Benton, T.; Cameron, D.D.; Carmel, Y.; Coomes, D.A.; Coulson, T.; Emmerson, M.C.; et al. Identification of 100 fundamental ecological questions. *J. Ecol.* **2013**, *101*, 58–67. [CrossRef]
41. Berdugo, M.; Kéfi, S.; Soliveres, S.; Maestre, F.T. Plant spatial patterns identify alternative ecosystem multifunctionality states in global drylands. *Nat. Ecol. Evol.* **2017**, *1*, 0003. [CrossRef]
42. Liautaud, K.; van Nes, E.H.; Barbier, M.; Scheffer, M.; Loreau, M. Superorganisms or loose collections of species? A unifying theory of community patterns along environmental gradients. *Ecol. Lett.* **2019**, *22*, 1243–1252. [CrossRef]
43. Rosindell, J.; Hubbell, S.P.; Etienne, R.S. The Unified Neutral Theory of Biodiversity and Biogeography at Age Ten. *Trends Ecol. Evol.* **2011**, *26*, 340–348. [CrossRef]
44. Beck, R.F.; Tober, D.A. Vegetational changes on creosotebush sites after removal of shrubs, cattle and rabbits. *Agric. Exp. Stn. Bull.* **1985**, *717*, 1–22.
45. Allen, E.B. *The Reconstruction of Disturbed Arid Lands: An Ecological Approach*; Routledge Press: New York, NY, USA, 2019.
46. Costello, D.F.; Turner, G.T. *Judging Condition and Utilization of Short-Grass Ranges on the Central Great Plains*; US Department of Agriculture: Washington, DC, USA, 1944.
47. Proffitt, A.; Bendotti, S.; Eastham, J. The effect of sheep trampling and grazing on soil physical properties and pasture growth for a red-brown earth. *Aust. J. Agric. Res.* **1993**, *44*, 317–331. [CrossRef]
48. Scheffer, M.; Carpenter, S.R. Catastrophic regime shifts in ecosystems: Linking theory to observation. *Trends Ecol. Evol.* **2003**, *18*, 648–656. [CrossRef]
49. Mäler, K.G. Development ecological resources and their management: A study of complex dynamic systems. *Eur. Econ. Rev.* **2009**, *44*, 645–665. [CrossRef]
50. Ellison, L. Influence of grazing on plant succession of rangelands. *Bot. Rev.* **1960**, *26*, 1–78. [CrossRef]
51. Rohde, K.; Stauffer, D. Simulation of geographical trends in Chowdhury ecosystem model. *Adv. Complex Syst.* **2005**, *8*, 4. [CrossRef]
52. Hu, Z.M.; Guo, Q.; Li, S.G.; Piao, S.L.; Knapp, A.K.; Ciais, P.; Li, X.R.; Yu, G.R. Shifts in the dynamics of productivity signal ecosystem state transitions at the biome-scale. *Ecol. Lett.* **2018**, *21*, 1457–1466. [CrossRef] [PubMed]
53. Wang, R.E.; He, B. Concept's Representation and Extension Nature Based on Matter-Element Models. *J. Petrochem. Univ.* **1997**, *10*, 83–87.

agronomy

MDPI

Article

The Microbiomes of Various Types of Abandoned Fallow Soils of South Taiga (Novgorod Region, Russian North-West)

Evgeny V. Abakumov [1,*], Grigory V. Gladkov [1,2], Anastasiia K. Kimeklis [1,2] and Evgeny E. Andronov [2,3]

1 Department of Applied Ecology, St. Petersburg State University, 199034 Saint-Petersburg, Russia; g.gladkov@spbu.ru (G.V.G.); a.kimeklis@spbu.ru (A.K.K.)
2 Laboratory of Microbiological Monitoring and Bioremediation of Soils, All-Russian Research Institute for Agricultural Microbiology, 196608 Saint-Petersburg, Russia; eeandr@gmail.com
3 V.V. Dokuchaev Soil Science Institute, 119017 Moscow, Russia
* Correspondence: e.abakumov@spbu.ru or e_abakumov@mail.ru

Abstract: More than 30 years have passed after the collapse of the Soviet Union, and huge areas of soil were left in a fallow state. The study of the microbiological status of fallow soils is an extremely urgent task because fallow soils represent the "hidden" food basket of Eurasia. In this context, we studied the influence of land use type (pasture, vegetable garden, hayfield, or secondary afforestation) on key agrochemical parameters and parameters of soil microbial biodiversity. All anthropogenically transformed soils included in the analysis showed increased humus content and pH shift to a more neutral side compared to the mature soil; the same seemed to be the case for all nutrient elements. It was established that the key factor regulating soil microbiome composition shift was the duration and degree of irreversibility of an agrogenic impact. The key phyla of soil microorganisms were Pseudomonadota, Acidobacteriota, Verrucomicrobiota, Bacteroidota, and Actinobacteriota. The proportion of other phyla was quite variative in soils of different land use. At the same time, all the 30-year-old abandoned soils were more similar to each other than to mature reference soil and 130-year-old soils of monoculture vegetable gardens. Thus, the first factor, regulating soil microbiome composition, is a continuation of soil agrogenic transformation. The second factor is the type of land use if the soil age was equal for fallow territory in the case of one initial podzol soil and one type of landscape. Thus, 30-year-old abandoned soils are intermediate in terms of microbial biodiversity between pristine natural podzols and plaggic podzol. It could be suggested that in the case of secondary involvement of soils in agriculture, the composition of the microbiome may turn to mature soil or to plaggic soil under intensive amelioration.

Keywords: abandoned soils; podzols; DNA sequencing; soil microbiome; soil fertility; south taiga

Citation: Abakumov, E.V.; Gladkov, G.V.; Kimeklis, A.K.; Andronov, E.E. The Microbiomes of Various Types of Abandoned Fallow Soils of South Taiga (Novgorod Region, Russian North-West). *Agronomy* **2023**, *13*, 2592. https://doi.org/10.3390/agronomy13102592

Academic Editors: Kesi Liu and Xinqing Shao

Received: 19 September 2023
Revised: 3 October 2023
Accepted: 7 October 2023
Published: 10 October 2023

1. Introduction

Soil conversion to fallow in the post-Soviet period affected huge areas of land previously used for arable farming for hundreds of years because its history in Russia is very long. Fallow lands play an important role in the sequestration of carbon dioxide in soils. Abandoned post-agricultural soils may accumulate or release carbon, depending on the type and age of abandoning and the type of land use [1]. Even though Russia ranks first in the world in the availability of land resources and is among the top five in terms of arable land, currently up to 40 out of 120 million hectares have been extracted from agricultural practices. This land was converted to fallow (abandoned lands) and is being transformed by natural and anthropogenic processes. The easiest and least expensive way to increase productive arable land and thereby dramatically increase the agricultural potential of the country is to return these 40 million hectares into circulation [2,3]. The area of fallow land in Northwest Russia is not decreasing, while the share of arable land is characterized by a steady decline, leading to the development of different scenarios of succession and soil formation. The accumulation of organic matter in fallow soils and an increase in the diversity of the molecular composition of humic acids in the profile vertical scale were observed during the postagrogenic successions. At the same time, a decrease in the share of aromatic components in the composition of organic matter is observed [4]. The last may lead to ambiguous effects on the emission of carbon dioxide.

Some of the soils of fallow lands, such as the outer islands of the Gulf of Finland, cannot be involved in the agricultural turnover and will remain untouched for a long time, as the islands are currently not inhabited [5]. These fallow agro-soils are unique natural monitoring models that can be used to study temporal trends in the soil-succession series. Some fallow lands will not be returned to the agricultural turnover because specially protected natural areas at regional and federal levels have been created on them. In the North-West of the Russian Federation, these are the Nizhnesvirsky State Reserve and the Valday National Park. The chronoseries of soils in such areas have been partially studied in terms of vegetation changes [6] and soil microbiome [7]. At the same time, the investigation of soil microbiome for benchmark, current agricultural, and fallow soils are few across the whole world and the Russian part of Eurasia [8]. Fragmentary data are available on the microbiome of Russian soils, including agrogenic soils [9–11], but these data are still insufficient to understand how the course of succession and land use affect the taxonomic composition and functional organization of the microbiome.

Since the soil microbiome is responsible for key biochemical processes in soils as well as their fertility, the study of the taxonomic composition of the microbiome is a priority for fallow soils. The relevance of this is reinforced by the recent interest in the re-engagement of agro-soils in agricultural practices. Studying the microbiome of fallow soils is also important in the context of elucidating their role in climate change and estimating carbon sequestration rates. Soils formed in the territory of the last glaciation as the youngest on the East European Plain can be one of the informative models for such studies. The end-moraine (glacial till) zone of the Valdai glaciation in the Novgorod region is the end zone of the distribution of the last glaciation [12]. Moraines, including local ones, zanders, and fluvioglacial sands, as well as ancient Pra-Msta river alluvial deposits, overlain by thin rewashed glacial deposits, are widespread here. The history of land use and agricultural practice here is well known and documented, as soils became abandoned from an arable form in 1993; thus, we have a good opportunity to compare various ways of abandoned soil development: afforestation of 30 and 70 years, pasturing, growing of meadow vegetation, current and abandoned vegetable gardens with one type of initial soil—podzol in fine sands, sublayered by the moraine loams of the Valday age.

That is why this work aimed to analyze the diversification of soil microbiome under the effect of different land use on the same type of soil and parent material in the southeastern part of the North-West Russian plain (Valday-Krestsy last glaciation moraine line). The following objectives were formulated: (1) to analyze soil morphology and chemical characteristics of the fine earth; (2) to investigate the taxonomy composition of soil microbiome; and (3) to analyze the parameters of alpha- and beta biodiversity of soil microorganisms in relation with type of land use and soil chemical characteristics.

2. Materials and Methods

2.1. The Study Sites

The study site's location is given in Figure 1. The research site is located in the territory of North-West Russia, in the Borovichi district of Novgorod region, in the vicinity of Velikiy Porog village. The sites under study are no more than 700 m apart, all of which is a single agro-landscape of the former state farm "Opechensky", which ceased to exist after the collapse of the Soviet Union.

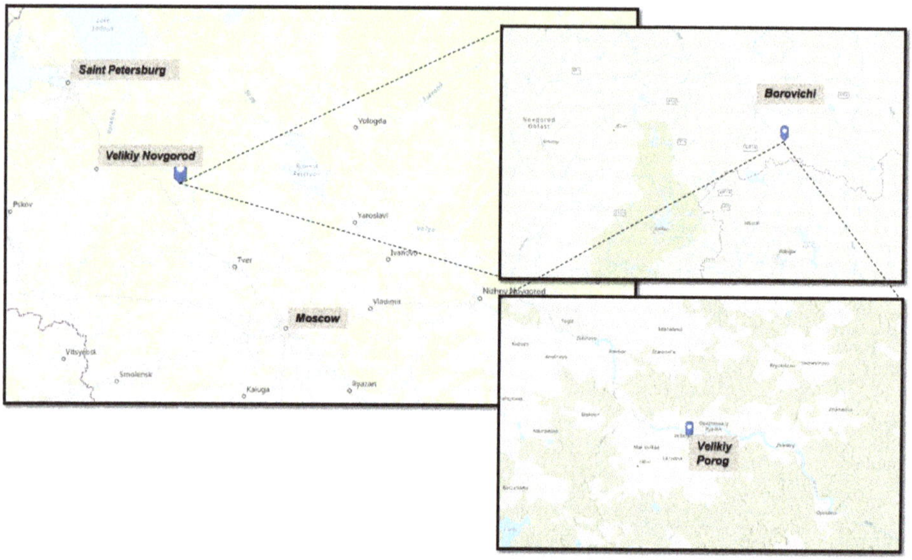

Figure 1. The location of the study area. Soil samples were collected near the Velikiy Porog village in the Borovichi district of Novgorod region (North-West Russia).

The average precipitation amount is 587 mm per year with evaporation about 430 mm. The annual mean temperature is 4.3 °C. The duration of the frost-free period is up to 130 days per year. The plot belongs to the south taiga bioclimatic zone with domination of podzols and retisols on drained watersheds and histic/gleyic soils in overmoisted landscape positions.

The history of farming in this region dates back a few centuries, but in this case, we studied the objects of 30-year-old fallow fields. The lands became fallow immediately after the collapse of the Soviet Union, i.e., in 1993 they were no longer plowed and fertilized. Further, a part of the fallow land was overgrown with secondary forest, a part was used continuously for haying, a part was subjected to permanent pasture digression (grazing of small cattle), and a part was used as a vegetable garden. Thus, the divergence of the soil formation process during 30 years after the land transition to fallow land is possible. We also studied the soil that reliably existed in arable condition for more than 130 years—a thick plaggen soil on podzol. All studied soils were formed on one type of parent material: sandy loam of water-glacial origin, underlain at the depth of 70–80 cm by red-brown moraine loams. The soil profiles are given in Figure 2. The study sites are located on the marginal part of the last, Valdai (Vyurm) glaciation, in the southern taiga subzone.

The natural (benchmark) vegetation of the sampling plots is Norway Spruce forest stands. The area of plowed land in this area during the Soviet era exceeded 80%, and even those spruce forests that appear to be primary underwent very significant anthropogenic impact. The sample plots, used for the analysis, are described in Figure 2. The mature soil (Figure 2a) was a podzol with a well-expressed iron-illuviated horizon, sublayed by a gleyification layer—G. Soils of 30-year old vegetable gardens (Figure 2b), secondary forest (Figure 2c), and pasture digression (Figure 2d) have loosed initial E and BF horizons; thus, they were involved to the arable horizon during Soviet times, and show no signs of secondary podzolization during the subsequent thirty-year stay in the deposit. Only the hayfield soil (Figure 2e) shows some signs of secondary podzolization in the upper part of the old plow horizon. As for the oldest arable soil, plaggen (Figure 2f), that has a humus A horizon with a total depth of about 40 cm, these soils were used for the monoculture of potatoes for at least 130 years. The coordinates of plots are the following: a—58.265268 N,

34.091997 E, b—58.269883 N, 34.083628 E, c—58.268745 N, 34.085468 E, d—58.269535 N, 34.083580 E, e—58.268671 N, 34.083279 E, and f—58.270034 N, 34.081949 E.

Figure 2. The studied soil profiles. (**a**) Mature soil (Mature soil), (**b**) 30-year-old abandoned vegetable garden (Garden), (**c**) 30-year-old secondary forest (Forest), (**d**) 30-year-old pasture digression (Pasture), (**e**) 30-year-old haymaking field (Hayfield), (**f**) 130-year-old plaggen (Plaggen) (soils formed under the long-term fertilization of bedding manure material applied in spring and fall from the farm. At the same time, horse, cow, and sheep manure was applied until 1940, and cow and sheep manure until 1970, after which chicken and goat manure dominated). Numbers to the right of the photos indicate the soil profile depth in cm.

2.2. Laboratory Analysis

The soil samples were collected on 10 May 2023 from the soil section wall at a depth of 10 cm using a special soil knife. The soil samples were taken in triplicate in each location. For the topsoil horizons, an average sample weighing 500 g was formed for routine soil analyses. For microbiological analysis, 2 g samples were taken in sterile tubes (in 4 replications). The samples for microbiological analysis were transported at +4 °C and stored at −20 °C.

For agrochemical analyses, the samples were air-dried, ground, and passed through a 2 mm sieve. The pH of the soil solution was measured using a Milwaukee Mi106 (Milwaukee Electronics (USA)) pH meter. A soil solution was prepared in a ratio of 1:2.5 with water [13]. The soil organic carbon (SOC) content was determined via the Tyurin method (comparable to Walkey and Blake's procedure) based on the oxidation of soil organic matter with a mixture of potassium dichromate and concentrated sulfuric acid [14]. The basal respiration (BR) and substrate-induced respiration (SIR) were measured according to the standard procedure in a laboratory in closed chambers under 60% soil humidity of whole water capacity. The content of available forms of ammonium nitrogen ($N\text{-}NH_4^+$)

and nitrate nitrogen (N-NO_3^-) was determined using a potassium chloride solution. The amounts of free potassium and phosphorus were determined via the Kirsanov method using phosphorus and potassium extraction with a 0.2 N hydrochloric acid solution [15,16]. The particle size distribution was measured using the classical sedimentometry method in 1 L volume cylinders with gravimetrical counting of each particle size fraction.

The total soil DNA was isolated in four replicates for each soil site by using the MN NucleoSpin Soil Kit (Macherey-Nagel, Dueren, Germany) using a Precellys 24 homogenizer (Bertin, Montigny-le-Bretonneux, France) according to the manufacturer's protocol. Quality control was carried out via PCR and agarose gel electrophoresis. The paired-end sequencing of the V4 variable region of the 16 S rRNA gene for a total of 24 samples was performed on the Illumina MiSEQ platform (Illumina, San Diego, CA, USA), using the primers 515 f (GTGCCAGCMGCCGCGGTAA) and 806 r (GGACTACVSGGGTATCTAAT) [17].

2.3. Data Processing and Statistical Analysis

The general processing of sequences was carried out in R 4.3 [18] using dada2 v. 1.28 (trimming at 200/180, error rate 2.5, truncation quality score 2) [19] and phyloseq v. 1.42 [20] packages, as described earlier [21]. The taxonomic annotation was performed with the SILVA 138.1 database [22] used as the training set; the phyla names were corrected according to LPSN [23]. The alpha-diversity (observed ASV and Simpson indices), beta-diversity (non-metric multidimensional scaling (NMDS) ordination of Bray-Curtis distances) metrics, and canonical correspondence analysis (CCA) were calculated using vegan v. 2.6-4 package [24] and visualized in phyloseq. The PERMANOVA analysis [25] was carried out using the adonis2 function in the vegan package.

3. Results and Discussion

Data on the soil chemical composition are given in Table 1. All the investigated soils were characterized by slightly acid pH values. The mature soil was more acidic than agrogenic ones. The highest pH was in plaggen soils; this is because the local population uses ash from stove heating to fertilize soils annually, as well as periodically applying purchased lime fertilizers. Thus, 30 years of being in the fallow state was not enough for a sharp change in the acid-base balance of soils. These data well correspond to the recently published results [26], which demonstrate that only after 25–30 years of a fallow state, a rapid change in soil acidity was detected. Thus, during the first three decades, abandoned soils act as a buffer in terms of acidity even in the case of cessation of fertilization, when the soil may not acidify for a long time, although reverse trends are possible [2]. According to the measurements of basal respiration, the respiration rate was the lowest in the mature soil and the highest in the secondary forest. T well corresponds with the previous data that soil organic matter is more stabilized in the end stage of pedogenesis, while the mineralization of a huge amount of fresh litter organic material may take place in the stage of active ecogenesis and soil formation [27]. The lowest values of p and K were in the mature podzol soils, which is typical for this type of soil with pronounced eluviation of any cations [28]. The highest level of these elements' accumulation was detected in the long-term amended plaggen soils. Other soils of our experiment were in the intermediate position in terms of p and K concentration between mature and plaggen soils. The pasture soil was the only one to demonstrate some increase in the mentioned nutrients, which could be a result of soil fertilization due to grazing livestock manure, or high spatial variability of soil agrochemical state, typical for the soil of post-Soviet agroecosystems [28,29] The content of ammonia nitrogen was low and that nitrate nitrogen was very low [30] in all soils investigated, with some variations among land use type. Thus, data on soil agrochemical composition may play a role as a predictor of soil microbial community composition and dynamics, which are analyzed below.

Table 1. Soil chemical analyses.

Sample ID	pH	TOC, %	BR	SIR	$N\text{-}NH_4^+$	$N\text{-}NO_3^-$	*p*	K
			$\mu g\ CO_2^-C/g \times hour$		Exchangeable forms, $mg\ kg^{-1}$			
Mature Soil	5.73	0.83 ± 0.06	0.82 ± 0.05	2.88 ± 0.08	4.08 ± 0.24	0.27 ± 0.09	25 ± 5	10 ± 2
Garden	6.11	1.40 ± 0.06	0.50 ± 0.09	0.87 ± 0.03	6.49 ± 0.13	0.49 ± 0.05	59 ± 4	13 ± 3
Forest	6.67	2.55 ± 0.05	2.03 ± 0.05	7.31 ± 0.22	11.78 ± 0.22	1.06 ± 0.08	70 ± 3	69 ± 4
Pasture	6.37	2.42 ± 0.09	1.60 ± 0.15	8.17 ± 0.72	11.52 ± 0.24	2.31 ± 0.09	161 ± 7	195 ± 6
Hayfield	6.23	2.21 ± 0.07	1.45 ± 0.12	3.48 ± 0.09	7.91 ± 0.31	0.70 ± 0.06	93 ± 6	93 ± 8
Plaggen	6.87	3.77 ± 0.03	1.27 ± 0.02	4.93 ± 0.25	5.92 ± 0.16	3.61 ± 0.14	499 ± 12	505 ± 19

Data on soil particle size distribution are given in Table 2. All the soils investigated are classified as sandy-textured with the dominance of the fraction of coarse sand inherited from the sandy-textured parent material. This is quite typical for podzol soils [31]. The physical clay particles were very low, which may predict the low absorption ability of soils both for chemical compounds and for microorganisms [32]; nevertheless, this influence was shown as not so direct and driven by particle size, especially in the case of the fungal community [33].

Table 2. Particle size distribution, fraction content, to fine earth (diameter of granulometric fractions).

Sample ID	1–0.25	0.25–0.05	0.05–0.01	0.01–0.005	0.005–0.001	<0.001	<0.01
Mature soil	50	38	3	1	1	7	9
Garden	45	30	16	1	1	7	9
Forest	29	48	11	1	4	7	12
Pasture	40	46	4	3	4	3	10
Hayfield	43	42	6	1	3	5	9
Plaggen	55	28	5	2	4	6	12

Soil Microbiome

Libraries of the 16 S rRNA gene for five sites of abandoned soils and one site of control soil were obtained. A total of 428,559 reads were classified into 5115 phylotypes. All investigated plots differed in their composition and representation of microorganisms. The most pronounced differences were detected between fallow soils, plaggen soils, and mature soil.

The parameters of alpha diversity of soil microorganisms are given in Figure 3. The number of phylotypes (ASVs) was different among the sites investigated. The lowest values of observed ASV, indicating community richness, were detected in the microbiomes of both forest soils—mature and 30-year-old forest on fallow land. The opposite data were obtained for the podzol soil microbiome under the urbanization effect [34]. It was previously shown that the composition of the microbiome of podzols was strongly influenced by the duration of the soil in a disturbed state [35]; in our case, several objects had one age of accumulation—30 years, which means that the microbiome is determined not by the age of the soil, but by the type of its use. The Shannon index characterizes evenness in the community structure and shows the degree of dominance of certain species in the community structure [36]. The lowest value of this index was in the mature forest soil and plaggen (Figure 3). While the richness of the plaggen microbiome did not reliably differ from fallow soils, its Simpson index was significantly lower than that of other fallow soils, which suggests that there was a decrease in evenness under long-term soil use for potato planting during more than 130 years. The values of all alpha indices were the highest in the soil under the pasture digression; thus, we can suggest that high soil biodiversity may be a result of intensive soil degradation. In addition, the soil of the hayfield had increased Shannon and Simpson indices values. Thus, both ecosystems of grassland were characterized by increased biodiversity. To conclude, forest ecosystems had less developed microbial biodiversity, but in the case of pasture, hayfield, garden, and plaggen soils, there was an increase in the diversity of the soil microbiome.

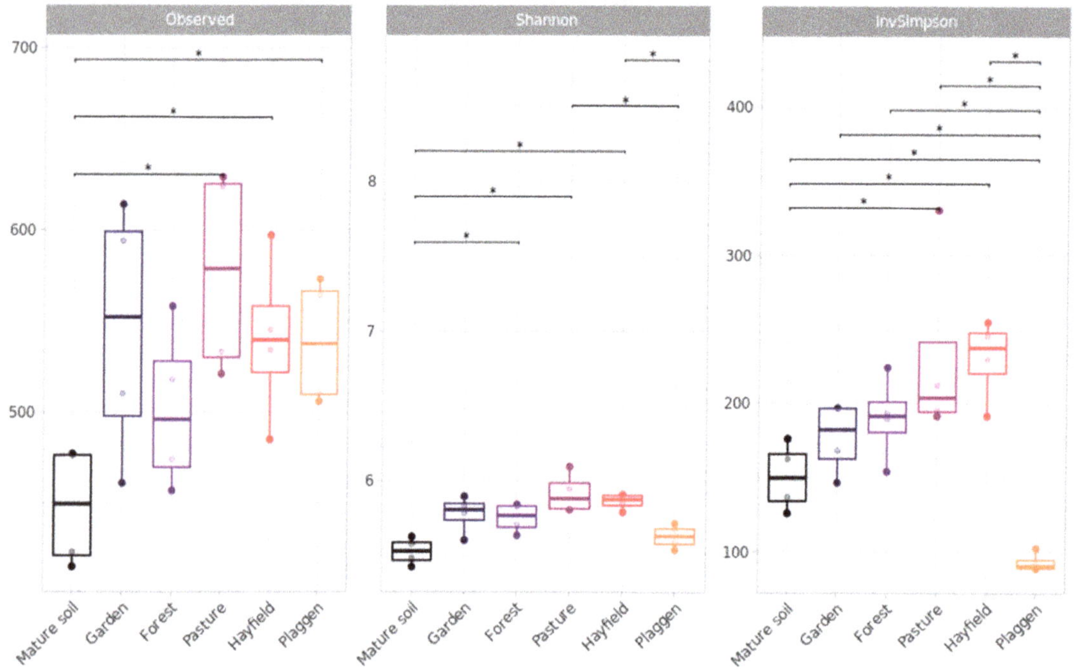

Figure 3. Parameters of alpha diversity (observed, Shannon, InvSimpson) of the microbial communities, each sample site in four replicates. Significant differences above the boxplots were assessed using the Mann–Whitney test, (*) p-value < 0.05.

An analysis of beta-biodiversity parameters (Figure 4) showed that the background soil and 130-year-old plaggen were the most different from each other and other fallow soils in terms of microbiome composition. In this case, the longest history of soil development in the course of agricultural impact was observed. In general, time is a very serious factor in determining the composition and function of the soil microbiome [36,37].

Data on the CCA plot (Figure 5) demonstrated that microbiomes of plaggen and mature soil reacted very differently to chemical composition than fallow soils. For plaggen soils, the increased amounts of available phosphorus, potassium, and nitrate nitrogen were the most important factors for explaining microbiome diversification. Total organic carbon and pH also contributed to the specificity of the plaggen microbiome. The basal and substrate-induced respiration values as well as ammonium nitrogen were the most important factors in the clusterization of all abandoned soils (hayfield, pasture, secondary forest, vegetable garden). Consequently, the mature soil microbiome was characterized by a negative correlation with the level of soil respiration and ammonium nitrogen content, which may demonstrate a stable state in mature soil in terms of biochemistry and microorganism activity in comparison with other ones.

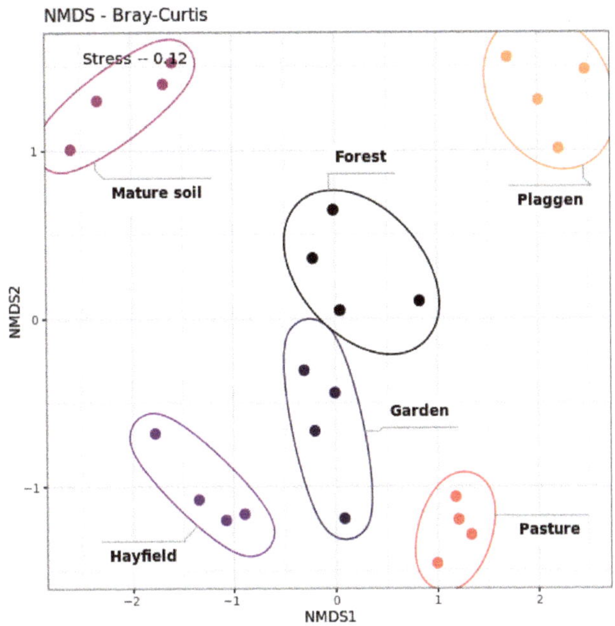

Figure 4. NMDS-plot of beta-biodiversity (Bray–Curtis distances), which demonstrates local differences between samples of abandoned (Plaggen, Garden, Pasture, Forest, Hayfield) and control (Mature soil) soils. Four replicate samples from one site are surrounded by ellipses.

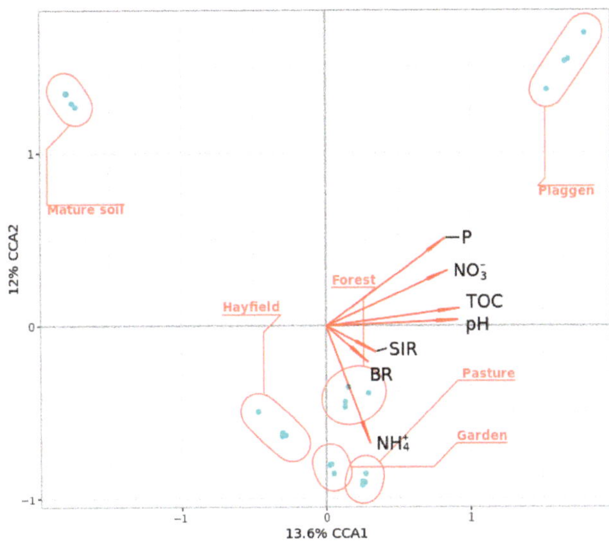

Figure 5. CCA plot demonstrating linear relationships between chemical characteristics and soil microbial communities of different abandoned soils. TOC—total organic carbon, P—phosphorus, $NH_4{}^+$—ammonium, $NO_3{}^-$—nitrates, SIR—soil-induced respiration, BR—basal respiration. K (potassium) is aliased with P because they were collinear (redundant). Percentages on the axis are the proportion explained values.

The relative abundances of major phyla are provided in Figure 6. The most represented phylotypes belonged to Pseudomonadota, Acidobacteriota, Verrucomicrobiota, Bacteroidota, and Actinobacteriota. All studied soils in terms of microbiome composition can be characterized as poorly anthropogenically transformed developed soils. This was indicated by a large number of representatives from the slightly cultured phyla Acidobacteriota and Verrucomicrobiota and bacteria that are more characteristic of the complex metabolically autochthonous microbiome (representatives of Planctomycetota, Chloroflexota, *Puia*, and *Haliangium*) (Figure 7). This well corresponds with the data of another fallow soil experiment, which show that in a 25-year-old fallow plot of albic retisol, the intensity of microbiological processes was close to those in virgin soddy-podzolic soils (Retisols), which were used as a control [38].

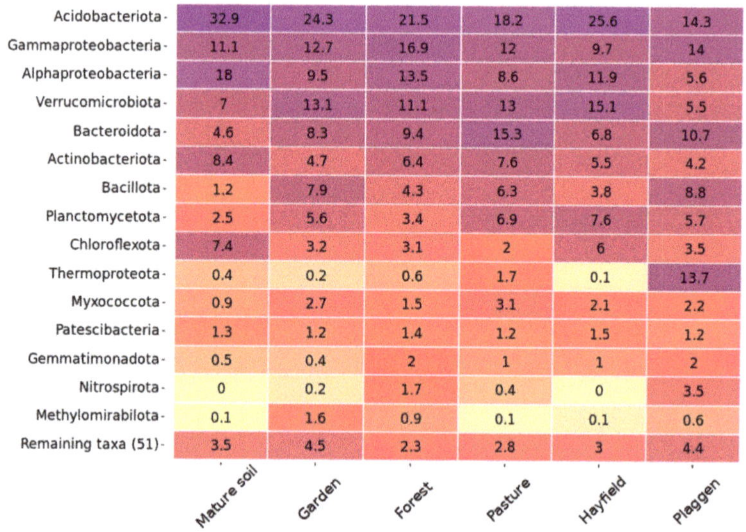

Figure 6. The relative abundance of major phyla in the microbiomes of abandoned and background soils. Pseudomonadota members are presented by two major classes. A darker color shows higher values, and a lighter one, lower.

The dominance of Acidobacteriota is known to be typical for organogenic and organomineral soil horizons [34]. Also, the dominance of Alpha- and Gammaproteobacteria, Actinobacteriota, Verrucomicrobiota, and Planctomycetota is known as typical for superficial podzol soil organic and organo-mineral horizons with acid reaction [8,37]. The representatives of Chloroflexota were more abundant in the mature soil, which may be linked to the high amount of fresh organic matter in the developed forest ecosystem, as it was mentioned recently that this group of microorganisms plays an important role in organic substance decomposition [39]. The microbiome of natural soils combines with the microbiome of agrogenic soils at both taxonomic and functional levels. Thus, the microbiome has components of the microbiome of sandy podzols and different variations of agrogenic soils. Thus, there was an evident trend in the increasing of Bacillota in all anthropogenically affected soils in comparison with the mature one. This is consistent with the findings that Bacillota appeared after the initial stage of lignocellulose decomposition [40]. The highest portion of Thermoproteota was in plaggen soils, which well corresponds with data on a high content of this group of microorganisms in fertile black soils of central Europe [41]. The plaggen soil was similar to black soils in terms of humus content, but not in terms of soil acidity; thus, the organic matter content may be a key factor of specificity of the soil microbiome of old cultivated soil. The phylum (Thermoproteota) is also known as a predictor of soil aggregation [41], and the aggregation intensively depends on organic matter content. The portion of Myxococcota was increased in all altered soil in comparison with benchmark (initial) one. The representatives of the Myxococcota phylum are well-known as micro predators able to destroy bacteria and eukaryotic organisms as well as to degrade complex macromolecules. An increase in their share in agricultural soils compared to natural soil may indicate certain shifts in soil metabolism [42] and may be responsible for the intensification of carbon-containing substance turnover. The role of Paterscibateria in soils is linked with the transport of metal in soil [43]; thus, its portion was more or less equal in all soils investigated because they belong to one geochemical group of sandy-textured iron-enriched south-taiga soils. High levels of nitrifying Nitrososphaerota archaea and Nitrospirota representatives were present, indicating that the microbiome reacted sharply to the introduction of excess nitrogen into the soil. These data contribute to the hypothesis [34] that the age of soil disturbance may play the most important role in the diversification of soil microbial biodiversity. The representatives of Gemmatimonadota did not show any significant differences among the soils.

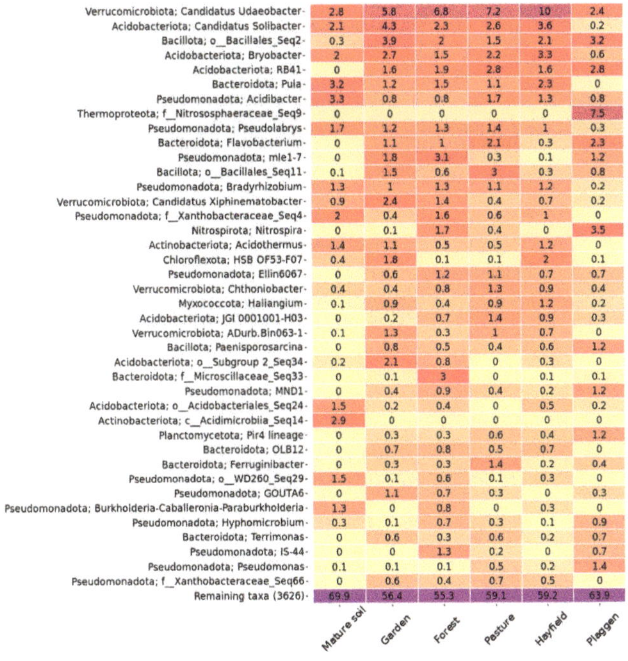

Figure 7. The relative abundance of the separate phylotypes in the microbiomes of abandoned and background soils. The row name includes phylum and the lowest determined taxon. A darker color shows higher values, and a lighter one, lower.

4. Conclusions

For the first time for the southern taiga subzone of the European territory of Russia, the development of soils and the formation of their microbiome 30 years after the transition to a fallow state were studied, as well as these parameters for the background soil and 130-year continuously used for potato cultivation plaggen podzol soil. All soils were formed from the same podzol formed on fluvioglacial sandy textured parent materials. Different fallow variants were studied: hayfield, vegetable garden, pasture digression, and secondary forest. It was found that the greatest differences in the chemical composition of soils and microbiome characteristics were characteristic of the pair of background podzol and 130-year-old plaggen. In the plaggen, the main factor affecting the microbiome was the content of organic matter, exchangeable phosphorus, and potassium compounds. The microbiome composition of 30-year-old fallow soils of all listed variants differed from both background soils and plaggen. All studied anthropogenically transformed soils showed increased humus content and pH shift to a more neutral side compared to the mature soil. The same seemed to be the case for all nutrient elements. It was established that the key factor of soil microbiome composition shift was the duration and degree of irreversibility of an agrogenic impact. The second factor was the content of nutrition elements—ammonium nitrogen in younger fallow soils and phosphorus and potassium in longer-developed soils.

Author Contributions: E.V.A.: writing—review and editing, project administration; funding acquisition; A.K.K.: laboratory procedures, data curation, investigation; G.V.G.: data curation, investigation, investigation, statistical treatment; E.E.A.: writing—review and editing. All authors have read and agreed to the published version of the manuscript.

Funding: This research was supported by the Russian Science Foundation, project No. 23-16-20003, dated 20 April 2023, and the Saint Petersburg Scientific Foundation, agreement No. 23-16-20003, dated 5 May 2023.

Institutional Review Board Statement: Not applicable.

Informed Consent Statement: Not applicable.

Data Availability Statement: Data are available at the NCBI Project PRJNA1023167, BioSample accessions SAMN37642152-7.

Acknowledgments: We thank the Centre for Genomic Technologies, Proteomics and Cell Biology (ARRIAM, Russia) for performing the preparation and sequencing of 16 S rRNA gene libraries. The authors are grateful to Timur Nizamutdinov, Roman Dyachkovsky, and Vyacheslav Polyakov, researchers of the Department of Applied Ecology, for their assistance in performing general chemical analyses of soils. This article is dedicated to the 300th anniversary of St. Petersburg State University.

Conflicts of Interest: The authors declare no conflict of interest.

References

1. Lyuri, D.I.; Nekrich, A.S.; Karelin, D.V. Cropland dynamics in Russia in 1990–2015 and soil emission of carbon dioxide. *Vestn. Mosc. Univ.* **2018**, *3*, 70–76.
2. Dymov, A.; Dubrovskiy, Y.; Startsev, V. Postagrogenic development of Retisols in the middle taiga subzone of European Russia (Komi Republic). *Land Degrad. Dev.* **2018**, *29*, 495–505. [CrossRef]
3. Orlova, O.I. Struggle for the ground: Restoration of laylands. *Karelian Sci. J.* **2015**, *2*, 130–133.
4. Kechaikina, I.O.; Ryumin, A.G.; Chukov, S.N. Postagrogenic transformation of organic matter in soddy-Podzolic soils. *Eurasian Soil Sci.* **2011**, *44*, 1077–1089. [CrossRef]
5. Iurmanov, A.; Romanov, M.; Marika, G.; Volodina, A.; Baikova, I.; Popov, I.; Markovets, M. Seagrass Zostera in the Russian Section of the Baltic Sea. *Geogr. Environ. Sustain.* **2022**, *15*, 111–115. [CrossRef]
6. Ipatov, V.S.; Lebedeva, V.K. Some aspects of the study of the functional structure of meadow phytocenosys. *Izv. Samara Sci. Cent. Russ. Acad. Sci.* **2012**, *36*, 1252–1256.
7. Frouz, J.; Nováková, A. Development of soil microbial properties in topsoil layer during spontaneous succession in heaps after brown coal mining in relation to humus microstructure development. *Geoderma* **2005**, *129*, 54–64. [CrossRef]
8. Pershina, E.; Ivanova, E.; Korvigo, I.; Chirak, E.; Sergaliev, N.; Abakumov, E.; Provorov, N.; Andronov, E. Investigation of the core microbiome in main soil types from the East European plain. *Sci. Total Environ.* **2018**, *631–632*, 1421–1430. [CrossRef]
9. Kaiser, K.; Wemheuer, B.; Korolkow, V.; Wemheuer, F.; Nacke, H.; Schöning, I. Driving forces of soil bacterial community structure, diversity, and function in temperate grasslands and forests. *Sci. Rep.* **2016**, *6*, 33696. [CrossRef]
10. Zverev, A.O.; Kichko, A.A.; Pinaev, A.G.; Provorov, N.A.; Andronov, E.E. Diversity Indices of Plant Communities and Their Rhizosphere Microbiomes: An Attempt to Find the Connection. *Microorganisms* **2021**, *9*, 2339. [CrossRef]
11. Semenov, M.V.; Ksenofontova, N.A.; Nikitin, D.A.; Tkhakakhova, A.K.; Lukin, S.M. Microbiological Parameters of Soddy-Podzolic Soil and Its Rhizosphere in a Half-Century Field Experiment with Different Fertilizer Systems. *Eurasian Soil Sci.* **2023**, *56*, 756–768. [CrossRef]
12. Malakhovsky, D.B.; Markov, K.K. *Geomorphology and Quaternary Deposits of the North-West of the European Part of the USSR*; Nauka: Leningrad, Russia, 1969; p. 256.
13. FAO. *Standard Operating Procedure for Soil pH Determination*; FAO: Rome, Italy, 2021.
14. FAO. *Standard Operating Procedure for Soil Organic Carbon: Tyurin Spectrophotometric Method*; FAO: Rome, Italy, 2021.
15. FAO. *Standard Operating Procedure for Soil Available Phosphorus—Olsen Method*; FAO: Rome, Italy, 2021.

16. *GOST R 54650–2011*; Soils: Determination of Mobile Phosphorus and Potassium Compounds by Kirsanov Method Modified by CINAO. Standartinform: Moscow, Russia, 2019; p. 8.
17. Caporaso, J.; Lauber, C.; Walters, W.; Berg-Lyons, D.; Lozupone, C.; Turnbaugh, P.; Fierer, N.; Knight, R. Global Patterns of 16S RRNA Diversity at a Depth of Millions of Sequences per Sample. *Proc. Natl. Acad. Sci. USA* **2011**, *108* (Suppl. S1), 4516–4522. [CrossRef] [PubMed]
18. R Core Team. *R: A Language and Environment for Statistical Computing*; R Foundation for Statistical Computing: Vienna, Austria, 2021.
19. Callahan, B.; Mcmurdie, P.; Rosen, M.; Han, A.; Johnson, A.J.; Holmes, S. DADA2: High-Resolution Sample Inference from Illumina Amplicon Data. *Nat. Methods* **2016**, *13*, 581–583. [CrossRef] [PubMed]
20. Mcmurdie, P.; Holmes, S. Phyloseq: An R Package for Reproducible Interactive Analysis and Graphics of Microbiome Census Data. *PLoS ONE* **2013**, *8*, e61217. [CrossRef] [PubMed]
21. Gladkov, G.; Kimeklis, A.; Zverev, A.; Pershina, E.; Ivanova, E.; Kichko, A.; Andronov, E.; Abakumov, E. Soil Microbiome of the Postmining Areas in Polar Ecosystems in Surroundings of Nadym, Western Siberia, Russia. *Open Agric.* **2019**, *4*, 684–696. [CrossRef]
22. Quast, C.; Pruesse, E.; Yilmaz, P.; Gerken, J.; Schweer, T.; Yarza, P.; Peplies, J.; Glöckner, F. The SILVA Ribosomal RNA Gene Database Project: Improved Data Processing and Web-Based Tools. *Nucleic Acids Res.* **2013**, *41*, D590–D596. [CrossRef]
23. Parte, A.C.; Sardà Carbasse, J.; Meier-Kolthoff, J.P.; Reimer, L.C.; Göker, M. List of Prokaryotic Names with Standing in Nomenclature (LPSN) Moves to the DSMZ. *Int. J. Syst. Evol. Microbiol.* **2020**, *70*, 5607–5612. [CrossRef]
24. Oksanen, J.; Blanchet, F.G.; Kindt, R.; Legendre, P.; Minchin, P.R.; O'hara, R.B.; Simpson, G.L.; Solymos, P.; Stevens, M.H.H.; Wagner, H. Vegan: Community Ecology Package. R Package Version 2.4-3. 2017. Available online: https://cran.r-project.org/package=vegan (accessed on 9 December 2022).
25. Anderson, M.J. Permutational Multivariate Analysis of Variance (PERMANOVA). In *Wiley StatsRef: Statistics Reference Online*; Balakrishnan, N., Colton, T., Everitt, B., Piegorsch, W., Ruggeri, F., Teugels, J.L., Eds.; Wiley: New York, NY, USA, 2017; pp. 1–15.
26. Litvinovich, A.V. Postagrogenic Evolution of Well Cultivated Soddy-Podzolic Soils in the Northwestern Nonchernozemic Zone. *Agrochemistry* **2009**, *7*, 85–93.
27. Bujalský, L.; Kaneda, S.; Dvorščík, P.; Frouz, J. In situ soil respiration at reclaimed and unreclaimed post-mining sites: Responses to temperature and reclamation treatment. *Ecol. Eng.* **2014**, *68*, 53–59. [CrossRef]
28. Bogdevitch, I.M.; Putyatin, Y.V.; Stanilevitch, I.S.; Lomonos, O.L. Dynamic of potassium supply level of arable and meadow soils of Belarus. *Soils Sci. Agrochem.* **2020**, *62*, 104–116.
29. Suleymanov, A.; Nizamutdinov, T.; Morgun, E. Assessing and geostatistical mapping of metal contamination in the polar arable plot (Yamal-Nenets Autonomous District, Russia). *Int. J. Environ. Sci. Technol.* **2023**, 1–10. [CrossRef]
30. Gamzikov, G.P.; Suleimenov, S.Z. Nitrogen Mineralization Capacity of Gray Forest Soil of the Novosibirsk Ob River Region during Composting and Fallowing of Plant Residues. *Eurasian Soil Sci.* **2021**, *54*, 729–737. [CrossRef]
31. Ponomareva, V. *Theory of Podzol Formation*; Nauka: Moscow, Russia, 1964.
32. Zheng, M.; Zhu, P.; Zheng, J.; Xue, L.; Zhu, Q.; Cai, X.; Cheng, S.; Zhang, Z.; Kong, F.; Zhang, J. Effects of soil texture and nitrogen fertilisation on soil bacterial community structure and nitrogen uptake in flue-cured tobacco. *Sci. Rep.* **2021**, *11*, 22643. [CrossRef] [PubMed]
33. Seaton, F.; George, P.; Lebron, I.; Jones, L.; Creer, S.; Robinson, D. Soil textural heterogeneity impacts bacterial but not fungal diversity. *Soil Biol. Biochem.* **2020**, *144*, 107766. [CrossRef]
34. Korneykova, M.V.; Vasenev, V.I.; Nikitin, D.A.; Soshina, A.S.; Dolgikh, A.V.; Sotnikova, Y.L. Urbanization Affects Soil Microbiome Profile Distribution in the Russian Arctic Region. *Int. J. Environ. Res. Public Health* **2021**, *18*, 11665. [CrossRef]
35. Gelashvili, D.B.; Iudin, D.I.; Rozenberg, G.S.; Snegireva, M.S.; Solntsev, L.A.; Fedjunin, V.A.; Yakimov, V.N. Fractal characteristics of the species structure of ichneumon wasp communities in the middle urals. *Dokl. Biol. Sci.* **2010**, *434*, 351–354. [CrossRef]
36. Lupwayi, N.; Clayton, G.; O'Donovan, J.; Harker, K.; Turkington, T.; Rice, W. Soil microbiological properties during decomposition of crop residues under conventional and zero tillage. *Can. J. Soil Sci.* **2004**, *84*, 411–419. [CrossRef]
37. Choma, M.; Šamonil, P.; Kaštovská, E.; Bárta, J.; Tahovská, K.; Valtera, M.; Šantrůčková, H. Soil Microbiome Composition along the Natural Norway Spruce Forest Life Cycle. *Forests* **2021**, *12*, 410. [CrossRef]
38. Naliukhin, A.N.; Khamitova, S.M.; Glinushkin, A.P.; Avdeev, Y.M.; Snetilova, V.S.; Laktionov, Y.V.; Surov, V.V.; Siluyanova, O.V.; Belozerov, D.A. Changes in the Metagenome of Prokaryotic Community as an Indicator of Fertility of Arable Soddy-Podzolic Soils upon Fertilizer Application. *Eurasian Soil Sci.* **2018**, *51*, 321–326. [CrossRef]
39. Wang, L.; Ye, X.; Hu, H.; Du, J.; Xi, Y.; Shen, Z.; Lin, J.; Chen, D. Soil bacterial communities triggered by organic matter inputs associates with a high-yielding pear production. *Soil* **2022**, *8*, 337–348. [CrossRef]
40. Saubenova, M.; Oleinikova, Y.; Sadanov, A.; Yermekbay, A.; Bokenov, D.; Shorabaev, Y. The input of microorganisms to the cultivation of mushrooms on lignocellulosic waste. *AIMS Agric. Food* **2023**, *8*, 239–277. [CrossRef]
41. Nelkner, J.; Huang, L.; Lin, T.W.; Schulz, A.; Osterholz, B.; Henke, C.; Blom, J.; Pühler, A.; Sczyrba, A.; Schlüter, A. Abundance, classification and genetic potential of *Thaumarchaeota* in metagenomes of European agricultural soils: A meta-analysis. *Environ. Microbiome* **2023**, *18*, 26. [CrossRef] [PubMed]

42. Colette, M.; Guentas, L.; Patrona, L.D.; Ansquer, D.; Callac, N. Dynamic of active microbial diversity in rhizosphere sediments of halophytes used for bioremediation of earthen shrimp ponds. *Environ. Microbiome* **2023**, *18*, 58. [CrossRef]
43. Tian, Z.; Li, G.; Tang, W.; Zhu, Q.; Li, X.; Du, C.; Li, C.; Li, J.; Zhao, C.; Zhang, L. Role of Sedum alfredii and soil microbes in the remediation of ultra-high content heavy metals contaminated soil. *Agric. Ecosyst. Environ.* **2022**, *339*, 108090. [CrossRef]

agronomy

Review

Research Progress on Grassland Eco-Assets and Eco-Products and Its Implications for the Enhancement of Ecosystem Service Function of Karst Desertification Control

Kangning Xiong *, Cheng He and Yongkuan Chi

School of Karst Science, Engineering Laboratory for Karst Desertification Control and Eco-Industry of Guizhou Province, Guizhou Normal University, Guiyang 550001, China
* Correspondence: xiongkn@gznu.edu.cn

Abstract: Grasslands in karst ecological fragile areas can effectively mitigate climate change, conserve biodiversity, maintain human well-being, and play a significant role in improving the health of regional ecosystems and farmers' livelihoods. Thus, the study of grassland ecological assets and ecological products comprehensively examines their effects on grassland ecosystem services based on the traditional paradigm. This procedure is crucial from a strategic perspective for rebuilding damaged grassland ecosystems in karst regions, strengthening the efficacy of desertification control, and encouraging sustainable economic growth. In this review, 143 pertinent works on grassland ecological assets and ecological products are numerically and qualitatively analyzed. The findings demonstrated the following: (i) After 2012, there was an exponential increase in the number of studies. The most frequently researched topics were ecological assets, functional enhancement, and service management contents, accounting for 82.09% of the total literature; the research regions were primarily distributed in Asia and North America. (ii) The research patterns were slowly diversifying and becoming more interdisciplinary. (iii) There are five key scientific issues to be addressed in the research on grassland ecosystems and we summarize the main developments and landmark achievements. (iv) There is an intrinsic relationship between grassland ecological assets, ecological products, and desertification control, and we propose insights into the enhancement of karst grassland ecosystem service functions based on three perspectives: fragile environment, trade-off synergy, and service management. This study provides valuable insights for the development of regional ecological livestock and the scientific promotion of integrated desertification control.

Keywords: structure optimization; function enhancement; ecological assets; service management; ecological products; supply capacity

Citation: Xiong, K.; He, C.; Chi, Y. Research Progress on Grassland Eco-Assets and Eco-Products and Its Implications for the Enhancement of Ecosystem Service Function of Karst Desertification Control. *Agronomy* **2023**, *13*, 2394. https://doi.org/10.3390/agronomy13092394

Academic Editor: Steven R. Larson

Received: 6 August 2023
Revised: 9 September 2023
Accepted: 13 September 2023
Published: 15 September 2023

1. Introduction

The intensification of global climate change, depletion of non-renewable resources, population explosion, and food shortages are generating many problems, such as biodiversity loss, grassland degradation, desertification, and shrinking of human living spaces [1,2]. In the past decades, the Convention on Biological Diversity (CBD) and the United Nations Sustainable Development Goals (SDGs) have emphasized a recovery strategy for global degraded grasslands, highlighting the multiple functions of grasslands in delaying atmospheric warming and increasing soil carbon sinks and human livelihoods [3,4]. Grasslands, mainly consisting of the legume and the grass family, are one of the pioneer communities of plants that predate ecological restoration [5]. Meanwhile, grasslands are also one of the producers of ecosystem cycle processes and food chains, which can provide a variety of products and services for sustainable socio-economic development to guarantee human well-being [6,7]. However, globally, the grasslands at varying degrees of risk of degradation represent about 40% of the existing grasslands, mainly in tropical and subtropical humid climate zones [8–10]. The essence of the current contradiction between grassland ecosystem protection and human society is the contradiction between the basic research on grassland ecosystem service (GES) functions and the environmental fragility of the combination of grassland resource scarcity, value, and ownership of power [11]. Therefore, clarifying the connotation and extension of the concepts of ecological assets and ecological products is a necessary condition for promoting the restoration of ecologically fragile areas. Ecological assets (EAs) are ecosystems that provide welfare for human society within a certain time and space range and under technological conditions [12]. Ecological products (EPs) refer to the physical and intangible products produced through clean production, recycling, consumption reduction, and emission reduction [13]. Liu et al. [14] proposed that grassland EAs are all resources and ecological environments that can enhance human well-being and serve as a "stock" for the pastoralists' economic development. Additionally, Zhang et al. [15] proposed that the EPs of grasslands are the final services and products used and consumed by human beings as a kind of "flow" under the joint action of biological production and human labor and are divided into public and business products [16]. The EP supply capacity of grasslands is based on the relationship between EAs natural resource endowment stock (Figure 1). Assets and products are prerequisites for human societies to value natural resources and are also intermediates in exploring economic development and environmental protection in ecologically fragile areas, providing important insights for a proper understanding of the human–land relationship.

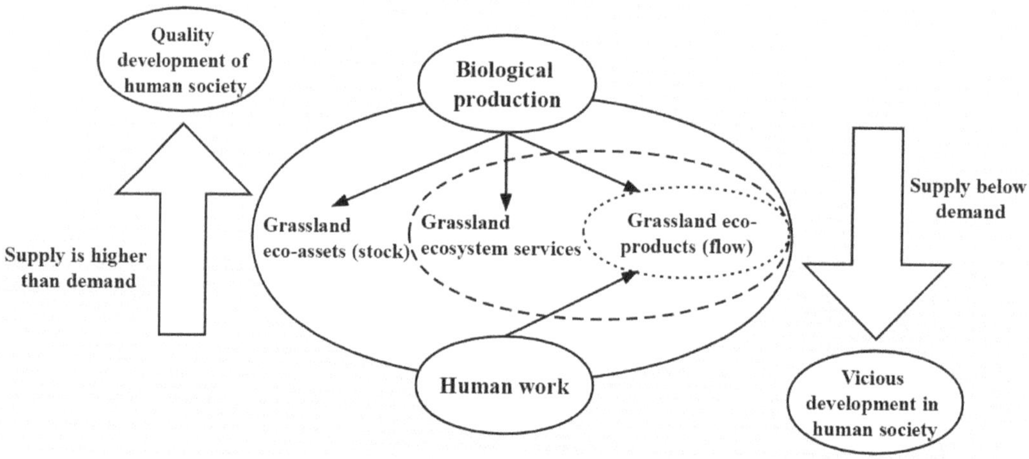

Figure 1. Relationship between grassland ecological assets, ecosystem services, and ecological products. Note: EAs contain ecosystem services, and ecosystem services contain EPs. When the supply and demand between natural ecosystem productivity and human society is an oversupply, society moves toward a virtuous cycle; on the contrary, society moves towards a vicious cycle.

Ecosystems and human health are being harmed by intensive fertilization and overgrazing practices in ecologically fragile areas [17]. As a result, the phenomenon of vulnerability is particularly common in karst ecosystems, which directly affects the restoration and conservation of degraded grasslands and sustainable socio-economic development [18]. Karst is due to the erosion process of soluble carbonate rocks by water flow, which forms a binary three-dimensional structure of surface–subsurface hydrological characteristics and a rich geomorphic landscape [19]. Global karst accounts for 1/5 of the land area, of which, the South China Karst (SCK) centered in Guizhou is the most widely distributed karst in a contiguous area, and its contribution to industry, agriculture, livestock, and safe water supplies for human life cannot be ignored [20,21]. However, the fragile natural environment and irrational human economic activities have led to social and environmental problems, such as soil erosion, vegetation degradation, increased rock outcrops, and increased poverty [22,23]. Curbing the expansion of rock desertification has quietly become a common problem in the world, and the Chinese government's plan to convert agricultural land into woodland and grassland provides a viable solution for the world [24]. SCK's desertification control has achieved significant results. At the same time, the process from "ecological cancer" (a special phenomenon of regional ecological imbalance caused by natural or human factors) to "green desert" (a phenomenon of large-scale and rapid artificial afforestation leading to a decrease in biodiversity and a decline in system function) is occuring. Although the rocky desertification control project has taken such measures as artificial afforestation and grassland planting, its species richness and ecosystem structure have suffered serious damage as a result of high-speed economic development. Examples include the emergence of the endangered *Primulina tabacum Hance*, the near-endangered *Castrodia elata*, the protected medicinal *Coptis chinensis*, and the invasive species *Ageratina adenophora* [25–27]. The fragmentation, and transferability of, and differences in, karst areas indirectly lead to the real dilemma of low yield, poor quality, and slow transformation of EPs [28]. Fundamentally, the relationship between EAs and EPs of grasslands under desertification control is that the supply is much smaller than the demand, and the high cost of farmers' input labor threatens their livelihoods [29]. Therefore, a clear understanding of the services and products that grasslands provide to human society is conducive to achieving ecological restoration. These services include the supply of raw materials, water purification, soil conservation, carbon sequestration, climate regulation, nutrient cycling, biodiversity conservation, and cultural aesthetics [30,31].

Currently, various forage planting combinations have been used to optimize the structure of grassland ecosystems and enrich the species diversity of soil communities [32,33]. By grazing at a reasonable level, soil erosion and farm poverty in karst areas can be effectively alleviated [34–36]. Effective trade-offs/synergies in the degree of human management are key to safeguarding the multifunctionality of grasslands and are very important to curbing land degradation and improving the quality of karst ecosystems, especially karst permanent grasslands (natural grasslands) [37,38]. However, there are few integrated studies on grassland ecosystem structure optimization, function enhancement, service management, EAs, and EPs. Therefore, this paper uses a systematic literature review approach. We aimed to (i) clarify the interrelationship between GES, EAs, and EPs, (ii) summarize the main developments and landmark achievements of grassland EAs and EPs, and (iii) extract five key scientific issues to be addressed at present. This paper helps to improve the scientific understanding of the components and functions of grassland ecosystems, so as to improve the quality and efficiency of grassland ecosystem ecological asset assessment and product supply capacity enhancement. It could provide decision-making guidelines for the scientific facilitation of integrated monitoring of desertification control and human welfare.

2. Materials and Methods

We used a systematic literature review (SLR) approach, which is a five-step cyclical process that incorporates program protocol, search, appraisal, synthesis, and report, with a characteristic methodology that is systematic, comprehensive, and reproducible [39]. This method has been utilized by researchers to both qualitatively and quantitatively analyze related research (Figure 2). The analytical approach of SLRs originated in the field of sanitation and nutrition evaluation and has proven to be applicable in a variety of fields such as mathematical modeling and geographical indications [40].

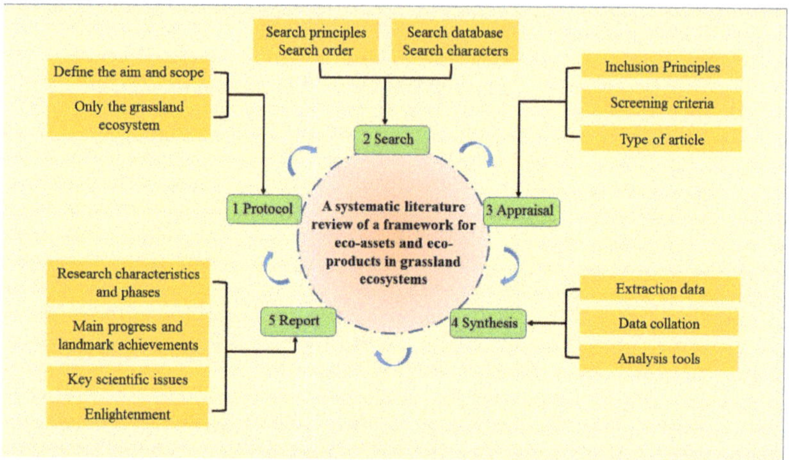

Figure 2. Technical route for the systematic literature review. The technical route is divided into a five-step, clockwise analysis cycle process from 1 to 5, including protocol, search, appraisal, synthesis, and report, and is completed by subroutines under each component.

2.1. Protocol

The protocol phase, using the China National Knowledge Infrastructure (CNKI) and Web of Science (WOS) databases for the literature search, developed a protocol to initiate a systematic literature review that is procedural, cross-disciplinary, and reproducible. The methodology's key aspect is the identification of study objectives, subjects, and boundaries, which helps researchers to validate scientific conjectures made prior to initiating the program [41]. The details of these conjectures were as follows: (1) the year of publication and the regional classification of the research; (2) the distribution of current hot topics; (3) the main developments and landmark achievements of the published literature; (4) the key scientific issues to be addressed in the future; and (5) the progress of research on EAs and EPs of grasslands for enhancing the ecosystem service function of karst desertification control.

2.2. Search

The search phase is shown in Table 1, and 4241 papers were obtained by entering the relevant search strings into the CNKI and WOS databases for primary searches according to the search order of title, abstract, and keywords. We inputted multiple terms related to grass, clicked the search window in the results for a secondary search, and excluded doctoral and master's theses by manual screening; a total of 488 papers were obtained.

2.3. Appraisal

During the evaluation phase, a full assessment of the 488 papers was performed to ensure the scientific validity and scope of the research. Clarification of the inclusion criteria, exclusion criteria, and type of evaluation allowed for further details and improved accuracy of the analysis. The inclusion criteria stipulated that (1) the search terms were present anywhere in the title, abstract, or keywords and (2) the content of the paper clarified the conceptual connotation of EAs, EPs, or services of grassland ecosystems.

The exclusion criteria included (1) duplicate papers; (2) publications with a weak impact; and (3) papers that could not be downloaded for reading.

Finally, based on the inclusion and exclusion criteria, 143 publications were used for SLR analysis. The content and number of all publications included service management (n = 28), structural optimization (n = 22), functional enhancement (n = 40), eco-assets (n = 51), and eco-products (n = 2). Qualitative and quantitative analysis of writings related to grassland ecological assets and ecological products. 85.31% of these literature were positive, 4.2% were negative, and 10.49% were neutral.

Table 1. Search databases and number of relevant papers on ecological assets and ecological products of grassland ecosystems. Retrieved 30 December 2022.

Literature Databases		Primary Search Terms (in Title, Abstract, and Keywords)	Search within Results	No. of Initially Acquired Publications	No. of Final Publications
CNKI and WOS	Retrieval string	"Ecosystem" AND "Structural optimization"	"Karst" and "Grassland" OR "Meadow" OR "Pasture" OR "Rangeland" OR "Steppe" OR "Prairie" OR "Savanna"	1658	181
		"Ecosystem" AND "Function enhancement"		817	84
		"Ecosystem" AND "Ecological assets" AND "Ecological products"		196	30
		"Ecosystem" AND "Ecological assets" AND "Services"		1199	129
		"Ecosystem" AND "Ecological products" AND "Services"		371	64
	Total			4241	488

2.4. Synthesis

During this phase, the analysis was carried out using tools such as Excel 2010, IBM SPSS Statistics 22.0, and Origin 2018, with a detailed delineation of themes, time period, and regional distribution. We concluded that grassland ecosystems are a source of abundant food and raw materials for human society. Our belief is that grassland ecosystem components, structures, and functions are crucial for human economic and social development. The focus on relationships, transformation mechanisms, and drivers of services, assets, and products is conducive to enhancing the systems' supply capacity and promoting the goals of the SLR.

2.5. Report

The main aim of the report phase is to analyze the results of the SLR and to establish the research framework for this article. Sections 3.1–3.3 discuss the trends over time, the proportion of research themes, and the distribution of the regions under study, respectively; Section 3.4 outlines the main developments and landmark achievements of the current research; Section 4.1 contains the key scientific issues to be addressed by future research; and Section 4.2 discusses the implications of the ecosystem service functions of grasslands for desertification control.

3. Results

3.1. Annual Distribution

Figure 3 shows the temporal distribution of the retrieved literature from 1994 to 2022 and examines the trend of the total grassland EAs and EPs literature volume according to the focus. Based on the distribution characteristics and landmark events, three research phases could be distinguished: the budding period, the development period, and the diversification period.

First, the budding period was from 1994 to 2004 and was characterized by an average of no more than two articles per year, with the ecosystem process of "plant–animal–soil" interaction as the main focus [42,43]. Global climate change, fire, overgrazing, and species invasion (e.g, *Ageratina adenophora, Amaranthus spinosus* L. and *Bidens pilosa* L.) disrupt the nutrient cycles of carbon, nitrogen, and phosphorus in soils. This affects the functions of microbial communities, alters species diversity, and accelerates the degradation of soil and grassland ecosystems [44,45]. In order to alleviate the contradiction between high-speed economic development and limited ecological resources, macro-control and environmental protection are necessary.

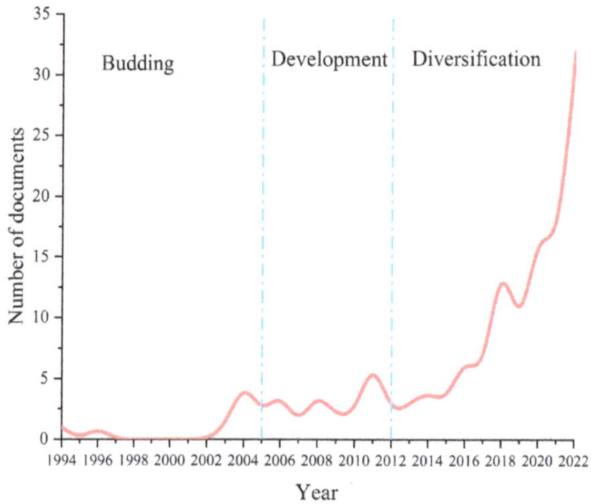

Figure 3. Annual distribution of published literature from 1994 to 2022.

The development period was from 2005 to 2011, with the UN-sponsored Millennium Ecosystem Assessment triggering a global scientific focus on restoring grassland ecosystems [46]. Structural optimization, functional enhancement, and ecological asset assessment of grassland ecosystems were the main concerns, with small fluctuations in the number of published papers. The researchers studied the various changes in microbial community structure, land use types, plant diversity conservation, and ecosystem service values. It has been shown that changing traditional grassland management patterns and improving grassland ecosystem services can help to ensure human food security, mitigate climate change, and enhance human welfare [47–49].

After 2012, the volume of published studies showed an exponential growth trend, and the United Nations focused on the restoration of different ecosystems in the next ten years [50]. In the context of global food security, biodiversity loss, and resource scarcity, aspects of service management, asset and product transformation, and ecological product supply capacity were the focus of the research in this period [51–54]. This period was characterized by studies on the diversity, intersectionality, and technological aspects of issues related to grassland ecosystems. Therefore, this period is known as the diversification phase. It was found that the sustainable development of animal husbandry was conducive to solving the livelihood problems of pastoralists and promoting the restoration of degraded grassland ecosystems.

In summary, the budding period focused on qualitative analyses of grassland ecosystem degradation. The data on system structure optimization and function enhancement support the shift from qualitative to quantitative assessments in the development period. The primary research in the diversification period highlighted the characteristics of the overall quality improvement path of regional ecosystems, the improvement of the supply capacity of high-quality EPs, and the management of ecosystem services.

3.2. Research Topic Distribution

This literature review classified and summarized the research contents of grassland EAs and EPs from the perspective of the structure, process, function, and services of ecosystems. We explored the research characteristics of structure optimization, function enhancement, service management, ecological asset assessment, and ecological product supply capacity improvement (Figure 4).

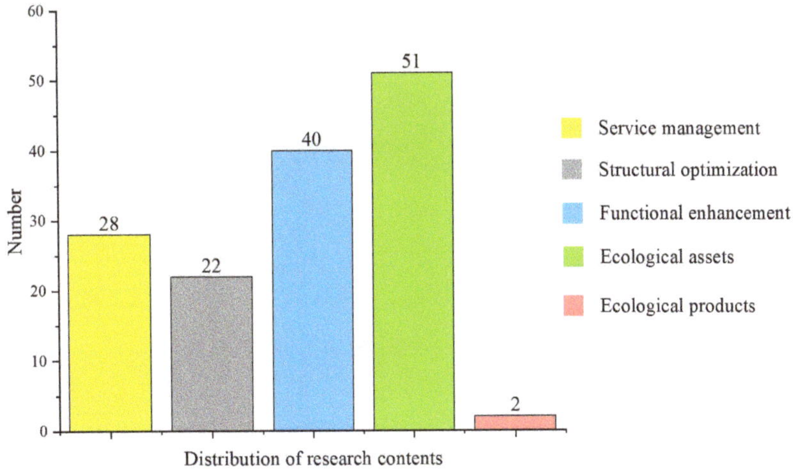

Figure 4. Thematic distribution of research literature.

The most researched themes were EAs, service management, and functional enhancement, which accounted for 82.09% of all papers. Among them, 34.93% were on EAs, 19.58% on service management, and 27.58% on functional enhancement. Because decision-makers and managers need to determine a list of regional ecological resources in order to provide data support for service management and improvement of degraded grassland ecosystems [55]. This shows that ecological asset assessment can change the status of grassland resources. It provides data monitoring for grazing time, technical regulation, and environmental protection red line delineation and regulates ecosystem service functional areas and management measures [56,57].

3.3. Study Area Distribution

Figure 5 shows the regional global distribution of grassland EAs and EPs studies. China had the highest number of published papers with 105 articles and 73.43 percent of the total. This reflects the importance of grassland EAs, EPs, and policy development in Asia. It also reflects the paradoxical phenomenon of changing dietary habits and inadequate regional supply in densely populated areas of Asia [58]. The United States of America followed China with 13 articles and 9.09 percent of the total. This reflects the significant downward trend in net primary productivity of grasslands in North America, which has the largest grassland area in the world. The phenomenon of grassland degradation has attracted the attention of researchers in the region [59]. Germany, France, and Italy accumulated more than three papers (two or less articles are not listed). The quantity here refers only to the regional distribution of the authors' affiliated countries.

3.4. Major Progress and Landmark Achievements

The world is facing the risk of the gradual degradation of grasslands. However, there are still challenges in the classification and value assessment of EAs, EPs, and ecosystem services. The mechanisms for grassland ecosystem structure optimization, function enhancement, service management, and product supply capacity enhancement have been seriously neglected. This review focused on the research themes of grassland ecosystem structure optimization, function enhancement, service management, EAs, and EPs. Figure 6 shows the framework for analyzing the literature on different topics.

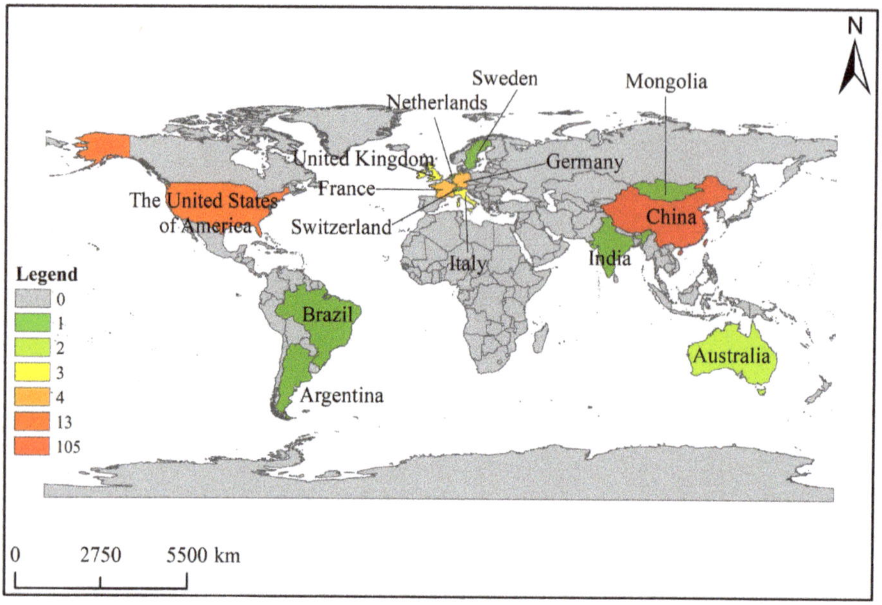

Figure 5. Regional distribution of the global number of published literature on eco-assets and eco-products of grassland ecosystems.

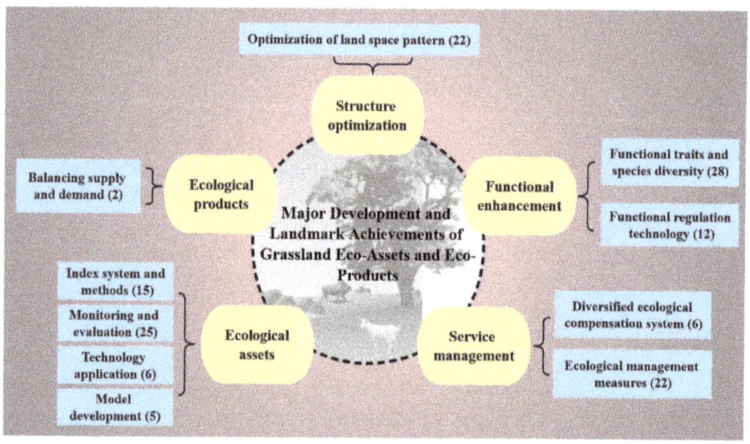

Figure 6. The frame diagram of eco-assets and eco-products of grassland ecosystems. (Note: The numbers in the graph indicate the amount of studies on that research topic).

3.4.1. Structure Optimization

The number of studies on the topic of structural optimization was 22. The earliest paper addressed the problem of diminished system productivity in the interlocking agricultural and pastoral areas of Altai, Xinjiang [60]. With the blurring of the boundaries of different ecosystems, analyzing the optimal allocation of EAs has gradually become the mainstream approach to ecosystem structural optimization. Currently, the spatial and temporal evolution of EAs is mediated by land use patterns, which reveal the indirect effects of climate change and human activities. This results in a general downward trend in the value of ecosystem services [61]. For example, to analyze the effects of population density on ecosystem structure and service values, Fei et al. [62] used a density estimation method. The results showed that population density was negatively correlated with ecosystem services, except for agricultural ecosystems. Thus, regional microhabitat differences and changes in population density show significant variability effects on ecosystems. Ecological restoration is the process of restoring dying ecosystems by returning farmland to forests and grasslands. This is a reasonable measure that will enable us to compensate for the environmental shortcomings and achieve the goal of optimizing the overall structural design. Li et al. [63] used spatial analysis to determine the spatial pattern of regional ecological assets. They clarified that the main drivers of asset change are the climate, population, land use, and urbanization process [64,65]. In summary, clarifying the main controlling factors of regional asset changes is conducive to the structural optimization of different ecosystems. It has provided data references and a scientific basis for ecological environment reconstruction.

3.4.2. Functional Enhancement

(1) Functional traits and species diversity

The theme of plant functional traits and species diversity contains 28 articles. In ecology, targeting the "soil microbial community–plant functional traits–species diversity" framework has been shown to be effective in improving ecosystem services. For example, a meta-analysis based on the exploration of mixed legume plantings in grasslands showed that net primary productivity (NPP) was significantly increased in grasslands with different soil environments. Studies have shown that legume forages may replace inorganic N fertilizers in the future, reducing damage to soil microbial communities from industrial fertilizers, and providing high-quality forage for livestock [42,66]. A study on the effect of optimal mowing time on the functional traits of community plants indicated that appropriate mowing frequency, time, and stubble retention height are beneficial to improving grass yield and quality [67]. Grazing and construction of vegetation barriers in sandy areas directly altered the contribution of the vegetation community and the physicochemical properties of the soil and protected biodiversity in ecologically fragile areas. For example, the intercropping of *Elymus nutans*, *Avena sativa*, and *Salix cupularis* [45,68]. Thus, the above three studies have demonstrated approaches for the improvement of the ecological service functions of grasslands and the effective restoration of ecosystem quality.

(2) Functional regulation technology

There were 12 papers on the topic of functional regulation techniques. Ecological problems have increased in recent years, with frequent fires and overgrazing activities destabilizing the structure of local grassland ecosystems [69], reducing plant community richness, and altering regional microhabitats. The decline in forage quality and the decrease in yield has led to a reduction in forage competitiveness. Therefore, the introduction of new species is an essential tool to maintain biodiversity and ecosystem stability [70]. For example, in studies of new species introductions, *Miscanthus* was found to be a relatively harmless forage grass in European regions, with high ecological and economic benefit in response to different hydrothermal conditions [71]. Thus, for future biodiversity conservation, this grass could be an effective option. At the same time, in order to prevent the disappearance of effective species in the future, it is necessary to establish a germplasm repository based on the requirements of genetic diversity. Thus, we need to clarify the different required materials, principles, and technical indicators, predict the current technical construction difficulties, and determine how to reasonably reduce input costs [72]. Using a quantitative analysis model based on Near-Infrared Spectroscopy (NIRS) technology to detect crude protein (CP) [73], high-quality grass species for breeding programs could be quickly and efficiently screened. However, it is important to combat the threat of invasive species (e.g., *Eupatorium odoratum* L., *Bidens pilosa* L., and *Avena fatua* L.) to grassland ecosystems and develop long-term monitoring and removal [74]. In addition, we should pay attention to the seasonal scarcity of feed and choose appropriate additives to improve the storage technology of silage feed. In order to obtain stable, safe, and high-quality feed [75].

3.4.3. Service Management

(1) Ecological management measures

The theme of ecological management measures was covered in 6 articles. The focus was on the macroscopic perspective of the relationship between regional EAs and ecological product inputs and outputs. The supply of natural ecosystems cannot meet the higher consumption needs of humans; therefore, in the future, an ecological product supply mechanism based on the public–private partnership (PPP) model should be established [76]. Studying the relationship between the economy and ecological environment in different regions is characterized by analyzing the degree of coupling and coordination between ecosystem service functions and supply capacity. At the same time, long-term integrated experimental network sites and forest–shrub–grass grazing crossover systems need to be established to improve ecosystem stability and supply capacity [77,78]. The politics of grass-based agriculture and livestock production, and the stability and resilience of the system vary from region to region. However, from the perspective of market analysis of ecological product supply and operation mechanisms, environmental management measures based on integration can play a significant role in alleviating the contradiction between the ecological environment and rapid economic development.

(2) Diversified ecological compensation system

A diversified ecological compensation system was the theme of 22 papers. Currently, ecosystem degradation and biodiversity loss are common. The "ecological protection red line" delineates ecologically essential and environmentally sensitive areas and seeks a long-term mechanism for simultaneous socio-economic development and ecological protection [79]. GES and local farmers' livelihoods are essential components of the sustainable management of socio-ecological systems [80]. Currently, it is necessary to pay more attention to the ecological and production functions of ecosystems, while also not neglecting the livelihood functions of farmers and vulnerable environmental areas. Therefore, grassland ecological compensation mechanism research should be based on environmental performance, income impact, and policy satisfaction as the indices to develop the grassland ecological compensation policy [81,82]. Only then, can we change grassland utilization and the ecological environment.

3.4.4. Ecological Assets

(1) Index system and methods

There were 15 articles on ecosystem service value assessment index systems and methods. In the past, ecological asset valuation was aimed at constructing an index system for the evaluation of ecosystem quality and area, and at monitoring biodiversity loss and the sustainable provision of ecosystem services [83]. For example, a set of 13 valuation indicators was initially developed using the forest indicator system and model to account for the value of grassland ecosystem services. With technological innovation, spatial analysis methods with a range of costs and computation times have been generated based on differences in scale, the difficulty of data acquisition, and research objectives [84]. Spatial analysis was used to study the spatial and temporal changes in landscape patterns for alpine, semi-arid, or ecologically hostile areas and to explore the impact mechanisms for improving ecosystem service functions and managing the development of agro-pastoral industries [85,86].

At present, the statement "green water and green mountains are the silver mountain of gold" points out that natural ecosystems contain enormous ecological and economic benefits. Incorporating the value of EAs into the social and market economic assessment system is necessary for society to attach more importance to the construction of an ecological civilization. In order to motivate institutional innovation in the ecological civilization construction, accounting methods such as green national economic accounting, national balance sheets, and account inventories of natural resource asset assessment frameworks have been developed [87–89]. For example, the integrated modeling framework integrates ecosystem services and landscape patterns into element, model, and data layers to explore practical solutions for regional ecosystem management [90]. The conceptual framework of "ecological assets–gross ecosystem product–green appreciation transformation–ecological compensation policy formulation" has made outstanding contributions to the process of ecological civilization construction. The gross ecosystem product (GEP) and ecological asset accounting are the basis for converting ecological benefits into economic benefits [91].

Simple and efficient valuation accounting methods were first identified by analyzing the ecological asset valuation objects and value categories [92]. The valuation methods were divided into three categories: direct market, alternative, and simulated market methods [93]. As shown in Tables 2 and 3, ecosystem service value assessment methods have their advantages and disadvantages. When comparing the scientific accuracy, objectivity, applicability, and credibility of the methods, the ranking is direct market method > alternative market method > simulated market method > energy valuation method.

Table 2. Advantages and disadvantages of different methods of accounting for ecosystem service value.

Evaluation Indicator	Method Type	Method	Equation	Formula Meaning	Advantages	Disadvantages	Reference
Organic matter production function	Direct market method	Market value method	$V_1 = \sum(m_i \times p_i)$	V_1 is the product value in CNY; m_i is the output of product category *i* in *t*; and P1 is the unit price of product category i in CNY·t^{-1}.	Evaluation is objective, with real-time data providing value and credibility.	Market prices follow the "supply and demand" law of economic development and are highly volatile; the data must be comprehensive and large in size.	[94]
Entertainment and leisure functions		Expense method	$V = V_1 + V_2$ $V_1 = \sum\limits_{i=1} P_i \times Q_i$ $V_2 = F + T + Q$	V is the cultural recreation value of the study area; V_1 is the research and cultural value; P_i is the average research value per unit area of ecosystem; and Q_i is the unit ecosystem area. V_2 is the recreational leisure value; F is the travel cost expenditure replaced by tourism income; T is the travel time value; and Q is other expenses.	The ecological value of cultural tourism areas can be directly quantified.	The method is limited and can only estimate the value of ecotourism areas.	[95]
Environmental purification function		Recovery and protection cost method	$V_a = \sum\limits_{i=1}^{n} X_i \times A_i \times P_i$	V_a is the value of the ecosystem's ability to absorb or deter pollutants; X_i is the capacity of the ecosystem to absorb pollutant i; A_i is the area of the ecosystem to absorb pollutants; P_i is the cost of treatment of pollutant *i*.	Ecological value can be quantified in terms of ecological restoration costs or protection costs.	Estimates are low, and the loss in value of ecosystem services that have been destroyed cannot be accurately estimated.	[96]

Table 2. *Cont.*

Evaluation Indicator	Method Type	Method	Equation	Formula Meaning	Advantages	Disadvantages	Reference
Soil conservation function	Alternative market approach	Opportunity cost method	$V = \sum_{i=1}^{n} S_i \times F_i$ $F = \sum_{i-1} S_i \div T$	In the first formula, V is the value of ecosystem soil fixation; i is the different ecosystems; S_i is the opportunity cost of soil erosion per unit area of ecosystem type i; and F_i is the area of abandoned land of ecosystem type i. In the second formula, F is the area of abandoned land; L is the total reduction in soil erosion for ecosystem type i; and T is the average thickness of surface soil in the study area.	A more comprehensive way to calculate the ecological home value of ecological resources.	Cannot account for the value of scarce resources.	[97]
Water-supporting function		Alternative engineering method	$V = P \times K \times R \times C$	V is the value of cultured water, in CNY·m^{-2}·a^{-1}; P is the amount of precipitation, in mm; K is the ratio of flow-producing precipitation to total precipitation, a constant; R is the coefficient of runoff reduction benefit, a constant; and C is the cost invested in the construction of 1 m^{-3} reservoir construction.	Approximates the value of damaged ecosystems.	The cost of alternative work makes it difficult to fully value the multiple functional benefits of ecosystems.	[98]
Carbon sequestration		Shadow engineering method	$V = NPP \times (1.2 \times P + 1.62 \times Q)$	V is the value of carbon sequestration, in CNY; P is the price of industrial oxygen production, in CNY·g^{-1}·c^{-1}; and Q is the fixed price of CO_2, in CNY·g^{-1}·c^{-1}.	Use alternative market techniques to find market prices for common public goods.	The selection of shadow prices is highly subjective and uncertain, and the prices of the selected commodities fluctuate significantly.	[99]
Biodiversity maintenance function	Simulated market method	Conditional value method (CVM)	$V = P \times P_a$	V is the value of biodiversity, in USD; P is the number of people that are willing to pay, in units; and P_a is the average price, in CNY.	For non-use (non-market) types of ecosystem services, the economic value of consulting public goods from a consumer's perspective is estimated scientifically and rationally.	The ambiguity in consumer awareness creates preference problems that are difficult to solve.	[100]

Table 3. Energy value theory method.

Scope of Application	Method	Advantages	Disadvantages	Reference
Calculation of the value of various types of services	Energy value theory method	Calculation of different categories and different natures of EAs; the establishment of a unified standard solar radiation energy calculation formula; the flexible use of different energy value conversion rates; the elimination of objective factors; and thus, it has broader application prospects.	Current scientific and technological developments limit the widespread use of this method.	[101]

(2) Monitoring and evaluation

There were 25 papers on monitoring and assessment topics, including 18 ecosystem-wide studies and 7 single-system or single-service function studies. These studies showed that the scope of research on EA value monitoring and assessment is divided into four main categories: global scale, regional scale, watershed scale, and sample plot scale. This classification only characterizes the relative size relationship of the different problems and objects.

Overall ecosystem: London et al. [6] initiated a study of global and regional ecological asset valuation. Focusing on climate change issues and the impact of large-scale human production activities on ecology can help alleviate conflicts between humans and the environment and alleviate environmental degradation. The research has focused on the effects of different types of ecosystem structure, quality, and area on ecosystem function. Srikanta et al. [102] used the elasticity coefficient (CE) and sensitivity coefficient (CS) to analyze the spatio-temporal variation in ecosystem service value (ESV). The results indicated that anthropogenic and rational landscape planning and optimization of territorial spatial resources benefit sustainable future economic development. For example, spatio-temporal variation in dryland ecosystems resulted in regional hot and cold spots of ecosystem services [103,104]. Accurately tracking changes in forest, cropland, and grassland areas can provide data to support decisions by ecological conservation managers.

Single-ecosystem and single-service functions: The existing research has increasingly focused on the synergistic relationship between the functional composition of regional single ecosystems, production management practices, and the value transfer and trade-offs of service functions [105,106]. This is a key insight for the maximization of the value of a single ecosystem. For a single grassland ecosystem, for example, Chinese scientists conducted ecological asset accounting. Through numerous expert questionnaires, analytic hierarchy processes (AHP), and induction methods, they established equivalent factors for different service functions and roughly derived ecological asset values for different grassland types across the country [107]. Using local farm yields, a revised model based on biomass calculations allowed for the estimation of ecosystem service values of forests, grasslands, and croplands. However, this accounting approach ignores the heterogeneous influence of geographically differentiated characteristics and climatic conditions on the value of ecological services. The range of static values of a single ecosystem function for water purification, soil conservation, and biodiversity as criteria for good or low ecosystem quality is questionable [108–110]. Wu et al. [111] used the precipitation storage method to estimate the interannual variation in water content in forest and grassland ecosystems in the Three Rivers Source Area. They clarified that the contribution of the improvement of water purification service function after the implementation of ecological projects was about 23.98%. Xiao et al. [112] used the soil loss equation model (USLE) to measure the soil conservation function of woodland, grassland, swamp, and farmland ecosystems on the Tibetan Plateau. They classified the soil conservation function into soil nutrient retention, wasteland reduction, and sedimentation reduction. The focus of more recent research has been on the simple summation of the values of various ecosystems. The mechanism of ecological asset transfer between different regions, landscape types, and functions is unclear, which directly affects the supply capacity of ecosystem services.

The conversion of ecological assets from quantified to unquantified is carried out to assess the overall index or grade of government performance and to provide data support for ecological compensation policies. Recent research has focused on the expanded index and ranking of total ecological assets to accurately compensate for weaknesses in ecologically vulnerable areas and to establish a complete compensation mechanism [113]. For example, Bo et al. [114] conducted an accounting of the comprehensive index of ecological assets in the Hinggan League, Inner Mongolia. They created a physical volume change table and a profit and loss table that showed that improving the quality of forests, grasslands, and wetlands promoted a virtuous cycle of ecosystem development. You et al. [55] examined a comprehensive index of county EAs, a liability account, and a profit and loss account to effectively reflect changes in ecosystem quality and delineate ecological asset classes. Huang et al. [115] used GIS spatial analysis to assess the change in EAs area and comprehensive index in key ecological functional areas, which showed an increasing trend in both quantity and quality of EAs. The results indicated that ecological engineering was the main driving force. This provides data support for the implementation of effective ecological engineering and the formulation of ecological compensation policies.

(3) Technology application

The theme of 6 articles was technology application. In order to comply with the development trend of the information age, we adhere to the principles of reducing evaluation costs, simplifying cumbersome procedures, and improving accounting efficiency. We use "3S" technology to evaluate, monitor, and analyze the spatiotemporal evolution of ecological assets, and effectively clarify their temporal trends, spatial distribution patterns, and influencing factors. For example, Chen et al. [38] showed that the value of ecosystem services in karst areas of China fluctuates upward over time, with higher ESVs in the southwest. Changes in wetlands, woodlands, and watersheds are the main reasons for the increase in ESV in this region. Xu et al. [116] pointed out that the overall trend of EAs in the Yangtze River Delta region has been decreasing over the past 12 years. The spatial distribution is characterized by a pattern of high in the south and low in the north, suggesting that the most effective way to increase regional EAs is to promote ecological restoration projects. In the regulation of ecological functions, the EPs of nature reserves play a crucial role. Through the assessment of water purification, soil conservation, and carbon sequestration, it was found that various ecosystem types supply different EPs [117]. There is a trend to use different types of data such as Landsat TM, MODIS, land use, NPP, and vegetation cover data to monitor and assess the value of EAs. This is particularly important for the scientific and systematic assessment of regional economic development levels.

(4) Model development

The theme of five articles was model development. Ecosystem services are assessed in terms of "physical quantity" and "value quantity". Based on the calibration of the ESV equivalence scale compiled by previous studies, this model can assess the value of regional ecosystem services with the advantages of efficiency, accuracy, and classification refinement [118]. On the one hand, the InVEST model was used to assess the regional ecosystem service capacity of typical desertification environments. It can accurately grasp the vital role of the three modules of water purification, soil conservation, and carbon storage for regional ecological management and the enhancement of ecosystem service functions [119]. On the other hand, the equilibrium factor and the yield factor, based on the ecological footprint model, show the differences in the capacity of environmental services provided by different ecosystems and can better express the carrying capacity of various ecosystems [120]. In summary, to solve the problem of slow policy formulation and accounting results, the ecological asset optimization assessment model maintains the balance between regional socio-economic development and environmental protection.

Quantitative prediction of future scenario model: The ecosystem service value prediction model scientifically predicts past, present, and future scenarios that can ensure the sustainable provision of quality services by ecosystems in the future [121]. For example, the CA-Markov model can accurately represent future vegetation succession trends, and it showed that controlling the leading drivers during stochastic changes in landscape patterns is conducive to improving future habitat quality and biodiversity service functions [122]. The CLUE-S model can more accurately simulate small-scale areas that drive changes in land use patterns [123]. In summary, the analysis of the spatio-temporal evolution patterns of ESV in the past, present, and future, combined with the future economic development direction of the region, will produce sustainable solutions for land policy makers and managers.

3.4.5. Ecological Products

The theme of two articles was the balance between supply and demand of EPs. The research is gradually moving toward the balanced development of the "three living functions". The ecological function of grasslands is the basis for maintaining ecological security; the production function of grasslands is the extension of eco-industry; and the living function of grasslands is the carrier of people's welfare. The "three functions" form an interlocking and coupled mechanism that directly affects the sustainable development of grassland ecosystems [124–126]. Balancing the supply of ecosystems with the consumption needs of socio-economic development is a fundamental issue in accelerating the overall enhancement of grassland ecosystem quality and the scientific promotion of integrated desertification control. Thus, we elucidate the relationship between ecosystems (producers, consumers) and humans (individual preferences, willingness to pay). Following the research idea of "qualitative-quantitative-orientation-strategy", the basic laws and framework systems of spatial transfer, management practices, value assessment mechanisms, and differences between supply and demand of ecosystem services can be explored [127,128]. With technological innovation, the introduction of "3S" technology and ecological footprint models to quantify the supply and demand of EAs [129], we can identify an appropriate ecological footprint that can help alleviate the conflict between supply and demand in grasslands. These cases show that grassland degradation, the introduction of high-quality grass species and production management methods, and the transfer of surplus animal husbandry laborers are important initiatives to improve the service capacity of grasslands, laying the foundation for a comprehensive improvement of regional ecological quality.

4. Discussion

We comprehensively reviewed the research process of grassland EAs and EPs through an overview of annual changes and stages in the literature and the distribution of research themes. The landmark results of recent research in structure optimization, function enhancement, service management, EAs, and EPs were summarized. On the basis of the above studies, we propose five key scientific issues to be addressed to apply the research on enhancing ecosystem service functions in desertification control, which we will systematically elaborate on in the following sections.

4.1. Key Scientific Issues That Need to Be Addressed

4.1.1. There Is an Urgent Need to Identify the Drivers of Change in Ecosystem EAs and EPs of Grasslands and to Find the Mechanisms of the Effects from Climate Change and Soil Characteristics

Natural and human factors, such as climate change, population density, urbanization, and industrial upgrading, combine to drive land use change and alter the supply of ecosystem services [61,130]. However, the drivers of grassland ecosystems require further analysis. The drivers of grassland ecosystems require a spatio-temporal perspective to highlight changes in the individual ecosystem service functions of grasslands. Remote sensing was used to analyze the seasonal spatio-temporal evolution of EAs of grassland ecosystems and to assess the carrying capacity of primary ecosystem functions. Similarly, individual ecosystem services are influenced to varying degrees by the soil biology of the sample site [131,132]. Therefore, the degree of influence of plant functional traits and soil characteristics on ecosystem services is essential to understanding the interrelationships and weights of different functions and to improve the quality of regional ecosystems.

4.1.2. Assessment Models and Techniques Should Focus on Enhancing Multifunctionality, Breaking Down Supply and Demand Barriers between Karst Grassland Ecosystems and Human Society, and Promoting a Balance between the Three Regional Functions

Grassland ecosystem assessment models and techniques are key to problem solving for the current state of land production functions only. Ecological footprint models and ecosystem service techniques can be used to assess ecological and livelihood functions, thus revealing the balancing mechanism of multiple ecosystem functions [133]. The analysis of the capacity and ecological footprint of grasslands is the basis for life–production–ecological functions. For example, some scholars have used the CASA and STARFM models to re-evaluate the ecological and environmental capacity of grassland ecosystems to produce a sustainable supply [134]. In particular, the preliminary assessment aimed to improve the ecological function of grassland ecosystems, to change grazing behavior and grassland distribution, and thus to provide sustainable and healthy management practices [135]. It also elucidated the state of grassland overload caused by the spatial mismatch between the rapid growth of livestock and the supply and demand of grassland resources. Therefore, scenario prediction for optimizing the future land, is a key factor in clarifying the supply and demand relationship between ecosystems and human society and the balance of living–production–ecological functions [64], which helps to optimize and enhance the grassland ecosystem assessment models and techniques.

4.1.3. In Response to the Inefficient Transformation of Grassland EAs and EPs, the Input–Output Ratio Should Be Improved to Increase the Supply Capacity of Grassland EPs in Karst Areas

To address the need for more research on the input–output relationship of EAs in grassland ecosystems, we proposed an optimization scheme for grassland ecosystem services by analyzing the factors affecting the supply of EPs. The transfer of EAs in grassland ecosystems currently occurs between regions and individual ecosystems, and preliminary knowledge has been developed on the mechanisms of transfer media, regions, and drivers [31,136]. However, the key to optimizing and controlling a single ecosystem lies in the positive transfer between different service functions. This measure should accelerate the efficiency of the conversion of EAs of grassland ecosystems into EPs, thus increasing the supply capacity of the ecosystem. At the same time, consideration should also be given to reducing the artificial input costs of ecological assets and increasing the output volume of EPs.

4.1.4. To Address the Issue of Trade-Off/Synergy Mechanisms for Grassland Ecosystem Services and Based on the Cascade Framework of "Ecosystem Structure–Process–Function–Services", a Holistic Approach to Improving Ecosystem Quality in Karst Areas Is Proposed

Using a qualitative and quantitative approach to identify GES trade-offs/synergies, the relevance of the different indicators needs to be clarified. Scholars have argued for enhancing the influence of internal and external factors on ecosystems based on the ecosystem "structure–process–function–services" cascade framework. Attention needs to be paid not only to the external factors that balance the multifunctional demand for grassland productive potential, but also to the complex trade-offs and synergies within different ecosystem functions [137,138]. As a result, rationalization grassland ecological management can help to balance and synergize the complex relationships between the different ecosystem services [139]. Especially in ecologically fragile karst areas, we can clarify the trade-offs and synergies between grassland ecosystem services in desertification control, to explore their impacts on human well-being.

4.1.5. In Response to the Uneven Spatio-Temporal Distribution of Grassland Resources in Different Regions, Different Functional Areas of Use Should Be Identified to Enhance the Supply of EPs in Karst Areas

The recent research has focused on cross-sectional comparisons of ecological asset values, establishing dimensionless metrics. However, based on the assessment of the status of EAs in regional grassland ecosystems, the composite indices and rankings vary significantly and only provide a basis for resource managers and policy makers to delineate different functional areas at the macroscopic level [54]. At the micro level, however, there is an urgent need for technology development, policy refinement and risk prevention awareness to optimize grassland management. For example, ecological intensification of grasslands (e.g., intercropping between agriculture and grass) maintains high levels of ecosystem services and biodiversity. Simultaneously increasing forage availability and the area of grass sown is key to improving the supply capacity of grassland ecosystems [140]. Some scholars have pointed out that the design of forest and grassland conservation models is aimed at optimizing management strategies from the perspective of farmers' food needs. However, the advantages and disadvantages of species invasions on ecosystems should considered, whether in terms of artificial management practices or in terms of improving service functions from the perspective of ecosystem interactions [141,142].

4.2. The Intrinsic Relationship between Grassland EAs, Products, and Desertification Control

Grassland EAs and EPs connote the conditions and utility that natural resources and ecological environments provide to humans [143]. An accurate analysis of the virtuous cycle and malignant development impulse of grassland ecosystems facilitates the formation of a cascading framework for a long-term mechanism of improving ecological livestock husbandry and human well-being under stone desertification management (Figure 7). This framework, based on the interplay between grassland ecosystem structures, processes, and functions, clarifies the ownership and scarcity of natural ecological resources. Through the value-added system of EA investment and operations, we can explore the market-based supply method and value realization mechanism of EPs. The ecological compensation standard, scope, and mode should be clarified, and weighing the pros and cons of the ecosystem is conducive to providing a research framework for the economic development of karst regions.

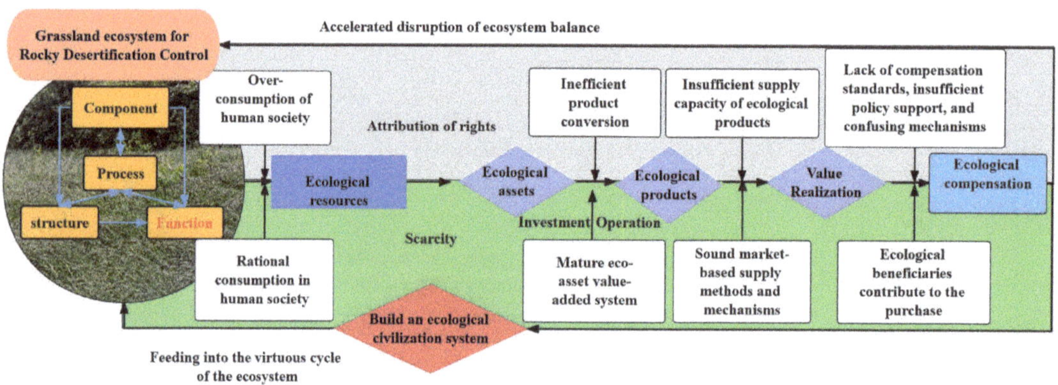

Figure 7. Framework diagram for forming final eco-products from grassland ecosystems for desertification control.

4.3. Insights into the Enhancement of Grassland Ecosystem Services in Desertification Control

In the 1990s, China carried out large-scale ecological restoration projects in response to the spread of desertification, with measures such as improving degraded grasslands and artificial planting in grassland ecosystems, which played a pivotal role in curbing the growth of desertification [144,145]. In this context, we present the findings of the recent grassland ecosystem research on karst desertification control from the perspective of preconditions, intermediate links, and important initiatives. We aim to provide directions for promoting the restoration of system quality in ecologically fragile areas.

4.3.1. A Deeper Understanding of the Fragility of the Karst Natural Environment Is a Prerequisite for Optimizing the Structure of Grassland EAs

Through a systematic review and compilation of the research on EAs and EPs of grassland ecosystems, the analysis focused on optimizing the allocation of grassland EAs to achieve the revitalization objectives under the natural conditions of the karst ecosystems. Based on the natural environment, karst ecosystems are located in a binary, three-dimensional landscape structure above and below the ground, with shallow and fragmented surface soils, which are mostly alkaline soils rich in calcium [146]. The sloping land has well-developed subsurface pore (fissure) spaces, which can easily cause soil nutrient and water loss in areas with heavy rainfall and high slopes. The rugged and steep topography, significant elevation differences due to the heterogeneity of water and thermal conditions, and vegetation diversity, growth, and development form a non-zonal distribution [147], resulting in considerable differences in the supply of GES. As a result, karst environmental ecosystems are prone to reverse succession of deterioration. Only plant populations with physiological characteristics of calcium liking, drought resistance, and lithophytic growth, as well as species with roots that adapt to rock and survive in crevices to extract nutrients, can grow and develop in this special environment.

4.3.2. Identifying the Relationship between the Supply and Demand of Grassland Ecosystems Is an Intermediate Step in the Functional Enhancement of Ecological Assets

The critical point in the development of organic livestock is the balance between human social consumption demands and the supply capacity of grassland ecosystems. The classification survey of EPs of karst grasslands and assessment of their supply capacity and influencing factors can prevent seasonal fodder shortage for livestock. This can reduce the cost of ecological asset valuation and ecological product distribution, accelerate improvements in input–output efficiency, and provide evidence to support decisions for optimizing grassland ecosystem functions.

4.3.3. Grassland Ecological Management Practices Are an Important Initiative to Enhance the Supply of Grassland Ecological Products

The measures to improve grassland ecosystem services are as follows: (1) A mixture of leguminous and herbaceous forages can be used to take advantage of their complementary benefits of increasing the net photosynthetic rate and water use rate. The nitrogen fixation properties of leguminous forages have a facilitating effect on the growth and development of herbaceous forages [148,149]. (2) Grasslands can use withered mulch to address the problem of accelerated surface water evaporation thanks to the reflection of solar radiation from rocky outcrops and to mitigate the effects of harsh environmental factors on the water-holding capacity of ecosystems [150]. Their planting structures and configurations should be optimized to modify the soil water–fertilizer exchange interactions and, to some extent, improve soil physicochemical properties. These studies will guide us in restoring degraded karst grassland ecosystems.

Therefore, to achieve the goal of a sustainable supply of high-quality EPs in the grassland ecosystem for karst desertification control, we need to optimize the allocation of grassland EAs and adjust the spatial distribution of water and fertilizer conditions in grassland ecosystems (Figure 8). We also need to facilitate the enhancement of the EPs supply capacity of grasslands, thereby promoting the improvement of the overall ecosystem service function.

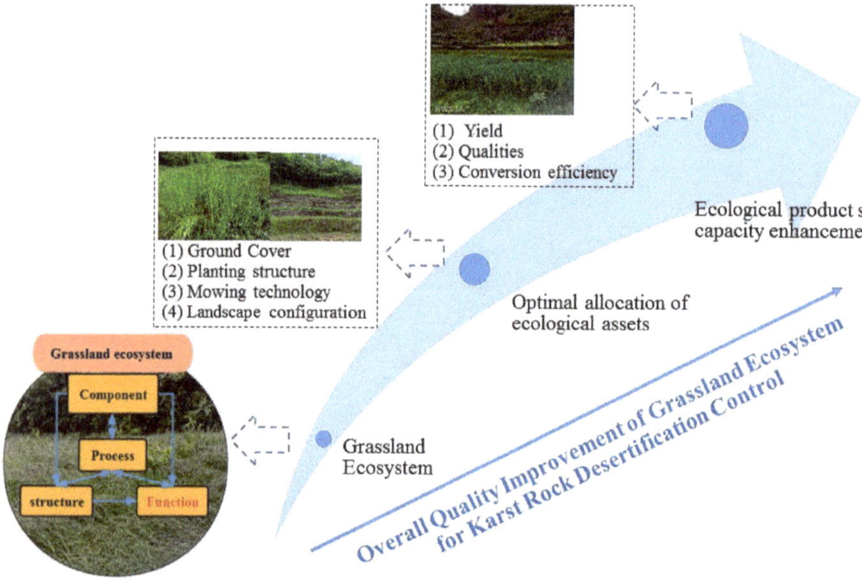

Figure 8. Roadmap for the optimal allocation of eco-assets in grassland ecosystems and the mechanism for eco-products supply capacity enhancement.

4.4. Limitations

The only databases used in this paper were WOS and CNKI, not Springs, Scoup, or other databases, and the search results may be biased. The summary of the research topic classification method and the key scientific issues to be addressed were also somewhat subjective.

5. Conclusions

In this paper, we conducted a systematic literature review of 143 articles retrieved through the CNKI and WOS foreign language journal resource service system, and the main conclusions were as follows: (1) The annual number of papers on EAs and EPs of grassland ecosystems has increased rapidly since 2012; the research stages can be divided into budding (1994–2004), development (2005–2011), and diversification periods (2012–2022). (2) The most common research themes were EAs, function enhancement, and service management, which accounted for 82.09% of the literature. The research regions were mainly located in Asia and North America. (3) The main progress and landmark achievements were in structural optimization, functional enhancement (functional traits and diversity and regulation techniques), service management (ecological management measures and diversified ecological compensation techniques), EAs (construction of index systems and selection of appropriate methods, monitoring and evaluation, application of technology, and model development), and EPs. (4) The five key scientific issues to be addressed in the future are identifying spatial and temporal changes in ecosystem assets and products; developing assessment models and technologies focused on enhancing multifunctionality; grassland EAs and products should focus on improving the efficiency of input–output ratios; establishing a synergistic mechanism for grassland ecosystem service trade-offs; and strengthening grassland ecological management in order to improve the supply of EPs. In the future, we should focus on the relationship between the optimal landscape allocation of EAs and the improvement of ecological product supply capacity; we also need to change the way decision makers and managers use grasslands, and give full play to the important role of grasslands in karst desertification control in improving ecosystem quality.

Author Contributions: K.X. proposed the main innovations and the construction of the overall writing framework of the paper, conducted the methodological research, reviewed and edited, provided funding, and wrote parts of it as well as the touch-up revisions in English; C.H. collected the data and organized the literature, designed the full-text images, and wrote part of the manuscript; Y.C. provided revisions and reviewed the final manuscript. All authors have read and agreed to the published version of the manuscript.

Funding: This research was financially supported by the Key Project of Science and Technology Program of Guizhou Province (No. 5411 2017 Qiankehe Pingtai Rencai); the World Top Discipline Program of Guizhou Province (No. 125 2019 Qianjiao Keyan Fa); and the China Overseas Expertise Introduction Program for Discipline Innovation (No. D17016).

Institutional Review Board Statement: Not applicable.

Data Availability Statement: The data presented in this study are available upon request from the corresponding author.

Acknowledgments: We appreciate the anonymous reviewers for their invaluable comments and suggestions on this manuscript.

Conflicts of Interest: The authors declare no conflict of interest.

References

1. Wu, G.; Cheng, Z.; Alatalo, J.M.; Zhao, J.; Liu, Y. Climate Warming Consistently Reduces Grassland Ecosystem Productivity. *Earth's Futur.* **2021**, *9*, e2020EF001837. [CrossRef]
2. Zhang, G.; Biradar, C.M.; Xiao, X.; Dong, J.; Zhou, Y.; Qin, Y.; Zhang, Y.; Liu, F.; Ding, M.; Thomas, R.J. Exacerbated grassland degradation and desertification in Central Asia during 2000–2014. *Ecol. Appl.* **2018**, *28*, 442–456. [CrossRef] [PubMed]
3. Soliveres, S.; Manning, P.; Prati, D.; Gossner, M.M.; Alt, F.; Arndt, H.; Allan, E. Locally rare species influence grassland ecosystem multifunctionality. *Philos. Trans. R. Soc. B Biol. Sci.* **2016**, *371*, 20150269. [CrossRef] [PubMed]
4. Bardgett, R.D.; Bullock, J.M.; Lavorel, S.; Manning, P.; Schaffner, U.; Ostle, N.; Chomel, M.; Durigan, G.; Fry, E.L.; Johnson, D.; et al. Combatting global grassland degradation. *Nat. Rev. Earth Environ.* **2021**, *2*, 720–735. [CrossRef]
5. Peeters, A. Importance, evolution, environmental impact and future challenges of grasslands and grassland-based systems in Europe. *Grassl. Sci.* **2009**, *55*, 113–125. [CrossRef]
6. London, J.; Park, J. *Man's Impact on the Global Environment: Assessment and Recommendation for Action*; Report of the Study of Critical Environmental Problems; MIT Press: Cambridge, MA, USA, 1970.
7. Daily, G. *Nature's Services: Societal Dependence on Natural Ecosystems*; Island Press: Washington, DC, USA, 1997.
8. White, R.; Murray, S.; Rohweder, M. *Pilot Analysis of Global Ecosystems: Grassland Ecosystems*; World Resources Institute: Washington, DC, USA, 2000.
9. Wilson, S. Grasses and grassland ecology. *Ann. Bot.* **2009**, *6*, 104. [CrossRef]
10. Buisson, E.; Le Stradic, S.; Silveira, F.A.O.; Durigan, G.; Overbeck, G.E.; Fidelis, A.; Fernandes, G.W.; Bond, W.J.; Hermann, J.-M.; Mahy, G.; et al. Resilience and restoration of tropical and subtropical grasslands, savannas, and grassy woodlands. *Biol. Rev.* **2018**, *94*, 590–609. [CrossRef]
11. Lemaire, G.; Hodgson, J.; Chabbi, A. Introduction: Food security and environmental impacts—Challenge for grassland sciences. In *Grassland Productivity and Ecosystem Services*; Lemaire, G., Hodgson, J., Chabbi, A., Eds.; CABI: Wallingford, UK, 2011.
12. Dong, T.; Zhang, L.; Xiao, Y.; Zheng, H.; Huang, B.; Ouyang, Z. Assessment of ecological assets and gross ecosystem product value in Ordos City. *Acta Ecol. Sin.* **2019**, *39*, 3062–3074.
13. Zeng, X.G.; Yu, H.Y.; Xie, F. Concept, classification and market supply mechanism of ecological products. *China Popul. Resour. Env.* **2014**, *24*, 12–17.
14. Liu, Y.X.; Fu, B.J.; Zhao, W.; Wang, S. Ecological asset accounting and ecosystem services evaluation: Concept intersection and key research priorities. *Acta Ecol. Sin.* **2018**, *38*, 8267–8276.
15. Zhang, L.B.; Yu, H.Y.; Hao, C.Z.; Wang, H.; Luo, R.J. Redefinition and connotation analysis of ecosystem product. *Res. Environ. Sci.* **2021**, *34*, 655–660.
16. Zhou, J.Y.; Xiong, K.N.; Wang, Q.; Tang, J.H.; Lin, L. A Review of Ecological Assets and Ecological Products Supply: Implications for the Karst Rocky Desertification Control. *Int. J. Environ. Res. Public Health* **2022**, *19*, 10168. [CrossRef] [PubMed]
17. Li, L.; Fan, Z.H.; Xiong, K.N.; Shen, H.; Guo, Q.Q.; Dan, W.H.; Li, R. Current situation and prospects of the studies of ecological industries and ecological products in eco-fragile areas. *Environ. Res.* **2021**, *201*, 111613. [CrossRef] [PubMed]
18. Dong, S.; Shang, Z.; Gao, J.; Boone, R.B. Enhancing sustainability of grassland ecosystems through ecological restoration and grazing management in an era of climate change on Qinghai-Tibetan Plateau. *Agric. Ecosyst. Environ.* **2020**, *287*, 106684. [CrossRef]

19. Ravbar, N.; Šebela, S. The effectiveness of protection policies and legislative framework with special regard to karst landscapes: Insights from Slovenia. *Environ. Sci. Policy* **2015**, *51*, 106–116. [CrossRef]
20. Ford, D.C.; Williams, P.W. *Karst Hydrogeology and Emorphology*; Wiley: Chichester, UK, 2007.
21. Gunn, J. Karst groundwater in UNESCO protected areas: A global overview. *Hydrogeol. J.* **2020**, *29*, 297–314. [CrossRef]
22. Xiong, K.N.; Li, J.; Long, M.Z. Features of soil and water loss and key issues in demonstration areas for combating karst rocky desertification. *Acta Geogr. Sin.* **2012**, *67*, 878–888.
23. Yang, Q.; Jiang, Z.; Yuan, D.; Ma, Z.; Xie, Y. Temporal and spatial changes of karst rocky desertification in ecological reconstruction region of Southwest China. *Environ. Earth Sci.* **2014**, *72*, 4483–4489. [CrossRef]
24. Trac, C.J.; Harrell, S.; Hinckley, T.M.; Henck, A.C. Reforestation programs in Southwest China: Reported success, observed failure, and the reasons why. *J. Mt. Sci.* **2007**, *4*, 275–292. [CrossRef]
25. Wu, J.; Shen, M.Y. Medicinal plant resources and eco—Industry development strategy in karst area of southwest China. *World For. Res.* **2020**, *33*, 66–74.
26. Su, W.C. Rare and endangered plants in Guizhou karst regions with the consideration of their conservation. *Resour. Environ. Yangtze Basin* **2002**, *11*, 111–116.
27. Yan, H.X.; Huang, C.Y.; Zhang, Z.B.; Cui, X.Q.; Deng, J.L.; Guan, S.K.; Pu, C.Y. Tissue culture and plant regeneration of leaves of Primulina guigangensis. *Chin. J. Trop. Crops* **2019**, *40*, 98–106.
28. Gao, J.X.; Fan, X.S. Connotation, Traits and Research Trends of Eco-Assets. *Res. Environ. Sci.* **2007**, *20*, 137–143.
29. Zhang, L.B.; Yu, H.Y.; Li, D.Q.; Jia, Z.Y.; Wu, F.C.; Liu, X. Connotation and value implementation mechanism of ecological products. *Trans. Chin. Soc. Agric. Mach.* **2019**, *50*, 173–183.
30. Costanza, R. Economic growth, carrying capacity, and the environment. *Ecol. Econ.* **1995**, *15*, 89–90. [CrossRef]
31. He, C.; Xing, K.; Chi, Y.; Song, S.; Fang, J.; He, S. Effect of landscape type change on spatial and temporal evolution of ecological assets in a karst plateau-mountain area. *Int. J. Environ. Res. Public Health* **2022**, *19*, 4477. [CrossRef]
32. Xiao, D.; Tan, Y.; Liu, X.; Yang, R.; Zhang, W.; He, X.; Wang, K. Effects of different legume species and densities on arbuscular mycorrhizal fungal communities in a karst grassland ecosystem. *Sci. Total Environ.* **2019**, *678*, 551–558. [CrossRef]
33. Song, S.Z.; Xiong, K.N.; Chi, Y.K.; He, C.; Fang, J.Z.; He, S.Y. Effect of Cultivated Pastures on Soil Bacterial Communities in the Karst Rocky Desertification Area. *Front. Microbiol.* **2022**, *13*, 922989. [CrossRef]
34. Sun, Q.Z.; Liu, R.L.; Chen, J.Y.; Zhang, Y.P. Effects of planting grass on soil erosion in karst demonstration areas of rock desertification integrated rehabilitation in Guizhou Province. *J. Soil Water Conserv.* **2013**, *27*, 67–72.
35. Xiong, K.N.; Xu, L.X.; Liu, K.X.; Guo, W.; Yang, S.M.; Liu, C.M.; Zhang, J.H. Coupling relationship between the development of ecological animal husbandry and the control of rocky desertification in karst mountain area. *J. Domest. Anim. Ecol.* **2016**, *37*, 72–79.
36. Wang, Z.; Deng, X.; Song, W.; Li, Z. What is the main cause of grassland degradation? A case study of grassland ecosystem service in the middle-south Inner Mongolia. *Catena* **2017**, *150*, 100–107. [CrossRef]
37. Schils, R.L.; Bufe, C.; Rhymer, C.M.; Francksen, R.M.; Klaus, V.H.; Abdalla, M.; Price, J.P.N. Permanent grasslands in Europe: Land use change and intensification decrease their multifunctionality. *Agric. Ecosyst. Environ.* **2022**, *330*, 107891. [CrossRef]
38. Chen, W.; Zhang, X.; Huang, Y. Spatial and temporal changes in ecosystem service values in karst areas in southwestern China based on land use changes. *Environ. Sci. Pollut. Res.* **2021**, *28*, 45724–45738. [CrossRef] [PubMed]
39. Grant, M.J.; Booth, A. A typology of reviews: An analysis of 14 review types and associated methodologies. *Health Inf. Libr. J.* **2009**, *26*, 91–108. [CrossRef] [PubMed]
40. Taklo, S.K.; Tooranloo, H.S.; Parizi, Z.S. Green Innovation: A Systematic Literature Review. *J. Clean. Prod.* **2020**, *279*, 122474. [CrossRef]
41. Xiao, J.; Xiong, K.N. A review of agroforestry ecosystem services and its enlightenment on the ecosystem improvement of rocky desertification control. *Sci. Total Environ.* **2022**, *852*, 158538. [CrossRef]
42. Rogers, H.H.; Runion, G.B.; Krupa, S.V. Plant responses to atmospheric CO_2 enrichment with emphasis on roots and the rhizosphere. *Environ. Pollut.* **1994**, *83*, 155–189. [CrossRef]
43. Asner, G.P.; Beatty, S.W. Effects of an African grass invasion on Hawaiian shrubland nitrogen biogeochemistry. *Plant Soil* **1996**, *186*, 205–211. [CrossRef]
44. Decaëns, T.; Jiménez, J.J.; Barros, E.; Chauvel, A.; Blanchart, E.; Fragoso, C.; Lavelle, P. Soil macrofaunal communities in permanent pastures derived from tropical forest or savanna. *Agric. Ecosyst. Environ.* **2004**, *103*, 301–312. [CrossRef]
45. Hickman, K.R.; Hartnett, D.C.; Cochran, R.C.; Owensby, C.E. Grazing Management Effects on Plant Species Diversity in Tallgrass Prairie. *J. Range Manag.* **2004**, *57*, 58. [CrossRef]
46. Millennium Ecosystem Assessment (MEA). *Ecosystems and Human Well-Being: The Assessment Series (Four Volumes and Summary)*; Island Press: Washington, DC, USA, 2005; Volume 10.
47. Patra, A.K.; Abbadie, L.; Clays-Josserand, A.; Degrange, V.; Grayston, S.J.; Loiseau, P.; Le Roux, X. Effects of grazing on microbial functional groups involved in soil ndynamics. *Ecol. Monogr.* **2005**, *75*, 65–80. [CrossRef]
48. Golodets, C.; Kigel, J.; Sternberg, M. Plant diversity partitioning in grazed Mediterranean grassland at multiple spatial and temporal scales. *J. Appl. Ecol.* **2011**, *48*, 1260–1268. [CrossRef]
49. Wrage, N.; Strodthoff, J.; Cuchillo, H.M.; Isselstein, J.; Kayser, M. Phytodiversity of temperate permanent grasslands: Ecosystem services for agriculture and livestock management for diversity conservation. *Biodivers. Conserv.* **2011**, *20*, 3317–3339. [CrossRef]

50. Yu, Y.; Zhao, W.W.; Martinez-Murillo, J.F.; Pereira, P. Loess Plateau: From degradation to restoration. *Sci. Total Environ.* **2020**, *738*, 104206. [CrossRef] [PubMed]
51. Porqueddu, C.; Ates, S.; Louhaichi, M.; Kyriazopoulos, A.P.; Moreno, G.; del Pozo, A.; Nichols, P.G.H. Grasslands in "Old World" and "New World" Mediterranean-climate zones: Past trends, current status and future research priorities. *Grass Forage Sci.* **2016**, *71*, 1–35. [CrossRef]
52. Coyle, C.; Creamer, R.E.; Schulte, R.P.O.; O'Sullivan, L.; Jordan, P. A Functional Land Management conceptual framework under soil drainage and land use scenarios. *Environ. Sci. Policy* **2016**, *56*, 39–48. [CrossRef]
53. Vargas, L.; Hein, L.; Remme, R.P. Accounting for ecosystem assets using remote sensing in the Colombian Orinoco River basin lowlands. *Earth Resour. Environ. Remote Sens./GIS Appl. VII.* **2016**, *11*, 26008. [CrossRef]
54. Wei, J.X.; Hu, A.; Gan, X.Y.; Zhao, X.D.; Huang, Y. Spatial and Temporal Characteristics of Ecosystem Service Trade-Off and Synergy Relationships in the Western Sichuan Plateau, China. *Forests* **2022**, *13*, 1845. [CrossRef]
55. You, X.; He, D.J.; Xiao, Y.; Fu, W.J.; Song, C.S.; Ouyang, Z.Y. Assessment of Eco-assets in a county area: A case of Pingbian County. *Acta Ecol. Sin.* **2020**, *40*, 5220–5229.
56. Zhang, X.Z.; Wang, X.D.; Gao, Q.Z.; Hou, T.P.; Shen, Z.X.; Fang, J.P. Research in ecological restoration and reconstruction technology for degraded alpine ecosystem, boosting the protection and construction of ecological security barrier in Tibet. *Acta Ecol. Sin.* **2016**, *36*, 7083–7087.
57. Zhao, X.J.; Su, J.D. Study on ecological protection redline in Qilian Mountains based on remote sensing and GIS. *Soil Water Conserv. China* **2019**, *8*, 33–37+69.
58. Qi, J.; Xin, X.; John, R.; Groisman, P.; Chen, J. Understanding livestock production and sustainability of grassland ecosystems in the Asian Dryland Belt. *Ecol. Process.* **2017**, *6*, 22. [CrossRef]
59. Gang, C.; Zhou, W.; Wang, Z.; Chen, Y.; Li, J.; Chen, J.; Qi, J.; Odeh, I.; Groisman, P.Y. Comparative Assessment of Grassland NPP Dynamics in Response to Climate Change in China, North America, Europe and Australia from 1981 to 2010. *J. Agron. Crop. Sci.* **2014**, *201*, 57–68. [CrossRef]
60. Sun, L.Q. The application of farming systems theory to ecological constructions in pasture area—An example in aletai city Xinjiang China. *Grass Feed. Livest.* **2004**, *4*, 5–8.
61. Wang, H.; Zhou, S.; Li, X.; Liu, H.; Chi, D.; Xu, K. The influence of climate change and human activities on ecosystem service value. *Ecol. Eng.* **2016**, *87*, 224–239. [CrossRef]
62. Fei, L.; Shuwen, Z.; Jiuchun, Y.; Kun, B.; Qing, W.; Junmei, T.; Liping, C. The effects of population density changes on ecosystem services value: A case study in Western Jilin, China. *Ecol. Indic.* **2016**, *61*, 328–337. [CrossRef]
63. Li, Y.; Yang, R.B.; Bi, J.P. Analysis on the pattern of ecological capital in CZT region. *Econ. Geogr.* **2015**, *35*, 184–188.
64. Zhang, S.Q.; Yang, P.; Xia, J.; Wang, W.Y.; Cai, W.; Chen, N.C.; Hu, S.; Luo, X.A.; Li, J.; Zhan, C.S. Land use/land cover prediction and analysis of the middle reaches of the Yangtze River under different scenarios. *Sci. Total Environ.* **2022**, *833*, 155238. [CrossRef]
65. Ferreira, L.M.R.; Esteves, L.S.; Souza, E.P.D.; de Souza, E.P.; dos Santos, C.A.C. Impact of the Urbanisation Process in the Availability of Ecosystem Services in a Tropical Ecotone Area. *Ecosystems* **2019**, *22*, 266–282. [CrossRef]
66. Ashworth, A.J.; Toler, H.D.; Allen, F.L.; Augé, R.M. Correction: Global meta-analysis reveals agro-grassland productivity varies based on species diversity over time. *PLoS ONE* **2020**, *15*, e0233402. [CrossRef]
67. Zhang, X.J.; Ma, W.; Wang, Z.W. Effects of mowing regime on community characteristics and forage yield and quality in Hulun Buir, China. *Chin. J. Appl. Ecol.* **2022**, *33*, 1555–1562.
68. Hu, J.J.; Zhou, Q.P.; Cao, Q.H.; Hu, J. Effects of ecological restoration measures on vegetation and soil properties in semi-humid sandy land on the southeast Qinghai-Tibetan Plateau, China. *Glob. Ecol. Conserv.* **2022**, *33*, e02000. [CrossRef]
69. Beal-Neves, M.; Vogel Ely, C.; Westerhofer Esteves, M.; Blochtein, B.; Lahm, R.A.; Quadros, E.L.; Abreu Ferreira, P.M. The Influence of Urbanization and Fire Disturbance on Plant-Floral Visitor Mutualistic Networks. *Diversity* **2020**, *12*, 14. [CrossRef]
70. Schaub, S.; Finger, R.; Buchmann, N.; Steiner, V.; Klaus, V.H. The costs of diversity: Higher prices for more diverse grassland seed mixtures. *Environ. Res. Lett.* **2021**, *16*, 094011. [CrossRef]
71. Lewandowski, I.; Clifton-Brown, J.C.; Scurlock, J.M.O.; Huisman, W. Miscanthus: European experience with a novel energy crop. *Biomass Bioenergy* **2000**, *19*, 209–227. [CrossRef]
72. Lu, X.X. Design and Construction of Plant Germplasm Banks. *Chin. Bull. Bot.* **2006**, *23*, 119–125.
73. Xiao-Fei, J.; Ming-hong, Y.; Shi-qie, B.; Da-xu, L.; Xiong, L.; Qi, W.; Yu, Z. Establishment of Quantitative Model for Analyzing Crude Protein in Phalaris arundinacea L. by Near Infrared Spectroscopy (NIRS). *Spectrosc. Spectr. Anal.* **2019**, *39*, 1731–1735.
74. Yang, C. Species, Harm, prevention and Control status and Future Development Trend of alien species Invasion in Guizhou. *J. Agric. Catastrophology* **2020**, *10*, 144–148+150.
75. Hou, M.L.; Ge, G.T.; Sun, L.; Zhou, T.R.; Zhang, Y.C.; Jia, Y.S. Effects of Formic Acid, Cellulose and Lactic Acid Bacteria on Silage Quality of Natural Forage of Typical Steppe. *Chin. J. Anim. Nutr.* **2015**, *27*, 2977–2986.
76. Qin, Y.; Zeng, X.G.; Xu, Z.H. Promoting the supply side reform of ecological products based on PPP model. *J. Arid Land Resour. Environ.* **2018**, *32*, 7–12.
77. Lemaire, G. Research priorities for grassland science: The need of long term integrated experiments networks. *Rev. Bras. Zootec.* **2007**, *36*, 93–100. [CrossRef]
78. Plieninger, T.; Huntsinger, L. Complex Rangeland Systems: Integrated Social-Ecological Approaches to Silvopastoralism. *Rangel. Ecol. Manag.* **2018**, *71*, 519–525. [CrossRef]

79. Zhao, T. A Study on the Delineation of Ecological Protection Red Lines for Grasslands in Six Northern Provinces. Master's Thesis, Northwest A&F University, Xianyang, China, 2018.
80. Wei, Y.; He, S.; Li, G.; Chen, X.; Shi, L.; Lei, G.; Su, Y. Identifying Nature–Community Nexuses for Sustainably Managing Social and Ecological Systems: A Case Study of the Qianjiangyuan National Park Pilot Area. *Sustainability* **2019**, *11*, 6182. [CrossRef]
81. Hu, Z.T.; Liu, D.; Qi, L.S. Grassland Eco-compensation: Ecological Performance, Income Effect and Policy Satisfaction. *China Popul. Resour. Environ.* **2016**, *26*, 165–176.
82. Yu, Y.; Wu, Y.; Wang, P.; Zhang, Y.L.; Yang, L.E.; Cheng, X.; Yan, J.Z. Grassland Subsidies Increase the Number of Livestock on the Tibetan Plateau: Why Does the "Payment for Ecosystem Services" Policy Have the Opposite Outcome? *Sustainability* **2021**, *13*, 6208. [CrossRef]
83. Yu, D.; Lu, N.; Fu, B. Establishment of a comprehensive indicator system for the assessment of biodiversity and ecosystem services. *Landsc. Ecol.* **2017**, *32*, 1563–1579. [CrossRef]
84. Zhao, T.Q.; Ouyang, Z.Y.; Zhen, H.; Wang, X.K.; Miao, H. Analyses on grassland ecosystem services and its indexes for assessment. *Chin. J. Ecol.* **2004**, *6*, 155–160.
85. Song, P.F.; Hao, Z.Q. Some ideas about ecological assets assessment. *Chin. J. Ecol.* **2007**, *18*, 2367–2373.
86. Nagy, R.K.; Bell, L.W.; Schellhorn, N.A.; Zalucki, M.P. Role of grasslands in pest suppressive landscapes: How green are my pastures? *Austral Entomol.* **2020**, *59*, 227–237. [CrossRef]
87. Ye, D.; Zhang, Y.F.; Li, Q.L.; Zhang, X.; Chu, C.L.; Ju, M.T. Assessing the Spatiotemporal Development of Ecological Civilization for China's Sustainable Development. *Sustainability* **2022**, *14*, 8779. [CrossRef]
88. Huang, R.B.; Zhao, Q. Discussion of preparation and audit on natural resource balance sheet. *Audit. Res.* **2015**, *1*, 37–43.
89. Dong, Z.F.; Qing, K.Y.; Liu, J.Y. Framework and method of natural resource assets evaluation in national key ecological functional areas. *Ecol. Econ.* **2022**, *38*, 13–21.
90. Liang, Y.J.; Liu, L.J. Integration of ecosystem services and landscape pattern: A review. *Acta Ecol. Sin.* **2018**, *38*, 7159–7167.
91. Ouyang, Z.Y.; Song, C.S.; Zheng, H.; Polasky, S.; Xiao, Y.; Bateman, I.J.; Liu, J.G.; Ruckelshaus, M.; Shi, F.Q.; Xiao, Y.; et al. Using gross ecosystem product (GEP) to value nature in decision making. *Proc. Natl. Acad. Sci. USA* **2020**, *117*, 14593–14601. [CrossRef]
92. Boland, J.J.; Freeman, A.M. *The Benefits of Environmental Improvement: Theory and Practice*; The Johns Hopkins University Press: Baltimore, MD, USA, 1979.
93. Zhou, C.; Li, G.P. A review of evaluation methods of ecosystem services: Also on the theoretical progress of contingent valuation method. *Ecol. Econ.* **2018**, *34*, 207–214.
94. Daily, G.C.; Söderqvist, T.; Aniyar, S.; Arrow, K.; Dasgupta, P.; Ehrlich, P.R.; Walker, B. The value of nature and the nature of value. *Science* **2000**, *289*, 395–396. [CrossRef] [PubMed]
95. Benson, J.F.; Willis, K.G. Valuing informal recreation on the Forestry Commission estate. *Q. J. For.* **1992**, *16*, 63–65.
96. Muscutt, A.; Harris, G.; Bailey, S.; Davies, D. Buffer zones to improve water quality: A review of their potential use in UK agriculture. *Agric. Ecosyst. Environ.* **1993**, *45*, 59–77. [CrossRef]
97. Zhou, C.X. Construction and Application of Ecological Asset Valuation Methods in Pingjiang County. Master's Thesis, Hunan Agriculture University, Changsha, China, 2016.
98. Gao, Y.M.; Wu, W.J.; Jiang, H.Q.; Duan, Y.; Zhou, X.F.; Ma, G.X. Spatiotemporal changes of water conservation services value based on global terrestrial ecosystem. *Res. Environ. Sci.* **2021**, *34*, 2696–2705.
99. Ma, C.X.; Liu, J.J.; Kang, W.B.; Sun, S.H.; Ren, J.H. Evaluation of forest ecosystem carbon fixation and oxygen release services in Shaanxi Province from 1999 to 2003. *Acta Ecol. Sin.* **2010**, *30*, 1412–1422.
100. Costanza, R.; d'Arge, R.; De Groot, R.; Farber, S.; Grasso, M.; Hannon, B.; Van Den Belt, M. The value of the world's ecosystem services and natural capital. *Nature* **1997**, *387*, 253–260. [CrossRef]
101. Odum, H.T. *Environmental Accounting: Energy and Environ-Mental Decision-Making*; John and Wiley and Sons: Hoboken, NJ, USA, 1996.
102. Sannigrahi, S.; Bhatt, S.; Rahmat, S.; Paul, S.K.; Sen, S. Estimating global ecosystem service values and its response to land surface dynamics during 1995–2015. *J. Environ. Manag.* **2018**, *223*, 115–131. [CrossRef]
103. Akhtar, M.; Zhao, Y.Y.; Gao, G.L.; Gulzar, Q.; Hussain, A. Assessment of spatiotemporal variations of ecosystem service values and hotspots in a dryland: A case-study in Pakistan. *Land Degrad. Dev.* **2022**, *33*, 1383–1397. [CrossRef]
104. Li, J.Y.; Chen, H.X.; Zhang, C.; Pan, T. Variations in ecosystem service value in response to land use/land cover changes in Central Asia from 1995–2035. *PeerJ* **2019**, *7*, e7665. [CrossRef] [PubMed]
105. Mbaabu, P.R.; Olago, D.; Gichaba, M.; Eckert, S.; Eschen, R.; Oriaso, S.; Choge, S.K.; Linders, T.E.; Schaffner, U. Restoration of degraded grasslands, but not invasion by Prosopis juliflora, avoids trade-offs between climate change mitigation and other ecosystem services. *Sci. Rep.* **2020**, *10*, 20391. [CrossRef]
106. Tang, Y.K.; Wu, Y.T.; Wu, K.; Guo, Z.W.; Liang, C.Z.; Wang, M.J.; Chang, P.J. Changes in trade-offs of grassland ecosystem services and functions under different grazing intensities. *Chin. J. Plant Ecol.* **2019**, *43*, 408–417. [CrossRef]
107. Xie, G.D.; Zhang, Y.L.; Lu, C.X.; Zheng, D.; Cheng, S.K. Study on valuation of rangeland ecosystem services of China. *J. Nat. Resour.* **2001**, *16*, 47–53.
108. Xiao, Q.; Hu, D.; Xiao, Y. Assessing changes in soil conservation ecosystem services and causal factors in the Three Gorges Reservoir region of China. *J. Clean. Prod.* **2017**, *163*, S172–S180. [CrossRef]

109. Grizzetti, B.; Lanzanova, D.; Liquete, C.; Reynaud, A.; Cardoso, A.C. Assessing water ecosystem services for water resource management. *Environ. Sci. Policy* **2016**, *61*, 194–203. [CrossRef]
110. Bartkowski, B. Are diverse ecosystems more valuable? Economic value of biodiversity as result of uncertainty and spatial interactions in ecosystem service provision. *Ecosyst. Serv.* **2017**, *24*, 50–57. [CrossRef]
111. Wu, D.; Shao, Q.Q.; Liu, J.Y. Assessment of Water Regulation Service of Forest and Grassland Ecosystems in Three-River Headwaters Region. *Bull. Soil Water Conserv.* **2016**, *36*, 206–210.
112. Yu, X.; Gaodi, X.; Kai, A.N. The function and economic value of soil conservation of ecosystems in Qinghai-Tibet Plateau. *Acta Ecol. Sin.* **2003**, *23*, 2367–2378.
113. Jing, X.D.; Tian, G.L.; Li, M.R.; Javeed, S.A. Research on the Spatial and Temporal Differences of China's Provincial Carbon Emissions and Ecological Compensation Based on Land Carbon Budget Accounting. *Int. J. Environ. Res. Public Health* **2022**, *18*, 12892. [CrossRef] [PubMed]
114. Bo, W.J.; Xiao, Y.; Wang, L.Y.; Wang, X.K.; Ouyang, Z.Y. Assessment of the status of ecological assets and variation of its characteristics: A case study of Hinggan League, Inner Mongolia. *Acta Ecol. Sin.* **2019**, *39*, 5425–5432.
115. Huang, B.B.; Zheng, H.; Xiao, Y.; Kong, L.Q.; Ouyang, Z.Y.; Wang, X.K. Effectiveness and Driving Forces of Ecological Asset Protection in National Key Ecological Function Regions. *Environ. Conform. Assess.* **2019**, *11*, 14–23.
116. Xu, X.B.; Chen, S.; Yang, G.S. Spatial and temporal change in ecological assets in the Yangtze River Delta of China 1995–2007. *Acta Ecol. Sin.* **2012**, *32*, 7667–7675.
117. Pang, L.H.; Chen, Y.M.; Fen, C.Y. Assessment of ecological products supplying capacities of natural reserve-A case of Hulun Buir Hui River Reserve. *J. Arid Land Resour. Environ.* **2014**, *28*, 110–116.
118. Xie, G.; Zhang, C.; Zhen, L.; Zhang, L. Dynamic changes in the value of China's ecosystem services. *Ecosyst. Serv.* **2017**, *26*, 146–154. [CrossRef]
119. Zhang, S.Y.; Bai, X.Y.; Wang, S.J.; Qing, L.Y.; Tian, Y.C.; Luo, G.J.; Li, Y. Ecosystem services evaluation of typical rocky desertiifcation areas based on InVEST model—A case study at Qinglong Country, Guizhou Provice. *J. Earth Environ.* **2014**, *5*, 328–338.
120. Guo, H.; Dong, S.W.; Wu, D.; Pei, S.X.; Xin, X.B. Calculation and analysis of equivalence factor and yield factor of ecological footprint based on ecosystem services value. *Acta Ecol. Sin.* **2020**, *40*, 1405–1412.
121. Li, B.W.; Yang, Z.F.; Cai, Y.P.; Xie, Y.L.; Guo, H.J.; Wang, Y.Y.; Zhang, P.; Li, B.; Jia, Q.P.; Huang, Y.P.; et al. Prediction and valuation of ecosystem service based on land use/land cover change: A case study of the Pearl River Delta. *Ecol. Eng.* **2022**, *179*, 106612. [CrossRef]
122. Hulst, R. On the dynamics of vegetation: Markov chains as models of succession. *Plant Ecol.* **1979**, *40*, 3–14. [CrossRef]
123. Zhao, G.L.; Hu, Y.C. Study on Ecosystem Service Value Changes Based on CLUE-S Models in Guangxi Karst Mountainous Area. *Res. Soil Water Conserv.* **2014**, *21*, 198–203+210+345.
124. Liu, X.Y.; Mou, Y.T. Research progress in the ecosystem services function and value of grasslands. *Acta Prataculturae Sin.* **2012**, *21*, 286–295.
125. Suttie, J.M.; Reynolds, S.G.; BaTello, C. *Grasslands of the World*; Food and Agriculture Organization of the United Nation: Rome, Italy, 2005; pp. 1–21.
126. Yan, Y.; Zhu, J.Y.; Wu, G.; Zhan, Y.J. Review and prospective applications of demand, supply, and consumption of ecosystem services. *Acta Ecol. Sin.* **2017**, *37*, 2489–2496.
127. Ma, L.; Liu, H.; Peng, J.; Wu, J.S. A review of ecosystem services supply and demand. *Acta Geogr. Sin.* **2017**, *72*, 1277–1289.
128. Yi, D.; Xiao, S.C.; Han, Y.; Ou, M.H. Review on supply and demand of ecosystem service and the construction of systematic frame-work. *Chin. J. Appl. Ecol.* **2021**, *32*, 3942–3952.
129. Xu, Y.; He, Z.W. Supply and demand balance analysis of ecological assets in Shenza County, northern Tibet based on RS and GIS. *Comput. Tech. Geophys. Geochem. Explor.* **2014**, *36*, 375–379.
130. Msofe, N.K.; Sheng, L.; Lyimo, J. Land Use Change Trends and Their Driving Forces in the Kilombero Valley Floodplain, Southeastern Tanzania. *Sustainability* **2019**, *11*, 505. [CrossRef]
131. Pang, Y.; Yu, C.Q.; Tu, Y.L.; Sun, W.; Luo, L.M.; Miao, Y.J.; Wu, J.X. The relationship between plant functional traits and multiple ecosystem services in a Tibetan grassland ecosystem. *Acta Ecol. Sin.* **2015**, *35*, 6821–6828.
132. Li, M.H.; Wang, X.Y.; Chen, J.C. Assessment of Grassland Ecosystem Services and Analysis on Its Driving Factors: A Case Study in Hulunbuir Grassland. *Front. Ecol. Evol.* **2022**, *10*, 841943. [CrossRef]
133. Yang, W.N.; Zhen, L.; Wei, Y.J. Food consumption and its local dependence: A case study in the Xilin Gol Grassland, China. *Environ. Dev.* **2019**, *34*, 100470. [CrossRef]
134. Bernhardt, J.R.; O'Connor, M.I. Aquatic biodiversity enhances multiple nutritional benefits to humans. *Proc. Natl. Acad. Sci. USA* **2021**, *118*, e1917487118. [CrossRef] [PubMed]
135. Bohnert, D.W.; Stephenson, M.B. Supplementation and sustainable grazing systems. *J. Anim. Sci.* **2016**, *94*, 15–25. [CrossRef]
136. Fan, X.S.; Gao, J.X.; Weng, W. Exploratory Study on Eco-Assets Transferring and the Valuating Models. *Res. Environ. Sci.* **2007**, *20*, 160–164.
137. Bisseleua, D.H.B.; Missoup, A.D.; Vidal, S. Biodiversity conservation, ecosystem functioning, and economic incentives under cocoa agroforestry intensification. *Conserv. Biol.* **2009**, *23*, 1176–1184. [CrossRef]
138. Huang, L.; Ning, J.; Zhu, P.; Zheng, Y.; Zhai, J. The conservation patterns of grassland ecosystem in response to the forage-livestock balance in North China. *J. Geogr. Sci.* **2021**, *31*, 518–534. [CrossRef]

139. Paudel, S.; Cobb, A.B.; Boughton, E.H.; Spiegal, S.; Boughton, R.K.; Silveira, M.L.; Swain, H.M.; Reuter, R.; Goodman, L.E.; Steiner, J.L. A framework for sustainable management of ecosystem services and disservices in perennial grassland agroecosystems. *Ecosphere* **2021**, *12*, e03837. [CrossRef]
140. Loucougaray, G.; Dobremez, L.; Gos, P.; Pauthenet, Y.; Nettier, B.; Lavorel, S. Assessing the Effects of Grassland Management on Forage Production and Environmental Quality to Identify Paths to Ecological Intensification in Mountain Grasslands. *Environ. Manag.* **2015**, *56*, 1039–1052. [CrossRef]
141. Shackleton, S.E.; Shackleton, R.T. Local knowledge regarding ecosystem services and disservices from invasive alien plants in the arid Kalahari, South Africa. *J. Arid Environ.* **2017**, *159*, 22–33. [CrossRef]
142. Sánchez-Romero, R.; Balvanera, P.; Castillo, A.; Mora, F.; García-Barrios, L.E.; González-Esquivel, C.E. Management strategies, silvopastoral practices and socioecological drivers in traditional livestock systems in tropical dry forests: An integrated analysis. *For. Ecol. Manag.* **2021**, *479*, 118506. [CrossRef]
143. Yu, G.R.; Yang, M. Ecological Economics Foundation Research on Ecological Values, Ecological Asset Management, and Value Realization: Scientific Concepts, Basic Theories, and Realization Paths. *Chin. J. Appl. Ecol.* **2022**, *33*, 1153–1165.
144. Zhu, X.-C.; Ma, M.-G.; Tateno, R.; He, X.-H.; Shi, W.-Y. Effects of Vegetation Restoration on Soil Carbon Dynamics in Karst and Non-Karst Regions in Southwest China: A Synthesis of Multi-Source Data. *Plant Soil* **2021**, *475*, 45–59. [CrossRef]
145. Song, S.Z.; Xiong, K.N.; Chi, Y.K.; Guo, W.; Liao, J.J. Study on Improvement of Degraded Grassland in Rocky Desertification Control in the Karst in Southern China. *J. Domest. Anim. Ecol.* **2019**, *40*, 82–87+96.
146. Yang, M.D. On the fragility karst environment. *Yunnan Geogr. Environ. Res.* **1990**, *1*, 21–29.
147. Li, C.L.; Dai, H.Q.; Peng, X.D.; Yuan, Y.F. Characteristics of Nutrient Loss in Runoff of Underground Pore (Fissure) on the Karst Bare Slopes. *J. Soil Water Conserv.* **2016**, *30*, 19–23+114.
148. Chi, Y.K.; Xiong, K.N.; Song, S.Z.; Xiao, H.; Zhang, Y.; Xu, L.X. Study on photosynthetic differences between monoculture and mixed sowing forage grass in rocky desertification area. *J. Domest. Anim. Ecol.* **2016**, *37*, 53–59.
149. Li, L.; Wang, Y.S.; Wang, K. Interspecific competition and co-existence in permanent grasses + Trifolium repens mixed pasture in Karst region. *Pratacultural Sci.* **2014**, *31*, 1943–1950.
150. He, F.Y.; Xiong, K.N.; Zhu, D.Y. Research progress on moisture effects of agroforestry in karst mountains. *China Forage* **2020**, *7*, 22–27.

agronomy

MDPI

Article

Applying Trait-Based Modeling to Achieve Functional Targets during the Ecological Restoration of an Arid Mine Area

Jian Hou *, Menghan Wu and Haobo Feng

Jixian National Forest Ecosystem Observation and Research Station, National Ecosystem Research Network of China, School of Soil and Water Conservation, Beijing Forestry University, Beijing 100083, China
* Correspondence: houjian@bjfu.edu.cn

Abstract: (1) Background: Because ecosystem degradation has become a global phenomenon which seriously affects the health of natural ecosystems and human well-being, restoration of degraded ecosystems has attracted increasing attention. However, many of the methods used in current ecological restoration work have rarely combined ecological restoration practices with the quantitative goal of restoring ecosystem function. (2) Methods: In this study, based on the conceptual framework of response-effect traits and Community Assembly by Trait Selection model (CATS model), a restoration strategy for a degraded abandoned mine in Wuhai City, China has been provided. This restoration strategy connected the ecosystem function targets to the appropriate recovery species and their required abundances. (3) Results: The results showed that a relative abundance ratio of 8:2 for S. grandis to B. dasyphylla was best for a shady slope, while a 6:4 ratio of K. tragus to B. dasyphylla was best for repair on a sunny slope of the degraded mine area. (4) Conclusion: This study provides a typical example of applying ecological theory in practice that will be useful for current and future studies and applications. This approach will ensure that governance efforts to restore degraded ecosystems are effective and efficient.

Keywords: plant functional trait; degraded ecosystem; mine area; trait-based model; ecological restoration

Citation: Hou, J.; Wu, M.; Feng, H. Applying Trait-Based Modeling to Achieve Functional Targets during the Ecological Restoration of an Arid Mine Area. *Agronomy* **2022**, *12*, 2833. https://doi.org/10.3390/agronomy12112833

Academic Editors: Kesi Liu and Xinqing Shao

Received: 3 October 2022
Accepted: 11 November 2022
Published: 13 November 2022

1. Introduction

Due to climate change and human disturbance, ecosystem degradation has become a global phenomenon which seriously affects the health of natural ecosystems and human well-being [1,2]. In this context, the restoration or reconstruction of degraded ecosystems has attracted increasing attention [3]. However, ecological restoration projects have rarely combined ecological restoration practices with the quantitative goal of restoring ecosystem function [4]. There is a considerable gap between the practitioners who carry out ecological restoration projects and the scientists who carry out theoretical research on ecological restoration [4]. Therefore, the practical application of ecological restoration theory and research is necessary to provide demonstrative examples that can inform and benefit the management and restoration of degraded ecosystems.

Abandoned mine areas are typical degraded ecosystems [5–7]. Because mineral resources are required for social and economic development, the demand for mines is increasing, which motivates mining efforts for these increasingly scarce materials. In arid areas, due to low rainfall, vegetation coverage is low and ecosystems are fragile [8]. Mines not only cause damage to the surface ecosystem, but also produce a large amount of sand and dust which represent a great environmental burden [6]. Atmospheric particulate matter as a source of air pollution not only harms the health of surrounding environments, but also aggravates climate change [9]. Furthermore, it can have a variety of adverse effects on humans and other living organisms, so it has attracted widespread attention from all countries of the world [9,10].

Due to the serious environmental degradation of abandoned mine areas, the natural recovery of damaged ecosystems is a slow and difficult process [6]. Therefore, ecological restoration measures should be adopted by management agencies to speed the recovery of abandoned mine areas. However, the improper selection of plant species for the restoration of arid mine areas, or the lack of systematic research on the environments of abandoned mine areas, will lead to differences in the ecosystem function of the restored ecosystem compared with the natural ecosystem. In practice, this would not only mean that the targets of ecological restoration were not achieved, but also that a lot of manpower and material resources had been wasted [11]. Recently, trait-based models have been developed and applied to translate ecosystem functional targets into required abundances of ecologically restorative species, providing targets for achieving the desired ecological restoration effect [12].

At present, theoretical studies on plant functional traits have begun to receive more and more attention among restoration ecologists [12–14]. Many studies have shown that the coupling of response and effect traits is an effective method for studying ecosystem function, but very few relevant experimental studies exist [4,15]. The concept of traits in reference to species functional traits was proposed in the 1990s [16] and is defined as the morphological, physiological, and phenological characteristics of individual plants [17]. Response traits refer to species functional traits that can respond to environmental change, and effect traits refer to traits that can influence ecosystem function [4]. Response and effect traits can be divided into different components according to the responses of different species to environmental change or their effects on ecosystem function. However, there is no clear defining boundary between them [18]. For example, Lavorel and Garnier [19] found that plant response traits were affected by soil nutrients, but these almost completely overlapped with the effect traits determining primary productivity (for example, specific leaf area appears in both components at the same time). However, there was almost no overlap between the response traits affected by fire (such as plant height) and the effect traits that determined the fire resistance of the community (such as plant water content). To date, studies have focused on plant roots [20], leaves [21,22], and various other organs and examined trait responses to the environment [23] as well as trait effects on ecosystem function [24,25]. However, few studies have combined the two aspects of response and effect traits. In order to obtain species compositions that meet specific restoration targets, Laughlin et al. [14] proposed the Community Assembly by Trait Selection model (CATS model) based on the concept framework of response-effect traits. Through this model, based on specific ecosystem function goals for specific restoration areas, suitable species and their required relative abundance ranges can be obtained.

In this study, we first investigated soil factors and plant functional traits, such as leaf area and root length, in Wuhai, a city in China's arid western region. Then, we investigated mine area environments in Wuhai and set appropriate ecosystem function targets for the ecological restoration of the mine area. Finally, the CATS model was used to connect the ecosystem function targets to the appropriate recovery species and their required abundances. Based on the concept framework of response-effect traits, this study has provided a useful example that advances the study of restoration ecology.

2. Materials and Methods

2.1. Study Area

The study area, including the abandoned mine and its surrounding areas, was located in Wuhai in the arid western region of China (Figure 1A,B). The study area has a typical continental climate with an average annual temperature of 9.8 °C. The annual precipitation in Wuhai ranges from 80 mm in the west to 250 mm in the east, and the average evaporation is about 3249 mm [26]. Brown calcic soil, gray desert soil, and chestnut soil are the main soil types in the area [26]. The study area biome changes from steppe in the east to desert in the west.

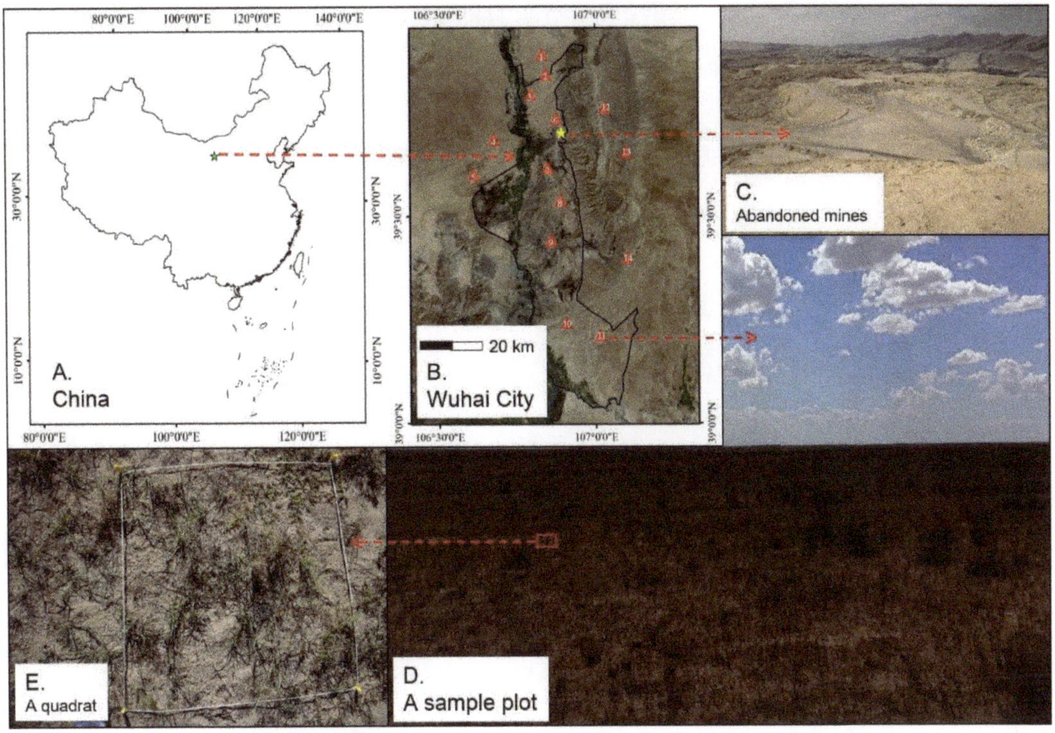

Figure 1. Research site and the plot locations of this study. (**A**): Location of Wuhai City in China; (**B**): Locations of abandoned mines and sample plots in Wuhai City; (**C**): Abandoned mines; (**D**): A sample plot in Wuhai City; (**E**): A quadrat in a sample plot.

The abandoned opencast mine is located in the center of the study area (Figure 1B,C). The terrain of the mining area has changed dramatically, and the wind erosion is serious. The mining work was started in 2011 and continues to this day. Before mining work, the local vegetation type was consistent with the surrounding areas. Now, due to intense human disturbance, there is almost no vegetation coverage in the abandoned mining area.

2.2. Plots Selection

From July to August of each year from 2018 to 2020, field surveys were conducted on the abandoned mine, which was selected as restoration area in this research, and its surrounding areas, which is the site outside of the restoration area, in Wuhai (Figure 1C). To consider the effect of topography on vegetation restoration in the abandoned mine, a shady slope sample plot and a sunny slope sample plot in the mine area were selected as environmental survey plots in restoration area. Furthermore, around the restoration area, 14 flat terrain plots were selected as native plant survey plots. In total, 16 survey plots were established. To reduce the effects of rainfall on the concentration of total suspended particulate matter in air, dust retention of plant leaves, and soil moisture, plot monitoring was conducted on the 8th day after rainfall.

2.3. Vegetation Investigation

Three 1×1 m quadrats were randomly selected within each plot, making a total of 48 quadrats (Figure 1D,E). Plant species names, community richness, number of plants per species, and coverage were recorded or estimated for each quadrat. Within the plot, 10 healthy plants with good growth and no disease were randomly selected as plant samples for each species. The whole plant was harvested with a shovel, and then soil was backfill. Leaves and intact roots of each plant sample were obtained. Root length, root diameter, root surface area, and root volume were analyzed with Winrhizo-Pro root analysis system. We soaked the collected leaves in distilled water and washed them. All the cleaning solution was then filtered through a filter membrane with a 0.15 μm pore size. The filter membrane with filtrate was then put into a 60 °C oven (Binder FED53) to dry to a constant volume, and then weighed with a 1/10,000 balance. The difference between the mass of filter membrane before and after filtration was taken as the dust retention of plant leaves. The fresh leaves were weighted and scanned (Epson 600), and the leaf area was calculated. We then put the leaves in the oven at 60 °C (Binder FED53) to dry to a constant weight, and weighed the dry leaves. The ratio of leaf area to dry weight was taken as the specific leaf area. The ratio of leaf dry weight to fresh weight was taken as leaf dry matter content. The ratio of leaf dust retention to leaf area was calculated as the dust retention per unit leaf area within 8 days.

2.4. *Soil Sampling*

In the center of each quadrat, a soil drill was used to obtain 30 g top soil samples (0–30 cm). Soil total nitrogen and phosphorus were determined using an automatic chemical analyzer (SmartCHM200). Soil water content was measured using the drying method.

2.5. *Total Suspended Particulate Matter (TSP) Concentration Monitoring*

A portable dust monitor (Dustmate) was used to monitor the concentration of TSP around the sample site from July to August 2020. Measurements were taken at each monitoring site once in the morning and once in the evening, and the average value was calculated as the background value of atmospheric dust concentration in the plot.

2.6. *Data Analysis*

2.6.1. Correction of Dust Retention by Plant Leaves

Since the background value of atmospheric dust concentration affects the dust retention process of plants, it is necessary to correct the dust retention amount per unit leaf area of different plots for their local atmospheric dust quantities so as to ensure that the dust retention amount per unit leaf area of all plants can be compared under similar environmental conditions. Since the mine area was selected as the restoration area in this study, we assessed the dust retention amount per unit leaf area in the mine area, based on the actual dust retention data per unit area in different plots, and the measured atmospheric TSP values in different regions. The correction formula is:

$$\mathrm{DR} = DR_{measure} \times \left(\frac{TSP_{restoration}}{TSP_{plot}} \right) \tag{1}$$

where DR is the corrected dust retention per unit leaf area of plant leaves; $DR_{measure}$ is plant dust retention per unit area measured in the quadrat; $TSP_{restoration}$ is the measured TSP in the mine area; and TSP_{plot} is the TSP measured in the plot.

2.6.2. Calculation of Community-Weighted Mean (CWM)

We calculated CWM to obtain the average values of species traits within plant communities.

$$\mathrm{CWM} = \sum_{i=1}^{n} p_i \times trait_i \tag{2}$$

where CWM is the average value of a trait within a plant community; p_i is the relative abundance of species i in the community, which can be calculated by the ratio of the abundance of species i to the abundance of all species in the community; and $trait_i$ is plant functional trait i.

2.6.3. Linear Model Fitting

We first calculated the value of CWMs and soil factors in each of the 14 herb plots outside of the abandoned mine. Then, a generalized linear model was used to analyze the relationship between soil factors and CWMs from the 14 herb plots, and the significant relationship, marked by the correlation curves with the high value of R^2, were selected to build the CATS model.

2.6.4. CATS Model

Firstly, based on the linear relationship between the investigated plant functional traits and soil factors, and the value of soil factors in the restoration area, the response trait target values in the restoration community were chosen to keep the restoration species in the restoration area alive. Secondly, based on the distribution of dust retention amount per unit leaf area of different plants, the maximum value was set as the effect trait target for the restoration area. This was selected based on the goal of maximizing the restoration community's dust retention ecosystem function. Finally, the CATS model was used to predict the required relative abundances of restoration species under the environmental conditions of the restoration area and based on the response-effect trait target values of the restoration community. The mathematical formula of this model is as follows:

$$\sum_{i=1}^{S} t_{ik} p_i = \overline{T_k} \quad (3)$$

$$\sum_{i=1}^{S} p_i = 1 \quad (4)$$

$$p_i > 0 \quad (5)$$

where, t_{ik} is trait k of species i; p_i is the relative abundance of species i in the restoration area; T_k is the trait target value in the repair community; and S is the total number of selected plant species. The model was solved in R-4.1.1 with package "limSolve".

3. Results

3.1. Plant Distribution

Plants in this study area are dominated by herbs and include 52 herb species. Among the 14 plant survey plots, the most widely distributed species included: *Bassia dasyphylla*, *Artemisia frigida*, and *Stipa gradis*. The *S. gradis* community in plot 7 had the highest coverage (76.8%), while the *A. fridgida* community in plot 3 had the lowest coverage (25.2%). The highest richness occurred in plot 8, which contained 8 herbs, while the lowest richness occurred in plot 4, which contained only 1 herb (Table 1).

Table 1. Vegetation and soil conditions of the plots.

Sample Plot	Vegetation Coverage (%)	Species Richness	Soil Total Nitrogen (g/kg)	Soil Total Phosphorus (g/kg)	Soil Water Content (%)
Mine area (sunny)	—	—	0.40	0.17	3.03
Mine area (shady)	—	—	0.39	0.21	4.50
1	34.6	2	0.16	0.17	2.34
2	27.3	4	0.20	0.27	3.20
3	25.2	5	0.28	0.26	2.66
4	54.7	1	0.13	0.16	1.42
5	67.4	2	0.25	0.23	1.72
6	29.4	4	0.42	0.36	4.53
7	76.8	7	0.55	0.39	7.63
8	68.1	8	0.40	0.33	6.12
9	44.9	6	0.58	0.28	8.19
10	25.5	3	0.23	0.21	2.98
11	28.3	2	0.27	0.21	2.94
12	43.9	4	0.30	0.35	3.38
13	39.1	4	0.30	0.36	4.27
14	41.6	3	0.36	0.35	3.56

3.2. Plant Functional Traits

The results showed that *Oxytropis racemosa* had the largest specific leaf area. Correspondingly, its leaf dry matter content was the smallest. The specific leaf area of *Asparagus gobicus* was smallest and, correspondingly, its leaf dry matter content was the largest. In terms of dust retention per unit leaf area of plants, among the 22 plants investigated, the species with the strongest dust retention ability was *Salsola collina*. The species with weakest dust retention ability was *Psammochloa villosa* (Table 2). *Heteropappus altaicus* had the largest total root length of 279.81 cm, while the total root length of *Saposhnikovia divaricata* was only 9.00 cm, which was the smallest. *Artemisia desertorum* had the largest root surface area of 64.55 cm^2, while *Asparagus gobicus* had the smallest largest root surface area of 5.67 cm^2. In addition, *Saposhnikovia divaricata* had the largest average root diameter of 8.05 mm, and the largest root volume of 4.58 cm^3 (Table 2).

Table 2. Plant functional traits.

	DR (g/m^2)	SLA (cm^2/g)	LDMC (g/g)	RL (cm)	RSA (cm^2)	RD (mm)	RV (cm^3)
CWM aim in sunny area	14.80	37.07	0.34	59.91	22.07	0.40	0.11
CWM aim in shady area	15.90	43.15	0.29	89.88	22.82	0.40	0.22
Allium mongolicum	1.81	56.94	0.28	60.35	20.82	1.10	0.57
Artemisia desertorum	2.45	23.73	0.26	120.95	64.55	1.70	2.74
Artemisia frigida	6.94	35.81	0.28	200.14	34.14	0.54	0.46
Asparagus gobicus	1.29	18.36	0.47	47.59	5.67	0.38	0.05
Astragalus grubovii	1.74	42.53	0.34	59.33	12.75	0.68	0.22
Bassia dasyphylla	17.79	61.76	0.21	131.14	14.79	0.36	0.13
Cleistogenes serotina	7.34	44.74	0.34	186.30	12.46	0.21	0.07
Echinops latifolius	4.09	31.13	0.39	65.21	8.98	0.44	0.10
Hedysarum scoparium	3.07	29.87	0.38	102.69	16.97	0.53	0.22
Heteropappus altaicus	5.48	46.79	0.24	279.81	52.60	0.60	0.79
Kali tragus	9.54	72.01	0.22	60.03	14.11	0.75	0.26
Olgaea leucophylla	5.26	59.12	0.25	17.10	17.05	3.17	1.35
Oxytropis racemosa	6.39	76.28	0.25	88.80	17.36	0.62	0.27
Peganum harmala	3.84	21.69	0.41	59.33	22.24	1.19	0.66
Peganum nigellastrum	1.34	45.62	0.21	92.07	24.90	0.86	0.54
Psammochloa villosa	0.47	32.19	0.37	54.74	51.34	2.99	3.83
Ptilotrichum cancscens	3.84	55.86	0.29	61.37	40.60	2.11	2.14
Salsola collina	18.03	32.26	0.22	172.76	22.87	0.42	0.24
Saposhnikovia divaricata	6.17	57.49	0.23	9.00	22.77	8.05	4.58
Sophora alopecuroides	7.35	41.66	0.28	78.52	28.10	1.14	0.80
Stipa grandis	2.56	22.26	0.43	38.87	6.43	0.53	0.09
Zygophyllum mucronatum	7.13	45.58	0.27	50.79	11.70	0.73	0.22

DR, dust retention per unit leaf area; SLA, specific leaf area; LDMC, Leaf dry-matter content; RL, root length; RSA, root surface Area; RD, root diameter; RV, root Volume. CWM aim in sunny/shady area, the trait aim of community-weighted mean in repair area of sunny/shady slope.

3.3. Concentration Characteristics of Total Suspended Particulate Matter

The total suspended particulate matter concentration in the study plots ranged from 74.62 $\mu g/m^3$ to 336.78 $\mu g/m^3$. Previously, Kriging interpolation has shown that the overall distribution of total suspended particulate matter concentration in Wuhai has zonal and radioactive characteristics. The high values were mainly concentrated in the western and north-eastern areas of Wuhai, while concentrations were low in the central and southern areas (Figure 2).

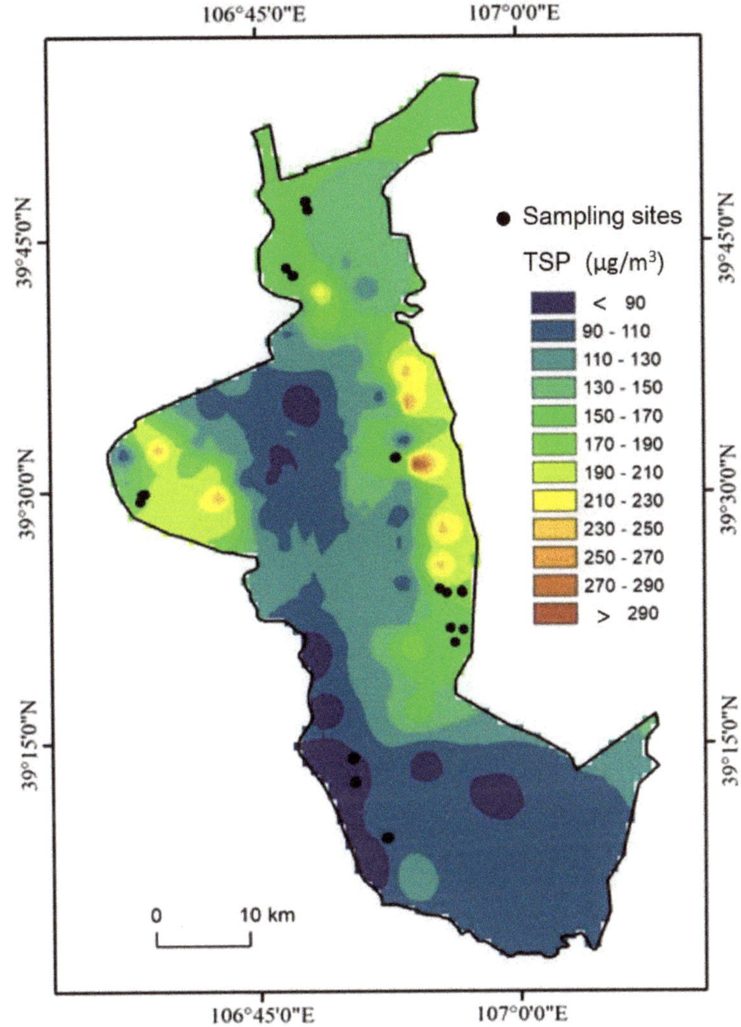

Figure 2. Distribution of total suspended particulate matter in city of Wuhai.

3.4. Soil Factors

Because the study area was located in an arid area, the soil nutrients and water contents were generally low. Soil total nitrogen content was between 0.13–0.58 g/kg. Soil total phosphorus content was between 0.17–0.39 g/kg, with slightly lower content in the restoration area (mine area) than in the other plots. Soil water content ranged from 1.42 to 8.19% (Table 1).

3.5. Linear Model Relating CWMs and Environmental Factors

After comparing different types of correlation curves, fitted by linear model, six curves with the highest value of R^2 were selected to build the CATS model. The results showed that the specific leaf area increased first and then decreased with increasing soil water content, and there was a significant correlation between them. The relationship between soil water content and leaf dry matter content also exhibited a significant correlation, but in a U-shaped curve. Root surface area had a significant negative correlation with increasing soil total nitrogen content. Root diameter increased initially and then decreased with increasing soil total nitrogen content and, overall, there was a significant positive correlation between them. The average root volume had a significant positive correlation with soil total phosphorus content. Root length had a significant positive correlation with soil total phosphorus content (Figure 3).

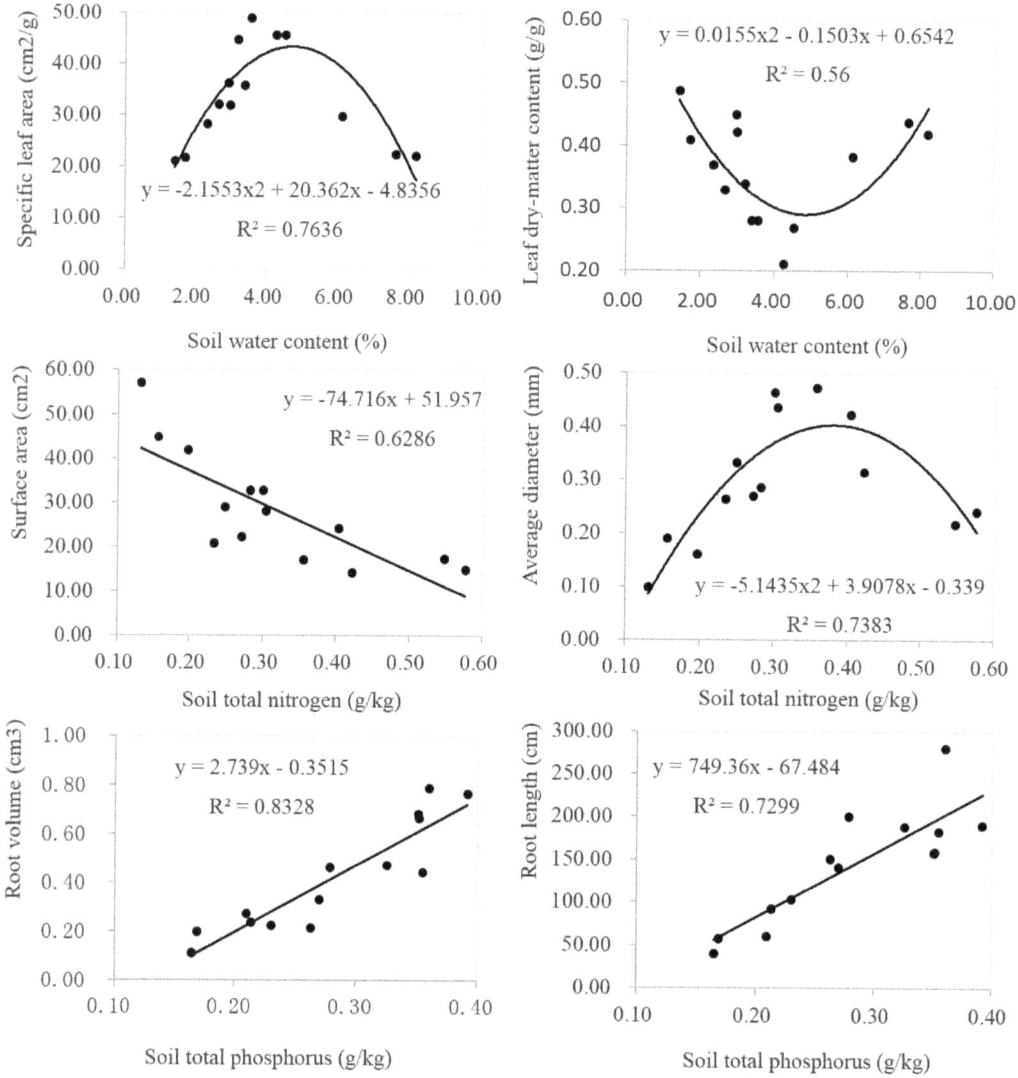

Figure 3. Fitting results of linear model for specific leaf area and soil water content, soil water content and leaf dry matter content, root surface area and soil total nitrogen, root diameter and soil total nitrogen content, average root volume and soil total phosphorus content, and root length and soil total phosphorus content.

3.6. Fitting Results of CATS Model

For the shady slope of the restoration area, the relative abundances of the species obtained by the CATS model were as follows: the average relative abundances of *Stipa grandis* and *Bassia dasyphylla* should be relatively high, with a median values of 0.46 and

0.13, respectively. This indicated that under these environmental conditions, *S. grandis* was suitable to be the dominant species in the future repair community, and *B. dasyphylla* was suitable to be the subdominant species in the future repair community. Therefore, an 8:2 ratio of *S. grandis* to *B. dasyphylla* would be the appropriate ratio in the future target community to achieve the target function in the shady slope ecosystem of the mine area (Figure 4). For the sunny slope of the restoration area, the relative abundances of the species obtained by the CATS model were as follows: the average relative abundances of *Kali tragus* and *B. dasyphylla* should be higher than all others. This indicated that the typical xerophytic herbs *K. tragus* and *B. dasyphylla* in a ratio of 6:4 are suitable for the repair community (Figure 4).

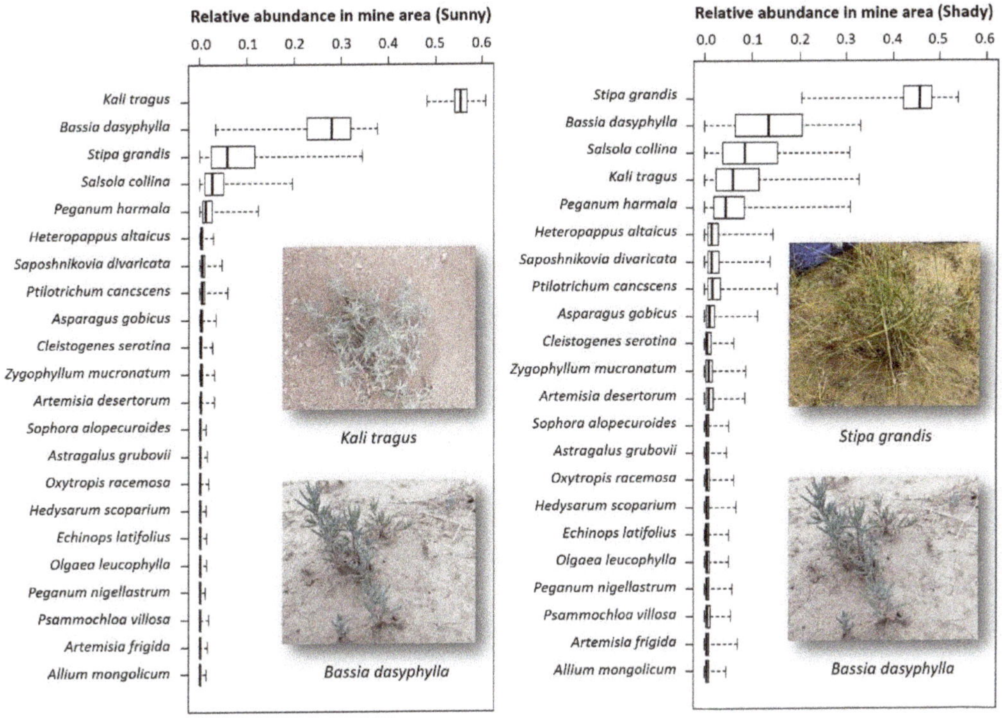

Figure 4. The relative abundances of the species in sunny or shady slope of mine area obtained by the CATS model.

4. Discussion

4.1. Relationships between Plants and Environmental Factors

Due to the low soil moisture and nutrient content in arid regions, the success of vegetation reconstruction in arid desert mine areas will depend largely on the selection of drought-tolerant plants [8]. Through natural selection, native species in arid areas have adapted to local conditions, and these adaptations have manifested as response traits. Xerophytes were the main plant type in the study area. They adopt various strategies to cope with environmental drought. For example, when the soil water content is low, plants can reduce their specific leaf area and increase leaf dry-matter content in response (Figure 3). In contrast, when soil water content is high, plants may use strategies opposite to those for drought stress (Figure 3). Soil nitrogen in the study area was low because it is mainly derived from decomposition of soil organic from litter [27,28], and there is relatively little litter in arid areas. In these conditions, it has been suggested that plants can increase the contact area between roots and soil by increasing root surface area and decreasing the average root diameter, both of which increase their nitrogen absorption ability (Figure 3). Compared with soil nitrogen, soil phosphorus content is most strongly affected by the soil matrix [27]. This indicated that soil phosphorus was less affected by the decomposition of litter than nitrogen in the arid study area. It could also be seen that the volume and length of plant roots developed well in areas rich in soil phosphorus (Figure 3). In addition, the dust retention per unit leaf area of herbaceous plants was significantly different among plots, which may have been due to differences in the leaf surface microstructures of different plants. Plant leaf surface morphology plays an important role in the retention of atmospheric particles. Future study is necessary to better understand the relationship between plant foliar microstructure characteristics and dust retention ability, and such a study would be conducive to the scientific and reasonable selection of dominant dust retaining plant species.

4.2. Prediction of Relative Abundance of Species Based on CATS Model

Two methods are commonly used in vegetation restoration work. One method is based on history. This method references past ecosystem conditions to determine the optimal plant community structure configuration [29]. However, this method may be disadvantageous because the past and future environmental conditions may be different. Therefore, vegetation restoration strategies based on the past cannot be guaranteed to apply to environmental conditions in the future. The other repair community determination method uses comparative tests conducted on different plant species [30]. When a species is shown to be suitable for the current environment, it can be included in the vegetation restoration community. In contrast to traditional methods, the CATS model goes one step further and predicts the required community assemblages for future environmental conditions based on their response and effect traits.

In the restoration area, the main species in the restoration community were *S. grandis*, *B. dasyphylla*, and *K. tragus*, which are suitable species for restoration in this area. The adaptation strategy when using *S. grandis*, which is a typical grassland plant species, is to expand the range of phosphorus absorption and improve phosphorus acquisition ability by increasing the underground biomass in low nutrient environments [31]. On the other hand, when the habitat conditions are poor, *S. grandis* can also cope with an environment with a smaller specific leaf area [31]. *S. grandis* is a pioneer plant in arid areas, which indicates that *S. grandis* has a high stress tolerance [31]. In the face of harsh habitats, *S. grandis* has unique adaptation strategies: it can quickly complete germination, excavation, and growth in areas with extremely rare rainfall. The root system of *S. grandis* occupies a wide space and can make full use of underground nutrients and water [32]. In addition, *S. grandis* has a small seed size so its seeds can be spread easily by the action of the wind. This can help speed the development of plant communities. *K. tragus* can reduce the pH of the soil near its rhizosphere, and change strong alkali environments [33]. This helps facilitate the absorption of soil nutrients and water by the plant community. These physiological characteristics are effective adaptations that help *K. tragus* survive in arid areas. The most typical morphological feature of *B. dasyphylla* is its pubescence covering [34]. This helps protect the internal tissues of the plant from harsh environmental effects [35]. In an arid environment, plant surface hairs can also help to adsorb condensed water [36]. Dense growth of *B. dasyphylla* increases the surface coverage of bare ground and plays a positive role in the fixation of aeolian soil [36].

In the process of vegetation restoration in the mine area, the selected plant species should be suitable for growth in the local environment. The study area was centered around a restoration area. The selected plant species were native plants that were present after long-term natural succession and were suitable for growth and development in the harsh environment. Therefore, the model results can be used to guide the vegetation restoration strategy for this restoration area. However, the results obtained in this study still need to be tested in practical applications in the future.

11. Damigos, D.; Kaliampakos, D. Assessing the benefits of reclaiming urban quarries: A CVM analysis. *Landsc. Urban Plan* **2003**, *64*, 249–258. [CrossRef]
12. Laughlin, D.C.; Strahan, R.T.; Huffman, D.W.; Sánchez Meador, A.J. Using trait-based ecology to restore resilient ecosystems: Historical conditions and the future of montane forests in western North America. *Restor. Ecol.* **2017**, *25*, S135–S146. [CrossRef]
13. Laughlin, D.C.; Joshi, C.; van Bodegom, P.M.; Bastow, Z.A.; Fule, P.Z. A predictive model of community assembly that incorporates intraspecific trait variation. *Ecol. Lett.* **2012**, *15*, 1291–1299. [CrossRef] [PubMed]
14. Laughlin, D.C.; Laughlin, D.E. Advances in modeling trait-based plant community assembly. *Trends Plant Sci.* **2013**, *18*, 584–593. [CrossRef] [PubMed]
15. Mensens, C.; De Laender, F.; Janssen, C.R.; Sabbe, K.; De Troch, M. Different response-effect trait relationships underlie contrasting responses to two chemical stressors. *J. Ecol.* **2017**, *105*, 1598–1609. [CrossRef]
16. Leps, J.; de Bello, F.; Lavorel, S.; Berman, S. Quantifying and interpreting functional diversity of natural communities: Practical considerations matter. *Preslia* **2006**, *78*, 481–501.
17. Violle, C.; Navas, M.-L.; Vile, D.; Kazakou, E.; Fortunel, C.; Hummel, I.; Garnier, E. Let the concept of trait be functional! *Oikos* **2007**, *116*, 882–892. [CrossRef]
18. Suding, K.N.; Lavorel, S.; Chapin, F.S.; Cornelissen, J.H.C.; DÍAz, S.; Garnier, E.; Goldberg, D.; Hooper, D.U.; Jackson, S.T.; Navas, M.-L. Scaling environmental change through the community-level: A trait-based response-and-effect framework for plants. *Glob. Change Biol.* **2008**, *14*, 1125–1140. [CrossRef]
19. Lavorel, S.; Garnier, E. Predicting changes in community composition and ecosystem functioning from plant traits: Revisiting the Holy Grail. *Funct. Ecol.* **2002**, *16*, 545–556. [CrossRef]
20. Ma, Z.; Guo, D.; Xu, X.; Lu, M.; Bardgett, R.D.; Eissenstat, D.M.; McCormack, M.L.; Hedin, L.O. Evolutionary history resolves global organization of root functional traits. *Nature* **2018**, *555*, 94–97. [CrossRef]
21. Keenan, T.F.; Niinemets, Ü. Global leaf trait estimates biased due to plasticity in the shade. *Nat. Plants* **2016**, *3*, 1–6. [CrossRef] [PubMed]
22. Wright, I.J.; Dong, N.; Maire, V.; Prentice, I.C.; Westoby, M.; Diaz, S.; Gallagher, R.V.; Jacobs, B.F.; Kooyman, R.; Law, E.A.; et al. Global climatic drivers of leaf size. *Science* **2017**, *357*, 917–921. [CrossRef] [PubMed]
23. Butler, E.E.; Datta, A.; Flores-Moreno, H.; Chen, M.; Wythers, K.R.; Fazayeli, F.; Banerjee, A.; Atkin, O.K.; Kattge, J.; Amiaud, B.; et al. Mapping local and global variability in plant trait distributions. *Proc. Natl. Acad. Sci. USA* **2017**, *114*, E10937–E10946. [CrossRef] [PubMed]
24. Gross, N.; Bagousse-Pinguet, Y.L.; Liancourt, P.; Berdugo, M.; Gotelli, N.J.; Maestre, F.T. Functional trait diversity maximizes ecosystem multifunctionality. *Nat. Ecol. Evol.* **2017**, *1*, 1300–1302. [CrossRef]
25. Kunstler, G.; Falster, D.; Coomes, D.A.; Hui, F.; Kooyman, R.M.; Laughlin, D.C.; Poorter, L.; Vanderwel, M.; Vieilledent, G.; Wright, S.J.; et al. Plant functional traits have globally consistent effects on competition. *Nature* **2016**, *529*, 204–207. [CrossRef]
26. Wang, W.F.; Hao, W.D.; Sian, Z.F.; Lei, S.G.; Wang, X.S.; Sang, S.X.; Xu, S.C. Effect of coal mining activities on the environment of *Tetraena mongolica* in Wuhai, Inner Mongolia, China—A geochemical perspective. *Int. J. Coal. Geol.* **2014**, *132*, 94–102. [CrossRef]
27. Rodríguez, A.; Durán, J.; Fernández-Palacios, J.M.; Gallardo, A. Spatial pattern and scale of soil N and P fractions under the influence of a leguminous shrub in a *Pinus canariensis* forest. *Geoderma* **2009**, *151*, 303–310. [CrossRef]
28. Hirobe, M.; Koba, K.; Tokuchi, N. Dynamics of the internal soil nitrogen cycles under moder and mull forest floor types on a slope in a *Cryptomeria japonica* D. Don plantation. *Ecol. Res.* **2003**, *18*, 53–64. [CrossRef]
29. Swetnam, T.W.; Allen, C.D.; Betancourt, J.L. Applied historical ecology: Using the past to manage for the future. *Ecol. Appl.* **1999**, *9*, 1189–1206. [CrossRef]
30. Zhang, Y.; Yang, J.Y.; Wu, H.L.; Shi, C.Q.; Zhang, C.L.; Li, D.X.; Feng, M.M. Dynamic changes in soil and vegetation during varying ecological-recovery conditions of abandoned mines in Beijing. *Ecol. Eng.* **2014**, *73*, 676–683. [CrossRef]
31. Liu, N.; Tian, Q.; Zhang, W. Comparison of adaptive strategies to phosphorus-deficient soil between dominant species *Artemisia frigida* and *Stipa krylovii* in typical steppe of Nei Mongol. *Chin. J. Plant Ecol.* **2014**, *38*, 905–915.
32. Ren, J.; Zhao, C.; Zhao, X.; Ma, J.; Li, Z.; Gou, F. Fractal characteristics of *Agriophyllum squarrosum* roots in desert-oasis ecotone in Jinta County, Gansu Province. *Acta Ecol. Sin.* **2020**, *40*, 5298–5305.
33. Li, C.; Ma, J.; Li, Y.; Fan, L. Effects of pH value on the morphological features and activity of roots of three life-form plant species. *Arid Zone Res.* **2010**, *27*, 915–920.
34. Grammatikopoulus, G.; Manetas, Y. Direct absorption of water by hairy leaves of *Phlomis fruticosaand* its contribution to drought avoidance. *Can. J. Bot.* **1994**, *72*, 1805–1811. [CrossRef]
35. Johnson, H.B. Plant pubescence: An ecological perspective. *Bot. Rev.* **1975**, *41*, 233–258. [CrossRef]
36. Zhuang, Y.; Zhao, W. Experimental Study of Effects of Artificial Dew on *Bassia dasyphylla* and *Agriophyllum squarrosum*. *J. Desert Res.* **2010**, *30*, 1068–1074.

agronomy

MDPI

Article

Effects of Long-Term Exclosure on Main Plant Functional Groups and Their Biochemical Properties in a Patchily Degraded Alpine Meadow in the Source Zone of the Yellow River, West China

Pengnian Yang [1], Xilai Li [1,2,*], Chenyi Li [1] and Jing Zhang [1]

[1] College of Agriculture and Animal Husbandry, Qinghai University, Xining 810016, China; 18093559658@163.com (P.Y.); chengyi_li0801@163.com (C.L.)

[2] State Key Laboratory of Plateau Ecology and Agriculture, Qinghai University, Xining 810016, China

* Correspondence: xilai-li@163.com

Abstract: This study aimed to understand the response of vegetation community characteristics in the degraded alpine meadow of the Source Zone of the Yellow River to exclosure of various lengths. Artificial fences were erected to prevent livestock grazing and let the degraded meadow recover naturally as a means of restoration. The research focused on a typical degraded alpine meadow in which four plots were fenced off for three periods of 1 year (E1), 4 years (E4), and 10 years (E10), plus a freely grazed plot as the control. The study compared and analyzed the differences in plant community characteristics, carbon (C), nitrogen (N), and phosphorus (P) reserves, as well as the stoichiometric characteristics of main functional groups in the alpine meadow over different exclosure durations. The results indicated that E10 long-term exclosure significantly increased the aboveground biomass of gramineous plants but reduced the aboveground biomass of miscellaneous grasses. However, when compared to E4 short-term exclosure, E10 resulted in a reduction in the aboveground biomass of Cyperaceae plants. On the other hand, E4 medium-term exclosure significantly increased the aboveground biomass of Gramineae and Cyperaceae. Exclosure significantly increased the nitrogen (N) and phosphorus (P) reserves of the aboveground plant communities. Among these communities, the plant communities in the E10 long-term exclosure had the highest N and P reserves. However, this exclosure length also led to a significant reduction in plant diversity. Furthermore, except for Cyperaceae, all functional groups were observed in E10 and E4 plots. The carbon–nitrogen ratio and carbon–phosphorus ratio of these groups were significantly lower than those of groups G and E1. Medium-term exclosure (E4) has a positive impact on the aboveground biomass as well as plants' nitrogen and phosphorus reserves. However, long-term exclosure (E10) has been observed to decrease species diversity and nutrient utilization efficiency of alpine meadow vegetation, which can be detrimental to the sustainable development of the alpine meadow ecosystem. Therefore, it is not recommended to implement long-term exclosure. Instead, a moderate level of grazing should be adopted after 4 years of exclosure.

Keywords: degraded patches; alpine meadow; exclosure duration; plant community characteristics; environmental factors

Citation: Yang, P.; Li, X.; Li, C.; Zhang, J. Effects of Long-Term Exclosure on Main Plant Functional Groups and Their Biochemical Properties in a Patchily Degraded Alpine Meadow in the Source Zone of the Yellow River, West China. *Agronomy* **2023**, *13*, 2781. https://doi.org/10.3390/agronomy13112781

Academic Editors: Xiaobo Qin, Kesi Liu and Xinqing Shao

Received: 19 September 2023
Revised: 12 October 2023
Accepted: 6 November 2023
Published: 9 November 2023

1. Introduction

Grassland is a predominant terrestrial ecosystem covering approximately 40% of the world's land area [1]. It serves as a crucial resource for animal husbandry and plays a vital role in biodiversity conservation, windbreak and sand fixation, and water conservation [2]. China boasts abundant grassland resources in a vast area of 3.92×10^8 ha. Notably, the grassland area in the Source Zone of the Yellow River amounts to 1.04×10^7 ha, serving as a significant carbon sink and water conservation region [3]. Situated in the ecologically sensitive Qinghai–Tibet Plateau, the Source Region of the Yellow River acts as a crucial nature reserve in China [4]. It also serves as the core area for the ecological functioning and water conservation of alpine meadows in the Qinghai–Tibet Plateau, possessing substantial ecological and economic values. Furthermore, this region is globally recognized as one of the high-altitude areas with rich biodiversity, characterized by a cold climate and fragile ecology [5]. The grassland ecosystem in the Source Region of the Yellow River has been significantly disturbed by human activities. The grassland in this area has been exploited beyond its sustainable carrying capacity, leading to serious overgrazing. Consequently, serious ecological problems such as grassland degradation, wetland shrinkage, and soil erosion have emerged frequently in recent decades [6]. Large-scale degradation of grassland has resulted in frequent ecological disasters, which not only hampers the sustainable development of grassland animal husbandry but also greatly impacts the ecological security and social stability of the local area, as well as the surrounding countries or regions. Therefore, how to curb grassland degradation and effectively restore the degraded grassland, and strengthen the management and restoration of the degraded grassland, has become an urgent and challenging issue worldwide.

Ecological restoration is the process of returning degraded grassland ecosystems to a healthy and stable state [7]. It can be achieved either naturally or via human intervention. The natural restoration cycle of degraded grassland ecosystems is lengthy, often taking decades or even centuries. To expedite the restoration process, artificial measures are necessary. China, on the other hand, began its efforts to restore and explore degraded grassland relatively late [8]. Currently, the main restoration measures employed in China include fenced exclosure, no-tillage reseeding, and fertilization, all of which have significantly improved the health of degraded grassland [9]. Fenced exclosure is a method of dividing grassland into smaller areas via wire fences to prevent them from being grazed by livestock and disturbed by other animals. Exclosure ensures that the grassland inside the fenced area is free from livestock grazing, trampling, and excretion, allowing it to undergo natural succession and natural recovery [10]. Compared with other manual recovery methods, the advantages of fenced exclosure include its low cost, wide applicability, and ease of implementation [11]. Extensive research has been conducted worldwide on various effects of fenced exclosure, such as plant community diversity; aboveground and underground biomass; community structure; plant carbon, nitrogen, phosphorus, and stoichiometric characteristics; soil seed bank; soil physical and chemical properties; as well as microbial community structure and diversity [12–18]. The findings indicated that fencing has the potential to enhance grassland biomass, improve soil structure, and increase water use efficiency [19]. Moreover, fencing proves beneficial for the restoration of degraded grassland and supports secondary succession, leading to the gradual recovery to healthy grassland [20]. Nevertheless, it is important to note that the impact of exclosure varies across different types of grasslands due to variations in climate conditions and the duration of exclosure.

Wu et al. [21] discovered that fencing has a significant positive impact on aboveground biomass in the northwest of the Qinghai–Tibet Plateau. This improvement benefits forage functional groups and restricts the growth of harmful weed functional groups. However, it does lead to a decrease in plant density and species diversity. Similarly, Xu et al. [22] found that fencing greatly enhances the coverage, height, and aboveground and underground biomass of plant communities in degraded temperate meadow steppe in the Hulunbeier steppe of North China. Furthermore, Asteken et al. [23] observed that 10 years of exclosure positively influenced the community diversity of temperate grassland and mountain meadow vegetation. It is worth noting that previous studies have focused primarily on the positive effects of fence exclosure on grassland ecosystems, neglecting its potential negative impacts on wildlife. After analyzing 208 literature sources worldwide on fencing, Smith et al. [24] discovered that only 7% of them examined both positive and negative effects of fencing. In a separate study, Boone et al. [25] found that fencing can impede the migration routes of herbivores, diminish the diversity of forage grass, and ultimately reduce the carrying capacity of grasslands. This indicates that long-term exclosure and grazing prohibition represent an extreme grassland management approach that has both protective and negative effects on the grassland ecosystem. In spite of the large number of studies on fencing, several knowledge gaps still remain. First, it remains unknown how the biomass of plant functional groups responds to exclosure duration. Additionally, different plant functional groups respond differently to exclosure in patchily degraded grassland. So far, there is a lack of research on how exclosure length can effectively balance forage productivity and plant diversity in the restorative process of patchily degraded alpine meadows, while minimizing the negative impact of exclosure.

This study focuses on the patchily degraded alpine meadow in the Source Zone of the Yellow River in the Qinghai–Tibet Plateau. This research explores how plant biophysical properties respond to different durations of exclosure in comparison with the control. The aim of this study is to investigate the long-term effects of exclosure on the patchily degraded alpine meadow and offer insights for the long-term restoration of alpine meadow ecosystems.

2. Materials and Methods

2.1. Overview of the Study Area

The study area is located in Henan Mongolian Autonomous County, Huangnan Prefecture, Qinghai Province. It is situated on the northeastern margin of the Qingnan Plateau at an average altitude of 3600–3800 m. The region experiences a typical plateau continental climate, characterized by short springs and autumns, indistinct seasons (the four seasons are not clear), and no absolute frost-free period. From May to October each year, the weather is warm and rainy, while from November to April of the following year, it becomes cold, dry, and windy. The average annual temperature ranges from 9.2 to 14.6 °C, and the annual precipitation ranges between 597.1 and 615.5 mm. The predominant soil type in the area is subalpine meadow soil.

The grassland area in Henan County is 661,500 ha, with 645,000 ha suitable for grazing. By 2021, the county had successfully and comprehensively restored the most severely degraded (e.g., denudated) alpine meadow known as "heitutan" and implemented measures to control grassland pests such as plateau pika. The restored areas included 30,600 ha of "heitutan", 20,000 ha sloped black soil, while 95,700 ha of grassland was treated with pest prevention and control measures, 81,300 ha fenced for long-term grazing exclosure, 46,700 ha for rotational grazing exclosure, 226,700 ha for short-term exclosure, and 27,300 ha of degraded grassland had partially recovered. Additionally, the county has established a grazing prohibited area of 435,600 ha and set 154,300 ha of grassland for balanced grazing (e.g., stocking rate does not exceed the grassland carrying capacity). However, there are still several challenges facing the subsequent utilization of the once enclosed grassland.

2.2. Experimental Design

In order to explore the mechanism of plant community construction in the alpine meadow under different exclosure years, and the influence of exclosure years on plant community diversity and productivity, this study adopts the philosophy of "using space as a proxy of time". In this study, 4 experimental plots with similar topography, vegetation composition, community characteristics, and degradation degree were selected in the enclosed alpine meadow in Henan County. The plots had an area of 300 m × 250 m, totaling 7.50 ha. They were fenced off in July 2011, 2017, and 2020, respectively. By August 2021, the experimental plots had been fenced for four durations: 10 years of long-term exclosure (E10), 4 years of medium-term exclosure (E4), 1 year of short-term exclosure (E1), and free grazing (G) without fencing. The location, altitude, and main composition of plant communities in the selected plots are presented in Table 1. It should be noted that the exclosure plot in this study refers to the continuous exclosure plot that has not involved grazing since the establishment of the fence.

Table 1. Location, elevation, and main composition of plant community of each plot.

Exclosure Duration	Plot Location	Elevation/m	Patch Numbers	Patch Density (Area Ratio)/%	The Main Composition of Plant Community
G	E 101°46′06″, N 34°41′02″	3610	1086.45	41.64	*Elymus nutans* Griseb., *Koeleria cristata* L., *Kobresia pygmaea* Willd., *Ligularia virgaurea*, *Gentiana scabra* Bunge, *Pedicularis spicata*, *Oxytropis ochrocephala*, *Carun buriaticum*, *Gentiana macrophylla* Pall., and *Taraxacum officinalis*
E1	E 101°46′08″, N 34°41′04″	3610	989.52	36.41	*Poa annua* L., *Elymus nutans* Griseb., *Kobresia pygmaea* Willd., *Koeleria cristata* L., *Potentilla anserina*, *Pedicularis spicata*, *Saussurea pulchra*, *Gentiana scabra* Bunge, and *Ligularia virgaurea*
E4	E 101°46′10″, N 34°41′06″	3610	536.71	18.63	*Elymus nutans* Griseb., *Poa annua* L., *Koeleria cristata* L., *Potentilla anserina*, *Ajania tenuifolia*, *Astragalus* sp., and *Gentiana scabra* Bunge
E10	E 101°26′34″, N 34°18′36″	3720	331.81	6.67	*Elymus nutans* Griseb., *Poa annua* L., *Koeleria cristata* L., *Carex* L., *Potentilla anserina*, and *Gentiana scabra* Bunge

The table above shows the status of the plot at the time of sampling in 2021, i.e., the recovery status of the plot at the end of the experiment. Patches refer to a state in which the meadow is fragmented due to degradation. G represents free grazing (control), E1 represents exclosure for 1a, E4 represents exclosure for 4a, and E10 represents exclosure for 10a (Figure 1) (the same below).

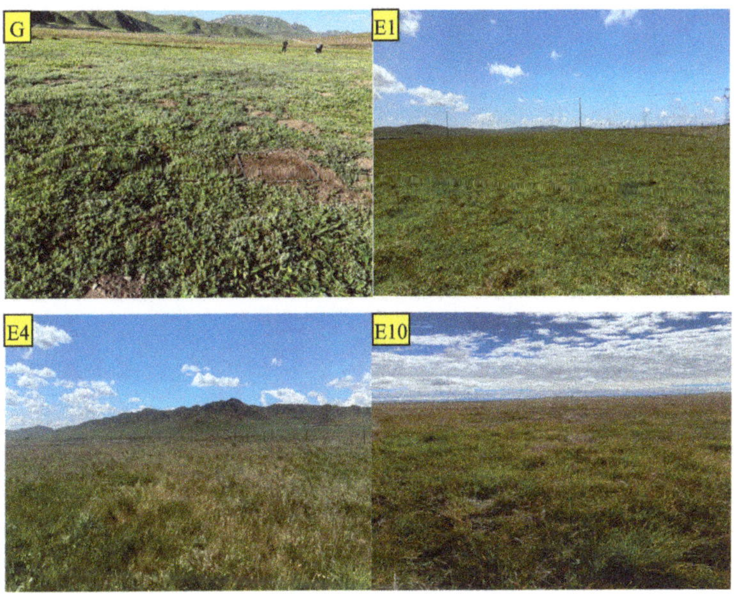

Figure 1. Sample landscape display.

2.3. Sample Collection and Analysis

2.3.1. Determination of Plant Community Dynamics, Population Characteristics, and Aboveground and Belowground Biomass

During the vigorous growth period of plants in August 2021, we investigated the characteristics of plant communities in the field. Six quadrants of 50 cm × 50 cm were randomly located in the plots of varying exclosure years. Within these quadrants, we surveyed plant species and total coverage, and measured height and density of various plants by functional groups. Additionally, we also sampled biomass for four types of plants (grasses, sedges, legumes, and forbs) by cutting them off the ground, and carefully packing them in plastic bags. Each bag was marked with the respective plant's name and transported to the laboratory. In the lab, the freshly collected plants were weighted and then placed in an oven at 90 °C for 30 min to kill them. Subsequently, we dried the plants in an oven at 65 °C until a constant weight. This dry weight represented the aboveground biomass of the plants during the peak month of the growing season. Additionally, we collected cylindrical soil samples with a height of 20 cm and a diameter of 8 cm from the middle of each sampling quadrant via root drilling. These soil samples were filtered using a 2 mm sieve in the lab to find plant roots, either alive or dead. The collected roots were then dried in an oven at 65 °C until a constant weight to measure the belowground biomass [26].

2.3.2. Plant Community Structure Indices

In the field, the plant properties within the selected quadrants were surveyed at the community level and expressed in three indices: the Shannon–Wiener index, H; Simpson index (D); and Pielous index (E). They are calculated as

$$\text{Shannon–Wiener diversity index}: \ H = -\sum_{i=1}^{S} P_i \ln P_i;$$

$$\text{Simpson diversity index}: \ D = 1 - \sum_{i=1}^{S} P_i^{\,2};$$

$$\text{Pielous evenness index}: \ E = H/\ln S;$$

where P_i represents the ratio of the number of individuals of the i-th species to the total number of individuals in the community. It can be calculated as $P_i = N_i/N$, where N_i is the number of individuals of the i-th species and N is the total number of individuals of all species. S denotes the total number of species in the community. The species diversity index, H, is calculated using the Shannon–Wiener index. The maximum species diversity index, H_{max}, is calculated as the natural logarithm of S, where S represents the total number of individuals of all species in the community [27].

2.3.3. Determination of Plant Carbon, Nitrogen, and Phosphorus

We determined the element content of the functional group of vegetation by separating all the aboveground parts of plants by functional group, crushing, and mixing them all with six replicates. Each composite sample represents the overall level of carbon, nitrogen, and phosphorus contents of the functional group of plants under study. Plant total carbon was determined using the potassium dichromate-H_2SO_4 oxidation external heating method. Total nitrogen and total phosphorus were determined using the AA3 continuous flow analyzer (Auto Analyzer 3-HR). The specific procedure was as follows: 0.2 g of plant samples were weighed and placed in a digester tube, to which 3.3 g of a catalyst (K_2SO_4:$CuSO_4$ = 10:1) and 10 mL of concentrated sulfuric acid were added. The mixture was placed in a 420 °C digester for 180 min. Afterwards, the resultant solution was transferred to a 100 mL flask and the volume was adjusted to the required level. Finally, the sample determination was carried out using the continuous flow analyzer.

2.4. Data Analysis and Representation

All the obtained original data were pre-processed using Microsoft Excel 2010 (Microsoft, Washington, DC, USA), and statistically analyzed using SPSS 19.0 (IBM, New York,

NY, USA). A one-way analysis of variance ($p < 0.05$) was used to examine the differences in plant biomass, diversity, and carbon, nitrogen, and phosphorus contents among different functional groups over different exclosure years. Histograms of plant carbon, nitrogen, and phosphorus content, and the stoichiometric ratio, were generated using Origin 2022. The correlation between plant carbon, nitrogen, and phosphorus contents and biomass was calculated and visualized using the R 4.0 (LUCENT, Murray Hill, NJ, USA). Canoco 5.0 (Microcomputer Power, Washington, DC, USA) was utilized for a redundancy analysis (RDA) and for plotting vegetation and soil characteristics.

3. Results

3.1. Plant Characteristics by Functional Group

Table 2 indicates significant or extremely significant differences in various indicators, with the exception of the total carbon content of plant roots, the nitrogen and phosphorus contents of sedge plants, and the aboveground biomass of leguminous plants. Notably, the carbon–nitrogen ratio and carbon–phosphorus ratio of gramineous plants, the nitrogen–phosphorus ratio of sedge plants, and the carbon–nitrogen ratio, carbon–phosphorus ratio, and nitrogen–phosphorus ratio of leguminous plants all exhibited significant or extremely significant differences. Similarly, the carbon–nitrogen ratio and carbon–phosphorus ratio of forbs also showed significant differences (Table 3).

Table 2. Results of ANOVA on the characteristics of main functional groups of plants.

Group/Type	df	F (C)	F (N)	F (P)	F (AGB)	F (Index)
Grasses	3	17.52 **	12.45 **	8.64 **	8.92 **	
Sedges	3	4.60 *	2.75 ns	2.34 ns	3.80 *	
Legumes	3	12.64 **	21.26 **	39.20 **	2.47 ns	
Forbs	3	14.50 **	11.92 **	12.15 **	10.3 **	
Plant litter	3	20.01 **	3.95 *	5.11 **	82.94 **	
Root	3	1.87 ns	6.79 **	3.96 *	7.16 **	
D	3					37.63 **
H	3					32.70 **
E	3					38.80 **

** Extremely significant difference ($p < 0.01$), * significant difference ($p < 0.05$), ns means no significant difference. In the table, grasses mean Poaceae family, sedges mean Cyperaceae family, legumes mean Fabaceae family, and forbs mean herbaceous plants from different families.

Table 3. Results of analysis of variance of carbon, nitrogen, and phosphorus stoichiometry of main functional groups of plants.

Group/Type	df	F (C/N)	F (C/P)	F (N/P)
Grasses	3	9.61 **	11.05 **	1.61 ns
Sedges	3	2.69 ns	2.96 ns	3.85 *
Legumes	3	10.32 **	43.60 **	65.73 **
Forbs	3	14.07 **	13.57 **	1.24 ns
Plant litter	3	1.64 ns	1.65 ns	1.80 ns
Root	3	0.56 ns	0.40 ns	1.06 ns

** Extremely significant difference ($p < 0.01$), * significant difference ($p < 0.05$), ns means no significant difference. In the table, grasses mean Poaceae family, sedges mean Cyperaceae family, legumes mean Fabaceae family, and forbs mean herbaceous plants from different families.

3.2. Main Plant Characteristics by Functional Group

Significant differences were observed in the total carbon content of gramineous plants in different exclosure durations. The total carbon content of long-term exclosure (E10) and medium-term exclosure (E4) was significantly lower than that of the control (G) and 1-year exclosure (E1) plots. Similarly, the total carbon content of Cyperaceae plants showed significant variations, with E4 and E10 being significantly lower than G. Furthermore, the total carbon content of Leguminosae plants also exhibited significant differences, with E10 being significantly lower than G, E1, and E4. For forb plants, E4 showed a significant decrease in the total carbon content compared to G, E1, and E10 (Figure 2a). Additionally, significant differences were observed in the total carbon content of plant litter, with E4 and E10 showing significantly lower values than E1 and G. However, no significant difference was found in the total carbon content of the belowground parts of the plant community among different exclosure durations (Figure 2b).

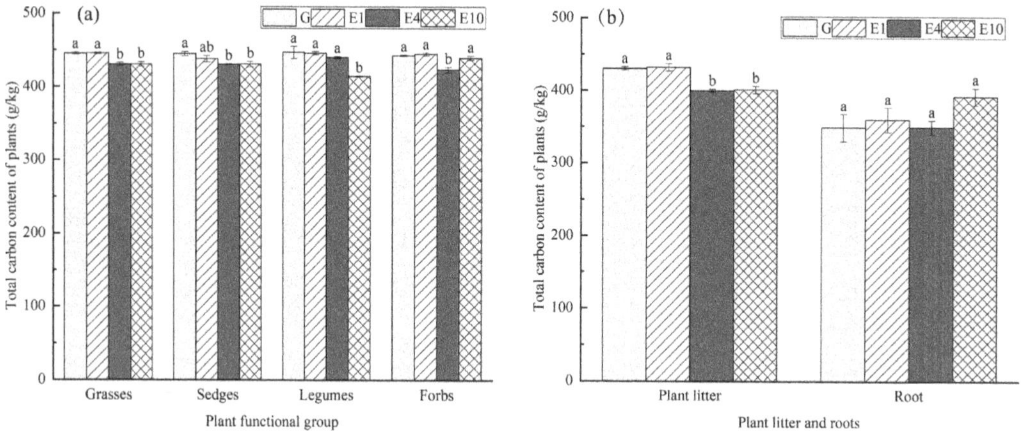

Figure 2. Total carbon content of different functional groups in plants. (**a**) Total carbon content of grasses, sedges, legumes, and forbs. (**b**) Total carbon content of plant litter and plant roots. Different lowercase letters in the same column show significant differences ($p < 0.05$).

Significant differences existed in the total nitrogen content of different plant groups over different exclosure years. E10 had a significantly higher nitrogen content in gramineous plants than E4, E1, and G. Similarly, E4 had a significantly higher nitrogen content than E1 and G. Among Cyperaceae plants, E10 had the highest nitrogen content that was significantly higher than E4. In the case of legumes, E10 had a significantly lower nitrogen content than E4, E1, and G. For forbs, E10 had a significantly higher nitrogen content than E4, E1, and G (Figure 3a). Significant differences were also observed in the total nitrogen content of plant litter, with E4 having a significantly lower nitrogen content than E1 and G. Furthermore, the roots of the plant community were significantly different in their nitrogen content, with E10 being significantly higher than E4, E1, and G (Figure 3b).

At different exclosure lengths, the total phosphorus content of gramineous plants showed significant variations. Both E10 and E4 had a significantly higher phosphorus content than E1 and G. Similarly, the total phosphorus content of Cyperaceae plants also showed significant differences between different exclosure lengths, with E4 having the highest content, significantly higher than E1. The total phosphorus content of leguminous plants also exhibited significant differences. E10 had a significantly higher phosphorus content than all three other durations, while E4 was significantly higher than E1 and G. For forbs, their total phosphorus content showed significant differences, with E10 being significantly higher than E1 and G, and E4 being significantly higher than E1 (Figure 4a). Furthermore, the total phosphorus content of plant litter also exhibited significant differences, with G being significantly higher than E1, E4, and E10. The roots of the plant community also had significant differences in their total phosphorus content, with E10 being significantly higher than E4, E1, and G (Figure 4b).

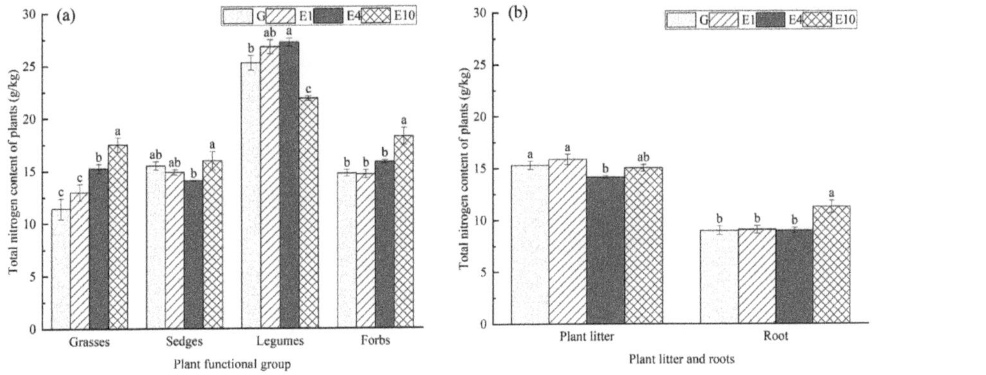

Figure 3. Total nitrogen content of different functional groups in plants. (**a**) Total nitrogen contents of grasses, sedges, legumes, and forbs. (**b**) Total nitrogen contents of plant litter and plant roots. Different lowercase letters in the same column show significant differences ($p < 0.05$).

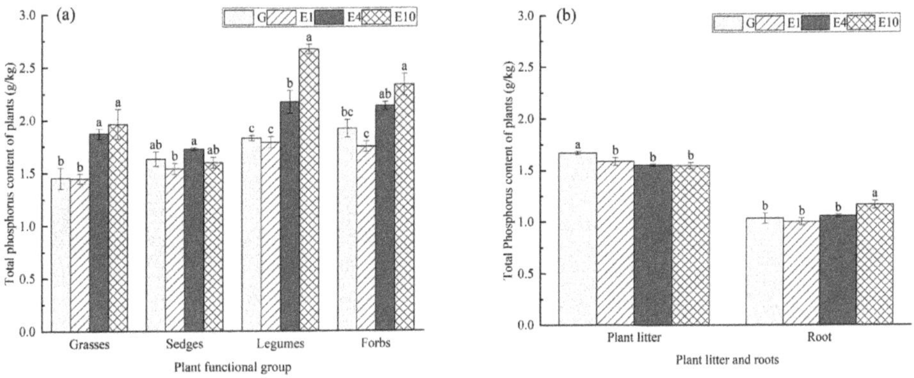

Figure 4. Total phosphorus content of different functional groups in plants. (**a**) Total phosphorus content of grasses, sedges, legumes, and forbs. (**b**) Total phosphorus contents of plant litter and plant roots. Different lowercase letters in the same column show significant differences ($p < 0.05$).

The community diversity indices (Shannon–Wiener index, H; Simpson index, D; and Pielous index, E) in E10 were notably lower than those of E4, E1, and G (Figure 5a). The aboveground biomass of gramineous plants exhibited significant variations among different exclosure years, with E10 and E4 showing significantly higher values than E1 and G. Similarly, the aboveground biomass of Cyperaceae plants also displayed significant differences, with E4 having the highest value, significantly surpassing E10, E1, and G. For leguminous plants, E10 had the lowest aboveground biomass, which was significantly lower than E1. The aboveground biomass of forbs also exhibited significant differences, with E10 being significantly lower than E4, E1, and G (Figure 5b). The biomass of plant litter showed significant differences among the treatments. E10 had the highest biomass, which was significantly higher than E4, E1, and G. Additionally, E4 had a significantly higher biomass than E1 and G (Figure 5c). The belowground biomass of the plants also exhibited significant differences among the treatments. E10 had the lowest biomass, which was significantly lower than E4, E1, and G. Furthermore, the three indexes of the E10 treatment were significantly lower than those of the other treatments (Figure 5d).

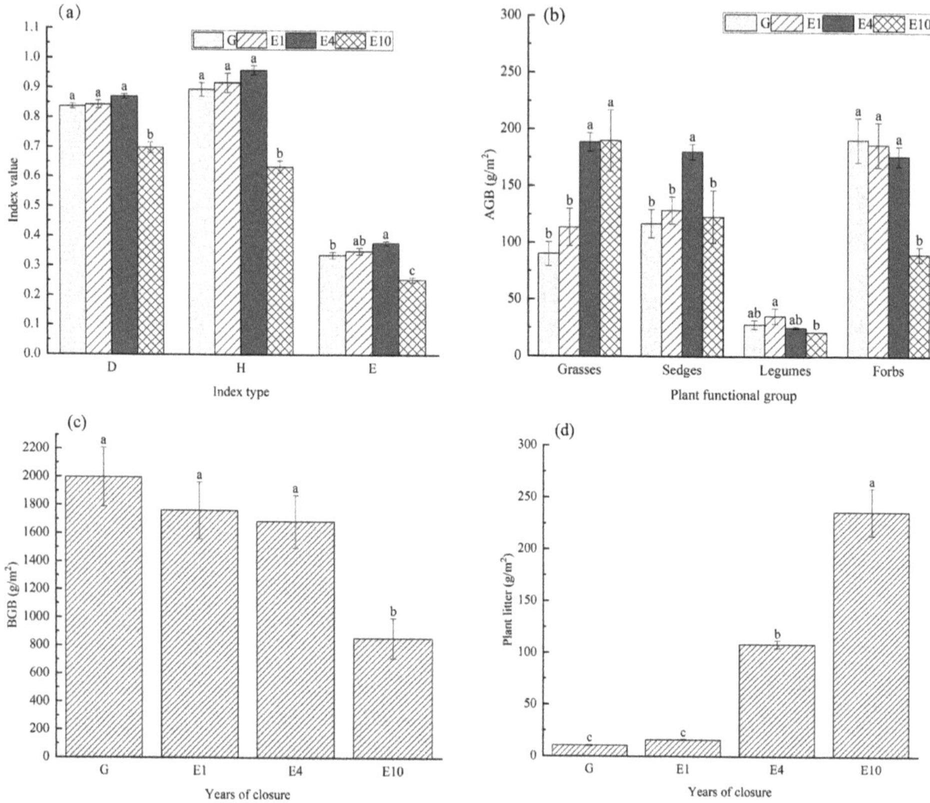

Figure 5. Plant community diversity index and biomass of different functional groups. (**a**) Plant community diversity index, (**b**) Aboveground biomass of grasses, sedges, legumes, and forbs, (**c**) Belowground biomass of plant communities, and (**d**) Plant litter content (AGB = aboveground biomass, and BGB = belowground biomass). Different lowercase letters in the same column show significant differences ($p < 0.05$).

There was a significant difference in the carbon–nitrogen ratio among gramineous plants of different exclosure durations ($p < 0.05$). The carbon–nitrogen ratio of long-term exclosure (E10) and medium-term exclosure (E4) was significantly lower than that of the control (G) and 1-year exclosure (E1). However, there was no significant difference in the carbon–nitrogen ratio of Cyperaceae plants. On the other hand, the C/N ratio of leguminous plants showed significant differences. E10 had a significantly higher ratio than G, E1, and E4, while E4 had a significantly lower ratio than G. The C/N ratio of forbs also exhibited significant differences, with E10 and E4 having a significantly lower ratio than G and E1. Interestingly, there was no significant difference in the carbon–nitrogen ratio between plant litter and the belowground parts of the plant community among different exclosure years (Table 4).

There was a significant difference in the carbon–phosphorus ratio of gramineous plants among different exclosure years ($p < 0.05$). Both long-term exclosure (E10) and medium-term exclosure (E4) had a significantly lower carbon–phosphorus ratio than the control group (G) and 1-year exclosure (E1). No significant difference was found in the carbon–phosphorus ratio of Cyperaceae plants. Significant differences were observed in the C/P ratio of legumes. E10 and E4 had significantly lower ratios than E1 and G, and E10 had a significantly lower ratio than E4. The C/P ratio of forbs also showed significant differences, with E10 and E4 having significantly lower ratios than E1 and G. However, no significant difference existed in the carbon–phosphorus ratio between plant litter and the roots of the plant community (Table 5).

Table 4. Carbon–nitrogen ratio of plants in different functional groups.

Exclosure Duration	Grasses	Sedges	Legumes	Forbs	Plant Litter	Root
G	40.50 ± 3.43 a	28.85 ± 0.77 a	17.78 ± 0.59 b	30.14 ± 0.64 a	28.21 ± 0.79 a	39.22 ± 2.88 a
E1	35.11 ± 2.59 a	29.50 ± 0.40 a	16.72 ± 0.40 bc	30.52 ± 0.83 a	27.32 ± 0.66 a	40.24 ± 3.19 a
E4	28.41 ± 0.79 b	30.71 ± 0.10 a	16.19 ± 0.20 c	26.80 ± 0.16 b	28.30 ± 0.42 a	38.96 ± 1.59 a
E10	24.84 ± 1.07 b	27.42 ± 1.43 a	18.98 ± 0.21 a	24.40 ± 1.13 c	26.74 ± 0.35 a	35.61 ± 2.85 a

Different lowercase letters in the same column show significant differences ($p < 0.05$).

Table 5. Carbon–phosphorus ratio of different functional groups of plants.

Exclosure Duration	Grasses	Sedges	Legumes	Forbs	Plant Litter	Root
G	313.89 ± 19.97 a	275.68 ± 12.33 a	244.64 ± 5.12 a	233.21 ± 10.25 a	257.71 ± 2.53 a	344.54 ± 30.73 a
E1	309.91 ± 9.49 a	286.21 ± 7.78 a	250.81 ± 7.72 a	255.60 ± 6.68 a	272.99 ± 6.95 a	363.32 ± 24.97 a
E4	231.19 ± 6.17 b	250.16 ± 2.01 a	205.20 ± 9.16 b	199.17 ± 4.90 b	258.39 ± 0.42 a	331.79 ± 8.11 a
E10	225.76 ± 17.76 b	271.89 ± 9.65 a	155.60 ± 2.69 c	190.83 ± 9.70 b	260.59 ± 8.33 a	338.33 ± 14.65 a

Different lowercase letters in the same column show significant differences ($p < 0.05$).

The nitrogen and phosphorus ratio of gramineous plants did not show significant differences among different exclosure years ($p > 0.05$). However, there was a significant difference in the ratio of nitrogen to phosphorus in Cyperaceae plants ($p < 0.05$). Among the treatments, E4 had the lowest ratio that was significantly lower than the other treatments. In the case of legumes, both E10 and E4 had significantly lower ratios than E1 and G. Additionally, E10 had a significantly lower ratio than E4, while E1 had a significantly higher ratio than G. On the other hand, there was no significant difference in the nitrogen to phosphorus ratio among the belowground parts of forbs, plant litters, and plant communities (Table 6).

Table 6. Ratios of nitrogen to phosphorus in different functional groups of plants.

Exclosure Duration	Grasses	Sedges	Legumes	Forbs	Grasses	Root
G	7.87 ± 0.46 a	9.62 ± 0.59 a	13.80 ± 0.25 b	7.74 ± 0.32 a	9.17 ± 0.30 a	8.82 ± 0.63 a
E1	9.00 ± 0.52 a	9.71 ± 0.30 a	15.02 ± 0.48 a	8.39 ± 0.22 a	10.04 ± 0.43 a	9.09 ± 0.33 a
E4	8.15 ± 0.16 a	8.14 ± 0.05 b	12.66 ± 0.48 c	7.43 ± 0.18 a	9.14 ± 0.13 a	8.55 ± 0.21 a
E10	9.12 ± 0.66 a	10.02 ± 0.54 a	8.20 ± 0.13 d	7.91 ± 0.58 a	9.76 ± 0.39 a	9.72 ± 0.63 a

Different lowercase letters in the same column show significant differences ($p < 0.05$).

3.3. *Correlation between Plant Carbon, Nitrogen, and Phosphorus Content and Biomass*

The correlation analysis of the carbon, nitrogen, and phosphorus contents and aboveground and belowground biomass of plants across different exclosure years revealed significant correlations between the carbon content of the plant community and the aboveground biomass of gramineous plants, aboveground biomass of miscellaneous plants, plant litter, and underground biomass of the plant community. The nitrogen content of the plant community is significantly related to the aboveground biomass of Cyperaceae, aboveground biomass of miscellaneous plants, plant litter, and underground biomass of the plant community. The phosphorus content of the plant community is significantly related to the aboveground biomass of Cyperaceae plants, aboveground biomass of miscellaneous plants, plant litter, and underground biomass of the plant community (Figure 6).

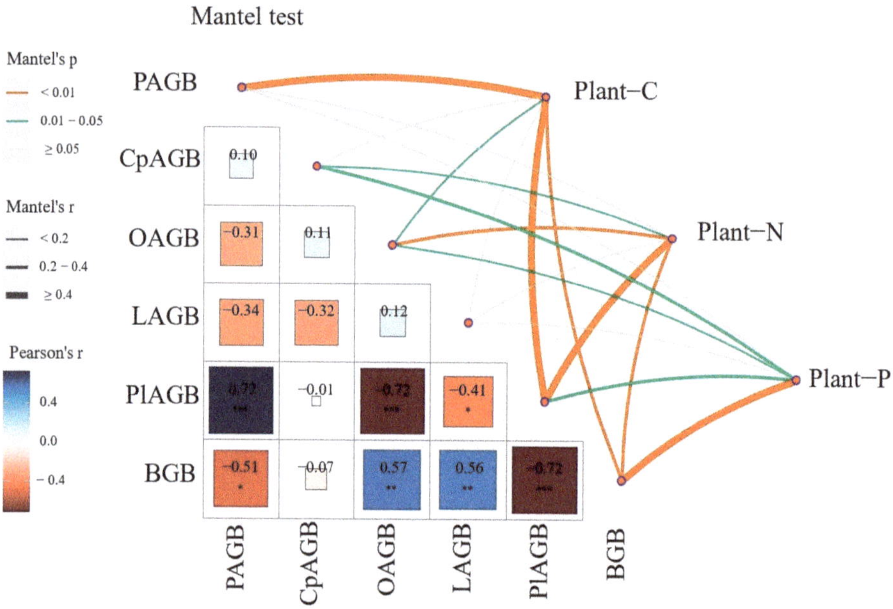

Figure 6. Correlation between plant carbon, nitrogen, and phosphorus contents and aboveground and belowground biomass with different exclosure years. PAGB, CpAGB, LAGB, OAGB, and PlAGB represent the aboveground biomass of grasses, sedges, legumes, forbs, and plant litters, respectively. BGB represents belowground biomass. * indicates $p < 0.05$, ** indicates $p < 0.01$, *** indicates $p < 0.001$.

3.4. Relationship between Plant Carbon, Nitrogen, Phosphorus Content and the Environment

The RDA analysis of soil physical and chemical properties (the data for soil physical and chemical properties can be found in Appendix A) revealed significant positive correlations between the total nitrogen and total phosphorus contents of gramineous plants and soil pH, available phosphorus (AP), organic carbon (SOC), water content (SWC), and total nitrogen (TN). On the other hand, the total carbon content showed a significant negative correlation with soil pH, AP, SOC, SWC, and TN. Additionally, the total phosphorus content of sedge plants exhibited a significant positive correlation with soil AP, while the total carbon and total nitrogen contents bore a significant negative correlation with soil AP. Moreover, the total carbon and total nitrogen contents of legumes were significantly and positively correlated with SWC and TN, whereas the total phosphorus content showed a significant negative correlation with SWC and TN. Significant correlations also existed between the total nitrogen and total phosphorus contents of forbs and soil AP, pH, SWC, and SOC. However, the total carbon content showed a significant negative correlation with soil AP, pH, SWC, and SOC. Additionally, significant negative correlations were observed between the contents of total carbon, total nitrogen, and total phosphorus in plant litter and soil pH, AP, SOC, and SWC. On the other hand, the contents of total carbon, total nitrogen, and total phosphorus in plant roots showed a significant positive correlation with soil SWC, TN, and TP. The biomass of gramineae, Cyperaceae, and plant litter was significantly and positively correlated with soil SWC, pH, TN, SOC, and AP. Conversely, the biomass of leguminosae, forbs, and plant roots showed a significant negative correlation with soil SWC, pH, TN, SOC, and AP. Overall, the response of carbon, nitrogen, and phosphorus in each functional group of plants to soil physical and chemical properties varied with the exclosure duration. However, there was a significant correlation with soil pH, organic carbon content, water content, total nitrogen content, and available phosphorus content (Figure 7).

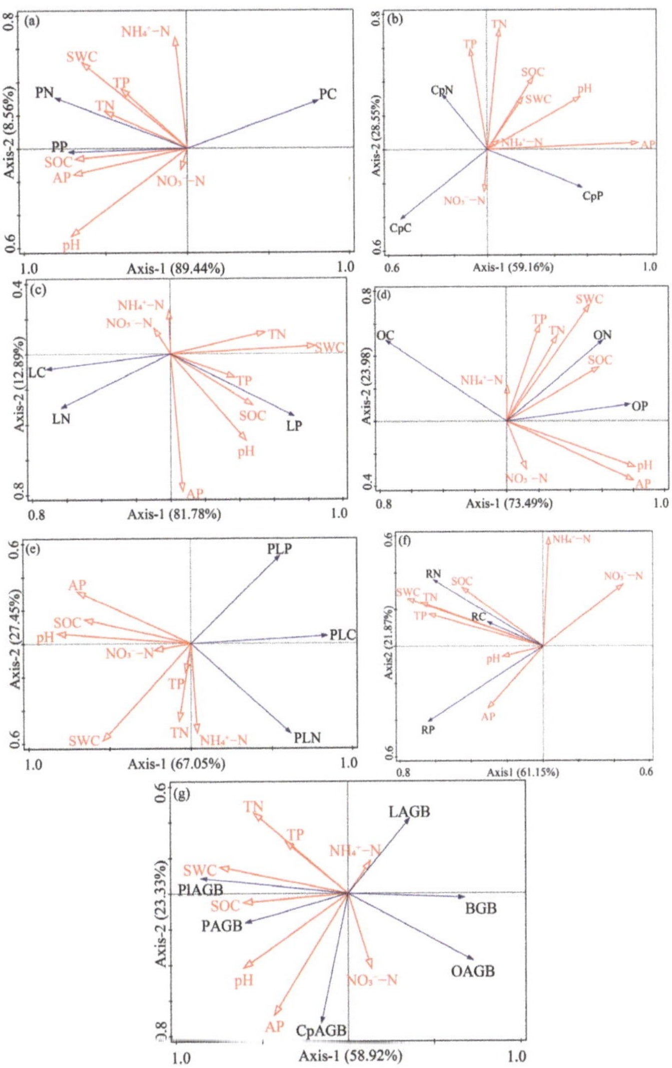

Figure 7. RDA analysis of soil physical and chemical properties and vegetation characteristics at different exclosure durations. Figure (**a–f**) represent the RDA analysis of carbon, nitrogen and phosphorus content and soil physicochemical properties of grasses, sedges, legumes, forbs, plant litter and root, respectively, and (**g**) represents the RDA analysis of soil physicochemical properties of plant biomass. The carbon, nitrogen, and phosphorus contents of gramineous plants are represented by PC, PN, and PP, respectively. Similarly, the carbon, nitrogen, and phosphorus contents of Cyperaceae plants are represented by CpC, CpN, and CpP, respectively. The carbon, nitrogen, and phosphorus contents of legumes are represented by LC, LN, and LP, respectively. The contents of carbon, nitrogen, and phosphorus in miscellaneous plants are represented by OC, ON, and OP, respectively. PAGB, CpAGB, LAGB, OAGB, and PlAGB represent the aboveground biomass of grasses, sedges, legumes, forbs, and plant litters, respectively. BGB represents belowground biomass. The contents of carbon, nitrogen, and phosphorus in plant litter are represented by plC, PlN, and PlP, respectively. RC, RN, and RP represent the content of carbon, nitrogen, and phosphorus in plant roots, respectively. NH_4^+−N represents ammonium nitrogen content, and NO_3^-−N represents nitrate nitrogen content.

4. Discussion

4.1. Effects of Long-Term Exclosure on the Diversity and Biomass of Main Plant Functional Groups

The restoration effect of fencing exclosure on the degraded alpine meadow primarily manifests in an increase in aboveground biomass, indicating cumulative productivity. However, the impact of long-term exclosure on plant diversity is not significant. Zhu et al. [28] conducted a study in the Three Parallel Rivers region of Tibet, and revealed that a 2-year short-term exclosure led to an increase in both aboveground and belowground biomass, but resulted in a reduced plant diversity. Similarly, Fenetahun et al. [29] found that exclosure in southern Ethiopia significantly increased aboveground vegetation biomass, but led to a decrease in plant diversity as the exclosure time extended. In our study, the plant diversity of E10 was significantly lower than that of E4, E1, and G. This can be attributed to the disturbance caused by livestock grazing in the grazed grassland of G, which weakens the competitiveness of dominant species from the Gramineae and Cyperaceae families, known for their favorable palatability in the alpine meadow plant community. This disturbance also promotes forbs and interferes with the formation of various habitat patches, resulting in the coexistence of populations at different succession stages. Long-term exclosure of grassland eliminates livestock interference, promotes rapid growth of palatable grasses and sedge plants, and creates a competitive advantage. However, this also leads to a decrease in biodiversity and a tendency towards uniformity in the habitat. Numerous studies have substantiated these findings. For instance, in a series of exclosure experiments conducted over 6, 11, and 12 years on the alpine grassland of the Qinghai–Tibet Plateau, Li et al. [30] observed that exclosure significantly reduced the Shannon–Wiener diversity index and Margalef richness index of vegetation. Similarly, research conducted in Arizona, southwestern United States, demonstrated that grazing exclusion resulted in a decrease in native species and reduced the overall species richness compared to moderate grazing [31]. Ebrahimi et al. [32] found that after 2, 4, and 6 years of fencing on dry pastures in southeastern Iran, the plant species richness was the highest after 6 years, due possibly to increased interspecific competition for resources such as light and/or nutrients. In this study, the richness index and diversity index of degraded alpine meadow plants after 4 years of exclosure were found to be the highest and significantly higher than those of E10. This can be attributed to the significant enhancement of Gramineae, a dominant species, in the fenced plots. As a result, the proportion of Gramineae biomass increased significantly with prolonged exclosure. The alpine meadow grasses, such as *Elymus dahuricus* and *Poa pratensis*, not only exhibit a strong tillering ability but also occupy the upper canopy, inhibiting the growth of sedges, legumes, and forbs. This finding is consistent with the significant decrease in the biomass of sedges observed in E10 in this study. Consequently, the lack of light and nutrients leads to a decrease in or even disappearance of some dwarf plants with weak competitiveness in the lower layer of the plant community. This ultimately results in a decrease in plant community diversity in alpine meadows [33].

Mechanistically, exclosure has both promoting and inhibiting effects on grassland productivity [34]. In this study, the aboveground biomass of gramineous plants increased with the increase in exclosure years, while the aboveground biomass of forbs decreased. Significant differences were observed between E10 and E4, as well as between E1 and G. These differences can be attributed to the low forage productivity of the degraded grassland induced by overgrazing prior to fencing. Exclosure of the degraded grassland allows the recovery of the original palatable grass and sedge grass communities, leading to short-term compensatory growth, seed germination, seedling recruitment, and bud tillering, ultimately improving grassland productivity [35]. E10 of sedge plants is lower than E4 due to the super compensatory growth mechanism of forage grass in the absence of interference from livestock. This is because a large amount of litter is decomposed at a slower rate than the turnover rate of plant production, which affects the efficiency of resource utilization [36]. This finding aligns with the observation that the biomass of plant litter of E10 is the highest, significantly surpassing E4, E1, and G in this study. Furthermore, E4 is significantly higher than E1 and G. The productivity of a community can also be influenced by changes in plant functional groups during the succession of enclosed communities. Generally, the climax community of restoration succession is more productive than the degraded community [37]. Therefore, the response of plants to exclosure is determined with the net effect of promotion and inhibition, which is closely tied to environmental conditions and grassland management strategies.

4.2. Effects of Long-Term Exclosure on Carbon, Nitrogen, and Phosphorus Contents of Plants

Carbon (C), nitrogen (N), and phosphorus (P) are fundamental elements that make up plants and influence their growth and development [38]. Carbon serves as the structural element of plants, forming the basis of all organic macromolecules. Nitrogen is the primary component of biological macromolecules such as proteins, while phosphorus is essential for the formation of genetic materials such as nucleic acids [39]. The ecological stoichiometric ratios of C, N, and P vary with the growing environment. In this study, except for forbs, the total carbon content of all functional groups was significantly lower in 10-year long-term and 4-year medium-term exclosures than 1-year exclosure and free grazing. This difference may be attributed to the compensatory growth of plants with high palatability that exhibit a strong carbon assimilation ability under the influence of livestock grazing and trampling. Consequently, the carbon content of grasses, sedges, and leguminous plants in the grazed grasslands is higher than in the long-term enclosed grassland. There were significant variations in the total nitrogen content among different plant families, namely Gramineae, Cyperaceae, and forbs. Specifically, the total nitrogen content of E10 was significantly higher than that of E4, E1, and G. Conversely, the total nitrogen content of legumes showed an opposite trend. On the one hand, the total nitrogen content of perennial legumes in long-term exclosure was lower than that in 1-year exclosure and free grazing treatments due to the decreased nitrogen availability. On the other hand, long-term enclosed perennial grasses and sedges were the dominant species in the secondary succession of alpine meadow vegetation communities. These species were at their peak growth stage, with young leaves comprising a relatively high proportion of the aboveground biomass, resulting in a high total nitrogen content. The total phosphorus content of each plant functional group was significantly higher in the 10-year long-term exclosure and 4-year medium-term exclosure than the 1-year exclosure and free grazing treatment. The exclusion of livestock in the alpine meadow vegetation community owing to long-term exclosure resulted in the formation of more seeds or reproductive branches. Additionally, the reproductive organs of the plants exhibit a higher phosphorus content.

4.3. Effects of Long-Term Exclosure on Plant Stoichiometric Characteristics

The ecological stoichiometric characteristics of carbon, nitrogen, and phosphorus vary among different plant communities in various grassland management schemes [40]. The carbon–nitrogen ratio and carbon–phosphorus ratio of Gramineae and Cyperaceae in alpine meadow communities indicate the carbon assimilation capacity of plants, which largely reflects their nutrient utilization efficiency [41]. In this study, except for Cyperaceae plants, the C/N ratio was significantly lower in long-term exclosure (E10) and medium-term exclosure (E4) than the control (G) and 1-year exclosure (E1). This can be attributed to the slow growth and low nutrient utilization efficiency of the plant community in long-term exclosure, while the vegetation community in free grazing and short-term exclosure exhibits opposite characteristics. The carbon–phosphorus ratio and carbon–nitrogen ratio were similar, indicating that free grazing and short-term exclosure have a strong vegetation carbon assimilation ability and a higher nutrient utilization efficiency than long-term exclosure. The content of the N/P ratio in plants remained relatively stable, except for Cyperaceae plants in E4 plots and Leguminosae plants in E10 and E4 treatments. In these cases, the N/P ratio was significantly lower than the G and E1 treatments. However, there were no significant differences in the N/P ratio among the other treatments, which aligns with the findings of Yu et al. [42].

4.4. Relationship between Enclosed Grassland and Environmental Factors

Exclosure has been found to have a significant rehabilitative effect on degraded grassland soil, with plant recovery following soil recovery [43]. Alpine meadow vegetation communities have the ability to influence soil hydrological processes and surface soil temperature through the input of light compounds and organic matter. Additionally, they can impact soil properties by providing habitats and resources for microscopic and macroscopic organisms [40–45]. However, changes in biotic and abiotic soil properties caused by vegetation may have counterproductive effects on aboveground vegetation biomass, which relies on soil to grow and develop [46]. Plant carbon (C), nitrogen (N), and phosphorus (P) content, as well as their stoichiometric ratios, vary with the type and content of N and P in the soil under different soil nutrient conditions [47]. In this study, the physical and chemical properties of enclosed soil such as total carbon, total nitrogen, and total phosphorus contents of Gramineous plants, sedge plants, Leguminous plants, forbs, plant litter, and plant roots did not respond to exclosure duration in a completely consistent manner. However, their responses were significantly correlated with soil pH, organic carbon content, water content, total nitrogen content, and available phosphorus content. The relationship between soil pH and nitrogen and phosphorus contents in Gramineous plants, forbs, and plant litter is significantly positive. A moderate increase in soil pH can enhance soil microbial activity, promote the decomposition of active organic substances in soil, and provide more nutrients such as nitrogen and phosphorus for plant growth. Additionally, maintaining a suitably acidic environment supports soil parent material development, improves soil fertility, and facilitates nutrient absorption by plants [48]. The nitrogen and phosphorus contents of Gramineous plants, forbs, and plant litters show a significant positive correlation with soil organic carbon content. This could be attributed to the impact of grazing and exclosure measures on Gramineous plants and forbs, as their roots respond to changes in the growing environment. Rapidly adapted growth strategies contribute to a substantial increase in organic carbon in the rhizosphere soil [49]. Previous studies have shown that soil water content plays a significant role in determining the nitrogen and phosphorus contents of various plant species. This finding is consistent with other findings obtained in ecologically fragile areas such as the Qinghai–Tibet Plateau, where soil water content is considered a key factor influencing plant nutrient levels [50]. Additionally, a strong correlation exists between the soil total nitrogen content and the total carbon, nitrogen, and phosphorus contents of the aboveground part. This correlation can be attributed to the direct influence of soil total nitrogen content on nutrient availability, soil and water conservation, aboveground and belowground biomass of vegetation, as well as the carbon, nitrogen, and phosphorus contents [51].

5. Conclusions

Short-term exclosure for a period up to 4 years has been found to have a significant positive impact on the biomass and diversity of the vegetation community in the patchily degraded alpine meadows in the study area. The 10-year long-term exclosure significantly reduces the species diversity of alpine meadows, which will lead to intensified interspecific competition, reduced ecological niche complementarity, and reduced resource use efficiency by species. From the perspective of functional groups, the biomass of grasses gradually increased, the aboveground biomass of sedge and legumes showed a trend of first increasing and then decreasing, and the biomass of forbs gradually decreased. Exclosure regulates the net primary productivity of a community by changing the proportion of different functional groups of plants in the community. On the other hand, long-term exclosure has been observed to enhance the storage of nitrogen and phosphorus and improve the nutrient utilization efficiency of gramineous and sedge plants. However, it also leads to a reduction in plant diversity within the alpine meadow vegetation community, ultimately compromising its stability and hindering the sustainable development of the degraded alpine meadow. To ensure the sustainability of such restored meadows, it is recommended to implement moderate grazing or mowing after a period of 4-year exclosure. More research is needed to assess whether this period can be further extended by a few more years.

Author Contributions: Conceptualization, P.Y.; Methodology, P.Y.; Software, P.Y.; Validation, X.L. and C.L.; Formal analysis, P.Y., C.L. and J.Z.; Resources, C.L.; Data curation, P.Y.; Writing—original draft, P.Y.; Writing—review & editing, P.Y., X.L. and J.Z.; Project administration, X.L.; Funding acquisition, X.L. All authors have read and agreed to the published version of the manuscript.

Funding: This study was financially supported by the Qinghai Science and Technology Department (Grant No. 2023-QY-210), the National Natural Science Foundation of China (U21A20191), the 111 Project of China (D18013), and Qinghai Science and Technology Innovation and Entrepreneurship Team Project titled 'Sanjiangyuan Ecological Evolution and Management Innovation Team' and the project of ecosystem succession and management direction in the world-class discipline of ecology at Qinghai University.

Data Availability Statement: The data presented in this study are available on request from the corresponding author. The data are not publicly available due to privacy.

Conflicts of Interest: The authors declare no conflict of interest.

Appendix A

Table A1. Differences in soil physical and chemical properties with different exclosure years [52].

Exclosure Duration	pH	SWC/%	ω (SOC)/g·kg^{-1}	ω (TN)/g·kg^{-1}	ω (TP)/g·kg^{-1}	ω (NH_4^+−N)/ mg·kg^{-1}	ω (NO_3^-−N)/ mg·kg^{-1}	ω (AP)/mg·kg^{-1}
G	6.46 ± 0.06 c	18.91 ± 0.01 b	93.38 ± 8.63 ab	3.85 ± 0.07 c	0.67 ± 0.01 b	9.00 ± 0.79 a	17.14 ± 1.14 ab	7.12 ± 0.38 b
E1	6.30 ± 0.06 d	21.65 ± 0.02 b	83.56 ± 9.52 b	4.37 ± 0.10 ab	0.75 ± 0.03 a	9.65 ± 0.93 a	14.78 ± 0.76 b	7.05 ± 0.18 b
E4	6.91 ± 0.06 a	21.88 ± 0.01 b	115.01 ± 0.60 a	4.00 ± 0.24 bc	0.72 ± 0.01 ab	9.17 ± 0.49 a	18.65 ± 0.05 a	9.51 ± 0.47 a
E10	6.68 ± 0.03 b	26.15 ± 0.01 a	116.75 ± 10.09 a	4.66 ± 0.15 a	0.78 ± 0.03 a	8.86 ± 0.26 a	14.95 ± 0.90 b	7.84 ± 0.43 b

Different lowercase letters in the same column show significant differences ($p < 0.05$).

References

1. Fu, B.; Li, S.; Yu, X.; Yang, P.; Yu, G.; Feng, R.; Zhuang, X. Chinese ecosystem research network: Progress and perspectives. *Ecol. Complex.* **2010**, *7*, 225–233. [CrossRef]
2. Bardgett, R.D.; Bullock, J.M.; Lavorel, S.; Manning, P.; Schaffner, U.; Ostle, N.; Chomel, M.; Durigan, G.; Fry, E.L.; Johnson, D.; et al. Combatting global grassland degradation. *Nat. Rev. Earth Environ.* **2021**, *2*, 720–735. [CrossRef]
3. Shao, Q.; Cao, W.; Fan, J.; Huang, L.; Xu, X. Effects of an ecological conservation and restoration project in the Three-River Source Region, China. *J. Geogr.* **2017**, *27*, 183–204. [CrossRef]
4. Zhang, L.; Xiao, P.; Yu, H.; Zhao, T.; Liu, S.; Yang, L.; He, Y.; Luo, Y.L.; Wang, X.; Dong, W.; et al. Effects of Climate Changes on the Pasture Productivity From 1961 to 2016 in Sichuan Yellow River Source, Qinghai-Tibet Plateau, China. Front. *Ecol. Evol.* **2022**, *10*, 908924. [CrossRef]
5. Liu, L.; Zhang, Y.; Bai, W.; Yan, J.; Ding, M.; Shen, Z.; Li, S.; Zheng, D. Characteristics of grassland degradation and driving forces in the source region of the Yellow River from 1985 to 2000. *J. Geogr.* **2006**, *16*, 131–142. [CrossRef]
6. Dong, S.C.; Zhou, C.J.; Wang, H.Y. Ecological crisis and countermeasures of the Three Rivers' Headstream Regions. *J. Nat. Resour.* **2002**, *17*, 713–720.
7. Martin, D.M. Ecological restoration should be redefined for the twenty-first century. *Restor. Ecol.* **2017**, *25*, 668–673. [CrossRef]
8. Ren, Y.; Lü, Y.; Fu, B. Quantifying the impacts of grassland restoration on biodiversity and ecosystem services in China: A meta-analysis. *Ecol. Eng.* **2016**, *95*, 542–550. [CrossRef]
9. Zhang, Z.H.; Zhou, H.K.; Zhao, X.Q.; Yao, B.Q.; Ma, Z.; Dong, Q.M.; Dong, Q.; Zhang, Z.H.; Wang, W.Y.; Yang, Y.W. Relationship between biodiversity and ecosystem functioning in alpine meadows of the Qinghai-Tibet Plateau. *Biodivers. Sci.* **2018**, *26*, 111. [CrossRef]
10. Xue, Y.F.; Zong, N.; He, N.P.; Tian, J.; Zhang, Y.Q. Influence of long-term enclosure and free grazing on soil microbial community structure and carbon metabolic diversity of alpine meadow. *J. Appl. Ecol.* **2018**, *29*, 2705–2712.
11. Jachowski, D.S.; Slotow, R.; Millspaugh, J.J. Good virtual fences make good neighbors: Opportunities for conservation. *Anim. Conserv.* **2014**, *17*, 187–196. [CrossRef]
12. Chen, Q.; Shang, Y.T.; Zhu, R.; Bao, Q.L.; Lin, S. Long-term enclosure at heavy grazing grassland affects soil nitrification via ammonia-oxidizing bacteria in Inner Mongolia. *Sci. Rep.* **2022**, *12*, 21464. [CrossRef]
13. Zhang, Z.W.; Han, J.H.; Yin, H.Y.; Xue, J.; Jia, L.Z.; Zhen, X.; Chang, J.J.; Wang, S.K.; Yu, B. Assessing the effects of different long-term ecological engineering enclosures on soil quality in an alpine desert grassland area. *Ecol. Indic.* **2022**, *143*, 109426. [CrossRef]

14. Shang, Z.H.; Cao, J.J.; Guo, R.Y.; Henkin, Z.; Ding, L.M.; Long, R.J.; Deng, B. Effect of enclosure on soil carbon, nitrogen and phosphorus of alpine desert rangeland. *Land Degrad. Dev.* **2017**, *28*, 1166–1177. [CrossRef]
15. Asadian, G.; Javadi, S.A.; Jafary, M.; Arzani, H.; Akbarzade, M. Relationships between environmental factors and plant communities in enclosure rangelands (case study: Gonbad, Hamadan). *J. Rangel. Sci.* **2016**, *7*, 20–34.
16. Bode, M.; Wintle, B. How to build an efficient conservation fence. *Conserv. Biol.* **2010**, *24*, 182–188. [CrossRef]
17. Zuo, W.Q.; Wang, Y.H.; Wang, F.Y.; Shi, G.X. Effects of enclosure on the community characteristics of Leymus chinensis in degenerated steppe. *Acta Prataculturae Sin.* **2009**, *18*, 12.
18. Yao, X.; Wu, J.; Gong, X.; Lang, X.; Wang, C.; Song, S.; Ahmad, A.A. Effects of long term fencing on biomass, coverage, density, biodiversity and nutritional values of vegetation community in an alpine meadow of the Qinghai-Tibet Plateau. *Ecol. Eng.* **2019**, *130*, 80–93. [CrossRef]
19. Babur, E.; Kara, O.; Fathi, R.A.; Susam, Y.E.; Riaz, M.; Arif, M.; Akhtar, K. Wattle fencing improved soil aggregate stability, organic carbon stocks and biochemical quality by restoring highly eroded mountain region soil. *J. Environ. Manag.* **2021**, *288*, 112489. [CrossRef]
20. Zhang, C.; Yan, R.R.; Liang, Q.W.; Na, R.S.; Li, T.; Yang, X.F.; Bao, Y.H.; Xin, X.P. Study on soil physical and chemical properties and carbon and nitrogen sequestration of grassland under different utilization modes. *Acta Prataculturae Sin.* **2021**, *30*, 90–98.
21. Wu, G.L.; Du, G.Z.; Liu, Z.H.; Thirgood, S. Effect of fencing and grazing on a Kobresia-dominated meadow in the Qinghai-Tibetan Plateau. *Plant Soil* **2009**, *319*, 115–126. [CrossRef]
22. Xu, L.; Nie, Y.; Chen, B.; Xin, X.; Yang, G.; Xu, D.; Ye, L. Effects of fence enclosure on vegetation community characteristics and productivity of a degraded temperate meadow steppe in northern China. *Appl. Sci.* **2020**, *10*, 2952. [CrossRef]
23. Astaiken, G.; Dong, Y.Q.; Zhou, S.J.; Nie, T.T.; Jiang, A.J.; An, S.Z. Effects of enclosure on plant community diversity and niche characteristics of different grassland types-Taking different types of grassland in Xinjiang as an example. *Pratacultural Sci.* **2023**, *40*, 1168–1185.
24. Smith, D.; King, R.; Allen, B.L. Impacts of exclusion fencing on target and non-target fauna: A global review. *Biol. Rev.* **2020**, *95*, 1590–1606. [CrossRef] [PubMed]
25. Boone, R.B.; Hobbs, N.T. Lines around fragments: Effects of fencing on large herbivores. *Afr. J. Range Forage Sci.* **2004**, *21*, 147–158. [CrossRef]
26. Lu, R.K. *Soil Agrochemical Analysis Method*; China Agricultural Science and Technology Press: Beijing, China, 2000; pp. 266–292.
27. Nie, Y.Y.; Xu, L.J.; Xin, X.P.; Chen, B.R.; Zhang, B.H. Effects of fence enclosure on the plant community composition and niche characteristics in a temperate meadow steppe. *Acta Prataculturae Sin.* **2020**, *29*, 11–22.
28. Zhu, N.; Sun, J.; Shi, N.; Wang, J.N.; Zhang, L.; Luo, D.L.; Shen, C.; Gai, A.H. Effects of Short-term Fence Enclosing on Plant Community and the Physical and Chemica Properties of Alpine Meadow Soils. *Acta Agrestia Sin.* **2023**, *31*, 834–843.
29. Fenetahun, Y.; Yuan, Y.; Xinwen, X.; Yongdong, W. Effects of grazing enclosures on species diversity, phenology, biomass, and carrying capacity in Borana Rangeland, Southern Ethiopia. *Front. Ecol. Evol.* **2021**, *8*, 623627. [CrossRef]
30. Li, R.F.; Niu, H.S.; Kong, Q.; Liu, Q. Effects of exclosure on plant and soil nutrients in an alpine grassland. *Pratacultural Sci.* **2021**, *38*, 399–409.
31. Loeser, M.R.R.; Sisk, T.D.; Crews, T.E. Impact of grazing intensity during drought in an Arizona grassland. *Conserv. Biol.* **2007**, *21*, 87–97. [CrossRef]
32. Ebrahimi, M.; Khosravi, H.; Rigi, M. Short-term grazing exclusion from heavy livestock rangelands affects vegetation cover and soil properties in natural ecosystems of southeastern Iran. *Ecol. Eng.* **2016**, *95*, 10–18. [CrossRef]
33. Inderjit. Plant invasions: Habitat invasibility and dominance of invasive plant species. *Plant Soil* **2005**, *277*, 1–5. [CrossRef]
34. Liu, M.; Liu, G.; Wu, X.; Wang, H.; Chen, L. Vegetation traits and soil properties in response to utilization patterns of grassland in Hulun Buir City, Inner Mongolia, China. *Chin. Geogr. Sci.* **2014**, *24*, 471–478. [CrossRef]
35. Yuan, J.; Li, H.; Yang, Y. The compensatory tillering in the forage grass Hordeum brevisubulatum after simulated grazing of different severity. *Front. Plant Sci.* **2020**, *11*, 792. [CrossRef] [PubMed]
36. Qin, R.; Wang, X. Effects of crown height on the compensatory growth of Italian ryegrass based on combined effects of stored organic matter and cytokinin. *Grassl. Sci.* **2020**, *66*, 29–39. [CrossRef]
37. Wang, C.T.; Long, R.J.; Wang, Q.L.; Jing, Z.C.; Shi, J.J. Changes in plant diversity, biomass and soil C, in alpine meadows at different degradation stages in the headwater region of three rivers, China. *Land Degrad. Dev.* **2009**, *20*, 187–198. [CrossRef]
38. Zechmeister-Boltenstern, S.; Keiblinger, K.M.; Mooshammer, M.; Peñuelas, J.; Richter, A.; Sardans, J.; Wanek, W. The application of ecological stoichiometry to plant–microbial–soil organic matter transformations. *Ecol. Monogr.* **2015**, *85*, 133–155. [CrossRef]
39. Malhotra, H.; Vandana; Sharma, S.; Pandey, R. Phosphorus nutrition: Plant growth in response to deficiency and excess. In *Plant Nutrients and Abiotic Stress Tolerance*; Hasanuzzaman, M., Fujita, M., Oku, H., Nahar, K., Hawrylak-Nowak, B., Eds.; Springer: Singapore, 2018; pp. 171–190.
40. Elser, J.J.; Fagan, W.F.; Kerkhoff, A.J.; Swenson, N.G.; Enquist, B.J. Biological stoichiometry of plant production: Metabolism, scaling and ecological response to global change. *New Phytol.* **2010**, *186*, 593–608. [CrossRef]
41. Song, S.; Wang, X.; He, C.; Chi, Y. Effects of Utilization Methods on C, N, P Rate and Enzyme Activity of Artificial Grassland in Karst Desertification Area. *Agronomy* **2023**, *13*, 1368. [CrossRef]
42. Yu, Q.; Chen, Q.S.; Elser, J.J.; He, N.P.; Wu, H.H.; Zhang, G.M.; Wu, J.G.; Bai, Y.F.; Han, X.G. Linking stoichiometric homoeostasis with ecosystem structure, functioning and stability. *Ecol. Lett.* **2010**, *13*, 1390–1399. [CrossRef]

43. Cheng, J.; Jing, G.; Wei, L.; Jing, Z. Long-term grazing exclusion effects on vegetation characteristics, soil properties and bacterial communities in the semi-arid grasslands of China. *Ecol. Eng.* **2016**, *97*, 170–178. [CrossRef]
44. Ayres, E.; Steltzer, H.; Simmons, B.L.; Simpson, R.T.; Steinweg, J.M.; Wallenstein, M.D.; Mellor, N.; Parton, W.J.; Moor, J.C.; Wall, D.H. Home-field advantage accelerates leaf litter decomposition in forests. *Soil Biol. Biochem.* **2009**, *41*, 606–610. [CrossRef]
45. Wardle, D.A.; Bardgett, R.D.; Klironomos, J.N.; Setala, H.; Van Der Putten, W.H.; Wall, D.H. Ecological linkages between aboveground and belowground biota. *Science* **2004**, *304*, 1629–1633. [CrossRef] [PubMed]
46. Van der Putten, W.H.; Bardgett, R.D.; Bever, J.D.; Bezemer, T.M.; Casper, B.B.; Fukami, T.; Wardle, D.A. Plant–soil feedbacks: The past, the present and future challenges. *J. Ecol.* **2013**, *101*, 265–276. [CrossRef]
47. McGroddy, M.E.; Daufresne, T.; Hedin, L.O. Scaling of C:N:P stoichiometry in forests worldwide: Implications of terrestrial redfield-type ratios. *Ecology* **2004**, *85*, 2390–2401. [CrossRef]
48. Yu, Y.C.; Yang, J.Y.; Zeng, S.C.; Wu, D.M.; Jacobs, D.F.; Sloan, J.L. Soil pH, organic matter, and nutrient content change with the continuous cropping of Cunninghamia lanceolata plantations in South China. *J. Soils Sediments* **2017**, *17*, 2230–2238. [CrossRef]
49. Filser, J.; Faber, J.H.; Tiunov, A.V.; Brussaard, L.; Frouz, J.; De Deyn, G.; Uvarov, A.V.; Berg, M.P.; Lavelle, P.; Loreau, M.; et al. Soil fauna: Key to new carbon models. *Soil* **2016**, *2*, 565–582. [CrossRef]
50. Liu, T.Y.; Zhou, T.C.; Sun, J.; Wang, Y.; Ye, C.C. Distribution and coupling characteristics of plant nitrogen and phosphorus along desertification gradients in alpine meadows, eastern Tibet Plateau. *Prataculturál. Sci.* **2021**, *38*, 209–220.
51. Hati, K.M.; Swarup, A.; Mishra, B.; Manna, M.C.; Wanjari, R.H.; Mandal, K.G.; Misra, A.K. Impact of long-term application of fertilizer, manure and lime under intensive cropping on physical properties and organic carbon content of an Alfisol. *Geoderma* **2008**, *148*, 173–179. [CrossRef]
52. Yang, P.N.; Li, X.L.; Li, C.Y.; Duan, C.W. Response of Soil Microbial Diversity to Long-term Enclosure in Degraded Patches of Alpine Meadow in the Source Zone of the Yellow River. *Huan Jing Ke Xue = Huanjing Kexue* **2023**, *44*, 2293–2303.

agronomy

Article

Vegetation Characteristics of the Main Grassland Types in China Respond Differently to the Duration of Enclosure: A Meta-Analysis

Cheng Liu [1], Hui Li [1], Kesi Liu [1,2,3,*], Xinqing Shao [1,3], Jing Huang [1], Muji Siri [1], Changliang Feng [1] and Xiaomeng Yang [1]

1 College of Grassland Science and Technology, China Agricultural University, Beijing 100193, China
2 National Field Station of Grassland Ecosystem, Guyuan 076550, China
3 Key Laboratory of Restoration Ecology of Cold Area in Qinghai Province, Northwest Institute of Plateau Biology, Chinese Academy of Sciences, Xining 810008, China
* Correspondence: kliu@cau.edu.cn

Abstract: Enclosure is one of the useful measures to protect and restore degraded grasslands, and it is widely used around the world. The vegetation characteristics of grasslands directly reflect the recovery status of degraded grasslands; however, conflicting results of plant traits were continually achieved in the numerous on-site studies of enclosure in the last two decades. It is necessary to conduct a systematic assessment to find a general conclusion for the effects of enclosure on different grasslands. Studies on the enclosure grasslands in China were taken as the objects to refine the relationships between grassland vegetation characteristics and enclosure measures using meta-analysis. Enclosure had positive effects on the restoration of vegetation coverage, aboveground and belowground biomass, and diversity of degraded grasslands. Different vegetation characteristics and grassland types showed different responses to enclosure duration. The vegetation productivity reached a maximum in the 11–15 years of enclosure for alpine grasslands and typical steppe grasslands, 6–10 years for desert grasslands, and more than 15 years of enclosure for meadow grasslands. Plant species diversity reached the peak values when alpine grasslands and typical steppe grasslands were enclosed approximately 10 years, desert grasslands approximately 11–15 years, and meadow grasslands approximately 5 years. These results indicated that the management strategies of enclosed grasslands should be adjusted reasonably according to the types and the management objectives of grasslands in order to maintain or even improve the condition and services of grassland ecosystems.

Keywords: degraded grassland; coverage; biomass; diversity; growth rate

Citation: Liu, C.; Li, H.; Liu, K.; Shao, X.; Huang, J.; Siri, M.; Feng, C.; Yang, X. Vegetation Characteristics of the Main Grassland Types in China Respond Differently to the Duration of Enclosure: A Meta-Analysis. *Agronomy* **2023**, *13*, 854. https://doi.org/10.3390/agronomy13030854

Academic Editor: Fujiang Hou

Received: 1 February 2023
Revised: 9 March 2023
Accepted: 10 March 2023
Published: 15 March 2023

1. Introduction

Grasslands, covering approximately 40% of the global terrestrial area [1], not only provide food production but also play essential roles in ecosystem diversity, carbon accumulation, and global climate change [2]. Unfortunately, large-scale climate change and excessive human activity disruption [3], such as livestock overgrazing, lead to severe degradation of grassland ecosystems, even desertification [4–7]. Grasslands in China account for close to 41% of the land area [8], and there are many types of grasslands, including alpine grasslands, desert grasslands, meadow grasslands, and typical steppe grasslands; however, approximately 90% of the grasslands were estimated to be degraded [9], resulting in the decline of grassland vegetation coverage, grassland vegetation productivity, plant diversity, and the deterioration of the entire system structure. For the sake of easing grassland degradation's negative influence and enhancing ecosystem functions and tolerance, many methods have been adopted, including plant rebuilding, reseeding, fertilizing, irrigation, and fencing enclosure [10,11]. Among these methods, enclosure by fence is a relatively low-cost measure and is currently commonly used in the restitution of degraded grasslands [12–14].

Enclosure can protect degraded grasslands from disturbance within a certain period of time and allow the natural restoration of degraded grasslands. Many studies have reported that enclosure has indeed reversed the negative effects and significantly improved vegetation composition, grassland productivity, soil structure, and soil organic matter accumulation [5,15,16]. Enclosing the severely degraded grassland can not only restore the degraded grassland to a certain extent but also avoid the re-degradation caused by non-use during the restoration process [17]. However, some studies have also shown that the enclosure has significant inhibitory effects on plant traits [18,19], as well as a drop in the number of vegetative species and species richness [20]; namely, the species richness of grasslands (5.4 m^{-2}) enclosed for 12 years compared with those grazed (16 m^{-2}) was significantly declined [21]. These controversial results reflect on grassland types and regions [22].

For instance, alpine grasslands' plant species richness and diversity reached the highest value after 5–7 years of enclosure [23,24], and the aboveground biomass (AGB) reached the highest value after 5–6 years of enclosure [25]. For typical steppe grasslands, species diversity decreased significantly in 6 years' enclosure [26], while a study noted that in an enclosure duration of 30 years, plant community diversity was still seen in higher levels [27]. Typical grasslands' vegetation coverage, biomass, and diversity increased significantly from 0 to 15 years and gradually decreased from 16 to 30 years [28]. Five-year duration in meadows, species abundance, and diversity have sustainable benefits [19], but in subalpine meadows of the Qinghai, enclosure had little sustainable benefit after four–five years [29,30]. In addition, others reported that the diversity of alpine meadows gradually decreased in the sixth year [31,32] or in the ninth year [33]. For desert grasslands, this fencing effect occurred only in the first 6 years [8,34]. These contradictory findings from on-site investigations impede our ability to make managerial decisions. Therefore, it is essential to make a comprehensive evaluation of the impact of duration enclosure on vegetation coverage, AGB, and plant diversity of different grassland types in different regions of China. In this study, we focus on the effect of vegetation coverage, biomass, and diversity, which can reflect plant photosynthetic capacity [35,36], energy flow, material circulation [37–41], and ecosystem stability and functions [42–45]. Specifically, we aimed to address the following questions: (1) Whether the enclosure of various types of grasslands resulted in a consistent response from the vegetation characteristics, and (2) How the enclosure duration affected the grassland vegetation characteristics. The policy of grassland enclosure has been implemented in China for several decades, and many scholars have conducted numerous studies related to enclosed grasslands in China, which can be used as good samples to elucidate these questions and achieve a general conclusion. Therefore, we considered the enclosed grasslands in China as the research object, collected the currently published articles on "enclosure" and "plant traits", and used meta-analysis to comprehensively explore the responses to enclosure of plant traits of four main grassland types. Through this study, corresponding theoretical support could be provided for the rational conservation and utilization of enclosed degraded grasslands.

2. Materials and Methods

2.1. Data Collection

This research targeted the responses to enclosure effects of four grassland types. To gather records that quantified the results of enclosure, we investigated peer-reviewed journal articles published from 1999 to 2021 using the Web of Science and China National Knowledge Infrastructure (CNKIT). The combinations of the following search term were used, along with "fence enclosure" OR "fence" OR "enclosure" OR "grazing exclusion" OR "grazing removal". Then, articles were refined by countries/regions "PEOPLES R CHINA". Some literature published as dissertations were also collected. More than 240 articles were preliminary obtained. To decrease the bias induced through the display literatures, the following criteria were used for the study selection:

(1) The study sites were grasslands in China, and the types of grassland could be judged through the text of the articles.
(2) The experiment was set up with the enclosure group and the grazing control group at the same time;
(3) There have been no different practices (e.g., fertilization or seeding) performed in the fenced sites;
(4) The grassland type, soil texture, and climate characteristics of the experimental group and the control group were the same;
(5) Each parameter in the chosen article had the records of common values, standard deviations or standard errors, and sample sizes;
(6) The means, standard deviations or standard errors, and pattern sizes of treatment and control have been immediately reported or could, in any other case, be determined from the chosen articles;
(7) Duration of enclosure was at least one year. When more than one article posted data from the same site, the contemporary publication with the most current data was given priority.

Finally, 91 eligible articles from specific study sites were chosen (Figure 1), and a list of date source used in the study are provided in the supplementary material section. The collected data related to vegetation characteristics included plant aboveground biomass (AGB), belowground biomass (BGB), plant coverage, and plant species diversity (Shannon–Wiener index). In addition, the information related to enclosure and grasslands was collected, including grassland types and the duration of enclosure. Data from tables were directly extracted, and data from figures were obtained using GetData Graph Digitizer 2.24. In this database, if the studies were in contrast regarding a number of controls (grazed plots with exceptional stocking rate) in opposition to an equal treatment, we calculated effect sizes for each assessment first and then pooled these effect sizes by using a separate meta-analysis. Furthermore, on the basis of the Chinese vegetation classification system, grassland was divided into four main types, including alpine grassland (MAP 100–500 mm; plants mainly consisting of *Kobresia pygmaea, Potentilla saundersiana, Poa crymophila*, and *Thymelaeaceae*), typical steppe grasslands (MAP 300–400 mm; plants mainly consisting of *Leymus chinesis, Stipa grandis, Artemisia sacrorum, and Thymus mongolicus*), meadow grassland (MAP > 400 mm; plants mainly consisting of *Kobresia humilis, Kobresia pygmaea, Poaceae, Asteraceae*, and *Farbaceae*), and desert grassland (MAP $\leq$ 200 mm; plants mainly consisting of *Asteraceae, Liliaceae*, and *Polygonaceae*). Duration of grazing exclusion was divided into five time intervals, namely $\leq$5 years, 6–10 years, 11–15 years, 16–20 years, and $\geq$20 years.

2.2. Data Analysis

The AGB effect of enclosure and vegetation coverage was calculated in the entire enclosed grasslands and the same grassland type to evaluate the responses to enclosure measures of plant traits of grassland ecosystems. The growth rate was calculated as follows:

$$\text{Growth rate} = \frac{(\text{values in the enclosed grassland} - \text{values in the control})}{\text{values in the control}}$$

To quantify the distinction of chosen variables between grazed and enclosed grasslands, the herbal log-transformed response ratio (ln*RR*) was used as the effect value to estimate the impact size [46,47]. The formula is as follows:

$$\ln RR = \ln\left(\frac{X_E}{X_C}\right) = \ln(X_E) - \ln(X_C)$$

where X_E and X_C are the average values of the indicators in the treatment (exclusion) and control (grazing) observations, respectively.

The variance (V) of each log response ratio's formula is as follows:

$$V_{\ln RR} = \frac{S_E^2}{N_E(X_E)^2} + \frac{Sc^2}{N_c(X_c)^2}$$

where S_E and S_C are the standard deviations of the treatment and the control, respectively; and N_E and N_C are the sample sizes of the treatment and the control, respectively. The lnRR is a standardized metric that approves assessment of data between treatment (exclusion) and control (grazing) in different units [46].

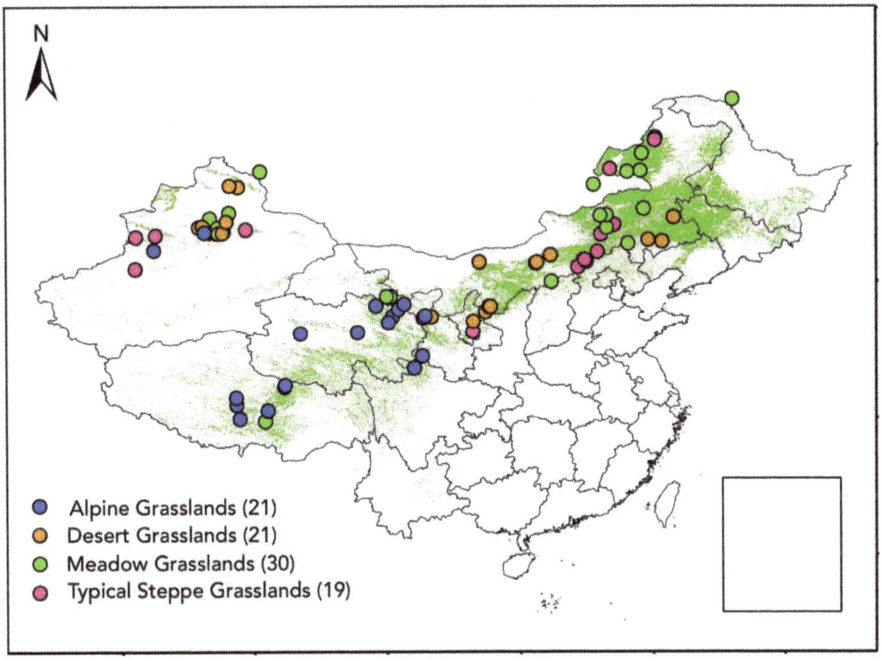

Figure 1. Study sites included in this meta-analysis (colored dots). Numbers represent the number of articles for the grassland types.

To better illustrate the effect of the treatment group on the indicator, the values of lnRR have been modified to estimate the proportion change in treatment and other variables relative to the control (%):

$$E = \left(e^{lnRR} - 1\right) \times 100\%$$

We used a random effect meta-analysis model to test the mean effect size for each study, which assumed that all perceived variation was due not only to sampling error but also to an authentic random error. Mean effect sizes and the 95% confidence intervals (CIs) were generated by a bootstrapping procedure based on 4999 iterations permutations using MetaWin 2.1 [48,49]. Using this method, the effects of vegetation characteristics to enclosures were considered as significant if the 95% CIs did not overlap zero, and significantly different from zero if the 95% CIs did not overlap zero. Microsoft Excel software was used for data processing, ArcGIS 10.2 software was used for data visualization, and figures were plotted in Sigmaplot.

3. Results

3.1. Effects of Enclosure on Vegetation Coverage

The growth rate of vegetation coverage across the enclosed grasslands was greater than 0% (Figure 2a) and showed varied dynamic changes at different enclosure duration. The coverage increased slightly in the short duration of the enclosure (less than five years), then reached a maximum at 10–15 year, approximately 80–100%, and then dropped sharply in the following enclosure years. This implies that the grassland coverage reached saturation in the 10–15 years following the enclosure and that the utilization rate of grassland resources is also the highest. For different types of grasslands (Figure 2b–e), vegetation coverage of alpine grassland did not significantly change in enclosure duration, always approximately 20–40% (Figure 2b). However, growth rate of desert and meadow grasslands both significantly increased with the duration of enclosure, and the growth rate of vegetation coverage after enclosure generally showed an upward trend year after year, with the growth rate being positively correlated with the enclosure period (Figure 2c,d). For typical steppe grasslands, when the enclosure period was less than ten years, the growth rate varied greatly, from 4.63% in the eighth year to 70.42% in the ninth year, but the overall growth rate did not exceed 100%. The highest growth rate occurred at the enclosure duration of 12 to 13 years, approximately 83.60–96.02%. When the enclosure period was more than 20 years, the growth rate of the coverage began to decrease (Figure 2e). This indicated that enclosure increased vegetation coverage regardless of grassland types, and the growth rate fluctuated with the grassland types and the duration of enclosure.

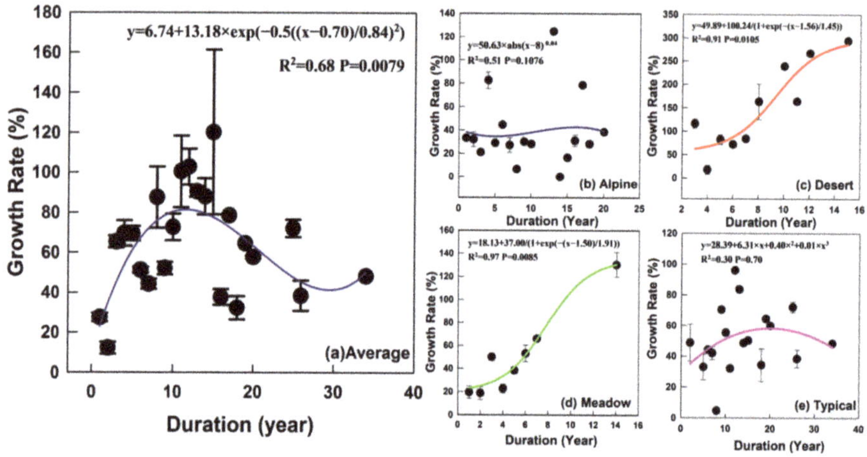

Figure 2. The growth rate of vegetation coverage across the enclosed grasslands (**a**) and in specific grassland types (**b**–**e**). Error bars are 95% bootstrapped confidence intervals.

The meta-analysis showed that the enclosure duration had noticeable effects on coverage of all grasslands (Figure 3). Vegetation coverage significantly increased during 1–15 years, and the highest effect value was 11–15 years; however, the coverage did not significantly change after the enclosure duration was 16–20 years (Figure 3a). To summarize, short-term enclosure can effectively increase vegetation coverage, whereas long-term enclosure is not conducive to increasing vegetation coverage and even has negative effects, and the best enclosure period is 11–15 years. However, regarding different grasslands, vegetation coverage had different responses to enclosure duration. Alpine grasslands' coverage significantly increased with the increase of enclosure duration (Figure 3b). The coverage of desert grasslands continuously increased in the first 1–10 years, then decreased, and the positive effect began to fade after the 15-year enclosure (Figure 3c). The mean effect size of meadow and typical steppe grasslands increased significantly in the enclosure years, and the vegetation coverage reached its highest level in 11–15 years (Figure 3d,e). As a result, it can be assumed that for a typical grassland, the best enclosure period is between 11 and 15 years. After that, as the length of the enclosure time is extended, the vegetation cover of the grassland may exhibit increasingly negative effects.

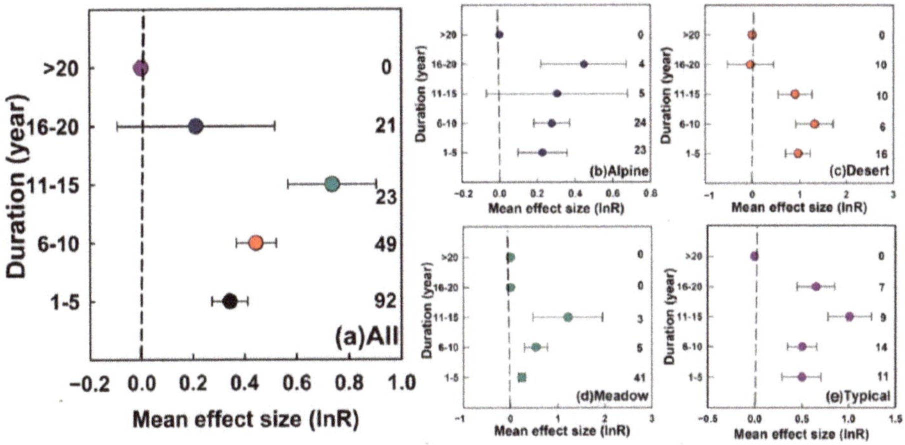

Figure 3. The effect value of vegetation coverage throughout different enclosure years (**a**), as well as the reaction of different types of grassland vegetation coverage to enclosure (**b**–**e**). The values are calculated effect sizes and 95% confidence intervals. Sample sizes are given to the right of the graph.

3.2. *Effects of Enclosure on Vegetation Biomass*

The average AGB effect of enclosure showed varied dynamic tendencies with enclosed duration, but the value of growth rate was more than 0% regardless of grassland type (Figure 4a). When the enclosure years exceeded 10, the aboveground biomass exhibited a clear upward trend, but as the enclosure time was continuously extended, the growth rate rose. When the years limit was less than 10, the growth rate ranged from 64.58 to 148.31%, with the maximum value emerging in the seventh year. At approximately 20 years of encirclement, it rapidly deteriorated, and the growth rate fell to 34.10%. The AGB effect of enclosure in the alpine grassland increased slightly in the first five years, then reached the maximum value of 376.46% in the 7th year. After the 10-year enclosure, the growth rate steeply decreased. The duration of enclosure increased, although not significantly (Figure 4b). The AGB effect of enclosure in desert grasslands showed some volatility, but the overall trend was upward along with increase of enclosure duration. The growth rate was 29.30% at 15 years of enclosure; it was 271.36% at 26 years (Figure 4c). For meadow grasslands, the AGB effect of enclosure had no obvious changes and retained a higher level when the duration of enclosure was within 10–15 years (Figure 4d). The growth rate of typical steppe grasslands' AGB reached the highest value in the 4th year, up to 249.26%; after that, the growth rate went down year by year, then quickly went up, and in the 15th year, it reached a high of 462.94% (Figure 4e).

AGB and BGB of all grasslands showed different responses to the enclosure effect (Figures 5 and 6). During the enclosure, the AGB increased, while the BGB had no significant change (Figure 6a). Under the enclosure period of 1–15 years, the response of aboveground biomass of vegetation to enclosure kept going up. During the enclosure period of 11–15 years, the effect of aboveground biomass was the largest value, and the positive effect was the strongest. The positive effect of sealing time on aboveground biomass was lessened (Figure 5a). In conclusion, as the enclosure period increased, the overall tendency of the influence on the above-ground biomass of grasslands was to first increase then weaken, and it was estimated that the maximum above-ground biomass of degraded grasslands would be reached in 11 to 15 years. Regarding different grasslands, alpine grasslands reached their peak value in 11–15 years, while the benefits of enclosure on AGB began to decrease in 16–20 years (Figure 5b). In contrast, their BGB had no significant change (Figure 6b). The enclosure of desert grasslands had a significant impact on AGB and BGB (Figures 5c and 6c). Desert grasslands' AGB reached the highest level in the 6–10 years of enclosure duration, then dropped or even returned to its initial level during the following 16–20 years (Figure 5c). The AGB of meadow maintained continuous growth during the 15-year enclosure period (Figure 5c), and the BGB had significant increase in the first 5 years, then decreased (Figure 6d). The AGB of typical steppe grasslands fluctuated during the enclosure years, and the maximum value occurred in the 11–15 years of enclosure (Figure 5e). Nevertheless, the BGB reached the highest value in the 6th to 10th year, then its beneficial effects gradually lessened. (Figure 6e).

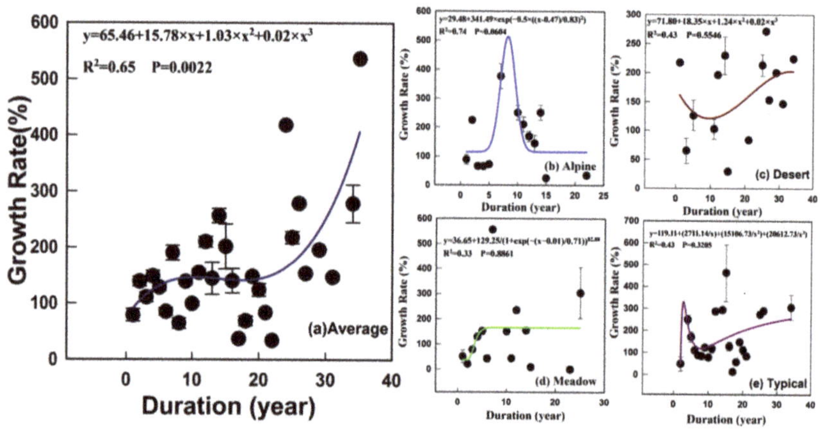

Figure 4. Average growth rate of aboveground biomass (**a**) and growth rate of aboveground biomass in different grassland types (**b–e**). Error bars are 95% confidence intervals.

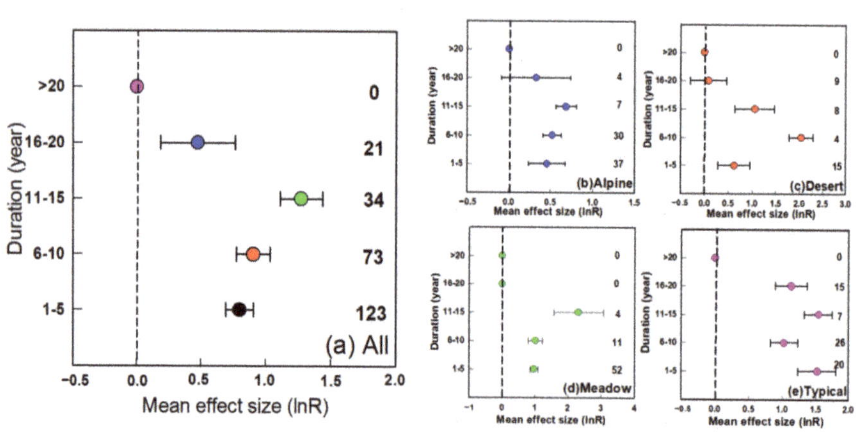

Figure 5. The effect value of aboveground biomass under different enclosure years (**a**), as well as the reaction of different types of grassland aboveground biomass to enclosure, were determined. (**b–e**). The values are calculated effect sizes and 95% confidence intervals. Sample sizes are given to the right of the graph.

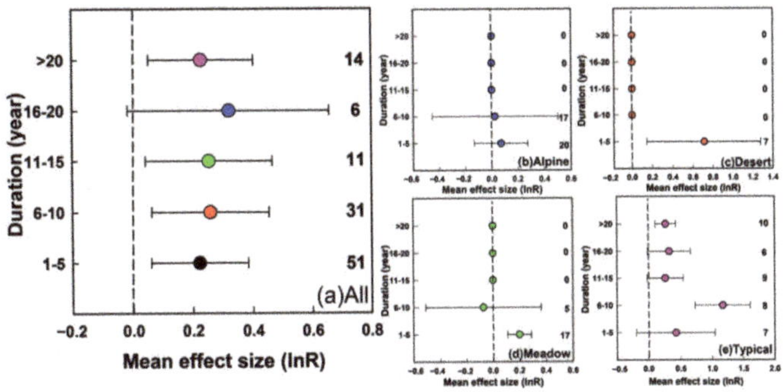

Figure 6. Effect value of underground biomass under different enclosure years (**a**) and response of different grassland underground biomass to enclosure (**b–e**). The values are calculated effect sizes and 95% confidence intervals. Sample sizes are given to the right of the graph.

3.3. *The Impact of Enclosure on Plant Species Diversity*

The grassland plant species diversity increased firstly and then decreased, and the effect value of plant species diversity reached the highest in the 11–15 years of enclosure (Figure 7a), indicating that the positive effect on grassland plant species diversity is the greatest between 11 and 15 years after enclosure. There is a certain degree of detrimental influence on the diversity of grassland vegetation when the enclosure duration reaches 15 years, and it may be increased as the enclosure term is continuously extended. This negative effect will become increasingly more substantial. As for specific grassland types, plant species diversity in the alpine grasslands increased significantly in the first 1–5 years and then decreased after 6–10 years of enclosure. When the enclosure was more than 20 years, the positive effects on plant species diversity ceased to be significant (Figure 7b). Plant species diversity in the enclosed desert grasslands generally increased within 15 years of enclosure—specifically, the enclosure raised diversity over 1–5 years, then it fell for 6–10 years and reached a peak during the 11th–15th year, and at 16 to 20 years, it started to decline once more, but not considerably (Figure 7c). Within the 10 years of enclosed, the plant species diversity in the meadow grasslands showed an increase in the first 1–5 years and then decreased (Figure 7d). The plant species diversity of typical steppe grasslands peaked in the 6th–10th year, then decreased after 11–15 years or even more than 20 years of enclosure. It showed negative effect values when the enclosure duration was more than 20 years, revealing that after more than ten years of enclosure, the variety of typical steppe grasslands started to steadily decline (Figure 7e).

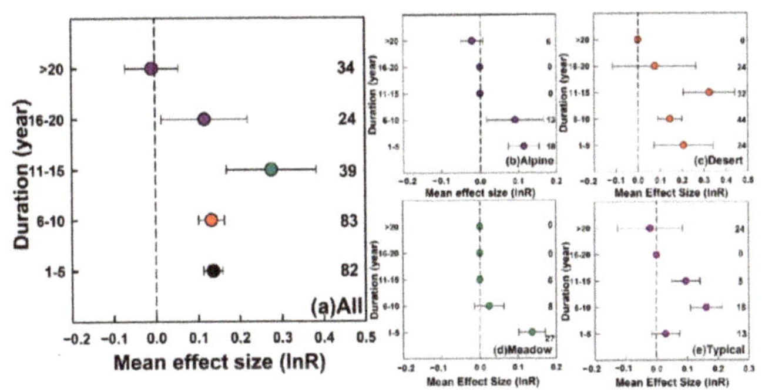

Figure 7. The effect value of plant species diversity under different enclosure years (**a**) and the response of different grassland plant species diversity to enclosure (**b–e**). The values are calculated effect sizes and 95% confidence intervals. Sample sizes are given to the right.

4. Discussion

Vegetation biomass and plant species diversity are the main indicators reflecting community structure characteristics and growth status of vegetation [50]. AGB can reflect the status of grassland restorations, and BGB has a central impact on the steadiness of grassland ecosystems [51]. Plant species diversity reflects the persistence and stability of grassland ecosystems [52]. Enclosure has a direct impact on vegetation coverage, biomass, and diversity [25,53]. Our meta-analysis showed a significantly positive effect of short-term enclosure on grassland coverage, biomass, and species diversity. Furthermore, vegetation coverage and biomass of enclosed grasslands to the enclosure years were essentially comparable, and after the duration of enclosure, all types of grasslands had a similar increased tendency in vegetation coverage, AGB, and BGB. It was well observed that the vegetation coverage, AGB, and BGB of different grasslands significantly increased in a certain enclosure period. These results are in accord with previous studies [54–57] which have verified that enclosure plays a positive role in restoration of degraded grasslands, and the accumulation of litter due to enclosure increases the organic matter into soil and promotes the growth of vegetation [58–60]. On the basis of the data analysis of coverage and biomass of all grasslands, it was indicated that the 11–15 years of enclosure duration had the strongest positive effects. However, specific to different types of grassland, their coverage and plant biomass reached peak values after different numbers of enclosure years. The desert grasslands required 6–10 years, the typical steppe grasslands required 11–15 years, and the AGB of alpine grasslands reached the highest value after 15 years, with their coverage continuing to increase within 20 years. The coverage and biomass of meadow grasslands showed a trend growth within the 15 years. The above results were in line with some on-site studies. For example, Yang et al. [61] found that the coverage and biomass of grassland enclosed for 15 years were greater than in non-enclosed grassland. Enclosing for 6–8 years significantly improved the vegetation coverage and AGB of alpine grasslands [37]. In less than 10 years of enclosed desert grasslands, significant increases in grassland coverage and biomass have been reported with the increase of enclosure duration [62–64]. However, the coverage and biomass of grassland generally changed from an increasing to decreasing trend after a certain enclosure period, likely because the increase of enclosure duration caused the habitats to become gradually homogenized; the diversity decreased, finally resulting in neither increase nor decrease in biomass [65]. Jin et al. [66] found that after the desert grasslands were enclosed more than 10 years, the structure of the plant community began to change, and perennial herbs and small shrubs gradually replaced the previously vigorous dominant species, resulting in a decrease in grassland coverage and biomass. Shan et al. [67] found that the vegetation coverage and biomass of typical steppe grasslands reached a maximum after the 11–15 years of enclosure and then decreased. All of these results elucidated that the types of grasslands should be considered when considering the optimal duration of enclosure through plant traits.

Values of effect size in our meta-analysis revealed that plant diversity of different grasslands had varied responses to the enclosure. The highest diversity of alpine grasslands appeared within 5 years of enclosure, in accordance with other studies which showed that enclosures of 6 years [25] or 4 years [39,68] in the alpine grasslands is the optimal duration. Longer enclosure duration marked decreases in alpine grasslands' species richness and diversity [11,33,69,70]. Nine-year enclosures have been noted to exhibit obvious reduction of plant diversity [71–73]. Desert grassland ecosystems are fragile and vulnerable, vegetation depends mainly on plant propagation and establishment [74,75]. Yang et al. [76] observed the species diversity of enclosed desert grasslands in 3, 5, and 8 years and found that diversity obviously increased with the duration of enclosure. Pan et al. [77] suggested that desert grasslands in short-term enclosure (less than 5 years) and long-term enclosure (lasting 15 years) both can allow species diversity to maintain a high level. The positive effects of enclosure on species diversity were the most significant in the 11–15 years, followed by enclosure for 1–5 years, and our results are in agreement with finding of Wu et al. [8] that 16-year enclosed desert grasslands had more plant species and greater stable ecosystems than 6-year enclosed desert grasslands. Although species diversity data of enclosed meadow grasslands was available only for enclosures of up to 10 years, our meta-analysis found that 1–5 years of enclosure had a significant positive effect on its diversity, which is consistent with the findings of Huang et al. [78], who noted that the species diversity increased within 5 years of enclosure in the meadow grasslands. This meta-analysis showed that species diversity of typical steppe grasslands reached a maximum after enclosure of 6–10 years, and the result is in line with Wu et al. [71], who found that species diversity reached a peak value after enclosure of 9 years, whereas the negative impact occurred when the enclosure duration was prolonged. However, there are also some reports that it takes 14 years [79] or 15 years for the diversity of typical steppe grasslands to reach its peak [80]. This difference could be due to the type and condition of the typical steppe grasslands at the time of enclosure. A natural succession process also includes enclosure and disturbing the grassland. The plant species diversity usually goes up as time passes, peaks in the middle, and then goes down. The inflection point will be different for different types of grassland and the conditions of each typical grassland at the start [68,81,82]. After the plant diversity increases significantly, the competition among species will increase, and the competitive exclusion effect will lead to the decline of community diversity [83], which will decrease the coverage or biomass of grasslands, and ultimately affect the stability of the system [23,84]. Therefore, we propose that the optimal enclosure period is vital and necessary for reestablishing degraded grasslands and is determined by the grassland types.

5. Conclusions

Enclosure had positive effects on vegetation restoration in degraded grassland ecosystems. The coverage of alpine grassland maintained continuous growth within 20-year enclosure, the AGB decreased after 15 years, and the species diversity began to decline after 10 years. If the objective is to increase alpine grassland yield to its maximum potential, the fence should be removed for approximately 15 years. On the other hand, if the objective is to restore the stability of the alpine grassland ecosystem, enclosure should be discontinued approximately 10 years after it has been established. The coverage and AGB of meadow grasslands continued to increase within 15 years, while BGB and plant species diversity declined after 5 years. Within 15 years, when grassland yields are at their optimum, fencing in meadows should be removed. In addition, at approximately the 5th year, plant species variety should be at its peak. Desert grasslands' coverage and AGB decreased after 10 years, while the species diversity decreased after 15 years. Removal of the fence at approximately 15 years after enclosure has begun for stabilizing the desert grassland ecosystem; when desert grassland production is at its peak, there must be no fence after the 10th year of enclosure. For typical steppe grasslands, the vegetation coverage and AGB decreased after 15 years, and the BGB and diversity decreased after 10 years. To maximize typical steppe grassland productivity, the fence should be removed after roughly 15 years of enclosure, and optimal plant species diversity of typical steppe grassland should be required after the 10th year of enclosure. These results showed that it is necessary to adjust the enclosure management strategies reasonably according to the grassland types and the management objectives of grassland vegetation to ensure the plant traits' health of grassland ecosystems.

Supplementary Materials: The following supporting information can be downloaded at: https://www.mdpi.com/article/10.3390/agronomy13030854/s1, Data sources.

Author Contributions: C.L., H.L., J.H., M.S., C.F. and X.Y., data curation, methodology, formal analysis, and writing—original draft preparation; C.L., manuscript writing; K.L. and X.S., conceptualization, methodology, validation, supervision, funding acquisition, and writing—review and editing. All authors discussed the results and contributed to the final manuscript. All authors have read and agreed to the published version of the manuscript.

Funding: This research was funded by Kesi Liu grant number 32271764 and the earmarked fund for CARS (CARS-3). And The APC was funded by Grant No. 32271764.

Data Availability Statement: The data sources presented in this study are available in supplementary material here.

Acknowledgments: This work was supported by the National Natural Science Foundation of China (Grant No. 32271764) and the earmarked fund for CARS (CARS-3).

Conflicts of Interest: The authors declare that they have no known competing financial interest or personal relationship that could have appeared to influence the work reported in this paper.

References

1. Ebrahimi, M.; Khosravi, H.; Rigi, M. Short-term grazing exclusion from heavy livestock rangelands affects vegetation cover and soil properties in natural ecosystems of southeastern Iran. *Ecol. Eng.* **2016**, *95*, 10–18. [CrossRef]
2. Carlier, L.; Rotar, I.; Vlahova, M.; Vidican, R. Importance and functions of grasslands. *Not. Bot. Horti. Agrobot.* **2009**, *37*, 25–30. [CrossRef]
3. Akiyama, T.; Kawamura, K. Grassland degradation in China: Methods of monitoring, management and restoration. *Grassl. Sci.* **2007**, *53*, 1–17. [CrossRef]
4. Zhou, H.; Zhao, X.; Tang, Y.; Gu, S.; Zhou, L. Alpine grassland degradation and its control in the source region of the Yangtze and Yellow Rivers, China. *Grassl. Sci.* **2005**, *51*, 191–203. [CrossRef]
5. Mofidi, M.; Jafari, M.; Tavili, A.; Rashtbari, M.; Alijanpour, A. Grazing exclusion effect on soil and vegetation properties in imam kandi rangelands, iran. *Arid Land Res. Manag.* **2013**, *27*, 32–40. [CrossRef]
6. Shang, Z.H.; Gibb, M.J.; Leiber, F.; Ismail, M.; Ding, L.M.; Guo, X.S.; Long, R.J. The sustainable development of grassland-livestock systems on the Tibetan plateau: Problems, strategies and prospects. *Rangel. J.* **2014**, *36*, 267. [CrossRef]
7. Feyisa, K.; Beyene, S.; Angassa, A.; Said, M.Y.; de Leeuw, J.; Abebe, A.; Megersa, B. Effects of enclosure management on carbon sequestration, soil properties and vegetation attributes in East African rangelands. *Catena* **2017**, *159*, 9–19. [CrossRef]
8. Wu, K.; Xu, W.; Yang, W. Short-term grazing exclusion does not effectively restore degraded rangeland in the Junggar desert of Xinjiang, China. *Grassl. Sci.* **2021**, *67*, 118–127. [CrossRef]
9. Liu, G.; Ma, C. Problems and countermeasures of grassland resources in China. *Shanxi Agric. Econ.* **2020**, *277*, 84–86. [CrossRef]
10. Wei, D.; Ri, X.; Wang, Y.; Wang, Y.; Liu, Y.; Yao, T. Responses of CO_2, CH_4 and N_2O fluxes to livestock exclosure in an alpine steppe on the Tibetan Plateau, China. *Plant Soil* **2012**, *359*, 45–55. [CrossRef]
11. Shi, X.; Li, X.G.; Li, C.T.; Zhao, Y.; Shang, Z.H.; Ma, Q. Grazing exclusion decreases soil organic C storage at an alpine grassland of the Qinghai–Tibetan Plateau. *Ecol. Eng.* **2013**, *57*, 183–187. [CrossRef]
12. Golodets, C.; Kigel, J.; Sternberg, M. Recovery of plant species composition and ecosystem function after cessation of grazing in a Mediterranean grassland. *Plant Soil* **2010**, *329*, 365–378. [CrossRef]
13. Al-Rowaily, S.L.; El-Bana, M.I.; Al-Bakre, D.A.; Assaeed, A.M.; Hegazy, A.K.; Ali, M.B. Effects of open grazing and livestock exclusion on floristic composition and diversity in natural ecosystem of Western Saudi Arabia. *Saudi J. Biol. Sci.* **2015**, *22*, 430–437. [CrossRef] [PubMed]
14. Fedrigo, J.K.; Ataide, P.F.; Filho, J.A.; Oliveira, L.V.; Jaurena, M.; Laca, E.A.; Overbeck, G.E.; Nabinger, C. Temporary grazing exclusion promotes rapid recovery of species richness and productivity in a long-term overgrazed Campos grassland. *Restor. Ecol.* **2018**, *26*, 677–685. [CrossRef]
15. Mata-González, R.; Figueroa-Sandoval, B.; Clemente, F.; Manzano, M. Vegetation changes after livestock grazing exclusion and shrub control in the southern chihuahuan desert. *West. N. Am. Naturalist* **2007**, *67*, 63–70. [CrossRef]
16. Zhou, X.; Chen, C.; Wang, Y. Long-Term exclusion of grazing increases soil microbial biomass but not diversity in a temperate grassland. *Open J. Soil Sci.* **2012**, *02*, 364–371. [CrossRef]
17. Wang, L.; Gan, Y.; Wiesmeier, M.; Zhao, G.; Zhang, R.; Han, G.; Siddique, K.H.M.; Hou, F. Grazing exclusion—An effective approach for naturally restoring degraded grasslands in Northern China. *Land Degrad. Dev.* **2018**, *29*, 4439–4456. [CrossRef]
18. Reeder, J.D.; Schuman, G.E. Influence of livestock grazing on C sequestration in semi-arid mixed-grass and short-grass rangelands. *Environ. Pollut.* **2002**, *116*, 457–463. [CrossRef]
19. Yan, Y.; Tang, H.; Xin, X.; Wang, X. Research progress on the effects of enclosure on grassland. *Acta Ecol. Sin.* **2009**, *29*, 5039–5046.
20. Schultz, N.L.; Morgan, J.W.; Lunt, I.D. Effects of grazing exclusion on plant species richness and phytomass accumulation vary across a regional productivity gradient. *J. Veg. Sci.* **2011**, *22*, 130–142. [CrossRef]
21. Maccherini, S.; Santi, E. Long-term experimental restoration in a calcareous grassland: Identifying the most effective restoration strategies. *Biol. Conserv.* **2012**, *146*, 123–135. [CrossRef]
22. Chen, T.; Sun, S.; Wang, Z.; Wu, H.; Zhang, J. Research progress on the effects of fence enclosure on vegetation community characteristics. *Agric. Tech. Serv.* **2020**, *37*, 42–43.
23. Fan, Y.; Hu, Y.; Li, K.; Yu, J.; Wang, X. Effects of different disturbances on species diversity and biomass in alpine grassland communities. *Arid. Zone Stud.* **2008**, 531–536. [CrossRef]
24. Dan, Z.; Bai, M.; Duo, J.; La, B. Effects of enclosure age on vegetation characteristics and soil nutrients in alpine meadows in Tibet. *Pratacultural Sci.* **2018**, *35*, 10–17.
25. Li, W.; Liu, Y.; Wang, J.; Shi, S.; Cao, W. Six years of grazing exclusion is the optimum duration in the alpine meadow-steppe of the north-eastern Qinghai-Tibetan Plateau. *Sci. Rep. UK* **2018**, *8*, 17269. [CrossRef] [PubMed]

26. Xu, L.; Gao, Q.; Wang, Y. Effects of 6 years of enclosure on species diversity and its relationship with aboveground biomass on slopes of typical steppe in temperate zones. *J. Ecol. Environ.* **2014**, *23*, 398–405.
27. Yan, Y.; Tang, H. Effects of fenced grazing prohibition on the characteristics of typical grassland communities in Inner Mongolia. *Northwest Bot. J.* **2007**, 1225–1232. Available online: https://kns.cnki.net/kcms/detail/detail.aspx?FileName=DNYX200706025&DbName=CJFQ2007 (accessed on 31 January 2023).
28. Jing, Z.; Cheng, J.; Chen, A. Assessment of vegetative ecological characteristics and the succession process during three decades of grazing exclusion in a continental steppe grassland. *Ecol. Eng.* **2013**, *57*, 162–169. [CrossRef]
29. Chu, X.; Xie, Y.; Shan, G.; Yuan, F.; Chen, G.; Yin, H. Effects of grazing and captivation on the community structure and species diversity of subalpine meadows in the southern margin of the Tibetan Plateau. *J. Grassl.* **2017**, *25*, 939–945.
30. Li, H.; Qian, J.; Chen, H.; Dong, J.; Wang, Y. Effects of Enclosure on Community Characteristics and Species Diversity in Hongsongwa Natural Reserve. *Res. Soil Water Conserv.* **2017**, *24*, 274–278, 283.
31. Xiong, D.; Shi, P.; Zhang, X.; Zou, C.B. Effects of grazing exclusion on carbon sequestration and plant diversity in grasslands of China—A meta-analysis. *Ecol. Eng.* **2016**, *94*, 647–655. [CrossRef]
32. Li, J.; Luo, Z.; Zhang, S.; Fan, T. Study on the effect of grassland fence enclosure on the improvement of degraded grassland in the zhongshan belt of the Ili River Valley. *Chin. Herbiv. Sci.* **2017**, *37*, 36–40.
33. Su, J.; Jing, G.; Jin, J.; Wei, L.; Liu, J.; Cheng, J. Identifying drivers of root community compositional changes in semiarid grassland on the Loess plateau after long-term grazing exclusion. *Ecol. Eng.* **2017**, *99*, 13–21. [CrossRef]
34. Dong, Y.; Sun, Z.; An, S.; Jing, C.; Wei, P. Community characteristics and carbon and nitrogen storage in arid and semiarid sagebrush deserts in Xinjiang, China: Effects of grazing exclusion. *Arid Land Res. Manag.* **2020**, *34*, 419–434. [CrossRef]
35. Yan, Y.; Lu, X. Is grazing exclusion effective in restoring vegetation in degraded alpine grasslands in Tibet, China? *PeerJ* **2015**, *3*, e1020. [CrossRef]
36. Li, W.; Cao, W.; Wang, J.; Li, X.; Xu, C.; Shi, S. Effects of grazing regime on vegetation structure, productivity, soil quality, carbon and nitrogen storage of alpine meadow on the Qinghai-Tibetan Plateau. *Ecol. Eng.* **2017**, *98*, 123–133. [CrossRef]
37. Sun, J.; Liu, M.; Fu, B.; Kemp, D.; Zhao, W.; Liu, G.; Han, G.; Wilkes, A.; Lu, X.; Chen, Y.; et al. Reconsidering the efficiency of grazing exclusion using fences on the Tibetan Plateau. *Sci. Bull.* **2020**, *65*, 1405–1414. [CrossRef] [PubMed]
38. Li, J.; Zhu, J.; Nayi, R.T.; Liu, H. Effects of fence enclosure on vegetation restoration in the spring and autumn grassland of Zhaosu Racecourse. *Prairie Lawn* **2007**, 45–48. Available online: https://kns.cnki.net/kcms/detail/detail.aspx?FileName=CYCP200706012&DbName=CJFQ2007 (accessed on 31 January 2023).
39. Li, J.; Liu, D.; Nayi, R.T.; Zhu, J. Changes in the composition of enclosed grassland communities and plant diversity in the Ili River Valley. *Prataculatural Sci.* **2013**, *30*, 736–742.
40. Li, J.; Cao, Q.; Nayi, R.T.; Zhu, J. Effects of enclosure on soil physicochemical properties and enzyme activities in spring and autumn grasslands of the Ili River Valley. *Chin. J. Grassl.* **2014**, *36*, 84–89.
41. Wen, D.; He, N.; Zhang, J. Dynamics of soil organic carbon and aggregate stability with grazing exclusion in the inner mongolian grasslands. *PLoS ONE* **2016**, *11*, e146757. [CrossRef] [PubMed]
42. Hooper, D.U.; Chapin, F.S.; Ewel, J.J.; Hector, A.; Inchausti, P.; Lavorel, S.; Lawton, J.H.; Lodge, D.M.; Loreau, M.; Naeem, S.; et al. Effects of biodiversity on ecosystem functioning: A consensus of current knowledge. *Ecol. Monogr.* **2005**, *75*, 3–35. [CrossRef]
43. Oliver, T.H.; Isaac, N.J.B.; August, T.A.; Woodcock, B.A.; Roy, D.B.; Bullock, J.M. Declining resilience of ecosystem functions under biodiversity loss. *Nat. Commun.* **2015**, *6*, 10122. [CrossRef]
44. Ren, Y.; Lü, Y.; Fu, B. Quantifying the impacts of grassland restoration on biodiversity and ecosystem services in China: A meta-analysis. *Ecol. Eng.* **2016**, *95*, 542–550. [CrossRef]
45. Tian, L.; Bai, Y.; Wang, W.; Qu, G.; Deng, Z.; Li, R.; Zhao, J. Warm- and cold- season grazing affect plant diversity and soil carbon and nitrogen sequestration differently in Tibetan alpine swamp meadows. *Plant Soil* **2021**, *458*, 151–164. [CrossRef]
46. Hedges, L.V.; Gurevitch, J.; Curtis, P.S. The meta-analysis of response ratios in experimenta ecology. *Ecology* **1999**, *80*, 1150–1156. [CrossRef]
47. Wu, Z.; Dijkstra, P.; Koch, G.W.; Peñuelas, J.; Hungate, B.A. Responses of terrestrial ecosystems to temperature and precipitation change: A meta-analysis of experimental manipulation. *Glob. Chang. Biol.* **2011**, *17*, 927–942. [CrossRef]
48. Adams, D.C.; Gurevitch, J.; Rosenberg, M.S. Resampling tests for meta-analysis of ecological data. *Ecology* **1997**, *78*, 1277–1283. [CrossRef]
49. Rosenberg, M.S.; Adams, D.C.; Gurevitch, J. *MetaWin: Statistical Software for Meta-Analysis, Version 2.0*; Sinauer: Sunderland, MA, USA, 2000.
50. Wang, Z.; Zhang, Q.; Staley, C.; Gao, H.; Ishii, S.; Wei, X.; Liu, J.; Cheng, J.; Hao, M.; Sadowsky, M.J. Impact of long-term grazing exclusion on soil microbial community composition and nutrient availability. *Biol. Fert. Soils* **2019**, *55*, 121–134. [CrossRef]
51. Song, X.; Wang, Y.; Wang, Z.; Kang, H.; Liu, C.; Li, Z.; Qu, Z.; Han, G.; Wang, Z. Relationship between soil respiration and community biomass in desert grasslands under different grazing intensities and water treatments. *J. Grassl.* **2019**, *27*, 962–968.
52. Nishizawa, K.; Tatsumi, S.; Kitagawa, R.; Mori, A.S. Deer herbivory affects the functional diversity of forest floor plants via changes in competition-mediated assembly rules. *Ecol. Res.* **2016**, *31*, 569–578. [CrossRef]
53. Li, G. Effects of Enclosure and Grazing on Biomass and CO_2 Exchange Rate of Grassland. M.A. Shanxi Agricultural University. 2014. Available online: https://kns.cnki.net/kcms2/article/abstract?v=z-Ap8PPs-htaUg7xTUW5

ucKRxyrNSJOBMGmvnpSoEXJbZgYMHYnXeh8NRwqC7k5U0O365ms9uQajxUkwra12dGY8tL0v7O4sp76EpZ3q4aBZxSK4KCw-bQ==&uniplatform=NZKPT&language=CHS (accessed on 15 June 2014).
54. Wu, J.; Zhang, X.; Shen, Z.; Shi, P.; Xu, X.; Li, X. Grazing-Exclusion Effects on Aboveground Biomass and Water-Use Efficiency of Alpine Grasslands on the Northern Tibetan Plateau. *Rangel. Ecol. Manag.* **2013**, *66*, 454–461. [CrossRef]
55. Li, X.; Chen, L.; Fan, R.; Wu, X.; Xie, Y. Effect of four typical plant community litter input on soil physical and chemical properties under the fenced condition in desert steppe. *J. Zhejiang Univ.* **2015**, *41*, 101–110.
56. Zhang, W.N.; Ganjurjav, H.; Liang, Y.; Gao, Q.Z.; Wan, Y.F.; Li, Y.; Baima, Y.Z.; Xirao, Z.M. Effect of a grazing ban on restoring the degraded alpine meadows of Northern Tibet, China. *Rangel. J.* **2015**, *37*, 89. [CrossRef]
57. Zhang, J.; Zuo, X.; Zhou, X.; Lv, P.; Lian, J.; Yue, X. Long-term grazing effects on vegetation characteristics and soil properties in a semiarid grassland, northern China. *Environ. Monit. Assess.* **2017**, *189*, 216. [CrossRef]
58. Cheng, J.; Wu, G.L.; Zhao, L.P.; Li, Y.; Li, W.; Cheng, J.M. Cumulative effects of 20-year exclusion of livestock grazing on above- and belowground biomass of typical steppe communities in arid areas of the Loess Plateau, China. *Plant Soil Environ.* **2011**, *57*, 40–44. [CrossRef]
59. Cheng, J.; Jing, G.; Wei, L.; Jing, Z. Long-term grazing exclusion effects on vegetation characteristics, soil properties and bacterial communities in the semi-arid grasslands of China. *Ecol. Eng.* **2016**, *97*, 170–178. [CrossRef]
60. Chen, Z.; Xie, Y.; Liu, M. Responses of aboveground biomass and species diversity of plants in penned alpine grasslands to key regulators. *Pratacultural Sci.* **2019**, *36*, 1000–1009.
61. Yang, J.; Zhan, W.; Wang, X. Effects of 10-year enclosure on the characteristics of plant communities in degraded alpine meadows in northern Tibet. *Chin. J. Grassl.* **2020**, *42*, 44–49.
62. Qu, W.; Pei, S.; Zhou, Z.; Zhang, B.; Fu, H. Effects of grazing and enclosure on soil organic carbon and vegetation characteristics of Alxa Desert Grassland. *Gansu For. Sci. Technol.* **2004**, 4–6. Available online: https://kns.cnki.net/kcms/detail/detail.aspx?FileName=GSLK200402002&DbName=CJFQ2004 (accessed on 31 January 2023).
63. Zhao, S.; Zuo, X.; Zhang, T.; Lü, P.; Yue, P.; Zhang, J. Responses of species diversity and biomass relationships to grazing intensity in the Urat desert grassland community. *Arid. Zone Stud.* **2020**, *37*, 168–177.
64. Li, N.; Yang, J.; Jia, Z.; Yan, R.; Chen, Q. Effects of different grazing methods on community characteristics and soil nutrients of the main grassland types in Alxa Zuoqi. *J. Grassl.* **2021**, *29*, 991–1003.
65. Yang, X.; Zhang, K.; Hou, R. Effects of captivity measures on vegetation community characteristics and aboveground biomass in semi-arid sandy grasslands. *Ecol. Environ.* **2005**, 730–734. Available online: https://kns.cnki.net/kcms/detail/detail.aspx?FileName=TRYJ200505025&DbName=CJFQ2005 (accessed on 31 January 2023).
66. Jin, C.; Li, G.; Zhao, P.; Song, L.; Gong, S. Effects of enclosure on community diversity and soil microorganisms of bitter bean. *North. Hortic.* **2019**, 118–124. Available online: https://kns.cnki.net/kcms/detail/detail.aspx?FileName=BFYY201902023&DbName=CJFQ2019 (accessed on 31 January 2023).
67. Shan, G.; Xue, S.; Chen, G.; Kuang, C.; Liu, Z.; Chu, X. Effects of seasonal enclosure on vegetation restoration in typical grasslands of Inner Mongolia. *J. Grassl.* **2012**, *20*, 812–818.
68. Miao, F.; Guo, Y.; Miao, P.; Guo, Z.; Shen, Y. Responses of alpine meadow community characteristics to captivity in the northeastern marginal region of the Qinghai-Tibet Plateau. *Pratacultural Sci.* **2012**, *21*, 11–16.
69. Zhou, H.; Zhou, L.; Liu, W.; Wang, Q.; Zhao, W. Effects of captivity measures on degraded and undegraded dwarf grass meadows. *Chin. Meadows* **2003**, 16–23. Available online: https://kns.cnki.net/kcms/detail/detail.aspx?FileName=ZGCD200305003&DbName=CJFQ2003 (accessed on 31 January 2023).
70. Zou, J.; Luo, C.; Xu, X.; Zhao, N.; Zhao, L.; Zhao, X. Relationship of plant diversity with litter and soil available nitrogen in an alpine meadow under a 9-year grazing exclusion. *Ecol. Res.* **2016**, *31*, 841–851. [CrossRef]
71. Wu, G.; Du, G.; Liu, Z.; Thirgood, S. Effect of fencing and grazing on a Kobresia-dominated meadow in the Qinghai-Tibetan Plateau. *Plant Soil* **2009**, *319*, 115–126. [CrossRef]
72. Wu, J.; Bao, X.; Li, J.; Zhao, N.; Gao, Y. Effects of different enclosure years on typical grassland communities and population characteristics of *E. macrophylla*. *J. Grassl.* **2010**, *18*, 490–495.
73. Gao, Y.; An, Y.; Qi, B.; Liu, J.; Yu, H.; Wang, D. Grazing exclusion mediates the trade-off between plant diversity and productivity in Leymus chinensis meadows along a chronosequence on the Songnen Plain, China. *Ecol. Indic.* **2021**, *126*, 107655. [CrossRef]
74. Miao, R.; Liu, Y.; Wu, L.; Wang, D.; Liu, Y.; Miao, Y.; Yang, Z.; Guo, M.; Ma, J. Effects of long-term grazing exclusion on plant and soil properties vary with position in dune systems in the Horqin Sandy Land. *Catena* **2022**, *209*, 105860. [CrossRef]
75. Miao, R.; Jiang, D.; Musa, A.; Zhou, Q.; Guo, M.; Wang, Y. Effectiveness of shrub planting and grazing exclusion on degraded sandy grassland restoration in Horqin sandy land in Inner Mongolia. *Ecol. Eng.* **2015**, *74*, 164–173. [CrossRef]
76. Yang, J.; Sun, Z.; Ba, D.; Ma, H.; Dong, Y. Effects of captivity on the diversity of functional groups of grassland vegetation and the characteristics of soil nutrients. *Chin. J. Grassl.* **2018**, *40*, 102–110.
77. Pan, X.; Liu, B.; Niu, S.; Wang, L.; Han, C. Effects of enclosure age on community structure and species diversity of desert grasslands. *Guizhou Agric. Sci.* **2018**, *46*, 95–99.
78. Huang, G.; Xi, Y.; Zhao, C.; Liu, R.; Yang, J.; Li, N.; Li, W. Effects of enclosure on the structure and biomass of subalpine grassland communities in Qilian Mountains. *J. Lanzhou Univ. (Nat. Sci. Ed.)* **2020**, *56*, 718–723.
79. Shan, G.; Xu, Z.; Ning, F.; Ma, Y.; Li, L. Effects of enclosure age on typical grassland community structure and species diversity. *Pratacultural Sci.* **2008**, *17*, 1–8.

80. Zhao, L.; Bai, X.; Tan, S.; Fan, W.; Wang, Z.; Wang, Q. Effects of different years of confinement on vegetation on typical grassland vegetation of the Loess Plateau. *Prataculturral Sci.* **2018**, *35*, 27–35.
81. Jiang, X.; Zhang, W.; Yang, Z.; Wang, G. Effects of different disturbance types on community structure and plant diversity in alpine meadows. *Northwest Bot. J.* **2003**, 1479–1485. Available online: https://kns.cnki.net/kcms/detail/detail.aspx?FileName=DNYX200309001&DbName=CJFQ2003 (accessed on 31 January 2023).
82. Shao, X.; Wang, K.; Wang, Y.; Liu, G. Dynamic changes in plant communities during natural restoration succession in typical grasslands. *Acta Ecol. Sin.* **2008**, 855–861. Available online: https://kns.cnki.net/kcms/detail/detail.aspx?FileName=STXB200802050&DbName=CJFQ2008 (accessed on 31 January 2023).
83. Loydi, A. Effects of grazing exclusion on vegetation and seed bank composition in a mesic mountain grassland in Argentina. *Plant Ecol. Divers.* **2019**, *12*, 127–138. [CrossRef]
84. Li, J.; Zheng, Z.; Zhao, N.; Gao, Y. The relationship between the versatility of grassland ecosystems and plant species diversity under the three utilization modes of cutting, enclosure and grazing. *Chin. J. Plant Ecol.* **2016**, *40*, 735–747.

agronomy

MDPI

Article

The Impact of Root-Invasive Fungi on Dominant and Invasive Plant Species in Degraded Grassland at Nanshan Pasture

Yanxia Zhang [1], Jiechao Chang [1], Jiayao Xie [1], Liquan Yang [1], Mohamed S. Sheteiwy [2], Abdel-Raouf A. Moustafa [3], Mohamed S. Zaghloul [3] and Haiyan Ren [1,*]

[1] College of Agro-Grassland Science, Nanjing Agricultural University, Nanjing 210095, China; 2021820026@stu.njau.edu.cn (Y.Z.); 2020120006@stu.njau.edu.cn (J.C.); 2022120010@stu.njau.edu.cn (J.X.); 2022820035@stu.njau.edu.cn (L.Y.)

[2] Department of Agronomy, Faculty of Agriculture, Mansoura University, Mansoura 35516, Egypt; salahco_2010@yahoo.com

[3] Botany Department, Faculty of Science, Suez Canal University, Ismailia 41522, Egypt; raoufmoustafa2@hotmail.com (A.-R.A.M.); zaghoul_mohamed@yahoo.com (M.S.Z.)

* Correspondence: hren@njau.edu.cn

Abstract: Overgrazing leads to the degradation of grazing lands, which seriously threatens the stability of grassland ecosystems. Root-invading fungi, as one of the main influencing factors, can cause plant diseases in grasslands, reduce the proportion of dominant plant species, increase the proportion of invasive poisonous weeds, and further aggravate degradation. In order to predict and improve the effects of root-invading fungi on grassland degradation, we conducted an in situ soil indoor control experiment using soils collected from non-degraded, moderately degraded, and severely degraded areas of Nanshan pasture in Hunan Province, China. We used monoculture or mixed grasslands of dominant plant species, including *Lolium perenne*, *Trifolium repens*, and the invasive weed *Persicaria hydropiper*, and inoculated them with local strains of pathogenic *Fusarium* species (*Fusarium boothii* and *Fusarium circinatum*) and beneficial fungi Arbuscular Mycorrhizal Fungi (AMF) and *Trichoderma hamatum* to explore how different strains of fungi affect plant growth and community dynamics. The results showed that *Fusarium* species (*Fusarium boothii* and *Fusarium circinatum*), as a major pathogenic fungus, inhibited the growth of the dominant grass *Lolium perenne* in moderately and severely degraded soils, which provided growth space and resources for invasive weeds *Persicaria hydropiper* and further aggravated the degree of grassland degradation. However, the collaborative effect of beneficial fungi (AMF and *Trichoderma*) and their inhibitory effect on *Fusarium* species (*Fusarium boothii* and *Fusarium circinatum*) could promote the growth of dominant plants and weeds in soils with varying degrees of degradation, which is beneficial to maintaining the stability and diversity of grassland plant communities. The collaborative effect of beneficial fungi could also increase the availability of nutrients in severely degraded soils. Therefore, using beneficial fungi (AMF and *Trichoderma*) for soil improvement and reducing the harm of pathogenic *Fusarium* species (*Fusarium boothii* and *Fusarium circinatum*) to plant growth is of great significance for promoting the protection and management of grassland ecosystems, as well as for the restoration and recovery of grasslands.

Keywords: degraded grassland; *Fusarium*; plant community; *Trichoderma*; arbuscular mycorrhizal fungi

Citation: Zhang, Y.; Chang, J.; Xie, J.; Yang, L.; Sheteiwy, M.S.; Moustafa, A.-R.A.; Zaghloul, M.S.; Ren, H. The Impact of Root-Invasive Fungi on Dominant and Invasive Plant Species in Degraded Grassland at Nanshan Pasture. *Agronomy* **2023**, *13*, 1666. https://doi.org/10.3390/agronomy13071666

Academic Editors: Kesi Liu and Xinqing Shao

Received: 19 May 2023
Revised: 15 June 2023
Accepted: 19 June 2023
Published: 21 June 2023

1. Introduction

Grasslands have played a significant role in the development of animal husbandry [1]. However, they are facing severe degradation issues due to overgrazing, leading to a massive invasion of weeds and a dramatic decrease in productivity [2,3]. It is necessary to conduct in-depth research on the mechanisms of weed invasion and take measures to prevent the degradation of grasslands.

Microorganisms are closely associated with weed invasion and may serve as important drivers for successful invasions. Studies have shown that after the invasion of alien weeds, there is a lack of corresponding pathogenic microorganisms in the new habitat, allowing the successful invasion of the alien weeds [4]. Additionally, when invasive plants enter new habitats, they can affect the growth of native plants. For example, the invasive plant *Chromolaena odorata* can increase the spore population of *Fusarium* in the soil, thus suppressing the growth of two native plants [5]. This phenomenon suggests that invasive plants can become a pathogen reservoir for native plants, severely impacting the ecosystem of native plants. Therefore, studying the interaction between microorganisms and weed invasion is crucial for understanding ecosystem changes during the invasion process. Root-invasive fungi are a group of internal microorganisms that spread through the soil and invade plant roots [6,7]. They have significant impacts on plant growth and development and can parasitize plant roots, resulting in substantial effects [8]. Some of these root-invasive fungi can invade weeds and parasitize their roots, making them alternative hosts for pathogenic fungi and promoting the spread of these pathogens [9,10].

However, current research on the interaction between microorganisms and plants after weed invasion mostly focuses on soil fungi, with relatively little study on root fungi [11–13]. Therefore, further research on the effects of root-invasive fungi on both invasive weeds and native plants, as well as their mechanisms of interaction with invasive weeds and native plants, is of great significance for maintaining grassland ecological balance. In this study, we have selected the dominant local plant *Lolium perenne*, *Trifolium repens*, and the invasive weed *Persicaria hydropiper* as research subjects to investigate the impact of root-invasive fungi on dominant and invasive plants in soils with varying degrees of degradation. This study aims to explore the role of root-invasive fungi in grassland degradation at both the species and community levels, helping us better understand the role and effects of root-invasive fungi in the invasion process and providing a scientific basis for grassland conservation and management.

2. Materials and Methods

2.1. Study Area

The Nanshan Pasture is located 80 km southwest of Chengbu Miao Autonomous County, on the southern edge of western Hunan Province, at 26°12′ N and 109°56′ E. It is a typical high-mountain moss grassland with a total area of 15,300 hectares, an average altitude of 1760 m, and a maximum altitude of 1940 m. The area belongs to the subtropical monsoon humid climate zone of central subtropical mountainous climate, with an average annual temperature of 11 °C, an average annual precipitation of 1218.5 mm, and an average relative humidity of 75% to 83%. The dominant plants in this grassland are *Stipa baicalensis*, *Lolium perenne*, and *Trifolium repens*, while *Potentilla reptans*, *Polygonum hydropiper*, and *Rumex acetosa* L. are subdominant species.

2.2. Preparation of Soil and Tested Fungi

This experiment collected two invasive plant specimens, namely *Rumex acetosa* and *Polygonum hydropiper*, from the degraded grassland of Nanshan pasture for the isolation of root-inhabiting fungi (Unpublished data). Subsequently, the isolated fungi were subjected to pathogenicity testing, and two species of Fusarium, namely *Fusarium boothii* and *Fusarium circinatum*, were selected as pathogenic fungi for testing. These two fungi can cause diseases in dominant grasses *Lolium perenne* and *Trifolium repens* in grasslands but have no significant effect on invasive plants such as *R. acetosa* and *P. hydropiper*. In addition, the experiment selected *Trichoderma hamatum*, a beneficial fungus isolated from *R. acetosa*, as a tested beneficial fungus.

According to the national standard "Classification Criteria for Natural Grassland Degradation, Desertification, and Salinization" (GB19377-2003), we assessed the degree of grassland degradation in Nanshan Pasture, Hunan Province. The assessment involved calculating vegetation coverage, aboveground biomass, and the rate of organic matter reduction in the 0–20 cm soil layer. We defined areas where the indicators decreased by 10% as non-degraded (ND) grassland, areas with a decrease of 20–40% as moderately degraded (MD) grassland, and areas with a decrease of over 50% as severely degraded (SD) grassland. Corresponding soil samples were collected from each degradation zone. During the sampling process, we randomly selected 10 sampling points in each degradation zone with a distance of 10 m between adjacent points. At each sampling point, we used a shovel to collect 10 kg of topsoil (0–10 cm), and in each degradation zone, we collected a total of 100 kg of soil samples. In total, we collected 900 kg of soil samples (100 kg per degradation zone × 3 degradation types × 3 repetitions). The collected soil samples were mixed and sieved (<0.5 cm) in the field to remove visible debris, such as large stones and plant remains, and then transported back to the laboratory for sterilization in a high-pressure sterilizer at 120 °C for 1 h and stored for further analysis.

2.3. Seed and Inoculum Preparation

First, the seeds were disinfected with a 1% NaClO solution for 2 min, then placed on sterilized filter paper in a culture dish and germinated in a 20 °C incubator for later use. Two weeks after seed germination, 5 consistent seedlings from each plant species were randomly selected and transplanted into prepared pots. Then, the pots were placed in the greenhouse of Nanjing Agricultural University according to a completely randomized design.

To prepare the fungal inoculum, the tested fungi (*Fusarium boothii*, *Fusarium circinatum*, and *Trichoderma hamatum*) were cultured on PDA for one week. Three 5 mm diameter fungal discs were taken from the edge of the colony using a sterilized puncher and cultured in LB medium while shaking on a shaker for 3 days. After filtering out the culture medium, the remaining fungal suspension was diluted with sterile water to prepare a fungal suspension with an absorbance of 2, which was used for inoculation.

Propagating and inoculating arbuscular mycorrhizal fungal spores (provided by the School of Resources and Environmental Sciences, Nanjing Agricultural University, Nanjing, China): Spores of *Glomus caledonium* are placed in a pot containing sterilized sandy soil with corn as the host for propagation. For inoculation, 20 g of the propagated sandy soil is weighed.

2.4. Greenhouse Experiment

At the species level, the experiment involves 162 pots with 3 different soil degradation levels (non-degraded, moderately degraded, severely degraded), 6 different fungal treatments (Fb, Fc, Fbc, AM, Fbc + AM, and CK), and 3 different plants (*Lolium perenne* L., *Trifolium repens* L., and *Persicaria hydropiper*), each with 3 replicates. Fb represents *Fusarium boothii*, and Fc represents *Fusarium circinatum*. Fbc represents a mixture of *Fusarium boothii* and *Fusarium circinatum*, AM represents *arbuscular mycorrhizal fungi*, AM + Fbc represents a mixture of *arbuscular mycorrhizal fungi* and two *Fusarium* species, and CK is the control.

At the community level, the experiment involves 72 pots with mixed plants (*Lolium perenne* L., *Trifolium repens* L., and *Persicaria hydropiper*), 3 different soil degradation levels (non-degraded, moderately degraded, severely degraded), and 8 different inoculation treatments (FM, TH, AMF, FM + TH, FM + AMF, AMF + TH, FM + TH + AMF, and CON), each with 3 replicates. FM represents a mixture of *Fusarium boothii* and *Fusarium circinatum*, TH represents *Trichoderma hamatum*, AMF represents *arbuscular mycorrhizal fungi*, FM + TH represents a mixture of *Trichoderma hamatum* and two *Fusarium* species, FM + AMF represents a mixture of *arbuscular mycorrhizal fungi* and two *Fusarium* species, AMF + TH represents a mixture of *Trichoderma hamatum* and *arbuscular mycorrhizal fungi*, FM + TH + AMF represents a mixture of *Trichoderma hamatum*, *arbuscular mycorrhizal fungi*, and two *Fusarium* species, and CON is the control.

The population experiment selected plastic pots with a diameter of 15 cm and a height of 15 cm, with each pot containing 1500 g of sterilized soil with varying degrees of degradation, for later use. The community experiment selected plastic pots with a diameter of 15 cm and a height of 15 cm, with each pot containing 3000 g of sterilized soil with varying degrees of degradation, for later use. The two experiments were consistent in all other experimental conditions.

Randomly select Pre-germinated seeds were randomly selected, and 5 seedlings with consistent growth were transplanted into each prepared pot. If a seedling died within the first week of the experiment, it was replaced immediately. The pots were placed in the Nanjing Agricultural University greenhouse according to a completely randomized design, and their positions were changed periodically to ensure even light exposure. The relative humidity was maintained at 70%. Plants received 16 h of light at 21 °C during the day and 8 h of light at 16 °C during the night, and were watered every other day. The initial soil moisture was measured twice a week (calculated as 17% of soil dry weight).

2.5. Harvesting and Measurements

After 18 weeks, the aboveground plants of all pots in both the species-level and community-level experiments were collected, dried in an oven (70 °C for 3 days), and weighed. The soil samples from the community-level experiment were air-dried in a ventilated area after passing through a 2 mm sieve for the determination of basic soil physical and chemical properties. The soil samples were extracted using 2 mol/L KCl, and nitrate nitrogen (NO_3^--N) was determined using colorimetry (at 220 nm and 275 nm) [14]. Ammonium nitrogen (NH_4^+-N) was determined using colorimetry (at 625 nm) after extraction with indophenol blue-potassium chloride [14]. The available phosphorus content of the soil samples was measured using the Olsen method (mg/g) [15]. Finally, the organic carbon content (SOC) of the soil samples was directly determined using a TOC analyzer [16].

2.6. Data Analysis and Statistical Analysis

All data were processed using Excel, and statistical analyses were performed after calculating the mean values of different observations within the same replication. A three-way ANOVA was conducted to evaluate the effects of plant species, degradation level, fungal treatment, and their interactions on aboveground biomass. A two-way ANOVA was used to examine the effects of degradation level, fungal treatment, and their interactions on total biomass and soil nutrients (organic carbon, available phosphorus, ammonium nitrogen, and nitrate nitrogen). The statistical analyses were conducted using IBM SPSS 22. In this paper, all significance levels were indicated at the 0.05 level. Graphs were generated using Origin2021.

3. Results

3.1. Effects of Pathogenic Fungi on Plant Biomass

Degradation level and fungal treatment had significant effects on the aboveground biomass of different plants (Table 1). The aboveground biomass of all three plants showed a decreasing trend with increasing degradation level (Figure 1). There was an interaction between degradation level and fungal treatment on the aboveground biomass of *Lolium perenne*, *Persicaria hydropiper*, and *Trifolium repens* L. (Table 1). Compared with the CK control, Fb, Fc, Fbc, and Fbc + AM decreased the aboveground biomass of *Lolium perenne* and *Trifolium repens* L. but had no significant effect on the aboveground biomass of *Persicaria hydropiper* in non-degraded and moderately degraded soils (Figure 1); among them, the aboveground biomass of *Lolium perenne* was lowest under the Fbc fungal treatment, and the aboveground biomass of *Trifolium repens* L. was lowest under the Fbc, Fb, and Fc fungal treatments. In severely degraded soils, Fb, Fc, Fbc, and AM + Fbc reduced the aboveground biomass of all three plants.

Table 1. Results from three-way ANOVA Table for the effects of plant species (P), degradation level (D), and fungal treatment (F) on aboveground biomass of species.

Treatment	DF	F	*p*
Plant species (P)	2	208.0	<0.001
Degree of degradation (D)	2	50,746.9	<0.001
Fungal treatment (F)	5	6696.7	<0.001
P × D	4	32.3	<0.001
P × F	10	501.2	<0.001
D × F	10	4.0	<0.001
P × D × F	20	3.6	0.08

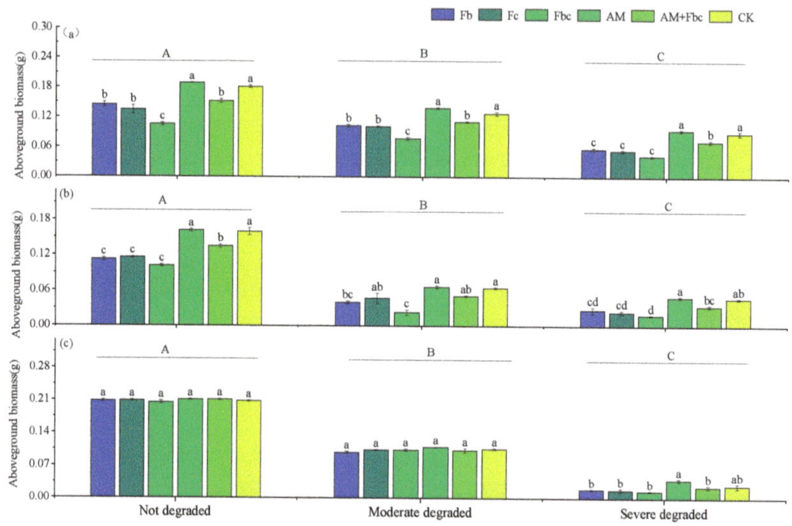

Figure 1. Changes in aboveground biomass of three plant species ((**a**) *Lolium perenne*, (**b**) *Trifolium repens* L., and (**c**) *Persicaria hydropiper*) after inoculation with fungi. Fb: *Fusarium boothii*, Fc: *Fusarium circinatum*, AM: Arbuscular mycorrhizal fungi. Different uppercase letters indicate significant differences in different degrees of degradation ($p < 0.05$), and different lowercase letters indicate significant differences in different fungal inoculation treatments under the same degree of degradation ($p < 0.05$).

3.2. Community Total Biomass

Both degradation level and fungal treatment have significant effects on the aboveground biomass of the community (Table 2). The aboveground biomass of the community decreases with increasing degradation level (Figure 2). In moderately degraded and severely degraded soils, the community aboveground biomass treated with pathogenic fungi (FM) is significantly lower than other treatments (TH, AMF, FM + TH, FM + AMF, FM + TH + AMF, and CON). In all three degraded soils, fungal treatment (TH, AMF, FM + TH, FM + AMF, and FM + TH + AMF) increases the community biomass, and the highest increase is observed under mixed treatment with pathogenic and beneficial fungi (FM + AMF and FM + TH + AMF) (Figure 2).

Table 2. The effects of different levels of soil degradation and fungal treatments on aboveground biomass (AB), soil organic carbon (SOC), soil inorganic nitrogen (NH_4^+-N and NO_3^--N), and available phosphorus (AP) in the soil community were studied using a two-way analysis of variance (ANOVA). The results provide F-values (F) and *p*-values (*p*), where ($p < 0.05$) indicates a significant difference and ($p < 0.01$) indicates a highly significant difference.

Treatment	DF	AB		SOC		NO_3^--N		NH_4^+-N		AP	
		F	*p*	F	*p*	F	*p*	F	*p*	F	*p*
Degree of degradation (D)	2	170,582.298	<0.001	3492.557	<0.001	73.577	<0.01	11.169	<0.001	770.432	<0.001
Fungal treatment (F)	7	221.834	<0.001	16.955	<0.001	4.595	<0.001	3.252	<0.01	7.571	<0.001
D × F	14	8.840	<0.001	6.121	<0.001	0.469	0.938	0.056	0.885	0.735	0.730

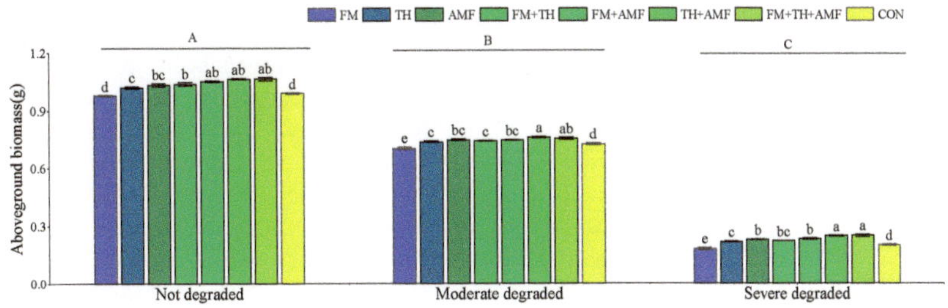

Figure 2. Changes in aboveground biomass of plant community after inoculation with fungi in different degrees of degradation. Different uppercase letters indicate significant differences in different degrees of degradation ($p < 0.05$), and different lowercase letters indicate significant differences in different fungal inoculation treatments under the same degree of degradation ($p < 0.05$).

3.3. The Biomass of Each Plant Species in the Community

The biomass of three plant species, namely *Lolium perenne, Trifolium repens,* and *Polygonum hydropiper*, decreased with increasing degradation degree under different fungal treatments, consistent with the results at the species level (Figure 3). In the same degraded soil, different fungal inoculations had different effects on the biomass of different plants. In contrast to the results in the previous chapter, the pathogenic fungal treatment (FM) significantly reduced the aboveground biomass of *L. perenne* in moderately and severely degraded soils but had no significant effect on *T. repens* and *P. hydropiper*. In different degraded soils, inoculation with different fungi (TH, AMF, FM + TH, FM + AMF, TH + AMF, and FM + TH + AMF) increased the biomass of the three plant species to varying degrees (Figure 3), but the TH + AMF treatment resulted in the most significant increase in biomass for the three plant species across different degraded soils. Furthermore, the aboveground biomass of *L. perenne, T. repens,* and *P. hydropiper* was significantly higher under the FM + TH + AMF treatment than under the FM treatment, indicating that the promotion of AMF and *Trichoderma* on the three plant species was greater than the inhibitory effect of *Fusarium*.

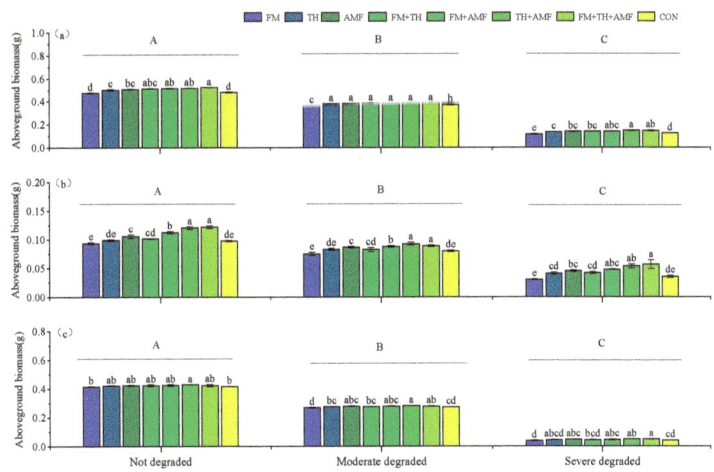

Figure 3. Changes in aboveground biomass of plant species in different degraded soils after inoculation with fungi. *Lolium Perenne* (**a**), *Trifolium repens* (**b**), and *Polygonum hydropiper* (**c**); FM: *Fusarium boothi* + *Fusarium circinatum*, Th: *Trichoderma hamatum*, AMF: arbuscular mycorrhizal fungi. Different uppercase letters indicate significant differences in different degrees of degradation ($p < 0.05$), and different lowercase letters indicate significant differences in different fungal inoculation treatments under the same degree of degradation ($p < 0.05$).

3.4. Soil Nutrients

Degradation and fungal treatments both significantly affect soil organic carbon, ammonium nitrogen, and available phosphorus content but only have an interactive effect on soil organic carbon content (Table 2). Under different fungal treatments, soil organic carbon content decreased with increasing degradation degree (Figure 4). In soils with different degrees of degradation, the beneficial fungus (AMF + TH) significantly increased soil organic carbon, ammonium nitrogen, and available phosphorus content compared to the untreated control (CON), while the pathogenic fungus (FM) only significantly decreased soil organic carbon content in moderately degraded soil, without significant effect on soil nitrate nitrogen and available phosphorus content (Figure 4).

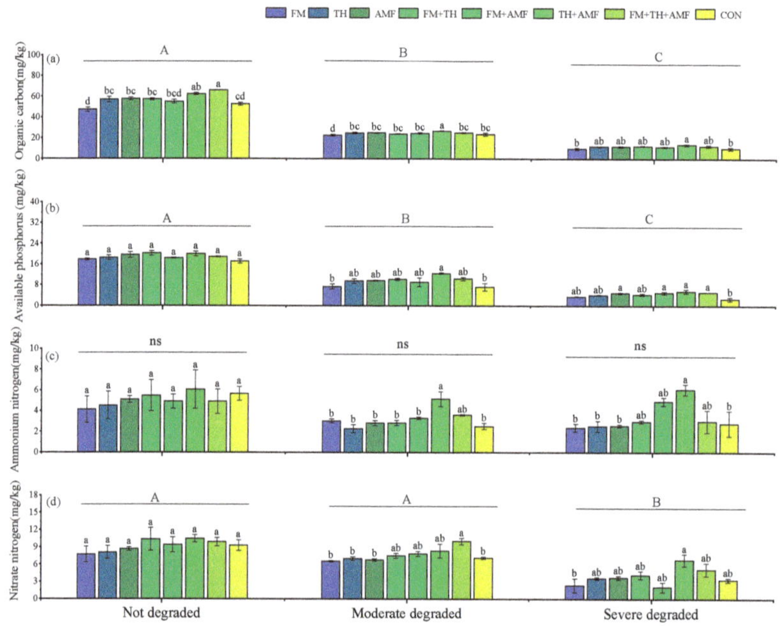

Figure 4. Changes in soil nutrients of plant communities under different degradation degrees after inoculation with fungi. (**a**) Organic carbon, (**b**) available phosphorus, (**c**) ammonium nitrogen, and (**d**) nitrate nitrogen; FM: *Fusarium boothi* + *Fusarium circinatum*, TH: *Trichoderma hamatum*, AMF: arbuscular mycorrhizal fungi. ns indicates no significant difference among degraded soils. Different uppercase letters indicate significant differences in different degrees of degradation ($p < 0.05$), and different lowercase letters indicate significant differences in different fungal inoculation treatments under the same degree of degradation ($p < 0.05$).

4. Discussion

In our study, we found that the parasitic effects of the invading weed fungi Persicaria hydropiper and Rumex acetosa-derived Fusarium on plants vary at different ecological levels, including the species level and the community level. At the species level, Fusarium can inhibit the growth of multiple plant species, while at the community level, it significantly affects the growth of Lolium perenne. Regardless of the level, Fusarium significantly inhibits the growth of the dominant plant species, Lolium perenne. Further research indicates that the combined action of symbiotic fungi such as AMF or Trichoderma can suppress the harmful effects of Fusarium on both dominant plant species and invading weeds while improving the availability of nutrients in degraded grassland soils. Therefore, we believe that root fungi from invading weeds exacerbate grassland degradation by suppressing the growth of native plants, and introducing symbiotic fungi can be an effective biocontrol measure to aid in the restoration of degraded grassland ecosystems.

4.1. Effects of Pathogenic Fungi on Plants

Pathogenic strains of *Fusarium* can parasitize multiple plant hosts, leading to a reduction in plant diversity [5,17]. In this study, we found that in non-degraded and moderately degraded soils, pathogenic strains of *Fusarium* isolated from the invading weeds *Persicaria hydropiper* and *Rumex acetosa*—when inoculated alone or in combination—reduced aboveground biomass of the dominant native plants *Lolium perenne* and *Trifolium repens*. The inhibitory effect was more significant with mixed inoculation but had no significant impact on the invading weeds. However, in severely degraded soils, *Fusarium* exhibited varying degrees of inhibition on both dominant native plants (such as *Lolium perenne* and *Trifolium repens*) and invading plants (*Polygonum hydropiper*). This indicates that the pathogenic strains of *Fusarium* isolated from the invading weeds *Persicaria hydropiper* and *Rumex acetosa* infect host plants in response to soil conditions, and the combined infection of *Fusarium* has a more pronounced inhibitory effect on plant growth [18].

According to the disease triangle theory, the parasitic relationship between pathogens and plants is closely related to soil conditions [19]. The results of this study further demonstrate that in severely degraded soils, pathogenic *Fusarium* infected a greater number of plant species, indicating a stronger inhibitory effect on plants in heavily degraded soils. Furthermore, the study results also indicate that in severely degraded soils, *Fusarium* has a significant impact on the aboveground biomass of *Lolium perenne* and *Trifolium repens* but has no significant effect on the invading plants. This suggests that the effect of *Fusarium* on plants varies depending on the plant species. Previous research has shown that invading weeds can thrive in nutrient-poor soils and possess strong adaptability [20,21]. Consistent with previous studies, our findings indicate that in degraded grasslands, invading weeds exhibit a stronger resistance to *Fusarium*. In contrast, invading weeds are better adapted to degraded soil conditions and can withstand pathogen infection more effectively.

4.2. The Role of Pathogenic Fungi in Plant Communities

A high nutrient supply can counterbalance the negative impact of soil pathogens on plants and maintain the stability of plant communities [22]. In this study, inoculation with two strains of *Fusarium boothi* and *Fusarium circinatum* resulted in a decrease in biomass in plant communities of different degraded soils. Compared to the control group without inoculation, the biomass was reduced by 1.14%, 2.8%, and 9.6% in non-degraded, moderately degraded, and severely degraded soil, respectively. These findings suggest that the root fungi *Fusarium boothi* and *Fusarium circinatum*, found in the roots of invasive weeds, may exhibit a more pronounced inhibitory effect on plant communities in severely degraded soils. Consistent with previous research, the lower nutrient content in degraded soils may enhance the negative impact of the root fungi *Fusarium boothi* and *Fusarium circinatum* on plants.

In degraded grasslands, the competitive ability of native dominant grasses is reduced, leading to a significant decrease in their proportion. The proportion of invasive weeds increases significantly, replacing native dominant grasses as the dominant plants in degraded grasslands [23,24]. In this study, co-inoculation with *Fusarium boothi* and *Fusarium circinatum* in different degraded soils resulted in a decrease in the total biomass of the plant community. In moderately and severely degraded soils, the biomass of the dominant grass *Lolium perenne* was significantly reduced, while the biomass of *Trifolium repens* and *Polygonum hydropiper* was unaffected. This indicates that the decrease in total biomass of the community is due to the inhibitory effect of *Fusarium boothi* and *Fusarium circinatum* on *Lolium perenne* growth. The inhibition of *Lolium perenne* by *Fusarium boothi* and *Fusarium circinatum* reduces its competitive ability in the grassland, further diminishing the proportion of native dominant grasses and providing growing space for weed invasion.

In plant communities, the interactions between different plants alter the relationship between pathogenic fungi and hosts [25]. Our study also found differences in the parasitic effects of the root fungi *Fusarium boothi* and *Fusarium circinatum* on plants at the species and community levels. At the species level, *Fusarium boothi* and *Fusarium circinatum* can inhibit the growth of multiple plant species, while at the community level, they only significantly inhibit the growth of *Lolium perenne*. This indicates that the interactions between *Lolium perenne*, *Trifolium repens*, and *Polygonum hydropiper* affect the relationship between *Fusarium boothi* and *Fusarium circinatum* and their hosts. However, whether at the species level or the community level, *Fusarium boothi* and *Fusarium circinatum* have a significant inhibitory effect on the growth of the dominant plant *Lolium perenne*. Therefore, we conclude that the inhibition of *Lolium perenne* by *Fusarium boothi* and *Fusarium circinatum* is one of the important factors contributing to the degradation of the Nanshan pasture [26]. Specifically, in the undegraded grassland of the Nanshan pasture, *Lolium perenne* is the dominant plant and is usually able to occupy more resources and space, while the weed *Polygonum hydropiper* is a subdominant plant that needs to grow in intense competition. *Fusarium boothi* and *Fusarium circinatum* reduce the competitive advantage of *Lolium perenne*, providing growing space and resources for invasive weeds such as *Polygonum hydropiper*, which favors the survival and reproduction of invasive weeds and ultimately leads to grassland degradation.

4.3. The Role of Symbiotic Fungi in the Community

Symbiotic fungi are closely related to plant growth and can influence plant nutrient absorption, compete for soil nutrients, decompose soil organic matter, and ultimately determine plant yield [27–29]. Our research shows that under different levels of degradation, inoculating AMF and *Trichoderma* increased the total biomass of the community by 5.7%, 4.8%, and 18% in non-degraded, moderately degraded, and severely degraded soil, respectively. This indicates that AM fungi and *Trichoderma* had a positive effect on plant communities in different degraded soils, and the promoting effect on plant communities was strongest in severely degraded soils [30]. This is consistent with previous studies that showed symbiotic fungi (AMF and *Trichoderma*) can promote plant growth, and this growth-promoting effect is stronger in nutrient-poor soils, indicating that the combined action of symbiotic fungi (AMF and *Trichoderma*) can promote the restoration of degraded grasslands and maintain community stability [31,32]. At the same time, symbiotic fungi can suppress the harmful effects of pathogenic microorganisms on plants. In this study, compared to inoculating only with *Fusarium boothi* and *Fusarium circinatum*, inoculating with MX (two *Fusarium* species + AMF + *Trichoderma*) significantly increased the community biomass, which may be because symbiotic fungi (AMF and *Trichoderma*) offset the negative effects of pathogenic fungi on plants. In addition, previous studies have shown that AMF and *Trichoderma* can increase the availability of soil nutrients by decomposing soluble or insoluble organic matter. In this experiment, AMF or *Trichoderma* increased the content of soil organic carbon, available phosphorus, and nitrate nitrogen, and the mixed treatment of symbiotic fungi (AMF and *Trichoderma*) significantly increased the availability of soil nutrients, indicating that the combined action of AMF and *Trichoderma* can produce a synergistic effect that improves soil nutrient limitations.

5. Conclusions

In conclusion, the results showed that the inhibition of the dominant plant *Lolium perenne* by two *Fusarium* species (*Fusarium boothi* and *Fusarium circinatum*) was one of the important causes of grassland degradation in Nanshan, and this inhibition is not conducive to the regeneration of degraded grassland *Lolium perenne*. However, the mixed application of AMF and *Trichoderma hamatum* can promote the growth of dominant plants and weed plants, reduce the harm of *fusarium* to plants, and increase the availability of soil nutrients; thus, grassland restoration and reconstruction can be achieved.

Author Contributions: H.R. designed the experiment, H.R. and Y.Z. performed the experiment, and all the authors contributed to writing the manuscript. All authors have read and agreed to the published version of the manuscript.

Funding: This project was supported by the National Natural Science Foundation of China (32271615), the Fundamental Research Funds for the Central Universities (KYCYXT2022008), and XUEKEN2022020.

Data Availability Statement: Please contact the first author with requests for data.

Acknowledgments: We are grateful to all the students who assisted with the collection and processing of data. Many thanks are expressed to the anonymous reviewers for their helpful suggestions.

Conflicts of Interest: The authors declare no conflict of interest.

References

1. Sun, H.; Zhou, H.; Wang, P. Weed invasion and grassland quality during degraded succession of mixed artificial grassland. *Grassl. Lawn* **2001**, *1*, 33–37.
2. Duan, D. Research on the establishment of artificial grassland in mountainous areas of southern China. *Sci. Bull.* **1992**, *8*, 180–183.
3. Hong, F.; Wang, Y. Review and thinking on animal husbandry of artificial grassland in South China. *J. Grassl. Sci. China* **2006**, *28*, 71–75.
4. Maron, J.L.; Klironomos, J.; Waller, L.; Callaway, R.M.; Austin, A. Invasive Plants Escape from Suppressive Soil Biota at Regional Scales. *J. Ecol.* **2014**, *102*, 19–27. [CrossRef]
5. Mangla, S.; Callaway, R.M. Exotic invasive plant accumulates native soil pathogens which inhibit native plants. *J. Ecol.* **2007**, *96*, 58–67. [CrossRef]
6. Callaway, R.M.; Thelen, G.C.; Barth, S.; Ramsey, P.W.; Gannon, J.E. Soil Fungi Alter Interactions between the Invader Centaurea Maculosa and North American Natives. *Ecology* **2004**, *85*, 1062–1071. [CrossRef]
7. Lambers, H.; Mougel, C.; Jaillard, B.; Hinsinger, P. Plant-Microbe-Soil Interactions in the Rhizosphere: An Evolutionary Perspective. *Plant Soil* **2009**, *321*, 83–115.
8. Lagueux, D.; Jumpponen, A.; Porras-Alfaro, A.; Herrera, J.; Chung, Y.A.; Baur, L.E.; Smith, M.D.; Knapp, A.K.; Collins, S.L.; Rudgers, J.A. Experimental Drought Re-Ordered Assemblages of Root-Associated Fungi across North American Grasslands. *J. Ecol.* **2021**, *109*, 776–792. [CrossRef]
9. Shiragane, H.; Usami, T.; Shishido, M. Weed Roots Facilitate the Spread of Rosellinia Necatrix, the Causal Agent of White Root Rot. *Microbes Environ.* **2019**, *34*, 340–343. [CrossRef]
10. Pearson, K.A.; Taylor, A.F.S.R.; Fuchs, M.E.; Woodward, S. Characterisation and Pathogenicity of Fusarium Taxa Isolated from Ragwort (*Jacobaea Vulgaris*) Roots. *Fungal Ecol.* **2016**, *20*, 186–192.
11. Soti, P.G.; Purcell, M.; Jayachandran, K. Soil Biotic and Abiotic Conditions Negate Invasive Species Performance in Native Habitat. *Ecol. Process.* **2020**, *9*, 18. [CrossRef]
12. Zhang, P.; Li, B.; Wu, J.; Hu, S. Invasive plants differentially affect soil biota through litter and rhizosphere pathways: A meta-analysis. *Ecol. Lett.* **2018**, *22*, 200–210. [CrossRef]
13. Ravichandran, K.R.; Thangavelu, M. Role and Influence of Soil Microbial Communities on Plant Invasion. *Ecol. Quest.* **2017**, *27*, 9–23. [CrossRef]
14. Bao, S. *Soil Agricultural Chemical Elements Analysis*; China Agriculture Press: Beijing, China, 2000; pp. 50–57.
15. Song, C.; Han, J.; Gao, C. Research progress on soil available phosphorus and its chemical testing methods. *Agric. Syst. Sci. Integr. Res.* **2010**, *26*, 283–287.
16. Nelson, D.W.; Sommers, L.E. Total Carbon, Organic Carbon, and Organic Matter. *Methods Soil Anal.* **1982**, *9*, 961–1010.
17. Bever, J.D.; Mangan, S.A.; Alexander, H.M. Maintenance of Plant Species Diversity by Pathogens. *Annu. Rev. Ecol. Evol. Syst.* **2015**, *46*, 305–325. [CrossRef]
18. Ampt, E.A.; van Ruijven, J.; Raaijmakers, J.M.; Termorshuizen, A.J.; Mommer, L. Linking ecology and plant pathology to unravel the importance of soil-borne fungal pathogens in species-rich grasslands. *Eur. J. Plant Pathol.* **2018**, *154*, 141–156. [CrossRef]
19. Jarosz, A.M.; Davelos, A.L. Effects of disease in wild plant populations and the evolution of pathogen aggressiveness. *New Phytol.* **1995**, *129*, 371–387. [CrossRef]
20. Tao, Z.W.; Bu, H.; Li, J.; Jia, P.; Qi, W.; Liu, K.; Du, G.Z. Effects of different artificial planting schemes on invasive weeds. *Glob. Ecol. Conserv.* **2021**, *28*, e01651. [CrossRef]
21. Ke, W.; Pan, Y.; Chen, L.; Huang, J.; Zhang, J.; Long, X.; Cai, M.; Peng, C. Adaptive photosynthetic strategies of the invasive plant Sphagneticola trilobata and its hybrid to a low-light environment. *Photosynthetica* **2022**, *60*, 549–561. [CrossRef]
22. Klironomos, J.N. Feedback with soil biota contributes to plant rarity and invasiveness in communities. *Nature* **2002**, *417*, 67–70. [CrossRef] [PubMed]
23. Kardol, P.; De Deyn, G.; Laliberté, E.; Mariotte, P.; Hawkes, C.V. Biotic plant-soil feedbacks across temporal scales. *J. Ecol.* **2013**, *101*, 309–315. [CrossRef]
24. Grman, E.; Robinson, T.M.P. Resource Availability and Imbalance Affect Plant-Mycorrhizal Interactions: A Field Test of Three Hypotheses. *Ecology* **2013**, *94*, 62–71. [CrossRef] [PubMed]
25. Ampt, E.A.; Francioli, D.; van Ruijven, J.; Gomes, S.I.F.; Maciá-Vicente, J.G.; Termorshuizen, A.J.; Bakker, L.M.; Mommer, L. Deciphering the interactions between plant species and their main fungal root pathogens in mixed grassland communities. *J. Ecol.* **2022**, *110*, 3039–3052. [CrossRef]

26. Maron, J.L.; Smith, A.L.; Ortega, Y.K.; Pearson, D.E.; Callaway, R.M. Negative Plant-Soil Feedbacks Increase with Plant Abundance, and Are Unchanged by Competition. *Ecology* **2016**, *97*, 2055–2063. [CrossRef]
27. Delgado-Baquerizo, M.; Reich, P.B.; Trivedi, C.; Eldridge, D.J.; Abades, S.; Alfaro, F.D.; Bastida, F.; Berhe, A.A.; Cutler, N.A.; Gallardo, A.; et al. Multiple elements of soil biodiversity drive ecosystem functions across biomes. *Nat. Ecol. Evol.* **2020**, *4*, 210–220. [CrossRef] [PubMed]
28. Fanin, N.; Gundale, M.J.; Farrell, M.; Ciobanu, M.; Baldock, J.A.; Nilsson, M.-C.; Kardol, P.; Wardle, D.A. Consistent effects of biodiversity loss on multifunctionality across contrasting ecosystems. *Nat. Ecol. Evol.* **2017**, *2*, 269–278. [CrossRef]
29. Guerra, C.A.; Bardgett, R.D.; Caon, L.; Crowther, T.W.; Delgado-Baquerizo, M.; Montanarella, L.; Navarro, L.M.; Orgiazzi, A.; Singh, B.K.; Tedersoo, L.; et al. Tracking, targeting, and conserving soil biodiversity. *Science* **2021**, *371*, 239–241. [CrossRef]
30. Reynolds, H.L.; Packer, A.; Bever, J.D.; Clay, K. Grassroots Ecology: Plant-Microbe-Soil Interactions as Drivers of Plant Community Structure and Dynamics. *Ecology* **2003**, *84*, 2281–2291. [CrossRef]
31. Shinichi, N.; Schielzeth, H. A General and Simple Method for Obtaining R2 from Generalized Linear Mixed-Effects Models. *Methods Ecol. Evol.* **2013**, *4*, 133–142.
32. Harman, G.E.; Uphoff, N. Symbiotic Root-Endophytic Soil Microbes Improve Crop Productivity and Provide Environmental Benefits. *Scientifica* **2019**, *2019*, 9106395. [CrossRef] [PubMed]

agronomy

Article

The Synergistic Effect of Biochar and Microorganisms Greatly Improves Vegetation and Microbial Structure of Degraded Alpine Grassland on Qinghai–Tibet Plateau

Jinsheng Li [1,2], Hui Li [2], Jianying Shang [3], Kesi Liu [2,4,*], Yixuan He [2] and Xinqing Shao [2,4,*]

1 School of Resources and Environment, Anhui Agricultural University, Hefei 230036, China; l628js@126.com
2 College of Grassland Science and Technology, China Agricultural University, Beijing 100193, China
3 Department of Water and soil Science, China Agricultural University, Beijing 100193, China
4 Key Laboratory of Restoration Ecology of Cold Area in Qinghai Province, Northwest Institute of Plateau Biology, Chinese Academy of Sciences, Xining 810008, China
* Correspondence: kliu@cau.edu.cn (K.L.); shaoxinqing@163.com (X.S.)

Abstract: The attenuation of soil organic carbon and the destruction of soil microbial structure are common manifestations of grassland degradation. The addition of exogenous organic carbon and microorganisms may be an effective way to quickly restore degraded grassland, but corresponding evaluations are still rare. We investigated the effects of effective microorganisms (EM) and biochar addition on vegetation biomass, microorganisms and soil properties in degraded alpine grassland. The treatments included a control (no biochar or EM addition, CK), EM addition (250 mL m^{-2} EM, M), biochar addition (4.00 kg m^{-2} biochar, C) and a mixture of biochar and EM (4.00 kg m^{-2} biochar and 250 mL m^{-2} EM, C+M). C, M and C+M rapidly increased vegetation biomass, soil organic carbon (TOC), total nitrogen (TN), available nitrogen (NH_4^+-N, NO_3^--N), available phosphorus (AP), total microbial biomass (MB), bacteria and fungus biomass in the soil, and also altered the microbial community structure. The content of soil nutrients in the C treatment was the highest, followed by C+M. The vegetation biomass and microbial biomass were the greatest in the C+M treatment, and increased by 101.04~198.52% and 22.14~45.41%, respectively. C+M can also enhance the presence of saprotrophic fungi, thereby facilitating the augmentation of both plant and soil nutrients. Overall, the biochar combined with EM addition had a synergistic effect on the restoration of degraded alpine grasslands.

Keywords: alpine grassland; degradation; effective microorganisms; biochar; soil physicochemical properties; soil microorganisms

Citation: Li, J.; Li, H.; Shang, J.; Liu, K.; He, Y.; Shao, X. The Synergistic Effect of Biochar and Microorganisms Greatly Improves Vegetation and Microbial Structure of Degraded Alpine Grassland on Qinghai–Tibet Plateau. *Agronomy* **2023**, *13*, 2203. https://doi.org/10.3390/agronomy13092203

Academic Editor: Wanting Ling

Received: 27 July 2023
Revised: 18 August 2023
Accepted: 21 August 2023
Published: 23 August 2023

1. Introduction

As one of the important terrestrial ecosystems, the world's widespread grassland ecosystems have unique ecosystem services and functions. Grasslands distributed on the Qinghai–Tibet Plateau are one of the four basic pastoral areas in China, and play a pivotal role in providing the material basis for local animal husbandry production [1]. Additionally, Alpine grasslands have the crucial ecological functions of climate regulation, carbon sequestration, nutrient cycling, water conservation and the maintenance of biodiversity [2–4]. Due to its unique natural environment, the ecosystem of alpine grassland is relatively fragile [5]. In recent years, the combined effects of unreasonable human disturbance, climate change and low ecological resilience have caused serious degradation in many alpine grasslands [5,6]. In response to grassland degradation, scientists have used numerous methods to restore damaged vegetation and improve degraded soil, such as grazing prohibition, reseeding, fertilization, etc. [7,8].

Many studies reveal that biochar enhances microbial biomass but has conflicting impacts on microbial diversity [9,10]. The deterioration of the structure and function of soil microbial communities and the attenuation of soil organic carbon are the most obvious characteristics of soil degradation [11,12]. The loss and destruction of soil microorganisms can not only interfere with the normal cycle of soil nutrient elements (C, N, P), but also limit plant growth by hindering plant nutrient uptake and photosynthesis [13–15]. Generally, the effects of biochar on microbial biomass are dependent on biochar properties, while those on microbial diversity are dependent on soil properties [16–18]. In agricultural production, the addition of microbial inoculum, especially effective microorganisms (EM), has been verified as an effective way to resolve the issue of the reduction in or inactivation of beneficial bacteria after soil degradation [19–21]. Compared with single-strain microbial inoculum, EM contains a variety of microbial communities (e.g., photosynthetic bacteria, lactic acid bacteria, saccharomyces), so it has a complex composition, a stable structure, and a wide range of functions [22,23]. Some studies have shown that EM addition can inhibit soil diseases, stimulate the reproduction of soil beneficial microorganisms, promote the decomposition of soil organic matter to increase soil nutrients, and improve the plant's absorption of nutrients [22,24,25]. However, most EM studies and utilization currently focus on arable lands, and there are few studies on degraded grasslands, especially degraded alpine grasslands.

The loss of organic carbon in soil seriously affects the healthy operation and maintenance of different terrestrial ecosystems, and even leads to ecosystem degradation [7,26]. Biochar has been widely used as a soil amendment in agricultural production to compensate for the massive loss of soil organic carbon due to soil degradation [27]. Biochar typically contains 50–80% carbon and is produced through the pyrolysis of animal and plant residues under complete or partial anoxic conditions [28]. Biochar enhances soil not only through the addition of nutrients and organic carbon [29], but also by leveraging the inherent structural properties that biochar possesses [30]. Biochar has a special microporous structure and strong adsorption ability, and can adsorb and immobilize mineral elements for plant growth and effectively regulate nutrient cycling in soil [31,32]. The organic molecules adsorbed on the surface of biochar can form organic matter through complex physical, chemical and biological reactions [33,34]. Based on its unique traits, biochar directly or indirectly improves soil physicochemical properties and nutrient availability, and regulates soil microbial biomass and soil microbial community structure [35,36]. For example, biochar can improve soil permeability and water retention capacity through its large specific surface area and porous structure, thereby providing a better living environment for soil microorganisms [36].

Currently, the application of biochar or EM mostly focuses on monoculture cropland, and is mixed fully with soil through plowing [37–39]. And the increases in total microbial diversity with biochar addition vary in acidic and sandy soils with low soil organic carbon content [40,41]. Compared to cropland, the complexity of grassland vegetation species and underground root systems is much higher. Furthermore, for the remediation of natural grassland soil in China, it is forbidden to use completely destructive methods on cropland such as plowing and mixing exogenous additives. So, the effects of biochar and EM on soil properties and microbial activities in grasslands may be different from those described in previous studies on farmland ecosystems. However, few studies have reported on the restoration of degraded alpine grasslands using a combination of biochar and EM, and a combined experiment on fragile degraded grasslands would provide a unique opportunity to explore the restoration effects on the vegetation biomass, soil physicochemical properties and microorganisms of degraded alpine grassland. Therefore, we investigated the improvement effect of the addition of biochar or EM separately or their mixture on degraded alpine grassland, and tested the following hypotheses: (1) Although degraded alpine grassland is different from cropland, adding biochar or EM could significantly increase the vegetation biomass and microorganisms and improve the soil physicochemical properties of degraded grassland. (2) In comparison to biochar or EM addition alone, a synergistic effect of the combined addition of biochar and EM might exist.

2. Materials and Methods

2.1. Study Site and Experimental Materials

This experiment was conducted in a moderately–severely degraded alpine grassland, defined according to the national standard (GB19377-2003) [42], in Senduo Town (36°35′ N, 101°42′ E) in the northeast part of the Qinghai–Tibetan Plateau. The site before the experiment was grazing land. The elevation of the experimental site is 3220 m a.s.l (above sea level). The climate of the study region belongs to a typical plateau continental climate, with mean annual precipitation of 403.80 mm and a mean annual temperature of 2.3 °C. Meanwhile, the annual evaporation is 1378.5 mm, and the annual sunshine hours are 2738 h [43]. The soil type is chernozem. The dominant plants across the whole experimental grassland include *Cleistogenes squarrosa, Poa crymophila, Carex tristachya, Elymus nutans.* Griseb, *Ligularia Cass, Stipa krylovii Roshev. Oxytropis* DC and *Stellera chamaejasme.* L.

Biochar was made using corn straw at 550 °C, which contained 10.2 g kg^{-1} total nitrogen, 508.9 g kg^{-1} organic carbon, 80.95 g kg^{-1} total phosphorus, 8.96 pH and 1595 μs cm^{-1} electrical conductivity. Effective microorganisms (EM) were selected as the microbial inoculum in this experiment in view of the good performance of EM in previous studies. EM was purchased from Beijing Baofeng Biological Technology Co. Ltd. (China), mainly including photosynthetic bacteria, lactic acid bacteria and yeast and other bacteria, and the number of effective viable bacteria contained was $\geq 10 \times 10^8$ cfu·mL^{-1}.

2.2. Experimental Design and Sampling

The experiment comprised four treatments, including a control (no biochar or EM, CK), biochar addition alone (2%, 4.00 kg m^{-2}, C), EM addition alone (250 mL m^{-2} EM, M), and the mixed addition of biochar and EM (C+M). A complete random block design was used with three repetitions. Within each block, experimental plots (2 m length × 2 m width) for each treatment were established with a 1.5 m buffer strip between each plot. The amount of biochar added was based on the ratio of its weight to the dry weight of 0–20 cm soil depth at the experimental site (initial soil bulk density was 1 g cm^{-3}). The dosage of EM was twice the maximum applied in the field, and the EM solution was diluted to 1:4 (EM: water, v/v) before its addition. In order to reduce the destruction to the grassland, the biochar and EM were applied to the soil via hole application and surface application. Details of the addition procedure and method have been described in the article of Li et al. [43]. In short, sixteen soil cores (20 cm depth, 3.5 cm diameter) were drilled in each plot with approximately 50 cm intervals between them. Firstly, half of the additives were added into each hole, then the rest were evenly sprayed on the surface of the plot. In order to prevent disturbance due to grazing, the entire field area was fenced during the experiment. The experiment was set up in May 2017.

Plant and soil samples were collected in late July in 2017, 2018 and 2019. For plant samples, three 50 cm × 50 cm quadrats were randomly selected in each plot to collect aboveground plants, and the collected plant samples were oven-dried at 65 °C to a constant weight and weighted. After harvesting the aboveground plants, soil samples were collected from the same quadrats. Three soil sampling sites were randomly selected and soil samples were separately taken from 0–10 cm and 10–20 cm soil layers using a soil auger (3.5 cm diameter). When collecting the soil samples, if the obtained sample coincided with the hole application, we abandoned that drilling sample and collect a new one.

The same layer of the soil sample was composted, and then, separated into three parts: one part was dried at 65°C for soil physiochemical analysis, one part was placed into an aluminum box to test the soil water content (SWC) and the third part was placed in a plastic bag and stored at −20°C for soil microorganism analysis.

The fresh soil was dried at 105 °C for 24 h to determine the soil water content (SWC), which was determined as follows:

$$SWC = (A2 - A3) \times 100/(A3 - GA1)$$

where A1 is the weight of the aluminum box, A2 is the weight of the aluminum box plus the original soil sample and A3 is the weight of the aluminum box plus the dried soil sample.

2.3. Soil Physicochemical Properties

Using a Fisher 2000 elemental analyzer, we measured soil total nitrogen (TN) and soil organic carbon (TOC) (Thermo Fisher Scientific, Rome, Italy). Using an acidity meter and conductivity meter, we measured soil pH and soil electric conductivity (EC) (METTLER TOLEDO, Zurich, Switzerland). The sodium bicarbonate extraction molybdenum antimony anti-colorimetric method was used to measure soil available phosphorus (AP). A flow autoanalyzer was used to determine NH_4^+-N and NO_3^--N (FIA Compact, Berlin, Germany).

2.4. Soil Microorganisms

The microbial communities and structures were measured using phospholipid fatty acid (PLFA), which was modified by White et al. [44]. Briefly, we added a single-phase mixture of chloroform, methanol and phosphate buffer (1:2:0.8) to 8 g of freeze-dried soil and extract lipids. Silicic acid column chromatography was used to isolate and concentrate the crude extracts. Then, they were saponified and methylated to obtain phospholipid fatty acid methyl esters (FAMEs).

Gas chromatography was used to analyze the FAMEs (Agilent 6850, New York, NY, USA). Then, we used Sherlock MIDI software to define the PLFAs (Newark, NJ, USA). The PLFA classification is shown in Table S1.

2.5. Statistical Analysis

We tested the differences in plant biomass, soil physicochemical properties and microorganisms between treatments and years, respectively, using ANOVA and Tukey's test (with a confidence of 95%) (using the multcomp package). The soil microbe community and structures indicated by PLFAs were evaluated and analyzed using principal component analysis (PCA) with the vegan package. Redundant analysis (RDA) was used to test the relationships between the microorganisms and environmental variables (using the R vegan packages). The angles between the arrows indicating PLFAs and environmental factors indicate the relationship between microorganisms and environmental factors. The correlation coefficients between soil physical and chemical properties were determined using Pearson correlation in R software. The correlations that were not significant were deleted. The relationships between the environmental variables and microbial communities (PLFAs) of different treatments were compared using the mantel test with the vegan package. To determine the compatibility of microorganisms and environmental factors in different treatments, Procrustes analysis was performed using the PCA (Bray–Curtis) results of different treatments with the vegan package. All statistical analyses were performed using R 4.0.2. The significance level for all statistical tests was $p < 0.05$.

3. Results

3.1. Changes in Aboveground Vegetation Biomass

Compared with CK, biochar or EM addition (C, M and C+M) significantly increased the aboveground biomass of degraded alpine grassland (Figure 1). During the three years, EM addition alone increased the above-ground biomass of degraded alpine grassland by 81.02~149.10%, biochar-only addition increased the aboveground biomass of degraded alpine grassland by 65.82~151.07%, and the addition of C+M increased the aboveground biomass of degraded alpine grassland by 101.04~198.52%. Overall, the aboveground biomass with the C+M treatment was significantly higher than with the other treatments. In addition, the aboveground biomass increased significantly with the increase in years, with values of 170.72 g m^{-2}, 183.83 g m^{-2} and 212.86 g m^{-2} in 2017, 2018 and 2019, respectively.

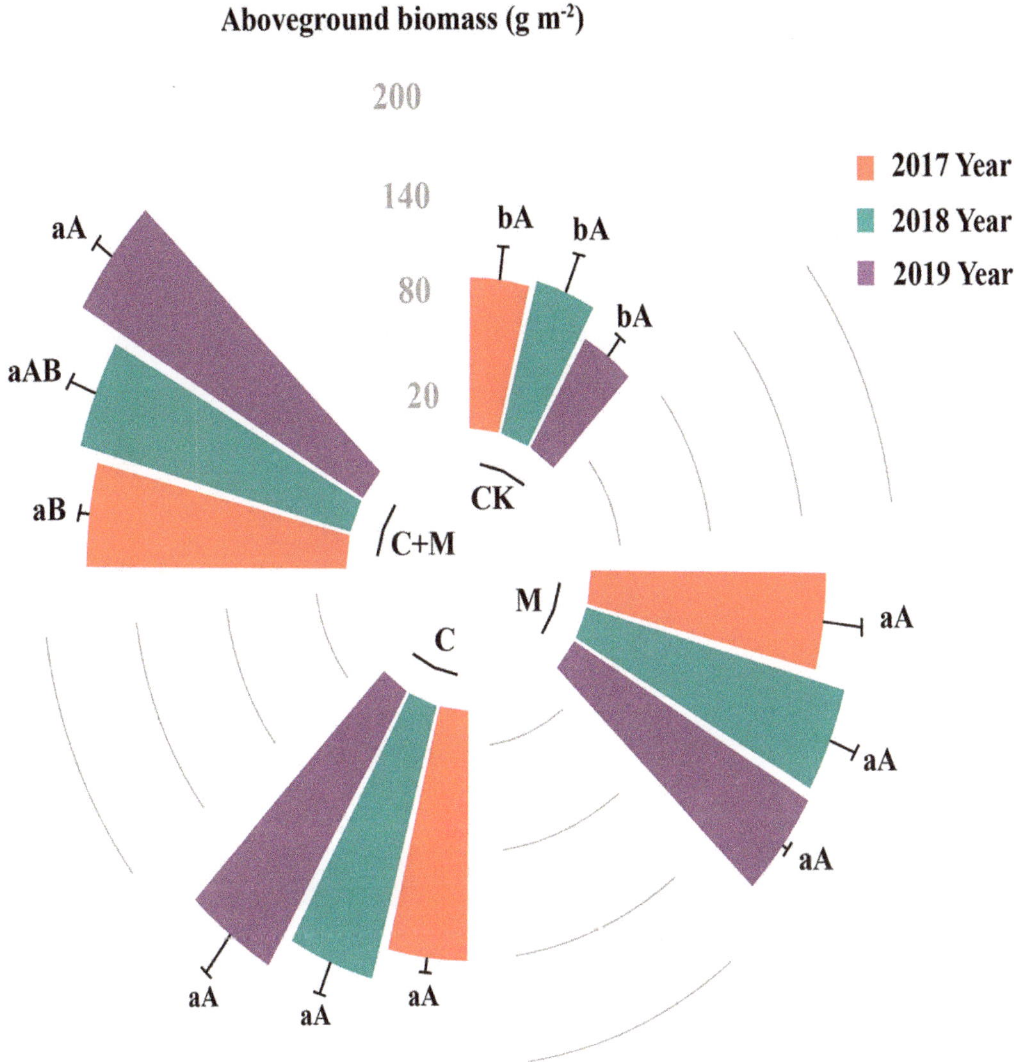

Figure 1. The variations in aboveground biomass (g m^{-1}) with different treatments in 2017, 2018 and 2019. Values are mean ± SE (n = 3). Different lowercase letters indicate significant differences among different treatments in the same year. Different capital letters indicate significant differences between years under the same treatments at $p < 0.05$.

3.2. Changes in Soil Physicochemical Properties

Biochar addition (C, C+M) increased the pH of the 0–20 cm soil layer, but the pH in each treatment first increased, and then, decreased with the increase in years (Table 1). Compared with CK, EM and biochar addition, especially C+M, significantly decreased soil EC and increased the SWC of the whole soil sample (0–10 cm and 10–20 cm). In addition, SWC had a greater increase with the C and C+M treatments compared with the EM-only addition, especially in the top soil (0–10 cm) (Table 1).

Table 1. The properties of pH, EC, SWC and C/N under different treatments in the 0–20 cm soil layers in 2017, 2018 and 2019.

Parameter	Year	Layer (cm)	CK	M	C	C+M
pH	2017	0–10	7.12 ± 0.05 bcB	6.99 ± 0.08 cB	7.36 ± 0.12 abC	7.46 ± 0.18 aB
	2018		7.55 ± 0.03 bA	7.63 ± 0.09 bA	8.22 ± 0.22 aA	8.32 ± 0.16 aA
	2019		7.28 ± 0.22 bAB	7.77 ± 0.13 aA	7.78 ± 0.05 aB	8.05 ± 0.08 aA
	2017	10–20	7.24 ± 0.11 bB	7.28 ± 0.12 bB	7.31 ± 0.08 aC	7.33 ± 0.11 aC
	2018		7.48 ± 0.09 bA	7.65 ± 0.07 bA	8.44 ± 0.08 aA	8.31 ± 0.08 aA
	2019		7.49 ± 0.05 cB	7.78 ± 0.01 bA	7.81 ± 0.11 bC	7.99 ± 0.06 aB
EC	2017	0–10	362.76 ± 46.31 aA	163.87 ± 27.02 cA	312.01 ± 20.85 bA	121.77 ± 11.93 cC
	2018		274.09 ± 28.54 bB	186.53 ± 16.66 bA	302.58 ± 31.56 aA	253.17 ± 14.39 bA
	2019		209.8 ± 21.20 aC	208.83 ± 32.09 aA	220.73 ± 26.97 aB	172.53 ± 17.28 aB
	2017	10–20	324.91 ± 11.21 aA	161.8 ± 28.45 bA	319.01 ± 16.06 aA	117.01 ± 11.6 cC
	2018		305.27 ± 99.51 aA	202.87 ± 26.05 cA	282.62 ± 15.51 bB	223.27 ± 13.27 cA
	2019		197.53 ± 14.41 aB	171.83 ± 25.92 aA	198.33 ± 23.1 aC	161.10 ± 15.5 aB
SWC (%)	2017	0–10	24.52 ± 0.93 bA	24.48 ± 1.90 bA	24.35 ± 0.83 bB	30.41 ± 2.82 aA
	2018		18.82 ± 0.38 cB	26.97 ± 2.41 bA	30.69 ± 1.81 aA	31.71 ± 1.42 aA
	2019		16.46 ± 1.01 cC	24.18 ± 0.77 abA	27.01 ± 2.66 aAB	22.92 ± 0.87 bB
	2017	10–20	21.14 ± 0.71 bA	21.71 ± 1.87 aA	21.89 ± 0.83 bB	25.68 ± 1.49 aB
	2018		19.85 ± 0.51 bA	25.43 ± 1.65 bB	27.31 ± 0.75 aA	29.70 ± 1.87 aA
	2019		16.08 ± 0.81 bA	16.86 ± 0.94 bC	20.55 ± 0.99 aB	19.83 ± 1.06 aC
C/N	2017	0–10	12.02 ± 0.49 bA	10.78 ± 1.42 bA	15.63 ± 1.35 aA	15.81 ± 0.09 aA
	2018		12.02 ± 1.38 abA	11.11 ± 0.31 bA	13.13 ± 0.58 aB	11.96 ± 1.37 aA
	2019		10.85 ± 0.57 bA	11.43 ± 0.77 bA	15.34 ± 0.58 aAB	11.69 ± 1.28 bA
	2017	10–20	12.68 ± 1.38 aA	12.33 ± 0.68 aA	12.73 ± 0.89 aA	11.77 ± 0.83 aA
	2018		10.08 ± 0.47 aB	9.70 ± 1.59 aA	9.22 ± 0.82 aB	10.33 ± 0.29 aA
	2019		11.28 ± 0.8 cA	11.34 ± 0.92 cA	13.67 ± 1.00 aA	12.56 ± 1.10 abA

Note: Values are mean ± standard error (n = 3). Different lowercase letters indicate significant differences among different treatments in the same year. Different capital letters indicate significant differences between years under the same treatments at $p < 0.05$. EC: electrical conductivity, SWC: soil water content.

Compared with CK, biochar and EM addition, especially C+M, significantly increased the TOC (Table 2). Moreover, the TOC in the deep soil (10–20 cm) increased significantly with the increase in the residence time of biochar or EM (Table 2). The TN also increased significantly after biochar or EM addition, but there was no significant difference between the C and M treatments (Table 2). The C/N had no significant change between the CK and M treatments, but increased in the treatments with biochar addition (C, C+M, Table 1). The NH_4^+-N in the treatments with biochar or EM addition was significantly higher in the top soil than that in CK, but in the deep soil, the NH_4^+-N in the C and C+M treatments was significantly higher than that in CK (Table 2). The NO_3^--N in the top soil had little change between the M and CK treatments in the first year (2017), but was significantly higher in the C and C+M treatments than in CK. In the third year (2019), the NO_3^--N in the biochar and EM addition treatments was higher than that in CK (Table 2). In the deep soil, the trend of NO_3^--N in the first year was similar to the top soil, but in the third year, the NO_3^--N decreased in the M treatment. However, there was no difference in the C and C+M treatments (Table 2). Interestingly, we found that TOC, NH_4^+-N and NO_3^--N in the C+M treatment were lower than those in C treatment, but higher than those in M treatment in 2018 and 2019. Compared with CK, AP in the deep soil was significantly increased in the biochar and EM addition treatments, especially in the C+M treatment (Table 2).

Table 2. Soil organic carbon (TOC), soil total nitrogen (TN), soil ammonia-nitrogen (NH_4^+-N), nitrate nitrogen (NO_3^--N) and available phosphorus (AP) in the 0–20 cm soil layers with different treatments in 2017, 2018 and 2019.

Parameter	Year	Layer (cm)	CK	M	C	C+M
TOC ($g\ kg^{-1}$)	2017	0–10	33.14 ± 1.63 bB	39.56 ± 4.49 bA	57.43 ± 5.43 aA	58.88 ± 0.75 aA
	2018		40.92 ± 1.62 cA	40.44 ± 1.05 cA	63.54 ± 2.55 aA	46.88 ± 0.46 bB
	2019		34.26 ± 1.04 dB	43.73 ± 0.87 cA	56.78 ± 0.43 aA	49.85 ± 2.05 bB
	2017	10–20	30.16 ± 1.4 bA	34.17 ± 1.91 aB	36.60 ± 0.87 aB	36.47 ± 0.73 aA
	2018		30.45 ± 0.68 bA	32.95 ± 0.55 bB	36.61 ± 0.7 aB	36.11 ± 2.26 aA
	2019		30.19 ± 0.21 cA	37.49 ± 1.1 bA	43.35 ± 4.12 aA	39.90 ± 1.92 abA
TN ($g\ kg^{-1}$)	2017	0–10	2.76 ± 0.12 bB	3.67 ± 0.27 aA	3.67 ± 0.12 aB	3.73 ± 0.10 aA
	2018		3.43 ± 0.26 bA	3.64 ± 0.02 bA	4.84 ± 0.17 aA	3.97 ± 0.45 bA
	2019		3.17 ± 0.24 bAB	3.84 ± 0.3 aA	3.71 ± 0.17 abB	3.80 ± 0.22 aA
	2017	10–20	2.40 ± 0.18 bB	2.78 ± 0.19 abA	2.89 ± 0.15 aB	3.11 ± 0.17 aB
	2018		3.03 ± 0.20 bA	3.51 ± 0.71 abA	4.00 ± 0.30 aA	3.49 ± 0.13 abA
	2019		2.69 ± 0.20 bB	3.33 ± 0.27 aA	3.17 ± 0.12 aB	3.31 ± 0.08 aAB
NH_4^+-N ($mg\ kg^{-1}$)	2017	0–10	18.33 ± 0.83 cC	23.13 ± 2.22 bC	11.74 ± 0.09 dC	29.45 ± 1.61 aB
	2018		35.05 ± 3.67 cA	43.15 ± 1.73 cA	70.37 ± 6.75 aA	58.74 ± 6.01 bA
	2019		26.13 ± 1.20 cB	28.39 ± 2.31 bB	48.35 ± 2.37 aB	36.97 ± 3.07 abB
	2017	10–20	16.71 ± 1.50 bB	23.96 ± 1.31 aA	9.65 ± 0.43 cB	19.36 ± 3.60 abB
	2018		22.17 ± 1.31 cA	16.42 ± 2.04 dB	34.56 ± 2.70 bA	40.44 ± 2.72 aA
	2019		14.64 ± 2.11 cB	9.27 ± 0.93 cC	36.06 ± 2.54 aB	22.19 ± 3.78 bB
NO_3^-N ($mg\ kg^{-1}$)	2017	0–10	7.57 ± 1.56 cAB	6.47 ± 0.47 cB	21.62 ± 1.25 aB	15.56 ± 0.33 bA
	2018		4.97 ± 0.70 bB	7.51 ± 0.33 aB	5.77 ± 1.08 bC	6.40 ± 0.33 abB
	2019		8.96 ± 1.72 cA	18.25 ± 4.49 bA	31.70 ± 5.22 aA	14.28 ± 2.08 bcA
	2017	10–20	7.18 ± 1.08 bAB	7.95 ± 0.79 bA	13.29 ± 1.11 aA	13.30 ± 0.26 aA
	2018		4.94 ± 0.79 bB	9.09 ± 0.56 aA	8.01 ± 1.31 aB	6.99 ± 1.22 abC
	2019		11.09 ± 2.99 aAB	9.90 ± 2.19 bA	13.79 ± 1.11 aA	10.83 ± 0.17 aB
AP ($mg\ kg^{-1}$)	2017	0–10	2.68 ± 0.24 bA	2.91 ± 0.36 bA	4.10 ± 0.41 aA	4.35 ± 0.20 aA
	2018		3.02 ± 0.24 bA	3.27 ± 0.95 abAB	4.37 ± 0.07 aA	3.43 ± 0.19 abB
	2019		1.57 ± 0.17 cB	3.16 ± 0.10 bA	3.31 ± 0.10 bB	4.02 ± 0.15 aA
	2017	10–20	1.79 ± 0.13 bA	2.83 ± 0.24 aAB	3.12 ± 0.24 aA	2.76 ± 0.10 aA
	2018		1.89 ± 0.36 bA	3.08 ± 0.35 aA	2.72 ± 0.12 aAB	2.79 ± 0.07 aA
	2019		1.49 ± 0.03 bA	2.24 ± 0.19 aB	2.37 ± 0.23 aA	2.60 ± 0.22 aA

Note: Values are mean ± standard error (n = 3). Different lowercase letters indicate significant differences among different treatments in the same year. Different capital letters indicate significant differences between years under the same treatments at $p < 0.05$.

3.3. Changes in Soil Microorganisms

In the 0–10 cm soil layer, biochar and EM addition caused a significant increase in the total biomass of microorganisms compared with CK (Figure 2a). With the increase in years, the microbial biomass in the biochar and EM addition treatments showed a significant increase, especially in the C+M treatment. The microbial biomass in the C+M treatment increased from 11.31 $\mu g\ g^{-1}$ in 2017 to 14.98 $\mu g\ g^{-1}$ in 2019 (Figure 2a). Similar to the top soil (0–10 cm soil depth), the microbial biomass in the deep soil (10–20 cm soil depth) increased significantly in the biochar and EM addition treatments with the increase in years (Figure 2b).

Figure 2. The variations in microbial biomass (**a**,**b**), bacteria (**c**,**d**) and fungi (**e**,**f**) with different treatments in 2017, 2018 and 2019 in 0–20 cm soil layers. Values are mean ± SE (n = 3). Different lowercase letters indicate significant differences among different treatments in the same year. Different capital letters indicate significant differences between years under the same treatments at $p < 0.05$.

Compared to CK, the content of bacteria in the biochar and EM addition was significantly higher in the top soil (Figure 2c). The bacteria in each treatment decreased first, and then, increased with the increase in years (2019 > 2017 > 2018). Fungus content in the M, C and C+M treatments was significantly higher than that in CK, and showed an increasing trend year by year in the top soil (Figure 2e). In the deep soil, bacteria and fungi had no significant changes among treatments, but increased with the increase in years in the C+M treatment (Figure 2d,f).

Biochar and EM addition significantly increased the F: B in the top soil compared with CK, especially in the C+M treatment (Table 3). GP:GN had no significant difference in the deep soil. However, as the years increased, GP:GN had a downward trend (Table 3). Meanwhile, the addition of biochar (C, C+M) significantly reduced the MB: TOC in the top cm soil (Table 3). The MB:TN in each treatment did not change significantly in the top soil in 2017 and 2018, but the MB:TN in the M, C and C+M treatments was higher in 2019 than in CK, and highest in the C+M treatment (3.94, Table 3).

Table 3. The properties of microorganisms in 0–20 cm soil layers with different treatments in 2017, 2018 and 2019.

Parameter	Year	Layer (cm)	CK	M	C	C+M
F:B	2017	0–10	0.18 ± 0.01 bB	0.24 ± 0.01 aA	0.22 ± 0.01 abB	0.25 ± 0.04 aA
	2018		0.21 ± 0.01 bA	0.25 ± 0.03 abA	0.27 ± 0.01 aA	0.29 ± 0.04 aA
	2019		0.20 ± 0.01 bA	0.27 ± 0.04 aA	0.19 ± 0.01 bC	0.30 ± 0.04 aA
	2017	10–20	0.17 ± 0.01 aB	0.19 ± 0.02 aB	0.21 ± 0.06 aA	0.19 ± 0.03 aB
	2018		0.22 ± 0.03 bA	0.27 ± 0.01 abA	0.23 ± 0.05 bA	0.32 ± 0.07 aA
	2019		0.19 ± 0.02 abC	0.16 ± 0.01 bB	0.24 ± 0.07 aA	0.19 ± 0.03 abB
GP:GN	2017	0–10	1.44 ± 0.11 aAB	1.51 ± 0.07 aA	1.37 ± 0.03 aB	1.45 ± 0.04 aA
	2018		1.55 ± 0.05 aA	1.47 ± 0.03 aA	1.41 ± 0.17 aA	1.41 ± 0.08 aA
	2019		1.29 ± 0.05 aB	1.34 ± 0.01 aB	1.13 ± 0.01 bB	1.13 ± 0.04 bB
	2017	10–20	1.12 ± 0.19 aA	1.24 ± 0.15 aA	1.24 ± 0.08 aA	1.00 ± 0.07 aB
	2018		1.22 ± 0.04 aA	1.14 ± 0.04 aA	1.24 ± 0.17 aA	1.22 ± 0.10 aA
	2019		1.11 ± 0.08 aA	1.12 ± 0.05 aA	1.21 ± 0.12 aA	1.04 ± 0.05 aAB
MB:TOC	2017	0–10	0.28 ± 0.01 aA	0.30 ± 0.04 aAB	0.21 ± 0.02 bA	0.19 ± 0.01 bB
	2018		0.22 ± 0.01 bB	0.25 ± 0.02 abB	0.21 ± 0.01 bA	0.27 ± 0.02 aA
	2019		0.30 ± 0.01 aA	0.34 ± 0.01 aA	0.23 ± 0.01 bA	0.30 ± 0.03 aA
	2017	10–20	0.27 ± 0.04 aB	0.24 ± 0.01 aB	0.23 ± 0.01 aB	0.23 ± 0.01 aB
	2018		0.28 ± 0.04 abB	0.23 ± 0.01 bB	0.30 ± 0.01 aAB	0.30 ± 0.04 aA
	2019		0.31 ± 0.01 aA	0.26 ± 0.03 aA	0.33 ± 0.05 aA	0.24 ± 0.02 aB
MB:TN	2017	0–10	3.36 ± 0.11 aA	3.16 ± 0.34 aB	3.20 ± 0.39 aAB	3.04 ± 0.13 aB
	2018		2.65 ± 0.19 bB	2.73 ± 0.10 bB	2.82 ± 0.05 abB	3.26 ± 0.35 aAB
	2019		3.27 ± 0.22 bA	3.85 ± 0.29 abA	3.57 ± 0.17 abA	3.96 ± 0.44 aA
	2017	10–20	3.34 ± 0.16 aA	2.95 ± 0.11 bAB	2.94 ± 0.20 bB	2.69 ± 0.08 bA
	2018		2.84 ± 0.45 abAB	2.21 ± 0.43 bB	2.78 ± 0.24 abB	3.14 ± 0.38 aA
	2019		3.50 ± 0.27 abA	3.58 ± 0.20 aA	3.74 ± 0.71 aA	2.85 ± 0.09 bA

Note: Values are mean ± standard error (n = 3). Different lowercase letters indicate significant differences among different treatments in the same year. Different capital letters indicate significant differences between years under the same treatments at $p < 0.05$. F/B—fungi: bacteria, GP/GN—Gram-positive: Gram-negative, MB: TOC—microbial biomass: total organic carbon, MB:TN—microbial biomass: total nitrogen.

In PCA, PC1 and PC2 explained the variation of 75.62% and 73.80% in the top and deep soil, respectively (Figure 3). With increasing years, the dispersion degree of the different treatments increased, and the degree of dispersion in the top soil was greater than that in the deep soil (Figure 3a,c). In the top soil, the relative contents of arbuscular mycorrhizal fungi (16.1w5c), saprotrophic fungi (18.2.w6,9c, 18.1.w9c) and methanotrophic bacteria (18.1.w7c) were higher and those of other microorganisms were lower in the biochar and EM addition treatments than in CK, especially in the C+M treatment (Figure 3a,b). In the top soil, the microbial community structure among the treatments was relatively similar in 2017; the relative contents of Gram-positive bacteria (a15.0, i16.0) and saprotrophic fungi (18.1.w9c) in the C and C+M treatments were higher than those in the M and CK treatments in 2018; and the relative contents of saprotrophic fungi (18.1.w9c, 18.2.w6,9c) and methanotrophic bacteria (18.1.w7c) were higher in the M treatment in 2019 (Figure 3c,d).

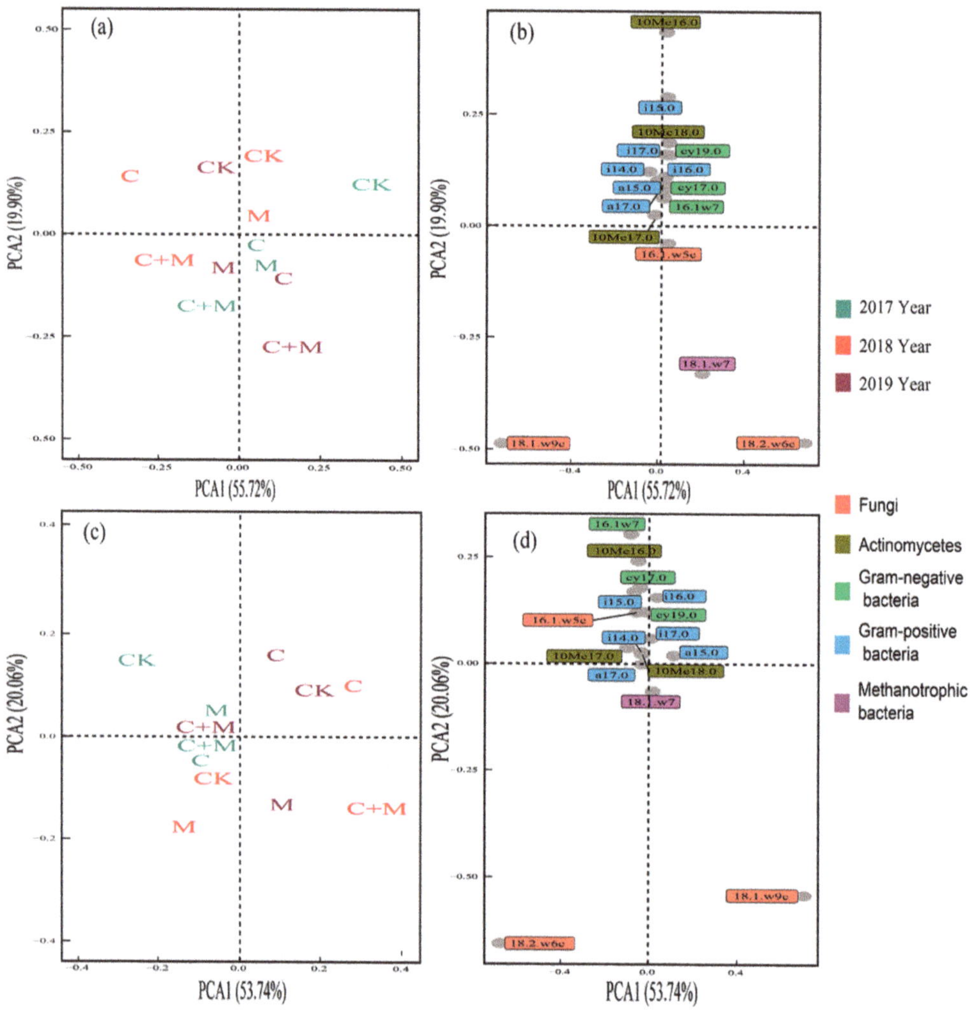

Figure 3. Principal component analysis (PCA) of microbial community in 0–20 cm soil layers: (**a**,**b**) 0–10 cm soil layer, (**c**,**d**) 10–20 cm soil layer.

3.4. The Interaction between Microorganisms and Soil Properties

Biochar and EM addition changed the interactions between soil microorganisms and soil properties. In CK, the interaction between different microbial communities and soil physicochemical properties was weak; only pH, EC and SWC had significant interactions with microbial communities. In the M treatment, different microbial communities had a significant relationship with TOC and TN, but the correlation between different soil physicochemical properties was weak. In the C treatment, fungi, actinomyces and Gram-positive bacteria had significant correlations with soil physicochemical properties. In the C+M treatment, TOC, TN and AP were significantly correlated with fungi, actinomyces and Gram-positive bacteria (Figure 4). Except for the M treatment, the *P* value of the Procrustes analysis in the other treatments was less than 0.05, and the M2 performance was C (M2 = 0.2891) < C+M (M2 = 0.4177) < CK (M2 = 0.4961) < M (M2 = 0.6081) (Figure 5). So, the relationship between the environmental variables and microbial community was better in the C treatment, followed by the C+M treatment (Figure 5b,d).

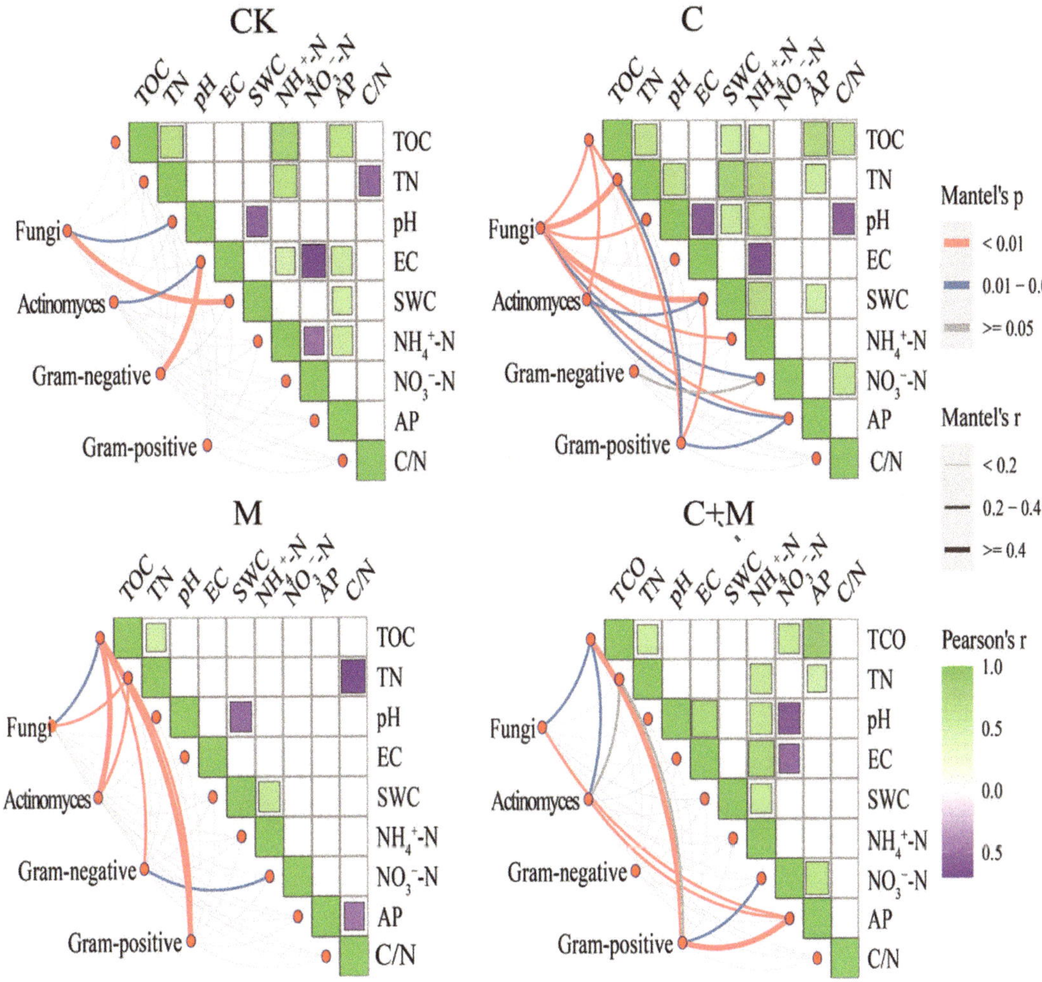

Figure 4. Pearson correlation and Mantel tests for different treatments. Statistical significance was set at $p < 0.05$.

In the RDA, the influence of C/N and SWC on the microbial community structure was smaller in the top soil than that of the other soil factors. Sf (Saprotrophic fungi) was positively correlated with most soil factors. The sensitivity of microorganisms to soil factors in the C and C+M treatments was higher than that in CK and M treatments, especially the C+M treatment (Figure 6a). In the deep soil, the influence of C/N and SWC on the microbial community structure was higher than that of the other soil factors (Figure 6b).

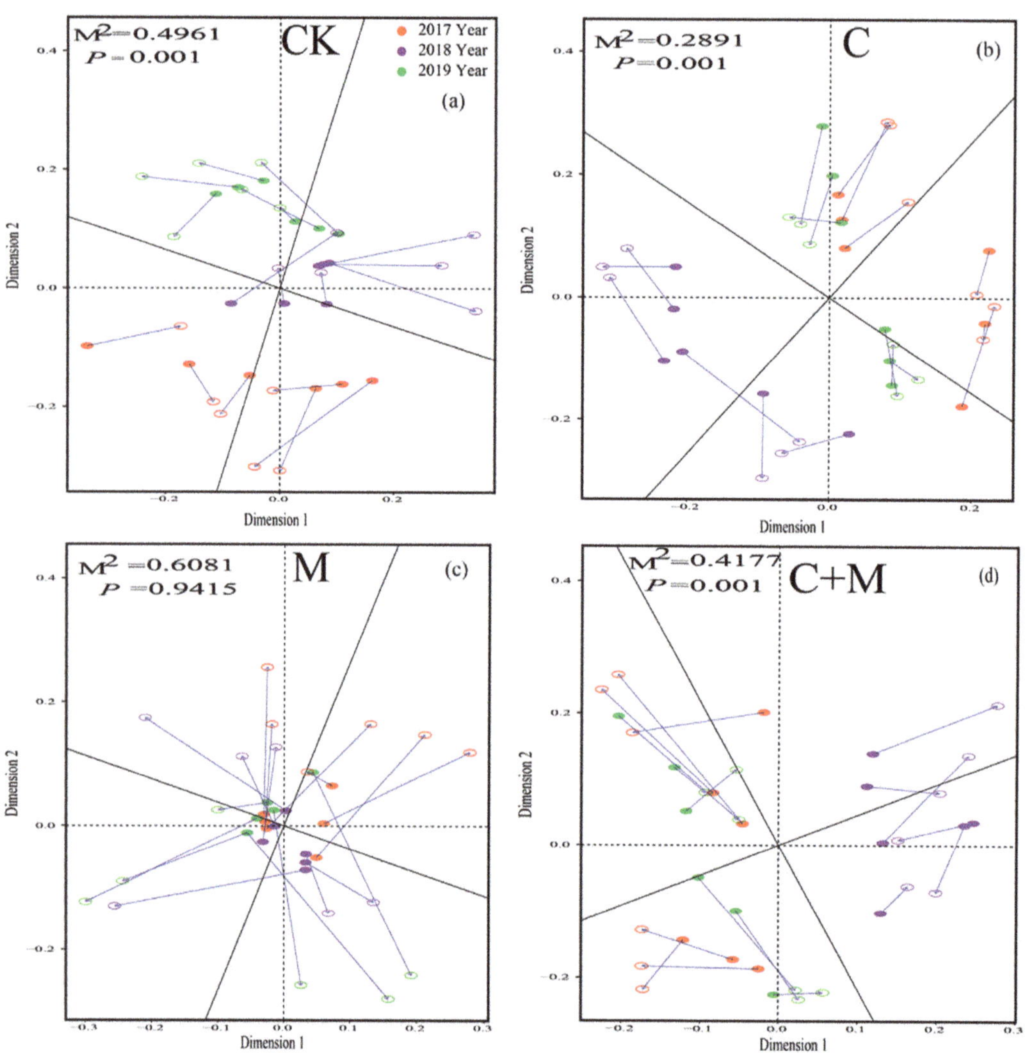

Figure 5. Procrustes analysis on the correlation between environmental variables and microbial community in different treatments. (**a**): CK, (**b**): C, (**c**): M, (**d**): C+M.

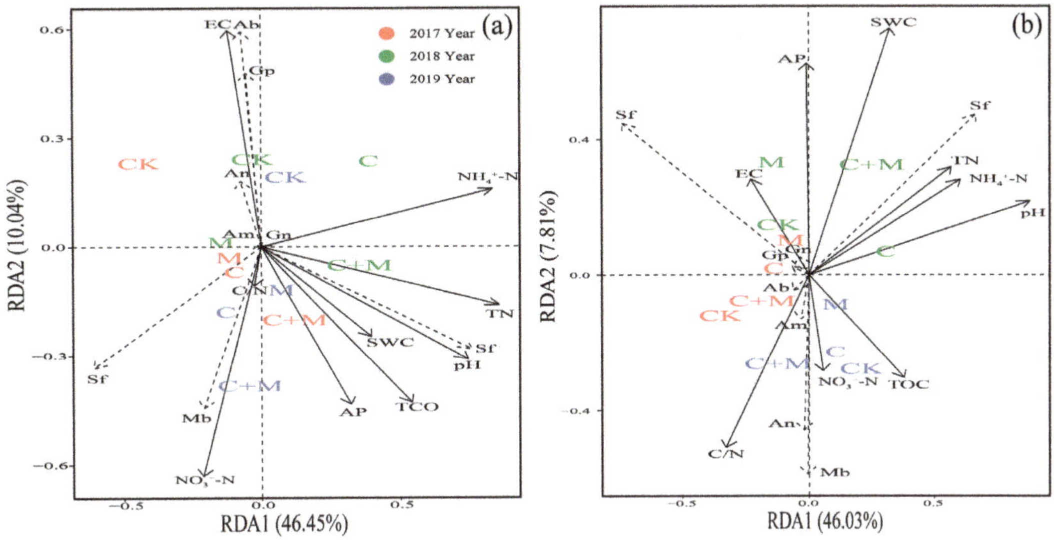

Figure 6. Redundancy analysis (RDA) of environmental variables and indicator PLFAs: (**a**) 0–10 cm soil layer, (**b**) 10–20 cm soil layer. Black lines represent environmental variables, and dashed line represents indicator PLFAs. See Tables 1, 2 and S1 for abbreviations.

4. Discussion

4.1. Responses of Aboveground Vegetation Biomass to the Addition of Biochar or EM

Biochar and EM addition significantly increased the biomass of aboveground vegetation (Figure 1). This study also found that the nutrient content in soil increased significantly after the addition of biochar or EM (Table 2). This might be the key reason for the significant increase in aboveground biomass. Studies have shown that biochar can change soil fertility and the availability of nutrients to plants [27,45]. This is because, on the one hand, biochar contains certain nutrients (N, P, K) that can promote the growth of plants [27,45]. On the other hand, biochar can increase the permeability of soil [46], change the physical properties of soil to increase the availability of soil nutrients [22] and improve plant nutrient absorption [27], thus promoting plant growth. And biochar improved soil proteobacteria abundance and most of the ammonia-oxidizing bacteria, including nitrogen-fixing bacteria, ammonia-oxidizing bacteria, cellulose-decomposing bacteria, nitrifying bacteria and denitrifying bacteria belonging to proteobacteria, meaning it plays a significant role in nitrogen recycling, which is beneficial for plant growth, yield and fruit seed quality [47]. Previous studies have shown that EM can be used as an activator of soil. The addition of EM can accelerate the decomposition of soil organic matter, improve soil fertility, increase soil nutrient elements and ultimately promote plant growth [48,49]. In addition, the beneficial flora in EM (such as photosynthetic bacteria and lactic acid bacteria) can not only increase plant photosynthesis and improve plant nutrient absorption, but also inhibit the growth of pathogenic bacteria [49]. We found that the biomass of aboveground vegetation was the largest in the C+M treatment (Figure 1). This indicated that biochar addition combined with EM had a strong synergistic effect. This might be because when biochar and EM were added together, EM compensated for the deficiency in microorganisms and provoked the reproduction of other beneficial bacteria, and biochar addition reduces nutrient loss and provides a good habitat environment for microorganisms [22,29,50]. Many studies confirm the above findings and biochar has been extensively studied as a soil amendment for carbon sequestration and for improving soil quality. The systematic understanding of the responses of soil microbial biomass and diversity to biochar addition shows that biochar increases microbial biomass but has variable effects on microbial diversity [9,51]. The application of biochar, particularly that produced under low temperatures and from nutrient-rich feedstocks, could better increase soil microbial biomass (based on phospholipid fatty acid analysis (MBCPLFA)) and diversity [52]. A study confirmed that the increases in total microbial diversity with biochar addition were greater in acidic and sandy soils with low soil organic carbon content [40]. These studies confirmed that the combination of biochar and EM was a more effective strategy to promote the productivity of degraded alpine grasslands.

4.2. Responses of Soil Physicochemical Properties to the Addition of Biochar or EM

Under the C and C+M treatments, the pH increased significantly (Table 1). Since biochar contains ash elements (K, Na, Mg), it can increase the soil salinity saturation and reduce the level of exchangeable hydrogen ions and aluminum ions through adsorption [53], thereby increasing soil pH. However, we found that the pH did not increase continuously with the increase in experimental time. This may be because the ash elements contained in biochar were leached out or absorbed by plants, preventing the continuous increase in pH. EM containing a lot of lactic acid bacteria not only makes the bacterial liquid acidic, but also produces organic acid during the decomposition of organic matter, therefore decreasing the soil pH [54]. The addition of EM and biochar decreased soil EC (Table 1). This is probably because EM can inhibit the accumulation of Na+ and increase the uptake of N, P, K^+, Fe, Zn and Cu by plants [22]. The decreasing effect of biochar on EC occurs mainly because biochar, on the one hand, adsorbs salt through its strong adsorption capacity and reduces soil salt ions [55], and on the other hand, can change the physical structure of the soil (increases its porosity) and accelerate the leaching of soil salts [56].

Soil TOC and TN under biochar and EM addition significantly increased, and showed an increasing trend with the increase in experimental years. The increasing intensity in the top soil was greater than that in the deep soil (Table 2). First, biochar generally contains a large amount of inert organic carbon, a small amount of variable organic carbon and abundant nitrogen. When biochar enters the soil, the organic carbon and nitrogen within the biochar will be added to the soil [57,58]. Second, biochar has a strong adsorption capacity, and can absorb small organic molecules in the soil and make them aggregate to form organic matter; then, it produces a negative priming effect with soil organic carbon mineralization, and finally, it increases soil carbon sequestration [59,60]. In addition, biochar can indirectly increase carbon and nitrogen in the soil by changing the soil physicochemical properties and biological characteristics. Biochar addition increases soil porosity and promotes the formation of soil aggregates [61], increases soil microorganisms to accelerate litter decomposition [62], and increases the formation of endogenous carbon and nitrogen and the input of exogenous carbon and nitrogen [56]. The beneficial microorganisms in the EM not only produce nutrients to improve soil fertility, but also improve soil permeability and aggregate structure to enhance the soil physicochemical properties and reduce soil particle loss, thereby increasing soil carbon and nitrogen [63]. In addition, EM can increase the activity of soil enzymes, promote the decomposition of soil organic matter, and then, increase the accumulation of soil carbon and nitrogen [64]. In this study, biochar-only addition significantly increased the C/N in the top soil, but EM addition (M, C+M) reduced the C/N, especially in the C+M treatment (Table 1). Biochar addition (C, C+M) significantly supplemented soil organic carbon. However, under the addition of M or C+M, some soil organic carbon was decomposed (Table 2), and thereby, the C/N was reduced. The C/N in the C+M treatment was maintained at the global average level of grasslands (11.8) [65]. This indicated that the biochar addition combined with EM could make soil carbon and nitrogen at a relatively balanced level, which was conducive to maintaining the health of the grassland soil system.

Biochar and EM addition also changed the condition of other nutrients in the soil. With the increase in years, NH_4^+-N first increased, and then, decreased, while the NO_3^--N was the opposite (Table 2). Jenkins et al. [66] found that the available nitrogen in biochar mostly existed in the form of nitrate nitrogen. So, when biochar entered the soil, nitrate nitrogen was rapidly supplemented (Table 2). In 2018, the soil at the study site had a relatively high pH and SWC. High soil water content may cause soil hypoxia. Together with high pH, this may inhibit nitrification and increase the accumulation of NH_4^+-N [67]. In addition, microorganisms prefer to use ammonium nitrogen rather than nitrate nitrogen [68]. Meanwhile, the MB: TN in each treatment was lower in 2018 than in 2017 and 2019 (Table 3). The combined effects of these factors might result in the accumulation of NH_4^+-N. However, with the decrease in SWC in 2019, nitrification was promoted (Table 2). As the residence time of biochar in soil increases, the adsorption capacity of soil to NH_4^+ is enhanced [69]. Therefore, even if the nitrification was enhanced, there was still high NH_4^+-N in the C and C+M treatments (Table 2). Although the inter-annual variation in ammonium nitrogen and nitrate nitrogen was different after the addition of biochar, biochar significantly increased the content of soil inorganic nitrogen. This may be because biochar can increase the activity of enzymes and nitrogen-fixing bacteria, thus increasing the mineralization of soil nitrogen [70,71]. Studies have shown that EM addition can promote the activity of soil microorganisms and enzymes, accelerate the decomposition of organic matter and increase the nitrogen fixation ability of beneficial microorganisms. Therefore, the addition of EM can increase the mineralization and fixation of nitrogen, which is consistent with the results of this experiment.

4.3. Responses of Soil Microorganisms to the Addition of Biochar or EM

The total amounts of microorganisms, bacteria and fungi increased in the soil of degraded alpine grassland when biochar or EM was added, and the microbial biomass showed an upward trend with the increase in experimental years (Figure 2). The high adsorption capacity and cation exchange capacity of biochar helps to hold nutrients in the soil, and provides a substrate for the growth and metabolism of soil microorganisms (Table 2) [60,68]. In addition, biochar provides a good habitat for the reproduction of microorganisms due to its porous structure and ability to change the porosity of soil [72]. EM can rapidly increase the activity and amount of microorganisms [25], and decompose soil organic matter to provide a substrate for the growth and metabolism of microorganisms [73].

Some studies have shown that biochar and EM can not only cause changes in microbial biomass but also change microbial community structure [72,74]. In our study, the relative content of saprotrophic fungi (18:1w9c, 18:2w6,9c) and F/B in the C, M and C+M treatments was greater than that in CK, while the relative content of actinobacteria was lower, especially in the C+M treatment (Figure 3). This indicated that the activity of saprotrophic fungi (18:1w9c, 18:2w6,9c) in degraded alpine grassland was promoted by biochar or EM addition. Saprotrophic fungi (18:1w9c, 18:2w6,9c) mainly grow on the surface of young roots, and mycelium can tightly bind around these young roots [75]. The pore size of biochar becomes larger with an increase in the time that biochar stays in soil, which provides a larger habitat for the growth of saprotrophic fungi [37]. EM can promote the growth of plant roots, which is beneficial to the growth of saprotrophic fungi. Therefore, the relative content of saprotrophic fungi in C+M was the highest (Figure 3). The increase in saprotrophic fungi (18:1w9c, 18:2w6,9c) can accelerate the decomposition of litter or dead roots, promote the increase in soil fertility, improve the absorption of mineral elements by vegetation roots and increase plant tolerance to harsh environments and disease resistance [75]. Overall, biochar combined with EM can not only promote the growth of plants on degraded grassland, but also increase soil nutrients and microbial biomass, and change the abundance of specific microorganisms on degraded grassland. In this paper, their joint positive effect was more beneficial to the rapid restoration of degraded alpine grasslands.

5. Conclusions

Biochar addition alone, as well as the combination of biochar and effective microorganisms (EM), demonstrated more pronounced enhancements in vegetation biomass, soil physicochemical properties and microorganisms in degraded alpine grassland, compared to EM addition alone. Soil parameters such as total organic carbon (TOC), total nitrogen

(TN), nitrate nitrogen (NO_3^--N), ammonium nitrogen (NH_4^+-N), available phosphorus (AP), soil water content (SWC) and microbial biomass exhibited significant increases. The addition of only biochar had the most pronounced impact on soil carbon, nitrogen and phosphorus. Conversely, when biochar was combined with EM, it exerted the strongest influence on above-ground vegetation biomass and microbial biomass. The addition of biochar or EM altered both the structure of the microbial community and its interaction with various soil parameters. The relative content of saprotrophic fungi (18:2w6,9c,18:1w9c) increased in the C, M and C+M treatments, and was the highest in C+M treatment. Consequently, the ecosystem of degraded alpine grassland experienced the most significant improvement when biochar was added in combination with EM.

Supplementary Materials: The following supporting information can be downloaded at: https://www.mdpi.com/article/10.3390/agronomy13092203/s1, Table S1: Microbial PLFA biomarkers and metrics [76–81].

Author Contributions: J.L., K.L. and X.S. contributed to the conception and design of this study. H.L. and Y.H. organized the database. J.L. performed the statistical analysis. J.L. and K.L. wrote the first draft of the manuscript. X.S., J.S. and K.L. wrote sections of the manuscript. All authors have read and agreed to the published version of the manuscript.

Funding: This project was supported by the National Key Research and Development Program of China (No. 2021YFD1300503) and the earmarked fund for CARS (CARS-34).

Data Availability Statement: The datasets generated and/or analyzed during the current study are available from the corresponding author on reasonable request.

Conflicts of Interest: The authors declare no conflict of interest.

References

1. Wei, Y.; Lu, H.; Wang, J.; Wang, X.; Sun, J. Dual Influence of Climate Change and Anthropogenic Activities on the Spatiotemporal Vegetation Dynamics Over the Qinghai-Tibetan Plateau From 1981 to 2015. *Earths Future* **2022**, *10*, e2021EF002566. [CrossRef]
2. Dong, S.; Zhang, J.; Li, Y.; Liu, S.; Dong, Q.; Zhou, H.; Yeomans, J.; Li, Y.; Li, S.; Gao, X. Effect of grassland degradation on aggregate-associated soil organic carbon of alpine grassland ecosystems in Qinghai-Tibetan Plateau. *Eur. J. Soil Sci.* **2019**, *71*, 69–79. [CrossRef]
3. Ran, Q.; Hao, Y.; Xia, A.; Liu, W.; Hu, R.; Cui, X.; Xue, K.; Song, X.; Xu, C.; Ding, B.; et al. Quantitative Assessment of the Impact of Physical and Anthropogenic Factors on Vegetation Spatial-Temporal Variation in Northern Tibet. *Remote Sens.* **2019**, *11*, 1183. [CrossRef]
4. Su, X.; Han, W.; Liu, G.; Zhang, Y.; Lu, H. Substantial gaps between the protection of biodiversity hotspots in alpine grasslands and the effectiveness of protected areas on the Qinghai-Tibetan Plateau, China. *Agric. Ecosyst. Environ.* **2019**, *278*, 15–23. [CrossRef]
5. Li, Y.; Dong, S.; Liu, S.; Zhou, H.; Gao, Q.; Cao, G.; Wang, X.; Su, X.; Zhang, Y.; Tang, L.; et al. Seasonal changes of CO_2, CH_4 and N_2O fluxes in different types of alpine grassland in the Qinghai-Tibetan Plateau of China. *Soil Biol. Biochem.* **2015**, *80*, 306–314. [CrossRef]
6. Han, W.; Lu, H.; Liu, G.; Wang, J.; Su, X. Quantifying Degradation Classifications on Alpine Grassland in the Lhasa River Basin, Qinghai-Tibetan Plateau. *Sustainability* **2019**, *11*, 7067. [CrossRef]
7. Bardgett, R.D.; Bullock, J.M.; Lavorel, S.; Manning, P.; Schaffner, U.; Ostle, N.; Chomel, M.; Durigan, G.; Fry, E.L.; Johnson, D.; et al. Combatting global grassland degradation. *Nat. Rev. Earth Environ.* **2021**, *2*, 720–735. [CrossRef]
8. Du, C.; Gao, Y. Grazing exclusion alters ecological stoichiometry of plant and soil in degraded alpine grassland. *Agric. Ecosyst. Environ.* **2021**, *308*, 107256. [CrossRef]
9. Zhang, G.; Zhao, Z.; Yin, X.A.; Zhu, Y. Impacts of biochars on bacterial community shifts and biodegradation of antibiotics in an agricultural soil during short-term incubation. *Sci. Total Environ.* **2021**, *771*, 144751. [CrossRef]
10. Yan, Y.; Jarvie, S.; Liu, Q.; Zhang, Q. Effects of fragmentation on grassland plant diversity depend on the habitat specialization of species. *Biol. Conserv.* **2022**, *275*, 109773. [CrossRef]
11. Che, R.; Wang, Y.; Lia, K.; Xu, Z.; Hua, J.; Wang, F.; Rui, Y.; Li, L.; Pang, Z.; Cui, X. Degraded patch formation significantly changed microbial community composition in alpine meadow soils. *Soil Tillage Res.* **2019**, *195*, 104426. [CrossRef]
12. Raiesi, F. A minimum data set and soil quality index to quantify the effect of land use conversion on soil quality and degradation in native rangelands of upland arid and semiarid regions. *Ecol. Indic.* **2017**, *75*, 307–320. [CrossRef]
13. Lei, L.; Xiao, W.; Zeng, L.; Frey, B.; Huang, Z.; Zhu, J.; Cheng, R.; Li, M.-H. Effects of thinning intensity and understory removal on soil microbial community in Pinus massoniana plantations of subtropical China. *Appl. Soil Ecol.* **2021**, *167*, 104055. [CrossRef]

14. Shen, J.; Tao, Q.; Dong, Q.; Luo, Y.; Luo, J.; He, Y.; Li, B.; Li, Q.; Xu, Q.; Li, M.; et al. Long-term conversion from rice-wheat to rice-vegetable rotations drives variation in soil microbial communities and shifts in nitrogen-cycling through soil profiles. *Geoderma* **2021**, *404*, 115299. [CrossRef]
15. Zhao, H.; Li, Q.; Jin, X.; Li, D.; Zhu, Z.; Li, Q.X. Chiral enantiomers of the plant growth regulator paclobutrazol selectively affect community structure and diversity of soil microorganisms. *Sci. Total Environ.* **2021**, *797*, 148942. [CrossRef]
16. Qiu, D.; Zhu, G.; Lin, X.; Jiao, Y.; Lu, S.; Liu, J.; Liu, J.; Zhang, W.; Ye, L.; Li, R.; et al. Dissipation and movement of soil water in artificial forest in arid oasis areas: Cognition based on stable isotopes. *Catena* **2023**, *228*, 107178. [CrossRef]
17. Ding, M.Q.; Yang, S.S.; Ding, J.; Zhang, Z.R.; Zhao, Y.L.; Dai, W.; Sun, H.J.; Zhao, L.; Xing, D.; Ren, N.; et al. Gut Microbiome Associating with Carbon and Nitrogen Metabolism during Biodegradation of Polyethene in Tenebrio larvae with Crop Residues as Co-Diets. *Sci. Total Environ.* **2023**, *57*, 3031–3041. [CrossRef]
18. Zhou, J.; Wang, L.; Zhong, X.; Yao, T.; Qi, J.; Wang, Y.; Xue, Y. Quantifying the major drivers for the expanding lakes in the interior Tibetan Plateau. *Sci. Bull.* **2022**, *67*, 474–478. [CrossRef]
19. Liang, J.-P.; Xue, Z.-Q.; Yang, Z.-Y.; Chai, Z.; Niu, J.-P.; Shi, Z.-Y. Effects of microbial organic fertilizers on Astragalus membranaceus growth and rhizosphere microbial community. *Ann. Microbiol.* **2021**, *71*, 11. [CrossRef]
20. Mayer, J.; Scheid, S.; Widmer, F.; Fließbach, A.; Oberholzer, H.-R. How effective are 'Effective microorganisms®(EM)'? Results from a field study in temperate climate. *Appl. Soil Ecol.* **2010**, *46*, 230–239. [CrossRef]
21. Mise, K.; Koyama, Y.; Matsumoto, A.; Fujita, K.; Kunito, T.; Senoo, K.; Otsuka, S. Pectin drives microbial phosphorus solubilization in soil: Evidence from isolation-based and community-scale approaches. *Eur. J. Soil Biol.* **2020**, *97*, 103169. [CrossRef]
22. Abd El-Mageed, T.A.; Rady, M.M.; Taha, R.S.; Abd El Azeam, S.; Simpson, C.R.; Semida, W.M. Effects of integrated use of residual sulfur-enhanced biochar with effective microorganisms on soil properties, plant growth and short-term productivity of Capsicum annuum under salt stress. *Sci. Hortic.* **2020**, *261*, 108930. [CrossRef]
23. Ortega, R.; Miralles, I.; Meca, D.E.; Gázquez, J.C.; Domene, M.Á. Effect of Organic and Synthetic Fertilizers on the Crop Yield and Macronutrients Contents in Soil and Pepper. *Commun. Soil Sci. Plant Anal.* **2016**, *47*, 1216–1226. [CrossRef]
24. Cui, Q.; Xia, J.; Yang, H.; Liu, J.; Shao, P. Biochar and effective microorganisms promote Sesbania cannabina growth and soil quality in the coastal saline-alkali soil of the Yellow River Delta, China. *Sci. Total Environ.* **2021**, *756*, 143801. [CrossRef] [PubMed]
25. Talaat, N.B. Effective microorganisms: An innovative tool for inducing common bean (*Phaseolus vulgaris* L.) salt-tolerance by regulating photosynthetic rate and endogenous phytohormones production. *Sci. Hortic.* **2019**, *250*, 254–265. [CrossRef]
26. Mitchell, E.; Scheer, C.; Rowlings, D.; Cotrufo, F.; Conant, R.T.; Grace, P. Important constraints on soil organic carbon formation efficiency in subtropical and tropical grasslands. *Glob. Change Biol.* **2021**, *27*, 5383–5391. [CrossRef] [PubMed]
27. Agegnehu, G.; Srivastava, A.K.; Bird, M.I. The role of biochar and biochar-compost in improving soil quality and crop performance: A review. *Appl. Soil Ecol.* **2017**, *119*, 156–170. [CrossRef]
28. Gao, S.; DeLuca, T.H.; Cleveland, C.C. Biochar additions alter phosphorus and nitrogen availability in agricultural ecosystems: A meta-analysis. *Sci. Total Environ.* **2019**, *654*, 463–472. [CrossRef]
29. El-Naggar, A.; Lee, S.S.; Rinklebe, J.; Farooq, M.; Song, H.; Sarmah, A.K.; Zimmerman, A.R.; Ahmad, M.; Shaheen, S.M.; Ok, Y.S. Biochar application to low fertility soils: A review of current status, and future prospects. *Geoderma* **2019**, *337*, 536–554. [CrossRef]
30. Gunarathne, V.; Senadeera, A.; Gunarathne, U.; Biswas, J.K.; Almaroai, Y.A.; Vithanage, M. Potential of biochar and organic amendments for reclamation of coastal acidic-salt affected soil. *Biochar* **2020**, *2*, 107–120. [CrossRef]
31. Kim, H.S.; Kim, K.R.; Yang, J.E.; Ok, Y.S.; Owens, G.; Nehls, T.; Wessolek, G.; Kim, K.H. Effect of biochar on reclaimed tidal land soil properties and maize (*Zea mays* L.) response. *Chemosphere* **2016**, *142*, 153–159. [CrossRef] [PubMed]
32. Mandal, S.; Thangarajan, R.; Bolan, N.S.; Sarkar, B.; Khan, N.; Ok, Y.S.; Naidu, R. Biochar-induced concomitant decrease in ammonia volatilization and increase in nitrogen use efficiency by wheat. *Chemosphere* **2016**, *142*, 120–127. [CrossRef]
33. Bhaduri, D.; Saha, A.; Desai, D.; Meena, H.N. Restoration of carbon and microbial activity in salt-induced soil by application of peanut shell biochar during short-term incubation study. *Chemosphere* **2016**, *148*, 86–98. [CrossRef]
34. El-Naggar, A.; Lee, S.S.; Awad, Y.M.; Yang, X.; Ryu, C.; Rizwan, M.; Rinklebe, J.; Tsang, D.C.W.; Ok, Y.S. Influence of soil properties and feedstocks on biochar potential for carbon mineralization and improvement of infertile soils. *Geoderma* **2018**, *332*, 100–108. [CrossRef]
35. Elzobair, K.A.; Stromberger, M.E.; Ippolito, J.A.; Lentz, R.D. Contrasting effects of biochar versus manure on soil microbial communities and enzyme activities in an Aridisol. *Chemosphere* **2016**, *142*, 145–152. [CrossRef]
36. Xu, W.; Whitman, W.B.; Gundale, M.J.; Chien, C.C.; Chiu, C.Y. Functional response of the soil microbial community to biochar applications. *GCB Bioenergy* **2020**, *13*, 269–281. [CrossRef]
37. Dangi, S.; Gao, S.; Duan, Y.; Wang, D. Soil microbial community structure affected by biochar and fertilizer sources. *Appl. Soil Ecol.* **2020**, *150*, 103452. [CrossRef]
38. Fernandez, J.M.; Nieto, M.A.; Lopez-de-Sa, E.G.; Gasco, G.; Mendez, A.; Plaza, C. Carbon dioxide emissions from semi-arid soils amended with biochar alone or combined with mineral and organic fertilizers. *Sci. Total Environ.* **2014**, *482–483*, 1–7. [CrossRef]
39. Rafique, M.; Ortas, I.; Ahmed, I.A.M.; Rizwan, M.; Afridi, M.S.; Sultan, T.; Chaudhary, H.J. Potential impact of biochar types and microbial inoculants on growth of onion plant in differently textured and phosphorus limited soils. *J. Environ. Manag.* **2019**, *247*, 672–680. [CrossRef]
40. Zhang, T.; Song, B.; Han, G.; Zhao, H.; Hu, Q.; Zhao, Y.; Liu, H. Effects of coastal wetland reclamation on soil organic carbon, total nitrogen, and total phosphorus in China: A meta-analysis. *Land Degrad. Dev.* **2023**, *34*, 3340–3349. [CrossRef]

41. Li, W.; Shi, Y.; Zhu, D.; Wang, W.; Liu, H.; Li, J.; Shi, N.; Ma, L.; Fu, S. Fine root biomass and morphology in a temperate forest are influenced more by the nitrogen treatment approach than the rate. *Ecol. Indic.* **2021**, *130*, 108031. [CrossRef]
42. *GB 19733-2003 [S]*; Classification Index of Natural Grassland Degradation, Desertification and Salinization. Ministry of Agriculture of the People's Republic of China, Standards Press of China: Beijing, China, 2003.
43. Li, J.; Zhao, Y.; Shao, X.; Huang, D.; Shang, J.; Li, H.; He, Y.; Liu, K. The Mixed Addition of Biochar and Nitrogen Improves Soil Properties and Microbial Structure of Moderate-Severe Degraded Alpine Grassland in Qinghai-Tibet Plateau. *Front. Plant Sci.* **2021**, *12*, 765041. [CrossRef] [PubMed]
44. White, D.C.; Davis, W.M.; Nickels, J.S.; King, J.D.; Bobbie, R.J. Determination of the sedimentary microbial biomass by extractible lipid phosphate. *Oecologia* **1979**, *40*, 51–62. [CrossRef] [PubMed]
45. Ding, Y.; Liu, Y.; Liu, S.; Li, Z.; Tan, X.; Huang, X.; Zeng, G.; Zhou, L.; Zheng, B. Biochar to improve soil fertility. A review. *Agron. Sustain. Dev.* **2016**, *36*, 36. [CrossRef]
46. Laghari, M.; Mirjat, M.S.; Hu, Z.; Fazal, S.; Xiao, B.; Hu, M.; Chen, Z.; Guo, D. Effects of biochar application rate on sandy desert soil properties and sorghum growth. *Catena* **2015**, *135*, 313–320. [CrossRef]
47. Zhang, H.; Ullah, F.; Ahmad, R.; Ali Shah, S.U.; Khan, A.; Adnan, M. Response of Soil Proteobacteria to Biochar Amendment in Sustainable Agriculture- A mini review. *J. Soil Plant Environ.* **2022**, *1*, 16–30. [CrossRef]
48. Hu, C.; Qi, Y. Long-term effective microorganisms application promote growth and increase yields and nutrition of wheat in China. *Eur. J. Agron.* **2013**, *46*, 63–67. [CrossRef]
49. Talaat, N.B.; Ghoniem, A.E.; Abdelhamid, M.T.; Shawky, B.T. Effective microorganisms improve growth performance, alter nutrients acquisition and induce compatible solutes accumulation in common bean (*Phaseolus vulgaris* L.) plants subjected to salinity stress. *Plant Growth Regul.* **2014**, *75*, 281–295. [CrossRef]
50. Lehmann, J.; Rillig, M.C.; Thies, J.; Masiello, C.A.; Hockaday, W.C.; Crowley, D. Biochar effects on soil biota—A review. *Soil Biol. Biochem.* **2011**, *43*, 1812–1836. [CrossRef]
51. Yang, Y.; Liu, L.; Zhang, P.; Wu, F.; Wang, Y.; Xu, C.; Zhang, L.; An, S.; Kuzyakov, Y. Large-scale ecosystem carbon stocks and their driving factors across Loess Plateau. *Carbon Neutrality* **2023**, *2*, 5. [CrossRef]
52. Cheng, M.; Cui, Y.; Yan, X.; Zhang, R.; Wang, J.; Wang, X. Effect of dual-modified cassava starches on intelligent packaging films containing red cabbage extracts. *Food Hydrocoll.* **2022**, *124*, 107225. [CrossRef]
53. He, X.; Xie, H.; Gao, D.; Khashi, U.R.M.; Zhou, X.; Wu, F. Biochar and Intercropping with Potato-Onion Enhanced the Growth and Yield Advantages of Tomato by Regulating the Soil Properties, Nutrient Uptake, and Soil Microbial Community. *Front. Microbiol.* **2021**, *12*, 695447. [CrossRef] [PubMed]
54. Schweinsberg-Mickan, M.S.Z.; Müller, T. Impact of effective microorganisms and other biofertilizers on soil microbial characteristics, organic-matter decomposition, and plant growth. *J. Plant Nutr. Soil Sci.* **2009**, *172*, 704–712. [CrossRef]
55. Thomas, S.C.; Frye, S.; Gale, N.; Garmon, M.; Launchbury, R.; Machado, N.; Melamed, S.; Murray, J.; Petroff, A.; Winsborough, C. Biochar mitigates negative effects of salt additions on two herbaceous plant species. *J. Environ. Manage.* **2013**, *129*, 62–68. [CrossRef]
56. Saifullah; Dahlawi, S.; Naeem, A.; Rengel, Z.; Naidu, R. Biochar application for the remediation of salt-affected soils: Challenges and opportunities. *Sci. Total Environ.* **2018**, *625*, 320–335.
57. Pan, S.-Y.; Dong, C.-D.; Su, J.-F.; Wang, P.-Y.; Chen, C.-W.; Chang, J.-S.; Kim, H.; Huang, C.-P.; Hung, C.-M. The Role of Biochar in Regulating the Carbon, Phosphorus, and Nitrogen Cycles Exemplified by Soil Systems. *Sustainability* **2021**, *13*, 5612. [CrossRef]
58. Wang, X.; Zhou, W.; Liang, G.; Song, D.; Zhang, X. Characteristics of maize biochar with different pyrolysis temperatures and its effects on organic carbon, nitrogen and enzymatic activities after addition to fluvo-aquic soil. *Sci. Total Environ.* **2015**, *538*, 137–144. [CrossRef]
59. Biederman, L.A.; Harpole, W.S. Biochar and its effects on plant productivity and nutrient cycling: A meta-analysis. *GCB Bioenergy* **2013**, *5*, 202–214. [CrossRef]
60. Jien, S.H.; Kuo, Y.L.; Liao, C.S.; Wu, Y.T.; Igalavithana, A.D.; Tsang, D.C.W.; Ok, Y.S. Effects of field scale in situ biochar incorporation on soil environment in a tropical highly weathered soil. *Environ. Pollut.* **2021**, *272*, 116009. [CrossRef]
61. Hu, F.; Xu, C.; Ma, R.; Tu, K.; Yang, J.; Zhao, S.; Yang, M.; Zhang, F. Biochar application driven change in soil internal forces improves aggregate stability: Based on a two-year field study. *Geoderma* **2021**, *403*, 115276. [CrossRef]
62. Palansooriya, K.N.; Wong, J.T.F.; Hashimoto, Y.; Huang, L.; Rinklebe, J.; Chang, S.X.; Bolan, N.; Wang, H.; Ok, Y.S. Response of microbial communities to biochar-amended soils: A critical review. *Biochar* **2019**, *1*, 3–22. [CrossRef]
63. Yang, W.; Gong, T.; Wang, J.; Li, G.; Liu, Y.; Zhen, J.; Ning, M.; Yue, D.; Du, Z.; Chen, G. Effects of Compound Microbial Fertilizer on Soil Characteristics and Yield of Wheat (*Triticum aestivum* L.). *J. Soil Sci. Plant Nutr.* **2020**, *20*, 2740–2748. [CrossRef]
64. Ren, H.; Qin, X.; Huang, B.; Fernandez-Garcia, V.; Lv, C. Responses of soil enzyme activities and plant growth in a eucalyptus seedling plantation amended with bacterial fertilizers. *Arch. Microbiol.* **2020**, *202*, 1381–1396. [CrossRef]
65. Wang, X.; Lü, X.; Zhang, H.; Dijkstrac, F.A.; Jiang, Y.; Wang, X.; Lu, J.; Wu, Y.; Wang, Z.; Han, X. Changes in soil C:N:P stoichiometry along an aridity gradient in drylands of northern China. *Geoderma* **2019**, *361*, 114087. [CrossRef]
66. Jenkins, J.R.; Viger, M.; Arnold, E.C.; Harris, Z.M.; Ventura, M.; Miglietta, F.; Girardin, C.; Edwards, R.J.; Rumpel, C.; Fornasier, F.; et al. Biochar alters the soil microbiome and soil function: Results of next-generation amplicon sequencing across Europe. *GCB Bioenergy* **2017**, *9*, 591–612. [CrossRef]

67. Ye, Z.; Liu, L.; Tan, Z.; Zhang, L.; Huang, Q. Effects of pyrolysis conditions on migration and distribution of biochar nitrogen in the soil-plant-atmosphere system. *Sci. Total Environ.* **2020**, *723*, 138006. [CrossRef]
68. Zhang, Y.; Zhao, H.; Hu, W.; Wang, Y.; Zhang, H.; Zhou, X.; Fei, J.; Luo, G. Understanding how reed-biochar application mitigates nitrogen losses in paddy soil: Insight into microbially-driven nitrogen dynamics. *Chemosphere* **2022**, *295*, 133904. [CrossRef]
69. Li, H.; Lu, X.; Xu, Y.; Liu, H. How close is artificial biochar aging to natural biochar aging in fields? A meta-analysis. *Geoderma* **2019**, *352*, 96–103. [CrossRef]
70. Sun, H.; Lu, H.; Chu, L.; Shao, H.; Shi, W. Biochar applied with appropriate rates can reduce N leaching, keep N retention and not increase NH3 volatilization in a coastal saline soil. *Sci. Total Environ.* **2017**, *575*, 820–825. [CrossRef]
71. Zhang, G.; Bai, J.; Zhao, Q.; Jia, J.; Wang, W.; Wang, X. Bacterial Succession in Salt Marsh Soils Along a Short-term Invasion Chronosequence of Spartina alterniflora in the Yellow River Estuary, China. *Microb. Ecol.* **2019**, *79*, 644–661. [CrossRef]
72. Zheng, J.; Chen, J.; Pan, G.; Liu, X.; Zhang, X.; Li, L.; Bian, R.; Cheng, K.; Jinwei, Z. Biochar decreased microbial metabolic quotient and shifted community composition four years after a single incorporation in a slightly acid rice paddy from southwest China. *Sci. Total Environ.* **2016**, *571*, 206–217. [CrossRef] [PubMed]
73. Jimenez-Gomez, A.; Garcia-Estevez, I.; Garcia-Fraile, P.; Escribano-Bailon, M.T.; Rivas, R. Increase in phenolic compounds of Coriandrum sativum L. after the application of a Bacillus halotolerans biofertilizer. *J. Sci. Food Agric.* **2020**, *100*, 2742–2749. [CrossRef] [PubMed]
74. Xu, Y.; Liu, C.; Bao, J.; Zhu, H.; Chen, Y.; Luo, Y.; Zhang, L. Microbial diversity and physicochemical properties in farmland soils amended by effective microorganisms and fulvic acid for cropping Asian ginseng. *Not. Bot. Horti Agrobot. Cluj-Napoca* **2022**, *50*, 12563. [CrossRef]
75. Cao, T.; Fang, Y.; Chen, Y.; Kong, X.; Yang, J.; Alharbi, H.; Kuzyakov, Y.; Tian, X. Synergy of saprotrophs with mycorrhiza for litter decomposition and hotspot formation depends on nutrient availability in the rhizosphere. *Geoderma* **2022**, *410*, 115662. [CrossRef]
76. Djukic, I.; Zehetner, F.; Mentler, A.; Gerzabek, M.H. Microbial community composition and activity in different Alpine vegetation zones. *Soil Biol. Biochem.* **2010**, *42*, 155–161. [CrossRef]
77. Evgrafova, S.Y.; Santruckova, H.; Shibistova, O.B.; Elhottova, D.; Cerna, B.; Zrazhevskaya, G.K.; Lloyd, D. Phospholipid fatty acid composition of microorganisms in pine forest soils of Central Siberia. *Biol. Bull.* **2008**, *35*, 452–458. [CrossRef]
78. Frostegard, A.; Baath, E. The use of phospholipid fatty acid analysis to estimate bacterial and fungal biomass in soil. *Biol Fert. Soils* **1996**, *22*, 59–65. [CrossRef]
79. Li, J.; Shao, X.; Huang, D.; Shang, J.; Liu, K.; Zhang, Q.; Yang, X.; Li, H.; He, Y. The addition of organic carbon and nitrogen accelerates the restoration of soil system of degraded alpine grassland in Qinghai-Tibet Plateau. *Ecol. Eng.* **2020**, *158*, 106084. [CrossRef]
80. Smith, A.P.; Marin-Spiotta, E.; Balser, T. Successional and seasonal variations in soil and litter microbial community structure and function during tropical postagricultural forest regeneration: A multiyear study. *Glob. Chang. Biol.* **2015**, *21*, 3532–3547. [CrossRef]
81. Sundh, I.; Borjesson, G.; Tunlid, A. Methane oxidation and phospholipid fatty acid composition in a podzolic soil profile. *Soil Biol. Biochem.* **2000**, *32*, 1025–1028. [CrossRef]

agronomy

MDPI

Article

Integrating Native Plant Mixtures and Arbuscular Mycorrhizal Fungi Inoculation Increases the Productivity of Degraded Grassland

Jiechao Chang [1], Kang Li [1], Jiayao Xie [1], Yanxia Zhang [1], Sitong Wang [1], Haiyan Ren [1,*] and Manqiang Liu [2]

1 College of Agro-Grassland Science, Nanjing Agricultural University, Nanjing 210095, China
2 Soil Ecology Lab, College of Resources and Environmental Sciences, Nanjing Agricultural University, Nanjing 210095, China
* Correspondence: hren@njau.edu.cn

Abstract: Intense human activities break the grassland–livestock balance and accelerate grassland degradation. We evaluated the use of native dominant species combined with arbuscular mycorrhizal fungi (AMF) in order to recover grassland and restrain grassland degradation. We conducted a full factorial greenhouse experiment to evaluate the interaction effects of native species of distinct traits grass *Lolium perenne* (L) and legume *Trifolium repens* (T) with arbuscular mycorrhizal fungi (AMF) inoculation on grass productivity and soil properties across non-degraded, lightly degraded, and severely degraded soils. The grass–legume mixture was manipulated with five ratios (T:L = 1:0, T:L = 1:1, T:L = 3:1, T:L = 1:3, T:L = 0:1). The results showed that *L. perenne* significantly increased grassland productivity at different grass–legume ratios, regardless of AMF presence or absence. AMF inoculation increased plant N and P content uptake and improved the productivity of degraded grasslands, especially in severely degraded grasslands. The NO_3^--N and available P concentrations increased in soil when the legume component increased from T:L = 0:1 (grass monoculture) to T:L = 1:0 (legume monoculture). This may be because the presence of *Lolium perenne* (L) can promote nitrogen fixation in legumes. Structural equation modeling indicated that grass–legume mixtures directly affected plant biomass, whereas AMF affected plant biomass via providing plant nutrients. A soil quality index based on minimum datasets indicated a significant positive effect of artificial grassland establishment on soil quality. We conclude that planting T:L = 0:1 and T:L = 1:3 combined with AMF inoculation can be used to recover degraded grassland production, and planting T:L = 1:1 and T:L = 1:3 plus AMF inoculation can be applied for grassland nutrient accumulation and stability maintenance.

Keywords: arbuscular mycorrhizal fungi; grassland restoration; grass–legume mixture; planting ratio; plant–soil interaction

Citation: Chang, J.; Li, K.; Xie, J.; Zhang, Y.; Wang, S.; Ren, H.; Liu, M. Integrating Native Plant Mixtures and Arbuscular Mycorrhizal Fungi Inoculation Increases the Productivity of Degraded Grassland. *Agronomy* **2023**, *13*, 7. https://doi.org/10.3390/agronomy13010007

Academic Editors: Kesi Liu and Xinqing Shao

Received: 1 November 2022
Revised: 5 December 2022
Accepted: 14 December 2022
Published: 20 December 2022

1. Introduction

Grassland degradation is an ecological process of reverse evolution caused by a series of human and natural events [1] At present, one-third of grasslands in China are degraded. Nanshan Pasture in Hunan Province is an important pastoral grassland area in southern China. However, due to excessive grazing and climate factors, most of the grassland area in Nanshan has seriously degraded, accompanied with plant biomass and species diversity loss, soil nutrients loss, and soil pH increase [2]. In response to these environmental disasters, various interventions have been implemented, such as fencing pastures, rodent and pest control, fertilization, and farming; however, these strategies require long recovery periods and have unstable effects [3].

The establishment of artificial grassland offers a sound approach for the restoration of degraded grasslands and to slow down the pace of grassland degradation [4,5]. Studies have shown that the establishment of artificial or semi-artificial grasslands by reseeding had a positive effect on productivity and soil structure in degraded grasslands. In addition, the soil microbial community plays a vital role in nutrient cycling, soil structure maintenance, and ecosystem productivity [6]. As the interactions between plant and microorganism are altered in degraded grassland ecosystems, the regulation of soil microorganisms is expected to improve the effect of sowing native grass species for the effective restoration of degraded grassland ecosystems.

Arbuscular mycorrhizal fungi (AMF), as substantial components of microbial communities in soils, form mutualistic associations with the roots of most terrestrial plants, helping them access soil nutrients (especially phosphorus), while accumulating plant carbohydrates [7,8]. Mycorrhizal symbionts can improve the absorption of soil mineral elements by the host plant, increase plant biomass, and improve soil quality and soil aggregate stability, thus affecting the stability of the plant–soil system. The role of AMF is most pronounced in vegetation restoration in degraded areas [9]. Rosales et al. (1997) illustrated that AMF should be considered for the restoration of degraded areas by reintroducing them or increasing their inoculum [10]. Therefore, the use of mycorrhizal biotechnology to restore degraded grassland ecosystems has been a main focus in research regarding the restoration of severely degraded grasslands [11]. However, the effects of AMF inoculation on plant species composition and soil–plant synergism in degraded grasslands remains unclear.

Grass–legume mixtures, which are widely used in mixed pastures and agricultural intercropping systems, can enhance community productivity and stability [12–14]. As two different functional groups, their characteristics of physiology and nutrient utilization benefit each other. The establishment of *Lolium perenne* and *Trifolium repens* mixture grassland has been widely used to reconstruct the grassland ecosystem in southern China [15]. The reseeding of *T. repens* and *L. perenne* in degraded grasslands is known to increase soil nitrogen availability, improve biomass yield, and maintain biodiversity [16]. However, the addition of AMF in leguminous and non-leguminous mixed seeding systems reduces nitrogen loss and improves nitrogen utilization efficiency [17]. Previous studies mostly focused on the complementary effects of grass–legume mixtures, and less research is related to the effects of exogenous beneficial microorganisms on vegetation restoration in mixed grass–legume planting grassland, thus restricting the recovery of degraded grasslands.

Interactions between above-ground and below-ground communities are vital in ecosystem functions, including improving ecosystem stability, maintaining biodiversity, and enhancing soil nutrient stability [18]. In this study, we introduced AMF combined with locally dominant grass species to regulate soil–plant interactions to accelerate grassland restoration. The objectives of this study were to investigate whether the addition of AMF was effective in restoring degraded grassland and to study the effects of the combination of two dominant species at different composition ratios and AMF inoculation on plant biomass, nutrient accumulation, and soil nutrients with the aim to aid the appropriate management schemes for the protection and restoration of degraded grasslands.

2. Materials and Methods

2.1. Study Area

Nanshan Pasture in Hunan province (Figure 1) is located in the southwest of Chengbu Miao Autonomous County in Shaoyang City, Hunan Province, China. It was established in March 1956. The soil of Nanshan Pasture is mainly mountain meadow soil and mountain yellow brown soil, and the pH is neutral to slightly acidic. Natural vegetation mainly includes *Arundinella yunnanensis Keng*, *Cathaya argyrophylla Chun et Kuang*, *Emmenopterys henryi Oliv*, and *Holcus lanatus*. The dominant grass species at this site are *Trifolium repens* (T) and *Lolium perenne* (L). The mean annual temperature is around 11 °C, and the average annual rainfall is approximately 1900 mm.

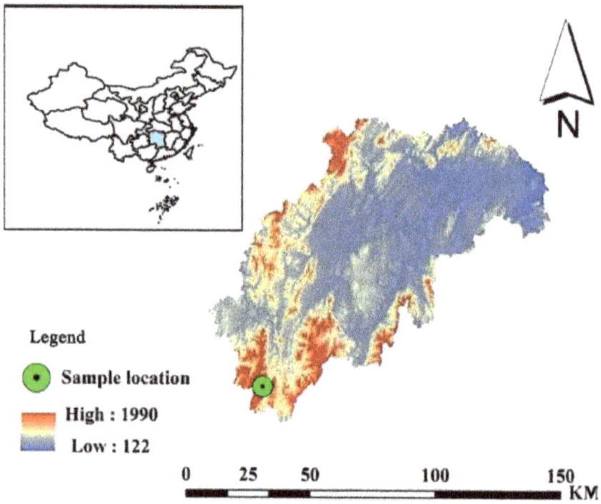

Figure 1. Soil sampling site at Nanshan Pasture, Hunan Province. The inset shows the location of Hunan Province in China. Data Center for Resource and Environmental Sciences, Chinese Academy of Sciences, 2021.

2.2. Soil and Inoculum Preparation

Based on the evaluation criteria of grassland degradation (BG19377-2003), non-degraded (NDG), lightly degraded (LDG), and severely degraded (SDG) grasslands were selected for sampling. Soil samples were collected on 30 July 2021. At each plot, we collected 60 kg of soil (5–20 cm deep) using shovels, and between samples, the shovels were sterilized with a 70% ethanol solution. The soil samples were transported to the laboratory, mixed, and passed through a 0.5 cm sieve to remove rocks and plant debris. The homogenized soil was divided into two equal parts. One part was stored at 4 °C. The other part was heat-sterilized (autoclaving at 121 °C for 60 min) and stored in a cool place for later use. A major disadvantage of sterilization is that the concentration of soil nutrients increases due to decomposition of the killed soil organisms [19]. Autoclaving increases the soil organic carbon (SOC) content. In general, the side effects of sterilization can be largely avoided with the addition of small amounts of living and sterilized soil to the sterilized background soil [20,21]. To minimize the effect of autoclaving on soil, we mixed the fresh soil or sterilized soil with each degradation degree and sterilized mixed soil (soil + sand) with the same degradation degrees at a ratio of 1:10, which were non-sterilized and sterilized soils.

Soils were inoculated with AMF strains purchased from the College of Resources and Environment Sciences, Nanjing Agricultural University, Nanjing, China. The five AMF strains included *Acaulospora scrobiculata Trappe*, *Glomus mosseas*, *Glomus caledonium*, *Glomus versiforme*, and *Glomus intraradices*. AMF spores contained in soil inoculation were propagated in autoclaved (121 °C for 120 min for one cycle) substrate (sand/soil, 1:2) with maize for two successive propagation cycles (three months for each cycle). The inoculum consisted of a mixture of the five strains at a dry weight ratio of 1:1:1:1:1.

2.3. Greenhouse Experiment

Seed was used of the ryegrass (*Lolium perenne*) Bond varieties, and of the legume (*Trifolium repens*) Sulky varieties. Mixtures are henceforth referred to by the first letters of their names. We carried out a full factorial greenhouse experiment, in which soil degradation degree (NDG, LDG, and SDG), planting ratio (T1L0, T3L1, T1L1, T0L1, and T1L3), and soil treatment (non-sterilized, sterilized, and sterilized + AMF) were the three factors (Table 1). The plant density in each pot remained unchanged, with 16 plants per pot and three replicates. One of five *T. repens* to *L. perenne* ratios (0:1, 1:3, 3:1, 1:0, and 1:1) was randomly assigned to each pot. All seeds were surface-disinfected (2 min in a 1% chloride solution), rinsed, and germinated on Petri dishes.

Table 1. A complete factorial design of 3 levels of soil degradation degree × 3 levels of soil biological treatment × 5 levels of plant mixtures.

Soil Degradation Degree	Soil Biological Treatment	Planting Ratios
Non-degraded/lightly/severely degraded	Sterilized soil	*Lolium perenne* monoculture *Trifolium repens* monoculture *L. perenne*:*T. repens* = 3:1 *L. perenne*:*T. repens* = 1:3 *L. perenne*:*T. repens* = 1:1
	Sterilized + AMF	*L. perenne* monoculture *T. repens* monoculture *L. perenne*:*T. repens* = 3:1 *L. perenne*:*T. repens* = 1:3 *L. perenne*:*T. repens* = 1:1
	Non-sterilized soil	*L. perenne* monoculture *T. repens* monoculture *L. perenne*:*T. repens* = 3:1 *L. perenne*:*T. repens* = 1:3 *L. perenne*:*T. repens* = 1:1

At the planting stage, the experiment included 135 pots (3 soil degradation degrees × 5 planting ratios × 3 soil treatments × 3 replicates). Plants were grown in plastic pots containing 500 g of steam-sterilized coarse sand covered with 2000 g of sterilized soil of different degradation levels and 200 g of fresh soil or 2200 g of sterilized soil mixture. For the AMF inoculation treatment, we added 250 g of AMF inoculum per pot, whereas for the non-inoculation treatment, 250 g of sterilized AMF inoculum was added per pot. Seedlings that died during the first week were immediately replaced, and pots were positioned randomly in greenhouses with 70% relative humidity. The plants were exposed to 16 h of light at 21 °C (day) and 8 h of darkness at 16 °C (night). The pots were watered every other day, and initial soil moisture (17% dry weight of soil) was measured twice a week.

2.4. Harvesting and Measurements

After 18 weeks, shoots were harvested, oven dried at 65 °C for 72 h to a constant weight, and weighed. The soil in the pots was collected through a 2 mm sieve and dried in a ventilated place to test the soil properties. Nitrate (NO_3^--N) was detected using a colorimetric method (220 nm minus 275 nm) after extraction with 2 mol/L of KCl [22], and ammonium (NH_4^+-N) was analyzed using a colorimetric method (625 nm) after indophenol blue-KCl extraction [23]. The available phosphorus (mg/g) was measured using the Olsen method [23].

2.5. Data Analysis and Statistical Analysis

A general linear model (GLM) was used to determine the effect of soil degradation degree, soil treatment, planting ratio, and their interactions on above-ground biomass, above-ground N and P contents, soil NH_4^+-N, soil NO_3^--N, and soil available P (Table 2). One-way ANOVA was used to evaluate the significant differences among treatments. All of the mentioned parameters were Box-Cox-transformed to ensure normality and homogeneity using a Shapiro–Wilk test. Duncan's multiple comparison was used to test the difference between the mean values of each treatment. Linear regression analysis was used to analyze the relationship between plant biomass and soil nutrients and plant N and P content. R 3.6.0 was used for statistical analysis (p-values of less than 0.05 were considered significant). The location map of the studied area was created by using ArcMap of ArcGIS software. SigmaPlot 12.5 was used to create the figures.

Table 2. Analysis of variance of the effects of soil degradation degree (S), planting ratio (P), and soil biological treatment (B) on plant and soil indicators.

Treatment	DF	Biomass	NH_4^+-N	NO_3^--N	Available P	Shoot N	Shoot P	Plant N/P
Soil degradation (S)	2	901.0 ***	73.25 ***	110.13 ***	498.13 *	426.19 *	909.14 *	108.26 ***
Planting ratio (P)	4	278.20 ***	5.39 ***	24.892 ***	1.97	82.96 *	90.02 *	4.70 *
Biological treatment (B)	2	10.99 ***	7.29 ***	19.648 *	0.05	5.83 *	10.19 *	13.99 ***
S × P	8	3.92 ***	2.80 ***	8.116 *	2.59 *	5.84 *	5.57 *	4.92 ***
S × B	4	15.24 ***	0.86	3.016 *	0.94	10.73 *	12.02 *	5.28 *
B × P	8	3.73 ***	1.42	1.188	1.83	2.67	1.53	1.41
S × P × B	16	2.48 ***	1.26	1.651	1.36	3.19	2.78 *	2.03 *

Note: * indicates significant difference ($p < 0.05$); *** indicates highly significant difference ($p < 0.01$).

A soil quality Index (SQI) was calculated using the minimum dataset (MDS) indicators that best represented soil functions. Using PCA, PCs with eigenvalues > 1 that explained >5% of the total variation were assumed to represent soil quality for MDS (Table 3). We then used the following sigmoidal curve to normalize and score the MDS indicators [24].

$$NL - SF(Y) = \frac{\mathrm{a}}{\left(1 + \left(\frac{\mathrm{x}}{x_0}\right)b\right)}$$

where *NL—SF* (*Y*) is the non-linear score of each indicator ranging from 0 to 1, a is the maximum value (a = 1), x is the value of the selected indicator, x_0 is the average value of each indicator, and b is the slope of the equation and was set to −2.5 for "more is better" functions and 2.5 for "less is better" functions.

Table 3. Principal component analysis of soil quality indicators.

Variable	PC 1	PC 2	PC 3
Available P	0.949	−0.092	−0.057
SOC	0.883	0.035	−0.127
Shoot P	0.855	0.025	0.159
Biomass	0.0747	−0.252	−0.449
Shoot N	−0.072	0.831	0.4
pH	−0.084	−0.587	0.599
NO_3^--N	0.613	−0.031	0.67
NH_4^+-N	0.466	0.543	−0.168
Eigenvalues	3.576	1.405	1.197
Cumulative contribution (%)	44.696	62.264	77.222

The final step for the soil quality assessment combined the selected indicators into an overall SQI using the following weighted additive equation [25].:

$$\mathrm{SQI} = \sum_{i=0}^{n} \mathrm{WiSi}$$

where W is the weighting factor for the soil property that equals the explanation of each PC divided by the total percentage of variation and S is a non-linear (NL− SQI) score. The SQI value is considered to represent overall soil quality, with higher values indicating better soil quality.

Structural equation models (SEMs) were used to analyze potential pathways which estimate AMF addition and planting ratio exert effects and indirect relationships between soil nutrients and plant shoot nutrient content on plant biomass. Soil nutrients (organic C, available N, and available P) were decreased separately through principal component analysis (PCA). The first principal component (PC1) in each group was used for the subsequent SEM analysis. The PCA results showed that in non-degraded grasslands, PC1 explained 73% and 80% of the total variance in soil nutrients and plant nutrients, respectively. In lightly degraded grasslands, PC1 explained 66% and 70% of the total variance of soil nutrients and plant nutrients, respectively, while in severely degraded grasslands, PC1 explained 77% and 66% of the total variance of soil nutrients and plant nutrients, respectively (Table 4).

Table 4. Partial correlation coefficients of principal components analysis (PCA) of variables in categories of plant nutrients and soil nutrients in soil degradation degree.

Variables	Component 1		
	Non-Degraded	Lightly Degraded	Severely Degraded
Soil nutrients			
AN (mg kg^{-1})	0.753	0.651	0.657
AP (mg kg^{-1})	0.752	0.822	0.775
SOC (mg kg^{-1})	0.388	−0.617	0.724
Cumulative (%)	72.75%	66.42%	77.36%
Plant nutrients			
Shoot N (mg g^{-1})	0.893	0.772	0.751
Shoot P (mg g^{-1})	0.893	−0.772	−0.751
Cumulative (%)	79.82%	69.60%	66.41%

3. Results

3.1. Changes in Plant Community Productivity under Different Treatments

All three treatments, i.e., the soil treatment (non-sterilized, sterilized, and sterilized + AMF), soil degradation degree (NDG, LDG, and SDG), and planting ratio (T1L0, T3L1, T1L1, T0L1, and T1L3), had significant effects on above-ground biomass, and there were significant interactions among them (Table 2). The highest plant biomass of 10.45 g DM/pot was observed under T1L3 in NDG non-sterilized soil, followed by 9.07 g DM/pot under L1T0 in NDG non-sterilized soil inoculated with AMF (Figure 2). Compared to the non-sterilized and sterilized treatments, AMF treatment of NDG soil induced a decrease in above-ground biomass (Figure 2). Compared with the non-sterilized and sterilized treatments, AMF inoculation significantly increased the above-ground biomass in both LDG and SDG soils, especially in communities with a high proportion of *L. perenne* (Figure 2). The aboveground biomass under different treatments was in the following order: T1L3 in NGD non-sterilized soil > T0L1 in NDG soil with AMF > T1L1 in NDG non-sterilized soil > T1L3 in NDG sterilized soil > T0L1 in NDG sterilized soil > T1L1 in NDG sterilized soil > T0L1 in LDG soil with AMF > T0L1 in LDG sterilized soil > T1L1 in LDG soil with AMF, and these above-ground biomass levels were significantly higher than those under the other treatments (Figure 2).

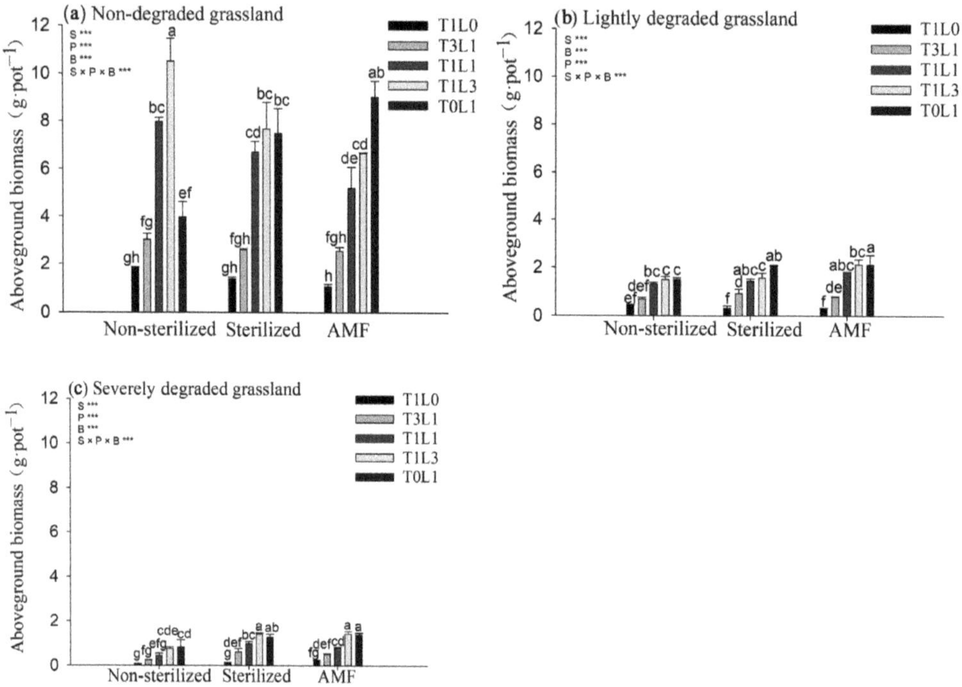

Figure 2. Effects of different treatments on plant community productivity. (**a**) Effects of soil type (S), planting ratio (P) , biological treatment (B) and their interactions (S × P × B) on aboveground biomass in non-degraded grassland soil (**b**) Effects of soil type (S), planting ratio (P) , biological treatment (B) and their interactions (S × P × B) on aboveground biomass in lightly degraded grassland soil. (**c**) Effects of soil type (S), planting ratio (P) , biological treatment (B) and their interactions (S × P × B) on aboveground biomass in severely degraded grassland soil. Different letters indicate the significant differences among treatments ($p < 0.05$). *** $p < 0.001$ indicate the significance level of each treatment. Note: soil degradation degree (S); planting ratio (P); soil biological treatment (B). T1L0, T3L1, T1L1, T1L3, T0L1 mean planting ratio of *T. repens* and *L. perenne*: 1:0, 3:1, 1:1, 1:3, 0:1, respectively.

3.2. Changes in Aboveground N and P Contents in Plants under Different Treatments

Arbuscular mycorrhizal fungi inoculation improved the plant nutrient contents (Figures 3b and 4b,c). In LDG and SDG soils, AMF inoculation increased the plant shoot N and P contents. However, AMF inoculation did not improve the plant shoot N and P contents in NDG soil (Figure 3b). The plant shoot N content was the highest under T1L0 in NDG soil. AMF inoculation increased the plant shoot P content in plant communities with a high proportion of *T. repens* in LDG and SDG soils (Figure 4). The plant N/P ratio in NDG soil was significantly lower than that in LDG and SDG soils (Figure 5). Overall, for NDG soil, the average plant N:P ratio was 3.66 ± 0.34 (mean ± standard error); for LDG soil, it was 18.36 ± 1.25; and for SDG soil, it was 15.42 ± 0.91.

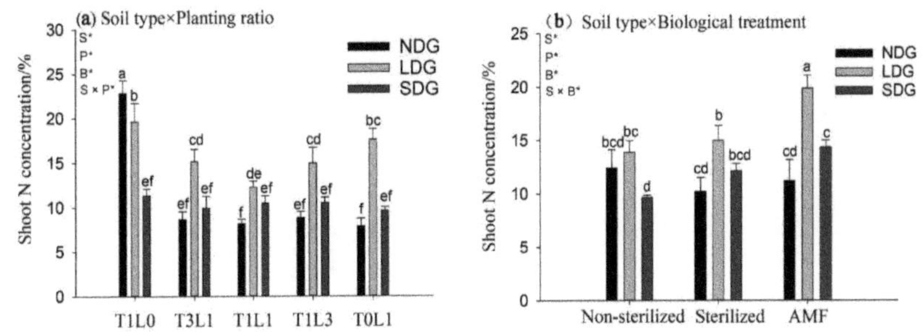

Figure 3. Effects of different treatments on shoot N content. (**a**) Effects of soil type (S), planting ratio (P) and their interactions (S × P) on shoot N concentration. (**b**) Effects of soil type (S), biological treatment (B) and their interactions (S × B) on shoot N concentration. Different letters indicate the significant differences among treatments ($p < 0.05$). * $p < 0.05$, $p < 0.1$ indicate the significance level of each treatment. Note: soil degradation degree (S); planting ratio (P); soil biological treatment (B); NDG: non-degraded grassland; LDG: lightly degraded grassland; SDG: severely degraded grassland. T1L0, T3L1, T1L1, T1L3, T0L1 mean planting ratio of *T. repens* and *L. perenne*: 1:0, 3:1, 1:1, 1:3, 0:1, respectively.

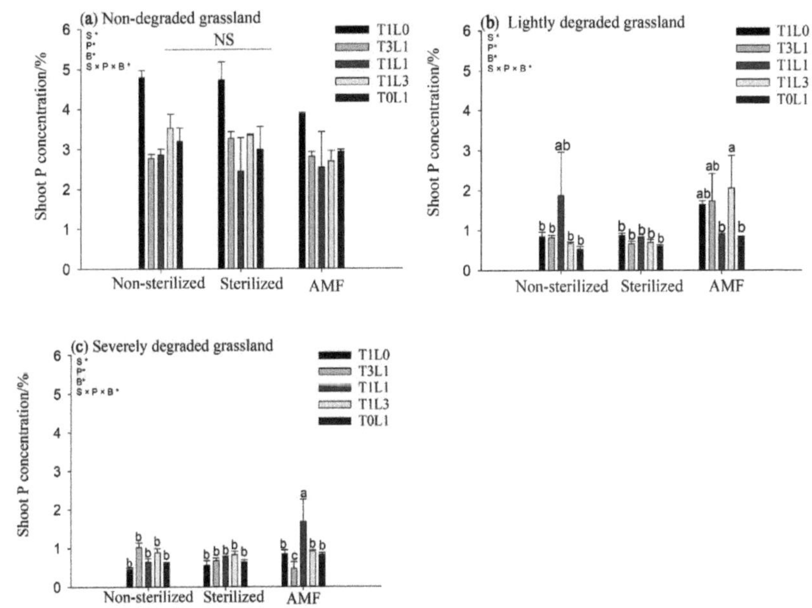

Figure 4. Effects of different treatments on shoot P content. (**a**) Effects of soil type (S), planting ratio (P) , biological treatment (B) and their interactions (S × P × B) on shoot P concentration in non-degraded grassland soil (**b**) Effects of soil type (S), planting ratio (P) , biological treatment (B) and their interactions (S × P × B) on shoot P concentration in lightly degraded grassland soil. (**c**) Effects of soil type (S), planting ratio (P) , biological treatment (B) and their interactions (S × P × B) on shoot P concentration in severely degraded grassland soil. Different letters indicate the significant differences among treatments ($p < 0.05$). * $p < 0.05$, $p < 0.1$ indicate the significance level of each treatment. Note: soil degradation degree (S); planting ratio (P); soil biological treatment (B). T1L0, T3L1, T1L1, T1L3, T0L1 mean planting ratio of *T. repens* and *L. perenne*: 1:0, 3:1, 1:1, 1:3, 0:1, respectively.

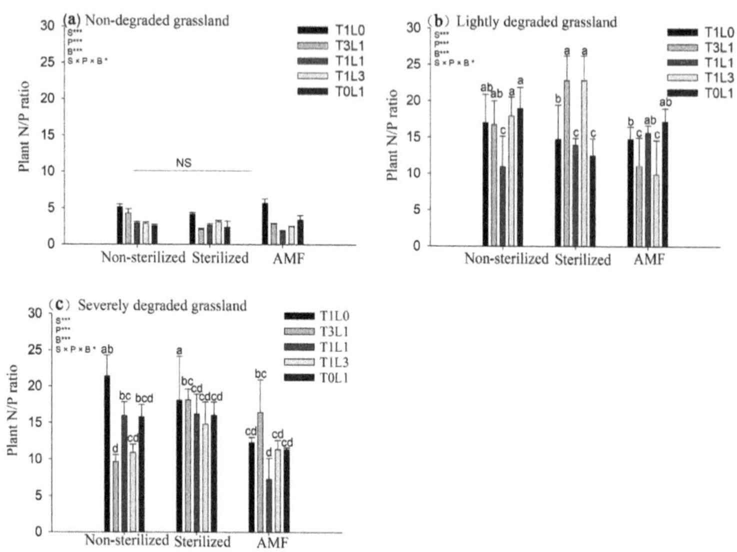

Figure 5. Effects of different treatments on plant N/P ratio. (a) Effects of soil type (S), planting ratio (P) , biological treatment (B) and their interactions (S × P × B) on plant N/P ratio in non-degraded grassland soil (b) Effects of soil type (S), planting ratio (P) , biological treatment (B) and their interactions (S × P × B) on plant N/P ratio in lightly degraded grassland soil. (c) Effects of soil type (S), planting ratio (P) , biological treatment (B) and their interactions (S × P × B) on plant N/P ratio in severely degraded grassland soil. Different letters indicate the significant differences among treatments ($p < 0.05$). *** $p < 0.001$, * $p < 0.05$, $p < 0.1$ indicate the significance level of each treatment. Note: soil degradation degree (S); planting ratio (P); soil biological treatment (B). T1L0, T3L1, T1L1, T1L3, T0L1 mean planting ratio of *T. repens* and *L. perenne*: 1:0, 3:1, 1:1, 1:3, 0:1, respectively.

3.3. Changes in Soil Nutrients under Different Treatments and SQI Values

The soil quality indexes after treatments are shown in Figure 6. The PCA based on soil physicochemical parameters showed that the eigenvalues of the first two PCs were >1, and PC1 explained 71.4% of the total variance (Table 3). Further, 77.22% of the total variance was explained by three PCs (Table 3). PC-1 was dominated by AP, SOC, and NO_3^--N, which accounted for 44.69% of the total variance. PC-2 was characterized by shoot N, pH, and NH_4^+-N, which accounted for 17.56% of the total variance. These variables were not correlated. Biomass was selected from PC-3 (explaining 14.95% of the variance).

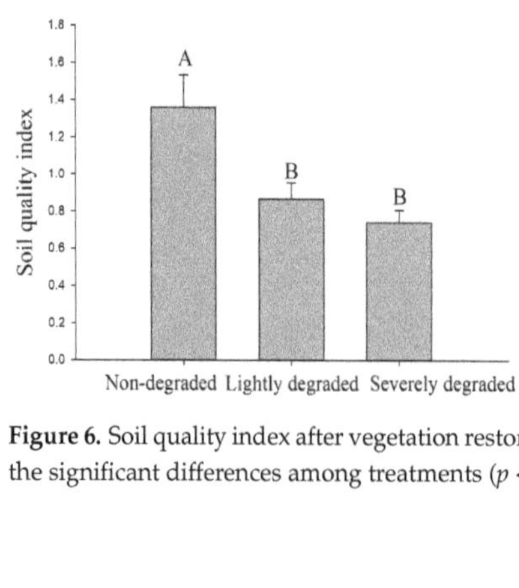

Figure 6. Soil quality index after vegetation restoration and AMF inoculation. Different letters indicate the significant differences among treatments ($p < 0.05$).

The soil treatment, planting ratio, and soil degradation degree significantly affected the soil nutrient contents (NH_4^+-N and NO_3^--N (Table 2). The soil degradation degree also significantly affected the soil's available P content (Table 2). The NH_4^+-N content was higher under T1L0, T0L1, and T1L1 in LDG soil (Figure 7a). The NO_3^--N and NH_4^+-N content were increased when the legume component increased (Figure 8a). The available nitrogen content in the three soil types decreased by 65%, 56.9%, and 16.8% after the treatments. There were also significant differences in the available P content under the different treatments. The available P content was higher in the communities with large proportions of legumes (Figure 9). The available P content in the three soil types decreased by 48%, 24%, and 13%. Soil pH generally increased by 77%. The SOC content significantly reduced by 66%, 40%, and 29%, and it reduced most in non-degraded soil. The soil treatment also significantly affected the soil NH_4^+-N and NO_3^--N contents. Compared with the sterilized and AMF treatments, the non-sterilized treatment reduced the soil NH_4^+-N content (Figure 7b). In NDG and LDG soils, AMF inoculation reduced the soil NO_3^--N content compared to the sterilized and non-sterilized treatments (Figure 8b).

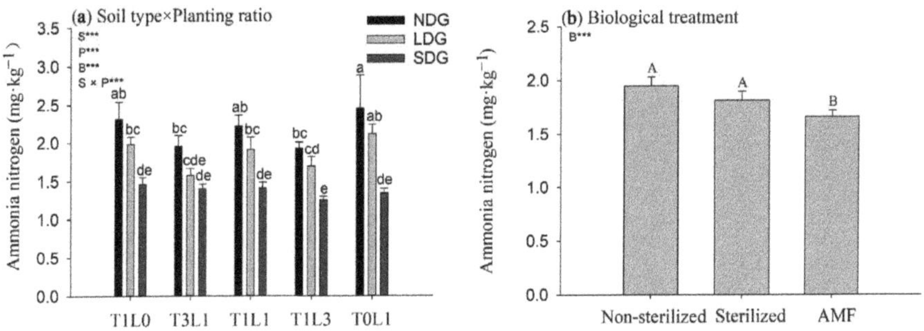

Figure 7. Effects of different treatments on soil NH_4^+-N content. (**a**) Effects of soil type (S), planting ratio (P) and their interactions (S × P) on soil ammonia nitrogen concentration. (**b**) Effects of biological treatment (B) on soil ammonia nitrogen concentration. Different capital and lowercase letters indicate significant differences between treatments. *** $p < 0.001$ indicate the significance level of each treatment. Note: soil degradation degree (S); planting ratio (P); soil biological treatment (B); NDG: non-degraded grassland; LDG: lightly degraded grassland; SDG: severely degraded grassland. T1L0, T3L1, T1L1, T1L3, T0L1 mean planting ratio of *T. repens* and *L. perenne*: 1:0, 3:1, 1:1, 1:3, 0:1, respectively.

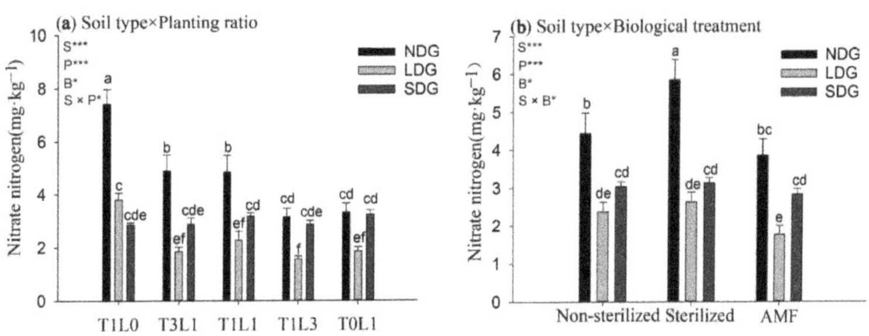

Figure 8. Effects of different treatments on soil NO_3^--N content. (**a**) Effects of soil type (S), planting ratio (P) and their interactions (S × P) on soil nitrate nitrogen concentration. (**b**) Effects of biological treatment (B) on soil nitrate nitrogen concentration. Different letters indicate the significant differences among treatments ($p < 0.05$). *** $p < 0.001$, * $p < 0.05$, $p < 0.1$ indicate the significance level of each treatment. Note: soil degradation degree (S); planting ratio (P); soil biological treatment (B); NDG: non-degraded grassland; LDG: lightly degraded grassland; SDG: severely degraded grassland. T1L0, T3L1, T1L1, T1L3, T0L1 mean planting ratio of *T. repens* and *L. perenne*: 1:0, 3:1, 1:1, 1:3, 0:1, respectively.

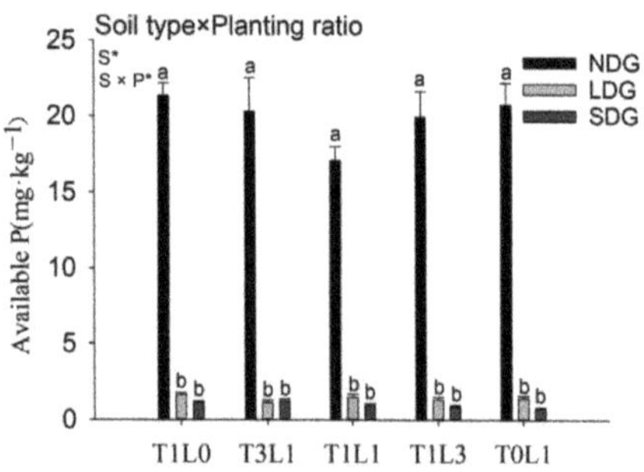

Figure 9. Effects of different treatments on soil available phosphorus content. Effects of soil type (S), planting ratio (P) and their interactions (S × P) on soil available P concentration. Different letters indicate the significant differences among treatments ($p < 0.05$). * $p < 0.05$, $p < 0.1$ indicate the significance level of each treatment. Note: soil degradation degree (S); planting ratio (P); soil biological treatment (B); NDG: non-degraded grassland; LDG: lightly degraded grassland; SDG: severely degraded grassland. T1L0, T3L1, T1L1, T1L3, T0L1 mean planting ratio of *T. repens* and *L. perenne*: 1:0, 3:1, 1:1, 1:3, 0:1, respectively.

3.4. Relationships between Plant Parameters and Soil Properties

The regression analysis showed that the community above-ground biomass increased with increasing NH_4^+-N and available P in the soil (Figure 10). The above-ground N content showed positive correlations with soil NH_4^+-N and NO_3^--N contents, but not with available P content (Figure 11). The above-ground P content showed positive correlations with the NH_4^+-N, NO_3^--N, and available P content in the soil (Figure 12).

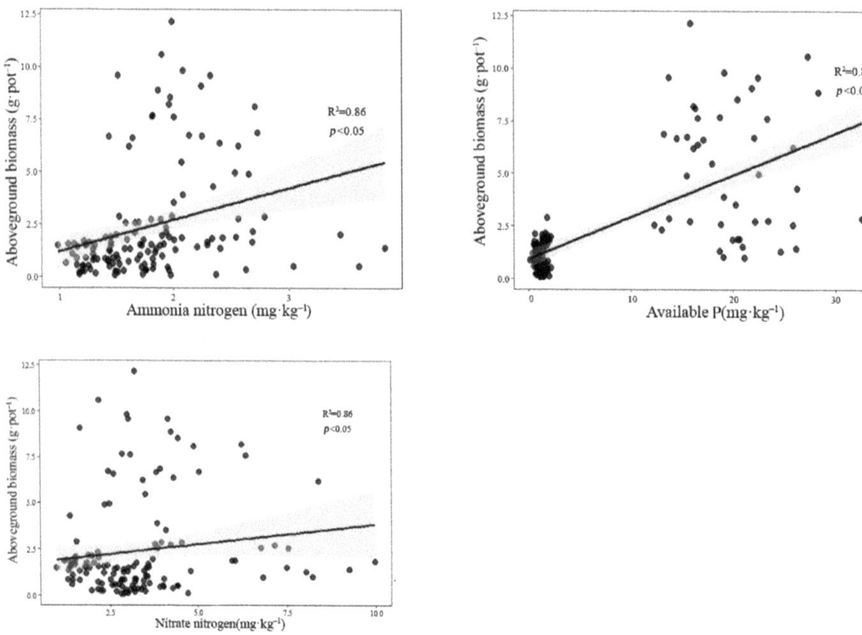

Figure 10. Relationships between available P, NH_4^+-N, and NO—-N contents in soil and plant community productivity.

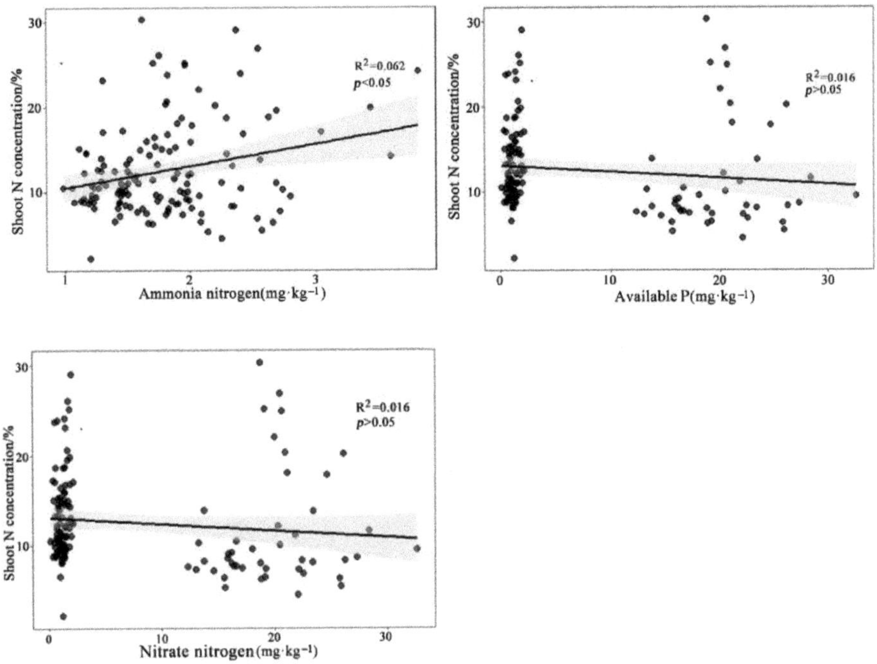

Figure 11. Relationships between available P, NH_4^+-N, and NO—-N contents in soil and plant N.

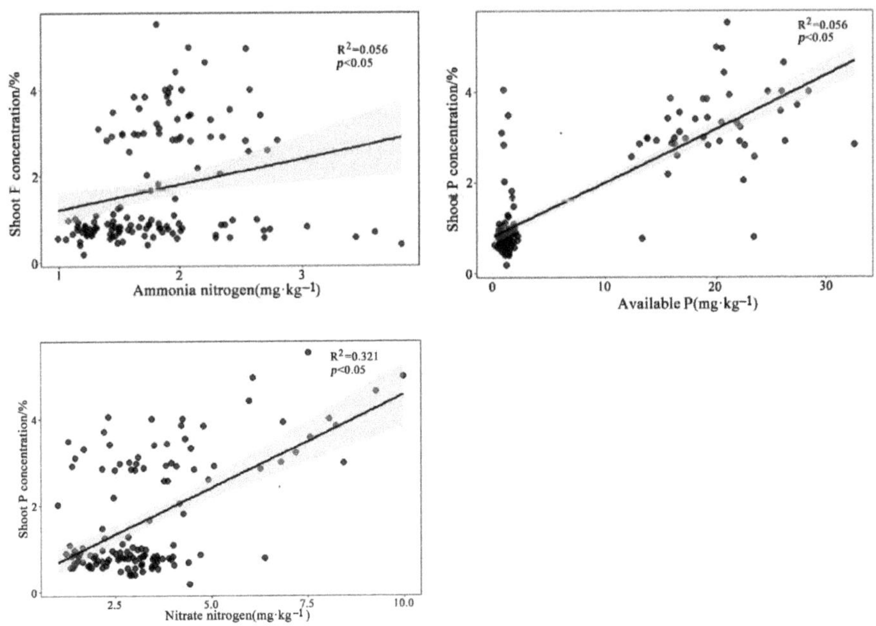

Figure 12. Relationships between available P, NH_4^+-N, and NO_3^--N contents in soil and plant P.

3.5. Pathways Determining Biomass

Two SEMs revealed that the planting ratio and AMF inoculation altered the plant biomass via different pathways in soils with different degrees of degradation, according to significant standardized path coefficients. Under different degrees of degradation, the planting ratio was the main pathway determining biomass, and it indirectly influenced the plant biomass through effects on shoot N and P contents. In LDG soil, AMF influenced the plant biomass through effects on soil nutrient contents. In SDG soil, AMF affected the plant biomass directly and indirectly by the affecting shoot N and P contents (Figure 13).

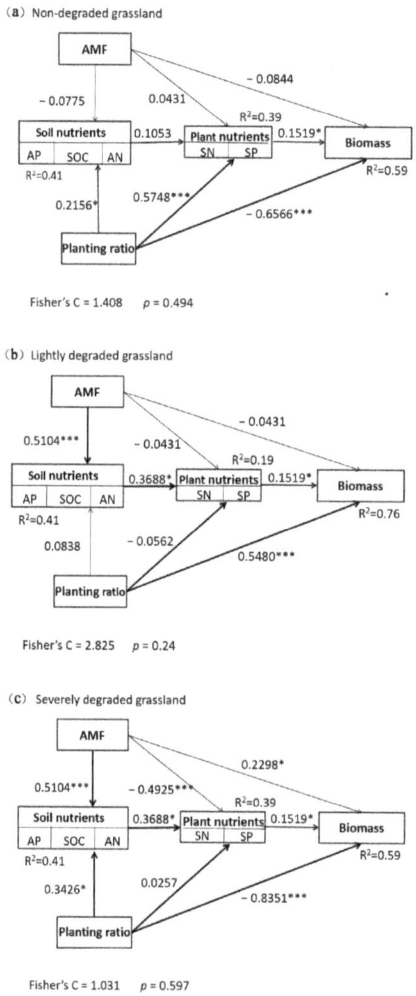

Figure 13. Structural equation models showing the effects of AMF addition and planting ratio on plant shoot biomass through soil nutrients and plant shoot nitrogen and phosphorus contents. (**a**) Non-degraded grassland. (**b**) Lightly degraded grassland. (**c**) Severely degraded grassland. Square boxes denote variables included in the models. Soil nutrient variables include soil NH_4^+-N, NO_3^--N, available P, and soil organic carbon. Soil nutrients are synthetic variables derived from the first axis of principal component analyses. The thickness of arrows is proportional to the standardized path coefficient. *** $p < 0.001$, * $p < 0.05$, $p < 0.1$ indicate the significance level of each effect. R^2 values represent the proportion of variance explained for each endogenous variable.

4. Discussion

Above-ground plant communities were affected by AMF inoculation, planting ratio, and grassland degradation state, as well as their interactions. The soil treatment and planting ratio regulated the plant shoot nutrient content and soil nutrients under different state of soil degradation. AMF inoculation increased the N and P contents in plant shoots, especially in nutrient-deficient environments, and enhanced the absorption of mineral elements by plants. One possible hypothesis is the strong nitrogen fixation ability of *T. repens*; communities with a large proportion of *T. repens* in the sowing mixture showed increased soil N accumulation.

4.1. Effects on Degraded Grassland Plant Communities

Soil microorganisms play an important role in plant growth [26]. Beneficial microorganisms, as ecosystem mediators, can be used to increase crop yields and modify microorganism communities to restore degraded grasslands [27]. Our findings revealed that plants grown in non-sterilized soil performed better than those grown in sterilized soil and sterilized soil inoculated with AMF in NDG soil. This difference could be attributable to the higher beneficial microbial diversity in non-sterilized soil and the antagonistic effects of dominant beneficial microorganisms against pathogens in the microbial community, which contributes to a better growth of the plants [28]. We detected no significant differences in plants growth between LDG non-sterilized and sterilized soils, attributing to the negative effects of soil-borne pathogens accumulated in the rhizosphere of the plant communities and the positive effects of AMF and other beneficial microorganisms [29]. Appropriate soil nutrients promoted the positive feedback effect of AMF on the plant community, whereas nutrient enrichment suppressed the beneficial effects of AMF on plant performance, negatively affecting the positive feedback of AMF on the plant community [30]. AMF have a more positive effect on plant growth in nutrient-poor soil [31]. In line with this, our study showed the AMF inoculation promoted plant community productivity more significantly in LDG and SDG soil than in NDG soil.

N and P are the main limiting nutrients for plant growth in terrestrial ecosystems [32]. The symbiotic relationships between plants and AM fungi can significantly promote N and P uptake and thus improve plant biomass [33] In degraded soil, AMF inoculation was found to promote the plant shoot N and P contents. AMF inoculation contributed to an increase in the above-ground biomass of *L. perenne*. There was no significant increase in the above-ground biomass of *T. repens*, which was partly attributed to the higher individual biomass of *L. perenne*, which reduced the competitive ability of *T. repens*. In addition, this differential growth response could be attributable to the fact that *T. repens* is not strongly dependent on AM fungi. *L. perenne* forms strong symbiotic relationships with AMF, which enables *L. perenne* to obtain and efficiently utilize mineral nutrients, thereby promoting growth.

The vegetation N:P ratios served as indicators of nutrient limitation, and it has been suggested that an N/P ratio <14 generally indicates N limitation, whereas a ratio >16 indicates P limitation, and some transitional states may exist when N/P ratios are between 14 and 16 [34]. In the present study, given that in the NDG soil in the study area, the mean N/P ratio was lower than 16 under all treatments, we established the NDG soil to be N-limited. However, the LDG and SDG soils were P-limited soils. In future efforts for the restoration of degraded grassland, inoculation with beneficial rhizosphere microorganisms combined with a higher proportion of legumes in the planting community or adding P fertilizer can contribute to the alleviation of N and P limitations, thereby accelerate the restoration of degraded grasslands.

4.2. Effects on Soil Properties of Degraded Grassland

Artificial grassland establishment increases soil's nutrient content by recruiting native vegetation, such as by sowing legumes [35]. Nitrogen is a main limiting factor for grassland productivity [36]. It has been widely reported that introducing legumes into grasslands leads to higher soil N availability [37,38]. Although NO_3^--N is highly mobile and can leach or denitrify, our study demonstrated that grass–legume mixtures at an appropriate proportion significantly increased the soil N content, regardless of the soil treatment. In addition, the soil'sNO_3^--N concentration decreased with the increase in the grass component. This may have been due to the large amount of NO_3^- absorbed by grass to maintain the soil mineral N at a low level and then to reduce the inhibition of legume N fixation. We concluded that the relative abundance of legumes had a great impact on the soil's N status in this grassland ecosystem. Furthermore, when the proportion of legumes increased in the mixed plant community, the soil's N mineralization and fixation increased, and the soil's N consumption decreased. The available phosphorus content in soil did not change obviously, possibly because soil's nutrient recovery requires more time. There is a clear need for further research to examine the effects of vegetation restoration on soil nutrients for longer periods of time if this approach is to be adopted to restore degraded grasslands.

Arbuscular mycorrhizal fungi (AMF) play a crucial role in promoting nutrient acquisition [39]. They can enhance a host plant's uptake of soil nutrients and transfer large amounts of inorganic N to the host plant. AM fungi can obtain substantial amounts of N from decomposing organic materials and directly utilize N from organic compounds [40], and then finally promote the efficiency of plant uptake and the utilization of soil N, thus reducing the residual N in the soil [41]. In this study, regardless of the planting ratio, the soil's NH_4^+-N and NO_3^--N contents decreased after AMF inoculation. This was because of the activation of soil nutrients via AMF inoculation, which improved soil nutrient availability and nutrient uptake, and thus reduced the soil's N content [42]. However, regardless of AMF addition, plants have a weak absorbability of available P, which may be determined by the nutritional requirements of plants themselves.

Plant–soil interaction is the most important feedback component in grassland restoration [43]. Previous research has indicated that soil nitrogen pools were strongly correlated with biomass [44]. In the present study, a regression analysis showed that above-ground biomass and plants' N and P contents were positively related with the soil's NH_4^+-N, NO_3^--N, and available P contents. Artificial vegetation establishment may increase the C and N storage by altering the quantity and quality of litter returned to the soil and further promote plant community recovery [45]. This research result was also reflected in the soil quality index. The soil quality index values showed that after vegetation restoration and AMF addition, there was little difference in the soil quality index between LDG and SDG. From a restoration perspective, the combination of AMF and grass–legume mixtures may enhance the uptake and utilization of soil nutrients and could be an effective strategy for degraded grassland restoration.

5. Conclusions

The restoration of vegetation and soil fertility reclamation is important for the recovery of degraded ecosystems. Grass–clover mixtures of *Lolium* (grass) and *Trifolium* (legume) on degraded grassland obviously improves soil quality. AMF increases plant community productivity by enhancing nutrient uptake, especially of P and N. SEM indicated that the planting ratio directly affects plant biomass, whereas AMF indirectly influences plant biomass via providing plant nutrients. Increasing the proportion of *L. perenne* can further improve plant community productivity. *T. repens* has a limited effect on community biomass but has a positive effect on forage quality and soil in mixed planting communities. We suggest that planting T:L = 0:1 and T:L = 1:3 combined with AMF inoculation can be used to recover degraded grassland production, and planting T1L1 and L3T1 with AMF inoculation can be applied for grassland nutrient accumulation and diversity conservation.

Author Contributions: H.R. designed the experiment; H.R. and J.C. Chang performed the experiment; and all the authors contributed to writing the manuscript. All authors have read and agreed to the published version of the manuscript.

Funding: This project was supported by the National Natural Science Foundation of China (32271615), the Fundamental Research Funds for the Central Universities (KYCYXT2022008), and XUEKEN2022020.

Data Availability Statement: Please contact the first author with requests for data.

Acknowledgments: We are grateful to all the students who assisted with the collection and processing of data. Many thanks are expressed to the anonymous reviewers for their helpful suggestions.

Conflicts of Interest: The authors declare no conflict of interest.

References

1. Peng, F.; Xue, X.; You, Q.; Huang, C.; Dong, S.; Liao, J.; Duan, H.; Tsunekawa, A.; Wang, T. Changes of soil properties regulate the soil organic carbon loss with grassland degradation on the Qinghai-Tibet Plateau. *Ecol. Indic.* **2018**, *93*, 572–580. [CrossRef]
2. Fayiah, M.; Dong, S.; Khomera, S.W.; Ur Rehman, S.A.; Yang, M.; Xiao, J. Status and challenges of Qinghai–Tibet Plateau's grasslands: An analysis of causes, mitigation measures, and way forward. *Sustainability* **2020**, *12*, 1099. [CrossRef]
3. Panakhyd, G.; Kotyash, U.; Yarmoluk, M.; Mizernyk, D.; Mashchak, Y. Improvement effect on the productivity of degraded grasslands. *Ştiinţa Agric.* **2014**, *2*, 3–8.
4. Carbon and nitrogen mineralization in soils under agro-pastoral systems in subtropical central Brazil. *Soil Sci. Plant Nutr.* **2002**, *48*, 179–184. [CrossRef]
5. Chen, K.; Zhou, H.; Lu, B.; Wu, Y.; Wang, J.; Zhao, Z.; Li, Y.; Wang, M.; Zhang, Y.; Chen, W.; et al. Single-Species Artificial Grasslands Decrease Soil Multifunctionality in a Temperate Steppe on the Qinghai–Tibet Plateau. *Agronomy* **2021**, *11*, 2092. [CrossRef]
6. Fisher, J.P.; Estop-Aragonés, C.; Thierry, A.; Charman, D.J.; Wolfe, S.A.; Hartley, I.P.; Murton, J.P.; Williams, M.; Phoenix, G.K. The influence of vegetation and soil characteristics on active-layer thickness of permafrost soils in boreal forest. *Glob. Change Biol.* **2016**, *22*, 3127–3140. [CrossRef]
7. Gonzalez-Chavez, C.; Harris, P.J.; Dodd, J.; Meharg, A.A. Arbuscular mycorrhizal fungi confer enhanced arsenate resistance on Holcus lanatus. *New Phytol.* **2002**, *155*, 163–171. [CrossRef]
8. Bao, X.; Zou, J.; Zhang, B.; Wu, L.; Yang, T.; Huang, Q. Arbuscular Mycorrhizal Fungi and Microbes Interaction in Rice Mycorrhizosphere. *Agronomy* **2022**, *12*, 1277. [CrossRef]
9. Tian, H.; Gai, J.P.; Zhang, J.L.; Christie, P.; Li, X.L. Arbuscular mycorrhizal fungi in degraded typical steppe of Inner Mongolia. *Land Degrad. Dev.* **2009**, *20*, 41–54. [CrossRef]
10. Rosales, J.; Cuenca, G.; Ramírez, N.; De Andrade, Z. Native colonizing species and degraded land restoration in La Gran Sabana, Venezuela. *Restor. Ecol.* **1997**, *5*, 147–155. [CrossRef]
11. Coutinho, E.S.; Barbosa, M.; Beiroz, W.; Mescolotti, D.L.; Bonfim, J.A.; Berbara, R.L.L.; Fernandes, G.W. Soil constraints for arbuscular mycorrhizal fungi spore community in degraded sites of rupestrian grassland: Implications for restoration. *Soil Biol.* **2019**, *90*, 51–57. [CrossRef]
12. Hebeisen, T.; LÜSCHER, A.; Zanetti, S.; Fischer, B.; Hartwig, U.; Frehner, M.; Hendrey, G.; Blum, H.; NÖSBERGER, J.O.S.E.F. Growth response of Trifolium repens L. and Lolium perenne L. as monocultures and bi-species mixture to free air CO_2 enrichment and management. *Glob. Chang. Biol.* **1997**, *3*, 149–160. [CrossRef]
13. Li, B.; Li, Y.Y.; Wu, H.M.; Zhang, F.F.; Li, C.J.; Li, X.X.; Lambers, H.; Li, L. Root exudates drive interspecific facilitation by enhancing nodulation and N2 fixation. *Proc. Natl. Acad. Sci. USA* **2016**, *113*, 6496–6501. [CrossRef] [PubMed]
14. Issah, G.; Schoenau, J.J.; Lardner, H.A.; Knight, J.D. Nitrogen fixation and resource partitioning in alfalfa (*Medicago sativa* L.), cicer milkvetch (*Astragalus cicer* L.) and sainfoin (Onobrychis viciifolia Scop.) using 15N enrichment under controlled environment conditions. *Agronomy* **2020**, *10*, 1438. [CrossRef]
15. Crème, A.; Rumpel, C.; Gastal, F.; de la Luz Mora Gil, M.; Chabbi, A. Effects of grasses and a legume grown in monoculture or mixture on soil organic matter and phosphorus forms. *Plant Soil* **2016**, *402*, 117–128. [CrossRef]
16. Ledgard, S.F.; Steele, K.W. Biological nitrogen fixation in mixed legume/grass pastures. *Plant Soil* **1992**, *141*, 137–153. [CrossRef]
17. Verzeaux, J.; Hirel, B.; Dubois, F.; Lea, P.J.; Tétu, T. Agricultural practices to improve nitrogen use efficiency through the use of arbuscular mycorrhizae: Basic and agronomic aspects. *Plant Sci.* **2017**, *264*, 48–56. [CrossRef]
18. He, J.S.; Dong, S.; Shang, Z.; Sundqvist, M.K.; Wu, G.; Yang, Y. Above-belowground interactions in alpine ecosystems on the roof of the world. *Plant Soil* **2021**, *458*, 1–6. [CrossRef]
19. Berns, A.E.; Philipp, H.; Narres, H.D.; Burauel, P.; Vereecken, H.; Tappe, W. Effect of gamma-sterilization and autoclaving on soil organic matter structure as studied by solid state NMR, UV and fluorescence spectroscopy. *Eur. J. Soil Sci.* **2008**, *59*, 540–550. [CrossRef]
20. De Long, J.R.; Kardol, P.; Sundqvist, M.K.; Veen, G.F.; Wardle, D.A. Plant growth response to direct and indirect temperature effects varies by vegetation type and elevation in a subarctic tundra. *Oikos* **2015**, *124*, 772–783. [CrossRef]
21. Gundale, M.J.; Kardol, P.; Nilsson, M.C.; Nilsson, U.; Lucas, R.W.; Wardle, D.A. Interactions with soil biota shift from negative to positive when a tree species is moved outside its native range. *New Phytol.* **2014**, *202*, 415–421. [CrossRef]
22. Zhang, S.; Huang, J.; Wang, Y.; Shen, Q.; Mu, L.; Liu, Z. Spatiotemporal heterogeneity of soil available nitrogen during crop growth stages on mollisol slopes of Northeast China. *Land Degrad. Dev.* **2017**, *28*, 856–869. [CrossRef]

23. BAO S, D. Soil and agricultural chemistry analysis. *Agric. Publ.* **2000**, 355–356.
24. Andrews, S.S.; Karlen, D.L.; Mitchell, J.P. A comparison of soil quality indexing methods for vegetable production systems in Northern California. *Agric. Ecosyst. Environ.* **2002**, *90*, 25–45. [CrossRef]
25. Andrews, S.S.; Mitchell, J.P.; Mancinelli, R.; Karlen, D.L.; Hartz, T.K.; Horwath, W.R.; Pettygrove, S.R.; Scow, K.M.; Munk, D.S. On-farm assessment of soil quality in California's Central Valley. *Agron. J.* **2002**, *94*, 12–23.
26. Hayat, R.; Ali, S.; Amara, U.; Khalid, R.; Ahmed, I. Soil beneficial bacteria and their role in plant growth promotion: A review. *Ann. Microbiol.* **2010**, *60*, 579–598. [CrossRef]
27. Coban, O.; De Deyn, G.B.; van der Ploeg, M. Soil microbiota as game-changers in restoration of degraded lands. *Science* **2018**, *375*, abe0725. [CrossRef] [PubMed]
28. Zhou, X.; Zhang, J.; Pan, D.; Ge, X.; Jin, X.; Chen, S.; Wu, F. p-Coumaric can alter the composition of cucumber rhizosphere microbial communities and induce negative plant-microbial interactions. *Biol. Fertil. Soils.* **2018**, *54*, 363–372. [CrossRef]
29. Wang, G.; Koziol, L.; Foster, B.L.; Bever, J.D. Microbial mediators of plant community response to long-term N and P fertilization: Evidence of a role of plant responsiveness to mycorrhizal fungi. *Glob. Chang. Biol.* **2022**, *28*, 2721–2735. [CrossRef]
30. Lin, G.; McCormack, M.L.; Guo, D. Arbuscular mycorrhizal fungal effects on plant competition and community structure. *J. Ecol.* **2015**, *103*, 1224–1232. [CrossRef]
31. Liang, J.F.; An, J.; Gao, J.Q.; Zhang, X.Y.; Yu, F.H. Effects of arbuscular mycorrhizal fungi and soil nutrient addition on the growth of Phragmites australis under different drying-rewetting cycles. *PLoS ONE* **2018**, *13*, e0191999. [CrossRef]
32. Ren, Z.; Niu, D.; Ma, P.; Wang, Y.; Fu, H.; Elser, J.J. Cascading influences of grassland degradation on nutrient limitation in a high mountain lake and its inflow streams. *Ecology* **2019**, *100*, e02755. [CrossRef] [PubMed]
33. Song, Z.; Bi, Y.; Zhang, J.; Gong, Y.; Yang, H. Arbuscular mycorrhizal fungi promote the growth of plants in the mining associated clay. *Sci. Rep.* **2020**, *10*, 1–9. [CrossRef] [PubMed]
34. van de Weg, M.J.; Meir, P.; Grace, J.; Atkin, O.K. Altitudinal variation in leaf mass per unit area, leaf tissue density and foliar nitrogen and phosphorus content along an Amazon-Andes gradient in Peru. *Plant Ecol. Divers.* **2009**, *2*, 243–254. [CrossRef]
35. Conant, R.T.; Paustian, K.; Elliott, E.T. Grassland management and conversion into grassland: Effects on soil carbon. *Ecol. Appl.* **2001**, *11*, 343–355. [CrossRef]
36. Liu, Y.; Men, M.; Peng, Z.; Houx, J.H., III; Peng, Y. Nitrogen availability determines ecosystem productivity in response to climate warming. *Ecology* **2022**, *103*, e3823. [CrossRef]
37. Mortenson, M.C.; Ingram, L.J. Carbon sequestration in rangelands interseeded with yellow-flowering alfalfa (*Medicago sativa* ssp. falcata). *Environ. Manag.* **2004**, *33*, S475–S481. [CrossRef]
38. Rao, S.C.; Northup, B.K.; Phillips, W.A.; Mayeux, H.S. Interseeding novel cool-season annual legumes to improve bermudagrass paddocks. *Crop. Sci.* **2007**, *47*, 168–173. [CrossRef]
39. Ramasamy, K.; Joe, M.M.; Kim, K.Y.; Lee, S.M.; Shagol, C.; Rangasamy, A.; Chung, J.B.; Sa, T.M. Synergistic effects of arbuscular mycorrhizal fungi and plant growth promoting rhizobacteria for sustainable agricultural production. *Korean J. Soil Sci. Fertil.* **2011**, *44*, 637–649. [CrossRef]
40. Hodge, A.; Fitter, A.H. Substantial nitrogen acquisition by arbuscular mycorrhizal fungi from organic material has implications for N cycling. *Proc. Natl. Acad. Sci. USA* **2010**, *107*, 13754–13759. [CrossRef]
41. Xu, H.; Shao, H.; Lu, Y. Arbuscular mycorrhiza fungi and related soil microbial activity drive carbon mineralization in the maize rhizosphere. *Ecotoxicol. Environ. Saf.* **2019**, *182*, 109476. [CrossRef] [PubMed]
42. Chen, S.; Zhao, H.; Zou, C.; Li, Y.; Chen, Y.; Wang, Z.; Jiang, Y.; Liu, A.; Zhao, P.; Wang, M. Combined inoculation with multiple arbuscular mycorrhizal fungi improves growth, nutrient uptake and photosynthesis in cucumber seedlings. *Front. Microbiol.* **2017**, *8*, 2516. [CrossRef] [PubMed]
43. Wang, C.T.; Wang, G.X.; Liu, W.; Wang, Y.; Hu, L.; Ma, L. Effects of establishing an artificial grassland on vegetation characteristics and soil quality in a degraded meadow. *Isr. J. Ecol. Evol.* **2013**, *59*, 141–153. [CrossRef]
44. Zak, D.R.; Grigal, D.F.; Gleeson, S.; Tilman, D. Carbon and nitrogen cycling during old-field succession: Constraints on plant and microbial biomass. *Biogeochemistry* **1990**, *11*, 111–129. [CrossRef]
45. Sun, D.S.; Wesche, K.; Chen, D.D.; Zhang, S.H.; Wu, G.L.; Du, G.Z.; Comerford, N.B. Grazing depresses soil carbon storage through changing plant biomass and composition in a Tibetan alpine meadow. *Plant Soil Environ.* **2011**, *57*, 271–278. [CrossRef]

agronomy

MDPI

Article

Sandy Habitats Play an Important Role in Shrub Encroachment in Grasslands

Lina Xie [1], Yuchen Li [2], Hongyu Guo [1], Chunwen Wang [1], Qing Chen [1], Peng He [1] and Chengcang Ma [1,*]

1 Tianjin Key Laboratory of Animal and Plant Resistance, College of Life Sciences, Tianjin Normal University, Tianjin 300387, China
2 College of Life Sciences, Nankai University, Tianjin 300071, China
* Correspondence: machengcang@163.com

Abstract: Shrub species have increased in density and cover in desertification areas, however, the role of sandy habitats in contributing to the expansion of shrubs is poorly understood. Although the effect of sandy habitats on plant growth and reproduction have been demonstrated, most existing studies lack either experimental demonstration or an integrated study during the whole shrub life cycle. We performed field and laboratory experiments to examine the responses of four stages in the life cycle of shrubs (seed germination, plant growth, seed reproduction, clonal reproduction) to sandy habitats (including sand substrate, sand burial and wind erosion) for *Caragana* shrubs. Results showed that both sand substrate and sand burial facilitated seed germination, seedling biomass, sapling establishment, plant growth, and root-shoot ratio of *Caragana*. Meanwhile, they both strongly increased seed number and seed preservation, and thus enhanced sexual reproduction. Sand burial favored clonal reproduction of *Caragana* by promoting the formation of branch-derived ramets, while wind erosion benefited clonal reproduction by facilitating the formation of root-derived ramets. These results suggested that sandy habitats facilitated seed germination, plant growth, sexual reproduction, and clonal reproduction of *Caragana*, which could explain why shrub abundance, shrub area and shrub height of *Caragana* in sandy areas was higher than in grasslands. Our study provided an experimental demonstration that sandy habitats promoted the population growth of *Caragana* shrubs during the whole life cycle and highlighted the significant role of sandy habitats in facilitating shrub encroachment in grasslands.

Keywords: sand burial; wind erosion; sand substrate; clonal reproduction; sexual reproduction; *Caragana* species

Citation: Xie, L.; Li, Y.; Guo, H.; Wang, C.; Chen, Q.; He, P.; Ma, C. Sandy Habitats Play an Important Role in Shrub Encroachment in Grasslands. *Agronomy* **2022**, *12*, 2858. https://doi.org/10.3390/agronomy12112858

Academic Editors: Kesi Liu and Xinqing Shao

Received: 16 October 2022
Accepted: 14 November 2022
Published: 16 November 2022

1. Introduction

Desertification, resulting from climatic changes (such as elevated levels of CO_2 and parallel increase in temperature) and human activities (such as overgrazing), is considered as one of the most critical ecological and environmental issues in grasslands worldwide [1,2]. Arid and semi-arid regions of northern China, characterized by low precipitation, loose soil structure and frequent strong winds, are especially vulnerable to desertification. The local herbaceous species are easily degraded under certain extreme conditions (e.g., drought, burial and erosion). However, such harsh sandy conditions favored the growth of the woody plants, most of which are psammophytes shrubs, due to their high windbreak and sand-fixation capacities [3]. Therefore, a large and rapidly increasing number of shrub plants encroach into grasslands, resulting in thicket grasslands with shrub patches [4,5]. A large number of studies have focused on the causes and consequences of shrub encroachment into grasslands [2], and some attention has also been paid to how shrubs adapt to the harsh conditions [6,7]. However, the contribution of sandy habitats in the process of shrub encroachment is poorly understood.

The relative proportions of sand, silt and clay particles, also known as soil texture, has a major role in influencing plants' growth [8]. Usually, a sandy soil substrate is specialized by a lower nutrient content and water retention capacity, but have a large overall volume of pore spaces and high permeability. Therefore, plants growing in sandy soil will allow their roots to extend more easily [9] and increase seed germination [10], but will reduce the cover and biomass of both shrubs and herbaceous species [11], and constrain the distribution of plants species [12]. Additionally, soils with high sand content would alter the relationship between mycorrhizal fungi and plants, thereby influencing plant biomass production in grasslands [13].

Sand burial and wind erosion are strong stresses for plants in sandy habitats. Previous studies have reported that sand burial or wind erosion affected the critical stages of plant life cycle. For example, sand burial affected seed germination [7,14–17], plant growth [18–20], clonal reproduction [21–23], and sexual reproduction [24], thus influencing community cover in sandy habitats [25]. Wind erosion also influenced seedling survival, plant growth [26,27], ramet numbers, survival rate and biomass [6,22,28], thereby influencing total community cover and ecosystem productivity [29].

Thus far, some studies have examined the effects of sandy habitats on plant growth and reproduction based on field observations [11,18–24]. However, such effects have never been examined by factor control experiments (except for seed germination) with psammophyte shrubs that commonly colonize and stabilize sandy land habitats. Moreover, studies assessing the effects of sandy habitats on shrub encroachment at population level, during the whole life cycle, are scarce.

Here, we conducted field and laboratory experiments to examine the response of four stages in the life cycle: seed germination, plant growth (plant height, plant biomass and root -shoot ratio), seed reproduction (reproduction effort and seed preservation), and clonal reproduction, in sandy habitats (including sand substrate, sand burial and wind erosion) for *Caragana* shrubs, which are the pioneer plants occurring naturally in the harsh, sandy conditions with frequent sand burial and wind erosion. It is the common observation that *Caragana* encroached into grassland in Inner Mongolia [4,5]. Previous studies have suggested that the causes of *Caragana* encroachment into grassland might be a result of their tolerance to drought and grazing [4,5]. However, limited empirical evidence exists on the role of sandy habitats in this ecological process. Therefore, the objective of this study was to examine the effects of sandy habitats on population growth of shrubs from the whole life cycle perspective, using *Caragana* as the model shrub, and to evaluate the role of sandy habitats in the process of shrub encroachment.

2. Materials and Methods

2.1. Study Species and Study Site

Caragana (Fabaceae) are xeromorphic, winter-deciduous shrubs with a high capability of trapping sand. They are widely distributed in sandy lands or in grasslands with high sand content. Therefore, *Caragana* shrubs are an ideal model for studying the value of sand context to the population growth of shrubs. There are 16 *Caragana* species distributed in the Inner Mongolia Steppe; *C. stenophylla* Pojark is one of the most important species due to its widespread distribution, forage value, and other ecological functions in the region [30]. *C. stenophylla* realizes population recruitment by both sexual and clonal reproduction. The clonal reproduction of *C. stenophylla* includes branch-derived and root-derived clonal reproduction. The branch-derived clonal reproduction produces adventitious roots from branches, forming new ramets. The root-derived clonal reproduction produces horizontal roots (spacers) from taproots of the mother plant or adult ramets; at the end or the middle of the horizontal roots, vertical roots grow deeper, and adventitious buds grow towards the soil surface, forming new ramets [31,32].

The Alxa region (37°24′~42°47′ N, 97°10′~106°52′ E; altitude ~1550 m; area ~270,000 km^2) in Inner Mongolia, China, is a hyper-arid area. The whole Alxa region is enclosed by the Helan Mountains to the east, the Qilian Mountains to the south, and the north-eastern part of the Tibetan Plateau to the south-west. The region has three deserts: the Tengger (Tengri) Desert in the south, the Badain Jaran (Baden Dzareng or Batan Tsalang) in the west, and the Ulan Buh (Wulanbuhe) in the northeast. We conducted the field study in the Alashan Left Banner, which is located in the northeast of the Tengger Desert (38°19′ N, 105°41′ E). In the study area, the mean annual precipitation is 110 mm; the mean annual temperature is 7.8 °C; the mean annual total sunshine time is 3200 h, with a mean daily solar radiation intensity of 1.71 kJ cm^{-2} d^{-1}. Driven by wind, sand burial with little vegetation cover (<20%) is dominated by xerophytic shrubs, such as the *Caragana* shrub species, accounting for ~60% of the total plant biomass.

2.2. *Caragana Population Survey in Grassland and Sandy Land*

In July 2019, we conducted a *Caragana* survey in two distinct microhabitats (sandy land and grassland). Here, we define "sandy land" as land covered by a sandy soil, with a vegetation cover of less than 5%, which includes areas of sandy desert [33]. Three plots were established on each of the two microhabitats. Each plot contained a 50 m × 50 m quadrat. For each quadrat, we first recorded the number of *C. stenophylla* shrubs in order to assess the shrub abundance (shrub/hectare). Then, we chose 10 shrubs at random, measured the length of the long axis and short axis of the shrub crown, the shrub height and nabkha height for each selected shrub, and calculated the shrub area (shrub area = π × semi-long axis × semi-short axis). Meanwhile, at the seed maturation stage, we cut three branches from each shrub, counted the number of seeds on branches, and calculated the reproductive effort (number of seeds/dry biomass (g)) after drying the branches at 60 °C for 72 h. Reproductive effort is traditionally defined as the proportion of total biomass allocation to seeds and other "obvious" reproductive structures [34].

2.3. *Seed Preservation Experiment for Sand Burial Treatments*

To assess the effect of sand burial on the preservation of the seeds, in July 2021, we performed seed preservation experiments in the three sandy land plots. In each plot, we established ten 50 cm × 50 cm quadrates, which were randomly assigned to two treatments: five quadrates for no sand burial treatment (seeds was totally exposed to the air, referred to "ground surface preservation" hereafter) and the other five quadrates for sand burial treatment (seeds was burial with 1 cm sandy soil, referred to "sand burial preservation" hereafter). We scattered 100 seeds, evenly, in each quadrate. We recorded the number of remaining seeds in each quadrate after 15 days.

2.4. *Seed Germination Experiment for Burial Treatments*

Washed, moist sand was placed in pots (10 cm deep × 20 cm diameter) with drainage holes on the bottom, to a depth of 3 cm; 100 *Caragana* seeds were placed in each of the pots and then assigned in the following treatments: buried by sand of 2 cm, 3 cm or 5 cm deep, or not buried (seeds lying on the sand surface). There were 5 replicates for each treatment. We conducted the germination trials in an incubator with 12 h light and dark cycles, and with temperature of 25 °C under light condition/15 °C under dark condition. Tap water was applied to keep the soil moist during the seed germination period. We recorded the number of seed germination after 30 days. All the seedlings were weighted after being oven-dried at 60 °C for 72 h. For the ungerminated seeds, we used 1% TTC stain to evaluate their vigor. If seeds showed viability but did not germinate, they were considered dormant seeds. If seeds showed no viability, they were considered dead seeds. We then calculated the seed germination rate and seed mortality rate.

2.5. *Planting Experiment with Grassland Soil and Sandy Land Soil*

Soils for this experiment were collected from the natural grassland (the soil is typical grey-brown desert soil, referred to hereafter as "grassland soil") and the sandy land (referred to hereafter as "sandy soil") in the field. After collection, soil was immediately sieved through a 5 mm mesh to remove stones and roots, thoroughly mixed, and brought back to the laboratory. A total of 15 plastic pots (17 cm diameter ×18 cm height) were used and divided into two treatments. Of these, 5 pots were randomly selected for the grassland soil treatment (G), and the remaining 10 pots were used for the sandy soil treatment (S). Each pot was filled with 3.5 kg of field soil and immediately watered with 100 mL Hoagland's solution; 50 *Caragana* seeds were sowed per pot at 2 cm burial depth. All pots were randomly kept in a greenhouse with a temperature regime of ~22 °C during daytime and ~18 °C during night and were watered regularly to keep the soil moist during the experiment. Forty-five days after sowing, we first calculated the sapling establishment rate (the number of saplings formed by 100 seeds). Then, we kept eight vigorous saplings in each pot and the others were removed. Five pots of S treatment were randomly selected for experimental sand burial treatment (SB) by artificially adding sand. Sand was added to each pot in the following 60 days with a total burial depth of 2 cm. The plant height, above and below ground biomass was measured 120 days after sowing, and the root-shoot ratio was calculated.

2.6. Probability of Branches Forming Adventitious Roots under Buried or Not Buried Conditions

In August 2019, we selected 10 *Caragana* shrubs in each of the three plots in sandy land. For each selected shrub, we marked four branches, of which two branches were buried in sand from the base up to 2/3 of their heights, the other two was completely exposed to the air. In August 2020, we counted the number of branches forming adventitious roots, and calculated their percentage.

2.7. The Number of Adventitious Buds on Horizontal Roots under Wind Eroded and Non-Wind Eroded Conditions

In August 2019, we selected five *Caragana* shrubs in each of the three plots in sandy land. For each selected shrub, we first carefully removed the sand dune and dug, 20 cm down from the surface, to expose the horizontal roots. Then, we chose two horizontal roots, one of which was buried in sand, and the other one was exposed to the air, with an exposure length of 100~120 cm. These selected roots were marked. In August 2020, we counted the number of adventitious buds on each horizontal root buried in sand and exposed to the air, respectively.

2.8. Statistical Analyses

We performed one-way ANOVA to examine the differences in seed germination rate, seed death rate and seedling biomass among burial depth treatments. If ANOVA showed significant differences, post hoc tests (Tukey HSD) were performed to compare these indices among the burial depth treatments. We performed *t*-tests to examine the differences in reproduction effort between grassland and sandy land, the differences in seed preservation between sand burial and ground surface, the differences in sapling establish rate and plant growth between grassland soil and sandy land soil, the differences in plant growth between sandy land soil and sandy land soil + sand burial, the differences in percentage of branches forming adventitious roots between sand burial and control, the differences in ramet number on horizontal roots between wind erosion and control, and the differences in shrub abundance between grassland and sandy land. We performed analyses, using GLMMs, with sampling shrubs within plot and plots within landscape (grassland and sandy land) as random variables (sampling shrubs were nested in plot; plots were nested in landscape) in order to examine the differences in shrub height, shrub area and nabkha height between grassland and sandy land. All analyses were performed with SPSS 21.0 (IBM, Armonk, NY, USA) and the significant level was $p < 0.05$.

3. Results

3.1. Effect of Sand Burial on Seed Germination

The burial depth significantly affected the seed germination rate ($F_{4,20} = 20.46$, $p < 0.05$), seed death rate ($F_{4,20} = 16.21$, $p < 0.05$) and seedling biomass ($F_{4,20} = 24.68$, $p < 0.05$) of *Caragana*. With an increased burial depth, the seed germination rate and seedling biomass first increased and then decreased. They were both highest at the intermediate burial depth of 2 cm (Figure 1). In contrast to the seed germination rate and seedling biomass, the seed death rate first decreased and then increased with the increase of burial depth, and it was the lowest when burial depth was 2 cm (Figure 1).

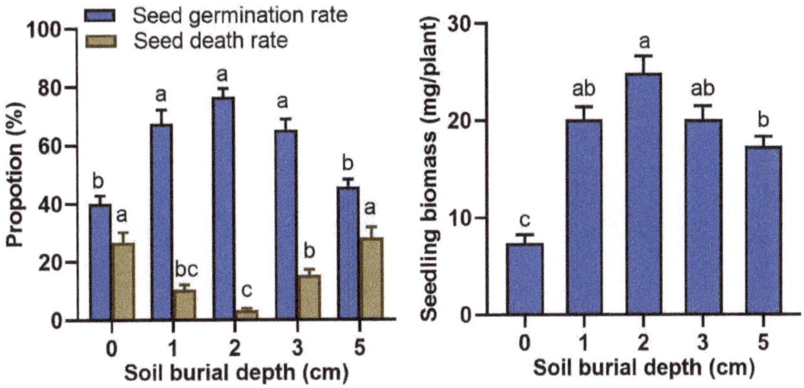

Figure 1. Effect of burial depth on seed germination rate, seed death rate and seedling biomass of *Caragana stenophylla* (mean ± SE). Different lowercase letters indicate significant differences among treatments, $p < 0.05$, Tukey HSD.

3.2. *Effects of Sandy Soil on Sapling Establishment and Plant Growth*

The sapling establish rate, plant height and biomass of *Caragana* was higher in sandy soil than in grassland soil (Figure 2a–c). Sand burial did not increase *Caragana* plant height, but significantly increased its biomass and root-shoot ratio (Figure 2b–d).

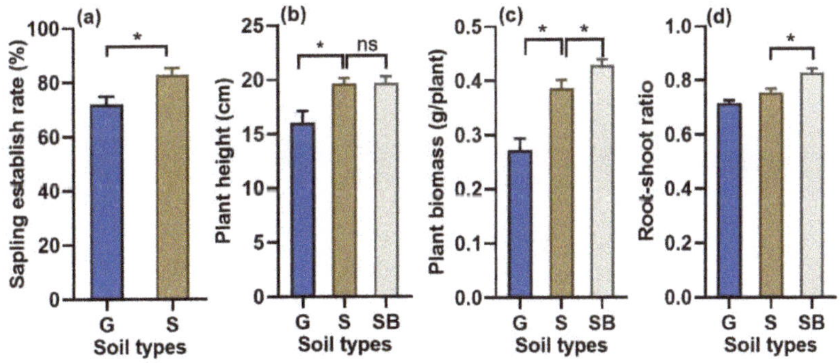

Figure 2. Effect of soil types on sapling establish rate (**a**), plant height (**b**), plant biomass (**c**) and root -shoot ratio (**d**) of *Caragana stenophylla* (mean ± SE). G, grassland soil; S, sandy land soil; SB, sandy land + sand burial. * indicate significant differences between G and S, and between S and SB, $p < 0.05$, ns indicates not significant difference, $p > 0.05$, *t*-test.

3.3. *Effect of Sandy Habitats on Seed Production and Seed Preservation*

The reproduction effort of *Caragana* increased in sandy land, as shown by the strongly increased biomass allocation to seeds. *Caragana* produced, on average, 4.63 seeds/g biomass in grassland, while 6.80 seeds/g biomass in sandy land (Figure 3a).

Sand burial significantly increased the seed preservation of *Caragana* (5.71 times higher than ground surface; Figure 3b). This indicated that sand habitat was beneficial to seed preservation in sandy land.

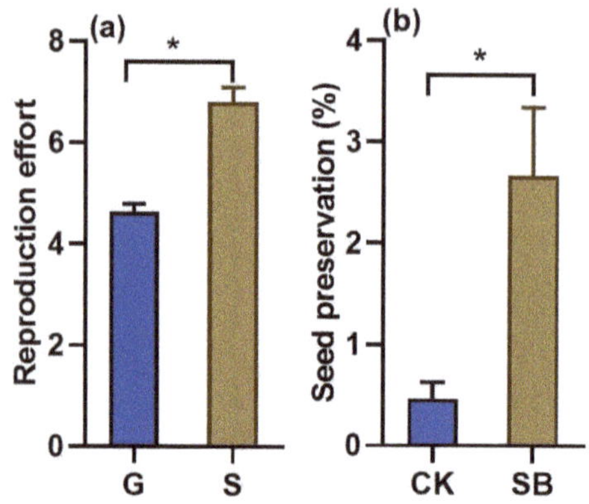

Figure 3. (**a**) Reproduction effort (seed/g biomass) of *Caragana stenophylla* in grassland and sandy land (mean ± SE). G, grassland; S, sandy land. * indicate significant differences between grassland and sandy land, $p < 0.05$, *t*-test. (**b**) Effect of sand burial on seed preservation of *Caragana stenophylla* (mean ± SE). CK, ground surface; SB, sand burial. * indicate significant differences between sand burial and ground surface, $p < 0.05$, *t*-test.

3.4. Effect of Sand Burial and Wind Erosion on Clonal Reproduction

Almost no adventitious roots were formed on the branches of *Caragana* under the control condition (no sand burial), and 70.0% of branches produced adventitious roots after soil burial, indicating that sand burial greatly facilitated branch-derived ramet formation (Figure 4a).

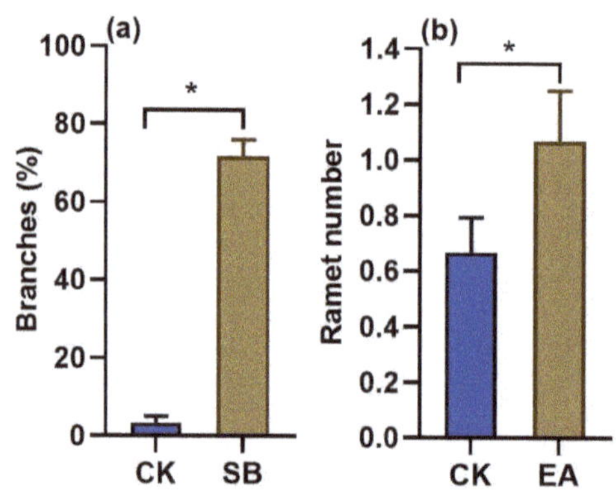

Figure 4. (**a**) Percentage of branches forming adventitious roots after *Caragana stenophylla* branches were sand buried (mean ± SE). CK, control; SB, sand burial. (**b**) Ramet establishment on the horizontal buried roots and the exposed roots for *Caragana stenophylla* (mean ± SE). CK, buried roots; EA, exposed roots. * indicate significant differences between treatments, $p < 0.05$, *t*-test.

Caragana produced more adventitious buds on horizontal exposed roots than buried roots. These results indicated that wind erosion could promote the ramet formation from root-derived clonal reproduction (Figure 4b).

3.5. Population Characteristics of Caragana in Grassland and Sandy Land

The *Caragana* population in grassland and sandy land had different quantity characteristics. The shrub abundance, area and height of *Caragana* in sandy land was higher than that in grassland (although marginally significant for shrub abundance; $p = 0.07$; Figure 5a–c), which indicates that sandy land enhances the population growth of *Caragana* in general. With an increasing shrub size and strong sand movement in sandy land, the nabkha height of *Caragana* in sandy land was 854% greater than that in grassland (Figure 5d).

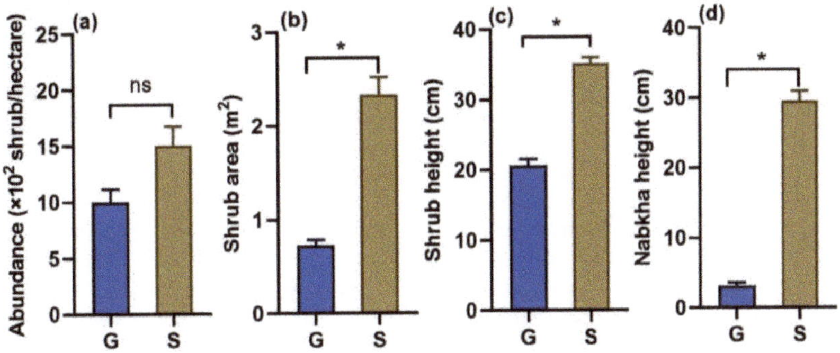

Figure 5. Shrub abundance (**a**), shrub area (**b**), shrub height (**c**) and nabkha height (**d**) of *Caragana stenophylla* in grassland and sandy land (mean ± SE). G, grassland; S, sandy land. * indicate significant differences between grassland and sandy land, $p < 0.05$, ns indicates not significant difference, $p > 0.05$, *t*-test or GLMMs.

4. Discussion

4.1. Sandy Habitats Facilitate Seed Germination and Seedling Growth of Shrubs

In the Alxa region, the wind is frequently strong and wind erosion occurs throughout the year. The bare land outside the shrub area is subjected to wind erosion, while the shrub area is frequently partially or completely buried by sand (Figure 5d). Some studies have shown that moderate burial improved seed germination, seedling survival and the growth of herbaceous plants, such as *Psammochloa villosa* [35], *Leymus secalinus* [15], as well as of desert shrubs, such as *Nitraria sphaerocarpa* and *Haloxylon ammodendron* [7], while others also found that seed germination of *Eremosparton songoricum* [14] and *Pinus thunbergii* [36] decreased after shallow partial burial treatment. We found that the seed germination and seedling biomass of *Caragana* reached the highest level at a sand burial depth of 1–2 cm. Similar results were observed for *Caragana korshinskii* [37]. Collectively, these results demonstrated that shallow sand burial stimulated germination and subsequent seedling recruitment. This is likely explained by the fact that sand burial could keep the seeds moist and stimulate seed germination. This led to higher shrub abundance in sandy habitats than in grassland habitats (Figure 5a).

4.2. Sandy Habitats Facilitate Plant Growth of Shrubs

Soil texture changes the water, air, heat, and nutrition status of land, thereby affecting the growth and development of plants growing in the soil. Our results showed that sandy soil promoted the sapling establishment of *Caragana* (Figure 2a). Sladonja et al. (2014) [10] also found that the percentage of seedling of pyrethrum was positively related to sand content. This might be because the sandy soil substrate had good air permeability with loosely packed matrixes, which would be beneficial for seed respiration, seed germination, thus facilitating seedling emergence and root expansion.

Previous studies found that the relative proportions of sand had strong negative effects on plant performance in general [38]. For example, soils with a higher sand content reduced cover and biomass for both shrub and herbaceous species [11] and constrained the distribution of species [12]. In contrast, our study found that sandy substrates favored *Caragana* growth (Figure 2b,c), which could explain why the height and area of the *Caragana* shrub in the sandy land was larger than that in the grassland (Figure 5b,c). This indicates that *Caragana* possesses an inherent ability to respond to sandy soil, which is consistent with its role as a primary colonizer in dry areas with high sand content.

Following the successful establishment of saplings of *Caragana* in the sandy land, they will continue to be buried by wind-deposited sand (Figure 5d). The sand burial significantly facilitated sapling growth (Figure 2c), thus forming large shrubs (Figure 5b,c). Similar results were also reported that partial sand burial promoted biomass in *Hedysarum laeve* [39], *Ulmus pumila* [18], and *Agriophyllum squarrosum* [17], while others found that when whole or partial shoots were buried by sand, biomass production decreased [20,24,35]. Those results suggested that the responses of sapling growth to sand burial might be species-specific.

The allocation and utilization of resources is a fundamental and vital activity of plants. Our study found that sand burial increased the root growth and root-shoot ratio (Figure 2d), indicating that resource allocation to the root increased after burial. Such results were also shown in *Nitraria sphaerocarpa* [40]. In the hyper-arid region (such as Alax with annal precipitation of 110 mm), water is the major limiting factor affecting plant growth and development. *Caragana* that withstand sand burial could increase biomass allocation to roots, thus acquiring more water from dry sandy soil to support its growth. However, contrasting results have been reported for *Artemisia ordosica*, which showed more biomass investment in shoots under burial [26,41].

4.3. Sandy Habitats Facilitate Seed Reproduction of Shrubs

The sexual reproductive efforts of *Caragana* in sandy land was higher than that in grassland (Figure 3a). Higher sexual reproductive efforts mean more seeds produced by *Caragana* in sandy land, thus forming more seedlings. This result was likely because sandy land enhanced the growth of *Caragana* (Figure 5b,c). The better the plant's growth, the stronger their fecundity capacity, and the more they invest in sexual reproductive organs, resulting in higher reproductive efforts in the sandy land.

In grassland, seed consumption by predators is very common, which immensely reduces seed quantity. Previous studies in the grassland of Loess Plateau showed that almost all *Caragana* seeds were eaten by seed predators within one week of the seeds falling on the ground [42]. However, in sandy land, shrubs can easily be buried by sand, thereby forming nabkhas of various sizes, from 20 to 40 cm in height, under shrub canopies due to wind-deposited sand (Figure 5d) [43]. In this case, *Caragana* seeds may fall under or next to shrub canopies where they were later buried at various depths in sand together with leaves of shrubs and other debris. Partial sand burial prevented seeds from predation and shielded dispersed seeds from wind, resulting in higher seed preservation (Figure 3b). Therefore, we hypothesized that sandy habitats facilitated the abundance of *Caragana* by improving seed preservation.

In summary, sandy soil substrate and sand burial are important environmental factors in sandy land, which increases seed number, seed preservation, seed germination, and subsequently improves sapling establishment and species fitness, and finally increases the individual numbers in population. This would explain why we found a higher abundance of *Caragana* in sandy land (Figure 5a). In addition, *Caragana*, cluster shrubs with a combined canopy, usually consists of many individuals of sexually reproduction [44]. An increase in sexual reproduction led to more individuals in each shrub cluster, resulting in an increase in shrub area (Figure 5b).

4.4. Sandy Habitats Facilitate Clonal Reproduction of Shrubs

Most studies found that sand burial had negative effects on clonal reproduction (number of ramets) for *Spartina alterniflora* [24] and *Calligonum arborescens* [6,23]. However, our results showed that sand burial promoted clonal reproduction of *Caragana* by producing more adventitious roots on their branches (Figure 4a). Such adventitious roots would be capable of forming more branch-derived ramets, thereby leading to increase of shrub area (Figure 5b). The positive effect of sand burial on formation of adventitious roots was also found for *Artemisia* species [45].

Severe erosion had negative effects on the clonal reproduction of xylophyta, such as *Calligonum arborescens* [6,23] and *Calligonum mongolicum* [22]. In our study, the horizontal roots, the reproductive organs of *Caragana*, could extend outward by more than 130 cm from the shrub [31]. Wind erosion exposed these horizontal roots to air, which promoted the formation of root-derived ramets (Figure 4b). Ramets close to the mother plant were interconnected with the mother plant to form larger shrubs (Figure 5b), while ramets away from the mother plant formed new shrubs that would increase the abundance of the *Caragana* population (Figure 5a). Our study suggested that spreading the horizontal roots may be an important adaptive strategy for shrub population persistence in sand habitats. Our results are contrary to Yu et al. (2008) [28], who showed that once rhizomes of *Psammochloa villosa* were exposed to the air, the associated ramets either died or became very weak, and ramet number decreased as erosion severity increased.

In summary, both concurrent branch-derived and root-derived clonal reproduction are important biological characteristics for *Caragana* to adapt to sandy environments. Sand burial favored the clonal reproduction of *Caragana* by promoting the formation of branch-derived ramets, while wind erosion benefited the clonal reproduction by facilitating the formation of root-derived ramets. Such mechanisms led to greater abundance and larger areas of *Caragana* shrubs in the sandy land.

5. Conclusions

Sandy habitats stimulated seed germination, promoted plant growth, increased sexual reproduction, and facilitated clonal reproduction of *Caragana* shrubs, which could explain why the abundance, area and height of the *Caragana* shrubs in sandy lands was higher than that in grasslands. Our results suggested that grassland degradation led to land desertification and such sandy habitats favored the population growth of *Caragana* shrubs, which was one of the important mechanisms for shrub encroachment in grasslands. In addition, the capability of fast growth and rapid population expansion in sandy land makes *Caragana* an ecological engineering species in ameliorate desertification.

Author Contributions: Conceptualization, L.X. and C.M.; methodology, L.X. and P.H.; formal analysis, Y.L.; investigation, L.X., Y.L. and C.W.; data curation, L.X. and P.H.; writing—original draft preparation, L.X. and C.M.; writing—review and editing, H.G., Q.C. and C.M.; visualization, L.X. and C.M. All authors have read and agreed to the published version of the manuscript.

Funding: This study was supported by the National Natural Science Foundation of China (31901140, 31570453, 31971437) (http://www.nsfc.gov.cn/; accessed on 1 January 2020).

Data Availability Statement: The data presented in this study are available on request from the corresponding author.

Conflicts of Interest: The authors declare no conflict of interest.

References

1. Reynolds, J.F.; Smith, D.M.S.; Lambin, E.F.; Turner, B.L.; Mortimore, M.; Batterbury, S.P.; Downing, T.E.; Dowlatabadi, H.; Fernández, R.J.; Herrick, J.E.; et al. Global desertification: Building a science for dryland development. *Science* **2007**, *316*, 847–851. [CrossRef] [PubMed]
2. Van Auken, O.W. Causes and consequences of woody plant encroachment into western North American grasslands. *J. Environ. Manag.* **2009**, *90*, 2931–2942. [CrossRef]
3. Li, S.L.; Yu, F.H.; Werger, M.J.; Dong, M.; During, H.J.; Zuidema, P.A. Mobile dune fixation by a fast-growing clonal plant: A full life-cycle analysis. *Sci. Rep.* **2015**, *5*, 8935. [CrossRef] [PubMed]
4. Xiong, X.; Han, X.; Bai, Y.; Pan, Q. Increased distribution of *Caragana microphylla* in rangelands and its causes and consequences in Xilin River Basin. *Acta Prataculturae Sin.* **2003**, *12*, 57–62.
5. Xiong, X.; Han, X.; Chen, Q.; Pan, Q. Increased abundance of woody plants in grasslands and savannas. *Acta Ecol. Sin.* **2003**, *23*, 2436–2443.
6. Luo, W.; Zhao, W. Effects of wind erosion and sand burial on growth and reproduction of a clonal shrub. *Flora* **2015**, *217*, 164–169. [CrossRef]
7. Wang, G.; Yu, K.; Gou, Q. Effects of sand burial disturbance on establishment of three desert shrub species in the margin of oasis in northwestern China. *Ecol. Res.* **2019**, *34*, 127–135. [CrossRef]
8. Tracy, S.R.; Black, C.R.; Roberts, J.A.; Mooney, S.J. Exploring the interacting effect of soil texture and bulk density on root system development in tomato (*Solanum lycopersicum* L.). *Environ. Exp. Bot.* **2013**, *91*, 38–47. [CrossRef]

9. White, R.E. *Principles and Practice of Soil Science: The Soil as a Natural Resource*, 4th ed.; Blackwell Publishing: Oxford, UK, 2005.
10. Sladonja, B.; Krapac, M.; Ban, D.; Uzila, Z.; Dudas, S.; Dorcic, D. Effect of soil type on pyrethrum seed germination. *J. Plant Prot. Res.* **2014**, *54*, 421–425. [CrossRef]
11. Li, X.R.; Zhang, Z.S.; Zhang, J.G.; Wang, X.P.; Jia, X.H. Association between vegetation patterns and soil properties in the southeastern Tengger Desert, China. *Arid Land Res. Manag.* **2004**, *18*, 369–383. [CrossRef]
12. Medinski, T.V.; Mills, A.J.; Esler, K.J.; Schmiedel, U.; Jürgens, N. Do soil properties constrain species richness? Insights from boundary line analysis across several biomes in south western Africa. *J. Arid Environ.* **2010**, *74*, 1052–1060. [CrossRef]
13. Zaller, J.G.; Frank, T.; Drapela, T. Soil sand content can alter effects of different taxa of mycorrhizal fungi on plant biomass production of grassland species. *Eur. J. Soil Biol.* **2011**, *47*, 175–181. [CrossRef] [PubMed]
14. Liu, H.L.; Shi, X.; Wang, J.C.; Yin, L.K.; Huang, Z.Y.; Zhang, D.Y. Effects of sand burial, soil water content and distribution pattern of seeds in sand on seed germination and seedling survival of *Eremosparton songoricum* (Fabaceae), a rare species inhabiting the moving sand dunes of the Gurbantunggut Desert of China. *Plant Soil* **2011**, *345*, 69–87. [CrossRef]
15. Zhu, Y.; Yang, X.; Baskin, C.C.; Baskin, J.M.; Dong, M.; Huang, Z. Effects of amount and frequency of precipitation and sand burial on seed germination, seedling emergence and survival of the dune grass *Leymus secalinus* in semiarid China. *Plant Soil* **2014**, *374*, 399–409. [CrossRef]
16. Ye, X.; Li, L.; Baskin, C.C.; Baskin, J.M.; Du, J.; Huang, Z. Sand burial helps regulate timing of seed germination of a dominant herb in an inland dune ecosystem with a semiarid temperate climate. *Sci. Total Environ.* **2019**, *680*, 44–50. [CrossRef]
17. Li, J.; Qu, H.; Zhao, H.; Zhou, R.; Yun, J.; Pan, C. Growth and physiological responses of *Agriophyllum squarrosum* to sand burial stress. *J. Arid Land* **2015**, *7*, 94–100. [CrossRef]
18. Shi, L.; Zhang, Z.J.; Zhang, C.Y.; Zhang, J.Z. Effects of sand burial on survival, growth, gas exchange and biomass allocation of *Ulmus pumila* seedlings in the Hunshandak Sandland, China. *Ann. Bot.* **2004**, *94*, 553–560. [CrossRef]
19. Zheng, M.; Lai, L.; Jiang, L.; An, P.; Yu, Y.; Zheng, Y.; Shimizue, H.; Baskin, J.M.; Baskin, C.C. Moderate water supply and partial sand burial increase relative growth rate of two *Artemisia* species in an inland sandy land. *J. Arid Environ.* **2012**, *8*, 105–113. [CrossRef]
20. Xu, L.; Huber, H.; During, H.J.; Dong, M.; Anten, N.P. Intraspecific variation of a desert shrub species in phenotypic plasticity in response to sand burial. *New Phytol.* **2013**, *199*, 991–1000. [CrossRef]
21. Li, F.; Xie, Y.; Zhu, L.; Jiang, L.; Chen, X.; Pan, B.; Deng, Z. Changed clonal growth form induced by sand burial facilitates the acclimation of *Carex brevicuspis* to competition. *PLoS ONE* **2015**, *10*, e0121270. [CrossRef]
22. Fan, B.; Zhao, C.; Zhang, X.; Sun, K. Impacts of sand burial and wind erosion on regeneration and growth of a desert clonal shrub. *Front. Plant Sci.* **2018**, *9*, 1696. [CrossRef] [PubMed]
23. Luo, W.; Zhao, W.; Zhuang, Y. Sand-burial and wind erosion promote oriented-growth and patchy distribution of a clonal shrub in dune ecosystems. *Catena* **2018**, *167*, 212–220. [CrossRef]
24. Xiao, Y.; Zhao, H.; Yang, W.; Qing, H.; Zhou, C.; Tang, J.; An, S. Variations in growth, clonal and sexual reproduction of *Spartina alterniflora* responding to changes in clonal integration and sand burial. *CLEAN–Soil Air Water* **2015**, *43*, 1100–1106. [CrossRef]
25. Ye, X.; Liu, Z.; Gao, S.; Cui, Q.; Liu, G.; Du, J.; Dong, M.; Huang, Z.; Cornelissen, J.H.C. Differential plant species responses to interactions of sand burial, precipitation enhancement and climatic variation promote co-existence in Chinese steppe vegetation. *J. Veg. Sci.* **2017**, *28*, 139–148. [CrossRef]
26. Li, S.L.; Werger, M.J.; Zuidema, P.A.; Yu, F.H.; Dong, M. Seedlings of the semi-shrub *Artemisia ordosica* are resistant to moderate wind denudation and sand burial in Mu Us sandland, China. *Trees* **2010**, *24*, 515–521. [CrossRef]
27. Liu, B.; Liu, Z.; Wang, L.; Wang, Z. Responses of rhizomatous grass *Phragmites communis* to wind erosion: Effects on biomass allocation. *Plant Soil* **2014**, *380*, 389–398. [CrossRef]
28. Yu, F.H.; Wang, N.; He, W.M.; Chu, Y.; Dong, M. Adaptation of rhizome connections in drylands: Increasing tolerance of clones to wind erosion. *Ann. Bot.* **2008**, *102*, 571–577. [CrossRef]
29. Zheng, M.; Song, J.; Ru, J.; Zhou, Z.; Zhong, M.; Jiang, L.; Hui, D.F.; Wan, S. Effects of grazing, wind erosion, and dust deposition on plant community composition and structure in a temperate steppe. *Ecosystems* **2021**, *24*, 403–420. [CrossRef]
30. Xie, L.N.; Ma, C.C.; Guo, H.Y.; Li, Q.F.; Gao, Y.B. Distribution pattern of *Caragana* species under the influence of climate gradient in the Inner Mongolia region, China. *J. Arid Land* **2014**, *6*, 311–323. [CrossRef]
31. Ma, C.C.; Zhang, J.H.; Guo, H.Y.; Li, Q.F.; Xie, L.N.; Gao, Y.B. Alterations in canopy size and reproduction of *Caragana stenophylla* along a climate gradient on the Inner Mongolian Plateau. *Flora* **2013**, *208*, 97–103. [CrossRef]
32. Wang, Z.W.; Xie, L.N.; Prather, M.C.; Guo, H.Y.; Han, G.D.; Ma, C.C. What drives the shift between sexual and clonal reproduction of *Caragana stenophylla* along a climatic aridity gradient? *BMC Plant Biol.* **2018**, *18*, 91. [CrossRef] [PubMed]
33. Yan, C.Z.; Wang, Y.M.; Feng, Y.S.; Wang, J.H. Macro-scale survey and dynamic studies of sandy land in Ningxia by remote sensing. *J. Desert Res.* **2003**, *23*, 34–37.
34. Reekie, E.G.; Bazzaz, F.A. Reproductive effort in plants. 1. Carbon allocation to reproduction. *Am. Nat.* **1987**, *129*, 876–896. [CrossRef]
35. Zhu, Y.; Dong, M.; Huang, Z. Response of seed germination and seedling growth to sand burial of two dominant perennial grasses in Mu-Us sandy grassland, semiarid China. *Rangel. Ecol. Manag.* **2009**, *62*, 337–344. [CrossRef]

36. Mao, P.L.; Mu, H.X.; Cao, B.H.; Liu, Y.H.; Fan, Z.F.; Wang, S.M. Effects of sand burial and overstory tree age on seedling establishment in coastal *Pinus thunbergii* forests in the northern Shandong Peninsula, China. *For. Chron.* **2016**, *92*, 357–365. [CrossRef]
37. Yang, H.L.; Liang, Z.L.; Zhu, X.W.; Mei, S.X.; Wang, H.Q.; Shen, Y.; Huang, Z.Y. Effects of sand burial and seed size on seed germination, seedling emergence and growth of *Caragana korshinskii* Kom (Fabaceae). *Acta Ecol. Sin.* **2012**, *32*, 7757–7763. [CrossRef]
38. Hook, P.B.; Burke, I.C. Biogeochemistry in a shortgrass landscape: Control by topography, soil texture, and microclimate. *Ecology* **2000**, *81*, 2686–2703. [CrossRef]
39. Zhang, C.Y.; Yu, F.H.; Dong, M. Effects of sand burial on the survival, growth, and biomass allocation in semi-shrub *Hedysarum laeve* seedlings. *Acta Bot. Sin.* **2002**, *44*, 337–343.
40. Zhao, W.Z.; Li, Q.Y.; Fang, H.Y. Effects of sand burial disturbance on seedling growth of *Nitraria sphaerocarpa*. *Plant Soil* **2007**, *295*, 95–102. [CrossRef]
41. Liu, B.; Liu, Z.; Guan, D. Seedling growth variation in response to sand burial in four *Artemisia* species from different habitats in the semi-arid dune field. *Trees* **2008**, *22*, 41–47. [CrossRef]
42. Zhao, X.; Ren, J. Influence of seed predation on regeneration of three *Caragana* species. *Biodivers. Sci.* **2005**, *13*, 514–519. [CrossRef]
43. Zhang, Y.Y.; Ma, C.C.; Han, L.; Gao, Y.B. Nabkha morphology and sand-fixing capability of four dominant *Caragana* species in the desert region of the Inner Mongolia Plateau. *Acta Ecol. Sin.* **2012**, *32*, 3343–3351. [CrossRef]
44. Xie, L.N.; Guo, H.Y.; Gabler, C.A.; Li, Q.F.; Ma, C.C. Changes in Spatial Patterns of *Caragana stenophylla* along a Climatic Drought Gradient on the Inner Mongolian Plateau. *PLoS ONE* **2015**, *10*, e0121234. [CrossRef] [PubMed]
45. Liu, B.; Liu, Z.; Lü, X.; Maestre, F.T.; Wang, L. Sand burial compensates for the negative effects of erosion on the dune-building shrub *Artemisia wudanica*. *Plant Soil* **2014**, *374*, 263–273. [CrossRef]

agronomy

MDPI

Article

Sexual Reproduction Is Not Responsible for *Caragana* Shrub Encroachment in Grasslands

Lina Xie [1], Yuchen Li [2], Mingyan Lin [1], Hongyu Guo [1], Yue Wang [1], Lihong Wang [1] and Chengcang Ma [1,*]

[1] Tianjin Key Laboratory of Animal and Plant Resistance, College of Life Sciences, Tianjin Normal University, Tianjin 300387, China; xielina1989@163.com (L.X.); mia200001@163.com (M.L.); hongyuguo88@163.com (H.G.); wangyue5182023@163.com (Y.W.); wanglihong5182023@163.com (L.W.)
[2] College of Life Sciences, Nankai University, Tianjin 300071, China; 2120211142@mail.nankai.edu.cn
* Correspondence: machengcang@163.com

Abstract: Shrubs tended to increase their abundance as climatic aridity and grazing intensity increased in the Inner Mongolian grassland. Increasing shrub abundance was believed to be due to enhanced reproduction. However, the effects of climatic aridity and grazing on the sexual reproduction of shrubs in grassland remain largely unclear. In this study, we conducted field experiments with *Caragana microphylla* to examine the variation of sexual reproduction aspects (seed production, seed vigor, and sapling establishment) along a climatic aridity gradient (subhumid, semiarid, arid, and dry arid zones) and a grazing intensity gradient (fenced, mildly grazed, and severely grazed). We then quantified the population growth rate based on seed production and sapling establishment rates. Our objective was to evaluate whether sexual reproduction is the main mechanism for *Caragana* encroachment into grasslands. We found that climatic aridity decreased seed quantity and seed vigor but increased the sapling establishment rate of *Caragana* shrubs. Under ungrazed conditions, climatic aridity did not affect population growth rates, while under grazing conditions, increased aridity stresses reduced population growth rates. Grazing reduced seed production, sapling establishment, and population growth rates. Climatic aridity enhanced the negative effects of grazing on sexual reproduction, while grazing intensified the negative effects of aridity on the population growth of *Caragana* shrubs. In conclusion, climatic aridity, grazing, and their combined effects had negative effects on the sexual reproduction of *Caragana* shrubs. Therefore, sexual reproduction could not fully explain the increased abundance of shrubs with increasing aridity and grazing. Clonal reproduction might be of considerable importance for understanding the mechanism of shrub encroachment in grasslands.

Keywords: shrub encroachment; population growth; sexual reproduction; grazing; climatic aridity gradient

Citation: Xie, L.; Li, Y.; Lin, M.; Guo, H.; Wang, Y.; Wang, L.; Ma, C. Sexual Reproduction Is Not Responsible for *Caragana* Shrub Encroachment in Grasslands. *Agronomy* **2023**, *13*, 1848. https://doi.org/10.3390/agronomy13071848

Academic Editors: Kesi Liu and Xinqing Shao

Received: 9 June 2023
Revised: 3 July 2023
Accepted: 11 July 2023
Published: 13 July 2023

1. Introduction

Woody encroachment is occurring in grasslands worldwide [1]. Alterations in the density of woody plants have been mainly attributed to environmental factors and the characteristics of encroaching shrubs. It is suggested that long-term overgrazing and climatic aridity led to grassland vegetation degradation with reductions in species richness, abundance, and cover, resulting in reductions in fire frequency and heterogeneous distributions of vegetation and soil resources, as well as land desertification. These conditions would favor woody plants over grasses [2–4]. It is also suggested that strong grazing resistance [2] and drought adaptability [5–8] of shrubs can confer on their adult plants a survival advantage under constant overgrazing and intensive climatic aridity. All of these characteristics are likely to alter the competition between grass and woody plants under biotic and abiotic stress. Generally, increasing shrub abundance was believed to be due to increased reproduction. So far, the role of reproduction in shrub encroachment is still unclear.

Caragana shrubs are the dominant native perennial woody plants in the Inner Mongolian grassland ecosystem. This region has a strong aridity gradient from the northeast to the southwest, along which their climatic aridity conditions range from subhumid, semi-arid, arid, dry arid, and intensively arid. Livestock grazing is the most important and pervasive anthropogenic disturbance in grasslands. Our previous studies documented that shrub encroachment by *Caragana* in the Inner Mongolian grassland had two characteristics [6,9]. First, *Caragana* shrubs tended to increase their encroachment as climatic aridity increased. Second, shrub encroachment became more serious as grazing intensity increased as well. Given that such a direct increase in abundance is attributed to an increase in reproduction, we hypothesized that climatic aridity and grazing would promote *Caragana* shrub reproduction, which would be responsible for the current shrub encroachment in the Inner Mongolian grassland.

There are few grasslands on earth with natural aridity gradients, so very few studies have been conducted to examine how the sexual reproduction of grassland plants varies across aridity gradients. Studies have shown that seed yield [10] and seed vigor [11–13] of herbaceous plants in grasslands decreased with increasing aridity. However, studies also showed that seed production was significantly greater in a semi-arid environment than in a comparable Mediterranean environment [14], and there were no clear trends in seed or pod-related traits along the aridity gradient from arid to humid zones [15]. Similarly, drought also affected seedling establishment and population growth rates. For example, drought severely limited tree regeneration, increased population mortality rates [16–18], and decreased seedling survival rates [19]. Other studies found that seed germination and seedling recruitment of grass were not strongly impacted by an increasing water deficit in the Patagonian steppe [20]. However, how seed production, sapling recruitment, and population growth of shrub species vary along a climatic aridity gradient remain largely unknown.

Contrary to climatic aridity effects, many studies offered insights into the effect of grazing on the sexual reproduction of grassland plants. First, grazing influences seed production for different plant species in grasslands in different ways. For example, grazing decreased seed production of *Atriplex vesicaria* [21], *Ruellia humilis* and *Amorpha canescens* [22], *Stipa grandis*, *Agropyron cristatum*, and *Cleistogenes squarrosa* [23], and decreased flower and fruit numbers of *Agrostis vinealis* and *Ranunculus bulbosussome* [24]. Whereas the reverse was also true for some unpalatable plant species, such as *Agrostis capillaris*, *Dianthus deltoides* [24], and *Ligularia narynensis* [25]. While grazing did not affect the reproduction of *Festuca gracillima* [26] and *Artemisia tridentata* [27]. Second, grazing directly affected the seedling establishment process. For example, grazing had a negative effect on plant recruitment and population growth [28–31] by decreasing seedling survival rate and increasing seedling mortality rate, thereby decreasing seedling number [32–35]. However, other studies found that grazing facilitated seedling establishment, especially for legume plants and non-grass herbs that are unpalatable [36,37]; moderate grazing increased the number of newborn seedlings [38], and the number of sexual offspring increased with increasing grazing intensity in grasslands [39,40]. Moreover, in a few cases, grazing did not affect seedling establishment in herbaceous and shrub species [41,42]. Given the above divergent observations, an improved understanding of how the sexual reproduction of shrubs varies along the grazing gradient would help predict the pattern of shrub encroachment.

Moreover, the joint effects of climatic aridity and grazing stress on population growth have rarely been addressed [43–45], even though these two stressors commonly co-occur in natural ecosystems and play important roles in shaping plant communities. Consequently, empirical studies are necessary to test their joint effects on the population growth of shrubs, which would advance our understanding of shrub encroachment into grasslands.

In this study, we conducted field experiments to test the sexual reproduction (seed production, seed vigor, and sapling establishment) of *Caragana microphylla* along a climatic aridity gradient (subhumid, semiarid, arid, and dry arid zones) and a grazing intensity gradient (fenced, mildly grazed, and severely grazed) in the Inner Mongolian grassland.

We then quantified population growth based on seed production and sapling establishment rates. Our objective was to evaluate whether sexual reproduction is the main factor in *Caragana* shrub encroachment into grasslands.

2. Study Area and Methods

2.1. Study Area and Grazing Treatment

We conducted the field experiment at four study sites in the Inner Mongolian Steppe in northern China. The study sites were distributed in a wide range of aridity gradients, including (1) the Xiwu site in the subhumid zone; (2) the Abaga site in the semiarid zone; (3) the Suniteyou site in the arid zone; and (4) the Siziwang site in the dry arid zone (Table 1). The mean annual precipitation (MAP) and mean annual temperature (MAT) of this region ranged from 210 mm to 354 mm and from 1.87 °C to 3.60 °C, respectively (Table 1). We obtained the MAP and MAT data from the ERA5-Land dataset (https://cds.climate.copernicus.eu/cdsapp#!/dataset/reanalysis-era5-land-monthly-means?tab=overview; accessed on 7 January 2023). The aridity level of each site was quantified as the Aridity Index (AI) = MAP/(10 + MAT), where a higher AI value indicates that a site is more mesic (less arid). The vegetation of the study sites is dominated by the shrub species *C. microphylla*.

Table 1. Location and environmental data of the study sites.

Site	Longitude (°E)	Latitude (°N)	Altitude (m)	Mean Annual Precipitation (mm)	Mean Annual Temperature (°C)	Aridity Index	Aridity Type
Xiwu	117°38′53″	44°36′50″	1024	354	2.42	28.51	Subhumid
Abaga	114°58′02″	44°04′00″	1126	245	1.87	20.64	Semi-arid
Suniteyou	112°56′39″	42°21′00″	1144	220	3.09	16.81	Arid
Siziwang	111°53′22″	41°47′28″	1492	210	3.60	15.44	Dry arid

We established three levels of grazing treatment within each site: fenced (F), mild grazing (MG), and severe grazing (SG). The grazing intensities were as follows: (1) 1.2 sheep/ha (MG), 2.5 sheep/ha (SG) at the Xiwu site; (2) 1.1 sheep/ha (MG), 2.2 sheep/ha (SG) at the Abaga and Suniteyou sites; (3) 1.0 sheep/ha (MG), 2.0 sheep/ha (SG) at the Siziwang site. Since vegetation cover decreases gradually with the increase in climatic aridity stress from the subhumid zone to the dry arid zone, the grazing intensity at each site was set according to the local vegetation condition for logistical reasons. The four study sites were all under long-term grazing treatments (over 14 years). Considering interannual climate variation, the experiments were conducted twice (seed production in 2017 and 2018; sapling establishment in 2018–2019 and 2019–2020).

2.2. Seed Production Experiment

In each grazing treatment, we established four subplots (the area of each subplot was more than 1 ha). During the pod ripening seasons in July–August 2017 and 2018, we first randomly chose 10 shrubs within each subplot. Then we collected three branches from each shrub, counted the number of mature pods and seeds on branches, and calculated the number of seeds/dry biomass (g) after drying the branches at 80 °C for 72 h.

2.3. Seed Vigor Experiment

In July–August 2017 and 2018, a large number of seeds were collected from each subplot, air dried, and stored at 4 °C. Then we conducted a seed germination experiment in November of the same year. 100 randomly selected seeds from each subplot were surface sterilized with a 1% NaClO solution for 10 min, followed by rinsing with distilled water several times. Sterilized seeds were placed in an incubator with 12 h light and 12 h dark cycles and a temperature of 25 °C under light conditions/15 °C under dark conditions. A seed with a germ length of 1 mm was defined as a germinated seed. We recorded the number of germinated seeds per day during the germination test until there was no seed germination for 5 consecutive days. Then, we calculated the seed germination index $GI = \sum Gt/Dt$, where Gt is the number of germinated seeds on day t and Dt is the time corresponding to Gt in days.

2.4. Sapling Establishment Experiment

The field-sowing experiments were conducted during 2018–2019 and 2019–2020, respectively. At the beginning of the growing season in 2018 and 2019, we established a transect of 100 m in each subplot and placed 1×1 m^2 quadrats at intervals of 10 m along each transect (in total, 10 quadrats per subplot). We sowed 100 *C. microphylla* seeds in each quadrat. The seeds sown at each site were collected from the same site the previous year. At the end of the growing season in the next year (18 months after sowing), we recorded the sapling number in each quadrat and then calculated the sapling establishment rate for each quadrat using the formula: sapling establishment rate = sapling number/100.

2.5. Population Growth Rate

We quantified the population growth rate by sexual recruitment using the formula: population growth rate = seed number (seed/g dry biomass) × sapling establishment rate (sapling/seed).

2.6. Statistical Analysis

We performed data analyses using GLMMs, with sampling shrubs within subplots and subplots within grazing treatments as random variables (sampling shrubs (n = 20) were nested in each subplot; subplots (n = 4) were nested in each grazing treatment) in order to examine the differences in seed production among climatic aridity zones and grazing intensities. We performed ANOVAs to analyze the differences in means of germination index and population growth rate among climatic aridity zones and grazing intensities. We used Tukey HSD post-hoc tests to identify the differences among individual treatments. We also performed GLMMs with quadrats within subplots and subplots within grazing treatments as random variables (sampling quadrats (n = 20) were nested in each subplot, and subplots (n = 4) were nested in each grazing treatment) in order to examine the differences in sapling establishment among climatic aridity zones and grazing intensities.

3. Results

3.1. Effects of Climatic Aridity and Grazing on Seed Production

C. microphylla shrubs at the wettest end of the arid gradient produced more seeds, and seed production gradually decreased as climatic aridity increased ($F_{3, 948}$ = 17.45, $p < 0.01$). There was a 64.6% decrease in seed production in the MG treatment across the aridity gradient, with the average seed production ranging from 1.27 seeds at the least arid site to 0.45 seeds at the most arid site. We also noticed that seed production in the F treatment did not change significantly along the aridity gradient (Figure 1).

We observed a similar trend along the grazing intensity, with a sharp decrease in seed production as grazing intensity increased ($F_{2, 948}$ = 67.45, $p < 0.01$). In the dry arid zone, seed production in the MG and SG treatments was 61.2% and 81% less than that in the F treatment, respectively (Figure 1).

As climatic aridity stress increased, the negative effect of grazing on seed production gradually increased, although the interaction effect between climatic aridity and grazing was not significant ($F_{6, 948}$ = 0.717, p = 0.64). Compared with the F treatment, seed production in the SG treatment decreased by 51% in the subhumid zone, 62% in the semiarid zone, 71% in the arid zone, and 81% in the dry arid zone, respectively (Figure 1).

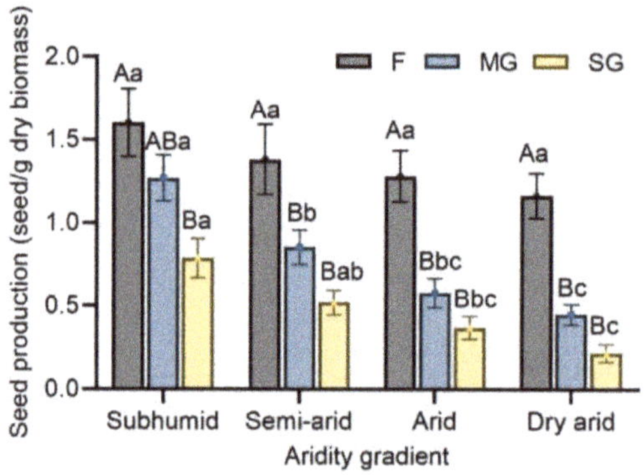

Figure 1. Effects of climatic aridity and grazing on seed production of *Caragana microphylla*. Error bars indicate standard errors. F, fenced; MG, mild grazing; SG, severe grazing. For the same climatic aridity, different uppercase letters indicate significant differences between grazing intensities; and for the same grazing intensity, different lowercase letters indicate significant differences between climatic aridity zones.

3.2. *Effects of Climatic Aridity and Grazing on Seed Vigor*

Seed germination indexes of *C. microphylla* showed a decreasing trend as climatic aridity stress increased ($F_{3,84} = 4.70$, $p < 0.01$), ranging from 17.5 in the subhumid zone to 14.8 in the dry arid zone in the fenced plots (Figure 2). Grazing had a significant effect on the seed germination index of *C. microphylla* ($F_{2,84} = 18.01$, $p < 0.01$), with the highest value in the F treatment and the lowest value in the MG treatment. Germination indexes in the arid zone were 63.0% lower in the MG treatment and 11.4% lower in the SG treatment, compared with those in the F treatment (Figure 2). The interaction between climatic aridity and grazing had no significant effects on the seed germination index of *C. microphylla* ($F_{6,84} = 1.37$, $p = 0.24$; Figure 2).

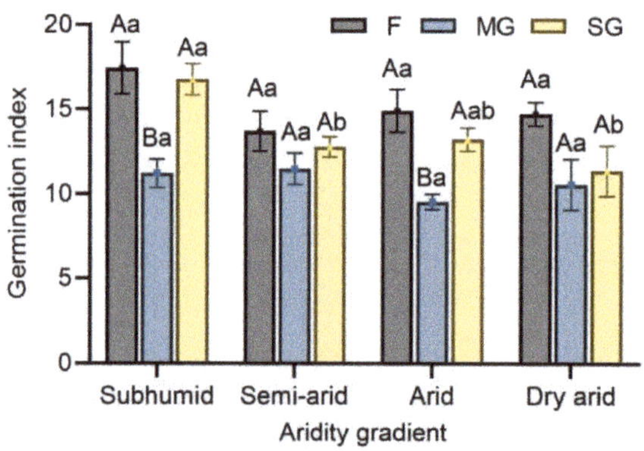

Figure 2. Effects of climatic aridity and grazing on seed germination index of *Caragana microphylla*. Error bars indicate standard errors. F, fenced; MG, mild grazing; SG, severe grazing. For the same climatic aridity, different uppercase letters indicate significant differences between grazing intensities; and for the same grazing intensity, different lowercase letters indicate significant differences between climatic aridity zones.

3.3. *Effects of Climatic Aridity and Grazing on Sapling Establishment*

The increase in the sapling establishment rate of *C. microphylla* was found from the subhumid zone to the dry arid zone ($F_{3,948} = 6.66$, $p < 0.01$). In the fenced plots, the sapling establishment rates in the semi-arid, arid, and dry arid zones were 5.3%, 18.8%, and 30.3%

higher than those in the subhumid zone, respectively (Figure 3). Sapling establishment rate decreased as grazing intensity increased ($F_{2, 948} = 55.28$, $p < 0.01$), and as a result, this rate was 56.5–73.6% lower in the SG treatment than that in the F treatment (Figure 3). The interaction between climatic aridity and grazing had no significant effect on the sapling establishment rate of *C. microphylla* ($F_{6, 948} = 0.18$, $p = 0.98$; Figure 3).

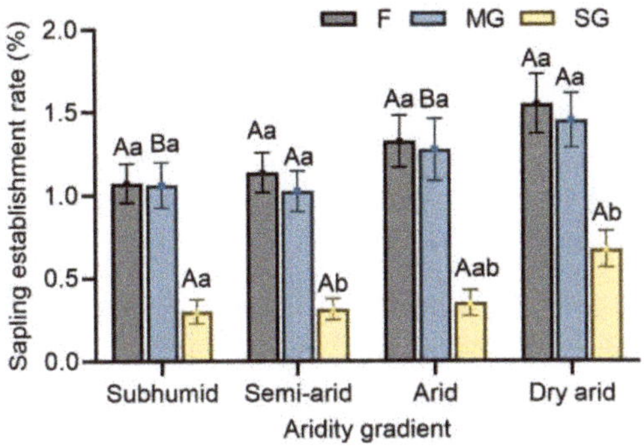

Figure 3. Effects of climatic aridity and grazing on sapling establishment rate of *Caragana microphylla*. Error bars indicate standard errors. F, fenced; MG, mild grazing; SG, severe grazing. For the same climatic aridity, different uppercase letters indicate significant differences between grazing intensities; and for the same grazing intensity, different lowercase letters indicate significant differences between climatic aridity zones.

3.4. *Effects of Climatic Aridity and Grazing on Population Growth*

The population growth rate of *C. microphylla* in the F treatment did not change significantly along the aridity gradient, whereas there was a strong decreasing trend of population growth rate in the MG treatment with increasing aridity. The population growth rate in the MG treatment was 50.1 saplings in the subhumid zone, 33.2 saplings in the semi-arid zone, 27.8 saplings in the arid zone, and 25.0 saplings in the dry arid zone, respectively. Thus, there was a significant interaction between climatic aridity and grazing, as grazing intensified the negative effect of climate aridity on the population growth rate of *C. microphylla* ($F_{6, 84} = 2.19$, $p = 0.05$; Figure 4).

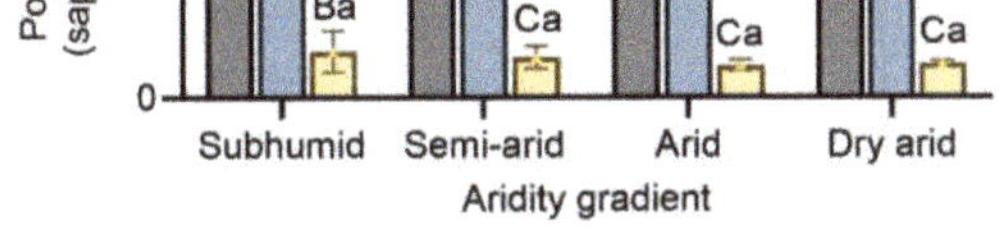

Figure 4. Effects of climatic aridity and grazing on population growth rate of *Caragana microphylla*. Error bars indicate standard errors. F, fenced; MG, mild grazing; SG, severe grazing. For the same climatic aridity, different uppercase letters indicate significant differences between grazing intensities; and for the same grazing intensity, different lowercase letters indicate significant differences between climatic aridity zones.

There was a sharp decrease in the population growth rate of *C. microphylla* as grazing intensity increased ($F_{2, 84} = 145.12$, $p < 0.01$). In the semi-arid zone, the population growth rate was 43.2% and 89.1% less in the MG and SG treatments than that in the F treatment (Figure 4).

4. Discussion

4.1. Effect of Climatic Aridity on Sexual Reproduction

Seed production of *C. microphylla* decreased across all four aridity levels. Patterns for seed production were consistent with those of some herbaceous plants and desert shrubs, in which plant populations from the most arid locations produced fewer seeds [46–49]. This was likely due to a reduced capacity to take up nutrients under aridity stresses [50], which would lead to a decrease in biomass allocation to sexual reproduction and thus decrease the fecundity of shrubs. In addition, aridity stress may cause the abortion of flowers and early fruit drops [51], which may account for hindering the seed development process. These results were contrary to patterns found for *C. stenophylla*, another shrub species of the same *Caragana* genus, whose seed number gradually increased from the semi-arid zone to the very arid zone [44].

The seed vigor of *C. microphylla* decreased with increasing aridity. Reduced seed vigor under the most arid conditions was also reported for herbs [11–13]. These may also be due to nutrient shortages caused by water deficit stress, which were not able to allow the production of more vigorous seeds [13,52]. This suggests that aridity could act as a selective pressure, unfavoring seed vigor. However, some other studies on seed germination showed that there were no significant differences in seed vigor for seeds from different arid zones [45].

Many studies have shown that aridity reduces seedling establishment in herbs [16–18] and shrubs [19,45]. In our study, the sapling establishment rate of *C. microphylla* increased from the subhumid zone to the dry arid zone. Such a response might enable *C. microphylla* to be more adaptive to arid conditions at seed germination and the seedling stage. This could be one of the important factors facilitating shrub encroachment into grasslands.

Under ungrazed conditions, the seed production and population growth rate of *C. microphylla* did not change significantly along the aridity gradient. This was probably due to the stronger drought adaptability of *Caragana* shrubs [6–8]. Moreover, *Caragana* shrubs are leguminous with nitrogen-fixing capability. In general, nitrogen-fixing plants would have higher water use efficiency and a stronger competitive advantage in more arid environments [53,54], which might enable them to survive, grow, and develop during prolonged periods of drought. In addition, leguminous plants tend to have higher levels of nitrogen-fixing capability in more arid areas (Armstrong 2019; Doctoral dissertation, San Jose State University), which would greatly increase soil nitrogen content in these areas [55]. Such increased soil nutrition conditions would also offset the negative effects of aridity stress on the sexual reproduction of *Caragana* shrubs. In contrast, under grazing conditions, increasing aridity stress resulted in a greater reduction in the population growth rate. This suggests that the strong drought tolerance of *C. microphylla* is insufficient for maintaining a relatively high population growth rate under hyper aridity conditions.

No consensus has been reached on whether sexual reproduction is the main factor in shrub encroachment into grasslands. The results of our study suggested that this is indeed possible. For example, we showed that shrub sapling establishment increased as climatic aridity stress increased. However, our study found no evidence that sexual reproduction could fully explain the phenomenon of a higher abundance of shrubs with increasing aridity. This could be attributed to the reduction in seed production along the aridity gradient, which led to no increase (under ungrazed conditions) and even a substantial decrease (under grazing conditions) in population growth rate with increasing aridity.

4.2. Effect of Grazing Intensity on Sexual Reproduction

Studies suggest that grazing has negative effects on seed production [56–59], sapling establishment [32–35], sexual recruitment [33,60,61], and thus population growth [28–31,62,63] of grassland plants. Indeed, our experiment showed that seed production and sapling establishment of *C. microphylla* were greatly reduced with the increase in grazing intensity, which reduced the population growth of this shrub plant. This could be explained by three potential mechanisms. First, many of the young shoots and leaves are prone to being consumed by livestock in grasslands under long-term grazing, potentially leading to reduced fertility in *Caragana* shrubs. There were studies on other plant species in arid steppes showing that grazing strongly reduced spike biomass, biomass allocation for sexual reproduction [64,65], and pollination efficiency [66]. The second potential mechanism is that livestock trampling could greatly increase soil compaction [67,68], resulting in reduced water retention and nutrient availability. The third potential mechanism could be that reproductive organs, flowers, fruits, seeds, and seedlings might be directly consumed by ungulates [67–69]. However, there were also some other studies showing that grazing did not affect fruit numbers [70] or even that grazing could promote seed production [25] and sapling establishment [36–38], indicating that grazing had no significant effect on population growth [33].

In our study, the seed vigor in the severe grazing plots was significantly higher than that in the mild grazing and fenced plots. This was probably because the increased grazing intensity significantly decreased the pod abundance of *C. microphylla* (Figure 1). *Caragana* has generally large and compact cushion-like canopies. Such shrub morphology could provide a shield for pods from herbivores, and thus only the pods hidden in shrub canopies could survive in the severe grazing plots, resulting in only a few mature *Caragana* pods. Specifically, the seed abundance in the severe grazing plots was 18.7–49.1% of that in the fenced plots. Thus, greater resources should be allocated to the survival seeds under severe grazing conditions, resulting in more vigorous seeds.

We found that *Caragana* population growth drastically decreased under long-term grazing, which did not support our hypothesis. Thus, sexual reproduction cannot fully explain the higher abundance of shrubs under increased grazing intensity. Whereas we noticed that the effect of grazing on *Caragana* shrubs would also vary with the duration of grazing. Under short-term grazing, livestock would mainly feed on palatable herbaceous plants rather than thorny *Caragana* shrubs, thereby reducing the competition intensity of herbaceous plants with shrubs, thus benefiting the growth and reproduction of *Caragana* shrubs [2,3], and facilitating shrub encroachment in grasslands. In contrast, under long-term grazing (as in the situation in this study), herbaceous plant populations would decline substantially, and livestock would mainly feed on less palatable *Caragana* shrubs, which would negatively affect the growth and reproduction of shrubs.

4.3. Interactive Effects between Climatic Aridity and Grazing on Sexual Reproduction

Our results showed that as climatic aridity increased, the negative effects of grazing on seed production of *Caragana* shrubs gradually increased. Herbaceous plants predominate at more mesic sites. Under such conditions, herbivores naturally prefer to feed on grasses rather than *Caragana* shrubs. Thus, *Caragana* shrubs were not significantly affected by grazing. However, increasing aridity would increase the cover of woody plants, potentially leading to an increase in the ratio of woody to herbaceous cover. Under such conditions, herbivores switched from a preference for grasses to a diet based mainly on *Caragana* plants [45]. Consequently, grazing has the strongest negative effects on the seed production of *Caragana* shrubs in the dry arid zone. These results suggested that grassland ecosystems would become more sensitive to grazing disturbance as aridity stress increased from the subhumid zone to the dry arid zone.

We also found that grazing intensified the negative effect of climatic aridity on the population growth rate of *Caragana*. This was probably due to the fact that young shoots, leaves, flowers, pods, seeds, and saplings of shrubs are likely consumed more by livestock under drier conditions.

Overall, our study demonstrated that the joint effect of climatic aridity and grazing stress greatly suppressed the sexual population growth of *Caragana* shrubs, suggesting that these two stress factors did not promote shrub encroachment through sexual reproduction. Thus, shrub encroachment may not be realized through sexual reproduction. Other forms of reproduction, such as clonal reproduction, may deserve further investigation for a better understanding of the mechanisms underlying shrub encroachment in grasslands.

5. Conclusions

Climatic aridity decreased seed quantity and seed vigor of *Caragana* shrubs, increased the sapling establishment rate, and reduced the population growth rate at the grazing sites, but did not affect the population growth rate at the ungrazed sites. Long-term grazing reduced seed production, sapling establishment, and population growth rates. Climatic aridity stress enhanced the negative effects of grazing on the sexual reproduction of *Caragana* shrubs while grazing intensified the negative effect of climatic aridity on the population growth rate of *Caragana* shrubs. Climatic aridity, grazing, and their combined effects had a negative influence on the sexual reproduction of *Caragana* shrubs. Sexual reproduction could not fully explain the increased abundance of *Caragana* shrubs with increasing aridity and grazing intensity. Clonal reproduction may be of considerable importance for understanding the mechanism of shrub encroachment in grasslands.

Author Contributions: Conceptualization, L.X. and C.M.; methodology, L.X.; formal analysis, Y.L. and M.L.; investigation, L.X., Y.L., Y.W. and L.W.; data curation, L.X.; writing—original draft preparation, L.X., H.G. and C.M.; writing—review and editing, H.G. and C.M.; visualization, L.X. and C.M. All authors have read and agreed to the published version of the manuscript.

Funding: This study was supported by the Natural Science Foundation of Tianjin (22JCQNJC00170) and the National Natural Science Foundation of China (31901140).

Data Availability Statement: The data presented in this study are available on request from the corresponding author.

Conflicts of Interest: The authors declare no conflict of interest.

References

1. Eldridge, D.J.; Bowker, M.A.; Maestre, F.T.; Roger, E.; Reynolds, J.F.; Whitford, W.G. Impacts of shrub encroachment on ecosystem structure and functioning: Towards a global synthesis. *Ecol. Lett.* **2011**, *14*, 709–722. [CrossRef] [PubMed]
2. Xiong, X.G. Increased distribution of *Caragana microphylla* in rangelands and its causes and consequences in Xilin River Basin. *Acta. Pratacult. Sin.* **2003**, *12*, 57–62.
3. van Auken, O.W. Causes and consequences of woody plant encroachment into western North American grasslands. *J. Environ. Manag.* **2009**, *90*, 2931–2942. [CrossRef] [PubMed]
4. Xie, L.N.; Li, Y.C.; Guo, H.Y.; Wang, C.W.; Chen, Q.; He, P.; Ma, C.C. Sandy habitats play an important role in shrub encroachment in grasslands. *Agronomy* **2022**, *12*, 2858. [CrossRef]
5. Ma, C.C.; Zhang, J.H.; Guo, H.Y.; Li, Q.F.; Xie, L.N.; Gao, Y.B. Alterations in canopy size and reproduction of *Caragana stenophylla* along a climate gradient on the Inner Mongolian Plateau. *Flora-Morphol. Distrib. Funct. Ecol. Plants* **2013**, *208*, 97–103. [CrossRef]
6. Xie, L.N.; Ma, C.C.; Guo, H.Y.; Li, Q.F.; Gao, Y.B. Distribution pattern of *Caragana* species under the influence of climate gradient in the Inner Mongolia region, China. *J. Arid Land* **2014**, *6*, 311–323. [CrossRef]
7. Ma, C.C.; Gao, Y.B.; Guo, H.Y.; Wang, J.L.; Wu, J.B.; Xu, J.S. Physiological adaptations of four dominant *Caragana* species in the desert region of the Inner Mongolia Plateau. *J. Arid Environ.* **2008**, *72*, 247–254. [CrossRef]
8. Ma, C.C.; Guo, H.Y.; Wu, J.B.; Wang, J.L.; Qi, S.X.; Wei, Y.R.; Gao, Y.B. Acclimation of photosynthetic traits of *Caragana* species to desert environment in Inner Mongolian Plateau. *Arid Land Res. Manag.* **2014**, *28*, 87–101. [CrossRef]
9. Guo, H.Y.; Guan, L.J.; Wang, Y.H.; Xie, L.N.; Prather, C.M.; Liu, C.G.; Ma, C.C. Grazing limits natural biological controls of woody encroachment in Inner Mongolia Steppe. *Biol. Open* **2017**, *6*, 1569–1574. [CrossRef]
10. May, F.; Giladi, I.; Ristow, M.; Ziv, Y.; Jeltsch, F. Plant functional traits and community assembly along interacting gradients of productivity and fragmentation. *Perspect Plant Ecol.* **2013**, *15*, 304–318. [CrossRef]

11. Volis, S.; Mendlinger, S.; Ward, D. Differentiation in populations of *Hordeum spontaneum* along a gradient of environmental productivity and predictability: Life history and local adaptation. *Biol. J. Linn. Soc.* **2002**, *77*, 479–490. [CrossRef]
12. Volis, S.; Mendlinger, S.; Ward, D. Demography and role of the seed bank in Mediterranean and desert populations of wild barley. *Basic Appl. Ecol* **2004**, *5*, 53–64. [CrossRef]
13. Volis, S. Correlated patterns of variation in phenology and seed production in populations of two annual grasses along an aridity gradient. *Evol. Ecol.* **2007**, *21*, 381–393. [CrossRef]
14. Lebrija-Trejos, E.; Lobato, M.C.C.; Sternberg, M. Reproductive traits and seed dynamics at two environmentally contrasting annual plant communities: From fieldwork to theoretical expectations. *Isr. J. Ecol. Evol.* **2011**, *57*, 73–90. [CrossRef]
15. Del Pozo, A.; Ovalle, C.; Aronson, J.; Avendano, J. Ecotypic differentiation in *Medicago polymorpha* L. along an environmental gradient in central Chile. I. Phenology, biomass production and reproductive patterns. *Plant Ecol.* **2002**, *159*, 119–130. [CrossRef]
16. Traveset, A.; Gulías, J.; Riera, N.; Mus, M. Transition probabilities from pollination to establishment in a rare dioecious shrub species (*Rhamnus ludovici-salvatoris*) in two habitats. *J. Ecol.* **2003**, *91*, 427–437. [CrossRef]
17. Castro, J.; Zamora, R.; Hódar, J.A.; Gómez, J.M. Seedling establishment of a boreal tree species (*Pinus sylvestris*) at its southernmost distribution limit: Consequences of being in a marginal Mediterranean habitat. *J. Ecol.* **2004**, *92*, 266–277. Available online: https://www.jstor.org/stable/3599591 (accessed on 7 January 2023). [CrossRef]
18. Castro, J.; Zamora, R.; Hódar, J.A.; Gómez, J.M. Alleviation of summer drought boosts establishment success of *Pinus sylvestris* in a Mediterranean mountain: An experimental approach. *Plant Ecol.* **2005**, *181*, 191–202. [CrossRef]
19. Rysavy, A.; Seifan, M.; Sternberg, M.; Tielbörger, K. Shrub seedling survival under climate change–comparing natural and experimental rainfall gradients. *J. Arid Environ.* **2014**, *111*, 14–21. [CrossRef]
20. Cipriotti, P.A.; Flombaum, P.; Sala, O.E.; Aguiar, M.R. Does drought control emergence and survival of grass seedlings in semi-arid rangelands?: An example with a Patagonian species. *J. Arid Environ.* **2008**, *72*, 162–174. [CrossRef]
21. Hunt, L.P. Low seed availability may limit recruitment in grazed *Atriplex vesicaria* and contribute to its local extinction. *Plant Ecol.* **2001**, *157*, 53–67. Available online: https://www.jstor.org/stable/20051160 (accessed on 7 January 2023). [CrossRef]
22. Hickman, K.R.; Hartnett, D.C. Effects of grazing intensity on growth, reproduction, and abundance of three palatable forbs in Kansas tallgrass prairie. *Plant Ecol.* **2002**, *159*, 23–33. [CrossRef]
23. Wan, H.; Bai, Y.; Schönbach, P.; Gierus, M.; Taube, F. Effects of grazing management system on plant community structure and functioning in a semiarid steppe: Scaling from species to community. *Plant Soil* **2011**, *340*, 215–226. [CrossRef]
24. Kratochwil, A.; Fock, S.; Remy, D.; Schwabe, A. Responses of flower phenology and seed production under cattle grazing impact in sandy grasslands. *Phytocoenologia* **2002**, *32*, 531–552. [CrossRef]
25. Liu, D.; An, S.Z.; Kong, Q.G.; Zhang, X.H. Population characters of *Ligularia narynensis* in Kalajun rangeland. *Pratacult. Sci.* **2010**, *27*, 25–29.
26. Oliva, G.; Collantes, M.; Humano, G. Reproductive effort and seed establishment in grazed tussock grass populations of Patagonia. *Rangeland Ecol. Manag.* **2013**, *66*, 164–173. [CrossRef]
27. Bates, J.D.; Davies, K.W. Cattle grazing and vegetation succession on burned sagebrush steppe. *Rangel. Ecol. Manag.* **2014**, *67*, 412–422. [CrossRef]
28. Farrington, S.J.; Muzika, R.M.; Drees, D.; Knight, T.M. Interactive effects of harvest and deer herbivory on the population dynamics of American ginseng. *Conserv. Biol.* **2009**, *23*, 719–728. [CrossRef]
29. Mayer, R.; Erschbamer, B. Seedling recruitment and seed-/microsite limitation in traditionally grazed plant communities of the alpine zone. *Basic. Appl. Ecol.* **2011**, *12*, 10–20. [CrossRef]
30. Mandle, L.; Ticktin, T. Interactions among fire, grazing, harvest and abiotic conditions shape palm demographic responses to disturbance. *J. Ecol.* **2012**, *100*, 997–1008. [CrossRef]
31. Mandle, L.; Ticktin, T.; Zuidema, P.A. Resilience of palm populations to disturbance is determined by interactive effects of fire, herbivory and harvest. *J. Ecol.* **2015**, *103*, 1032–1043. [CrossRef]
32. Buckley, Y.M.; Bolker, B.M.; Rees, M. Disturbance, invasion and re-invasion: Managing the weed-shaped hole in disturbed ecosystems. *Ecol. Lett.* **2007**, *10*, 809–817. [CrossRef]
33. Li, S.L.; Yu, F.H.; Werger, M.J.; Dong, M.; Ramula, S.; Zuidema, P.A. Understanding the effects of a new grazing policy: The impact of seasonal grazing on shrub demography in the Inner Mongolian steppe. *J. Appl. Ecol.* **2013**, *50*, 1377–1386. [CrossRef]
34. Macias, D.; Mazía, N.; Jacobo, E. Grazing and neighborhood interactions limit woody encroachment in wet subtropical savannas. *Basic Appl. Ecol.* **2014**, *15*, 661–668. [CrossRef]
35. Osem, Y.; Fogel, T.; Moshe, Y.; Brant, S. Managing cattle grazing and overstorey cover for the conversion of pine monocultures into mixed Mediterranean woodlands. *Appl. Veg. Sci.* **2015**, *8*, 261–271. [CrossRef]
36. Wu, G.L.; Du, G.Z.; Liu, Z.H.; Thirgood, S. Effect of fencing and grazing on a *Kobresia*-dominated meadow in the Qinghai-Tibetan Plateau. *Plant Soil* **2009**, *319*, 115–126. [CrossRef]
37. Khishigjargal, M.; Dulamsuren, C.; Lkhagvadorj, D.; Leuschner, C.; Hauck, M. Contrasting responses of seedling and sapling densities to livestock density in the Mongolian forest-steppe. *Plant Ecol.* **2013**, *214*, 1391–1403. [CrossRef]
38. Zimmermann, H.; Renison, D.; Leyer, I.; Hensen, I. Do we need livestock grazing to promote *Polylepis australis* tree recruitment in the Central Argentinean Mountains? *Ecol. Res.* **2009**, *24*, 1075–1081. [CrossRef]

39. Forbis, T.A. Seedling demography in an alpine ecosystem. *Am. J. Bot.* **2003**, *90*, 1197–1206. [CrossRef]
40. Wu, G.L.; Li, W.; Li, X.P.; Shi, Z.H. Grazing as a mediator for maintenance of offspring diversity: Sexual and clonal recruitment in alpine grassland communities. *Flora-Morphol. Distrib. Funct. Ecol. Plants* **2011**, *206*, 241–245. [CrossRef]
41. Bisigato, A.J.; Bertiller, M.B. Temporal and micro-spatial patterning of seedling establishment. Consequences for patch dynamics in the southern Monte, Argentina. *Plant Ecol.* **2004**, *174*, 235–246. [CrossRef]
42. Evju, M.; Halvorsen, R.; Rydgren, K.; Austrheim, G.; Mysterud, A. Effects of sheep grazing and temporal variability on population dynamics of the clonal herb *Geranium sylvaticum* in an alpine habitat. *Plant Ecol.* **2011**, *212*, 1299–1312. [CrossRef]
43. Faust, C.; Süss, K.; Storm, C.; Schwabe, A. Threatened inland sand vegetation in the temperate zone under different types of abiotic and biotic disturbances during a ten-year period. *Flora-Morphol. Distrib. Funct. Ecol. Plants* **2011**, *206*, 611–621. [CrossRef]
44. Xie, L.N.; Chen, W.Z.; Gabler, C.A.; Han, L.; Guo, H.Y.; Chen, Q.; Ma, C.C.; Gu, S. Effects of grazing intensity on seed production of *Caragana stenophylla* along a climatic aridity gradient in the Inner Mongolia Steppe, China. *J. Arid Land* **2016**, *8*, 890–898. [CrossRef]
45. Xie, L.N.; Guo, H.Y.; Chen, W.Z.; Liu, Z.; Gu, S.; Ma, C.C. Effects of grazing on population growth characteristics of *Caragana stenophylla* along a climatic aridity gradient. *Rangeland. Ecol. Manag.* **2018**, *71*, 98–105. [CrossRef]
46. Harel, D.; Holzapfel, C.; Sternberg, M. Seed mass and dormancy of annual plant populations and communities decreases with aridity and rainfall predictability. *Basic Appl. Ecol.* **2011**, *12*, 674–684. [CrossRef]
47. Wang, R.Z.; Gao, Q. Climate-driven changes in shoot density and shoot biomass in *Leymus chinensis* (Poaceae) on the North-east China Transect (NECT). *Global Ecol. Biogeogr.* **2003**, *12*, 249–259. [CrossRef]
48. Wang, R.Z.; Gao, Q. Morphological responses of *Leymus chinensis* (Poaceae) to the large-scale climatic gradient along the North-east China Transect (NECT). *Divers. Distrib.* **2004**, *10*, 65–73. [CrossRef]
49. Lotan, A.; Izhaki, I. Could abiotic environment shape fleshy fruit traits? A field study of the desert shrub *Ochradenus baccatus*. *J. Arid. Environ.* **2013**, *92*, 34–41. [CrossRef]
50. Zhou, H.Y.; Li, X.R.; Fan, H.W.; Wang, X.P.; Tan, H.J.; Li, A.X. Physiological characteristics of several *Caragana shrub* species under extreme conditions. *J. Desert Res.* **2005**, *25*, 182–190.
51. Fang, X.; Turner, N.C.; Yan, G.; Li, F.; Siddique, K.H. Flower numbers, pod production, pollen viability, and pistil function are reduced and flower and pod abortion increased in chickpea (*Cicer arietinum* L.) under terminal drought. *J. Exp. Bot.* **2010**, *61*, 335–345. [CrossRef]
52. Salazar-Tortosa, D.; Castro, J.; Saladin, B.; Zimmermann, N.E.; Rubio De Casas, R. Arid environments select for larger seeds in pines (*Pinus* spp.). *Evol. Ecol.* **2020**, *34*, 11–26. [CrossRef]
53. Sheffer, E.; Batterman, S.A.; Levin, S.A.; Hedin, L.O. Biome-scale nitrogen fixation strategies selected by climatic constraints on nitrogen cycle. *Nat. Plants* **2015**, *1*, 15182. [CrossRef]
54. Adams, M.A.; Turnbull, T.L.; Sprent, J.I.; Buchmann, N. Legumes are different: Leaf nitrogen, photosynthesis, and water use efficiency. *Proc. Natl. Acad. Sci. USA* **2016**, *113*, 4098–4103. [CrossRef]
55. Ouyang, S.; Tian, Y.; Liu, Q.; Zhang, L.; Sun, Y.; Xu, X.; Liu, Y. Symbiotic nitrogen fixation and interspecific transfer by *Caragana microphylla* in a temperate grassland with 15 N dilution technique. *Appl. Soil Ecol.* **2016**, *108*, 221–227. [CrossRef]
56. Piqueras, J. Herbivory and ramet performance in the clonal herb *Trientalis europaea* L. *J. Ecol.* **1999**, *87*, 450–460. [CrossRef]
57. Peco, B.; Borghi, C.E.; Malo, J.E.; Acebes, P.; Almirón, M.; Campos, C.M. Effects of bark damage by feral herbivores on columnar cactus *Echinopsis* (= *Trichocereus*) *terscheckii* reproductive output. *J. Arid. Environ.* **2011**, *75*, 981–985. [CrossRef]
58. Becklin, K.M.; Kirkpatrick, H.E. Compensation through rosette formation: The response of scarlet gilia (*Ipomopsis aggregata*: Polemoniaceae) to mammalian herbivory. *Botany* **2006**, *84*, 1298–1303. [CrossRef]
59. Hayes, C.L.; Talbot, W.A.; Wolf, B.O. Woodrat herbivory influences saguaro (*Carnegiea gigantea*) reproductive output. *J. Arid Environ.* **2013**, *89*, 110–115. [CrossRef]
60. Boulant, N.; Kunstler, G.; Rambal, S.; Lepart, J. Seed supply, drought, and grazing determine spatio-temporal patterns of recruitment for native and introduced invasive pines in grasslands. *Divers. Distrib.* **2008**, *14*, 862–874. [CrossRef]
61. López-Sánchez, A.; Schroeder, J.; Roig, S.; Sobral, M.; Dirzo, R. Effects of cattle management on oak regeneration in northern Californian Mediterranean oak woodlands. *PLoS ONE* **2014**, *9*, e105472. [CrossRef] [PubMed]
62. Bradley, B.A. Regional analysis of the impacts of climate change on cheatgrass invasion shows potential risk and opportunity. *Glob. Chang. Biol.* **2009**, *15*, 196–208. [CrossRef]
63. Concilio, A.L.; Loik, M.E.; Belnap, J. Global change effects on *Bromus tectorum* L.(Poaceae) at its high-elevation range margin. *Global Chang. Biol.* **2013**, *19*, 161–172. [CrossRef] [PubMed]
64. Zhao, P.P.; Shao, W.S.; Jin, C.Q.; Song, L.; Gong, S.; Li, G. Effects of enclosure on biomass allocation characteristics of *Agropyron mongolicum* population in desert steppe. *Ecol. Environ. Sci.* **2017**, *26*, 2024–2029. [CrossRef]
65. Wei, Z.J.; Yan, R.R.; Yun, X.J.; Chu, W.B.; Yang, J. Study on biomass and energy allocation of major plant species in desert steppe under different grazing systems. *J. Desert Res.* **2011**, *31*, 1124–1130. [CrossRef]
66. Devoto, M.; Medan, D. Effects of grazing disturbance on the reproduction of a perennial herb, *Cypella herbertii* (Lindl.) Herb.(Iridaceae). *Plant Syst. Evol.* **2004**, *243*, 165–173. [CrossRef]
67. Gómez, J.M. Long-term effects of ungulates on performance, abundance, and spatial distribution of two montane herbs. *Ecol. Monogr.* **2005**, *75*, 231–258. [CrossRef]

68. Heckel, C.D.; Bourg, N.A.; McShea, W.J.; Kalisz, S. Nonconsumptive effects of a generalist ungulate herbivore drive decline of unpalatable forest herbs. *Ecology* **2010**, *91*, 319–326. [CrossRef]
69. Warner, P.J.; Cushman, H.J. Influence of herbivores on a perennial plant: Variation with life history stage and herbivore species. *Oecologia* **2002**, *132*, 77–85. [CrossRef]
70. Perevolotsky, A.; Schwartz-Tzachor, R.; Yonathan, R.; Ne'eman, G. Geophytes–herbivore interactions: Reproduction and population dynamics of *Anemone coronaria* L. *Plant Ecol.* **2011**, *212*, 563–571. [CrossRef]

agronomy

MDPI

Article

The Potential for Endozoochorous Dispersal of *Vachellia nilotica* Seeds by Goats: Implications for Bush Encroachment

Fortune L. Manganyi [1,2,*], Julius Tjelele [2], Khanyisile R. Mbatha [1,3], Ngoako Letsoalo [2] and Francuois Müller [2]

1 Department of Agriculture and Animal Health, School of Agriculture and Life Sciences, Florida Campus, University of South Africa, Florida 1709, South Africa; mbathkr@unisa.ac.za

2 Agricultural Research Council—Animal Production, Irene 0062, South Africa; jtjelele@arc.agric.za (J.T.); letsoalonl@arc.agric.za (N.L.); mullerf@arc.agric.za (F.M.)

3 School of Interdisciplinary Research and Graduate Studies, College of Graduate Studies, Muckleneuk Campus, University of South Africa, Pretoria 0003, South Africa

* Correspondence: manganyif@arc.agric.za

Abstract: Seed dispersal has become an important component in understanding plant–animal interaction. Recently, there have been concerns about the role of ruminants, particularly browsers, in the dispersal of woody plant seeds. This study aimed to determine whether including *Vachellia nilotica* seeds in livestock, diets could reduce the spread of viable seeds in the rangelands and control bush encroachment. The shoots and seed pods of *Vachellia nilotica* were harvested and analyzed for fiber, protein, and mineral nutrients at different seed pods inclusion rates, with or without a feed additive. Six diets were selected for feeding 24 goats and quantifying seed recovery and germination after ingestion. Results indicated that including seed pods and feed additives to *Vachellia nilotica* shoots significantly improved the quality of the fodder. Chipping the seed pods prior to including them in the diet resulted in 13% intact seed recovery, and approximately 2% of these seeds were recovered after ingestion. These recovered seeds were mostly still viable but were still dormant as seed coats were not sufficiently damaged after ingestion. Therefore, viable seeds may still be dispersed in the rangeland, leading to further bush encroachment.

Keywords: chipping; dispersal; encroachment; feed additives; seed germination; seed pods

Citation: Manganyi, F.L.; Tjelele, J.; Mbatha, K.R.; Letsoalo, N.; Müller, F. The Potential for Endozoochorous Dispersal of *Vachellia nilotica* Seeds by Goats: Implications for Bush Encroachment. *Agronomy* **2023**, *13*, 1599. https://doi.org/10.3390/agronomy13061599

Academic Editors: Kesi Liu and Xinqing Shao

Received: 16 May 2023
Revised: 7 June 2023
Accepted: 9 June 2023
Published: 13 June 2023

1. Introduction

Many researchers consider bush encroachment as a significant rangeland management problem that negatively affects livestock production and the livelihoods of farmers [1–3]. This is because bush encroachment suppresses the growth of herbaceous vegetation and therefore reduces productivity and the quality of rangeland forage resources [3,4]. Furthermore, increasing tree densities in rangelands reduces forage accessibility to livestock, further negatively affecting the utilization of rangelands by livestock [1,3,5]. Although there has been significant financial investment into the eradication of the encroaching woody plant species, the interventions have not yielded conclusive management strategies due to the persistent and adaptive nature of the encroaching species [3]. In addition, extensively reared livestock often make use of the encroaching tree species, making them useful resources to farmers, which makes it difficult to motivate the eradication of the encroaching woody plant species [6].

Seed dispersal by herbivores has become an important issue in plant ecology, where both wild game animals and domestic livestock play an important role in endozoochorous seed dispersal [6]. Many reasons have been given for the increased rate and extent of encroaching species [3]. One among many drivers of this is the influence of endozoochory, i.e., seeds that are ingested and subsequently dispersed by livestock [7–9]. This is because the pods of many woody plants form an important part of the diet of livestock and wildlife during the dry season due to their high nutritive value [8] compared to grasses and herbaceous vegetation in this season. However, some of the seeds that remain intact after ingestion passes through the gut, where they are scarified, which in turn allows for greater efficiency in the dispersal and effective seedling establishment of the encroaching plant species [8,9]. Endozoochorous seed dispersal is influenced by several factors, such as seed size, hardness, number of seeds consumed, and animal species, to survive chewing and rumination [6,8–10]. Other studies show that small seeds (<2.5 mm in width) are most likely to escape chewing and rumination compared to large seeds, while hard-coated seeds have higher chances of passing through the gut without physical harm [9,11,12].

To limit the further spreading of encroaching tree species in rangelands and, at the same time, improve rangeland conditions and reduce dry season feed gaps, the active harvesting of the encroaching tree species as a feed source has been proposed [2,13–16]. There are positive effects that thinning and/or eradication of the encroaching trees have on rangeland conditions [3,14,15,17]. However, limited information exists to show how these trees could potentially be used as alternative feed sources by livestock, other than their use as leaf meal [2,6,17,18]. It has been observed that encroaching tree species are highly nutritious and have a high potential of being used as supplementary feed [13,19,20]. On the contrary, some of the woody plant species have very low dry matter digestibility and high levels of secondary compounds and, thus, have low potential to use as a feed source without additional processing [21]. The digestibility of these fodder trees can, however, be improved through different processing techniques. Furthermore, most studies focus on using the trees as a cut-and-carry resource without the addition of feed additives and mostly neglect the impacts of seed ingestion on the potential for contributing to further bush encroachment [17,19,22,23]. This, in turn, leaves a gap in our understanding of how these species could be used effectively without causing further encroachment through endozoochory.

This study, therefore, evaluated the quality of *Vachellia nilotica* fodders with and without seeds and the inclusion of feed additives. Thereafter, the best quality diet treatments were fed to livestock from where feed intake, seed intake, seed recovery, and the germination potential of the recovered seeds were determined. The study aimed to answer the questions (1) Will the inclusion of seed pods as well as feed additives improve the nutritional quality of the *Vachellia nilotica* fodders? and (2) whether the inclusion of the seed pods and feed additive could lead to dormancy breaking while passing through the gut and therefore improve or not seed germination.

2. Materials and Methods

2.1. Seed Collection, Preparation, and Initial Viability Screening

Vachellia nilotica seed pods were hand-picked at the Agricultural Research Council (ARC)—Roodeplaat Experimental Farm (28°19′ E, 25°35′ S) in Pretoria, South Africa. Seed pods were separated from the shoots and stored in a cool, dry area pending feed formulation. Five replicates of whole seed pods were weighed to a mass of 250 g, and the number of seed pods within each replicate was counted [9,24]. Thereafter, the seeds were removed from the seed pods, and the number of seeds was counted. Secondly, an additional ten replicates of 250 g seed pods were chipped using a Tandem 6.5 hp chipper, and the number of whole seeds (i.e., undamaged seeds) recovered was quantified after chipping. Thereafter, a representative number of chipped and un-chipped seeds were used to determine the initial viability of the seeds. Thirdly, the un-chipped seeds were scarified by clipping the seed coat with a clipper to expose the embryo. The seeds were immersed in a 1% Tetrazolium chloride solution (3,5-triphynyl chloride) for 18 h in a dark germination chamber and stored at room temperature. Thereafter, each seed was cut longitudinally through the endosperm to expose the embryo and evaluated for staining through a light microscope [25,26]. Seeds that stained red were regarded as viable, while unstained seeds were regarded as dead.

2.2. Feed Creation and Nutritional Quality Determination

Edible *Vachellia nilotica* shoots consisting of edible branches and leaves (30 cm long and approximately 1.0 cm diameter) were harvested using the tree pruner at the end of the wet season (March–April 2021) at the fruiting stage (i.e., plants with seed pods). During this time, all plant material still contained green leaves were harvested, but the seed pods were harvested towards the end of this period at maturity. The shoots and pods were collected from 70 different trees; seed pods were collected when matured and dry and were kept separate from shoots and merged differently to form a composite sample. The samples of shoots were chipped using a woodchipper (Tandem 6.5 hp chipper/shredder) and mixed thoroughly to obtain a uniform mixture. A uniform sample of 250 g chipped shoots was used as the base, and chipped seed pods were included in a 4:1, 4:2, 4:3, and 4:4 ratio. For each feed treatment created, the seed pods were chipped separately and included into the chipped shoots and mixed. Four replicates of each treatment were developed, as well as two control treatments which consisted of only 250 g chipped shoots and another consisting of only 250 g chipped seed pods. Additionally, a feed additive (Voermol LS33) at a recommended rate of 800 mL/10 kg for the small stock was added in all six treatments, resulting in a total of 12 feed treatments (i.e., six with the feed additives and six without the feed additives). A sub-sample of 150 g of each feed was collected, oven-dried at 60 °C until a constant mass was achieved, and milled to pass through a 3 mm mesh and stored for chemical analyses.

From the dried and milled feed samples, a 0.5 g sub-sample was digested using a technique described by Zasoski et al. [27]. After digestion, an aliquot of the digested solution was used for the determination of calcium (Ca), magnesium (Mg), phosphorus (P), potassium (K), sodium (Na), manganese (Mn), iron (Fe), zinc (Zn), and copper (Cu) using an ICP-OES (Inductively Coupled Plasma Optical Emission Spectrometer—Agilent 725 (700 Series), Agilent Technologies, Santa Clara, CA, USA). The ICP-OES can determine the quantity of each element in each sample simultaneously. Prior to analyses, the instrument was calibrated against a series of standard solutions containing all the elements of interest in alignment with the operating procedures of the manufacturer. Furthermore, 8–12 g of the plant samples were used to determine the total nitrogen (N) concentrations using the dry oxidation method [28,29] in a Flash 2000 CHNS-O Analyzer (Thero Scientific, Waltham, MA, USA). For each analysis, the instrument was calibrated against a known standard (Phenylalanine) which contained 8.48% N. Total N was converted to crude protein (CP) by multiplying %N with 6.25 [30]. Neutral detergent fiber (NDF) and acid detergent fiber (ADF) were determined using a Dosi fiber analyzer system (Labex (Pty) Ltd., Edenvale, South Africa) according to the methods of van Soest et al. [31]. The NDF and ADF values obtained were used to calculate the digestible dry matter (DDM) [32], metabolizable energy (ME) [33], total digestible nutrients (TDN) [34], digestible forage energy (DFE) [33], digestible organic matter (DOM) [33], net energy for lactation (NE_L) [32], net energy for maintenance (NE_M) [32], and net energy for gain/growth (NE_G) [32] using Equations (1)–(8).

$$\text{DDM (\%)} = 88.9 - (\text{ADF} \times 0.779) \quad (1)$$

$$\text{ME (Mcal/kg DM)} = (1.01 \times \text{DFE}) - 0.45 \quad (2)$$

$$\text{TDN (\%)} = 87.84 - (0.7 \times \text{ADF}) \quad (3)$$

$$\text{DFE (Mcal/kg DM)} = 0.04409 \times \text{TDN} \quad (4)$$

$$\text{DOM (\%)} = \text{TDN} \div 1.05 \quad (5)$$

$$NE_L \text{ (Mcal/kg DM)} = 1.044 - (0.0119 \times \% \text{ ADF}) \quad (6)$$

$$NE_M \text{ (Mcal/kg DM)} = ((1.37 \times \text{ME}) - (0.3042 \times \text{ME}) + (0.051 \times \text{ME})) - 0.508 \quad (7)$$

$$NE_G \text{ (Mcal/kg DM)} = ((1.42 \times \text{ME}) - (0.3836 \times \text{ME}) + (0.0593 \times \text{ME})) - 0.7484 \quad (8)$$

2.3. Feeding Trial and Seed Recovery

Based on the nutritional analyses, six experimental diets, each representing an experimental treatment, were used for the feeding trial (Table 1). The feeding and seed recovery trial was conducted at the ARC, Irene Experimental farm, Gauteng province, South Africa. A total of twenty-four female indigenous veld goats (South African veld goats) of approximately two years old with an average body weight of 29.6 ± 1.33 kg were used in the study. The twenty-four goats were divided into four groups of six per group. Each animal in a group was regarded as a replicate. The animals were acclimatized to the experimental conditions for 14 days prior to the start of data collection, during which time they were fed chipped *Vachellia nilotica* shoots and grass hay. The goats were housed in individual metabolic pens (2 m × 1 m) with slatted floors, each fitted with feed and water troughs in a well-ventilated covered area. After acclimatization, each animal was fed the experimental diet at 3% of their body weight and hay grass as a basal diet. All experimental animals were allowed to consume their assigned diets within 24 h, after which the remaining materials were collected [8]. Left-over feed was weighed, and feed intake was determined by calculating the difference between the feed offered to the animals and the remaining feed. The number of seeds in the feeds was also quantified. The fecal collection commenced immediately after the experimental feeding period started and continued until no seeds were recovered in the feces for three consecutive days. The number of seeds recovered per day per animal was recorded in order to calculate the seed recovery percentage. The collected feces were immersed in cold water until soft and then washed with tap water through a wire mesh until the water was clear. A cabinet with a light source below a glass surface was used to separate seeds from the fecal remains. Seeds recovered from each animal per day were counted and stored in brown paper bags prior to the germination trial. Using these counts, the number of days when the first seeds were recovered from the fecal matter after ingestion was recorded. At the end of the trial, the number of days to 10%, 50%, and 90% of the total number of seeds recovered was calculated. From these calculations, the time taken from 10% to 90% seed recovery was determined and used as an indication of seed recovery uniformity".

Table 1. Experimental diets used for feeding goats.

Experimental Diets	
1	100% *Vachellia nilotica* seed pods (un-chipped)
2	100% *Vachellia nilotica* seed pods (chipped)
3	100% *Vachellia nilotica* seed pods (un-chipped) + Feed Additive
4	100% *Vachellia nilotica* seed pods (chipped) + Feed Additive
5	4:4 (*Vachellia nilotica* Shoots: *Vachellia nilotica* chipped seed pods)
6	4:4 (*Vachellia nilotica* Shoots: *Vachellia nilotica* chipped seed pods) + Feed Additive

2.4. Germination of Recovered Seeds

Germination tests were performed at the ARC National Forage Genebank Seed Laboratory according to the International Seed Testing Association standards. The recovered seeds per animal per day were counted and stored in a brown paper bag resulting in six replicates per experimental feed treatment pending germination trial. The three control treatments, i.e., unscarified seeds, seeds that passed through the chipper, and mechanically scarified seeds by scarifying the seed coat with a clipper, were created. All seeds were germinated in 12 cm petri dishes on a single disk of Whatman No. 1 filter paper. The petri dishes were maintained in seed germination chambers at a temperature of 25 °C [8] for the duration of the trial. The seeds were watered with 5 mL of dH_2O, and watering was done as needed throughout the duration of the germination period. Seed germination was recorded daily for 28 days, and germinated seeds were removed from the petri dishes on a daily basis to minimize excessive water uptake by germinated seeds. The total germination percentage was calculated at the end of the germination period following the technique by Armke and Scott [35]. All seeds that did not germinate at the end of 28 days were counted and subjected to a tetrazolium chloride viability test [8]. Seeds were scarified and soaked in 1% tetrazolium solution (3,5-triphynyl chloride) for 18 h in an incubator at 25 °C. Thereafter, each seed was cut longitudinally through the endosperm to expose the embryo, and staining was recorded by viewing the seeds under a stereo microscope [25,26].

2.5. Statistical Analysis

SPSS Version 22 (SPSS Inc., Chicago, IL, USA, 2013) was used for all statistical analyses. Nutritional quality, seed recovery, and seed viability data were subjected to a one-way analysis of variance (ANOVA) using a Fishers' LSD post hoc test to separate means and identify statistically significant differences ($p < 0.05$) between the different treatments.

3. Results

3.1. Seed Pods and Seed Characteristics

The collected seed pods contained, on average, 10 ± 0.2 seeds per pod and had an average weight of 2.7 ± 0.1 g. After chipping ten replicates of 250 g intact seed pods, with each replicate containing approximately 1108 ± 24 seeds, approximately 141 ± 22 intact seeds ($13\% \pm 0.3$) were recovered. The viability of the intact seeds recovered after chipping was 97%, and the seeds directly tested after removal from the seeds pods had a viability of 96%.

3.2. Nutritional Quality of *Vachellia nilotica* Fodders

Significantly ($p < 0.05$) higher concentrations of N, K, Mg, P, Zn (Table 2), CP, and TDN (Table 3) were found in the 100% *Vachellia nilotica* seed pods compared to the 100% *Vachellia nilotica* shoots. Furthermore, the *Vachellia nilotica* fodders created from 100% *Vachellia nilotica* seed pods also had significantly ($p < 0.05$) higher digestibility, i.e., lower ADF and NDF, higher DDM, DOM, and higher energy content (ME, NE_L, and NE_G) compared to the fodders containing 100% *Vachellia nilotica* shoots (Table 3). When the seed pods were added to the shoots, N, P, K (Table 2), and CP (Table 3) concentrations significantly increased ($p < 0.05$) from the 100% pure shoot fodder concentrations. The DDM, TDN, DOM, ME, NE_L, and NE_G (Table 3) content only significantly increased ($p < 0.05$) in the 100% shoot fodders at the highest seed pod inclusion levels, i.e., 4:4 shoots/pods. No significant differences ($p \geq 0.05$) were found in any of the other mineral nutrients between the 100% shoot fodders and the shoots + seed pod fodders, irrespective of the seed pod inclusion levels (Table 2). Neutral detergent fiber content significantly decreased ($p < 0.05$) when seed pods were added to the shoots, irrespective of the seed pod inclusion levels, while ADF content (Table 3) only decreased at the highest seed pod inclusion levels (Table 3).

Table 2. Mean (±SEM) Mineral nutrient content in experimental diets created from *Vachellia nilotica* shoots and seed pods with or without the addition of a feed additive (Voermol LS33). Different letters for each variable measured indicate statistically significant differences ($p \geq 0.05$) between different experimental diets within a column. P = probability, F = ratio of statistics.

Experimental Diet	N%	K g/kg	Ca g/kg	Mg g/kg	P g/kg	Na g/kg	Fe g/kg	Mn g/kg	Zn g/kg	Cu g/kg
100% Seed pods	2.2 ± 0.1^{b}	16.1 ± 0.3^{f}	5.8 ± 0.3^{a}	16.7 ± 0.2^{cd}	1.7 ± 0.1^{f}	0.1 ± 0.003^{a}	0.06 ± 0.003^{a}	0.02 ± 0.001^{a}	0.03 ± 0.001^{b}	0.01 ± 0.0004^{a}
100% Shoots	1.1 ± 0.1^{a}	6.7 ± 0.1^{a}	6.4 ± 0.4^{a}	11.5 ± 1.6^{ab}	0.7 ± 0.1^{a}	0.2 ± 0.003^{a}	0.16 ± 0.02^{bc}	0.03 ± 0.005^{bc}	0.03 ± 0.001^{a}	0.01 ± 0.0004^{b}
4:1 (Shoots/Seed pods)	1.5 ± 0.01^{a}	8.6 ± 0.2^{b}	6.6 ± 0.2^{a}	14.3 ± 1.3^{bc}	0.8 ± 0.03^{b}	0.1 ± 0.004^{a}	0.18 ± 0.02^{bcd}	0.03 ± 0.004^{c}	0.02 ± 0.001^{a}	0.01 ± 0.0011^{b}
4:2 (Shoots/Seed pods)	1.4 ± 0.02^{a}	10.0 ± 0.2^{c}	6.2 ± 0.03^{a}	12.6 ± 0.1^{ab}	0.9 ± 0.01^{bc}	0.2 ± 0.01^{a}	0.19 ± 0.02^{cd}	0.03 ± 0.001^{bc}	0.03 ± 0.001^{ab}	0.01 ± 0.0003^{b}
4:3 (Shoots/Seed pods)	1.5 ± 0.03^{b}	10.7 ± 0.03^{cd}	5.8 ± 0.1^{a}	11.8 ± 0.3^{a}	1.0 ± 0.02^{de}	0.2 ± 0.01^{a}	0.16 ± 0.03^{bc}	0.03 ± 0.002^{ab}	0.03 ± 0.0004^{a}	0.01 ± 0.0005^{b}
4:4 (Shoots/Seed pods)	1.5 ± 0.04^{b}	10.7 ± 0.03^{cd}	5.1 ± 0.2^{a}	12.2 ± 0.5^{ab}	1.1 ± 0.01^{e}	0.2 ± 0.01^{a}	0.15 ± 0.004^{bc}	0.03 ± 0.001^{ab}	0.02 ± 0.001^{a}	0.01 ± 0.0005^{b}
100% Seed pods + LS33	2.5 ± 0.04^{c}	17.4 ± 0.3^{g}	5.7 ± 0.3^{a}	20.3 ± 0.2^{f}	1.6 ± 0.1^{f}	1.5 ± 0.2^{b}	0.08 ± 0.002^{a}	0.04 ± 0.001^{c}	0.04 ± 0.002^{c}	0.01 ± 0.0005^{b}
100% Shoots + LS33	2.3 ± 0.01^{b}	11.1 ± 0.1^{d}	8.3 ± 0.1^{a}	22.9 ± 1.0^{g}	0.9 ± 0.03^{cde}	2.1 ± 0.1^{c}	0.23 ± 0.01^{cd}	0.07 ± 0.004^{f}	0.05 ± 0.002^{c}	0.02 ± 0.0013^{d}
4:1 (Shoots/Seed pods) + LS33	2.1 ± 0.18^{b}	13.0 ± 0.6^{e}	6.5 ± 0.1^{a}	19.7 ± 1.2^{ef}	0.9 ± 0.02^{bc}	2.6 ± 0.4^{d}	0.19 ± 0.01^{cd}	0.06 ± 0.003^{e}	0.06 ± 0.004^{d}	0.02 ± 0.0003^{cd}
4:2 (Shoots/Seed pods) + LS33	1.9 ± 0.03^{b}	13.2 ± 0.3^{e}	5.7 ± 0.2^{a}	13.2 ± 0.3^{def}	1.0 ± 0.1^{e}	1.7 ± 0.1^{bc}	0.16 ± 0.01^{bc}	0.05 ± 0.001^{d}	0.05 ± 0.001^{c}	0.01 ± 0.0006^{c}
4:3 (Shoots/Seed pods) + LS33	2.0 ± 0.03^{b}	13.9 ± 0.4^{e}	5.7 ± 0.1^{a}	17.3 ± 0.3^{de}	1.1 ± 0.02^{e}	1.9 ± 0.1^{bc}	0.15 ± 0.01^{bc}	0.05 ± 0.001^{d}	0.05 ± 0.003^{c}	0.01 ± 0.0004^{c}
4:4 (Shoots/Seed pods) + LS33	2.0 ± 0.11^{b}	13.8 ± 0.2^{e}	5.5 ± 0.2^{a}	17.7 ± 0.8^{de}	1.0 ± 0.1^{e}	1.8 ± 0.4^{bc}	0.14 ± 0.01^{b}	0.05 ± 0.003^{d}	0.05 ± 0.004^{c}	0.01 ± 0.0005^{c}
Significance	$F_{(11,36)} = 9.5$ $p \leq 0.001$	$F_{(11,36)} = 118.2$ $p \leq 0.001$	$F_{(11,36)} = 1.0$ $p = 0.474$	$F_{(11,36)} = 21.1$ $p \leq 0.001$	$F_{(11,36)} = 54.0$ $p \leq 0.001$	$F_{(11,36)} = 38.1$ $p \leq 0.001$	$F_{(11,36)} = 9.5$ $p \leq 0.001$	$F_{(11,36)} = 32.9$ $p \leq 0.001$	$F_{(11,36)} = 32.3$ $p \leq 0.001$	$F_{(11,36)} = 16.5$ $p \leq 0.001$

N = Nitrogen, K = Potassium, Ca = Calcium, Mg = Magnesium, P = Phosphorus, Na = Sodium, Fe = Iron, Mn = Manganese, Zn = Zinc, Cu = Copper.

Table 3. Mean (±SEM) Crude protein, fiber, digestibility, and energy content in experimental diets created from *Vachellia nilotica* shoots and seed pods with or without the addition of a feed additive (Voermol LS33). Statistically significant differences ($p \geq 0.05$) between different experimental diets are indicated by different letters for each variable measured. P = probability, F = ratio of statistics.

Experimental Diet	CP%	ADF%	NDF%	DDM%	TDN%	DOM%	DFE Mcal/kg	ME Mcal/kg	NE_L Mcal/kg	NE_M Mcal/kg	NE_G Mcal/kg
100% Seed pods	13 ± 0.4 ^{de}	28.3 ± 0.9 ^{a}	29.3 ± 1.1 ^{ab}	66.1 ± 2.1 ^{de}	67.4 ± 0.6 ^{cd}	64.2 ± 0.6 ^{cd}	3.0 ± 0.03 ^{a}	2.6 ± 0.03 ^{b}	0.7 ± 0.01 ^{d}	2.3 ± 0.03 ^{a}	2.1 ± 0.03 ^{d}
100% Shoots	7 ± 0.8 ^{a}	39.6 ± 2.9 ^{b}	48.6 ± 3.1 ^{e}	58.0 ± 2.3 ^{a}	60.1 ± 2.0 ^{a}	57.2 ± 1.9 ^{a}	2.7 ± 0.09 ^{a}	2.2 ± 0.09 ^{a}	0.6 ± 0.03 ^{a}	2.0 ± 0.10 ^{a}	1.7 ± 0.10 ^{a}
4:1 (Shoots/Seed pods)	9 ± 0.1 ^{b}	35.9 ± 0.8 ^{b}	39.7 ± 2.7 ^{d}	60.9 ± 0.6 ^{ab}	62.7 ± 0.6 ^{ab}	59.7 ± 0.5 ^{ab}	2.8 ± 0.02 ^{a}	2.3 ± 0.03 ^{a}	0.6 ± 0.01 ^{ab}	2.1 ± 0.03 ^{a}	1.8 ± 0.03 ^{ab}
4:2 (Shoots/Seed pods)	9 ± 0.1 ^{b}	36.0 ± 1.1 ^{b}	39.5 ± 0.4 ^{d}	60.9 ± 0.9 ^{ab}	62.7 ± 0.8 ^{ab}	59.7 ± 0.7 ^{ab}	2.8 ± 0.03 ^{a}	2.3 ± 0.03 ^{a}	0.6 ± 0.01 ^{ab}	2.1 ± 0.04 ^{a}	1.8 ± 0.04 ^{ab}
4:3 (Shoots/Seed pods)	9 ± 0.2 ^{b}	31.9 ± 0.4 ^{b}	35.8 ± 1.6 ^{abc}	61.0 ± 0.3 ^{ab}	62.7 ± 0.3 ^{ab}	59.8 ± 0.3 ^{ab}	2.8 ± 0.01 ^{a}	2.3 ± 0.01 ^{a}	0.6 ± 0.01 ^{ab}	2.1 ± 0.01 ^{a}	1.8 ± 0.01 ^{ab}
4:4 (Shoots/Seed pods)	10 ± 0.2 ^{b}	29.3 ± 1.9 ^{a}	30.0 ± 1.1 ^{ab}	66.1 ± 1.5 ^{de}	67.3 ± 1.3 ^{cd}	64.1 ± 1.3 ^{cd}	3.0 ± 0.06 ^{a}	2.6 ± 0.06 ^{b}	0.7 ± 0.02 ^{d}	2.3 ± 0.07 ^{a}	2.0 ± 0.06 ^{d}
100% Seed pods + LS33	16 ± 0.3 ^{f}	25.6 ± 1.2 ^{a}	30.0 ± 1.6 ^{ab}	65.8 ± 0.9 ^{cde}	67.1 ± 0.8 ^{cd}	63.9 ± 0.8 ^{cd}	3.0 ± 0.04 ^{a}	2.5 ± 0.04 ^{b}	0.7 ± 0.01 ^{cd}	2.3 ± 0.04 ^{a}	2.0 ± 0.04 ^{cd}
100% Shoots + LS33	14 ± 0.1 ^{e}	30.1 ± 0.7 ^{a}	32.9 ± 0.9 ^{bc}	65.5 ± 0.5 ^{cd}	66.8 ± 0.5 ^{cd}	63.6 ± 0.5 ^{cd}	2.9 ± 0.02 ^{a}	2.5 ± 0.02 ^{ab}	0.7 ± 0.01 ^{cd}	2.3 ± 0.02 ^{a}	2.0 ± 0.02 ^{cd}
4:1 (Shoots/Seed pods) + LS33	12 ± 1.1 ^{c}	34.5 ± 1.4 ^{b}	48.1 ± 1.4 ^{bcd}	62.0 ± 1.1 ^{bc}	63.7 ± 1.0 ^{bc}	60.7 ± 1.0 ^{bc}	2.8 ± 0.04 ^{a}	2.4 ± 0.05 ^{a}	0.6 ± 0.02 ^{bc}	2.2 ± 0.05 ^{a}	1.9 ± 0.05 ^{bc}
4:2 (Shoots/Seed pods) + LS33	12 ± 0.2 ^{c}	29.7 ± 2.7 ^{a}	30.5 ± 1.0 ^{ab}	65.8 ± 2.1 ^{cde}	67.0 ± 1.9 ^{cd}	63.9 ± 1.8 ^{cd}	3.0 ± 0.08 ^{a}	2.5 ± 0.08 ^{b}	0.7 ± 0.03 ^{cd}	2.3 ± 0.09 ^{a}	2.0 ± 0.09 ^{cd}
4:3 (Shoots/Seed pods) + LS33	12 ± 0.2 ^{c}	27.2 ± 1.2 ^{a}	31.1 ± 1.0 ^{ab}	67.7 ± 1.0 ^{de}	68.8 ± 0.9 ^{d}	65.5 ± 0.8 ^{d}	3.0 ± 0.04 ^{a}	2.6 ± 0.04 ^{b}	0.7 ± 0.01 ^{d}	2.4 ± 0.04 ^{a}	2.1 ± 0.04 ^{d}
4:4 (Shoots/Seed pods) + LS33	12 ± 0.7 ^{c}	24.9 ± 1.1 ^{a}	37.9 ± 4.2 ^{cd}	69.5 ± 0.9 ^{e}	70.4 ± 0.8 ^{d}	67.1 ± 0.8 ^{d}	3.1 ± 0.03 ^{a}	2.7 ± 0.04 ^{b}	0.8 ± 0.01 ^{d}	2.5 ± 0.04 ^{a}	2.2 ± 0.04 ^{d}
Significance	$F_{(11,36)}$ = 29.0 $p \leq 0.001$	$F_{(11,36)}$ = 8.0 $p \leq 0.001$	$F_{(11,36)}$ = 10.1 $p \leq 0.001$	$F_{(11,36)}$ = 8.1 $p \leq 0.001$	$F_{(11,36)}$ = 7.9 $p \leq 0.001$	$F_{(11,36)}$ = 7.4 $p \leq 0.001$	$F_{(11,36)}$ = 1.0 $p = 0.474$	$F_{(11,36)}$ = 3.6 $p = 0.004$	$F_{(11,36)}$ = 8.3 $p \leq 0.001$	$F_{(11,36)}$ = 1.0 $p = 0.474$	$F_{(11,36)}$ = 8.0 $p \leq 0.001$

CP = Crude protein, ADF = Acid detergent fiber, NDF = Neutral detergent fiber, DDM = Digestible dry matter, TDN = Total digestible nutrients, DOM = Digestible organic matter, DFE = Digestible forage energy, ME = Metabolizable energy, NE_L = Net energy for lactation, NE_M = Net energy for maintenance, NE_G = net energy for gain/growth.

The addition of the feed additive (Voermol LS33- molasses-based protein, vitamin, and mineral supplement) to the 100% *Vachellia nilotica* seed pods and 100% *Vachellia nilotica* shoot fodders significantly increased ($p < 0.05$) the concentrations of N, K, Mg, Na, Mn, and Zn. Phosphorus concentrations were only significantly increased ($p < 0.05$) when the additive was added to the shoots (Table 2). Similarly, the addition of the feed additives to the 100% seed pod and 100% shoot fodders significantly ($p < 0.05$) increased the CP content (Table 3). Feed additive added to 100% seed pods significantly decreased ($p \geq 0.05$) ADF and NDF content; however, no significant differences ($p \geq 0.05$) were observed on DDM, TDN, DOM, NE_L, and NE_G, whereas there was a significant increase ($p < 0.05$) when feed additives were added to the 100% *Vachellia nilotica* shoot fodders (Table 3). The addition of feed additive to the fodders created from the pure *Vachellia nilotica* shoot + the different seed pod inclusion levels, irrespective of the inclusion levels, resulted in significantly higher ($p < 0.05$) concentrations of K, Mg, Na, Mn, Zn, and Cu (Table 2) and CP (Table 3), while N only increased at the 4:1 and 4:2 seed pod inclusion levels (Table 2) and DDM, TDN, DOM, NE_L, and NE_G only increased at the 4:2 and 4:3 seed pod inclusion levels (Table 3).

3.3. Selection of Diets, Feeding Trial, and Seed Recovery after Ingestion

Vachellia nilotica feeds were selected for the feeding trial as experimental diets (Table 3). Generally, the total amount of feed consumed and the remaining after the experimental period did not differ statistically among the six experimental diets ($p > 0.05$; Table 4). However, the number of seeds ingested was significantly higher ($p < 0.05$) when the seed pods were chipped, irrespective of the addition of the feed additives (Table 4). The number of seeds recovered was significantly lower ($p < 0.05$) in the diets containing chipped seed pods compared to those with the control, i.e., whole seed pods (Table 4). Less than 2% of the chipped seed pods were recovered after ingestion, while 3% and 6% of seeds in the diets containing whole seed pods with or without Voermol LS33 were recovered (Table 4). No differences ($p \geq 0.05$) were found between the experimental diets in regard to when the first seeds were recovered from the feces (Table 4), while significant differences ($p < 0.05$) were observed in the number of seeds recovered between experimental diets. However, it was evident that the seed retention rate generally remained longer ($p < 0.05$) in the digestive tract of goats when they were fed diets containing whole seed pods, irrespective of the addition of the feed additive (Table 4). Furthermore, uniformity in seed recovery, calculated as the time taken between 10% and 90% of seeds recovered, indicated that the diets containing chipped seed pods with the addition of the feed additives resulted in a significantly ($p < 0.05$) shorter retention period in the gut, i.e., two days, while the recovery of the seed from the fecal matter was spread over 5–6 days long (Table 4).

3.4. Germination Potential of Recovered Seeds

The unscarified seeds that were not fed to the goats were mostly dormant (88%), with only 3.2% of the seeds being able to germinate (Table 5). Mechanically scarified seeds had a germination percentage of 79% (Table 5). Although chipping of the seed pods significantly increased ($p < 0.05$) the germination potential compared to control (unscarified seeds), more than 65% of the recovered seeds remained dormant (Table 5). Approximately 80% of the recovered seeds from whole seed pods diets were dormant, which was similar to the control (unscarified seeds) treatment (Table 5).

Table 4. Mean (±SEM) Feed, seed ingested, and seed recovery from the goats. Statistically significant differences ($p \geq 0.05$) between different experimental diets are indicated by different letters for each variable measured. P = probability, F = ratio of statistics.

Treatments	Feed Ingested	Feed Remains	% Seed Ingested	% Seed Remains	% Recovery	1st Recovery	50% Recovery	90% Recovery	Uniformity
Whole Seed pods	64.7 ± 9.8^{a}	35.3 ± 9.8^{a}	58.8 ± 9.7^{a}	41.2 ± 9.7^{b}	3.5 ± 1.3^{b}	1.7 ± 0.2^{a}	3.8 ± 0.3^{c}	5.7 ± 0.3^{c}	3.0 ± 0.4^{cd}
Whole Seed pods + LS33	65.7 ± 9.7^{a}	34.3 ± 9.7^{a}	57.4 ± 9.9^{a}	42.6 ± 9.9^{b}	6.1 ± 2.2^{c}	1.0 ± 0.0^{a}	3.5 ± 0.2^{bc}	6.2 ± 0.6^{c}	4.0 ± 0.6^{d}
Chipped seed pods	68.9 ± 10.9^{a}	31.1 ± 10.9^{a}	90.1 ± 6.7^{b}	9.9 ± 6.7^{a}	0.6 ± 0.2^{a}	1.5 ± 0.2^{a}	2.3 ± 0.5^{a}	3.7 ± 1.0^{a}	2.2 ± 0.9^{bc}
Chipped seed pods + LS33	83.0 ± 8.6^{a}	17.0 ± 8.6^{a}	85.2 ± 6.7^{b}	14.8 ± 6.7^{a}	0.8 ± 0.3^{a}	1.5 ± 0.2^{a}	2.8 ± 0.4^{ab}	3.8 ± 0.3^{b}	1.8 ± 0.2^{a}
Chipped (shoots, seed pods)	57.9 ± 6.4^{a}	42.1 ± 6.4^{a}	90.8 ± 6.8^{b}	9.2 ± 6.8^{a}	0.8 ± 0.2^{a}	1.7 ± 0.3^{a}	3.0 ± 0.4^{bc}	4.3 ± 0.5^{b}	2.3 ± 0.6^{bc}
Chipped (shoots, seed pods) + LS33	78.1 ± 7.6^{a}	21.9 ± 7.6^{a}	71.4 ± 8.3^{b}	28.6 ± 8.3^{a}	1.9 ± 1.2^{a}	1.8 ± 0.3^{a}	3.2 ± 0.5^{bc}	4.0 ± 0.5^{b}	1.7 ± 0.4^{a}
Significance	$F_{(5,36)} = 1.10$ $p = 0.382$	$F_{(5,36)} = 1.08$ $p = 0.391$	$F_{(5,36)} = 3.58$ $p = 0.012$	$F_{(5,36)} = 3.58$ $p = 0.012$	$F_{(5,36)} = 3.55$ $p = 0.012$	$F_{(5,36)} = 0.61$ $p = 0.693$	$F_{(5,36)} = 3.60$ $p = 0.014$	$F_{(5,36)} = 10.31$ $p \leq 0.001$	$F_{(5,36)} = 6.13$ $p \leq 0.001$

Table 5. Mean (±SEM) Germination trial and uniformity for seed recovered from the feces. Statistically significant differences ($p \geq 0.05$) between different experimental diets are indicated by different letters for each variable measured. P = probability, F = ratio of statistics.

Treatments	Germination (%)	Dormant Seed (%)	Dead Seed (%)
Seed pods	14.1 ± 2.8 [c]	83.0 ± 3.6 [d]	3.0 ± 1.3 [a]
Seed pods + LS33	8.7 ± 1.1 [b]	87.6 ± 1.7 [d]	3.7 ± 0.7 [a]
Chipped seed pods	16.8 ± 1.1 [c]	79.5 ± 3.0 [c]	3.7 ± 2.6 [a]
Chipped seed pods + LS33	17.1 ± 2.1 [c]	77.1 ± 2.4 [c]	5.7 ± 2.7 [a]
Chipped (shoots+ seed pods)	14.4 ± 2.2 [c]	77.8 ± 3.0 [c]	7.8 ± 1.0 [a]
Chipped (shoots+ seed pods) + LS33	13.1 ± 2.1 [c]	79.4 ± 3.2 [c]	7.4 ± 2.3 [a]
Unscarified	3.2 ± 1.0 [a]	88.4 ± 3.5 [d]	8.4 ± 2.9 [a]
Scarified	78.8 ± 2.2 [e]	0.0 ± 0.0 [a]	21.2 ± 2.2 [b]
Chipped	29.6 ± 2.9 [d]	65.2 ± 3.4 [b]	5.2 ± 1.0 [a]
Significance	$F_{(8,45)} = 121.3$ $p \leq 0.001$	$F_{(8,45)} = 89.6$ $p \leq 0.001$	$F_{(8,45)} = 7.8$ $p \leq 0.001$

4. Discussion

4.1. Nutritional Quality of *Vachellia nilotica* Fodders

Browse plants such as *Vachellia nilotica* are major sources of livestock feed during the dry season, partially due to their ability to retain their nutritional value during the dry season [36], contrary to grasses. This, along with their rate and extent of encroachment, have the potential to be a good alternative source of feed for livestock. At the end of the active growing period, the nutritional value of browse plants may not be sufficient to sustain livestock. An example of this is reported by Britz et al. [37], who indicated that maturing of browse plants resulted in a decline in the nutritional quality in terms of their mineral nutrients, digestibility, protein, and energy content. Thus, the best time to harvest the material for fodder is during the vegetative or early reproductive stages [20,37,38]. However, other studies have shown that some of the browse seed pods during the end of the wet season have higher nutritional value and could be used to improve the quality of the fodders created by these encroaching tree species [2,16,20,38,39]. These findings are in accordance with the findings of the current study, where results indicated that seed pods generally contained relatively higher quality mineral nutrients, crude protein, and lower amounts of fiber and therefore increased digestibility and energy content compared to the edible shoots. Both seed pods and shoots in the current study contained sufficient concentrations of K, Ca, Mg, Fe, Mn, Zn, Cu, and crude protein to meet the minimum requirements of 5–15 g/kg, 1.8–10 g/kg, 1 g/kg, 0.03–0.1 g/kg, 0.02–0.04 g/kg, 0.02–0.05 g/kg, 0.005–0.1 g/kg, and 7–8%, respectively, to maintain livestock condition [30,33]. However, only the 100% seed pod diet contained sufficient P concentrations to meet the maintenance requirements of (1.6–6 g/kg) ruminants. In addition, neither the seed pods nor the shoots contained sufficient Na concentrations to meet the minimum requirements of 0.4–1.8 g/kg to maintain livestock conditions [30]. The fiber content in the shoots of *Vachellia nilotica* had 28% (ADF) and 29% (NDF), and the seed pods had 39% (ADF) and 48% (NDF), which both fall within the adequate range of 19–40% ADF and 25–40% NDF for normal rumen functions [31,39–41]. Small ruminants such as goats and sheep require high concentrations of degradable fiber in their daily diets for rumen function [40]. However, a high level of fiber is often associated with decreased forage intake [40–42]. Furthermore, forages with a digestible dry matter (DDM) content of greater than 60% are regarded as high-quality forages as intake will not be impacted [31,40]. According to the study results, the *Vachellia nilotica* shoots alone contained 58% DDM while the seed pods had a DDM content of 66%, indicating the importance of the inclusion of the seed pods in livestock diets. The energy content of the *Vachellia nilotica* shoots and seed pods individually was sufficient to sustain the energy requirements for small ruminants (goats/sheep) during the dry season [30]. In addition, both shoots and seed pods had sufficient metabolizable energy (ME) content to meet the energy requirements of lambs up to 20 kg (3.9–10.5 MJ kg^{-1} DM) as well as those of 40–60 kg dry ewes (7.6–10.2 MJ kg^{-1} DM). However, neither shoots nor pods were found to have sufficient ME content (14.5–17.7 MJ kg^{-1} DM) to sustain pregnant and lactating (15.5–19.4 MJ kg^{-1} DM) ewes.

The nutritional value of shoots with the addition of seed pods was found to be higher compared to shoots alone, and it contained sufficient levels of mineral nutrients, CP, digestibility, and energy content to maintain livestock conditions during the dry season [30,33,42] Furthermore, the addition of seed pods to the shoots improved the mineral nutrient content of the forages and was found to meet the minimum requirements to maintain small stock conditions. This was true for all mineral nutrients, except for P and Na, which was below the minimum requirements levels of small stock [30,33]. Moreover, the addition of the feed additives to the pure seed pods and pure shoots further increased the nutritional quality of the *Vachellia nilotica* shoots, resulting in a CP content that was suitable for maintaining highly productive livestock herds, which have a minimum requirement of 13–14% CP.

4.2. Feed Intake, Seed Recovery, and Seed Germination Potential of Recovered Seeds

Hard-coated seeds tend to be protected against damage during ingestion and rumination, which in turn results in the recovery of more undamaged seeds in the feces [8,43,44]. Therefore, diets containing the chipped seed pods were found to be consumed in higher amounts than when seed pods were offered as a whole, irrespective of the addition of the feed additive in this study. Furthermore, the seed recovery was relatively high for whole seed pod diets and low for chipped seed pod diets. This might be because chipping the seed pods before feeding results in easy ingestion, making it easier to digest and thus reducing the recovery of intact seeds. The first seed recovery for all treatments ranged between 1–2 days; however, the time to 90% recovery was longer for diets that contained whole seed pods compared to chipped seed pods.

Seed recovery and survival after passage through the gut depends on factors such as the hardness of the seed coat, the size of the seeds, the associated diet fed with the seed pods as well as the number of seeds ingested [8,44,45]. Furthermore, seed recovery and germination after ingestion may be influenced by factors such as chewing and rumination [9,22]. Results from the current study showed that there was a low percentage of intact seeds that passed through the digestive tract of goats, especially from chipped seed pods diets. However, those seeds remained viable and had substantial germination potential. Although the relative viability of seeds that passed through the rumen was lower than those that were mechanically scarified and chipped seeds, it was significantly higher than untreated seeds. The relatively low loss in viability of ingested seeds is a good trade-off for the likelihood of these seeds being dispersed by animals away from the parent tree. Therefore, the study partially supported the hypothesis that feed additives will improve the digestibility of seed pods consumed by goats, thus reducing seed dispersals and viability.

5. Conclusions

Edible *Vachellia nilotica* shoots in this study were found to contain insufficient crude protein content for maintaining livestock conditions during the dry season. However, adding seed pods to the shoots significantly increased the nutritional quality. Chipping of the seed pods before inclusion in livestock diets resulted in significantly lower numbers of seeds recovered, with more than 85% of seeds being broken and damaged to the point where they did not germinate. Feeding the remaining seeds led to a further reduction in seeds recovered, with only 2% of the whole seeds fed to the livestock being recovered. Therefore, processing the seed pods prior to adding these to the shoot material in diets already significantly reduces the number of seeds that could potentially be dispersed throughout the rangeland. However, it is important to remember that the majority of the 2% of seeds recovered were dormant but still viable. These seeds could potentially still lead to further bush encroachment. Therefore, further research is required to determine whether these seeds passing through the gut of the livestock will survive the dry season to germinate and establish in the next wet season.

Author Contributions: Conceptualization, F.M. and J.T.; methodology, F.M., K.R.M., N.L. and F.L.M.; validation, F.M. and F.L.M.; formal analysis, F.M. and F.L.M.; investigation, F.M., J.T., K.R.M., N.L. and F.L.M.; validation, F.M.; data curation, F.M. and F.L.M.; writing—original draft preparation, F.L.M.; writing—review and editing, F.L.M., J.T., K.R.M., N.L. and F.M.; validation, visualization, F.L.M. and F.M.; supervision, J.T., K.R.M., N.L. and F.M.; project administration, F.M. and J.T.; funding acquisition, F.M. and J.T. All authors have read and agreed to the published version of the manuscript.

Funding: This research was funded by the Red Meat Research and Development Fund of South Africa (RMRD-SA) P02000229 and the National Research Foundation (NRF) of South Africa (149845 and 118595).

Institutional Review Board Statement: Not applicable.

Informed Consent Statement: Not applicable.

Data Availability Statement: All data presented in this manuscript are available from the corresponding author upon reasonable request.

Acknowledgments: We are thankful to the following people for assisting with data collection: Selokela Nathaniel, Tshabalala Josephine, Molefe Refimah, Molifi Francinah, Maluleka Thomas, Makhafola Martha, Malepe Kamogelo and Nxedlana Nontuthuko.

Conflicts of Interest: The authors declare no conflict of interest. The funders had no role in the design of the study; in the collection, analyses, or interpretation of data; in the writing of the manuscript; or in the decision to publish the results.

References

1. Smit, I.P.; Prins, H.H. Predicting the effects of woody encroachment on mammal 14 communities, grazing biomass and fire frequency in African savannas. *PLoS ONE* **2015**, *10*, e0137857.
2. Mlambo, V.; Smith, T.; Owen, E.; Mould, F.; Sikosana, J.; Mueller-Harvey, I. Tanniniferous *Dichrostachys cinerea* fruits do not require detoxification for goat nutrition: In sacco and in vivo evaluations. *Livest. Prod. Sci.* **2004**, *90*, 135–144. [CrossRef]
3. O'Connor, T.G.; Puttick, J.R.; Hoffman, M.T. Bush encroachment in southern Africa: Changes and causes. *Afr. J. Range Forage Sci.* **2014**, *31*, 67–88. [CrossRef]
4. Stevens, N.; Erasmus, B.F.N.; Archibald, S.; Bond, W.J. Woody encroachment over 70 years 24 in South African savannahs: Overgrazing, global change or extinction aftershock? *Philos. Trans. R. Soc. B Biol. Sci.* **2016**, *371*, 20150437. [CrossRef] [PubMed]
5. Lesoli, M.S.; Gxasheka, M.; Solomon, T.B.; Moyo, B. Integrated plant invasion and bush encroachment management on Southern African Rangelands. *Herbic. Curr. Res. Case Stud. Use* **2013**, *11*, 259–313. [CrossRef]
6. Abbas, S.; Nichol, J.E.; Zhang, J. The accumulation of species and recovery of species composition along a 70 years succession in a tropical secondary forest. *Ecol. Ind.* **2019**, *106*, 105524. [CrossRef]
7. Or, K.; Ward, D. Maternal effects on the life histories of bruchid beetles infesting *Acacia raddiana* in the Negev desert, Israel. *Entomol. Exp. Appl.* **2007**, *122*, 165–170. [CrossRef]
8. Tjelele, T.J.; Ward, D.; Dziba, L.E. Do Herbivores Facilitate Seed Germination and Seedling Recruitment of Woody Plants? Ph.D. Thesis, University of KwaZulu-Natal, Pietermaritzburg, South Africa, 2014.
9. Tjelele, T.; Ward, D.; Dziba, L.E. The effects of passage through the gut of goats and cattle, and the application of dung as a fertilizer on seedling establishment of *Dichrostachys cinerea* and Acacia nilotica. *Rangel. J.* **2015**, *37*, 147–156. [CrossRef]
10. Jaganathan, G.K.; Yule, K.; Liu, B. On the evolutionary and ecological value of breaking physical dormancy by endozoochory. *Perspect. Plant Ecol. Evol. Syst.* **2016**, *22*, 11–22. [CrossRef]
11. Frost, R.A.; Mosley, J.C.; Roeder, B.L. Recovery and viability of sulfur cinquefoil seeds from the feces of sheep and goats. *Rangel. Ecol. Manag.* **2013**, *66*, 51–55. [CrossRef]
12. Marchetto, K.M.; Heuschele, D.J.; Larkin, D.J.; Wolf, T.M. Goat digestion leads to low survival and viability of common buckthorn (*Rhamnus cathartica*) seeds. *Nat. Areas J.* **2020**, *40*, 150–154. [CrossRef]
13. Gebeyew, K.; Beriso, K.; Mohamed, A.G.; Silassie, G.; Melaku, S. Review on the Nutritive Value of Some Selected *Acacia* Species for Livestock Production in Dryland Areas. *Adv. Dairy Res.* **2015**, *3*, 139. [CrossRef]
14. Mnisi, C.M.; Mlambo, V. Influence of harvesting site on chemical composition and potential protein value of *Acacia erioloba*, *A. nilotica* and *Ziziphus mucronata* leaves for ruminants. *J. Anim. Physiol. Anim. Nutr.* **2016**, *101*, 994–1003. [CrossRef]
15. Mnisi, C.M.; Mlambo, V. Application of near infrared reflectance spectroscopy to the nutritional assessment of tree leaves as potential protein supplements for ruminants. *Trop. Agric.* **2017**, *94*, 9–19.

16. Mokoboki, H.K.; Sebola, A.N.; Ravhuhali, K.E.; Nhlane, L. Chemical composition, in vitro ruminal dry matter degradability and dry matter intake of some selected browse plants. *Cogent Food Agric.* **2019**, *5*, 1587811. [CrossRef]
17. Mapiye, C.; Chimonyo, M.; Dzama, K.; Strydom, P.E.; Marufu, M.C.; Muchenje, V. Nutritional status, growth performance and carcass characteristics of Nguni steers supplemented with *Acacia karroo* leaf-meal. *Livest. Sci.* **2019**, *126*, 206–214. [CrossRef]
18. Tefera, S.; Mlambo, V.; Dlamini, B.J.; Dlamini, A.M.; Koralagama, K.D.N.; Mould, F.L. Chemical composition and in vitro ruminal fermentation of common tree forages in the semi-arid rangelands of Swaziland. *Anim. Feed Sci. Technol.* **2008**, *142*, 99–110. [CrossRef]
19. Tanner, J.C.; Reed, J.D.; Owen, E. The nutritive value of fruits (pods with seeds) from four Acacia spp. compared with extracted Noug (*Guizotia abyssinica*) meal as supplements to maize stover for Ethiopian highland sheep. *Anim. Prod.* **1990**, *51*, 127–133. [CrossRef]
20. Ravhuhali, K.E.; Mlambo, V.; Beyene, T.S.; Palamuleni, L.G. Effects of soil type on density of trees and nutritive value of tree leaves in selected communal areas of South Africa. *S. Afr. J. Anim. Sci.* **2020**, *50*, 88–98. [CrossRef]
21. Mlambo, V.; Sikosana, J.L.N.; Smith, T.; Owen, E.; Mould, F.L.; Muesller-Harvey, I. An evaluation of NaOH and wood ash for the inactivation of tannins in *Acacia nilotica* and *Dichrostachys cinerea* fruits using an in vitro rumen fermentation technique. *Trop. Agric.* **2011**, *88*, 44–54.
22. Mlambo, V.; Smith, T.; Owen, E.; Mould, F.L.; Sikosana, J.L.N.; Mueller-Harvey, I. Wood ash inactivates tannins in *Acacia nilotica* and *Dichrostachys cinerea* fruits: An in vitro evaluation. *Anim. Feed Sci. Technol.* 2005, *in press*.
23. Mapiye, C.; Chimonyo, M.; Dzama, K.; Muchenje, V.; Strydom, P.E. Meat quality attributes of Nguni steers supplemented with *Acacia karroo* leaf-meal. *Meat Sci.* **2010**, *8*, 621–627. [CrossRef] [PubMed]
24. Lacey, J.R.; Wallander, R.; Olson-Rutz, K. Recovery, germinability, and viability of leafy spurge (*Euphorbia esula*) seed chewing rodents. *Afr. J. Ecol.* **1992**, *33*, 194–210.
25. International Seed Testing Association. *International Rules for Seed Testing*; ISTA: Bassersdorf, Switzerland, 2012.
26. International Seed Testing Association. International rules for seed testing. *Seed Sci. Technol.* **1985**, *13*, 299–355.
27. Zasoski, R.J.; Burau, R.G. A Rapid Nitric-Perchloric Acid Digestion Method for Multi-Element Tissue Analysis. *Commun. Soil Sci. Plant Anal.* **1977**, *8*, 425–436. [CrossRef]
28. Jimenez, R.R.; Ladha, J.K. Automated elemental analysis: A rapid and reliable but expensive measurement of total carbon and nitrogen in plant and soil samples. *Commun. Soil Sci. Plant Anal.* **1993**, *24*, 1897–1924. [CrossRef]
29. Matejovic, I. Total nitrogen in plant material determined by means of dry combustion: A possible alternative to determination by Kjeldahl digestion. *Commun. Soil Sci. Plant Anal.* **1995**, *26*, 2217–2229. [CrossRef]
30. Meissner, H.H. Nutrient supplementation of the grazing animal. In *Pasture Management in South Africa*; Tainton, N.M., Ed.; University of Natal Press: Pietermaritzburg, South Africa, 2000; pp. 96–115.
31. Van Soest, P.J.; Robertson, J.B.; Lewis, B.A. Methods of dietary fiber, neutral detergent fiber and non-starch poly-saccharides in relation to animal nutrition. *J. Dairy Sci.* **1991**, *74*, 3583–3597. [CrossRef]
32. Rasby, R.J.; Kononoff, P.J.; Anderson, B.E. Understanding and Using a Feed Analysis Report. University of Nebraska–Lincoln Extension, Institute of Agriculture and Natural Resources. 2008. Report G1892. Available online: https://extensionpublications.unl.edu/assets/pdf/g1892 (accessed on 21 August 2021).
33. Meissner, H.H.; Zacharias, P.J.K.; O'Reagain, P.J. Forage quality (feed value). In *Pasture Management in South Africa*; Tainton, N.M., Ed.; University of Natal Press: Pietermaritzburg, South Africa, 2000; pp. 66–88.
34. Schroeder, J.W. Forage Nutrition for Ruminants. NDSU Extension Publication AS-1250. Fargo: North Dakota State University Extension Service. 2009. Available online: https://www.ag.ndsu.edu/publica-tions/livestock/quality-forage-series-forage-nutrition-forruminants (accessed on 21 August 2021).
35. Armke, F.W.; Scott, C.B. Using cattle to disperse seeds for winter forage plants. *Texas. J. Agric. Nat. Resour.* **1999**, *12*, 28–38.
36. Aruwayo, A.; Adeleke, R.A. A Review of Chemical Constituents and Use of Browse Plants in the Tropics. FUD-MARec. *Chem. Sci.* **2020**, *1*, 72–81.
37. Britz, A.E.; Cyster, L.F.; Cupido, C.F.; Samuels, M.I.; Mpanza, T.D.E.; Müller, F.L. Nutritional quality of *Calobota sericea* fodders harvested at different phenological stages–opportunities for inclusion in fodder flow programs for extensive livestock farmers. *Afr. J. Range Forage Sci.* **2022**, 1–6. [CrossRef]
38. Mudzengi, I.P.; Murwira, A.; Halimani, T.; Fritz, H.; Murungweni, C. Relationships between farmer perceptions and temporal variation in nutritive value of browse species on savanna rangelands. *Afr. J. Range Forage Sci.* **2020**, *37*, 237–242. [CrossRef]
39. Uguru, C.; Lakpini, C.A.M.; Akpa, G.N.; Bawa, G.S. Nutritional potential of *Acacia nilotica* pods for growing Red Sokoto goats. *IOSR J. Agric. Vet. Sci.* **2014**, *7*, 43–49.
40. Spencer, R. Nutrient Requirements of Sheep and Goats, ANR-0812. Alabama Cooperative Extension System. 2018. Available online: https://www.google.com/url?sa=t&rct=j&q=&esrc=s&source=web&cd=&ved=2ahUKEwiHgsjsg7__AhW8gFYBHZogCBkQFnoECAoQAQ&url=https%3A%2F%2Fwww.aces.edu%2Fwp-content%2Fuploads%2F2018%2F11%2FANR-0812.pdf&usg=AOvVaw345TGcQxXFWyoQFd5G7IXU (accessed on 21 August 2021).
41. McDonald, P.; Edwards, R.A.; Greenhalgh, J.F.D.; Morgan, C.A.; Sinclair, L.A.; Wilkinson, R.G. *Animal Nutrition*, 7th ed.; Longman: New York, NY, USA, 2010.
42. Rinehart, L. Ruminant Nutrition for Graziers. ATTRA–National Sustainable Agriculture Information Service. 2008. Available online: https://extension.usu.edu/rangelands/ou-files/Ruminant_nutritiongrazing (accessed on 21 August 2021).

43. Abdel, D.A.R. Growth and Vegetative Propagation of Some of the Indigenous Tree Species of the Sudan. Master's Thesis, University of Khartoum, Khartoum, Sudan, 1997.
44. Whitacre, M.K.; Call, C.A. Recovery and germinability of native seed fed to cattle. *West. N. Am. Nat.* **2006**, *66*, 121–128. [CrossRef]
45. Bodmer, R.E.; Ward, D. Frugivory in large mammalian herbivores. In *Large Herbivore Ecology, Ecosystem Dynamics and Conservation*; Danell, K., Duncan, P., Bergstrom, R., Pastor, J., Eds.; Cambridge University Press: Cambridge, UK, 2006; pp. 232–260.

agronomy

MDPI

Article

Alfalfa (*Medicago sativa* L.) Nitrogen Utilization, Yield and Quality Respond to Nitrogen Application Level with Center Pivot Fertigation System

Yunling Wang [1,2,†], Maona Li [3,†], Jiali Guo [1] and Haijun Yan [1,4,*]

[1] College of Water Resources and Civil Engineering, China Agricultural University, Beijing 100083, China; wangyunling@cau.edu.cn (Y.W.); s20223091907@cau.edu.cn (J.G.)
[2] College of Animal Science and Technology, Hebei Agricultural University, Baoding 071001, China
[3] College of Grassland Science and Technology, China Agricultural University, Beijing 100083, China; limaona2020@cau.edu.cn
[4] Engineering Research Center of Agricultural Water-Saving and Water Resources, Ministry of Education, Beijing 100083, China
* Correspondence: yanhj@cau.edu.cn
† These authors contributed equally to this work.

Abstract: Nitrogen (N) application with a center-pivot fertigation system is commonly used for alfalfa production in the North China Plain, with its barren soil, but improper N application has resulted in lower N use efficiency and even negative effects on yield and quality. The effects of N application levels on soil NO_3^--N content, nitrogen utilization, yield and quality of alfalfa at various cuttings and established years were evaluated under sprinkler fertigation to optimize the N application schedule. Four N application levels of 0, 12.5, 25 and 37.5 kg N ha^{-1} (N0, N1, N2 and N3) for each cutting were applied to alfalfa at the early vegetative stage from the first to third year after establishment. The results showed that the variation in soil NO_3^--N content was mainly concentrated in the topsoil (0–40 cm) after N application via sprinkler fertigation. N uptake amount was dramatically improved through the N application for one-year-old alfalfa. Compared to the yield under the N0 treatment, the N application significantly improved the yield of alfalfa at the first two cuttings in the first year. The N use efficiency generally reached a high value with a low N application level. N application had an insignificant effect on the alfalfa quality classification during the three years. For obtaining high yield, quality and N use efficiency of alfalfa planted in the North China Plain, the N application level is recommended as 12.5 kg N ha^{-1} at the first cutting and 37.5 kg N ha^{-1} at the second cutting for one-year-old alfalfa.

Keywords: alfalfa; nitrogen fertilization; yield; crude protein content; relative feed values; nitrogen use efficiency

Citation: Wang, Y.; Li, M.; Guo, J.; Yan, H. Alfalfa (*Medicago sativa* L.) Nitrogen Utilization, Yield and Quality Respond to Nitrogen Application Level with Center Pivot Fertigation System. *Agronomy* **2024**, *14*, 48. https://doi.org/10.3390/agronomy14010048

Academic Editor: Koki Toyota

Received: 26 October 2023
Revised: 13 December 2023
Accepted: 20 December 2023
Published: 23 December 2023

1. Introduction

With the development of the dairy industry, the demand for forage with high quality is increasing. Alfalfa (*Medicago sativa* L.), a forage crop high in protein, is one of the major forage crops that are extensively planted in China. Based on the support of policies accelerating the development of alfalfa production, the alfalfa planting area has expanded further to 2206.67 thousand ha [1]. As a high-quality forage, alfalfa has high nitrogen (N) content. The critical concentration of N in alfalfa plants, the minimum N content in shoots required to produce the maximum aboveground biomass at a given time, ranges from 2.7% to 3.3% [2], with the highest concentration of nitrogen reaching 3.97% [3]. Being a leguminous crop, alfalfa can assimilate atmospheric N_2 through biological nitrogen fixation (BNF). Annual N_2 fixation can reach 50–450 kg ha^{-1}, and about 30–80% of plant N comes from fixed N [4]. However, alfalfa roots and nodules are not fully developed in the early stage, and the development of active nodules takes 2–4 weeks after planting, even when rhizobium is inoculated [5,6]. Therefore, a large N demand cannot be satisfied through N_2 fixation at the early stage [7]. Applying modest N fertilizer to alfalfa is beneficial to obtain a higher yield and quality in areas with low soil nitrate content and soil organic matter or in the seeding year [4,8].

The impacts of N application on alfalfa yield and quality vary in different growth stages, planting years and soil fertility [9–11]. Generally, N fertilization could improve alfalfa yield in the seeding year with an application level of 40–80 kg N ha^{-1} yr^{-1} [9]. Especially under the condition of low soil fertility, N application could increase alfalfa yield in the first year after established [4]. Some reports also show that the carbon requirements of the nodules could not be fulfilled after alfalfa shoot cutting, and N_2 fixation was reduced [12]. Nitrogen plays a dominant role in regulating carbon assimilation [13], and appropriate application of N fertilizer at this stage could improve N_2 fixation and alfalfa yield [7,14]. However, He et al. [15] reported that N application had no significant effect on the yield of alfalfa in the seeding year and would increase weed yield. Moreover, higher N application levels (225–300 kg N ha^{-1} yr^{-1}) could reduce the alfalfa yield in the first year of planting [16] because higher N application levels have an inhibitory effect on the root growth and development, nodulation and configuration [17–19]. Many studies have shown that N application had no significant effect on the yield of well-established alfalfa [11,18,20], as 70–80% of N demand could be satisfied through N_2 fixation [21]. Meanwhile, Fan et al. [10] found that N application could improve the yield of both young and old alfalfa stands owing to the absence of nodulation. Generally, N application has a significant effect on the alfalfa quality, which could increase the total nitrogen content, nitrate nitrogen and ammonium nitrogen content, and crude protein content (CP) but reduce the neutral detergent fiber (NDF) and acid detergent fiber (ADF) [22–24]. N fertilization could increase crude protein by improving non-protein nitrogen (NPN) that can be degraded to ammonia in the rumen [25,26]. Ammonia is a source for microbial protein synthesis, but excess ammonia is absorbed through the rumen wall and largely excreted as urea in the urine [27]. This could increase the metabolic burden on animals and have a negative impact on the environment. Therefore, animal scientists do not recommend abundant N application to increase the N content in alfalfa forage. However, some researchers have also reported that N application had no significant effect on the alfalfa quality during the first year of establishment [28] and even reduced CP and relative feed values (RFV) with high N application levels [16]. RFV is considered a crucial parameter for alfalfa quality evaluation [29]. Generally, RFV has a negative relationship with ADF and NDF [30]. Overall, the response of alfalfa yield and quality to N application is complicated. The optimal N application level for alfalfa is difficult to be consistent under different conditions.

Nitrogen fertilizer, as one of the most important agricultural inputs, plays an important role in improving crop yield. Annual N fertilizer consumption in China accounted for more than 35% of the total N fertilizer consumption of the world and about 60% of the whole fertilizer consumption in China. However, nitrogen use efficiency (NUE) of crop production in China ranged from 20% to 40% in 2017 [31], which was lower than that in Europe (52%) and the United States of America (68%) [32]. A large amount of N fertilizer has been severely lost due to improper fertilization methods. In alfalfa production, center-pivot irrigation systems have been widely used all over the world [33] because of their advantages, such as large coverage area, low labor costs, high irrigation efficiency, easy fertigation and high application uniformity [34,35]. In recent years, the fertigation method using center-pivot irrigation systems, particularly in the application of urea solutions, has been applied increasingly in China [17]. Compared with furrow irrigation, water could be saved 60–72% with a center-pivot sprinkler fertigation system [36]. Meanwhile, N fertilizer could be saved by 78% in sandy loam soil or 52% in sandy soil with fertigation, compared to that with the traditional fertilization method [37]. Since the end of the 20th century, the center-pivot sprinkler fertigation system has been popularized and applied in the United States and could reduce N leaching loss and improve N use efficiency [34,38]. Owing to the high degree of automation for sprinkler fertigation, N fertilizer could be supplied for alfalfa in each cutting rather than being totally applied at the beginning of the growing season. N demand for alfalfa during different cuttings might be various, but the same N application for each cutting was common in alfalfa production. Therefore, optimization of the N application schedule for alfalfa in different cuttings and established years should be further conducted.

The objectives of this study were (1) to evaluate the effects of four N application levels on the dry matter yield, CP content, RFV and N utilization of alfalfa in different cuttings and years with sprinkler fertigation and (2) to optimize the N application schedule for alfalfa planted in the North China Plain.

2. Materials and Methods

2.1. Experimental Site

The field experiments were conducted at the China Agricultural University Experimental Station from 2016 to 2018. The station is located in the North China Plain in Zhuozhou, Hebei Province (39°27′ N, 115°51′ E), and has a typical temperate sub-humid continental monsoon climate with a summer precipitation pattern (Figure 1). During the experimental period, the average total precipitation was 440.0 mm, 56.61% of which fell between July and August. The average monthly air temperature, wind speed, solar radiation and relative humidity during the three years recorded by a weather station (Watchdog 2000, Spectrum Technologies Inc., Haltom City, TX, USA) installed at the experimental site, are presented in Figure 1. Air temperature and solar radiation both increased first and then decreased with the increase of the month. The relative humidity reached a high level from July to October. Wind speed was faster during March and April (Figure 1). The soil type is classified as EutyicCambisols (World Reference Base (WRB) 2014) and is mainly sandy soil. Other physicochemical properties of the root zone soil (0–80 cm) are shown in Table 1. Overall, soil properties in the field were relatively homogeneous. Soil fertility was relativel y lower, especially the soil organic matter and nitrate nitrogen content. Seeds of the alfalfa cultivar WL363HQ were sown on 28 August 2015, with a row spacing of 30 cm and a sowing rate of 22.5 kg ha^{-1}. All the alfalfa seeds were purchased from the RYTWAY seed company (Beijing, China) and were produced in America. Meanwhile, all the seeds were coated and inoculated with rhizobium during production. Before alfalfa was established, the previous crops were two species of annual gramineous forage grasses (tall fescue (*Festuca arundinacea* Schreb.) and smooth bromegrass (*Bromus inermis* Leyss.)). Meanwhile, diammonium phosphate (N $\geq$ 18% and $P_2O_5 \geq$ 46%) of 150 kg ha^{-1} and potassium sulphate ($K_2O \geq$ 52%) of 100 kg ha^{-1} were applied before alfalfa seeding. During the period of experiments, imazethapyr aqueous solution (5%) was used in weed management. Lambda-cyhalothrin and acetamiprid were used in pest control. During the branching stage at the 1st cutting in 2017, superphosphate ($P_2O_5 \geq$ 14%) of 300 kg ha^{-1} and potassium sulphate ($K_2O \geq$ 52%) of 120 kg ha^{-1} were applied to all treatments. In 2016–2018, regrowth of alfalfa began on 29 March, 1 April and 5 April, respectively. Alfalfa was harvested at the early flowering stage (10% of alfalfa in the field bloomed) and four times each year. The harvest dates are shown in Table 2.

Table 1. Soil physicochemical properties at different soil depths in the field at the beginning of the experiment. (mean ± SD; n = 12).

Soil Depth (cm)	Bulk Density (g cm^{-3})	Field Capacity (cm^3 cm^{-3})	Soil pH	Available Phosphorus Content (mg kg^{-1})	Available Potassium Content (mg kg^{-1})	Organic Matter (g kg^{-1})	Nitrate Nitrogen Content (mg kg^{-1})	Ammonium Nitrogen Content (mg kg^{-1})
0–20	1.62 ± 0.06	0.25 ± 0.01	8.32 ± 0.11	45.2 ± 5.97	80.87 ± 8.65	9.84 ± 0.82	4.91 ± 0.68	2.70 ± 0.32
20–40	1.58 ± 0.09	0.20 ± 0.04	8.34 ± 0.10	56.9 ± 6.46	70.93 ± 7.08	10.37 ± 1.05	3.25 ± 0.46	1.68 ± 0.53
40–60	1.57 ± 0.04	0.23 ± 0.03	8.29 ± 0.11	3.53 ± 1.34	70.98 ± 10.94	8.29 ± 0.91	2.18 ± 0.65	1.67 ± 0.63
60–80	1.59 ± 0.05	0.16 ± 0.04	8.35 ± 0.09	11.50 ± 3.64	55.96 ± 6.07	8.35 ± 1.62	2.15 ± 0.37	1.44 ± 0.69

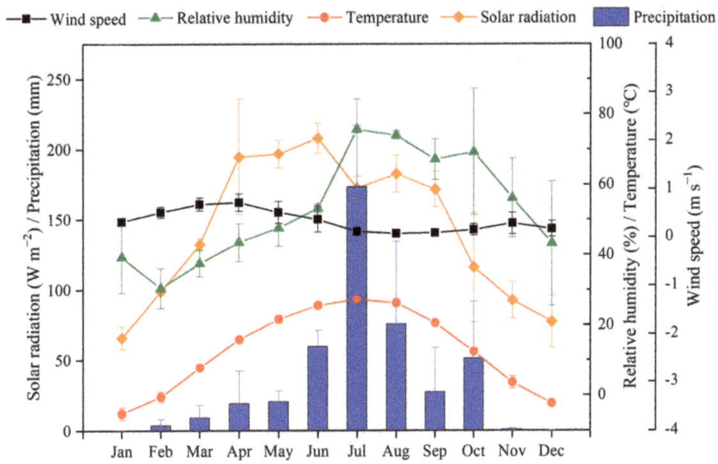

Figure 1. Average monthly weather data during 2016–2018 at the experimental station in the North China Plain.

Table 2. The irrigation amount, fertilizer application date and harvest date during the three years.

Year	Cuttings	Irrigation Amount (mm)	Fertilizer Application Date	Harvest Date
2016	1st cutting	192.0	4 April	26 May
	2nd cutting	68.0	4 June	6 July
	3rd cutting	51.0	18 July	17 August
	4th cutting	78.0	27 August	1 October
2017	1st cutting	165.0	1 April	19 May
	2nd cutting	129.0	4 June	5 July
	3rd cutting	39.0	19 July	14 August
	4th cutting	42.5	26 August	29 September
2018	1st cutting	95.0	12 April	24 May
	2nd cutting	126.0	5 June	4 July
	3rd cutting	12.0	19 July	13 August
	4th cutting	80.0	26 August	28 September

2.2. Experimental Design

To determine the effects of N application level on alfalfa yield, quality and N utilization, the field experiments were conducted based on a single-factor experimental design. Four N application levels, set at 0, 12.5, 25.0 and 37.5 kg N ha^{-1} for each cutting (N0, N1, N2 and N3), were applied to the alfalfa through a center-pivot sprinkler fertigation system. Particularly, no additional N fertilizer was applied, and only the initial soil N content was available for alfalfa under N0 treatment. Urea (N $\geq$ 46%) was used as nitrogen fertilizer in this study. During the three years, alfalfa was harvested four times during the growing season in this region. All of the treatments had three repetitions with a plot size of 7 m $\times$ 10 m, and the layout is shown in Figure 2. The fertilizer treatments were all conducted during the period from regrowth to branching of alfalfa during each cutting. The dates for fertilizer application during the three years are shown in Table 2.

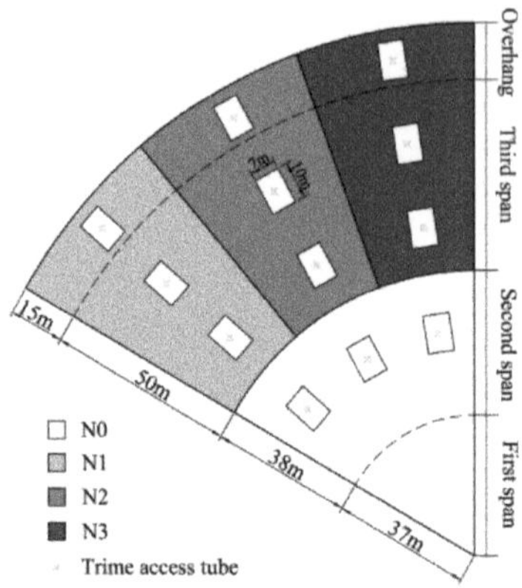

Figure 2. Distribution of experimental plots under the center pivot irrigation system. N0, N1, N2, and N3 represent the N application levels of 0.0, 12.5, 25.0, and 37.5 kg N ha^{-1} for each cutting, respectively.

2.3. *Fertigation System and Irrigation Management*

During the experimental period, water and fertilizer were supplied to alfalfa through a center-pivot irrigation system. The center pivot was composed of three spans and an overhang, with a total length of 140 m. The lengths of the first, second and third span and overhang were 37, 38, 50 and 15 m, respectively. The sprinklers used in this center pivot were the Nelson R3000, assembled with pressure regulators of 0.138 MPa (20 psi) (Nelson Irrigation Corp., Walla Walla, WA, USA). To realize the opening and closing of individual sprinklers, solenoid valves were installed at the connection between each drop hose and lateral pipe. Additionally, a fertigation system consisting of a piston injection pump and a fertilizer storage tank was equipped for the center pivot. The urea solution, mixed uniformly in the storage tank, was injected steadily into the center pivot through the pump and sprayed into the field with the irrigation water. The radial uniformity coefficients of water and fertilizer application with the whole center pivot were up to 89.84% and 88.77%, respectively [39]. During fertilization, sprinklers in the first and second spans were all closed, and only those in the third span and overhang were open. Different N application levels for the N1–N3 treatments were achieved by adjusting the fertilizer concentration in the storage tank, and the volume of fertilizer solution applied to the three N treatments was the same. Generally, fertilization for the three N treatments could be completed within 2 h. After fertilization, different irrigation amounts were immediately supplemented for N0 and the other N treatments to guarantee the same total irrigation amount for all treatments, which was determined based on crop water requirement. The desired water application in specific spans could be realized by changing the center pivot's travel speed and controlling the opening and closing of the sprinklers by turning the electric solenoid valves on or off. As previously mentioned, the irrigation amounts for different treatments were the same and calculated as the difference between the accumulated crop evapotranspiration (ET_c) and the accumulated effective precipitation ($\geq$5 mm). Crop evapotranspiration (ET_c) was calculated using reference evapotranspiration (ET_o) and crop coefficients (K_c). ET_o was calculated from the weather data using the Penman–Monteith equation. Daily K_c was first generated through FAO-56 and then adjusted with the actual lengths of the growing period and weather data in the experimental station [40]. During the experiment, weather data, including solar radiation, wind speed, temperature, relative humidity and precipitation at 2 m height, were collected every 0.5 h with a weather station (Watchdog 2000, Spectrum Technologies Inc., USA) installed at the experimental site. Irrigation was performed for all plots when the average soil water content at the soil depth of 0–40 cm reached as low as 60% of the field water capacity. An irrigation amount of 20 mm was applied for alfalfa in the first irrigation event of the first cutting during the three years to ensure the regrowth of alfalfa. The total irrigation amounts during the whole alfalfa growing season were 389.0, 375.5 and 313.0 mm for 2016, 2017 and 2018, respectively. Irrigation amounts for each cutting during the three years are shown in Table 2.

2.4. Measurement Parameters

2.4.1. Soil Water Content

Soil water content generally varied with irrigation, rainfall and crop water use. To determine the irrigation time based on the soil water content in this experiment, soil moisture at the depths of 0–20 cm, 20–40 cm, 40–60 cm and 60–80 cm were measured using a TDR Trime-tube system (Trime-T3 TDR, IMKO Ltd., Ettlingen, Baden-Württemberg, Germany) 1 d before and after irrigation during the entire growth period in the three years.

2.4.2. Soil Nitrate Nitrogen Content

N fertilization and plant nutrient uptake could change the soil nitrogen content. Particularly, soil NO_3^--N content is much higher than soil NH_4^+-N content in dryland. To assess the effects of N application on soil NO_3^--N content, soil samples were collected at soil depths of 0–20 cm, 20–40 cm, 40–60 cm and 60–80 cm once 1 or 2 days before fertigation, once within 10 days after fertigation, and once at the end of each cutting in 2016–2018. In 2017–2018, additional measurements of soil NO_3^--N content were conducted once at each growing stage. Soil nitrate-nitrogen content was determined using an AA3 continuous flow analyzer (Bran + Luebbe, Norderstedt, Schleswig-Holstein, Germany).

2.4.3. Total Nitrogen Content of Alfalfa

During the experiments, the total nitrogen content of alfalfa shoots for each cutting was measured using the semi-micro Kjeldahl method for nitrogen determination. All alfalfa shoots for each sample were dried at 105 °C for 0.5 h and then at 75 °C for 48 h. The dried shoots were ground, sifted and boiled in H_2SO_4-H_2O_2. Finally, the total nitrogen content of the plants was determined using a semi-automatic Kjeldahl nitrogen analyzer (K1305A, Shengsheng, Shanghai, China). Nitrogen use efficiency was calculated based on N application level, total nitrogen content and dry matter yield of alfalfa. The formulas are as follows [41]:

$$N_{uptake} = TN \times Y_{DM}, \quad (1)$$

$$\text{ANR} = \frac{(N_x - N_0)}{N_{level}} \quad (2)$$

and

$$\text{NAE} = \frac{(Y_x - Y_0)}{N_{level}}, \quad (3)$$

where N_{uptake} is the total N uptake of alfalfa, kg ha^{-1}; TN is the total nitrogen content, kg kg^{-1}; Y_{DM} is the dry matter yield of alfalfa, kg ha^{-1}; ANR is the apparent recovery of N fertilizer, kg kg^{-1}; N_x is the total N uptake of alfalfa shoot with N application, kg ha^{-1}; N_0 is the total N uptake of alfalfa shoot without N application, kg ha^{-1}; N_{level} is the N application level, kg ha^{-1}; NAE is the agronomic efficiency of N fertilizer, kg kg^{-1}; Y_x is the dry matter yield of alfalfa with N application, kg ha^{-1}; Y_0 is the dry matter yield of alfalfa without N application, kg ha^{-1}.

2.4.4. Yield and Quality of Alfalfa

During the measurements of alfalfa yield, fresh alfalfa plants from three squares (1 m × 1 m) in each plot were cut to a stubble height of 5 cm and weighed as fresh weight. Some samples randomly collected from the fresh alfalfa in each square were dried at 105 °C for 0.5 h and then at 75 °C for 48 h. The ratio of dry weight to fresh weight ($R_{d/w}$) of the sample was calculated. Finally, the dry matter yield of alfalfa per unit area was determined as the product of $R_{d/w}$ and the fresh weight of each square and used as the main yield parameter for further analysis. In addition, after drying, alfalfa samples from each square were further smashed to analyze the content of crude protein (CP) and the relative feed value (RFV). The CP content of alfalfa was measured through near-infrared reflectance spectroscopy (NIR-TR-3750, Foss NIR Systems, Inc., Stockholm, Sweden) based on the model proposed by NIRS Forage and Feed Testing Consortium (NIRSC) [42]. RFV was calculated using the following equations [29]:

$$\mathrm{RFV} = \frac{DDM(\%DM) \times DMI(\%BW)}{1.29}, \tag{4}$$

$$DMI(\%BW) = \frac{120}{NDF(\%DM)} \tag{5}$$

and

$$DDM(\%DM) = 88.9 - 0.779 \times ADF(\%DM), \tag{6}$$

where RFV is the relative feed value, and the unit is the percent of dry matter (%*DM*); *DDM* is the digestible dry matter, %*DM*; *DMI* is the dry matter intake, and the unit is the percent of body weight (%*BW*); *NDF* is the neutral detergent fiber, % *DM*; *ADF* is the acid detergent fiber, %*DM*. Meanwhile, the *NDF* and *ADF* of alfalfa were also measured through near-infrared reflectance spectroscopy. According to the hay grading standard for legumes proposed by the American Forage and Grassland Council [29], the alfalfa quality was graded.

2.5. Statistical Analysis

The linear and quadratic regression analyses of total N uptake, dry matter yield, ANR, NAE and quality of alfalfa with N application level were performed, and the significance of the regression function was investigated through an F test (evaluated as significant at the $p < 0.05$ level). The regression analysis and graphs were performed with Origin Pro2022 software (OriginLab, Northampton, MA, USA). In the figures, no regression curve and equation indicated that relationships between parameters and N application level were insignificant.

3. Results

3.1. Temporal and Spatial Variations of Soil Nitrate N Content

During the experimental period in 2016–2018, soil NO_3^--N content at a depth of 0–40 cm reached a high value and showed great changes over time (Figure 3). Differently, soil NO_3^--N content at depths of 40–80 cm remained relatively stable throughout the growing season. At the beginning of the growing season in 2016, soil NO_3^--N content under different treatments was relatively low, ranging from 0.74 to 5.18 mg kg^{-1}. After the first and third N applications, soil NO_3^--N content at different depths did not noticeably increase. After the second N application, soil NO_3^--N content at a depth of 0–20 cm significantly improved under all treatments, up to 13.88–39.43 mg kg^{-1}. Under the treatments of N2 and N3, soil NO_3^--N content at a depth of 20–40 cm also increased markedly. After the fourth N application, soil NO_3^--N contents at depths of 0–20 cm and 60–80 cm increased under N1–N3 treatments. However, soil NO_3^--N content was reduced to the initial level at the end of each cutting. During the years 2017 and 2018, soil NO_3^--N content was measured once at each growing stage. At the beginning of the experiments in 2017, soil NO_3^--N content at a depth of 0–20 cm increased up to 3.93–21.96 mg kg^{-1}. Under treatments with or without N application, soil NO_3^--N content at a depth of 0–20 cm increased first and then decreased after fertigation or irrigation in 2017 and 2018. In 2018, the soil NO_3^--N content at a depth of 0–20 cm even reached up to 73.29 mg kg^{-1}. Overall, the maximum increment in soil NO_3^--N content (0–20 cm) one week after fertigation reached 94.53, 118.50 and 114.26 kg ha^{-1} for N1, N2 and N3 treatments, respectively, which were much higher than the amount of N application. The increase in soil NO_3^--N content was insignificant after fertigation at the third cutting during the three years due to rainy weather. Meanwhile, soil NO_3^--N content at a depth of 60–80 cm under N2 and N3 treatments increased at the second cutting in 2016, the third cutting in 2018, and the fourth cutting during the three years, which might also be due to heavy rainfall. In general, a significant reduction in soil NO_3^--N content at a depth of 0–20 cm occurred during the late vegetative stage and bud stage of alfalfa. At the end of the growing season in 2018, the soil NO_3^--N content at a depth of 0–80 cm increased by 35.28, 22.03, 24.19 and 9.13 kg ha^{-1} under treatments of N0, N1, N2 and N3, respectively, compared to that at the beginning of the growing season in 2016.

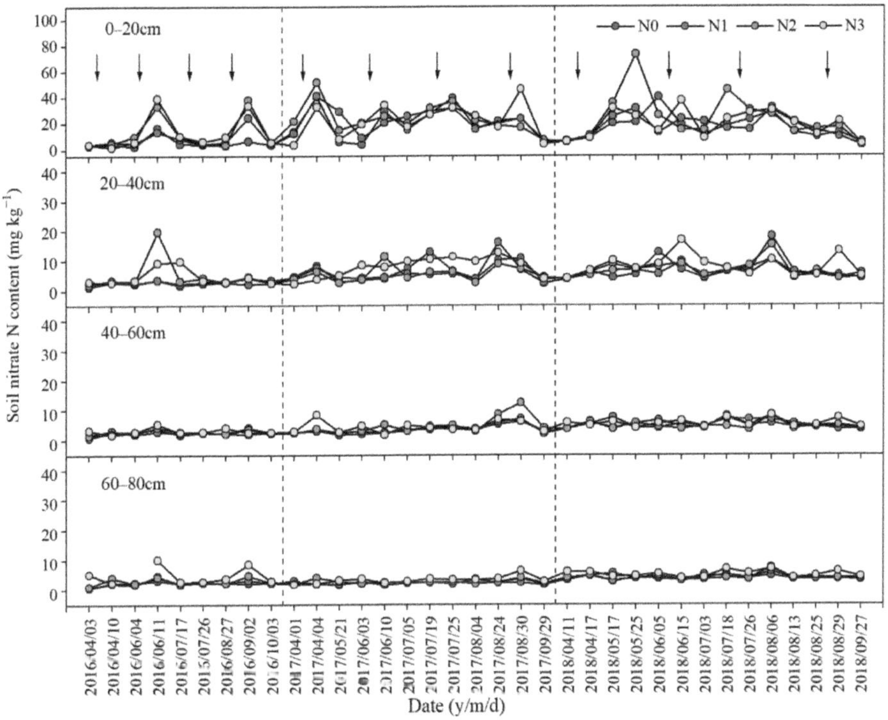

Figure 3. Spatiotemporal variation of soil NO_3^--N content under different treatments in 2016–2018. N0, N1, N2 and N3 represent N application levels of 0.0, 12.5, 25.0 and 37.5 kg N ha^{-1} for each cutting, respectively. Arrows represent the N application events.

3.2. Nitrogen Uptake by Alfalfa Shoot

N uptake by alfalfa shoot generally decreased with stand age and cutting (Figure 4). The average annual N uptake under different treatments for one-, two- and three-year-old alfalfa were 441, 413 and 314 kg ha^{-1}, respectively. The proportion of N uptake for each cutting to annual N uptake during the three years averaged 33.3%, 32.8%, 20.0% and 13.9% for the first to fourth cutting, respectively. In 2016, the total N uptake amount ranged from 386 to 497 kg ha^{-1}, which had a remarkable linear and positive relationship with the N application level (R^2 = 0.99). Especially in the first two cuttings, the N uptake amount had a prominent relationship with the N application level. A significant linear relationship was found between the N application level and the N uptake amount of the first cutting (R^2 = 0.96) (Figure 4b). N uptake amount of the first cutting ranged from 116 to 176 kg ha^{-1} and increased with the N application level. In the second cutting, regression analysis of N uptake amount and N application level showed a significant parabolic relationship with an upward opening (R^2 = 0.99). N uptake amount under N3 treatment reached the highest value of 191 kg ha^{-1}, which was 34.5% higher than that under N0 treatments. N uptake amount of the third and fourth cuttings had no evident relationship with the N application level. In the third cutting, the N uptake of alfalfa increased first and then decreased with the N application level. N uptake amount under the N2 treatment was the highest, with values of 87 kg ha^{-1} and 20.4% higher than that under the N0 treatment. In the fourth cutting, N uptake amounts under different treatments were very similar, ranging from 54 to 61 kg ha^{-1}. In 2017, no significant relationship existed between the total N uptake amount for the whole growing season or each cutting and N application level (Figure 4a,c). The total N uptake amount of alfalfa for the whole growing season with N application was 406–439 kg ha^{-1}, which was 3.9–12.4% higher than that under N0 treatment. In 2018, a significant parabolic relationship with a downward opening was observed between the total N uptake amount and N application level. The total N uptake amount under the N3 treatment was up to 327 kg ha^{-1}, which was 12.7% higher than that under the N0 treatment (Figure 4a). Meanwhile, the N application level only had a prominent relationship with the N uptake amount in the fourth cutting (Figure 4d). In the fourth cutting, a significant parabolic relationship with an upward opening existed between the N uptake amount and the N application level. N uptake amount under N3 treatment was the highest and 27.7% higher than that under N0 treatment.

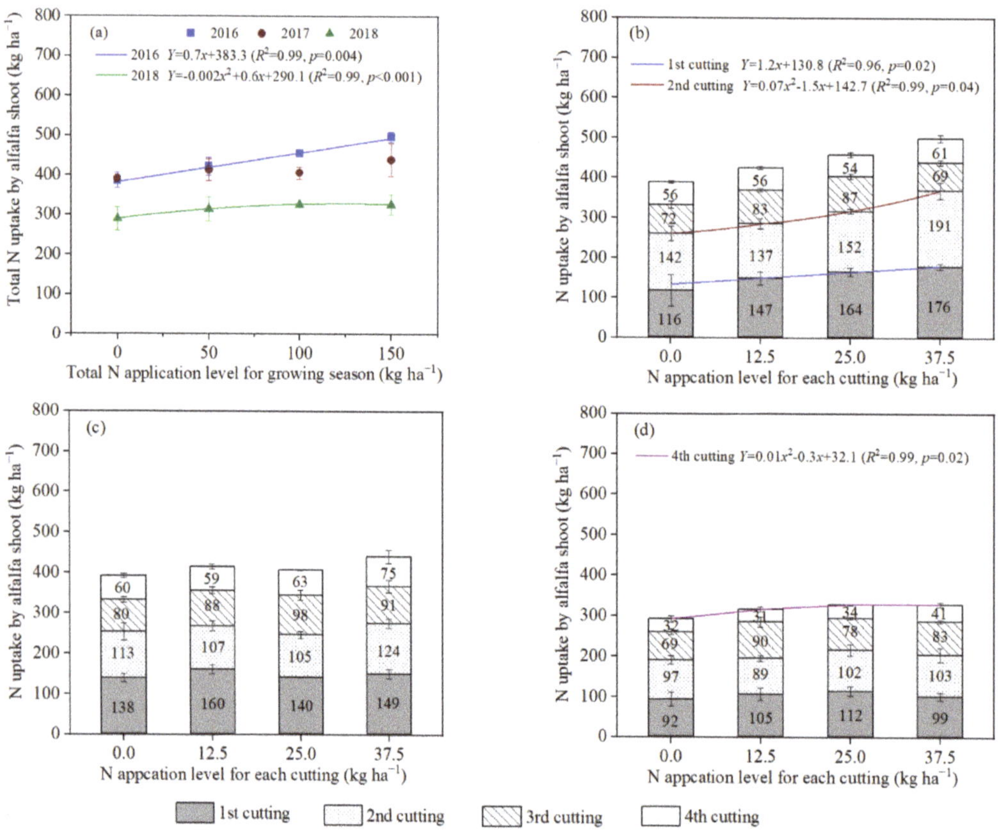

Figure 4. Total N uptake amount of alfalfa shoot for the whole growing season during the three years (**a**) and each cutting in 2016 (**b**), 2017 (**c**) and 2018 (**d**) under different N application levels. Linear or quadratic regression curves and equations indicated that the relationship between N uptake amount and N application level was significant.

3.3. Dry Matter Yield of Alfalfa

During the three years, alfalfa yield gradually reached a high value in the first two years and decreased with cutting (Figure 5). The alfalfa yield of the first and second cuttings was relatively higher, which could reach 63.16–70.33% of the annual yield. Overall, the N application level had a significant parabolic relationship with a downward opening with the annual yield for the three years (Figure 5a). Compared with the N0 treatment, total alfalfa yield could be dramatically increased through N application. In 2016, the average annual yield increased with N application and reached the highest value of 14,368.85 kg ha^{-1} with N3 treatment, which was 41.37% higher than that under treatment without N application. The annual yield was increased by 24.16% and 23.72% under the N1 and N2 treatments, respectively, compared to the yield of the N0 treatment. In 2017, compared to the alfalfa annual yield of the N0 treatment with an average value of 12,849.58 kg ha^{-1}, the alfalfa annual yield under N1, N2 and N3 treatments increased by 12.52%, 6.72% and 13.78%, respectively. In 2018, the annual yield with N application was 12.53–15.67% higher than that under N0 treatment. On the whole, the annual yield reached the highest value of 12,849.58–14,619.74 kg ha^{-1} in 2017. Compared with the annual yield in 2016 and 2017, it was reduced by 8.99–28.22% in 2018, which was mainly due to the emergence of low air temperature (average daily temperature < 8 °C, the minimum air temperature reaching −1.8 °C) during spring green-up stage and serious insect attacks in the third and fourth cuttings. Furthermore, N application level had a significant relationship with the alfalfa yield of the first, second and third cuttings in the first year (Figure 5b). In the first and third cutting, a significant parabolic relationship with a downward opening existed between alfalfa yield and N application level ($p < 0.01$). Alfalfa yield of the first cutting increased with N application level with a decreasing slope and reached the highest under N3 treatment. Compared with yield under N0 treatment, it was improved by 42.4–75.3% with N application. In the second cutting, alfalfa yield had a prominent linear and positive relationship with the N application level. Alfalfa yield could be increased by 44.9% under the N3 treatment compared to that under the N0 treatment. The alfalfa yield of the third cutting increased first and then decreased with the N application level. A similar variation tendency and the relationship between yield and N application level was also found at the third cutting in 2017 and 2018 (Figure 5c,d). Compared to that under N0 treatment, the alfalfa yield of the third cutting could be improved by N1 or N2 treatment in 2016–2018 with the maximum proportion of 19.0%, 15.3% and 7.4%, respectively. N application level had no remarkable relationship with the alfalfa yield of the fourth cutting. In 2017 and 2018, the N application level had no significant relationship with the alfalfa yield in the first and second cuttings. In contrast to that under N0 treatment, alfalfa yield increased by less than 15.9% with nitrogen application at the two cuttings of the two years (Figure 5c,d). Alfalfa yield of the fourth cutting had a significant parabolic relationship with the N application level over the two years. Meanwhile, alfalfa yield increased with the N application level and got the highest value under N3 treatment, but only 273.9–394.5 kg ha^{-1} (22.3–26.9%) higher than that under N0 treatment.

3.4. Nitrogen Utilization of Alfalfa

The relationships between the apparent recovery of N fertilizer (ANR) and agronomic efficiency of N fertilizer (NAE) for alfalfa with the N application level varied in different cuttings and years (Figure 6). In 2016, a significant linear and negative relationship between the N application level and ANR was observed at the third cutting (R^2 = 0.584). The ANR of alfalfa under different treatments was relatively higher, with an average value of 1.61–2.41 kg kg^{-1} in the first cutting, but the ANR of the fourth cutting could be negative under the treatments of N1 and N2. Additionally, the N application level only had a significant linear and negative relationship with the NAE of alfalfa in the first and third cuttings. In the two cuttings, the NAE decreased with the N application level and reached the highest with an average value of 121.15 kg kg^{-1} and 36.20 kg kg^{-1} under N1 treatment in the first and third cuttings, respectively. In the second and fourth cuttings, NAE decreased first and then increased with the N application level. The NAE of N2 treatment was generally low and even negative in the fourth cutting. In 2017, the N application level only had prominent linear and negative relationships with the ANR and NAE in the first cutting, and the ANR and NAE of the N1 treatment were significantly higher than those of other treatments. During the second and fourth cuttings, the negative ANR appeared with N1 and N2 treatments. The NAE of the N2 treatment was also the lowest, except for that of the third cutting. The NAE with N3 treatment was the lowest in the third cutting but the highest in the fourth cutting. In 2018, the N application level had linear and negative relationships with the ANR and NAE at the third cutting. The ANR and NAE of N1 treatment were generally higher in the first and third cuttings. Variously, a noteworthy linear and positive relationship between N application level and ANR existed in the fourth cutting. In the second and fourth cuttings, ANR under N1 treatment appeared to have a negative value. Overall, ANR and NAE were relatively higher in 2016 and were generally up to a high value in the first cutting. In addition, the ANR and NAE of N1 or N2 treatments appeared to have a negative value in the second and fourth cuttings, which revealed that the yield or N uptake of alfalfa was not improved through N application and was even reduced compared to that under treatment without N application.

3.5. Quality of Alfalfa

In the first established year of alfalfa (2016), N application generally had an insignificant relationship with the CP of alfalfa (Figure 7). In general, the CP decreased first and then increased with the N application level. The CP of alfalfa under N0 treatment remained at a high level at different cuttings. N application did not significantly improve the CP of alfalfa. N application level only had a significant parabolic relationship with the RFV in the third cutting in 2016. The RFV generally decreased first and then increased with the N application level. In the first, second and fourth cuttings, RFV was high under N0 and N2 treatments and higher than that under N1 and N3 treatments. In the third cutting, the RFV under N3 treatment reached a high level with a mean value of 174.82%, which was higher than that under other treatments. In 2017, relationships between the N application level and the RFV and CP of alfalfa were insignificant in the first three cuttings but remarkable in the fourth cutting. In the fourth cutting, the CP had a significant linear and positive relationship with the N application level. The CP reached the highest value, averaging 25.3% under N3 treatment, which was higher than that under other treatments. The average RFV under N3 treatment was up to 180.12%, which was 10.2%, 16.8% and 17.4% higher than those under N0, N1 and N2 treatments, respectively. In 2018, the N application level only had notable parabolic relationships with the RFV and the CP in the third cutting. In the third cutting, the CP and RFV increased first and then decreased with the N application level, reaching a high level under N1 and N2 treatments. Overall, the CP of alfalfa during the three years ranged from 19.3% to 28.7%. Based on the hay grading standard for legumes according to the CP content, alfalfa in this experiment could be classified as premium-grade grass. While classifying grass grade according to the RFV, alfalfa in 2016 could be generally graded as the premium, and only that under the N1 treatment in the third cutting as the first-grade grass. In 2017 and 2018, only alfalfa in the fourth cutting reached the premium grade standard. Except for the alfalfa of the second cutting in 2017, which was classified as the second-grade standard, alfalfa in the other cuttings in 2017 and 2018 could be classified as first-grade grass. Overall, alfalfa quality in 2016 was better than that in 2017 and 2018. Compared to the N0 treatment, the N application did not significantly improve the CP of alfalfa in 2016 and increased the RFV or CP only at the third and fourth cuttings in 2016–2018 but did not affect the quality classification of alfalfa.

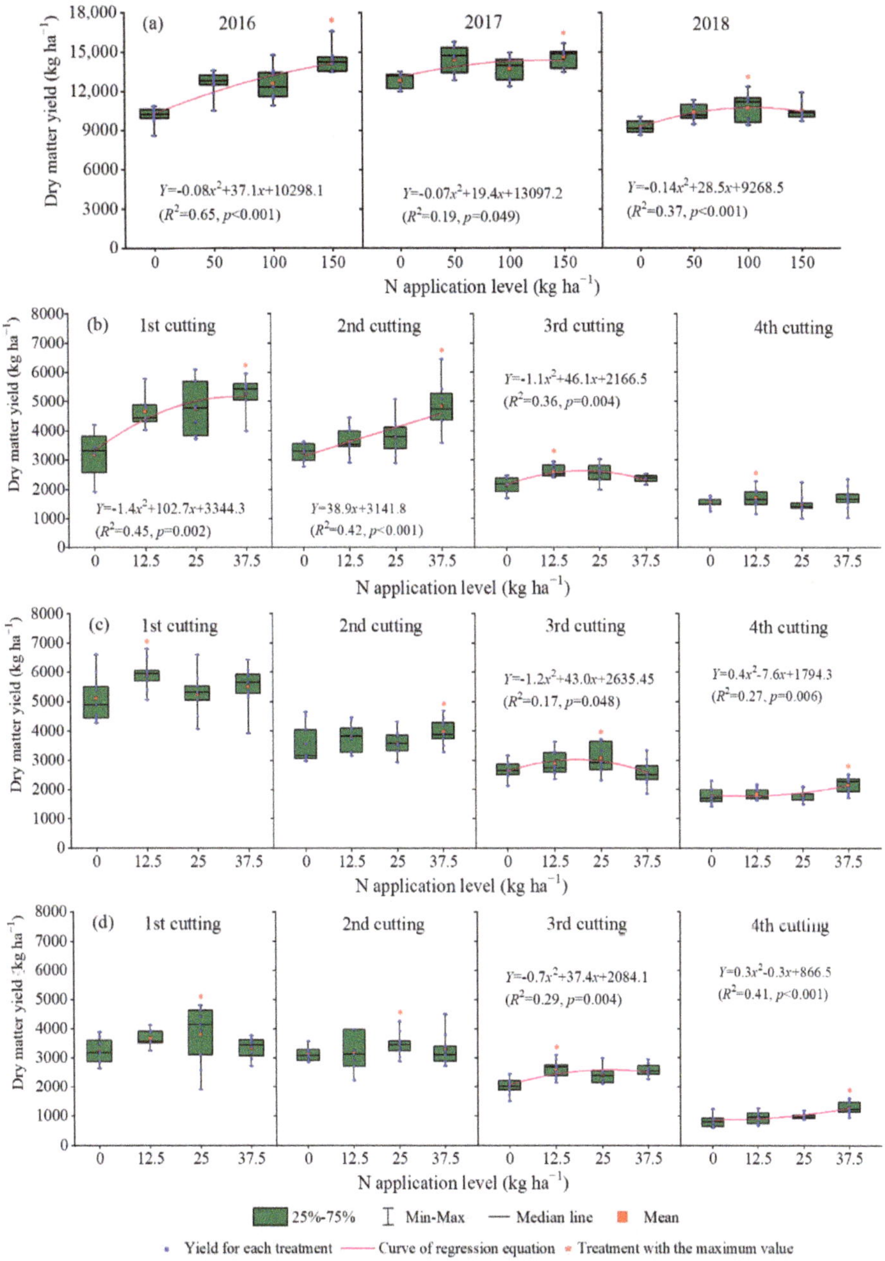

Figure 5. Dry matter yield of alfalfa under different N application levels for each cutting in 2016 (**a**), 2017 (**b**) and 2018 (**c**), as well as the annual yield of the three years (**d**).

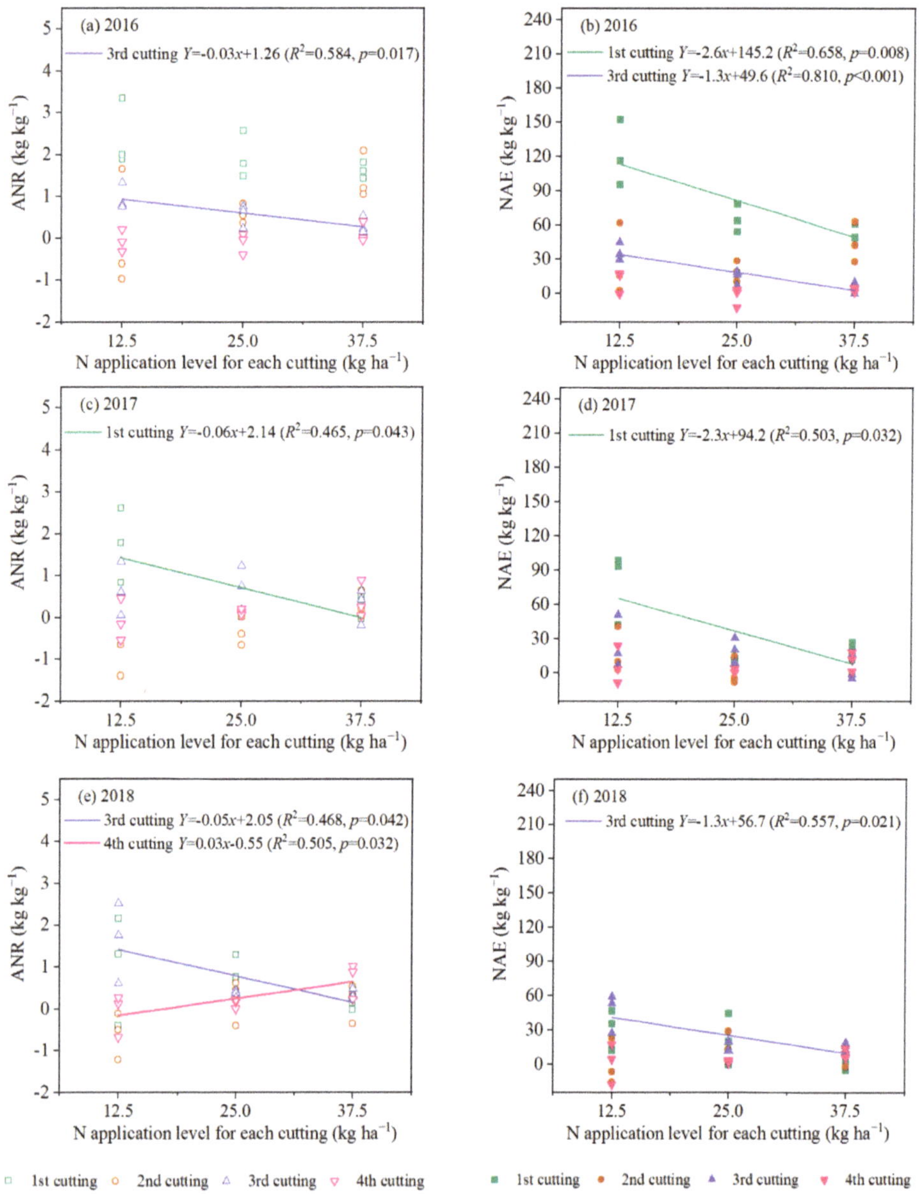

Figure 6. Apparent recovery of N fertilizer (ANR) (**a**,**c**,**e**) and agronomic efficiency of N fertilizer (NAE) (**b**,**d**,**f**) under different N application levels in 2016 (**a**,**b**), 2017 (**c**,**d**) and 2018 (**e**,**f**). Linear regression curves and equations indicated that the relationship of ANR and NAE with N application level was significant.

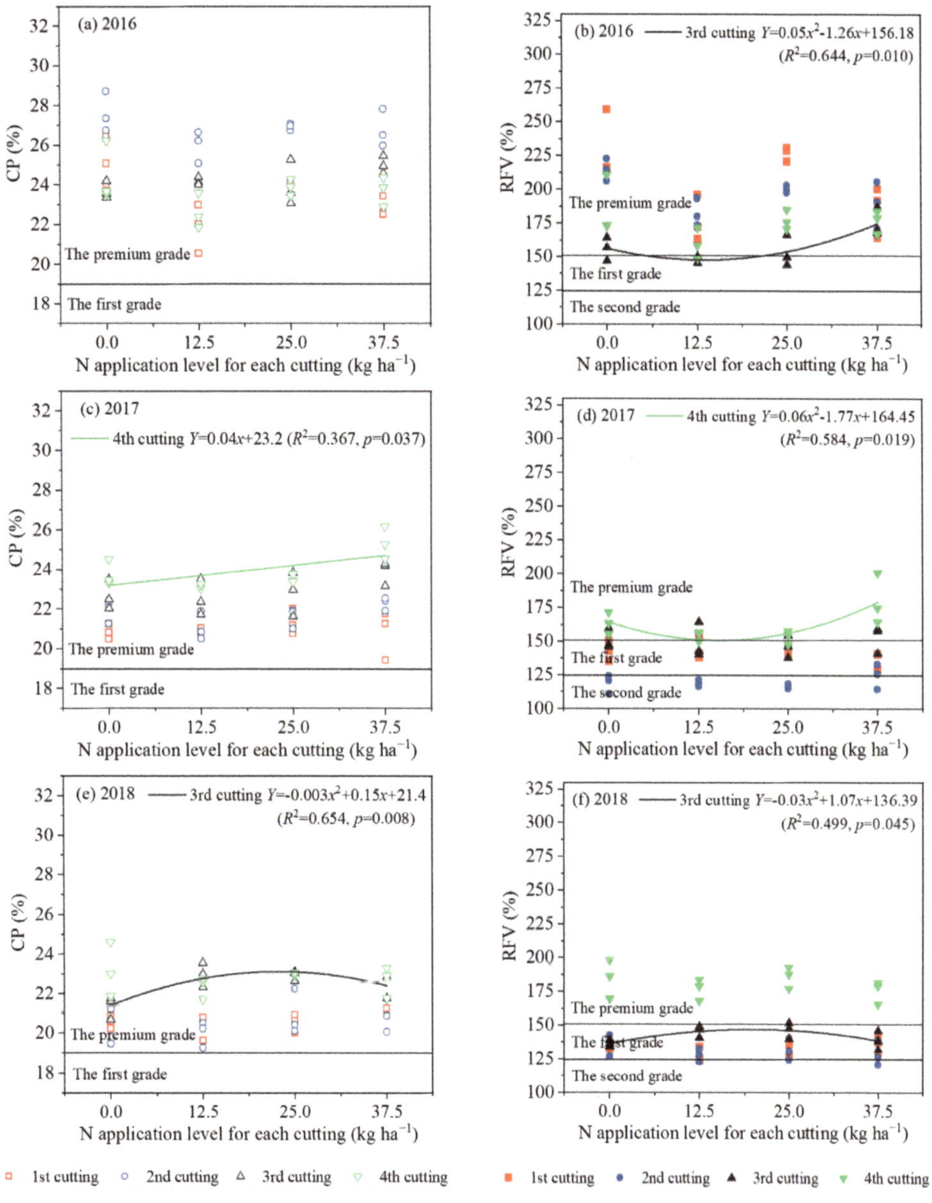

Figure 7. The crude protein (CP) (**a**,**c**,**e**) and the relative feed value (RFV) (**b**,**d**,**f**) of alfalfa under different N application levels in 2016 (**a**,**b**), 2017 (**c**,**d**) and 2018 (**e**,**f**). Linear or quadratic regression curves and equations indicated that the relationship of CP and RFV with N application level was significant.

4. Discussion

Changes in soil nitrogen content mainly appeared at a depth of 0–40 cm after N application. Soil NO_3^--N content at 40–80 cm only increased under N2 and N3 treatments during the rainy season (from June to September). The N applied using sprinkler fertigation did not result in serious N leaching. Similar to our results, many researchers have indicated that soil NO_3^--N content was higher at a depth of 0–40 cm, and N leaching could be reduced under sprinkler fertigation [43,44]. Meanwhile, N leaching could be increased through N application level [45,46], thus soil NO_3^--N content at the deep soil layer was improved under a higher N application level, and no significant N leaching occurred due to the comparatively lower N application level (12.5–37.5 kg N ha^{-1}) in our study. The increment of soil NO_3^--N accumulation was much higher than the N application amount, which might be due to the N input from N mineralization and N_2 fixation [47]. The gross N immobilization level could be 18–119 mg kg^{-1} per day in drying–rewetting soils under sprinkler irrigation [48,49]. The amount of N_2 fixation for alfalfa fields could reach the value of 200–500 kg N ha^{-1} yr^{-1} [50,51]. In addition, soil NO_3^--N accumulated amount for the soil layer of 0–80 cm was increased under all treatments, and the increment reached the highest value under the N0 treatment after the third year of alfalfa production. This indicates that planting alfalfa could improve soil nitrogen content due to N_2 fixation. During the three years, N uptake by alfalfa shoot ranged from 290 to 497 kg ha^{-1} yr^{-1} under different N application levels, which was similar to the results reported by Wu et al. [52]. In the first year after alfalfa was established, the total N uptake amount of alfalfa significantly increased with N application level due to the weak roots and lower ability to fix N, which was consistent with the results reported by Fan et al. [10]. However, in contrast to their results [10], N application level had no significant influence on N uptake for alfalfa established for two and three years in this study. This might be because no irrigation was applied in their experiments, but irrigation was applied based on the soil moisture in our study. Suitable water application with sprinkler irrigation could improve N_2 fixation and mineralization [47], which could satisfy N uptake and impair the effect of N application for older alfalfa.

The effects of N application on the yield and quality of alfalfa were complex and depended on soil conditions, stand ages and weather conditions [53]. In this study, the relationship between alfalfa yield and N application level could be mostly quantified through quadratic regression. Yield, in the first year, responded strongly to the N application. Compared to the yield under N0 treatment, N application could significantly improve alfalfa yield at the first and second cuttings, with the highest proportion of 44.9% and 75.3% under an N application level of 37.5 kg ha^{-1}, respectively, but only 8.7–19.0% at the third and fourth cuttings in the first year after planting. Similar to our study, results reported by Elgharably and Benes [7] also showed that the increase of yield through N application at the first cutting was higher than that at the third cutting. This might be because the lower nodule number restricted the N_2 fixation and total N derived from the atmosphere [7]. In the second and third years, the yield improved through nitrogen application, which was less than 15.9% of that under N0 treatment at the first and second cuttings but could reach approximately 30% at the third and fourth cuttings. The first two cuttings occurred in the spring and early summer in the North China Plain, where abundant nodules in two-year-old and three-year-old alfalfa, along with suitable temperature and soil moisture, likely enhanced N_2 fixation. However, the period of the third cutting entered the summer with hot and rainy weather, and the fourth cutting occurred during the autumn with low temperatures. Soil temperatures that are too high or too low and waterlogging can reduce N_2 fixation [5,54]. Consequently, the advantage of N application in improving yield was greater at the third and fourth cuttings of two-year- and three-year-old alfalfa in this study. It indicated that N application could improve alfalfa yield when the N_2 fixation was limited. N application might also be recommended for the legume in cold regions, where N_2 fixation could be restricted by cold stress [55,56]. The annual yield of alfalfa for the three years increased with N application compared to yields without N application. This aligns with findings from other studies, indicating that nitrogen application can enhance alfalfa yield during its initial establishment period [4,10]. Although many studies have indicated that N application is not recommended for old alfalfa stands [18,20], applying a small amount of N fertilizer to two- or three-year-old alfalfa stands could increase annual yield due to poor soil fertility in the experimental site. However, N application in the third cutting generally did not increase soil NO_3^- -N content and N leaching loss owing to heavy rain. Meanwhile, the improvement of yield through N application was relatively small at the fourth cutting, suggesting that N application might not be economical. Therefore, N application was not recommended for alfalfa at the third and fourth cuttings in the second and third years despite the active effect on alfalfa yield. The influence of N application level on N use efficiency was significant only in some cuttings in the three years, and N use efficiency generally decreased with N application level. The relationship between ANR/NAE and N application level could be quantified using a linear regression. This was consistent with the conclusions of other researchers [16]. The significantly higher ANR and NAE were only obtained with a nitrogen application level of 12.5 kg N ha^{-1} in the first cutting of the first and second years and the third cutting of the first and third years. Different from the yield, the negative effect of N application on the CP content and RFV of alfalfa appeared at the first and second cuttings of the first year after establishment, and CP and RFV under treatment without N application were generally up to the highest value. It accorded with the result proposed by Feng et al. [30] that alfalfa yield is inversely proportional to the quality. This might be because higher N application during the early growth stage of alfalfa could accelerate cell wall components and fiber quantities, resulting in a reduction in the CP and RFV of alfalfa [57,58]. In the second and third years, the N application improved the CP or RFV during the last two cuttings. The relationship between CP/RFV and N application level could be mostly quantified using quadratic regression. However, N application had an insignificant influence on alfalfa quality classification, which might be due to the fact that the alfalfa variety planted in this study had the advantages of large leaves, slim stems and good palatability [59]. Consequently, a nitrogen application level of 12.5 kg N ha^{-1} is recommended for alfalfa at first cutting in the first year to obtain a higher yield and N use efficiency. Meanwhile, due to the insignificant difference in N use efficiency under the three N application levels and the higher yield and good quality noticed under the N3 treatment, a nitrogen application level of 37.5 kg N ha^{-1} is recommended for alfalfa of the second cutting in the first year. For the two-year-old and three-year-old alfalfa, nitrogen application is not recommended for any cutting, considering economic and environmental factors.

5. Conclusions

After nitrogen fertilizer was applied through sprinkler fertigation, the variation in soil NO_3^--N content was mainly concentrated around the soil depth of 0–40 cm. Fluctuation in soil NO_3^--N content at a depth of 40–80 cm only appeared with an N application level of 25.0–37.5 kg N ha^{-1} during the rainy season. After three years of planting alfalfa, soil NO_3^--N content could be increased. During the three years, N uptake by alfalfa shoot for each year ranged from 290 to 497 kg ha^{-1}. N application only had a noteworthy positive influence on the N uptake amount of one-year-old alfalfa. N application could improve the annual yield of alfalfa during the three years, especially in the first year. Compared to the yield under the N0 treatment, the N application significantly improved the alfalfa yield at the first and second cuttings in the first year after establishment. In the second and third years, the N application level had a significant relationship with the yield at the third and fourth cuttings, and the yield could be increased through N application. The N use efficiency of alfalfa at the first and third cuttings showed significant linear and negative correlations with the N application level and reached the highest under an N application level of 12.5 kg N ha^{-1} during the three years. Mostly, the N application level had no remarkable relationship with the N use efficiency during the other cuttings. Compared to alfalfa quality under N0 treatment, N application had negatively affected the alfalfa quality in the first year but had a positive influence during the second and third years. However, the N application had minimal effect on the quality classification of alfalfa during the three years. Overall, the response to N application was strongest for alfalfa yield, followed by ANR and NAE, as well as quality. For obtaining the high yield, quality and N use efficiency of alfalfa planted in the North China Plain with sub-humid climate, N application levels of 12.5 and 37.5 kg N ha^{-1} are recommended for alfalfa of the first and second cuttings, respectively, and alfalfa at the third and fourth cuttings does not require N application in the first year after establishment. N application is not recommended for alfalfa in the second and third year after establishment.

Author Contributions: During this study, H.Y. and Y.W. conceived and designed the experiments; Y.W. and M.L. performed experiments; Y.W., M.L. and J.G. analyzed the data; Y.W., M.L. and H.Y. wrote and revised the manuscript. Overall, Y.W. and M.L. equally contributed to this paper and should be regarded as co-first authors. All authors have read and agreed to the published version of the manuscript.

Funding: This study was financially supported by the National Key R&D Program of China (2022YFD1300804), the China Agriculture Research System of MOF and MARA (CARS-34), and the National Natural Science Foundation of China (51939005).

Data Availability Statement: All data generated or analyzed during this study contained within this article and are available upon reasonable request.

Conflicts of Interest: The authors declare that they have no known competing financial interests or personal relationships that could have appeared to influence the work reported in this paper.

References

1. National Animal Husbandry Station of China. *Forage Statistics in China*; China Agriculture Press: Beijing, China, 2021; p. 70.
2. Darrell, A.M. *Forage Crops*; McGRAW-HILL Book Company: New York, NY, USA, 1984; p. 172.
3. Ntuall, W.F. Effect of N, P, and S fertilizers on alfalfa grown on three soil types in northeastern Saskatchewan. II. Nitrogen, P and S uptake and concentration in herbage. *Agron. J.* **1985**, *77*, 224–228. [CrossRef]
4. Hannaway, D.B.; Shuler, P.E. Nitrogen Fertilization in Alfalfa Production. *J. Prod. Agric.* **1993**, *6*, 80–85. [CrossRef]
5. Liu, Y.; Wu, L.; Baddeley, J.A.; Watson, C.A. Models of Biological Nitrogen Fixation of Legumes. A Review. *Agron. Sustain. Dev.* **2011**, *31*, 155–172. [CrossRef]
6. FAO. *Legume Inoculants and Their Use*; FAO Fertilizer and Plant Nutrition Service; FAO: Rome, Italy, 1984.
7. Elgharably, A.; Benes, S. Alfalfa biomass yield and nitrogen fixation in response to applied mineral nitrogen under saline soil conditions. *J. Soil Sci. Plant Nut.* **2021**, *21*, 744–755. [CrossRef]
8. Helalia, A.M.; Al-Tapir, O.A.; Al-Nabulsi, Y.A. The influence of irrigation water salinity and fertilizer management on the yield of Alfalfa (*Medicago sativa* L.). *Agric. Water Manag.* **1996**, *31*, 105–114. [CrossRef]
9. Bélanger, G.; Richards, J.E. Dynamics of biomass and N accumulation of alfalfa under three N fertilization rates. *Plant Soil* **2000**, *219*, 177–185. [CrossRef]
10. Fan, J.; Du, Y.; Wang, B.; Turner, N.C.; Wang, T.; Abbott, L.K.; Stefanova, K.; Siddique, K.H.M.; Li, F. Forage yield, soil water depletion, shoot nitrogen and phosphorus uptake and concentration, of young and old stands of alfalfa in response to nitrogen and phosphorus fertilisation in a semiarid environment. *Field Crops Res.* **2016**, *198*, 247–257. [CrossRef]
11. Pucek, T.R.; Pyš, J.B. The nutritive value and yield of alfalfa in relation to nitrogen fertilization on sulphur industry reclaimed lands. *J. Agron. Crops Sci.* **1996**, *177*, 295–303. [CrossRef]
12. Aranjuelo, I.; Molero, G.; Erice, G.; Aldasoro, J.; Arrese-Igor, C.; Nogués, S. Effect of shoot removal on remobilization of carbon and nitrogen during regrowth of nitrogen-fixing alfalfa. *Physiol. Plant* **2015**, *153*, 91–104. [CrossRef] [PubMed]
13. Gao, L.; Su, J.; Tian, Q.; Shen, Y. Contrasting strategies of nitrogen absorption and utilization in alfalfa plants under different water stress. *J. Soil Sci. Plant Nut.* **2020**, *20*, 1515–1523. [CrossRef]
14. Raun, W.R.; Johnson, G.V.; Phillips, S.B.; Thomason, W.E.; Dennis, J.L.; Cossey, D.A. Alfalfa yield response to nitrogen applied after each cutting. *Soil. Sci. Soc. Am. J.* **1999**, *63*, 1237–1243. [CrossRef]
15. He, F.; Xie, K.; Li, X. Effect of nitrogen fertilizer and seeding rate on yield of alfalfa and weeds. *Pol. J. Environ. Stud.* **2018**, *27*, 647–653. [CrossRef]
16. Kamran, M.; Yan, Z.; Jia, Q.; Chang, S.; Ahmad, I.; Ghani, M.U.; Hou, F. Irrigation and nitrogen fertilization influence on alfalfa yield, nutritive value, and resource use efficiency in an arid environment. *Field Crops Res.* **2022**, *284*, 108587. [CrossRef]
17. Li, M.; Wang, Y.; Adeli, A.; Yan, H. Effects of application methods and urea rates on ammonia volatilization, yields and fine root biomass of alfalfa. *Field Crops Res.* **2018**, *218*, 115–125. [CrossRef]
18. Oliveira, W.S.D.; Oliveira, P.P.A.; Corsi, M.; Rodrigo, F.; Duarte, S.; Tsai, S.M. Alfalfa yield and quality as function of nitrogen fertilization and symbiosis with *Sinorhizobium meliloti*. *Sci. Agric.* **2004**, *61*, 433–438. [CrossRef]
19. Zhu, Y.; Sheaffer, C.C.; Vance, C.P.; Graham, P.H.; Russelle, M.P.; Montealegre, C.M. Inoculation and nitrogen affect herbage and symbiotic properties of annual medicago species. *Agron. J.* **1998**, *90*, 781–786. [CrossRef]
20. Xie, K.; Li, X.; He, F.; Zhang, Y.; Wan, L.; David, B.H.; Wang, D.; Qin, Y.; Gamal, M.A.F. Effect of nitrogen fertilization on yield, N content, and nitrogen fixation of alfalfa and smooth bromegrass grown alone or in mixture in greenhouse pots. *J. Integr. Agric.* **2015**, *14*, 1864–1876. [CrossRef]
21. Wivstad, M.; Mårtensson, A.M.; Ljunggren, H.D. Field measurement of symbiotic nitrogen fixation in an established lucerne ley using 15N and an acetylene reduction method. *Plant Soil* **1987**, *97*, 93–104. [CrossRef]

22. Zhao, J.; Huang, R.; Wang, X.; Ma, C.; Li, M.; Zhang, Q. Effects of combined nitrogen and phosphorus application on protein fractions and nonstructural carbohydrate of alfalfa. *Front. Plant Sci.* **2023**, *14*, 1124664. [CrossRef] [PubMed]
23. Chase, L.E.; Long, T.A.; Washko, J.B.; Baumgardt, B.R. Effect of nitrogen fertilization on constituents of alfalfa. *J. Dairy. Sci.* **1976**, *59*, 170–174. [CrossRef]
24. Cherney, D.J.R.; Cherney, J.H.; Siciliano-Jones, J. Alfalfa composition and in sacco fiber and protein disappearance as influenced by nitrogen application. *J. Appl. Anim. Res.* **1995**, *8*, 105–120. [CrossRef]
25. Berça, A.S.; Cardoso, A.D.S.; Longhini, V.Z.; Tedeschi, L.O.; Boddey, R.M.; Reis, R.A.; Ruggieri, A.C. Protein and carbohydrate fractions in warm-season pastures: Effects of nitrogen management strategies. *Agronomy* **2021**, *11*, 847. [CrossRef]
26. Vanderwalt, J.G. Nitrogen-metabolism of the ruminant liver. *Aust. J. Agric. Res.* **1993**, *44*, 381–403. [CrossRef]
27. Lapierre, H.; Lobley, G.E. Nitrogen recycling in the ruminant: A review. *J. Dairy Sci.* **2001**, *84*, 223–236. [CrossRef]
28. Justes, E.; Thiebeau, P.; Avice, J.C.; Lemaire, G.; Volenec, J.J.; Ourry, A. Influence of summer sowing dates, N fertilization and irrigation on autumn VSP accumulation and dynamics of spring regrowth in alfalfa (*Medicago sativa* L.). *J. Exp. Bot.* **2002**, *53*, 111–121. [CrossRef]
29. Rohweder, D.A.; Barnes, R.F.; Jorgensen, N. Proposed hay grading standards based on laboratory analyses for evaluating quality. *J. Anim. Sci.* **1978**, *47*, 747–759. [CrossRef]
30. Feng, Y.; Shi, Y.; Zhao, M.; Shen, H.; Xu, L.; Luo, Y.; Liu, Y.; Xing, A.; Kang, J.; Jing, H.; et al. Yield and quality properties of alfalfa (*Medicago Sativa* L.) and their influencing factors in China. *Eur. J. Agron.* **2022**, *141*, 126637. [CrossRef]
31. Chen, B.; Ren, C.; Wang, C.; Duan, J.; Reis, S.; Gu, B. Driving forces of nitrogen use efficiency in Chinese croplands on county scale. *Environ. Pollut.* **2023**, *316*, 120610. [CrossRef]
32. Zhang, X.; Davidson, E.A.; Mauzerall, D.L.; Searchinger, T.D.; Dumas, P.; Shen, Y. Managing nitrogen for sustainable development. *Nature* **2015**, *528*, 51–59. [CrossRef]
33. Li, Y.; Su, D. Alfalfa water use and yield under different sprinkler irrigation regimes in north arid regions of China. *Sustainability* **2017**, *9*, 1380. [CrossRef]
34. Gascho, G.J.; Hook, J.E.; Mitchell, G.A. Sprinkler-applied and side-dressed nitrogen for irrigated corn grown on sand. *Agron. J.* **1984**, *76*, 77–81. [CrossRef]
35. O'Shaughnessy, S.A.; Evett, S.R.; Andrade, M.A.; Workneh, F.; Price, J.A.; Rush, C.M. Site-specific variable-rate irrigation as a means to enhance water use efficiency. *Trans. ASABE* **2016**, *59*, 239–249. [CrossRef]
36. Schepers, J.S.; Varvel, G.E.; Watts, D.G. Nitrogen and water management strategies to reduce nitrate leaching under irrigated maize. *J. Contam. Hydrol.* **1995**, *20*, 227–239. [CrossRef]
37. Walter, C.B. Impact of residual soil nitrate on in-season nitrogen applications to irrigated corn based on remotely sensed assessments of crop nitrogen status. *Precis. Agric.* **2005**, *6*, 509–519. [CrossRef]
38. Stone, K.C.; Camp, C.R.; Sadler, E.J. Corn yield response to nitrogen fertilizer and irrigation in the Southeastern Coastal Plain. *Appl. Eng. Agric.* **2010**, *26*, 429–438. [CrossRef]
39. Yan, H.; Ma, J.; Wang, Z. Design and field experiment on fertilizer injection device in center pivot irrigation system. *Trans. Chin. Soc. Agric. Mach.* **2015**, *46*, 100–106, (In Chinese with English Abstract).
40. Allen, R.G.; Pereira, L.S.; Raes, D.; Smith, M. Crop evapotranspiration: Guidelines for computing crop water requirements. In *Irrigation and Drainage Paper No. 56*; United Nations FAO: Rome, Italy, 1998; pp. 24–28.
41. Congreves, K.A.; Otchere, O.; Ferland, D.; Farzadfar, S.; Williams, S.; Arcand, M.M. Nitrogen use efficiency definitions of today and tomorrow. *Front. Plant Sci.* **2021**, *12*, 637108. [CrossRef]
42. Shenk, J.S.; Westerhaus, M.O. The application of near infrared reflectance spectroscopy (NIRS) to forage analysis. In *Forage Quality, Evaluation, and Utilization*; Fahey, G.C., Jr., Ed.; JohnWiley & Sons: Hoboken, NJ, USA, 1994; pp. 406–449. [CrossRef]
43. Yang, W.; Jiao, Y.; Yang, M.; Wen, H.; Gu, P.; Yang, J.; Liu, L.; Yu, J. Minimizing soil nitrogen leaching by changing furrow irrigation into sprinkler fertigation in potato fields in the Northwestern China Plain. *Water* **2020**, *12*, 2229. [CrossRef]
44. Lv, G.; Kang, Y.; Li, L.; Liu, S. Nutrient distribution, growth, and water use efficiency in maize following winter wheat irrigated by sprinklers or surface irrigation. *Irrig. Drain.* **2011**, *60*, 338–347. [CrossRef]
45. Li, Y.; Huang, G.; Chen, Z.; Xiong, Y.; Huang, Q.; Xu, X.; Huo, Z. Effects of irrigation and fertilization on grain yield, water and nitrogen dynamics and their use efficiency of spring wheat farmland in an arid agricultural watershed of Northwest China. *Agric. Water Manag.* **2022**, *260*, 107277. [CrossRef]
46. Liu, X.; Ju, X.; Zhang, F.; Pan, J.; Christie, P. Nitrogen dynamics and budgets in a winter wheat–maize cropping system in the North China Plain. *Field Crops Res.* **2003**, *83*, 111–124. [CrossRef]
47. Barakat, M.; Cheviron, B.; Angulo-Jaramillo, R. Influence of the irrigation technique and strategies on the nitrogen cycle and budget: A review. *Agric. Water. Manag.* **2016**, *178*, 225–238. [CrossRef]
48. Bengtsson, G.; Bengtson, P.; Månsson, K.F. Gross nitrogen mineralization-, immobilization-, and nitrification rates as a function of soil C/N ratio andmicrobial activity. *Soil Biol. Biochem.* **2003**, *35*, 143–154. [CrossRef]
49. Pulleman, M.; Tietema, A. Microbial C and N transformations during drying and rewetting of coniferous forest floor material. *Soil Biol. Biochem.* **1999**, *31*, 275–285. [CrossRef]
50. Issah, G.; Schoenau, J.J.; Lardner, H.A.; Knight, J.D. Nitrogen fixation and resource partitioning in alfalfa (*Medicago sativa* L.), cicer milkvetch (*Astragalus cicer* L.) and sainfoin (*Onobrychis viciifolia* Scop.) using ^{15}N enrichment under controlled environment conditions. *Agronomy* **2020**, *10*, 1438. [CrossRef]

51. Walley, F.L.; Tomm, G.O.; Matus, A.; Slinkard, A.E.; van Kessel, C. Allocation and cycling of nitrogen in an alfalfa-bromegrass sward. *Agron. J.* **1996**, *88*, 834–843. [CrossRef]
52. Wu, W.; Liu, M.; Wu, X.; Wang, Z.; Yang, H. Effects of deficit irrigation on nitrogen uptake and soil mineral nitrogen in alfalfa grasslands of the inland arid area of China. *Agric. Water Manag.* **2022**, *269*, 107724. [CrossRef]
53. Wang, L.; Xie, J.; Luo, Z.; Niu, Y.; Coulter, J.A.; Zhang, R.; Li, L. Forage yield, water use efficiency, and soil fertility response to alfalfa growing age in the semiarid Loess Plateau of China. *Agric. Water Manag.* **2021**, *243*, 106415. [CrossRef]
54. Havelka, U.D.; Boyle, M.G.; Hardy, R.W.F. Biological nitrogen fixation. In *Nitrogen in Agricultural Soils*; Stevenson, F.J., Ed.; ASA: Madison, WI, USA, 1982; pp. 365–422.
55. Thurston, C.L.; Grossman, J.M.; Fudge, R.; Maul, J.E.; Mirsky, S.; Wiering, N. Cold stress reduces nodulation and symbiotic nitrogen fixation in winter annual legume cover crops. *Plant Soil* **2022**, *481*, 661–676. [CrossRef]
56. Alexandre, A.; Oliveira, S. Response to temperature stress in rhizobia. *Crit. Rev. Microbiol.* **2012**, *39*, 219–228. [CrossRef] [PubMed]
57. Liu, M.; Wang, Z.; Mu, L.; Xu, R.; Yang, H. Effect of regulated deficit irrigation on alfalfa performance under two irrigation systems in the inland arid area of midwestern China. *Agric. Water Manag.* **2021**, *248*, 106764. [CrossRef]
58. Islam, M.R.; Garcia, S.C. Effects of sowing date and nitrogen fertilizer on forage yield, nitrogen- and water-use efficiency and nutritive value of an annual triple-crop complementary forage rotation. *Grass Forage Sci.* **2012**, *67*, 96–110. [CrossRef]
59. Li, Y.; Wu, F.; Shi, S.; Bai, X. Evaluation on production and nutritional value of 13 introduced alfalfa cultivars in Hexi corridor of Gansu Province. *Agric. Res. Arid Areas* **2019**, *37*, 119–129. (In Chinese with English Abstract)

agronomy

Article

Desertification Reversal Promotes the Complexity of Plant Community by Increasing Plant Species Diversity of Each Plant Functional Type

Kaiyang Qiu [1,2], Zhigang Li [1,*], Yingzhong Xie [1,*], Dongmei Xu [1], Chen He [3] and Richard Pott [2]

[1] College of Forestry and Prataculture, Ningxia University, Yinchuan 750021, China; kaiyangqiu@nxu.edu.cn (K.Q.)
[2] Institute of Geobotany, Leibniz Universität Hannover, 30167 Hannover, Germany
[3] College of Horticulture and Plant Protection, Henan University of Science and Technology, Luoyang 471000, China
* Correspondence: lizhigang@nxu.edu.cn (Z.L.); xieyz@nxu.edu.cn (Y.X.); Tel.: +86-9512061860 (Z.L. & Y.X.); Fax: +86-9512061697 (Z.L. & Y.X.)

Abstract: Desertification reversal is globally significant for the sustainable development of land resources. However, the mechanisms of desertification reversal at the level of plant community are still unclear. We hypothesized that desertification reversal has clear effects on plant community composition, plant functional types (PFTs), and other vegetation characteristics, including plant diversity and biomass, and their changes in the early stages of reversal are more dramatic than in later stages. We investigated the vegetation of four to five different stages of desertification reversal at each of seven large study sites in southwestern Mu Us Sandy Land, China. The results show that the dominant species in very severe desertification areas were replaced by perennial grasses in potential desertification areas. The importance values of annual forbs and perennial sub-shrubs decreased dramatically (from 42.59 and 32.98 to 22.13 and 5.54, respectively), whereas those of perennial grasses and perennial forbs increased prominently (from 13.26 and 2.71 to 53.94 and 11.79, respectively) with the reversal of desertification. Desertification reversal increased the complexity of plant community composition by increasing plant species in each PFT, and C_3 plants replaced C_4 plants to become the dominant PFT with reversal. Plant species richness and species diversity rose overall, and aboveground plant biomass significantly ($p < 0.05$) increased with the reversal of desertification. Most vegetation characteristics changed more strikingly in the early stages of desertification reversal than in later stages. Our results indicate that the type and composition of the plant community were dramatically affected by desertification reversal. Anthropogenic measures are more applicable to being employed in early stages than in later stages, and Amaranthaceae C_4 plants are suggested to be planted in mobile dunes for the acceleration of desertification reversal. This study is useful for designing strategies of land management and ecological restoration in arid and semiarid regions.

Keywords: C_4 plant; ecosystem restoration; plant functional type; restoration strategy; species importance value; species diversity

Citation: Qiu, K.; Li, Z.; Xie, Y.; Xu, D.; He, C.; Pott, R. Desertification Reversal Promotes the Complexity of Plant Community by Increasing Plant Species Diversity of Each Plant Functional Type. *Agronomy* **2024**, *14*, 96. https://doi.org/10.3390/agronomy14010096

Academic Editor: Shicheng Li

Received: 30 September 2023
Revised: 21 December 2023
Accepted: 27 December 2023
Published: 30 December 2023

1. Introduction

Desertification is a global phenomenon that affects a vast area and a substantial proportion of the earth's population [1–3], causing many problems such as land degradation and the reduction of ecosystem services [4]. Desertification is becoming a serious social–economic and environmental issue worldwide [5,6]. Desertification is, therefore, a matter that needs a global attention and is an important topic for intensive research. On the other hand, because of the reduction in anthropogenic pressure and the improvement in land use and management techniques, desertification in some regions is becoming actively or passively reversed. Desertification reversal is becoming an important field of sustainable development. The science of desertification reversal, especially the process and mechanism of the reversal, is helpful for ecological restoration in arid and semiarid regions. As one of the most prominent elements in terrestrial ecosystems, vegetation plays an important role in the process of desertification reversal. We selected the following variables as vegetation characteristics in the present study: plant community composition, plant functional types, plant biomass, plant species richness, and diversity.

Plant community composition plays a critical role in ecosystem functioning [7,8]. A plant functional type (PFT) is a group of plants that show similar responses to environmental circumstances and exert similar influence on the main ecosystem processes [9–11]. Revealing the changes in PFT is thus beneficial for understanding the process of desertification reversal and understanding useful implications for land management. Plant biomass is the best index to indicate the magnitude of the productivity in a grassland ecosystem and directly shows the quality of the structure and the function of the ecosystem. Biodiversity is another essential characteristic of vegetation. Plant species richness and diversity are indices for plant biodiversity. Previous studies have proposed that desertification contributes significantly to biodiversity loss [3,12], which is alleviated by combating desertification. Plant community composition, PFT, plant biomass, and biodiversity are consequently the prominent indices for the description of an ecosystem undergoing the reversal of desertification. Past studies have shown a close link between desertification reversal and plant responses. A large number of studies have documented that vegetation restoration can effectively promote the reversal process of desertification. In semi-arid regions, the involvement of shrubs increased plant richness, the biomass of fungi, actinomycetes, and other bacteria, and improved soil fertility and the N mineralization rate [13]. Shrubs promote biological soil crusts and have a positive impact on soil quality [14]. Plants can also improve soil properties and promote soil bacterial diversity by changing the soil microbial community structure [15,16]. Another study showed that plant cover in desert ecosystems can be significantly affected by environmental factors like soil nutrients [17]. However, there is relatively less research on plant community responses to desertification reversal at the level of plant functional type and plant diversity and their ecological implications. A study on this topic can contribute to understanding the process and management of ecosystem restoration in desertified grassland.

Plant community structure can be significantly changed by changes in environmental factors [17]. The measurement of changes in plant community composition serves as a good chance to monitor the land degradation [18]. Plant species have different responses to changes in the habitat as a result of plant community succession, and this leads to changes in plant species diversity and community composition [19,20]. But little attention has been paid to plant community succession during the reversal of sandy desertification. [21] found that plant species richness increased significantly under the effect of the hydraulic redistribution of *Populus euphratica*. But how plant diversity changes in a broader ecological context of desertification reversal is still not clear. Plant species diversity may develop more quickly in earlier stages during vegetation restoration [22]. How the magnitude of vegetation development differentiates between different restoration stages is still poorly understood. On the other hand, to expose the differences in the magnitude of vegetation restoration between different stages is useful for understanding the pattern and regularity of ecological restoration and is helpful for grasping and controlling vegetation succession in the process of restoration, which can in further serve land management in degenerated ecosystems.

To solve the problems mentioned above and better understand the process and mechanisms of the reversal of sandy desertification at the plant community level, we hypothesized a clear effect of desertification reversal on plant community composition, functional types, and other vegetation characteristics, including plant diversity and biomass; we also hypothesized that these vegetation characteristics develop more dramatically in the earlier stages of reversal than in the later stages. We carried out this study in different locations in Yanchi County, southern Mu Us Sandy Land, which is one of the largest anthropogenic deserts in China. The desertification reversal in the present study is a process of natural restoration without seedlings due to human measures such as livestock removal. This study will provide a scientific basis for vegetation development and land management in the process of desertification reversal.

2. Materials and Methods

2.1. Study Area

The study was conducted in Yanchi County (106°30′–107°47′ E, 37°04′–38°10′ N) in Ningxia Autonomous Region in northern China. Yanchi County is located in the southwestern part of Mu Us Sandy Land. The natural environment in this area is unfavorable, with strong wind, much sand, and a drought climate. The climate in this area belongs to the mid-temperate continental climate [23] and is profoundly influenced by dry and cold air from Mongolia and Siberia in the wintertime. The annual mean temperature is 8.1 °C [24]. The average temperature of the hottest month (July) is 22.3 °C and that of the coldest month (January) is −8.9 °C. The highest temperature measured was 38.1 °C and the lowest was −29.6 °C. The multi-year average frostless period is 165 days [24] and the annual precipitation is 296.4 mm, which falls mainly between July and October. The vegetation types in Yanchi County include shrubland, grassland, meadow, sandy land (dune), and desert vegetation [25]. The main types are typical steppe and desert steppe [26]. Vegetation in the typical steppe is composed mainly of xerophytic and mid-xerophytic plant species, such as *Agropyron cristatum*, *Stipa grandis*, *Stipa bungeana*, and *Thymus serpyllum* var. *mongolicus* [25]. Desert steppe includes species like *Oxytropis aciphylla*, *Agriophyllum squarrosum*, and *Salsola beticolor*. The main soil types in this area are sierozem and eolian sandy soil [24,27].

Desertification and its reversal have simultaneously taken place in different areas in Yanchi County during the past 50 years [24,28] because of the natural conditions and anthropogenic effects. This region underwent serious desertification between 1961 and 1989. Human activities, including overgrazing, over-cropping, deforestation, and over-excavation of medicinal plants, have played an important role in the progression of desertification in Yanchi County [29]. The desertification in this region started to reverse after 1989 [28] and especially after 2003 due to grazing prohibition [28]. On the basis of the local policy of the prohibition of grazing in the entire region and the national policy of the Grain for Green Project, the vegetation in this area began to recover naturally. Recent studies show that the policy of grazing prohibition played an important role in the reversal of desertification [30].

This region has been undergoing a marvelous desertification reversal in recent decades due to the application of a series of management measures. As these measures may also be significant for land management in other regions, especially for arid regions that are susceptible to land degradation, we summarize the management measures utilized in this region as follows. Firstly, several ecological projects have been implemented in Yanchi County, including the Three North-Shelter Forest Program, the Grain for Green Project, the Grazing Prohibition Program, and the Project for the Amelioration of Degraded Grassland by Re-seeding. These four projects were implemented in this region in 1979, 2001, 2002, and 2003, respectively, and have been significantly contributing to desertification reversal. For example, from 1979 to 2016, this region underwent five phases of the Three North-Shelter Forest Program. During this period, a large number of drought-tolerant shrubs, such as *Caragana korshinskii* and *Corethrodendron scoparium*, were planted on dunes to fix the mobile sand via other management measures such as setting a straw checkerboard barrier to prevent sand movement.

Secondly, the People's Congress of Ningxia has legislated several regulations, such as the grassland management regulations of Ningxia Autonomous Region and the grazing prohibition regulations of Ningxia Autonomous Region. The regulations provide a basis for land management during desertification reversal. Thirdly, the government of Yanchi County has launched many publicity campaigns to increase people's understanding of the significance of grazing prohibition for the restoration of this ecologically vulnerable area. Fourthly, livestock such as cattle and sheep are raised at farmers' homes during the period of grazing prohibition. Large quantities of grasses are planted in specific rangelands to provide plenty of forage for livestock bred at home. Fifthly, people in this region receive extra money from the government each year as compensation for the grazing prohibition. Sixthly, many rangelands are enclosed with a wire fence, especially in severely desertified areas. Seventhly, several penalty measures are employed when the regulations are violated, e.g., anyone who allows grazing during the period of grazing prohibition will have to pay a fine as a penalty.

Lastly, a series of management measures are applied to supervise the implementation of grazing prohibition. Each level of the government, including the village, the town, and the county, are designated relevant responsibilities for management. The government of the county establishes special police and executive teams to patrol the fields regularly for violation of the regulations. People are also encouraged to report to the police when they see an offense, and they will receive money as a reward if the violation is determined to be true. Additionally, the media are invited to participate in the supervision and expose violations.

2.2. Field Design and Laboratory Analyses

Based on a detailed survey of the whole Yanchi County, we selected seven large study sites located in Southern Mu Us Sandy Land. All selected study sites are located in the former typical desertification areas and are in the middle and northern parts of Yanchi County. At each study site, we used the method of space for time and randomly (in the locations) selected four to five sub-sites with different stages of desertification reversal based on a comprehensive criteria, mainly according to the proportion of bare (mobile) sand area to total ground area (SAND, %) and to vegetation cover (COV, %) [24,26,31,32]. SAND is a critical criterion because it relates to several indices, such as wind erosion intensity and aeolian activity. Due to the drought climate and strong wind, the desertification in this region has mainly been caused by wind erosion from the pressure of human activities. Therefore, we took SAND and COV as the main criteria for the classification of reversal stages. Areas with SAND and COV accounting for (1) more than 50% and less than 10%, (2) 30–50% and 10–30%, (3) 10–30% and 30–40%, (4) less than 10% and 40–50%, and (5) 0% and more than 50% of the total ground area, respectively, were classified as reversal stages of (1) very severe desertification (VSD), (2) severe desertification (SD), (3) moderate desertification (MD), (4) light desertification (LD), and (5) potential desertification (PD), respectively [24,32–35]. In addition to SAND and COV, we also took the intensity of wind erosion and aeolian activity [33,36] into account as subsidiary indices for the classification of stages of desertification reversal. There were only four study sites that had potential desertification areas. These sub-sites were undergoing different stages of desertification reversal due to the differences in factors such as restoration measures, micro-topography, and geographic location (whether directly against the prevailing wind direction).

The vegetation investigation was carried out in July and August during the optimal period for plant growth. Three plots were randomly established in each sub-site. At the center of each plot, we set a quadrat of 1 m × 1 m. Around each quadrat, in the range of 5 m × 5 m, we threw an iron circle with a diameter of 0.5 m ten times, and all the plant names enclosed within the circle were recorded each time. The total number of times that one species was enclosed in the circle was taken as the frequency [37]. For each quadrat, the abundance, height, and cover of each plant species was carefully investigated and recorded. The abundance was counted one by one. The height was measured using the highest plant of each species in the quadrat. The cover of each species was measured via the needling method, with 100 points homogeneously distributed in the quadrat [38]. The aboveground parts of all plants in the quadrat were then cut, collected, and weighed immediately.

In the laboratory, the aboveground parts of the plants were dried in an oven at 105 °C for 30 min and then at 80 °C for 24 h until the weight of the plants stopped changing. All of these dried plants were then weighed again after cooling to room temperature. The dried weight was taken as the aboveground plant biomass (hereafter, "plant biomass"). The plants were identified to be C_3 and C_4 species by observing their leaves and by compiling data from 23 published works [39–61]. The C_3 and C_4 plants were identified by measuring the stable carbon isotope ratio ($\delta^{13}C$) of plant tissues in the cited studies [45,60].

2.3. Statistical Analyses

The data were processed with Excel 2021 and SPSS 13.0. The importance value (IV) of a given plant species was calculated with the following formula [21] (Equation (1)):

$$IV = (RA + RH + RC + RF)/4 \times 100 \quad (1)$$

where RA, RH, RC, and RF represent the relative abundance, relative height, relative cover, and relative frequency, respectively. RA is the proportion of the number of individuals of this species to the sum of the individuals of all species in the same quadrat. RH, RC, and RF were calculated in the same way [62]. The formula was developed from the following equation [63] (Equation (2)):

$$DV = (RA + RH + RC)/3 \times 100 \quad (2)$$

where DV is the dominance value; RA, RH, and RC represent the same meanings as above.

The number of plant species in a quadrat was taken as the species richness. Species diversity was calculated with the Shannon–Wiener index [64–67] (Equation (3)):

$$H' = -\sum_{i=1}^{S} P_i \ln P_i \quad (3)$$

where H' is the Shannon–Wiener index, S is the number of plant species, and P_i is the proportion of the number of individuals of species i to the total number of individuals of all plant species (Equation (4)):

$$P_i = n_i/N \quad (4)$$

where n_i is the number of individuals of species i, and N is the total number of individuals of all plant species.

The Kolmogorov–Smirnov test (K-S test) for normality was implemented for each group of data. For data whose distributions were not normal, a logarithmic transformation of the data was conducted and then they were tested again with the K-S test. Variance analyses between different desertification reversal stages were conducted with SPSS 13.0 [68].

3. Results

3.1. Changes in Vegetation Conditions and Quantitative Plant Characteristics

Plant species importance values (IVs) changed dramatically with the reversal of sandy desertification (Table 1). In the very severe desertification (VSD) areas and severe desertification (SD) areas, *Artemisia ordosica* and *Agriophyllum squarrosum* were the dominant species of the plant communities. The total IVs of these two species accounted for 62.2% and 28.6% in VSD and SD areas, respectively. The IV of *A. ordosica* decreased gradually from VSD areas via the middle desertification (MD) areas to potential desertification (PD) areas. *A. squarrosum* disappeared in the MD and PD areas and almost in the light desertification (LD) areas. *A. ordosica* and *A. squarrosum* are pioneer plants for sand fixation in sandy desertification areas.

The IVs of perennial grass species, including *Leymus secalinus*, *Pennisetum flaccidum*, *Cleistogenes squarrosa*, *Agropyron mongolicum*, and *Stipa bungeana*, increased in general with the reversal of desertification (Table 1). The IV of *Artemisia scoparia* increased substantially from 0.41% in VSD to 15.80% in PD. However, the IVs of some other annual species, such as *Salsola beticolor* and *Setaria viridis*, decreased in general with the desertification reversal. It was more common for the IVs to change much more significantly in the early stages (from VSD to MD) of desertification reversal than in the later stages (from MD to PD) (Table 1).

Table 1. Effects of the reversal of sandy desertification on plant species importance values (IVs) in Southern Mu Us Sandy Land, China *.

Species	Life Form	Desertification Reversal Stage				
		VSD	SD	MD	LD	PD
Artemisia ordosica Krasch.	PS	32.89	12.25	14.72	10.54	4.31
Agriophyllum squarrosum (L.) Moq.	AF	29.27	16.35		0.47	
Leymus secalinus (Georgi) Tzvel.	PG	9.52	21.30	4.96	11.85	26.46
Salsola beticolor Iljin	AF	9.42	5.99	1.46	4.30	2.26
Setaria viridis (L.) Beauv.	AG	8.46	5.38	1.26	5.88	2.85
Pennisetum flaccidum Griseb.	PG	3.74	0.69	12.58	4.34	9.26
Bassia scoparia (L.) A.J.Scott	AF	2.85	3.43	0.81	2.29	
Ixeris chinensis subsp. *versicolor* (Fisch. ex Link) Kitam.	PF	2.18	3.68	6.62	9.94	4.11
Panzerina lanata var. *alaschanica* (Kuprian.) H. W. Li	PF	0.53	0.09	0.09		
Artemisia scoparia Waldst. et Kit.	AF	0.41	1.01	19.42	18.98	15.80
Euphorbia humifusa Willd.	AF	0.30	1.32	0.75	0.18	0.82
Chenopodium acuminatum Willd.	AF	0.16		5.21	1.84	0.34
Tribulus terrestris L.	AF	0.11	0.60			
Lespedeza potaninii Vass.	PS	0.09	1.81	1.37	2.32	0.10
Grubovia dasyphylla (Fisch. & C. A. Mey.) Freitag & G. Kadereit	AF	0.07	2.06	0.48	0.46	0.88
Oxytropis racemosa Turcz.	PL		8.06	0.25	0.31	
Sophora alopecuroides L.	PS		6.77	1.65	1.19	1.04
Cleistogenes squarrosa (Trin.) Keng	PG		1.89	5.20	2.13	7.09
Astragalus melilotoides Pall.	PL		1.85	0.98	2.82	0.58
Salix psammophila C.Wang et Ch.Y. Yang	SH		0.96			
Corethrodendron fruticosum var. *mongolicum* (Turczaninow) Turczaninow ex Kitagawa	PS		0.74			
Agropyron mongolicum Keng	PG		1.03	0.66	2.91	8.44
Oxytropis aciphylla Ledeb.	PS		0.69			
Cynanchum thesioides (Freyn) K. Schum	PF		0.54	0.13		
Gueldenstaedtia stenophylla Bge.	PL		0.53	0.72	0.19	
Inula salsoloides (Turcz.) Ostenf.	PF		0.32			
Vincetoxicum mongolicum Maxim.	PF		0.24	0.27		0.34
Olgaea leucophylla (Turcz.) Iljin	PF		0.13			
Euphorbia esula L.	PF		0.13	0.81	0.09	
Polygala tenuifolia Willd.	PF		0.09			
Echinops gmelinii Turcz.	AF		0.06	0.32	0.29	0.59
Chenopodium album L.	AF			4.13	1.14	0.94
Stipa breviflora Griseb.	PG			3.62	4.46	0.98
Thermopsis lanceolata R. Br.	PL			3.57	1.79	0.20
Aster altaicus Willd.	PF			2.25	1.88	4.66
Setaria arenaria Kitag.	AG			1.46	0.08	1.13
Cuscuta chinensis Lam.	AF			1.36		
Glycyrrhiza uralensis Fisch.	PL			1.14	1.74	1.93
Equisetum ramosissimum Desf.	PF			0.94		
Polygonum aviculare L.	AF			0.40		
Stipa bungeana Trin.	PG			0.16		1.71
Plantago asiatica L.	PF			0.12		0.11
Incarvillea sinensis Lam.	AF			0.11	0.19	
Kalidium foliatum (Pall.) Moq.	SH				1.27	
Artemisia sacrorum Ledeb.	PF				1.24	
Limonium aureum (L.) Hill. ex Kuntze	PF				1.00	
Peganum nigellastrum Bge.	PF				0.91	
Typha minima Funk.	PF				0.54	0.88
Sonchus arvensis L.	PF				0.34	
Lactuca tatarica (L.) C. A. Mey.	PF				0.09	
Artemisia dubia Wall. ex Bess. var. *subdigitata* (Mattf.) Y. R. Ling	PF					1.37
Artemisia mongolica Fisch. ex Bess.	PF					0.32
Silene conoidea L.	AF					0.27
Oxybasis glauca (L.) S. Fuentes, Uotila & Borsch	AF					0.23

* Desertification reversal stages: VSD: very severe desertification; SD: severe desertification; MD: moderate desertification; LD: light desertification; PD: potential desertification. Life forms: AF: annual forb (a forb is an herbaceous plant that is not a grass species); AG: annual grass; PF: perennial forb; PG: perennial grass; PL: perennial legume; PS: perennial sub-shrub; SH: shrub. The abbreviations in the following tables have the same meanings.

The number of plant species of annual forbs and perennial grasses increased from SD to MD, and that of perennial forbs and perennial legumes increased abruptly from VSD to SD and then maintained relative stability (Table 2). The IVs of life form groups noticeably changed with the reversal of sandy desertification. The IVs of annual forbs and perennial sub-shrubs decreased dramatically from 42.59% and 32.98% in VSD to 22.13% and 5.54% in PD, respectively. That of perennial grasses and forbs increased prominently from 15.97% in VSD to 65.83% in PD. Perennial grasses became the main life form in PD, with an IV value of 53.94%. In general, the number of plant species and IVs summarized in life forms changed more dramatically in the early stages (from VSD to MD) of desertification reversal than in the later stages (from MD to PD) (Table 2).

Table 2. Effects of the reversal of sandy desertification on plant life forms in Southern Mu Us Sandy Land, China *.

Life Form		Desertification Reversal Stage				
		VSD	SD	MD	LD	PD
AF	S	8	8	11	10	9
	IV	42.59	30.82	34.45	30.14	22.13
AG	S	1	1	2	2	2
	IV	8.46	5.38	2.72	5.96	3.98
PF	S	2	8	8	9	7
	IV	2.71	5.22	11.23	16.03	11.79
PG	S	2	4	6	5	6
	IV	13.26	24.91	27.18	25.69	53.94
PL	S	0	3	5	5	3
	IV	0.00	10.44	6.66	6.85	2.71
PS	S	2	5	3	3	3
	IV	32.98	22.26	17.74	14.05	5.54
SH	S	0	1	0	1	0
	IV	0.00	0.96	0.00	1.27	0.00

* S: the number of plant species. The other abbreviations are the same as in Table 1.

3.2. *Effects of Desertification Reversal on Plant Functional Types*

Six plant functional types were classified, including C_3 and C_4 plants, forbs, grasses, legumes, and sub-/shrubs (Table 3). We identified 35 species of C_3 plants and 17 species of C_4 plants. Amaranthaceae comprised the largest numbers of C_4 species (eight species), followed by Poaceae (with four C_4 species) (Table 3). Desertification reversal had clear effects on the composition of plant functional types (PFTs). In the early stages of reversal, PFTs were mainly forbs and sub-shrubs; in the middle and later stages, other PFTs such as grasses were included (Table 3). In VSD, C_4 plants accounted for 66.67% of the total number of plant species, and their importance value (IV) was much higher than that of C_3 plants (Table 3). Most of the C_4 plants found in VSD were also distributed in other stages, and only 4 of 17 C_4 species occurred just in one reversal stage, such as *Polygala tenuifolia* in SD (Table 3), suggesting an extensive range of adaptation in C_4 plants. The IV of C_3 plants increased steadily from 12.20 to 68.52 in the process of desertification reversal (Table 4). Although the percentages of the number of plant species (S%) in the later stages (from MD to PD) of desertification reversal remained relatively steady for legumes, grasses, and forbs, the IV of legumes decreased from 20.45 to 3.85 from VSD to PD, and that of grasses increased from 21.72 to 57.92 in the whole process of reversal (Table 4). On the other hand, most PFTs changed more dramatically in the early stages (from VSD to MD) in both S% and IVs than in the later stages, except for grasses in S% and IV and forbs in IV (Table 4).

Table 3. Effects of the reversal of sandy desertification on plant composition and the detailed information of plant functional types in Southern Mu Us Sandy Land, China *.

Desertification Reversal Stage		Species	Family	C_3/C_4	PFTs
VSD and SD	Early stages	*Tribulus terrestris* L.	Zygophyllaceae	C_4	Forb
SD		*Salix psammophila* C.Wang et Ch.Y. Yang	Salicaceae	C_3	Shrub
SD		*Corethrodendron fruticosum* var. *mongolicum* (Turczaninow) Turczaninow ex Kitagawa	Fabaceae	C_3	Legume/ sub-shrub
SD		*Oxytropis aciphylla* Ledeb.	Fabaceae	C_3	Legume/ sub-shrub
SD		*Inula salsoloides* (Turcz.) Ostenf.	Asteraceae	C_3	Forb
SD		*Olgaea leucophylla* (Turcz.) Iljin	Asteraceae	C_3	Forb
SD		*Polygala tenuifolia* Willd.	Polygalaceae	C_4	Forb
SD MD		*Cynanchum thesioides* (Freyn) K. Schum	Asclepiadaceae	C_3	Forb
VSD SD MD		*Panzerina lanata* var. *alaschanica* (Kuprian.) H. W. Li	Labiatae	C_3	Forb
VSD SD LD		*Agriophyllum squarrosum* (L.) Moq.	Amaranthaceae	C_4	Forb
MD	Middle stages	*Cuscuta chinensis* Lam.	Convolvulaceae	C_4	Forb
MD		*Equisetum ramosissimum* Desf.	Equisetaceae	C_3	Forb
MD		*Polygonum aviculare* L.	Polygonaceae	C_3	Forb
SD MD LD		*Oxytropis racemosa* Turcz.	Fabaceae	C_3	Legume
SD MD LD		*Gueldenstaedtia stenophylla* Bge.	Fabaceae	C_3	Legume
SD MD LD		*Euphorbia esula* L.	Euphorbiaceae	C_3	Forb
MD LD	Later stages	*Incarvillea sinensis* Lam.	Bignoniaceae	C_3	Forb
LD		*Kalidium foliatum* (Pall.) Moq.	Amaranthaceae	C_4	Shrub
LD		*Artemisia sacrorum* Ledeb.	Asteraceae	C_3	Forb
LD		*Limonium aureum* (L.) Hill. ex Kuntze	Plumbaginaceae	C_3	Forb
LD		*Peganum nigellastrum* Bge.	Zygophyllaceae	C_3	Forb
LD		*Sonchus arvensis* L.	Asteraceae	C_3	Forb
LD		*Lactuca tatarica* (L.) C. A. Mey.	Asteraceae	C_3	Forb
LD PD		*Typha minima* Funk.	Typhaceae	C_3	Forb
PD		*Artemisia dubia* Wall. ex Bess. var. *subdigitata* (Mattf.) Y. R. Ling	Asteraceae	—	Forb
PD		*Artemisia mongolica* Fisch. ex Bess.	Asteraceae	C_3	Forb
PD		*Silene conoidea* L.	Caryophyllaceae	C_3	Forb
PD		*Oxybasis glauca* (L.) S. Fuentes, Uotila & Borsch	Amaranthaceae	C_4	Forb
SD MD PD		*Vincetoxicum mongolicum* Maxim.	Asclepiadaceae	C_3	Forb
MD PD		*Stipa bungeana* Trin.	Poaceae	C_3	Grass
MD PD		*Plantago asiatica* L.	Plantaginaceae	C_3	Forb
MD LD PD		*Chenopodium album* L.	Amaranthaceae	C_4	Forb
MD LD PD		*Stipa breviflora* Griseb.	Poaceae	C_3	Grass
MD LD PD		*Thermopsis lanceolata* R. Br.	Fabaceae	C_3	Legume
MD LD PD		*Aster altaicus* Willd.	Asteraceae	C_3	Forb
MD LD PD		*Setaria arenaria* Kitag.	Poaceae	C_4	Grass
MD LD PD		*Glycyrrhiza uralensis* Fisch.	Fabaceae	C_3	Legume
VSD SD MD LD	Common stages	*Bassia scoparia* (L.) A.J.Scott	Amaranthaceae	C_4	Forb
SD MD LD PD		*Sophora alopecuroides* L.	Fabaceae	C_3	Legume/ Sub-shrub
SD MD LD PD		*Cleistogenes squarrosa* (Trin.) Keng	Poaceae	C_4	Grass
SD MD LD PD		*Astragalus melilotoides* Pall.	Fabaceae	C_3	Legume
SD MD LD PD		*Agropyron mongolicum* Keng	Poaceae	C_3	Grass
SD MD LD PD		*Echinops gmelinii* Turcz.	Asteraceae	C_3	Forb
VSD MD LD PD		*Chenopodium acuminatum* Willd.	Amaranthaceae	C_4	Forb
VSD SD MD LD PD		*Artemisia ordosica* Krasch.	Asteraceae	C_4	Sub-shrub
VSD SD MD LD PD		*Leymus secalinus* (Georgi) Tzvel.	Poaceae	C_3	Grass
VSD SD MD LD PD		*Salsola beticolor* Iljin	Amaranthaceae	C_4	Forb
VSD SD MD LD PD		*Setaria viridis* (L.) Beauv.	Poaceae	C_4	Grass
VSD SD MD LD PD		*Pennisetum flaccidum* Griseb.	Poaceae	C_4	Grass
VSD SD MD LD PD		*Ixeris chinensis* subsp. *versicolor* (Fisch. ex Link) Kitam.	Asteraceae	C_3	Forb
VSD SD MD LD PD		*Artemisia scoparia* Waldst. et Kit.	Asteraceae	C_3	Forb
VSD SD MD LD PD		*Euphorbia humifusa* Willd.	Euphorbiaceae	C_4	Forb
VSD SD MD LD PD		*Lespedeza potaninii* Vass.	Fabaceae	C_3	Legume/ sub-shrub
VSD SD MD LD PD		*Grubovia dasyphylla* (Fisch. & C. A. Mey.) Freitag & G. Kadereit	Amaranthaceae	C_4	Forb

* PTF: plant functional types. The other abbreviations are the same as in Table 1.

Table 4. Effects of the reversal of sandy desertification on plant species richness and importance values summarized in plant functional types in Southern Mu Us Sandy Land, China *.

Plant Functional Type		Desertification Reversal Stage				
		VSD	SD	MD	LD	PD
C_3 plant	S	4	18	22	22	18
	S%	26.67	60.00	62.86	62.86	60.00
	IV	12.20	49.85	50.47	65.07	68.52
C_4 plant	S	10	11	12	13	11
	S%	66.67	36.67	34.29	37.14	36.67
	IV	87.27	50.05	49.42	34.92	30.11
Legume [a]	S	1	7	7	7	5
	S%	6.67	23.33	20.00	20.00	16.67
	IV	0.09	20.45	9.68	10.36	3.85
Grass	S	3	5	8	7	8
	S%	20.00	16.67	22.86	20.00	26.67
	IV	21.72	30.29	29.90	31.65	57.92
Forb	S	10	16	19	19	16
	S%	66.67	53.33	54.29	54.29	53.33
	IV	45.30	36.04	45.68	46.17	33.92
Sub-shrub	S	2	6	3	5	3
	S%	13.33	20.00	8.57	14.29	10.00
	IV	32.98	23.22	17.74	16.51	5.45

* Legumes, including sub-shrubs, that also belong to *Fabaceae*. The included legume sub-shrubs were also included in the plant functional type of sub-shrub.

3.3. Responses of Plant Diversity and Biomass to the Reversal of Sandy Desertification

Species richness varied between 1.0 and 11.3; species diversity changed from 0.0 to 2.01 (Table 5). For each study site, species richness and diversity generally increased with the reversal of sandy desertification. There were significant differences ($p < 0.05$) in variables of species richness and diversity between different desertification reversal stages. In most study sites, plant species richness changed more dramatically in the early stages (from VSD to MD) than in the later stages (from MD to PD) of desertification reversal, whereas plant species diversity changed more obviously in the early stages in around half of all study sites, whereas in other study sites it changed more clearly in the later stages (Table 5). Taking all study sites as a whole, the trends in summary statistics of vegetation characteristics were clearer than those for single study sites. Mean species richness and diversity increased from 3.0 and 0.57 in VSD to 9.0 and 1.22 in PD, respectively, for the whole data set for all study sites.

Table 5. Effects of the reversal of sandy desertification on plant diversity in Southern Mu Us Sandy Land, China *.

Study Site	Reversal Stage	Species Richness	Species Diversity
NHZ	1	2.3 ± 0.6^{a}	0.67 ± 0.04^{a}
	2	7.0 ± 2.0^{b}	1.08 ± 0.51^{ab}
	3	9.3 ± 0.6^{b}	1.05 ± 0.16^{ab}
	4	8.0 ± 2.6^{b}	1.51 ± 0.13^{b}
HB	1	2.3 ± 0.6^{a}	0.56 ± 0.26^{a}
	2	4.0 ± 1.0^{b}	1.02 ± 0.19^{ab}
	3	7.7 ± 1.2^{c}	1.02 ± 0.59^{ab}
	4	7.7 ± 1.5^{c}	1.19 ± 0.21^{b}
	5	8.7 ± 2.9^{c}	1.01 ± 0.31^{ab}
HBN	1	2.7 ± 0.6^{a}	0.92 ± 0.25^{ab}
	2	4.3 ± 2.5^{ac}	0.82 ± 0.81^{a}
	3	9.7 ± 1.5^{bd}	2.01 ± 0.10^{b}
	4	5.7 ± 1.5^{c}	1.13 ± 0.28^{ab}
	5	11.3 ± 1.2^{d}	1.62 ± 0.66^{ab}
YZZ	1	6.3 ± 1.2^{a}	0.88 ± 0.25^{a}
	2	8.0 ± 1.0^{a}	0.50 ± 0.10^{a}
	3	7.7 ± 1.5^{a}	0.65 ± 0.33^{a}
	4	10.3 ± 1.2^{b}	1.66 ± 0.13^{b}

Table 5. *Cont.*

Study Site	Reversal Stage	Species Richness	Species Diversity
MC	1	4.3 ± 1.5 [a]	0.90 ± 0.23 [a]
	2	6.0 ± 1.0 [ac]	1.13 ± 0.29 [a]
	3	10.3 ± 1.2 [b]	1.28 ± 0.06 [a]
	4	7.3 ± 1.5 [c]	0.90 ± 0.33 [a]
WZZ	1	1.0 ± 0.0 [a]	0.00 ± 0.00 [a]
	2	5.7 ± 2.1 [b]	0.89 ± 0.37 [b]
	4	5.7 ± 2.1 [b]	0.71 ± 0.13 [b]
	5	9.0 ± 1.0 [c]	1.47 ± 0.30 [c]
LJHZ	1	2.3 ± 1.2 [a]	0.07 ± 0.07 [a]
	2	5.0 ± 1.0 [bc]	0.68 ± 0.24 [b]
	3	3.7 ± 1.5 [ab]	0.22 ± 0.26 [ac]
	4	5.3 ± 0.6 [bc]	1.42 ± 0.14 [bd]
	5	7.0 ± 1.7 [c]	0.80 ± 0.52 [bd]
All study sites	1	3.0 ± 1.8 [a]	0.57 ± 0.40 [a]
	2	5.7 ± 1.9 [b]	0.87 ± 0.41 [b]
	3	8.1 ± 2.5 [cd]	1.04 ± 0.62 [bc]
	4	7.1 ± 2.2 [c]	1.22 ± 0.37 [c]
	5	9.0 ± 2.3 [d]	1.22 ± 0.53 [c]

* Desertification reversal stages: (1) very severe desertification (VSD); (2) severe desertification (SD); (3) moderate desertification (MD); (4) light desertification (LD); (5) potential desertification (PD). All study sites: taking the data set for all study sites as a whole. Values are shown as mean ± standard deviation. Different letters within the same study site and variable indicate significant differences between the different desertification reversal stages at the $p < 0.05$ level.

The value of plant biomass was very small in very severe desertification areas and increased gradually with the reversal of desertification at each study site (Figure 1). There were significant differences between different reversal stages. However, there were also some exceptions. At the study site of HBN, the plant biomass in LD was greater than in PD; at the study site of YZZ and LJHZ, the plant biomass in MD was higher than in LD. In general, changes in plant biomass were larger in the early stages of desertification reversal than in the later stages (Figure 1).

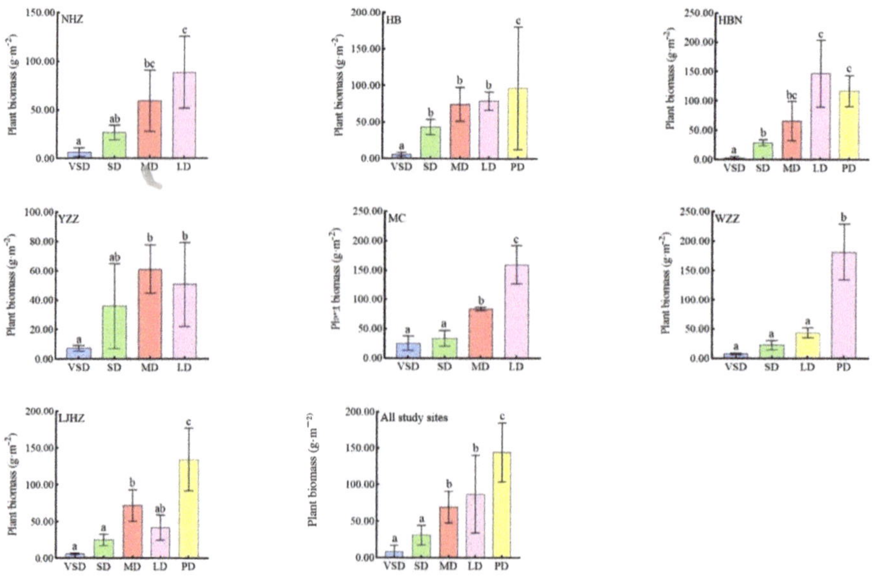

Figure 1. Effects of the reversal of sandy desertification on plant biomass in Southern Mu Us Sandy Land, China. Different letters within the same study site indicate significant differences between the different stages of desertification reversal at the $p < 0.05$ level. The length of the column is the mean value of plant biomass; the length of the bar on each column is the value of the standard deviation.

4. Discussion

4.1. Changes in Vegetation Conditions and Plant Community Characteristics with Desertification Reversal

Our hypotheses were supported by the results. For the first hypothesis, the species importance values (IVs) of both pioneer plant species (*Artemisia ordosica* and *Agriophyllum squarrosum*) for sand fixation decreased dramatically along the desertification reversal sequence, indicating a reversal process of sandy desertification. This result is consistent with the desertification classification in this study based on bare (mobile) sand area proportion, vegetation cover, the intensity of wind erosion, and aeolian activity. This agreement shows that plant community composition could also be an index for desertification reversal.

Changes in IVs of *A. squarrosum* and *Artemisia scoparia* occurred in opposite directions. *A. squarrosum* disappeared suddenly in moderate desertification (MD) areas, where its IV was 0, whereas the IV of *A. scoparia* increased abruptly in MD areas. These two plant species are both annual forbs. The reason for the observed change may be competition between plant species with the same life form. The environment in MD areas was not suitable for *A. squarrosum* anymore, and it was replaced by *A. scoparia* and some other plant species. *A. scoparia* is a species that does not survive well in an extremely harsh environment like the VSD areas, and [69] showed that the growth of *A. scoparia* significantly benefited from higher nitrogen concentrations.

Our results also show that sub-shrubs, together with shrubs, accounted for a significant IV in very severe desertification (VSD) areas, severe desertification (SD) areas, and MD areas, indicating the importance of the existence of sub-shrubs and shrubs in the process of desertification reversal. In accord with this study, ref. [13] showed that the establishment of shrubs was an important step for desertification reversal, and the number of vascular plant species increased due to shrub establishment.

This may be attributed to the positive effects of organic matter provided by perennial shrubs on microbial activity [70] and the effects of shrub rhizosphere components on microbial diversity [71], which both contribute to ameliorating soil conditions and may be beneficial to the re-establishment of perennial grasses. The establishment of shrubs also leads to substantial changes in plant community composition [13], as was also shown by this study. One related mechanism that changes plant species diversity and community composition may be the inhabitation of seed-spreading birds on shrubs; other comprehensive factors may include mycorrhizas and symbionts [13]. Moreover, ref. [72] showed that soil properties in shifting dunes were improved by shrub establishment and sand immobilization; this process facilitated the establishment of herbaceous plant species in the ecosystem [73,74]. Consequently, the growth of shrubs and sub shrubs in severely desertified grasslands may be an advantage for the initiation of the reversal of sandy desertification, although shrubs and sub-shrubs will be replaced gradually by perennial grasses and other herbaceous taxa during the process of desertification reversal.

The changes in IVs of *L. secalinus*, *P. flaccidum*, *C. squarrosa*, *A. mongolicum*, and *S. bungeana* in our study suggest that the importance of perennial grasses as a plant group gradually increased in communities with the reversal of sandy desertification. This plant group may play an important role in communities, especially in potential desertification (PD) areas. The changes in perennial grasses, perennial forbs, sub-shrubs, and annual forbs indicate that plant communities developed gradually from sub-shrub- and annual forb-dominated communities to perennial grass-dominated communities. Previous studies also suggested the re-establishment of perennial grasses as a consequence of desertification reversal [2], which is consistent with our results.

Perennial grasses are re-established with the reversal of desertification, which is probably affected by the interactions between three factors, including soil nutrients, compaction, and infiltration [2]. Soil compaction is a decrease in soil porosity or an increase in soil bulk density caused by internal or external loads [75]. Water infiltration is the capability of soil to absorb irrigation or rainfall. The accumulation of soil nutrients is affected by changes in soil infiltration and compaction [2]. Previous studies have suggested that increased infiltration is significantly related to decreased soil compaction [76]. An increase in infiltration and a decrease in soil compaction were reported as a consequence of long-term livestock removal [77], which supported the accumulation of soil nutrients over time [2]. This result will further affect the re-establishment of perennial grasses, which will provide a favorable environment for the further reversal of sandy desertification.

4.2. Relationships between Vegetation Characteristics and Desertification Reversal

In contrast to widely accepted knowledge, *Amaranthaceae*, rather than *Poaceae*, is the family that included the largest number of C_4 species in our study, suggesting a more complex situation of plant composition in desertified areas than in areas with normal conditions. This is mainly due to the fact that plant species of Amaranthaceae are more adaptive to the xerophytic environment in desert ecosystems. This result is in accordance with [78]. Our study found that C_4 plants accounted for the largest number of plant species and importance values in VSD areas, indicating that C_4 plants are more adaptive to the first stage of desertification reversal than C_3 plants. This further indicates that C_4 species may be considered for planting in mobile dunes as an anthropogenic measure to promote desertification reversal. We also found that C_4 plants could adapt well to habitats of different stages of desertification reversal. On the other hand, few C_3 plants were found in the early stages (Table 4), indicating the relatively narrow adaptation of C_3 plants. Our results are supported by another study that held that C_4 plants should be a useful measure for combating desertification [79]. All of these findings indicate that an application of selected Amaranthaceae C_4 plants in mobile dunes would be more efficient and practical than an application of C_3 or other C_4 plants and may also be beneficial for the maintenance of vegetation in later stages of desertification reversal.

Our study shows that species richness and diversity generally increased with the reversal of sandy desertification. Plant species diversity was found to be strongly related to soil chemical and physical properties, which resulted in different habitats [80]. The ameliorated soil conditions in less desertified grasslands can lead to higher species diversity due to the relatively adequate soil fertility [22] and other improved soil properties [65]. Plant diversity improved by shrub encroachment was also shown to be associated with higher soil fertility [13].

Plant species diversity increases not only in the natural process of desertification reversal described in our study but also during ecological restoration management by planting artificial shrubs in mobile dunes [72]. On the other hand, changes in plant species diversity also influence the reversal of sandy desertification. Higher plant species diversity yields more differences in litter components in both physical and chemical aspects, which is beneficial to soil formation and nutrient cycling [3]. Soils in areas with desertification reversal, therefore, have better conditions and higher rates of nutrient cycling due to higher species diversity. This, in turn, promotes the further reversal of sandy desertification. Moreover, ref. [81] showed that soil microbial communities responded rapidly to changes in plant species diversity in grassland ecosystems, and higher plant diversity resulted in higher bacterial activity and diversity, which is important for litter decomposition and other biogeochemical processes. In addition to plant species diversity, plant biomass also increased significantly with the reversal of sandy desertification in our study. Plant biomass is an important vegetation characteristic reflecting plant growth status. Plants grew better in the reversed grasslands than in the desertified grasslands, which might have been due to the improved soil conditions in the former. Increments in plant biomass, in turn, contribute to the further reversal of sandy desertification. Plant cover is of key importance for the accumulation of organic matter and fine particles in the soil, providing a physical barrier [70] that will prevent aeolian soil erosion. Both the organic matter accumulation and physical barrier function caused by plant cover will promote the continued process of desertification reversal.

4.3. *Differences in Changes in Vegetation Characteristics between Early and Later Stages of Desertification Reversal*

Our results also supported our second hypothesis. Most vegetation characteristics changed more dramatically in the early stages of desertification than in the later stages. These results indicate that applying engineering measures in the early stages is more practical than in the later stages, when these measures are considered. Engineering measures such as using straw checkerboards to arrest the sand and establishing sand barriers have been demonstrated to be successful methods in practice [72,82]. These methods are particularly suitable for vegetation restoration and dune stabilization [82] in the first two stages of sandy desertification reversal. Some other measures, such as leveling the ground of dunes [83], are advantageous for the reversal of very severe desertification areas. Additionally, to plant pioneer plant species in areas of the first two stages is helpful for vegetation recovery, especially Amaranthaceae C_4 plants in mobile dunes. It is particularly useful to protect the native perennial grasses that occur in the early stages, which will facilitate the community succession to the later stages of desertification reversal. On the other hand, engineering measures would help vegetation restoration by providing better physical conditions. But engineering measures need a significant input of energy and resources, and the physical conditions in later stages are normally much better than in early stages. Engineering measures in later stages may thus not generate as many effects as in early stages. Nevertheless, some basic human measures such as enclosure [84] are still necessary for the later stages of desertification reversal. In a word, anthropogenic and natural restoration should be combined, and the early stages, rather than the later stages, should apply appropriate anthropogenic measures when they are under consideration.

5. Conclusions and Implications

The plant community composition, functional types, and other vegetation characteristics (plant diversity, biomass, etc.) of different sites in different desertification reversal stages were analyzed. The natural dominant species in the sites evolved from annual forbs and sub-shrubs to perennial grasses and forbs. In the later stage of restoration, relatively stable near-natural sand-fixing vegetation was formed, showing a trend of succession to desertification grassland. Our results show that plant community composition, species importance value, and plant functional types changed significantly during the process of desertification reversal, and most of the vegetation characteristics changed more in the early stages of desertification reversal than in the later stages. In the early stages, the importance value of C_4 plants was higher than that of C_3 plants. With the reversal process, C_3 plants gradually replaced C_4 plants as the dominant species, indicating that C_4 plants have stronger adaptability than C_3 plants and can survive in harsher environments. Therefore, in the process of artificial sand fixation, C_4 plants or sub-shrubs should be considered for application to mobile sand dunes to change the soil quality and accelerate the reversal of desertification. Collectively, desertification reversal promotes the complexity of the plant community by increasing the plant species diversity of each plant functional type, especially in the early stages of desertification reversal.

Author Contributions: K.Q., Z.L. and Y.X. conceived the idea and designed the study; K.Q., Y.X. and D.X. carried out the field and laboratory work; K.Q. and C.H. performed the data analysis; K.Q. and R.P. wrote the paper. All authors contributed to the review and improvement of the manuscript. All authors have read and agreed to the published version of the manuscript.

Funding: This work was supported by the Ministry of Science and Technology of the People's Republic of China (grant number 2016YFC0500505), the National Natural Science Foundation of China (grant number 31160484), and the China Scholarship Council (CSC).

Data Availability Statement: The data presented in this study are available on request from the corresponding author. The data are not publicly available due to privacy restrictions.

Acknowledgments: We would like to thank Cory Matthew, Jenny Smith, and Tsu-Wei Chen for their critical reading. We also thank the members of Sidunzi Agricultural Research Center in Yanchi

County, Ningxia, for the accommodations during our field investigation. We are grateful to Xia Li, Chao Li, and Lijia Chen for their assistance with the field work.

Conflicts of Interest: The authors declare no conflict of interest.

Abbreviations

COV	Vegetation cover
IV	Importance value
DV	Dominance value
LD	Light desertification
MD	Moderate desertification
PD	Potential desertification
PFT	Plant functional type
RA	Relative abundance
RC	Relative cover
RF	Relative frequency
RH	Relative height
S%	Percentages of the number of plant species
SAND	Proportion of the bare (mobile) sand area to the total ground area
SD	Severe desertification
VSD	Very severe desertification

References

1. Adger, W.N.; Benjaminsen, T.A.; Brown, K.; Svarstad, H. Advancing a political ecology of global environmental discourses. *Dev. Chang.* **2001**, *32*, 681–715. [CrossRef]
2. Allington, G.R.H.; Valone, T.J. Reversal of desertification: The role of physical and chemical soil properties. *J. Arid. Environ.* **2010**, *74*, 973–977. [CrossRef]
3. MEA. *Millennium Ecosystem Assessment, Ecosystems and Human Well-Being: Desertification Synthesis*; World Resources Institute: Washington, DC, USA, 2005.
4. Li, S.; Xie, Y.C. Investigating coupled impacts of climate change and socioeconomic transformation on desertification by using multitemporal landsat images: A case study in central Xilingol, China. *IEEE Geosci. Remote Sens. Lett.* **2013**, *10*, 1244–1248. [CrossRef]
5. UNCCD. *United Nations Convention to Combat Desertification, Elaboration of an International Convention to Combat Desertification in Countries Experiencing Serious Drought and/or Desertification, Particularly in Africa*; United Nations: New York, NY, USA, 1994.
6. Zhou, W.; Sun, Z.G.; Li, J.L.; Gang, C.C.; Zhang, C.B. Desertification dynamic and the relative roles of climate change and human activities in desertification in the Heihe River Basin based on NPP. *J. Arid Land* **2013**, *5*, 465–479. [CrossRef]
7. Johnson, D.; Phoenix, G.K.; Grime, J.P. Plant community composition, not diversity, regulates soil respiration in grasslands. *Biol. Lett.* **2008**, *4*, 345–348. [CrossRef] [PubMed]
8. Thakur, M.P.; Eisenhauer, N. Plant community composition determines the strength of top-down control in a soil food web motif. *Sci. Rep.* **2015**, *5*, 9134. [CrossRef] [PubMed]
9. Díaz, S.; Cabido, M. Plant functional types and ecosystem function in relation to global change. *J. Veg. Sci.* **1997**, *8*, 463–474. [CrossRef]
10. Noble, I.R.; Gitay, H. A functional classification for predicting the dynamics of landscapes. *J. Veg. Sci.* **1996**, *7*, 329–336. [CrossRef]
11. Walker, B.H. Biodiversity and Ecological Redundancy. *Conserv. Biol.* **1992**, *6*, 18–23. [CrossRef]
12. Ricketts, T.H.; Dinerstein, E.; Olson, D.M.; Loucks, C.J.; Eichbaum, W.; DellaSala, D.A.; Kavanagh, K.; Hedao, P.; Hurley, P.; Carney, K.; et al. *Terrestrial ecoregions of North America: A Conservation Assessment*; Island Press: Washington, DC, USA, 1999.
13. Maestre, F.T.; Bowker, M.A.; Puche, M.D.; Belen Hinojosa, M.; Martinez, I.; Garcia-Palacios, P.; Castillo, A.P.; Soliveres, S.; Luzuriaga, A.L.; Sanchez, A.M.; et al. Shrub encroachment can reverse desertification in semi-arid Mediterranean grasslands. *Ecol. Lett.* **2009**, *12*, 930–941. [CrossRef]
14. Zhou, X.J.; An, X.L.; De Philippis, R.; Ye, C.R.; Ke, T.; Zhang, Y.R.; Chen, L.Z. The facilitative effects of shrub on induced biological soil crust development and soil properties. *Appl. Soil Ecol.* **2019**, *137*, 129–138. [CrossRef]
15. Liu, W.S.; Qiu, K.Y.; Xie, Y.Z.; Huang, Y.Y.; Wang, R.X.; Li, H.C.; Meng, W.F.; He, Y.; Li, Y.Y.; Li, H.Q.; et al. High-Throughput absolute quantification sequencing reveals that a combination of leguminous shrubs is effective in driving soil bacterial diversity during the process of desertification reversal. *Microb. Ecol.* **2023**, *86*, 1145–1163. [CrossRef] [PubMed]
16. Zhou, Z.Y.; Wang, G.Z.; Yu, M.H.; Gao, G.L.; Ding, G.D. The leguminous Hedysarum shrubs effectively drive the diversity and structural composition of soil bacterial community through rhizocompartments in the process of desertification reversal. *Land Degrad. Dev.* **2023**, *34*, 4833–4846. [CrossRef]
17. Zhao, X.F.; Xu, H.L.; Zhang, P.; Fu, J.Y.; Tu, W.X.; Zhang, Q.Q. The effects of nutrient addition on plant species diversity in desert grassland, Xinjiang, northwest China. *Quat. Int.* **2013**, *298*, 152–160. [CrossRef]

18. Böcük, H.; Türe, C.; Ketenoğlu, O. Plant diversity and conservation of the Northeast Phrygia region under the impact of land degradation and desertification (Central Anatolia, Turkey). *Pak. J. Bot.* **2009**, *41*, 2305–2321.
19. Elmberg, J.; Nummi, P.; Pöysä, H.; Sjöberg, K. Factors affecting species number and density of dabbling duck guilds in North Europe. *Ecography* **1993**, *16*, 251–260. [CrossRef]
20. Gill, A.H.; Ahmad, K.S.; Habib, S.; Hameed, M.; Ahmad, M.S.A.; Nawaz, T.; Ahmad, F.; Batool, R. Impact of highly saline wetland ecosystem on floral diversity of the Cholistan Desert. *Pak. J. Bot.* **2012**, *44*, 107–112.
21. Yang, X.D.; Zhang, X.N.; Lv, G.H.; Ali, A. Linking Populus euphratica hydraulic redistribution to diversity assembly in the arid desert zone of Xinjiang, China. *PLoS ONE* **2014**, *9*, e109071. [CrossRef]
22. Fang, W.; Peng, S.L. Development of species diversity in the restoration process of establishing a tropical man-made forest ecosystem in China. *For. Ecol. Manag.* **1997**, *99*, 185–196. [CrossRef]
23. Xie, Y.Z.; Qiu, K.Y.; Xu, D.M.; Shi, X.F.; Qi, T.Y.; Pott, R. Spatial heterogeneity of soil and vegetation characteristics and soil-vegetation relationships along an ecotone in Southern Mu Us Sandy Land, China. *J. Soils Sediments* **2015**, *15*, 1584–1601. [CrossRef]
24. Chen, X.H.; Duan, Z.H. Changes in soil physical and chemical properties during reversal of desertification in Yanchi County of Ningxia Hui autonomous region, China. *Environ. Geol.* **2009**, *57*, 975–985. [CrossRef]
25. Qiao, F.; Zhang, S.Y.; Zhang, K.B.; Yang, J.J.; Li, R.; Liu, G. The dynamic change of vegetation coverage in the Yanchi County monitoring based on remote sensing. *Res. Soil Water Conserv.* **2006**, *13*, 181–183, 186. (In Chinese with English Abstract)
26. Li, J.Y.; Xu, B.; Yang, X.C.; Jin, Y.X.; Gao, T.; Yu, H.D.; Ma, H.L.; Qin, Z.H.; Zhao, L.N. Temporal and Spatial Variations of Grassland Desertification in Lingwu and Yanchi of Ningxia, China). In Proceedings of the Second International Conference on Agro-Geoinformatics, Fairfax, VA, USA, 12–16 August 2013; pp. 356–360.
27. Institute of Soil Science; Chinese Academy of Sciences. *Chinese Soil Taxonomy*; USTC Press: Hefei, China, 2001. (In Chinese)
28. Zhou, T.J.; Zhao, T.N. Study on analysis of land use change in Yanchi County. *Res. Soil Water Conserv.* **2005**, *12*, 116–118. (In Chinese with English Abstract)
29. Zhou, L.H.; Yang, G.J.; Wang, T. Analysis on driving factors of desertification in northern China: A case study of Yanchi County). In Proceedings of the International Geoscience and Remote Sensing Symposium, Honolulu, HI, USA, 25–30 July 2010; pp. 2127–2130.
30. Zhou, L.; Zhu, Y.; Yang, G.; Luo, Y. Quantitative evaluation of the effect of prohibiting grazing policy on grassland desertification reversal in northern China. *Environ. Earth Sci.* **2013**, *68*, 2181–2188. [CrossRef]
31. Li, Y.; Zhao, H.; Zhao, X.; Zhang, T.; Chen, Y. Biomass energy, carbon and nitrogen stores in different habitats along a desertification gradient in the semiarid Horqin Sandy Land. *Arid Land Res. Manag.* **2006**, *20*, 43–60. [CrossRef]
32. Zhao, H.L.; Zhou, R.L.; Zhao, X.Y.; Zhang, T.H. Ground Discriminance on positive and negative processes of land desertification in Horqin Sand Land. *J. Desert Res.* **2008**, *28*, 8–15. (In Chinese with English Abstract)
33. Li, S.; Yang, P.; Gao, S.; Chen, H.; Yao, F. Dynamic changes and developmental trends of the land desertification in Tibetan Plateau over the past 10 years. *Adv. Earth Sci.* **2004**, *19*, 63–70. (In Chinese with English Abstract)
34. Li, X.R.; Jia, X.H.; Dong, G.R. Influence of desertification on vegetation pattern variations in the cold semi-arid grasslands of Qinghai-Tibet Plateau, North-west China. *J. Arid Environ.* **2006**, *64*, 505–522. [CrossRef]
35. Verón, S.R.; Paruelo, J.M.; Oesterheld, M. Assessing desertification. *J. Arid Environ.* **2006**, *66*, 751–763. [CrossRef]
36. Zhu, Z.D. Concept, cause and control of desertification in China. *Quat. Sci.* **1998**, *18*, 145–155, (In Chinese with English Abstract)
37. Wang, H.Z.; Liu, C.Q.; Zhang, Y.L. *Grassland Analysis and Production Design*; China Agriculture Press: Beijing, China, 1997. (In Chinese)
38. Song, Y.C. *Vegetation Ecology*; East China Normal University Press: Shanghai, China, 2001. (In Chinese)
39. An, H.; Li, G.Q. Effects of grazing on carbon and nitrogen in plants and soils in a semiarid desert grassland, China. *J. Arid Land* **2014**, *7*, 341–349. [CrossRef]
40. Cheng, J.M.; Jing, Z.B.; Jin, J.W.; Gao, Y. Restoration and utilization mechanism of degraded grassland in the semi-arid region of Loess Plateau. *Sci. Sin. Vitae* **2014**, *44*, 267–279. (In Chinese with English Abstract) [CrossRef]
41. Cui, Y.Q.; Ma, J.Y.; Sun, W.; Sun, J.H.; Duan, Z.H. A preliminary study of water use strategy of desert plants in Dunhuang, China. *J. Arid Land* **2015**, *7*, 73–81. [CrossRef]
42. Feng, Y.; Duan, S.M.; Mu, S.Y.; Zhao, L.; Zhao, X.H. Geographic distribution and ecology of C_4 plants in Xinjiang. *Arid Land Geogr.* **2012**, *35*, 145–153. (In Chinese with English Abstract)
43. Jiang, G.M.; He, W.M. Species- and habitat-variability of photosynthesis, transpiration and water use efficiency of different plant species in Maowusu Sand Area. *Acta Bot. Sin.* **1999**, *41*, 1114–1124.
44. Li, M.C.; Yi, X.F.; Zhang, X.A.; Li, L.X. The list of C_4 plants in alpine locality of Qinghai Plateau. *Acta Bot. Boreali-Occident. Sin.* **2005**, *25*, 1046–1050. (In Chinese with English Abstract)
45. Liu, X.Q.; Wang, R.Z.; Li, Y.Z. Photosynthetic pathway types in rangeland plant species from Inner Mongolia, North China. *Photosynthetica* **2004**, *42*, 339–344. [CrossRef]
46. Ma, J.Y.; Sun, W.; Liu, X.N.; Chen, F.H. Variation in the stable carbon and nitrogen isotope composition of plants and soil along a precipitation gradient in northern China. *PLoS ONE* **2012**, *7*, e51894. [CrossRef]
47. Niu, S.L.; Jiang, G.M.; Gao, L.M.; Li, Y.G.; Liu, M.Z. Comparison of gas exchange traits of different plant species in Hunshandak Sand Area. *Acta Phytoecol. Sin.* **2003**, *27*, 318–324.

48. Su, P.X.; Xie, T.T.; Zhou, Z.J. Geographical distribution of C_4 plant species in desert regions of China and its relation with climate factors. *J. Desert Res.* **2011**, *31*, 267–276. (In Chinese with English Abstract)
49. Wang, G.A.; Mao, H.J. $\delta^{13}C$ variations of C_3 plants in dry and rainy seasons. *Mar. Geol. Quat. Geol.* **2001**, *21*, 43–47. (In Chinese with English Abstract)
50. Wang, J. Along the 400 mm isopleth of mean annual precipitation in China. *Mendeley Data* **2017**. [CrossRef]
51. Wang, R.Z. Photosynthetic pathways and life forms in different grassland types from North China. *Photosynthetica* **2002**, *40*, 243–250. [CrossRef]
52. Wang, R.Z. Photosynthetic pathways, life forms, and reproductive types for forage species along the desertification gradient on Hunshandake desert, North China. *Photosynthetica* **2002**, *40*, 321–329. [CrossRef]
53. Wang, R.Z. Photosynthetic pathway and morphological functional types in the steppe vegetation from Inner Mongolia, North China. *Photosynthetica* **2003**, *41*, 143–150. [CrossRef]
54. Wang, R.Z. Photosynthetic and morphological functional types from different steppe communities in Inner Mongolia, North China. *Photosynthetica* **2004**, *42*, 493–503. [CrossRef]
55. Wang, R.Z. Photosynthetic pathways and life form types for native plant species from Hulunbeier Rangelands, Inner Mongolia, North China. *Photosynthetica* **2004**, *42*, 219–227. [CrossRef]
56. Werger, M.J.A.; van Staalduinen, M.A. Eurasian steppes. In *Ecological Problems and Livelihoods in a Changing World*; Springer: Berlin/Heidelberg, Germany, 2012.
57. Yin, L.J.; Zhu, L. A preliminary study on C_3 and C_4 plants in the Northeast steppes and their ecological distribution. *J. Appl. Ecol.* **1990**, *1*, 237–242. (In Chinese with English Abstract)
58. Zhang, B.; Ning, Y.F.; An, Z.S.; Liu, W.G. Abundance of C_4/C_3 plants in the Chinese Loess Plateau and their response to plant growing environment. *Quat. Sci.* **2015**, *35*, 801–808. (In Chinese with English Abstract)
59. Zhang, C.J.; Chen, F.H.; Ming, J. Study on modern plant C-13 in western China and its Significance. *Chin. J. Geochem.* **2003**, *22*, 97–106.
60. Zheng, S.X.; Shangguan, Z.P. Spatial patterns of foliar stable carbon isotope compositions of C_3 plant species in the Loess Plateau of China. *Ecol. Res.* **2007**, *22*, 342–353. [CrossRef]
61. Ziska, L.H. Evaluation of the growth response of six invasive species to past, present and future atmospheric carbon dioxide. *J. Exp. Bot.* **2003**, *54*, 395–404. [CrossRef]
62. Myers, J.H.; Bazely, D.R. *Ecology and Control of Introduced Plants*; Cambridge University Press: Cambridge, UK, 2003.
63. Zhang, J.; Zhao, H.; Zhang, T.; Zhao, X.; Drake, S. Community succession along a chronosequence of vegetation restoration on sand dunes in Horqin Sandy Land. *J. Arid Environ.* **2005**, *62*, 555–566. [CrossRef]
64. Belsky, A.J. Effects of grazing, competition, disturbance and fire on species composition and diversity in grassland communities. *J. Veg. Sci.* **1992**, *3*, 187–200. [CrossRef]
65. Chawla, A.; Kumar, A.; Rajkumar, S.; Singh, R.D.; Thukral, A.K.; Ahuja, P.S. Correlation of multispectral satellite data with plant species diversity vis-à-vis soil characteristics in a landscape of Western Himalayan Region, India. *Appl. Remote Sens. J.* **2010**, *1*, 1–13.
66. Magurran, A.E. *Ecological Diversity and Its Management*; Princeton Univ. Press: Princeton, NJ, USA, 1998.
67. Shannon, C.E. A mathematical theory of communication. *Bell Syst. Tech. J.* **1948**, *27*, 379–423, 623–656. [CrossRef]
68. Lu, W.D. *SPSS for Windows Statistics Analysis*, 3rd ed.; Press of Electronics Industry: Beijing, China, 2003. (In Chinese)
69. Du, F.; Sui, Y.Y.; Xu, X.X.; Zhang, X.C.; Hao, W.F.; Shao, M.G.; Hu, L.J.; Shan, L. Morphological responses of six plant species to patchy habitat with different nitrogen concentrations using in-growth core method. *Pol. J. Ecol.* **2013**, *61*, 257–269.
70. Berg, N.; Steinberger, Y. Are biological effects of desert shrubs more important than physical effects on soil microorganisms? *Microb. Ecol.* **2010**, *59*, 121–129. [CrossRef] [PubMed]
71. Diedhiou, S.; Dossa, E.L.; Badiane, A.N.; Diedhiou, I.; Sène, M.; Dick, R.P. Decomposition and spatial microbial heterogeneity associated with native shrubs in soils of agroecosystems in semi-arid Senegal. *Pedobiologia* **2009**, *52*, 273–286. [CrossRef]
72. Su, Y.Z.; Zhao, H.L. Soil properties and plant species in an age sequence of *Caragana microphylla* plantations in the Horqin Sandy Land, north China. *Ecol. Eng.* **2003**, *20*, 223–235. [CrossRef]
73. Su, Y.; Zhang, T.; Li, Y.; Wang, F. Changes in soil properties after establishment of *Artemisia halodendron* and *Caragana microphylla* on shifting sand dunes in semiarid Horqin Sandy Land, northern China. *Environ. Manag.* **2005**, *36*, 272–281. [CrossRef] [PubMed]
74. Zuo, X.A.; Zhao, X.Y.; Zhao, H.L.; Zhang, T.H.; Guo, Y.R.; Li, Y.Q.; Huang, Y.X. Spatial heterogeneity of soil properties and vegetation-soil relationships following vegetation restoration of mobile dunes in Horqin Sandy Land, Northern China. *Plant Soil* **2009**, *318*, 153–167. [CrossRef]
75. Alakukku, L. Soil compaction. In *Ecosystem Health & Sustainable Agriculture, Book 1. Sustainable Agriculture*; Jakobsson, C., Ed.; The Baltic University Programme, Uppsala University: Uppsala, Sweden, 2012.
76. Valone, T.J. Effects of long-term livestock removal on water infiltration rate in arid grasslands. In *Grasslands: Ecology, Management and Restoration*; Schroder, H.G., Ed.; Nova Science Publishers: New York, NY, USA, 2008; pp. 81–91.
77. Steffens, M.; Kölbl, A.; Totsche, K.U.; Kögel-Knabner, I. Grazing effects on soil chemical and physical properties in a semiarid steppe of Inner Mongolia (P.R. China). *Geoderma* **2008**, *143*, 63–72. [CrossRef]
78. Pyankov, V.I.; Gunin, P.D.; Tsoog, S.; Black, C.C. C_4 plants in the vegetation of Mongolia: Their natural occurrence and geographical distribution in relation to climate. *Oecologia* **2000**, *123*, 15–31. [CrossRef]

79. Zhang, J.W. C_4 plants, one of the ways of desertification control. *Arid Zone Res.* **2006**, *23*, 384–387. (In Chinese with English Abstract)
80. de Carvalho, F.; Godoy, E.L.; Lisboa, F.J.G.; Moreira, F.M.D.; de Souza, F.A.; Berbara, R.L.L.; Fernandes, G.W. Relationship between physical and chemical soil attributes and plant species diversity in tropical mountain ecosystems from Brazil. *J. Mt. Sci.* **2014**, *11*, 875–883. [CrossRef]
81. Loranger-Merciris, G.; Barthes, L.; Gastine, A.; Leadley, P. Rapid effects of plant species diversity and identity on soil microbial communities in experimental grassland ecosystems. *Soil Biol. Biochem.* **2006**, *38*, 2336–2343. [CrossRef]
82. Guo, Y.R.; Zhao, H.L.; Zuo, X.A.; Drake, S.; Zhao, X.Y. Biological soil crust development and its topsoil properties in the process of dune stabilization, Inner Mongolia, China. *Environ. Geol.* **2007**, *54*, 653–662. [CrossRef]
83. Su, Y.Z.; Zhao, W.Z.; Su, P.X.; Zhang, Z.H.; Wang, T.; Ram, R. Ecological effects of desertification control and desertified land reclamation in an oasis–desert ecotone in an arid region: A case study in Hexi Corridor, northwest China. *Ecol. Eng.* **2007**, *29*, 117–124. [CrossRef]
84. Li, Y.Q.; Zhao, H.L.; Zhao, X.Y.; Zhang, T.H.; Li, Y.L.; Cui, J.Y. Effects of grazing and livestock exclusion on soil physical and chemical properties in desertified sandy grassland, Inner Mongolia, northern China. *Environ. Earth Sci.* **2010**, *63*, 771–783. [CrossRef]

MDPI
St. Alban-Anlage 66
4052 Basel
Switzerland
www.mdpi.com

Agronomy Editorial Office
E-mail: agronomy@mdpi.com
www.mdpi.com/journal/agronomy